L: freigegeben 29.2.20..
trotz

Rechtsanwälte:
- Schmutztitel S... rechts und
  ...

Achtung: CD im Signal nicht
enthalten gewesen –
eingeklebt

HS: _nicht_ freigegeben,
wird nachgebessert
in Druckerei.

Computer Simulation of Materials at Atomic Level

edited by Péter Deák,
Thomas Frauenheim, and Mark R. Pederson

Berlin · Weinheim · New York · Chichester · Brisbane · Singapore · Toronto

Editors:

Professor Dr. Péter Deák, Technical University of Budapest, Department of Atomic Physics, Hungary
Professor Dr. Thomas Frauenheim, University of Paderborn, Department of Physics, Paderborn, Germany
Professor Dr. Mark R. Pederson, Naval Research Laboratories, Complex Systems Theory Branch, Washington, DC, USA

With 238 figures

1st edition

Library of Congress Card No.: applied for

Die Deutsche Bibliothek – CIP-Cataloguing-in-Publication-Data
A catalogue record for this publication is available from Die Deutsche Bibliothek

ISBN 3-527-40290-X

This book was carefully produced. Nevertheless, authors, editors, and publishers do not warrant the information contained therein to be free of errors. Readers are advised to keep in mind that statements, data, illustrations, procedural details, or other items may inadvertently be inaccurate.
All rights reserved (including those of translation into other languages). No part of this book may be reproduced in any form – by photoprinting, microfilm, or any other means – nor transmitted or translated into a machine language without written permission from the publishers. Registered names, trademarks, etc. used in this book, even when not specifically marked as such, are not to be considered unprotected by law.

© WILEY-VCH Verlag Berlin GmbH, Berlin (Federal Republic of Germany), 2000

Printed on non-acid paper.
The paper used corresponds to both the U. S. standard ANSI Z.39.48 – 1984 and the European standard ISO TC 46.

Printing and Bookbinding: Druckhaus „Thomas Müntzer", Bad Langensalza

Printed in the Federal Republic of Germany

WILEY-VCH Verlag Berlin GmbH
Bühringstraße 10
D-13086 Berlin
Federal Republic of Germany

Computer Simulation of Materials
Deák/Frauenheim/Pederson (eds.)

Preface

The idea for this compilation of papers grew out of two workshops organized by Thomas Frauenheim during the last four years. The first (Chemnitz, 1996) was entitled "First-Principles, Tight-Binding and Empirical Methods for Materials Simulation" and the second (Paderborn, 1998) was entitled "Massively Parallel and Superscalar Applications in Computational Materials Science". Many of the participants of these workshops had collaborations or cooperations with one another. Lectures from 26 prominent groups representing 9 countries were presented during these workshops. While many workshops revolve around a group of related methodologies, the "cohesive quality" that emerged from these two workshops was in fact due to the diversity of the methodologies discussed. Because of this diversity the materials applications exhibited a commensurately broader range than what would generally be observed in a more specialized workshop where the limitations of any given methodology tend to partially lead to problem preselection. Actual materials-based questions, of course, arise by practical needs and are indeed wide ranged. As such, we observed that a broad repertoire of materials methodologies at one's finger tips might indeed be the optimal way to avoid problem preselection and instead bridge some gaps and allow us to go beyond the conventional limits of computational materials science, condensed matter physics and chemistry. Indeed, the sheer complexity of the materials problems in the new millenium will require a new generation of computational scientists to be fluent in many, rather than one or two, computational methods and duely informed about the problem areas for which each and every computational methodology is applicable. Realization of this goal will be nontrivial since the complexity of computational materials science and condensed matter physics presently requires most groups to concentrate their efforts on one or two computational methodologies. Presently young researchers from a given group must spend several years learning a specific methodology and thereby become "hostage" to their own expertise. Exchanges, such as these workshops, then provide an excellent avenue for younger developing scientists, and even senior scientists, to broaden their understanding of methods outside their immediate scope and hopefully expand their computational arsenal so that higher compexity materials-related problems can be tackled.

We believe all participants enjoyed the excitement that was generated by the wide variety of topics covered and the very real prospect of theoretical prediction of technologically useful materials properties. Further, we expect that many researchers would benefit from the dissemination of this diverse set of tools. As such, we have attempted to compile this information in a way that would be useful to novice computational solid state physicists and materials scientists. The hope here is that as the field matures young scientists will more often decide on an interesting application and then determine the methodology or methodologies that will most beneficially impact their area of interest.

Of course, no group of theorists interested in real-life materials problems can work without close contact to experimental and technological research and development groups. Indeed almost all the groups represented herein have such contacts. The potential of *microscopic modeling on a physical basis* for helping to understand and solve problems could be more widely recognized and enhanced by the *experimental materials science and technology* community. Such interactions could be extended if the latter had a better overview of the possibilities of the former and *intensified* if the Edisonian community could fully appreciate the *limits and capabilities* of the various theoretical methods. *The other purpose* of this compilation is to demonstrate to the other fields of materials science and technology the wide range of problems for which atomistic modeling will provide insight and to enable this community to independently determine which methodology is most appropriate for reliably answering their particular questions.

With these goals in mind this compilation has been divided into two parts. The first part, *Methods*, contains indepth descriptions of many methods used in condensed matter and molecular physics. We attempted to include every major approach which is based on individual interatomic interactions. These approaches include classical empirical potentials, semi-empirical and non-empirical tight binding methods, and *ab initio* methods which include Hartree-Fock and Density-Functional based theories. We asked that the contributors be concise about the theoretical foundations, explaining only the basic concepts, but to elaborate on questions associated with implementation. We also asked them to explain the limits of the method and the critical parameters of the actual calculations which ultimately determine the quality of the output. Such requirements lead authors to primarily address implementations in their own area of expertise and to briefly mention other variants. The second part, *Applications*, presents a diversity of problems related to materials properties and phenomena where these methods can be fruitfully applied. We asked the authors of this part to emphasize issues related to accuracy of the calculated data but also to show how the raw data can be used to interpret and/or foretell experimental results.

For both parts of this volume, we strived to assemble a collection of individual non-overlapping papers that as a whole represented a broad range of topics of current interest in materials science. This necessitated going beyond the topics represented in the aforementioned workshops, and the result is a mix of contributions representing those topics as well as many other materials properties. In all cases we asked for new or updated work, and the end result was ultimately determined by the willingness of the invited authors to devote considerable time and effort in their contribution and to adhere to the restrictions arising from the broader aims of this volume. As with any volume of this nature size restrictions and author availability call for some qualifications. This compilation is in fact a snapshot of the present field of atomistic computational materials science. It does not and can not manage to cover every important method and all their possible aspects. Notably absent is a methodological discussion of plane-wave-based algorithms (which, however, is easily accessible in several recent reviews). However, several contributions in the applications section are based on such methods. The individual papers do not attempt to review their respective fields completely. While useful to a young scientist it can not be a full substitute since knowledge of advanced physics and chemistry is assumed rather than taught. Computational materials science is the most appropriate umbrella for this collection, but this field also contains a huge variety of problems and those which may be addressed atomistically are merely a subset. Within this subset our compilation represents problems related to semiconductors, dielectrics, molecular assembled materials and special transition metal systems.

While a photographic snapshot can record a portion of a large event, it generally neither features nor includes all participants but often captures the overall enthusiasm and excitement of the event. We think the snapshot of Computational Materials Science enclosed within this special volume of physica status solidi, together with the resulting book, achieve our specific aims and further illustrate the way in which the interactions between the fields of experimental and computational materials science are being expanded at this and other levels. In addition to the book we distribute a CD with a collection of demonstration versions of many of the computer codes used by the researchers. We encourage both experimentalists and theorists to play around with these codes to develop a greater idea of what is possible with each of these codes. We intend to continue the workshops on computational materials science as well.

We thank all authors who participated in this project. Finally we would like to thank the publishing house Wiley-VCH for endorsing this project and express our gratitude to the editorial office, Karin Müller (editor of the journal version), Gesine Reiher (editor of the book version), Dr. Michael Bär (publisher) and Professor Martin Stutzmann (editor-in-chief).

Budapest 1999

Péter Deák
Thomas Frauenheim
Mark R. Pederson

Contents

Methods

P. Deák — Choosing Models for Solids 9

D.W. Brenner — The Art and Science of an Analytic Potential 23

Th. Frauenheim, G. Seifert, M. Elstner, Z. Hajnal, G. Jungnickel, D. Porezag, S. Suhai, and R. Scholz
A Self-Consistent Charge Density-Functional Based Tigh-Binding Method for Predictive Materials Simulations in Physics, Chemistry and Biology. . . 41

R. Dovesi, R. Orlando, C. Roetti, C. Pisani, and V.R. Saunders
The Periodic Hartree-Fock Method and Its Implementation in the CRYSTAL Code. 63

R.W. Tank and C. Arcangeli
An Introduction to the Third-Generation LMTO Method 89

P.R. Bridden and R. Jones
LDA Calculations Using a Basis of Gaussian Orbitals 131

J.R. Chelikowsky, Y. Saad, S. Öğüt, I. Vasiliev, and A. Stathopoulos
Electronic Structure Methods for Predicting the Properties of Materials: Grids in Space. 173

M.R. Pederson, D.V. Porezag, J. Kortus, and D.C. Patton
Strategies for Massively Parallel Local-Orbital-Based Electronic Structure Methods 197

D. Porezag, M.R. Pederson, and A.Y. Liu
The Accuracy of the Pseudopotential Approximation within Density-Functional Theory. 219

G. Galli — Large-Scale Electronic Structure Calculations Using Linear Scaling Methods 231

R.E. Rudd and J.Q. Broughton
Concurrent Coupling of Length Scales in Solid State Systems 251

Applications

K. Jackson — Electric Fields in Electronic Structure Calculations: Electric Polarizabilities and IR and Raman Spectra from First Principles 293

S. Srinivas and J. Jellinek
Ab initio Monte Carlo Investigations of Small Lithium Clusters. 311

C. ASHMAN, S.N. KHANNA, and M.R. PEDERSON
   Structure and Isomerization in Alkali Halide Clusters 323

P. ORDEJÓN Linear Scaling ab initio Calculations in Nanoscale Materials with SIESTA 335

M. ELSTNER, TH. FRAUENHEIM, E. KAXIRAS, G. SEIFERT, and S. SUHAI
   A Self-Consistent Charge Density-Functional Based Tight-Binding Scheme for Large Biomolecules 357

P.D. TEPESCH and A.A. QUONG
   First-Principles Calculations of α-Alumina (0001) Surfaces Energies with and without Hydrogen 377

A. GROSS Ab initio Molecular Dynamics Simulations of Reactions at Surfaces 389

J. KOLLÁR, L. VITOS, B. JOHANSSON, and H.L. SKRIVER
   Metal Surfaces: Surface, Step and Kink Formation Energies 405

A.Y. LIU Linear-Response Studies of the Electron–Phonon Interaction in Metals . . 419

P. LEARY, C.P. EWELS, M.I. HEGGIE, R. JONES, and P.R. BRIDDON
   Modelling Carbon for Industry: Radiolytic Oxidation 429

R. SCHOLZ, J.-M. JANCU, F. BELTRAM, and F. BASSANI
   Calculation of Electronic States in Semiconductor Heterostructures with an Empirical spds* Tight-Binding Model 449

H.M. URBASSEK and P. KLEIN
   Constant-Pressure Molecular Dynamics of Amorphous Si 461

M. HAUGK, J. ELSNER, TH. FRAUENHEIM, T.E.M. STAAB, C.D. LATHAM, R. JONES, H.S. LEIPNER, T. HEINE, G. SEIFERT, and M. STERNBERG
   Structures, Energetics and Electronic Properties of Complex III–V Semiconductor Systems 473

S.K. ESTREICHER
   Structure and Dynamics of Point Defects in Crystalline Silicon 513

J.E. LOWTHER Superhard Materials 533

U.V. WAGHMARE, E. KAXIRAS, and M.S. DUESBERY
   Modeling Brittle and Ductile Behavior of Solids from First-Principles Calculations 545

F. DELLA SALA, J. WIDANY, and TH. FRAUENHEIM
   Comparison of Simulation Methods for Organic Molecular System: Porphyrin Stacks 565

Contents

F. CORÀ and C.R.A. CATLOW
    Quantum Mechanical Investigations on the Insertion Compounds of Early Transition Metal Oxides. 577

W.R.L. LAMBRECHT and S.N. RASHKEEV
    From Band Structures to Linear and Nonlinear Optical Spectra in Semiconductors 599

M.J. CALDAS   Si Nanoparticles as a Model for Porous Si 641

U. GERSTMANN, M. AMKREUTZ, and H. OVERHOF
    Paramagnetic Defects. 665

J. BERNHOLC, E.L. BRIGGS, C. BUNGARO, M. BUONGIORNO NARDELLI, J.-L. FATTEBERT,
K. RAPCEWICZ, C. ROLAND, W.G. SCHMIDT, and Q. ZHAO
    Large-Scale Applications of Real-Space Multigrid Methods to Surfaces, Nanotubes and Quantum Transport 685

A. DI CARLO   Semiconductor Nanostructures 703

Subject Index   723

phys. stat. sol. (b) **217**, 9 (2000)

Subject classification: 61.46.+w; 61.50.−f; 71.10.−w; 71.23.An

Choosing Models for Solids

P. Deák

*Surface Physics Laboratory, Department of Atomic Physics, TU Budapest,
Budafoki út 8., H-1111 Budapest, Hungary
(p.deak@eik.bme.hu)*

(Received August 10, 1999)

The atomistic simulation of properties and phenomena in solid materials requires a suitable model of the real system with a number of atoms which is still manageable at the chosen level of approximation. Since typical solids consist of atoms in the order of 10^{23}, only a very small fraction of them can be treated explicitly. The effect of the rest on the explicitly treated part has to be taken into account somehow. This paper attempts to categorize and explain the various tricks applied in modeling solids, showing their strength and weakness.

1. Introduction

Even though the term "material" is more general, when talking about "materials properties and phenomena", usually condensed matter is meant. This implicit distinction comes from the fact that materials science and technology evolved by dealing with structural and functional engineering materials, whereas fluids and gases were the working media only. The majority of the contributions in the present volume deals with an even more restricted class of materials: the solids. By that those pieces of condensed matter are meant which:

− are in the solid state under normal conditions and have characteristic relaxation times under stress $\tau > 10^{10}$ s ($\tau = \eta/E$; where η is the viscosity and E is the Young modulus),

− exhibit at least short-range order on the atomic scale (i.e., at least one of the atom types has the same first neighbor coordination everywhere),

− are larger than about 10 nm in diameter (consist roughly over 50000 atoms).

This last criterion separates solids from clusters which are also treated in this volume but − due to their size − do not need simplified models. The goal of the present paper is to give a short guide to those models which scale down the problem of real solids to a level tractable by present day atomistic computer simulations.

A special class of solids are crystals, i.e., solids with long-range order. The fact that the coordination of every atom type in any neighbor shell is identical, ensures translational symmetry in the bulk of the material. Considering the critical size of solids, the surface to volume ratio can be neglected and artificial boundary conditions can be applied to extrapolate the translational symmetry to infinity. As a consequence, (Newtonian) equations of motion or (Schrödinger) equation of state has to be solved only for one periodically repeated unit cell. Due to the translational symmetry, the calculations can be conveniently performed in momentum space. Indeed, between 1940 and 1970, the main task of solid state physics was to determine the properties of perfect crystals within this construction.

Even local deviations from the overall periodicity (i.e. bulk defects and surfaces), let alone the lack of long-range order, forfeit the principal basis for applying conventional momentum space description. Since these "imperfections" cannot be neglected, or rather they make most practical application of the material possible, the papers of this volume have to deal with this situation. The usual way is to handle the immediate environment of the critical part explicitly, taking into account somehow the effect of the rest of the solid. It is the purpose of my contribution to categorize and explain the tricks usually applied when facing the problem of having to solve equations – in principle – for a many-body system of particles well in excess of 10^5. The paper is intended to be a tutorial, rather than a review. Therefore, emphasis lies on the main ideas of modeling without any claim of completeness. In Section 2 the treatment used for perfect crystals is shortly given. Section 3 describes models of defective crystals, while Section 4 deals with surfaces and solids without long-range order.

2. The Perfect Crystal

A general assumption (almost always made) is the Born-Oppenheimer approximation. If the atomic vibrations in the crystal are strictly harmonic, a many-body Schrödinger equation for the electrons can be solved at fixed nuclei, and the total energy of the electrons can be added to the effective potential of the vibrating nuclei. The assumption is justified as long as the solid remains essentially elastic (Hook's law) and the temperature is sufficiently low (thermal expansion negligible). Of course, neither condition is satisfied exactly but the deviation can be taken into account by interaction between the (nominally independent) elementary excitations of both systems. This is also usually the case even for such defects (non-radiative recombination centers) where the two systems are definitely coupled. In Jahn-Teller unstable systems a vibronic rather than an electronic wavefunction should be used.

As mentioned above, the perfect crystal is assumed to be invariant to translations by the lattice vectors

$$\mathbf{R_l} = \sum_i l_i \mathbf{a}_i , \quad (1)$$

where $\mathbf{a}_i$ are the primitive unit vectors and $\mathbf{l} = [l_1, l_2, l_3]$ are arbitrary integers. As a consequence, a reciprocal lattice can be defined for the crystal in momentum space with lattice vectors

$$\mathbf{G_g} = \sum_j g_j \mathbf{b}_j , \quad (2)$$

where the primitive unit vectors $\mathbf{b}_j$ satisfy the condition

$$\mathbf{a}_i \mathbf{b}_j = 2\pi \delta_{ij} . \quad (3)$$

The symmetric unit cell of the reciprocal lattice (defined as points closer to one lattice point than to all others) is the Brillouin zone (BZ). Due to the translational symmetry, the elementary excitations of the many-electron system (Bloch electrons) or of the vibrating lattice (phonons) can be expressed by Bloch waves of the form

$$\varphi_{n\mathbf{k}}(\mathbf{r}) = u_{n\mathbf{k}}(\mathbf{r}) \exp(i\mathbf{kr}) \quad (4)$$

with wave vector $\mathbf{k}$ restricted to the BZ. The *microscopic periodicity* of the crystal is expressed by the fact that the Bloch waves satisfy Bloch's *periodicity condition*,

$$\varphi_{n\mathbf{k}}(\mathbf{r} + \mathbf{R_l}) = \varphi_{n\mathbf{k}}(\mathbf{r}) \exp(i\mathbf{k}\mathbf{R_l}) , \quad (5)$$

which makes the modulating function $u_{nk}(\mathbf{r})$ invariant with respect to translations by the lattice vectors. Since the real solid is finite (N_i unit cells in the direction of the unit vector $\mathbf{a}_i$), the perfect translational symmetry can only be ensured by imposing artificial *macroscopic periodicity*, i.e., applying the *cyclic boundary conditions* of Born and von Kármán,

$$\varphi_{n\mathbf{k}}(\mathbf{r} + N\mathbf{a}_i) = \varphi_{n\mathbf{k}}(\mathbf{r}), \qquad (6)$$

which makes the number of $\mathbf{k}$ vectors in the BZ equal to $N_1 N_2 N_3 = N$, and their density equal to $V/8\pi^3$ (where V is the volume of the solid).

Due to this construction, an identical, $\mathbf{k}$-dependent set of equations (size of the set given by the number of atoms in the unit cell) applies for all unit cells. In principle, the equations should be solved for all $\mathbf{k}$-points in the BZ, the more so, since they contain a sum (average) over all $\mathbf{k}$. Instead of doing the calculation on a dense mesh of points over the irreducible part of the BZ, the so-called *special $\mathbf{k}$-point theorem* [1] is often applied.

The average of the function $f(\mathbf{k})$,

$$\bar{f} = \frac{1}{V_{BZ}} \int_{V_{BZ}} f(\mathbf{k}) \, d^3k \qquad (7)$$

can be approximated by a weighted sum over special $\mathbf{k}$-points:

$$\tilde{f} = \sum_{\mathbf{q}} \omega_{\mathbf{q}} f(\mathbf{k}_{\mathbf{q}}). \qquad (8)$$

If $f(\mathbf{k})$ is invariant for translations by the reciprocal lattice vectors $\mathbf{G}_\mathbf{g}$, and to any symmetry operation $\hat{a}$ of the point group G of the lattice, then the Fourier expansion of $f(\mathbf{k})$ is

$$f(\mathbf{k}) = f_0 + \sum_{m=1}^{M} f_m A_m(\mathbf{k}) + \sum_{m=M+1}^{\infty} f_m A_m(\mathbf{k}), \qquad (9)$$

where

$$A_m(\mathbf{k}) = \frac{1}{n_G} \sum_{\hat{a} \in G} \exp[i\mathbf{k}(\hat{a}\mathbf{R}_\mathbf{m})] \qquad (10)$$

(n_G is the number of elements in the point group). Since for any m, the BZ-integral of $A_m(\mathbf{k})$ vanishes, $\bar{f} = f_0$. Substituting eq. (9) into eq. (8) gives

$$\tilde{f} = \sum_{\mathbf{q}} \omega_{\mathbf{q}} f_0 + \sum_{m=1}^{M} f_m \sum_{\mathbf{q}} \omega_{\mathbf{q}} A_m(\mathbf{k}_{\mathbf{q}}) + \sum_{m=M+1}^{\infty} f_m \sum_{\mathbf{q}} \omega_{\mathbf{q}} A_m(\mathbf{k}_{\mathbf{q}}), \qquad (11)$$

For high enough M the Fourier coefficients f_m become negligible, and the third term in eq. (11) vanishes. If the sum of the weighting factors is unity, the error of the approximation is determined by

$$\tilde{f} - \bar{f} \approx \sum_{m=1}^{M} f_m \sum_{\mathbf{q}} \omega_{\mathbf{q}} A_m(\mathbf{k}_{\mathbf{q}}). \qquad (12)$$

The special $\mathbf{k}$-points and their weighting factors should, therefore, be chosen to make $\sum_{\mathbf{q}} \omega_{\mathbf{q}} A_m(\mathbf{k}_{\mathbf{q}})$ vanish for all $m \leq M$. The error of the approximation diminishes almost exponentially with M.

Of course, the aim is to find a small enough set of special **k**-points with M as high as possible. A generally used procedure for generating such sets was given by Monkhorst and Pack [2]. For a given number Q, special **k**-points can be generated using the primitive unit vectors of the reciprocal lattice with the coefficients

$$q_i = \frac{2p_i - Q - 1}{2Q}; \qquad i = 1, 2, 3; \qquad p_i = 1 \ldots Q. \tag{13}$$

The appropriate weighting factors are $\omega_q = 1/Q^3$. The set generated this way is called a $Q \times Q \times Q$ MP set. (For lattices with symmetry lower than cubic, Q may be different in the three directions of the primitive reciprocal vectors $\mathbf{b}_i$.) Some of the resulting **k**-vectors may be equivalent with respect to the point group operations. Only one of each such group has to be involved in the calculation – with the weighting factor multiplied by the number of vectors in the group. In principle, a single **k**-point could also be applied [3] but usually 3 to 10 point sets are applied in case of small unit cells. As we will see, smaller **k**-point sets are sufficient for supercells.

3. Models of a Crystal with a Defect

If only a single point defect is introduced into an otherwise perfect crystal, the translational symmetry is, in principle, lost completely (see Fig. 1), and equations should be solved for all atoms of the solid. One could, in principle, take a smaller piece but the size must be chosen in such a way that there be at one least shell of atoms between the defect and the surface which remains undisturbed. The perturbations introduced by either source may have a range between 10 and 50 Å, so 2500 to 250000 atoms may be necessary. Obviously, the application of some kind of a simplified model is desirable. The obvious way is to separate the part of the crystal perturbed by the defect from the still crystalline background, and treat only the former in an explicit manner. On the one hand, the need for economizing calls for a model consisting of as few atoms as possible. On the other, the explicitly treated part of the crystal should, in principle, extend far enough for the effect of the defect ("effect defective": see Shakespeare: Hamlet!) to become negligible. That means, up to the point where

A: the amplitude of the localized wave functions

B: the deviations of charge density, and

C: the deviations of host atom positions relative to the unperturbed case

become close to zero. In practice, this is only assumed for a manageable group of atoms around the defect – usually called a CLUSTER. (The three criteria above are almost never satisfied to a necessary degree with clusters which can be squeezed into the computer. Therefore, at least, some kind of convergence test in all three respects are proper.) While the cluster is handled explicitly at any given level of calculation, the problem arises at its surface, i.e. with the representation of the rest of the material. As is shown schematically in Fig. 1, there are two basic possibilities. Section 3.1 deals with the supercell model, while Section 3.2 describes simple (classic) and more complicated (quantum) forms of embedding the cluster into the crystalline environment. Finally, Section 3.3 introduces the cyclic cluster model which is conceptually a supercell type approach but is technically more closely related to the simple molecular cluster model.

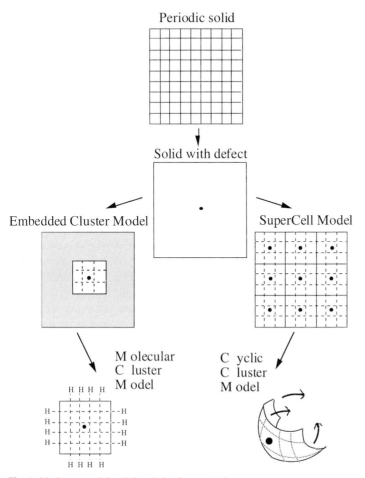

Fig. 1. Various models of the defective crystal

3.1 The supercell model (SCM)

The easy way out from loss of periodicity is to restore it in an artificial way, i.e., to construct a superlattice with the cluster as unit cell [4, 5]. A cluster chosen in shape of a unit cell (allowing repetition without gaps and overlaps) is called a *supercell*. In the supercell model (SCM), the momentum space description of the perfect crystals can be applied without modification as described in the previous section. It has to be considered, however, that the supercell – in terms of the perfect lattice – usually contains a multitude of primitive unit cells, so (see eq. (3)) the BZ corresponding to it in the reciprocal space is reduced (RBZ) relative to the primitive one (PBZ). In a "perfect" supercell (removing the defect from it), the vectors **K** of the RBZ represent a set of primitive **k**-vectors. If the superlattice has the same point group as the primitive one, the set $\{\mathbf{k}\}_0$ corresponding to the center of the RBZ, $\mathbf{K} = 0$, forms a special **k**-point set [6]. Therefore, with properly chosen SCM, a calculation for the sole point Γ ($\mathbf{K} = 0$) of the RBZ may give a good description of the extended (crystalline) states. For a diamond lattice, the sets represented by the center of the RZB in case of the 16 (f.c.c.), 32 (b.c.c.), 54 (f.c.c.), 64 (s.c.), 128 (f.c.c.), 216 (s.c.) atom supercells make the expression in eq. (12) vanish up to $M = 3, 5, 8, 7, 14$ and 16, respectively. However, if the defect is reintroduced into the supercell, the Γ-point is definitely no special point for the superlattice of defects and — unless the supercell is sufficiently big — the $\mathbf{K} = 0$ approximation might not be adequate. The SCM is often applied to defects in semiconductors. The neutral vacancy in silicon is a good test case for convergence. In a recent study [7] it was shown that the difference of the formation energy computed in the $\mathbf{K} = 0$ approximation and using a $2 \times 2 \times 2$ MP set changes in 32, 64, 128, and 216 atom supercell calculations by 3.05, 0.56, 0.30, and 0.04 eV, respectively. This shows the importance of testing the $\mathbf{K}$-point set in supercell calculations, especially for relatively small supercells. The necessity of using a good quality $\mathbf{K}$ set arises due to the interaction of defect wave functions in the superlattice (c.f. point A at the beginning of this section). This manifests itself also in the dispersion of the defect levels between various $\mathbf{K}$-points. It has to be noted, though, that the vacancy formation energy change less than 0.2 eV going from a 64 to a 216 atom supercell while the dispersion still remains substantial [8]. Usually, the level position of the isolated defect is determined in such cases by a tight binding fit.

The atoms on the boundary of the supercell should be kept fixed. This assumes that the criterion mentioned in point C at the beginning of the section, is assumed to be satisfied. In practice, however, the host atom relaxation is artificially limited by the size of the SCM, and this may influence the formation energy of defects directly. (Usually, only the relaxation of the first 1–2 neighbor shell influences the defect wave function considerably.) Again, for the neutral vacancy in silicon, a calculation using a $4 \times 4 \times 4$ MP set resulted in differences of 0.629, 0.385, 0.132, and 0.125 eV going from 16, 54, 64, 128 atom SCMs, respectively, to a 216 atom supercell [9]. The convergence with respect to allowed atomic relaxation is not always so favorable. Since even present day parallel supercomputers cannot handle more than about 200 to 300 atoms on a first-principles basis, the contribution of long-range relaxation beyond the SCM boundary to the total energy might be accounted for [10] by an independent term obtained from a Keating model [11]. This is essentially a set of imaginary host atoms around the supercell, connected by springs with force constants calculated for the perfect SCM. The relaxation of these atoms are determined self-consistently with the displacement of the atoms in the SCM. (In that case the SCM boundary atoms are also allowed to relax.) This approach should, however, be handled with caution. Since the space is, in principle, filled with the repetition of the supercell, the addition of an environment is conceptually unclean and, although the quantum mechanical and the classical forces are separated, the self-consistency between them may be problematic.

Criterion B is usually easy to meet by SCMs in the size range of 50 to 100 atoms in metallic or purely covalent solids, but may require very large SCMs in strongly ionic ones. Due to the long range of the Coulomb interaction, the defect may cause far reaching polarization effects which have to be accounted for in a self-consistent way. A "pia fraus", as with the long-range mechanical forces above, is not possible within the framework of the SCM. This constitutes the most severe limitation of its application. The SCM is most often applied to covalent and weakly ionic systems.

3.2 Embedded cluster models

When the periodicity is lost, the explicitly treated cluster around the defect should be embedded somehow into the undisturbed background. In case of a classical mechanics simulation (lattice or molecular dynamics), the application of a Keating model [11] – with atoms in the background arranged according to the perfect lattice, as described at the end of the previous subsection – is appropriate. In strongly ionic systems, where coupling between the cluster and the background is mainly by classical Coulomb interactions, the long-range forces from the background can be accounted for by a Madelung term. A higher level of approximation applies a shell model for the environment [12], i.e., places a set of ions with rigid positive cores and polarizable, negatively charged shells [13] around the cluster. Interaction between core and shell can be characterized by a spring constant. The parameters of this shell model, the charge and the spring constant can be determined, e.g., from self-consistent calculations on model systems [14]. If criterion B is not satisfied, however, charges in the embedding point charge model or in the shell model must be determined self-consistently with charges in the cluster [15].

Except for ionic solids, the interaction between the cluster and the background is quantum mechanical in nature, i.e., the electronic Schrödinger equation is to be solved. A rigorous mathematical basis for the separation of cluster and background is provided by Green-function techniques. There are essentially two possible ways: the perturbed crystal and the perturbed cluster approach.

3.2.1 Perturbed crystal approach

Let us assume that the defect introduces a perturbation U relative to the perfect crystal,

$$\hat{H} = \hat{H} + \hat{U} \tag{14}$$

and the perturbation is localized to the cluster. Using Green's operator of the perfect crystal,

$$\hat{G}_0(E) = (E - \hat{H}_0)^{-1} \tag{15}$$

in the Green's operator of the perturbed one

$$\hat{G}(E) = (E - \hat{H})^{-1} \tag{16}$$

Dyson's equation follows,

$$\hat{G}(E) = \hat{G}(E)_0 + \hat{G}_0(E)\,\hat{U}\hat{G}(E)\,, \tag{17}$$

which can be rewritten to

$$\hat{G}(E) = [1 - \hat{G}_0(E)\,\hat{U}]^{-1}\,\hat{G}_0(E)\,. \tag{18}$$

Therefore, if the Green function of the perfect crystal is known in any localized basis representation, and the matrix elements of the perturbation can be constructed on the same basis, the Green function of the perturbed crystal can be calculated. Density of states can be obtained directly from the trace of the imaginary part of the Green-function or, by rewriting the Schrödinger equation as

$$(H_0 + U)\,\psi = E\psi \rightarrow U\psi = (E - H_0)\,\psi \rightarrow [1 - \hat{G}_0(E)\,\hat{U}]\,\psi = 0 \tag{19}$$

the one-electron energies and wave functions can be determined after solving the secular equation of the matrix equation obtained from eq. (19) on the given basis [16].

The perturbed crystal approach looks like conceptually the cleanest and most sophisticated solution as far as the description of the crystalline background is concerned. The problems arise more with the cluster. The perturbation U should be constructed self-consistently and at the same level as the crystal potential. The perfect crystal problem can easily be solved on a plane wave basis but then the solutions must be transformed into Wannier functions in order to construct the perturbation. This is rather cumbersome. Alternatively, a localized basis set can be used for the perfect crystal in the first place. This may, however, limit the speed or accuracy of the calculation. The perturbed crystal approach becomes computationally exhaustive very quickly if the cluster has to be increased to fulfill criteria A and B or when the cluster undergoes reconstruction leading to substantial lowering of the symmetry [17]. Therefore, this method is most often used in metals where the screening effect of electrons is strong and the extent of atomic displacements is smaller.

3.2.2 Perturbed cluster approach

Since the use of a localized basis is advisable anyway, one could start with partitioning the matrix of the Hamiltonian in direct space according to basis functions in the cluster (C) and in the background (B)

$$H = \begin{pmatrix} H_{CC} & H_{CB} \\ H_{BC} & H_{BB} \end{pmatrix}. \tag{20}$$

Introducing the inverse of the Green function, Q by

$$QG = \begin{pmatrix} Q_{CC} & Q_{CB} \\ Q_{BC} & Q_{BB} \end{pmatrix} \begin{pmatrix} G_{CC} & G_{CB} \\ G_{BC} & G_{BB} \end{pmatrix} = \begin{pmatrix} I_{CC} & 0 \\ 0 & I_{BB} \end{pmatrix}. \tag{21}$$

It follows [18] that

$$Q_{CC} G_{CC} - Q_{CB} G_{BB} Q_{BC} Q_{CC}^{-1} = I_{CC} \tag{22}$$

from which the Green function of the perturbed cluster can be expressed as the sum of the Green function of the unembedded cluster and a corrective term

$$G_{CC} = Q_{CC}^{-1} + Q_{CC}^{-1} Q_{CB} G_{BB} Q_{BC} Q_{CC}^{-1} \equiv Q_{CC}^{-1} + \Delta G_{CC}. \tag{23}$$

The fundamental embedding assumption is that G_{BB} can be replaced by the matrix valid for the crystal without defect. Although this holds for Q_{BB}^{-1}, is not necessarily true for G_{BB}. By integration of the second term in eq. (23), the correction to the density can be obtained

$$\Delta \varrho_{CC} = \lim_{\eta \to 0^+} -\frac{1}{\pi} \operatorname{Im} \int_{-\infty}^{E_F} dE \, \Delta G_{CC}(E + i\eta). \tag{24}$$

From a practical point of view, an energy dependent embedding matrix [19]

$$M_{CC}(E) = \theta(E - E_F) \int_{-\infty}^{E_F} d\varepsilon \, \frac{\varrho_{CC}^0(\varepsilon)}{\varepsilon - E} - \theta(E_F - E) \int_{E_F}^{\infty} d\varepsilon \, \frac{\varrho_{CC}^0(\varepsilon)}{\varepsilon - E} \tag{25}$$

has to be calculated and stored, where $\varrho_{CC}^0(\varepsilon)$ is the projected density of states in the cluster region for the crystal without the defect

$$\varrho_{CC}^0(\varepsilon) = \frac{1}{V_{BZ}} \sum_l \int_{V_{BZ}} d\mathbf{k}\, a_{\mu l}^*(\mathbf{k})\, a_{\nu l}(\mathbf{k})\, \delta[\varepsilon - \varepsilon_l(\mathbf{k})]. \qquad (26)$$

The expansion coefficients $a_{\nu l}(\mathbf{k})$ have to be taken from subspace C.

In the perturbed cluster approach the fulfillment of criteria A, B and C is paramount but that still does not warrant the fundamental embedding assumption. This may be improved by a strictly variational partitioning between the C and B regions [20]. Although the perturbed cluster approach requires an elaborate and complicated computational scheme, for charged defects in moderately ionic solids, or for defects with substantial reconstruction in metals it represents a solution with perspective.

3.2.3 The molecular cluster model (MCM)

The simplest way of (quantum mechanical) embedding is based on the assumption that the electronic states of some solids can well be described by localized bonds. (The set of canonical one-electron orbitals may, in principle, be transformed into a set of bonds, and lone pair orbitals or dangling bonds which are, to a high degree, localized to two and one center, respectively.) This is mostly the case for semiconductors. Ideally, the quantum mechanical interaction of the cluster with the background is completely represented by the localized bonds crossing the interface between the two regions. If these bonds could be kept unchanged while cutting off the atoms of the background, the embedding would be still perfect (save for long range Coulomb interactions which can be treated as mentioned at the beginning of Section 3.2). To achieve that, the dangling bonds on the cluster surface have to be saturated in such a way that the new bonds become equivalent with the original bonds of the crystal. In principle, one can construct monovalent pseudoatoms with parameters fitted to restore the original bonding. Since the fitting is never perfect, in most cases simply hydrogen atoms are used. The hydrogen saturated cluster of atoms is a molecule in the chemical sense (e.g., five atoms of diamond saturated with hydrogen atoms form the molecule neopentane, C_5H_{12}), therefore this model is often called the molecular cluster model (MCM). The MCM evolved from the *defect molecule model* [21] as the first attempt for embedding the localized orbitals of the defect and its first neighbor shell into an atomistically treated background. Its popularity, ever since its introduction (to my knowledge by Messmer and Watkins [22]), is unbroken thanks to its absolute simplicity and to the fact that any standard computer code for solving the Schrödinger equation of a molecule can be applied. (In fact, up to now, it is the MCM only which allows the application of real many-body theories.)

The basic assumption of the MCM is reasonable for valence band states. Assuming that the size of the MCM is big enough for criteria A, B and C to be satisfied, the geometry and energy of formation for defects can, therefore, be calculated to a good accuracy. (In fact, as mentioned in Section 3.1, even without criterion A being entirely satisfied.) Recent studies [23] on the silicon vacancy show the energy of formation to be convergent within 0.15 eV at the MCM size $Si_{122}H_{100}$ (9 shells of Si neighbors). Comparing this with the difference of 0.125 eV, going from a 128 to a 216 atom supercell [9], see Section 3.1, it can be seen, that the convergence properties of the MCM are not worse in this respect than that of the SCM (surface/defect interaction versus defect/defect interaction). The problem of the MCM arises from the fact that conduction band states (or defect states derived from them) are inherently not of localized nature, i.e. cannot be localized to pairs of atoms. Therefore, the "embedding" idea of the MCM does not work for those states, and calculated quantities like the gap converge to the perfect crystal value very slowly [24], reaching convergence only as the size of the MCM approaches the limit between real clusters and solids (c.f. the definition given in the Introduction). Also the one-electron energies related to defect orbitals, which are dominated by conduction band admixture, converge rather poorly, allowing only estimations to be made for their position in the gap [25]. There are some additional disadvantages. The point group symmetry is reduced even without introducing the defect, and the uniformity of the sites is lost. Unless the size of the MCM is really big relative to the defect, the environment of an impurity in different sites may change considerably. This makes comparison of total energies in different configurations problematic. The bonds to hydrogen atoms create a dipole layer on the surface of the MCM, shifting ionization energies and heats of formations in a manner which is hard to account for [26]. Also, this dipole layer induces charge transfer between concentric shells of the cluster, even if it consists of nominally equivalent atoms. This effect can be somewhat relieved by tuning the X–H (X: host atom) distances [27] but eliminating it requires considerable increase in MCM size.

With all its limitation though, MCM is likely to stay for covalent solids, especially as the computer capacities increase.

3.3 The cyclic cluster model (CCM)

Getting rid of the problems (lack of appropriate point group symmetry, inequivalence of sites, surface dipole induced charge transfer) of the MCM while preserving its simple (direct space) approach, could easily be achieved by applying the *cyclic boundary condition* of Born and von Kármán directly *to the cluster*. This can be visualized as tying one end of the cluster to the other (see Fig. 1). In practice, this is achieved by modifying the interatomic distance and direction cosine matrices of each atom. Early attempts to introduce such "periodic" (a misnomer!) cluster models [28, 29] did not achieve better convergence properties, though. One has to realize that – despite of the direct (not momentum) space approach – such models should be related to the supercell model [30]. A cluster in the form of a unit cell (a requirement for cyclic boundary conditions) is namely a supercell, however, applying the cyclic boundary conditions (6) directly to the cluster (without assuming the periodicity condition (5)) means a crystal consisting of a single unit cell, i.e. $N_i = 1$. Consequently, the RBZ contains only one point: $\mathbf{K} = 0$. As mentioned in Section 3.1, to get a good model of the defective solid in a $\mathbf{K} = 0$ supercell approximation, the $\{\mathbf{k}\}_0$ set ($\mathbf{k}$-vectors in the PBZ corresponding to $\mathbf{K} = 0$ in the RBZ) should make a good quality special $\mathbf{k}$-point set and criterion A must be satisfied. The *cyclic cluster model* (CCM) is understood [31] as cyclic boundary conditions applied to a supercell chosen to meet these conditions. The CCM is, however, not identical with an SCM in $\mathbf{K} = 0$ approximation! The fact that the "crystal" consists of a single supercell restricts the interactions to within the supercell [31, 32]. On the one hand, of course, this limits the accuracy of reproducing extended states but, on the other, it excludes most of the artificial defect-defect interactions of the SCM. (I recall that in

Table 1
Comparison of results obtained on a 32 atom CCM to those of a full band structure calculation [33] for diamond

symm. points	full band struct. calc. [33] spd basis	C_{64} [32] sp basis
Γ_1	−21.68	−21.21
$\Gamma_{25'}$	0	0
$\Gamma_{15'}$	5.59	5.60
$\Gamma_{2'}$	13.21	13.55
X_{1v}	−12.90	−12.66
X_{4v}	− 6.43	− 6.36
X_{1c}	4.65	5.34
X_{4c}	16.87	16.49
$L_{2'}$	−15.79	−15.52
L_{1v}	−13.73	−13.25
$L_{3'}$	− 2.86	− 2.92
L_3	8.47	9.13
L_{1c}	8.90	9.45

SCM a set of **K**-points – instead of the sole **K** = 0 – is only required due to the interaction of the artificially repeated defects!) Table 1 compares the one-electron eigenvalues obtained for the perfect 32 atom CCM with the band structure of diamond using the same method for solving the Schrödinger equation, but with an sp basis versus an spd one. As can be seen, the agreement is good (the difference in the width of the indirect gap is mainly due to the lack of d orbitals), so the CCM is a well balanced approximation unifying the advantages of SCM (full point group symmetry, homogeneous environment for the defect, no surface states or dipole layer, good description of the extended states) and MCM (computational speed, simple direct space approach, no defect–defect interactions). It has to be noted that in the CCM there is no dispersion of defect levels to monitor criterion A. However, if the localization of the defect wave function does not decay smoothly toward the cluster boundary (showing an increase on the last shell), that is a sure sign of the cluster being too small.

4. Generalization

The models described in Section 3. were primarily developed for defects in bulk crystals. However, these are – with some modifications – also the ones most often used for surfaces and interfaces in crystals, as well as in disordered solids.

4.1 Surfaces and interfaces

The models used in present day atomistic simulations of surfaces and interfaces differ mainly in the amount of periodicity. In case of surface superstructures or pseudomorphic interfaces the two-dimensional periodicity can easily be extended artificially with a periodicity perpendicular to the interface. The periodically repeated N+M layer slab can be treated entirely in momentum space, i.e., it is essentially a supercell model. Care has to be taken to achieve convergence regarding N and M (in case of free surfaces M layers of vacuum has to be added). As described in Section 3.1 the SCM works also for local defects, i.e., a slab model can also applied if the parallel-to-surface periodicity is also artificial. In that case, of course, size convergence within the plane of the interface should also be considered. Naturally, a slab-CCM can also be applied.

Alternatively, a cluster with a free surface could be embedded into a semi-infinite crystalline solid. Green-function techniques can be used for the two-dimensionally periodic surface itself [38], however, the solutions for a semi-infinite crystal themselves are much too complicated for use as an embedding medium for local surface phenomena. Therefore, if classical embedding schemes (Keating model, set of point charges, shell model) will not do, usually an MCM with a surface is applied. If quantum mechanical embedding is unavoidable, the critical part of the surface is embedded in a larger MCM (cluster in cluster embedding). A better compromise is the usage of hydrogen saturators on a two-dimensionally periodic slab, i.e., a mixture of SCM and MCM tactics.

4.2 Disordered solids

In lack of long-range order, the traditional way of modeling was the embedding in an effective medium represented by an average potential or scattering matrix. Since these methods do not use any detailed structural information about the environment (beyond the radial distribution or pair correlation function), they cannot be regarded as "atomistic" simulations. A step in this direction is the embedding into a Bethe lattice (see e.g. ref. [34]). The Bethe lattice is an infinite tree-like sequence of branches and nodes without closed rings of atoms. It provides appropriate first neighbor coordination for any number of atoms in the environment.

It has to be noted that not only the structure but also the properties of disordered materials are determined by short-range order. Therefore, the criteria mentioned at the beginning of Section 3 are easier to satisfy with relatively small clusters than in the case of crystals. In recent years, the increase in computer capacity allowed the explicit treatment of sufficiently big systems without embedding. Only the surface states have to be gotten rid of and that may be achieved by the application of cyclic boundary conditions (as in CCM but with no consideration to $\{k_0\}$ selection). Using molecular dynamics, Monte Carlo or reverse Monte Carlo methods suitable models can be found which are in thermodynamic equilibrium and conform to measured structural information [35 to 37].

References

[1] D. J. CHADI and M. L. COHEN, Phys. Rev. B **8**, 5747 (1973).
[2] H. J. MONKHORST and J. D. PACK, Phys. Rev. B **13**, 5188 (1976).
[3] A. BALDERSCHI, Phys. Rev. B **7**, 5212 (1973).
[4] R. P. MESSMER and G. D. WATKINS, Inst. Phys. Conf. Ser. **16**, 255 (1973).
[5] S. G. LOUIE, M. SCHLÜTER, J. R. CHELIKOWSKY, and M. L. COHEN, Phys. Rev. B **13**, 1654 (1976).
[6] R. A. EVARESTOV and V. P. SMIRNOV, phys. stat. sol. (b) **99**, 463 (1980).
[7] M. J. PUSKA, S. PÖYKKÖ, M. PESOLA, and R. NIEMINEN, Phys. Rev. B **58**, 1318 (1998).
[8] S. J. CLARK and G. J. ACKLAND, Phys. Rev. B **56**, 47 (1997).
[9] A. ZYWIETZ, J. FURTHMÜLLER, and F. BECHSTEDT, phys. stat. sol. (b) **210**, 13 (1998).
[10] M. NEEDELS, J. D. JOANNOPOULOS, Y. BAR-YAM, and S. T. PANTELIDES, Phys. Rev. B **43**, 4208 (1991).
[11] P. N. KEATING, Phys. Rev. **145**, 637 (1966).
[12] A. B. KUNZ and D. L. KLEIN, Phys. Rev. B **17**, 4614 (1978).
[13] N. F. MOTT and M. J. LITTLETON, Trans. Faraday Soc. **34**, 485 (1938).
[14] YE LI, D. C. LANGRETH, and M.R. PEDERSON, Phys. Rev. B **55**, 16456 (1997).

[15] M. R. Hayns and L. Dissado, Theoret. Chim. Acta **37**, 147 (1975).
[16] S. T. Pantelides, Rev. Mod. Phys. **50**, 797 (1978).
[17] P. J. Kelly and R. Car, Phys. Rev. B **45**, 6543 (1992).
[18] A. J. Fisher, in: Qantum Mechanical Cluster Calculations in Solid State Studies, Eds. R. W. Grimes, C. R. A. Catlow and A. L. Schluger, World Scientific Publ. Co., Singapore 1992 (p. 47).
[19] C. Pisani, R. Orlando, and R. Nada, ibid (p. 117).
[20] U. Gutdeutsch, U. Birkenheuer, and N. Rösch, J. Chem. Phys. **109**, 2056 (1998).
[21] C. A. Coulson and M. J. Kearsley, Proc. Roy. Soc. A **241**, 433 (1957).
[22] R. P. Messmer and G. D. Watkins, Rad. Eff. Def. Solids **9**, 9 (1971).
[23] S. Ögüt, H. Ch. Kim, and J. R. Chelikowsky, Phys. Rev. B **56**, R 11353 (1997).
[24] A. Zunger, J. Phys. C **7**, 76 (1974).
[25] A. Resende, R. Jones, S. Öberg, and P. R. Briddon, Phys. Rev. Lett. **82**, 2111 (1999).
[26] P. Deák, L. C. Snyder, R. K. Singh, and J. W. Corbett, Phys. Rev. B **36**, 9612 (1987).
[27] S. Estreicher, Phys. Rev. B **37**, 858 (1988).
[28] A. J. Bennett, M. McCaroll, and R. P. Messmer, Phys. Rev. B **3**, 1397 (1971).
[29] A. Zunger, J. Phys. C **7**, 96 (1974).
[30] P. Deák, J. Kazsoki, and J. Giber, Phys. Rev. B **66**, 395 (1978).
[31] P. Deák and L. C. Snyder, Phys. Rev. B **36**, 9619 (1987).
[32] J. Miró, P. Deák, C. P. Ewels, and R. Jones, J. Phys. Condensed Matter **9**, 9555 (1997).
[33] G. B. Bachelet, H. S. Greenside, G. A. Baraff, and M. Schlüter, Phys. Rev. B **24**, 4745 (1981).
[34] M. F. Thorpe, in: Excitations in Disordered Systems, Ed. M. F. Thorpe, Plenum Press, New York, 1982 (pp. 85 to 107).
[35] S. R. Elliott, Physics of Amorphous Materials, 2nd ed., J. Wiley & Sons, New York 1990 (pp. 151 to 169).
[36] S. Kugler, L. Pusztai, L. Rosta, P. Chieux, and R. Bellisent, Phys. Rev. B **48**, 7685 (1993).
[37] B. R. Djordjevic, N. F. Thorpe, and F. Wooten, Phys. Rev. B **52**, 5685 (1995).
[38] H. L. Skriver and N. M. Rosengard, Phys. Rev. B **43**, 9538 (1991).

phys. stat. sol. (b) **217**, 23 (2000)

Subject classification: 71.15.–m; S5

The Art and Science of an Analytic Potential

D. W. Brenner

Department of Materials Science and Engineering, North Carolina State University, Raleigh, North Carolina, 27695-7907, USA

(Received August 10, 1999)

Two aspects of the development of an effective interatomic potential energy function are discussed. The first is the derivation of sound functional forms that are motivated by quantum-mechanical bonding principles. The second aspect is the development of empirical corrections and fitting parameters that are often necessary to make a potential function practical for specific applications. An analytic bond-order function for carbon is used as an example.

1. Introduction

Analytic potential energy functions (sometimes referred to as empirical or classical potentials) are simplified mathematical expressions that attempt to model interatomic forces arising from the quantum mechanical interaction of electrons and nuclei. Their use is generally necessitated either by the desire to model systems with sizes and/or timescales that exceed available computing resources required for quantum calculations, or to gain qualitative insight into things like bonding preferences that may not be immediately obvious from the results of numerical calculations.

Atomistic simulation utilizing analytic potential energy expressions has a long and successful history in many areas of science. The earliest example of this type of simulation is probably the seminal work of Hirschfelder, Eyring and Topley [1], who in 1936 modeled the hydrogen exchange reaction using an analytic interatomic force expression and classical trajectories. Simulations of condensed-phase phenomena such as ion–solid collisions [2] and liquid dynamics [3] using continuous (rather than impulsive) interatomic forces were first carried out in the early 1960's. Extensive biological applications were then developed over the following two decades, see e.g. [4]. Currently, analytic potentials are typically used for applications in which collective phenomena requiring many atoms or long times is to be studied with atomic resolution [5].

While several "standard" potential functions have emerged for particular classes of systems (e.g. the embedded-atom method for close-packed metals [6, 7]), at present there is no definitive functional form that adequately describes all types of multi-atom bonding. Instead, potentials are often developed for specific applications with functions and parameters determined on an *ad hoc* basis. This process leads to considerable and justified uncertainty with regard to the reliability of quantitative results produced by analytic potentials.

To be effective, an analytic potential energy function must possess the following critical properties.

1. *Flexibility:* A potential energy function must be sufficiently flexible that it accommodates as wide a range as possible of fitting data. For solid systems, this data might

include crystalline lattice constants, cohesive energies, elastic properties, vacancy formation energies, and surface energies.

2. *Accuracy:* A potential should be able to accurately reproduce an appropriate fitting data base.

3. *Transferability:* A potential function should be able to describe at least qualitatively, if not with quantitative accuracy, structures not included in a fitting data base.

4. *Computational efficiency:* Evaluation of the function should be relatively efficient depending on quantities such as system sizes and timescales of interest, as well as available computing resources.

Often times criteria 1 and 2 are emphasized in the development of an analytic potential, with the assumption that these will lead to transferability. Unfortunately, it is often the case, especially with *ad hoc* functional forms, that the opposite occurs. As more arbitrary fitting parameters are added, functions may lose significant transferability. Careful examination of the literature suggests that analytic potential functions with the highest degree of transferability are those based on sound quantum-mechanical bonding principles, and not necessarily those with the most parameters. The trick to producing an effective potential therefore is balancing sound functional forms with any necessary empirical parameter fitting.

The intent of this paper is to present two sides of the development of an analytic interatomic potential energy expression using a bond-order potential for carbon as an example. The first side, which is discussed in Section 3, is the development of a relatively simple functional form that captures the essence of quantum mechanical bonding. The second side of potential development, which is discussed in Section 4, centers on making the potential practical by incorporating additional empirically-derived functions and parameters. These are fit to specific bonding structures that may not be satisfactorily captured by the underlying formalism. This process often requires some level of intuitive chemical insight, considerable trial-and-error, and significant tenacity. Because there is no rigorous methodology in this process with which all systems can be fit, it is often referred to as an "art".

2. The Abell-Tersoff Analytic Bond-Order Potential

The analytic bond-order potential energy formalism discussed in this paper was originally introduced by Abell [8]. This expression models the local attractive electronic contribution to the binding energy E_i^{el} of an atom i using an interatomic bond-order that modulates a two-center interaction,

$$E_i^{\text{el}} = -\sum_{j(\neq i)} b_{ij} V^{\text{A}}(r_{ij}), \tag{1}$$

where the sum is over nearest neighbors j of atom i, b_{ij} is the bond-order function between atoms i and j, $V^{\text{A}}(r)$ is the pair term, and r is the scalar distance between the atoms. The function $V^{\text{A}}(r)$, which represents bonding from valence electrons, is assumed to be transferable between different atomic hybridizations. All many-body effects such as changes in the local density of states with varying local bonding topologies are included in the bond-order function. Abell suggested that the major contribution to the bond-order function is local coordination z, and using a Bethe lattice derived the approximation $b = z^{-1/2}$. By balancing the attractive local bonding contributions with a pair sum of repulsive interactions, Abell was able to show that the wide range of stable bonding configurations (e.g. close-packed versus molecular solids) can be rationalized by different ratios of slopes of the repulsive to attractive pair terms, while maintaining the approximate bonding universality suggested by Smith and coworkers [9]. Assuming exponentials for the repulsive and attractive pair interactions, the analytic interatomic potential energy form for the cohesive energy E_{coh} of a collection of atoms becomes

$$E_{\mathrm{coh}} = \sum_i E_i; \qquad E_i = \sum_{j(\neq i)} \left[A\, e^{-\alpha r_{ij}} - b_{ij} B\, e^{-\beta r_{ij}} \right], \qquad (2)$$

where E_i is the binding energy of individual atoms.

A practical implementation of Abell's bond-order formalism was developed by Tersoff [10, 11] for modeling group IV materials. He introduced an empirical functional form for the bond-order that incorporates angular interactions while still maintaining coordination as the dominant feature determining structure. With his empirical modification of Abell's functional form, Tersoff was not only able to stabilize the diamond lattice against shear, but he was also able to obtain a reasonable fit to elastic constants and phonon frequencies for silicon, germanium, carbon and their alloys with just a few parameters [10, 11]. Using data from density-functional calculations, Tersoff also showed that a single set of exponential functions for two-center attractive and repulsive terms can provide a reasonable quantitative fit to bond lengths and energies for silicon for coordinations ranging from one (the diatomic) to twelve (a face-centered cubic lattice). A subtle but crucial feature of Tersoff's bond-order function is that it did not assume different forms for the angular terms for different hybridizations (e.g. a lowest energy tetrahedral angle for fourfold coordination) as is done in more traditional valence-force expressions [12]. Instead, it uses an angular function that is determined by a global fit to structures with various coordinations. This feature, together with a physically-motivated functional form as discussed in the following section, provides the function with an extraordinary degree of transferability.

The Abell-Tersoff bond-order formalism has been analyzed in terms of the behavior of potential barriers for chemical reactions. It was concluded that it satisfies the correct trends relating barrier position and height relative to reaction exothermicity [13]. This feature was exploited in the development of simplified bond-order functions for modeling molecular solids that were used to simulate shock-induced phase transitions [14] and detonation propagation [15, 16].

A variety of other empirical bond-order expressions have been developed since the Tersoff form was introduced. These include improved forms for modeling silicon [17 to 20], potentials for other covalently-bonded systems such as GaAs [20] and SiN [21, 22], and forms for molecular structures that include hydrogen [23 to 27]. There has also been recent progress in incorporating weak non-bonded interactions within the bond-order formalism [28, 29]. All of these extensions of the original Abell-Tersoff form significantly increase the range of systems and types of processes that can be modeled with this approach.

3. The Science of an Analytic Potential

An analysis of the quantum-mechanical basis of the Abell-Tersoff bond-order potential is discussed in this section. The approach followed differs from the original derivation of Abell [8] by drawing heavily upon analyses from Foulkes and Haydock [30], Sutton and Balluffi [31], Finnis and Sinclair [32] and others [33]. While there are undoubtedly other routes to this bond-order expression, this analysis has the advantage that it starts with first-principles concepts, and passes through several other well-established potential energy expressions before arriving at a bond-order potential energy expression. For a similar, but somewhat more complete treatment of the quantum-mechanical basis of analytic potentials, the reader is referred to Ref. [33].

Density functional theory (DFT) shows that for an interacting system of electrons with a non-degenerate energy in an external potential, the ground-state electronic energy is a unique functional of the electron density [34]. For most cases of interest, the external potential is simply the Coulomb field due to the nuclei. The electronic energy E^{DF} in DFT is therefore given by

$$E^{DF}[\varrho(r)] = T[\varrho(r)] + \int [\varrho(r)\varrho(r')/(|r-r'|)]\,dr\,dr' + \int V_N(r)\,dr + E_{xc}[\varrho(r)], \tag{3}$$

where the first term of the right-hand side of Eq. (3) is the kinetic energy of a noninteracting electron gas with a density $\varrho(r)$, the second term is the classical electrostatic potential of the electron gas, $V_N(r)$ is the potential energy due to the positive nuclei, and E_{xc} is the exchange-correlation functional.

DFT also demonstrates that the electronic energy is minimized by the correct electron density for a given potential. This variational principle of DFT leads to a system of one-electron equations that can be self-consistently solved to obtain a ground-state energy [35]. These Kohn-Sham one-electron equations can be written as

$$[T + V_H(r) + V_N(r) + V_{xc}(r)]\,\phi_i^{K-S} = \varepsilon_i\,\phi_i^{K-S}, \tag{4}$$

where ϕ_i^{K-S} are the one-electron Kohn-Sham orbitals, ε_i are the eigenenergies of these orbitals, $V_H(r)$ is the Hartree potential, and $V_{xc}(r)$, called the exchange-correlation potential, is the functional derivative of the exchange-correlation energy. When solved self-consistently, the total electronic energy can be calculated from Eq. (3) using the electron densities obtained from Eq. (4). This expression takes into account exchange and correlation interactions, and requires no approximations other than the form of the density functional.

Rather than using Eq. (3) to obtain the electronic energy, the expression

$$E^{K-S}[\varrho(r)] = \sum_k \varepsilon_k - \int \varrho^{sc}[V_H(r)/2 + V_{xc}(r)]\,dr + E_{xc}[\varrho^{sc}(r)] \tag{5}$$

is usually used, where ϱ^{sc} is the self-consistent electron density and ε_k are the eigenvalues of the Kohn-Sham orbitals. The integral on the right-hand side of Eq. (5), corrects for the fact that the eigenvalue sum includes the exchange-correlation energy and double-counts the electron–electron Coulomb interactions.

One of the strengths of density functional theory is that the error in electronic energy is second order in the difference between any given electron density and the true ground-state density. Based on this property, it is expected in principle that reasonably good estimates for electronic energies can be obtained without having to iterate to a fully self-consistent solution. An obvious procedure would be to construct the Kohn-Sham orbitals from a given input charge density, and then sum the occupied Kohn-Sham orbital energies and calculate the double-counting terms in Eq. (5) from the output density given by these orbitals. This process, however, has two difficulties. First, while summing the orbital energies is straightforward, calculating the double-counting terms can be computationally intensive depending on the system size and basis set. Second, the upper limit of the energy value is often not sufficiently close to the self-consistent value to be useful.

Harris [36] as well as Foulkes and Haydock [30] demonstrated that the electronic energy calculated from a single iteration of the energy functional

$$E^{\rm H}[\varrho^{\rm in}(r)] = \sum_k \varepsilon_k^{\rm out} - \int \varrho^{\rm in}[V_{\rm H}(r)/2 + V_{\rm xc}(r)]\,{\rm d}r + E_{\rm xc}[\varrho^{\rm in}(r)] \qquad (6)$$

is also second-order in the error in change density. This expression is generally referred to as the Harris (or sometimes Harris-Foulkes) functional. It is important to note that while the Kohn-Sham orbital energies $\varepsilon_k^{\rm out}$ are still calculated, the double counting terms involve only the *input* charge density and not the density given by these orbitals. While the correct input electron density yields the correct ground-state energy, the Harris functional is not variational; it can give an energy either higher or lower than the true ground-state energy. In practice it has been found that with a judicious choice of input density this expression can produce results that match fully self-consistent calculations better than the variational upper bound produced by a full single step of the Kohn-Sham procedure [37 to 40].

With the Harris functional the input electron density can be chosen to simplify the calculation of the double counting terms, and the calculation does not require self-consistency. As demonstrated by Foulkes and Haydock [30], these features can be used to derive a typical tight-binding total-energy expression. Tight-binding expressions give the total energy $E_{\rm tot}$ for a system of atoms as a sum of eigenvalues of a set of occupied non-self-consistent one-electron molecular orbitals plus some additional analytic function of relative atomic distances. A pair additive sum over atomic distances is often assumed for the analytic function, leading to the total energy expression

$$E_{\rm tot} = \sum_i \sum_{j(\neq i)} \theta(r_{ij}) + \sum_k \varepsilon_k\,, \qquad (7)$$

where r_{ij} is the scalar distance between atoms i and j, $\theta(r)$ is the pair-additive interatomic interaction, and ε_k are the energies of the occupied orbitals. The simplest and most widespread tight-binding expressions use eigenenergies from a wavefunction expanded in an orthonormal minimal basis of short-range atom-centered orbitals with parameterized two-center Hamiltonian matrix elements.

Justification for the tight-binding energy expression above can be understood by analyzing the Harris functional [30]. First, the use of non-self-consistent one-electron molecular orbitals assumed in tight-binding expressions is justified if they correspond to the Kohn-Sham orbitals constructed from the input charge density in the Harris functional. Second, if the input electron density is approximated by a sum of overlapping, atom-centered spherical orbitals, then it can be shown that the double-counting terms in the Harris functional are given by

$$\sum C_{\rm a} + \tfrac{1}{2} \sum_i \sum_{j(\neq i)} U_{ij}(r_{ij}) + U_{\rm np}\,, \qquad (8)$$

where C_a is a constant intra-atomic energy, $U_{ij}(r_{ij})$ is a short-range pair-additive energy that depends on the scalar distance r_{ij} between atoms i and j, and U_{np} is a nonpair-additive contribution that comes from the exchange-correlation functional. If the regions where overlap of electron densities from three or more atoms are small, Foulkes and Haydock have shown that U_{np} is well approximated by a pair sum that can be added to U_{ij} [30]. Hence the assumption of pair-additivity for the analytic function in Eq. (7) is justified. Finally, spherical atomic orbitals lead to the simple form

$$V_{xc}(r) = \sum_i V_i(r) + U(r) \tag{9}$$

for the one-electron potential needed to calculate the orbital energies in the Harris functional. The function $V_i(r)$ is an additive atomic term that includes core electrons as well as Hartree and exchange-correlation potentials, and $U(r)$ arises from nonlinearities in the exchange-correlation functional. Although not two-centered, the contribution of the latter term is relatively small. Thus, the use of strictly two-center matrix elements in the tight-binding Hamiltonian is also justified.

The atomic binding energy arising from the orbital energies in a tight-binding expression can be further simplified using straightforward quantum-mechanical bonding ideas. In a simple linear combination of atomic orbitals/molecular orbital picture, the energy of atomic orbitals lose their degeneracy as atoms are brought together and molecular orbitals form. Roughly half of the molecular orbitals have energies lower than the atomic orbitals on isolated atoms, and roughly half have energies greater than the isolated atomic orbitals. Using Hund's rule, the orbitals with the lowest energies are occupied first, which, assuming that all states are not occupied, leads to a lowering of electronic energy. This energy lowering contributes to the formation of chemical bonds.

For solids, the broadening of atomic orbital energies as atoms are brought together results in energy bands. These bands can be described using a density of states $D(e)$ defined as

$$D(e) = \sum_k \delta(e - \varepsilon_k), \tag{10}$$

where ε_k are orbital energies, δ is the delta function, and the sum is over both occupied and unoccupied orbitals. The function $D(e)$ gives the number of states with energies in the range $e + de$. The electronic energy arising from the occupied orbitals that comprise an energy band can be written as the integral

$$E_{tot}^{el} = \int D(e)\, e\, de \tag{11}$$

whose upper limit is the Fermi energy.

Like any distribution, the shape of the density of states can be described by its moments about some energy value. These moments can be conveniently chosen about the energy of the atomic orbitals ε^{atomic} from which the molecular orbitals are formed so that the nth moment of the density of states is given by

$$\mu^n = \sum_k (e - \varepsilon^{atomic})^n D(e). \tag{12}$$

With this definition (and neglecting charge transfer), the first moment of the distribution μ^1 is zero. The second moment μ^2 describes the width of the density of states. The third and fourth moments are related to the degree of skewness about the center of the distribution and the tendency to form a gap in the middle of the distribution, respectively. Because the binding energy relative to the free atoms comes primarily from the spread in orbital energies, it is reasonable to assume that the energy should be most closely related to the second moment of the density of states. Indeed, it has been shown that for many simple systems (e.g. neglecting charge transfer) the orbital energy using Eq. (11) is roughly proportional to the square root of the second moment of the DOS [32, 33]. Therefore, for these systems it is possible to calculate the electronic bond energy without having to explicitly calculate either the orbitals or higher order moments of the density of states. Instead, only the second moment of the density of states and the slope of the line relating the square root of the second moment to the integral of the occupied orbital energies needs to be known.

The electronic energy of individual atoms E_i^{el} in a solid can be calculated by applying the moments idea to the local density of states $d_i(e)$ by defining this energy as

$$E_i^{el} = \int d_i(e)\, e\, \mathrm{d}e, \tag{13}$$

where the upper limit of the integral is the Fermi energy. With this definition, the global density of states is recovered by summing all of the local densities of states

$$D(e) = \sum_i d_i(e), \tag{14}$$

and the total electronic energy is the sum of the electronic energies associated with the individual atoms. According to the discussion above, an electronic bond energy can be determined for each individual atom if the second moment of the local density of states and the equation of the line giving the relationship between the square root of this value and the energy are known.

The moments concept would not be of much use if all of the eigenvalues and eigenvectors of the molecular orbitals in a tight-binding calculation had to be determined to obtain the expansion coefficients needed to find the local densities of states. Fortunately, the moments of the local density of states can be related to the bonding topology through the moments theorem without having to explicitly calculate molecular orbitals [41, 42]. This theorem states that the *n-th moment of the local density of states on an atom i is determined by the sum of all paths between n neighboring atoms that start and end at atom i.* To obtain an exact local density of states for a given atom with this theorem, all of the moments, and therefore all of the possible paths starting and ending at that atom, have to be determined. This quickly becomes a non-trivial exercise as higher moments are calculated. However, as discussed above, it has been demonstrated that knowing only the second moment can still often give a good estimate of the bond energy. The second moment is related to the loops beginning and ending on a given atom requiring only two "hops", which is simply the number of nearest neighbors z. Therefore, it can be concluded from this analysis that the local electronic bond energy for each atom arising from the molecular orbitals is approximately proportional to the square root of the number of neighbors

$$E_i^{el} \propto z^{1/2}. \tag{15}$$

This result is called the second-moment approximation. A resemblance between this expression and the relationship between the bond-order and coordination derived by Abell is beginning to emerge.

For regular solids nearest neighbors are determined unambiguously. For disordered systems and many defects, the range of possible bond lengths makes defining which atoms are nearest neighbors less clear. Therefore, to develop a practical analytic potential function from the second-moment approximation, a definition of neighboring atoms must be addressed. One of the first attempts to do this within the moments concept was made by Finnis and Sinclair [32] who assumed that the coordination could be replaced with a sum of functions that decay exponentially with distance. This formalism yields a bond energy for atom i in the second-moment approximation of

$$E_i^{\text{el}} \propto -\left[\sum_{j(\neq i)} e^{-\beta r_{ij}}\right]^{1/2}, \tag{16}$$

where β is a parameter. The total density of states is the sum of the local density of states, and so the total bond energy for a system of atoms is the sum of the bond energies of the individual atoms. Including a proportionality constant B and adding pair-additive interactions between atoms to balance the bond energy yields the Finnis-Sinclair analytic potential energy function [32]

$$E_{\text{coh}}^{\text{F-S}} = \sum_i \left[\sum_{j(\neq i)} [A\,e^{-\alpha r_{ij}}] - B\left[\sum_{j(\neq i)} e^{-\beta r_{ij}}\right]^{1/2}\right]. \tag{17}$$

This relatively simple expression captures the essence of quantum-mechanical bonding. Hence, it can be used in large-scale simulations to introduce basic elements of quantum mechanics into the interatomic forces.

Equation (16) can be rearranged to an expression similar to that of the bond-order potential described above by the following straightforward algebra [43]:

$$E_i^{\text{el}} = -B\left[\sum_{j(\neq i)} e^{-\beta r_{ij}}\right]^{1/2} = -\left[\sum_{j(\neq i)} e^{-\beta r_{ij}}\right]^{1/2} \cdot \frac{\left[\sum_{j(\neq i)} e^{-\beta r_{ij}}\right]^{1/2}}{\left[\sum_{k(\neq i)} e^{-\beta r_{ik}}\right]^{1/2}}$$

$$= -B\left[\sum_{j(\neq i)} e^{-\beta r_{ij}}\right]\left[\sum_{k(\neq i)} e^{-\beta r_{ik}}\right]^{-1/2} = -\sum_{j(\neq i)}\left[B\,e^{-\beta r_{ij}}\left[e^{-\beta r_{ij}} + \sum_{k(\neq i,j)} e^{-\beta r_{ik}}\right]^{-1/2}\right]$$

$$= -\sum_{j(\neq i)}\left[B\,e^{-\beta/2 r_{ij}}\left[1 + \sum_{k(\neq i,j)} e^{-\beta(r_{ik}-r_{ij})}\right]^{-1/2}\right]. \tag{18}$$

With this derivation the analytic bond-order b_{ij} is

$$b_{ij} = \left[1 + \sum_{k(\neq i,j)} e^{-\beta(r_{ik}-r_{ij})}\right]^{-1/2} \tag{19}$$

and the two-center pair potential is

$$V^{\text{A}}(r) = B\,e^{-(\beta/2)r}. \tag{20}$$

For a regular solid all of the distances are the same, and the bond-order function is proportional to the inverse root of the local coordination as derived by Abell using a Bethe lattice [8]. It is worth noting that a Bethe lattice has no closed loops, and therefore the second-moment approximation is exact. Adding angular interactions to the terms in the bond-order function and modifying the expression related to the difference in lengths for the *ij* and *ik* bonds leads to the Tersoff empirical bond-order function.

Pettifor and coworkers [44 to 46] have gone beyond the Tersoff empirical bond-order form by deriving an analytic form directly from tight-binding orbitals using the moments theorem. This expression includes explicit angular interactions, and therefore is somewhat less empirical than the Tersoff form (and the form discussed below).

4. The Art of an Empirical Potential

In 1990, an empirical bond-order expression was reported that is appropriate for describing both solid-state carbon and hydrocarbon molecules on an equal footing [23]. The form of this expression, which is based on previous ground-breaking work by Tersoff and others as discussed above in Section 2, allows for bond formation and breaking with appropriate changes in atomic hybridization. Although originally developed to model the chemical vapor deposition of diamond films, this expression found use in simulating a wide range of other chemical processes. These include fullerene formation from graphitic ribbons [47]; collision and subsequent reaction of C_{60} with a diamond surface [48]; compression of C_{60} between graphitic sheets [49]; the pick-up of a molecule from a surface by another gas-phase molecule (an Eley-Rideal reaction sequence) [50]; surface patterning via reactive force microscopy [51]; and compression, indentation, reaction, and friction at diamond surfaces [52 to 54].

The success of this potential for describing hydrocarbon bonding was due primarily to three factors. First, carbon is very well behaved. For example, Pauling-like relations exist between bond strengths, energies and force constants, and molecular atomization energies are largely bond additive. Second, there is a massive and relatively accurate database of information on properties of carbon-based species. This feature means that information entering the fitting database can be carefully chosen for a consistent level of accuracy, and to encompass pertinent structures and properties. Finally, as emphasized in the previous section, sound quantum mechanical principles rather than *ad hoc* equations motivate the form of the analytic expression.

The initial strategy in the 1990 hydrocarbon potential was to develop the carbon and the hydrogen interactions separately, and then to combine the two functions to model hydrocarbon molecules [23]. In this section the development of a potential function for the pure carbon part of a "second generation" hydrocarbon potential energy expression is outlined. The focus is on both the practical choice of functional forms and on the fitting procedure. Specific parameter and fitting data values will be given elsewhere. These aspects of potential function development are not obvious from the formalism development presented in the previous section. Instead, they require a combination of trial-and-error and a certain degree of chemical insight. The discussion in this section is therefore an example of the "art" of effective empirical potential development.

The first step in the fitting procedure is to recognize that despite the physical basis of the potential expression Eq. (2), a site-additive cohesive energy severely limits the ability of the functional form to adequately model carbon radicals and conjugation. π

bonds form between pairs of carbon atoms when there are one or more p orbitals that are not participating in σ bonding on *each of the two bonded atoms*. When only one of the atoms contains a free p orbital, a π bond cannot form. Instead, this bonding configuration is better described as a single σ bond plus a radical associated with the atom containing the free p orbital. In a simple site-additive cohesive energy scheme, the energy of a given atom depends only on the local environment of that atom, and not on the local environment of its neighboring atoms. Hence, the radical structure described above would be treated as a σ bond plus "half of a π bond", rather than a σ bond plus a radical. This results in overbinding of radical structures. This has serious consequences for describing surface reconstructions that are energetically driven by removing radicals and forming π bonds such as the (111) π-bonded chain reconstruction [55].

One straightforward way to correct for the overbinding of radicals associated with this formalism is to rewrite the site-additive cohesive energy Eq. (2) as a bond-additive energy

$$E_{\text{coh}} = \sum_i \sum_{j(\neq i)} [V^R(r_{ij}) - \bar{b}_{ij} V^A(r_{ij})], \qquad \bar{b}_{ij} = \left[\frac{b_{ij} + b_{ji}}{2} \right] \tag{21}$$

where $V^R(r)$ and $V^A(r)$ are repulsive and attractive pair terms, respectively. With the cohesive energy in this pair form, terms can be applied that take into account the number of neighbors for each atom in a pair of bonded atoms, hence correcting overbinding for the situation described above. Similarly, the pair-additive form of the cohesive energy can also be used to incorporate terms that account for energy barriers associated with rotation about dihedral angles for double bonds and conjugation effects, as discussed below.

The specific analytic forms for the pair terms and empirical bond-order function given below are relatively complicated, and the number of nonlinear parameters needed to accurately model carbon bonding for the wide range of atomic hybridizations considered is large. Therefore, to simplify the fitting procedure, a two-step process is used. In the first step, functional forms and parameters for the pair terms in Eq. (21) together with discrete values of the empirical bond-order are obtained. The data to which the potential is fit in this step consists of bond energies, lengths and force constants for triple, double, conjugated double (in graphite) and single bonds, as well as theoretical cohesive energies for solid-state simple cubic (s.c.) and face-centered cubic (f.c.c.) lattices. In the second step, a functional form and parameters for the bond-order function are fit to the discrete values of the bond-order determined in the first step, as well as additional properties such as vacancy formation energies, barriers for dihedral rotation about double bonds, and conjugation effects.

Minimum energy distances for the molecular bonds listed above can be taken from standard literature references. To derive a set of bond energies, heats of formation and zero-point energies for the molecules ethyne, ethene, ethane, cyclohexane and benzene relative to diamond can be used to determine molecular atomization energies. Assuming a constant value for the carbon–hydrogen bond energy in each of these molecules, energies can be derived for carbon–carbon single, double, conjugated double and triple bonds that sum to the atomization energies for each molecule. These atomization energies in turn provide a reasonable prediction of energies for molecules with a mix of single, double and triple bonds.

Values for the force constants K for the complete range of atomic hybridizations considered here can be obtained from Badger's rule

$$K = a(r_e - b)^{-3}, \tag{22}$$

where r_e is the minimum energy bond distance, and a and b are adjustable parameters. This approach has the advantage of providing a consistent and continuous set of force constants as a function of bond distance over attempting to use experimental vibrational frequencies. Values for the two parameters in Eq. (22) can be determined from force constants and minimum energy distances for two types of carbon–carbon bonds. Data for single and triple bonds were used in this case for two reasons. First, reliable data for these two types of bonds are available from the bulk modulus and lattice constant of diamond, and from the bond length and stretching vibrational frequency of acetylene. Second, these two bonding types represent boundaries for the molecular bonds of interest.

In the 1990 potential form, exponential functions were used both for the repulsive and attractive pair terms in Eq. (21) [23]. It was found that with these functional forms, bond lengths, energies and force constants could not be simultaneously fit. Therefore, two sets of parameters in the exponential functions were introduced: one that reproduces bond lengths and energies, and one that reproduces bond energies and force constants. The original exponential form has the further disadvantage that the repulsive term goes to a finite value as the distances between atoms go to zero, limiting the possibility of modeling processes involving highly energetic atomic collisions.

In developing a "second generation" function, it was concluded that the inability to simultaneously fit bond lengths, energies and force constants is not inherent to the formalism, but rather is a result of the restriction of using exponentials for the pair terms. After considerable trial and error, it was determined that a screened Coulomb potential with effective charge Q for the repulsive term

$$V^R(r) = (1 + Q/r) A e^{-\alpha r} \tag{23}$$

and the sum of three exponentials for the attractive term

$$V^A = \sum_n b_n e^{-\beta_n r} \tag{24}$$

have sufficient flexibility to fit the bond properties.

With the fitting data set described above, there are three properties, energy, length and force constant, for each of the four molecular bond types, resulting in a total of twelve fitting data values. The pair functions Eqs. (23) and (24) above contain a total of nine parameters. Discrete values of the bond-order for each of the molecular carbon–carbon bonds considered results in additional four parameters for the first fitting step. Assuming a bond-order value of one for the diatomic results in a total of twelve parameters to be determined. While the functional forms can be partly linearized to help simplify the fitting procedure, there is no obvious one-to-one correspondence between parameters and data values. Therefore, a somewhat complicated nonlinear fitting problem results that requires a combination of fitting methods (e.g. conjugate gradient and simulated annealing) to determine a reasonably optimized set of parameters. Once the pair interactions are determined, bond-orders for the s.c. and f.c.c. lattices can be determined by fitting to the respective cohesive energies.

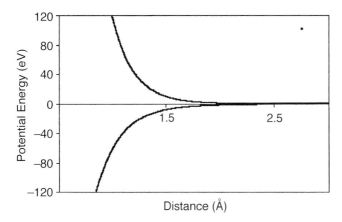

Fig. 1. Attractive and repulsive pair potentials

In Fig. 1 are shown the resulting repulsive and attractive pair terms. The effective pair interactions of three of the four types of molecular bonds mentioned plus bonds appropriate for the f.c.c. and s.c. lattices are plotted in Fig. 2 [56]. The curves in Fig. 2 were obtained by multiplying the attractive term by the discrete bond-order value appropriate for each of the bonds and adding the result to the repulsive term. The database bond energies and distances are indicated in Fig. 2 by the symbols. Overall the fit is generally quite good.

Worthy of note is that fitting the energy of the s.c. and f.c.c. lattices to density functional calculations results in predicted bond lengths within a few percent of those given by the first-principles calculations. The relative agreement between the bond distances, despite not being included in the fitting database, suggests that the function, which was fit to molecular bonding properties, is reasonably transferable to high-density structures. Similarly, the value of the bond-order needed to fit the energy of carbon–carbon bonds in benzene predicts a bond length within a few percent of experiment.

The data used in the second step of the fitting process consists of the six discrete values of the empirical bond-order function, vacancy formation energies for diamond and graphite, and barriers for rotation about double bonds. For full double bonds, the energy barrier for dihedral rotation in ethene can be used. The energy difference between graphite and a solid-state model consisting of sp^2 hybridized carbon in which all dihedral angles are 90° was used to determine an activation energy for rotation about a conjugated double bond [57].

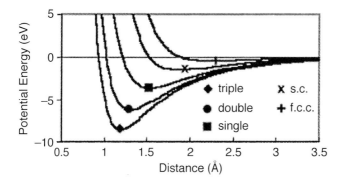

Fig. 2. Effective pair interactions for various bond types

Following the 1990 hydrocarbon bonding expression, the bond-order expression b_{ij} between atoms i and j is written as the sum of terms

$$\bar{b}_{ij} = \frac{[b_{ij}^{\sigma\pi} + b_{ji}^{\sigma\pi}]}{2} + b_{ij}^{\pi}, \qquad (25)$$

where values for $b_{ij}^{\sigma\pi}$ and $b_{ji}^{\sigma\pi}$ depend on the local bonding environment of atom i and atom j, respectively. The function b_{ij}^{π} is further written as a sum of the two terms

$$b_{ij}^{\pi} = \Pi_{ij}^{RC} + b_{ij}^{DH}. \qquad (26)$$

The value of the first term on the right-hand side of Eq. (26) depends on whether a bond between atoms i and j has radical character and is part of a conjugated system, and the value of the second term depends on the dihedral angle for carbon–carbon double bonds. For bonds that are not part of a conjugated system, and for planar double bonds, both terms are equal to zero. A further discussion of these forms is given below.

Following Tersoff, and after considerable trial-and-error, the following function was used for the first two terms in the bond-order Eq. (25):

$$b_{ij}^{\sigma\pi} = \left[1 + \sum_{k(\neq i,j)} G(\cos(\theta_{ijk}))\right]^{-1/2} \qquad (27)$$

where the subscripts refer to the atom identity (see Fig. 3). The function $G(\cos(\theta))$ in Eq. (27) modulates the contribution that each nearest-neighbor makes to the bond-order according to the cosine of the angle of the bonds between atoms i and k and atoms i and j. This function can be determined as follows. The diamond lattice and graphitic sheets contain only one angle each, 109.47° and 120°, respectively. Equation (27), together with the values of the bond-order yields values for $G(\cos(\theta))$ at each of these angles. The energy difference between the linear C_3 molecule and one bent at an angle of 120° (as given by a density functional calculation [58]) can be used to find a value for $G(\cos(\theta = 180°))$. Because in a s.c. lattice the bond angles among nearest neighbors are 90° and 180°, the value of $G(\cos(\theta = 180°))$ combined with the value of the bond-order for the s.c. lattice can be used to find a value of $G(\cos(\theta = 90°))$. Finally, the f.c.c. lattice contains angles of 60°, 120°, 180° and 90°. A value for $G(\cos(\theta = 90°))$ can therefore be determined from the values of $G(\cos(\theta))$ determined above and the value of the bond-order for the f.c.c. lattice. The values of $G(\cos(\theta))$ determined in this way are given by the squares in Fig. 4. This approach to fitting an angular interaction does not rely on the expansion about an equilibrium bond angle; instead, a single function is

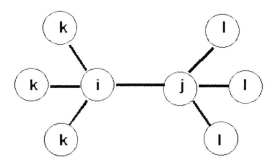

Fig. 3. Illustration of the atom numbering notation used in the text. The bond being considered is between atoms i and j

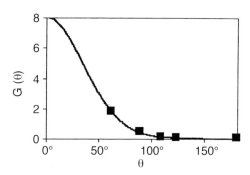

Fig. 4. Angular function entering the bond-order. Squares denote fitting values. The line represents a piece-wise spline fit to the data

used for all atomic hybridizations. The influence of bond angles on the potential energy varies in a continuous fashion as bonding environments change.

It is encouraging that fitting $G(\cos(\theta))$ in this way yields values that become monotonically smaller as the angle increases. This observation can be rationalized using valence shell electron pair repulsion theory which assumes that all molecules with covalent bonding want to have as large a set of bond angles as possible due to repulsion between pairs of electrons in separate bonds [59]. This behavior was not built into the potential, but rather is apparently a consequence of the fitting procedure and the quantum-mechanical basis of the functional form.

To complete the analytic form for the angular function $G(\cos(\theta))$, sixth order polynomial splines in $\cos(\theta)$ were used. For values of the angle θ between 109.47° and 120°, the polynomial was fit to the values of $G(\cos(\theta))$ determined above as well as the first and second derivatives of $G(\cos(\theta))$ with respect to $\cos(\theta)$ at these two angles. Values of the second derivatives of $G(\cos(\theta))$ at 109.47° and 120° were fit to the elastic constant c_{11} for diamond and the in-plane elastic constant c_{11} for graphite, respectively. The first derivatives were chosen to suppress spurious oscillations in the splines. The resulting splines are given as the solid line in Fig. 4.

The term Π_{ij}^{RC} in Eq. (26) represents the influence of radical energetics and π bond conjugation on the bond energies as discussed above. A tricubic spline F can be used for this function

$$\Pi_{ij}^{\mathrm{RC}} = F_{ij}(N_i^{\mathrm{t}}, N_j^{\mathrm{t}}, N_{ij}^{\mathrm{conj}}), \tag{28}$$

where N_i^{t} and N_j^{t} are the number of neighbors of atoms i and j, respectively, and N_{ij}^{conj} is a function that depends on local conjugation. To determine whether a bond is part of a conjugated system, the value of N_{ij}^{conj} is given by the function

$$N_{ij}^{\mathrm{conj}} = 1 + [F(X_{ik})]^2 + [F(X_{jl})]^2, \tag{29}$$

where

$$\begin{aligned} F(X_{ik}) &= 1; \quad X_{ik} < 2, \\ &= [1 + \cos(B(X_{ik} - 2))]/2; \quad 2 < X_{ik} < 3, \\ &= 0; \quad X_{ik} > 3, \end{aligned} \tag{30}$$

and

$$X_{ik} = N_k^{\mathrm{t}} - 1. \tag{31}$$

If all of the carbon atoms that are bonded to a pair of carbon atoms i and j have four or more neighbors, Eqs. (29) to (31) yield a value of unity for N_{ij}^{conj} and the bond between these atoms is not considered to be part of a conjugated system. As coordinations of the neighboring atoms decrease, N_{ij}^{conj} becomes greater than unity, indicating a conjugated bonding configuration. Furthermore, the form of these equations distinguishes between different configurations that can lead to conjugation. For example, the value of N_{ij}^{conj} for a carbon–carbon bond in graphite is nine, while that for a bond in benzene is three. This difference, not included in the 1990 analytic potential form [27], yields considerable extra flexibility for fitting the energies of conjugated systems.

Equations (28) to (31) were not derived from the second-moment approximation/bond-order formalism discussed in the previous section. Instead, they were empirically obtained by analyzing various bonding configurations and possible functional forms. These functions provide a straightforward way of incorporating conjugation effects without having to diagonalize a matrix or go beyond near neighbors, and they smoothly account for changes in conjugation as bonds break and form. However, they do not capture phenomena like ring versus chain formation that even simple Hückel approaches are able to capture. These types of limitations must be weighed when analyzing predictions from a potential such as this.

The term b_{ij}^{DH} in Eq. (25) is given by

$$b_{ij}^{\text{DH}} = T_{ij}(N_i^t, N_j^t, N_{ij}^{\text{conj}}) \left[\sum_{k(\neq i,j)} \sum_{l(\neq i,j)} (1 - \cos^2(\mathbf{T}_{ijkl})) \right], \tag{32}$$

where

$$\mathbf{T}_{ijkl} = \mathbf{e}_{jik} \times \mathbf{e}_{ijl}. \tag{33}$$

The function $T_{ij}(N_i^t, N_j^t, N_{ij}^{\text{conj}})$ is a tricubic spline, and the functions $\mathbf{e}_{jik}$ and $\mathbf{e}_{ijl}$ are unit vectors in the direction of the cross products $\mathbf{R}_{ji} \times \mathbf{R}_{ik}$ and $\mathbf{R}_{ij} \times \mathbf{R}_{jl}$, respectively, where the $\mathbf{R}$'s are vectors connecting the subscripted atoms. These equations incorporate barriers for dihedral angle rotation about carbon–carbon double bonds using a method similar to those employed in molecular force fields [12]. The value of this function is zero for a planar system, and unity for angles of 90°. Therefore, the function T determines the barrier for rotation about these bonds. This function can be parameterized so that carbon–carbon bonds that are not double bonds make no contribution to the bond-order. It can also distinguish differences in barriers between conjugated and nonconjugated double bonds. Its value for nonconjugated carbon–carbon bonds can be fit to data such as the barrier for rotation about the carbon–carbon bond in ethene. Similar to the term describing bond conjugation above, this expression for incorporating barriers for rotation about double bonds was not derived from quantum-mechanical principles, but rather was empirically derived through analysis of bonding configurations. Therefore, the same cautions apply with regard to structures for which this term has a strong influence.

Extensive testing of the analytic potential described here by a variety of researchers suggests that the potential function does a reasonable job of fulfilling the four criteria for an effective interatomic potential listed in the Introduction. This testing included evaluation of point defects and grain boundary energies in diamond [60 to 62], diamond surface reconstructions [63], properties of fullerenes [64 to 66] and the structure of carbon melts [67].

There is one aspect to the potential that remains unsatisfactory. In both deriving the quantum-mechanical basis for the bond-order potential and in developing empirical fitting to make it practical, it has been implicitly assumed that the bonding interaction extends only to nearest neighbors. Experience suggests that a straightforward extension to longer interaction distances both introduces severe complications into the fitting process and compromises much of the transferability of the potential arising from its derivation from quantum mechanics. From a practical viewpoint, using switching functions that introduce a narrow distance over which interactions are "turned-off" provides a solution, but at the cost of nonphysical and severe changes in interatomic forces over this range. A more thorough analysis of the quantum mechanics of these expressions, perhaps the non-self-consistency introduced by the Harris functional, needs to be completed to find a more satisfactory method of handling the difficulties associated with the range of the potential.

5. Conclusions

In this paper, two aspects of classical interatomic potential development have been presented, namely the development of a central set of analytic expressions that reflect the underlying principles of quantum-mechanical bond formation, and the empirical fitting of these expressions to a specific system, carbon. Experience suggests that both aspects are critical to developing transferable and reasonably accurate interatomic potential energy expressions that are suitable for large-scale atomistic simulation. It is rare that an analytic expression derived completely from quantum mechanics is exceptionally accurate, and it is the rare expression derived completely empirically that exhibits reasonable transferability.

Do analytic potential energy expressions have a future in atomistic simulation? While an unequivocal answer is of course impossible, it is reasonable to assume that at least in the near-term problems requiring atomic resolution will continue to be pursued for system sizes and timescales outside those accessible to first-principles methods. Therefore, the development of analytic bonding expressions, while perhaps less active than a decade ago, will likely continue to hold a central place in atomistic modeling into the next decade and even beyond.

Acknowledgements D. Areshkin, B. I. Dunlap, S. J. Frankland, B. J. Garrison, J. Glosli, J. A. Harrison, J. W. Mintmire, O. A. Shenderova, S. B. Sinnott, D. Srivastava, S. Stuart, and C. T. White are thanked for helpful discussions leading to some of the ideas presented here. Financial support from the Office of Naval Research, the National Science Foundation, Lawrence Livermore National Laboratory and NASA is gratefully acknowledged.

References

[1] J. Hirschfelder, H. Eyring, and B. Topley, J. Chem. Phys. **4**, 170 (1936).
[2] J. B. Gibson, A. N. Goland, M. Milgram, and G. H. Vineyard, Phys. Rev. **120**, 1229 (1960).
[3] A. Rahman, Phys. Rev. **136A**, 405 (1964).
[4] L. M. Balbes, S. W. Mascarella, and D. B. Boyd, in: Reviews in Computational Chemistry, Vol. 5, Eds. K. B. Lipkowitz and D. B. Boyd, VCH Publ., New York 1994 (pp. 337 to 379).
[5] P. Vashishta, R. K. Kalia, and A. Nakano, Comput. Sci. Engng. **1**, 56 (1999).

[6] M. S. Daw and M. I. Baskes, Phys. Rev. Lett. **50**, 285 (1983).
[7] F. Ercolessi, M. Parrinello, and E. Tossatti, Phil. Mag. A **58**, 213 (1988).
[8] G. C. Abell, Phys. Rev. B **31**, 6184 (1985).
[9] J. H. Rose, J. R. Smith, and J. Ferrante, Phys. Rev. B **28**, 1935 (1983).
[10] J. Tersoff, Phys. Rev. Lett. **56**, 632 (1986).
[11] J. Tersoff, Phys. Rev. B **39**, 5566 (1989).
[12] N. L. Allinger, Y. H. Yuh, and J. H. Li, J. Amer. Chem. Soc. **111**, 8551 (1989).
[13] D. W. Brenner, Mater. Res. Soc. Bull. **21**, 36 (1996).
[14] D. H. Robertson, D. W. Brenner, and C. T. White, Phys. Rev. Lett. **25**, 3231 (1991).
[15] D. W. Brenner, in: Shock Compression of Condensed Matter, Eds. S. C. Schmidt, R. D. Dick, J. W. Forbes, and D. G. Tasker, North-Holland Publ. Co., Amsterdam 1992 (p. 115).
[16] D. W. Brenner, D. H. Robertson, M. L. Elert, and C. T. White, Phys. Rev. Lett. **70**, 1821 (1992).
[17] K. E. Khor and S. Das Sarma, Phys. Rev. B **38**, 3318 (1988).
[18] E. Kaxiras and K. C. Pandey, Phys. Rev. B **38**, 12736 (1988).
[19] M. Z. Bazant, E. Kaxiras, and J. F. Justo, Phys. Rev. B **56**, 8542 (1997).
[20] C. D. Scheerschmidt, Phys. Rev. B **58**, 4538 (1998).
[21] F. D. Mota, J. F. Justo, and A. Fazzio, J. Appl. Phys. **86**, 1843 (1999).
[22] K. Alba and W. Moller, Comp. Mater. Sci. **10**, 111 (1998).
[23] D. W. Brenner, Phys. Rev. B **42**, 9458 (1990).
[24] A. J. Dyson and P. V. Smith, Surf. Sci. **355**, 140 (1996).
[25] A. J. Dyson and P. V. Smith, Surf. Sci. **375**, 45 (1997).
[26] M. V. R. Murty and H. A. Atwater, Phys. Rev. B **51**, 4889 (1995).
[27] K. Beardmore and R. Smith, Phil. Mag. A **74**, 1439 (1996).
[28] J. Che, T. Cagin, and W. H. Goddard III, Theo. Chem. Acc. **102**, 346 (1999).
[29] S. Stuart and J. A. Harrison, private communication.
[30] W. M. C. Foulkes and R. Haydock, Phys. Rev. B **39**, 12520 (1989).
[31] A. P. Sutton and R. W. Balluffi, in: Interfaces in Crystalline Materials, Clarendon Press, Oxford 1995 (pp. 150 to 239).
[32] M. W. Finnis and J. E. Sinclair, Phil. Mag. A **50**, 45 (1984).
[33] D. W. Brenner, O. A. Shenderova, and D. A. Areshkin, in: Reviews in Computational Chemistry, Vol. 5, Eds. K. B. Lipkowitz and D. B. Boyd, Wiley-VCH, New York 1998 (pp. 207 to 239).
[34] P. Hohenberg and W. Kohn, Phys. Rev. **136**, B864 (1964).
[35] W. Kohn and L. J. Sham, Phys. Rev. A **140**, 1133 (1965).
[36] J. Harris, Phys. Rev. B **31**, 1770 (1985).
[37] A. J. Read and R. J. Needs, J. Phys: Condensed Matter **1**, 7565 (1989).
[38] H. M. Polatoglu and M. Methfessel, Phys. Rev. B **37**, 10403 (1988).
[39] M. W. Finnis, J. Phys: Condensed Matter **2**, 331 (1990).
[40] I. J. Robertson, M. C. Payne, and V. Heine, J. Phys: Condensed Matter **3**, 8351 (1991).
[41] F. Cyrot-Lackmann, J. Phys. Chem. Solids **29**, 1235 (1968).
[42] A. P. Sutton, Electronic Structure of Materials, Clarendon Press, Oxford 1993.
[43] D. W. Brenner, Phys. Rev. Lett. **63**, 1022 (1989).
[44] D. G. Pettifor, Phys. Rev. Lett. **63**, 2480 (1989).
[45] D. Pettifor and I. I. Oleinik, Phys. Rev. B **59**, 8487 (1999).
[46] D. Pettifor and I. I. Oleinik, Phys. Rev. B **59**, 8500 (1999).
[47] D. H. Robertson, D. W. Brenner, and C. T. White, J. Phys. Chem. **96**, 6133 (1992).
[48] R. C. Mowrey, D. W. Brenner, B. I. Dunlap, J. W. Mintmire, and C. T. White, J. Phys. Chem. **95**, 7138 (1991).
[49] D. W. Brenner, J. A. Harrison, C. T. White, and R. J. Colton, Thin Solid Films **206**, 220 (1991).
[50] E. R. Williams, G. C. Jones, Jr., L. Fang, R. N. Zare, B. J. Garrison, and D. W. Brenner, J. Amer. Chem. Soc. **114**, 3207 (1992).
[51] S. B. Sinnott, R. J. Colton, C. T. White, and D. W. Brenner, Surf. Sci. **316**, L1055 (1994).
[52] J. A. Harrison, C. T. White, R. J. Colton, and D. W. Brenner, Surf. Sci. **271**, 57 (1992).
[53] J. A. Harrison and D. W. Brenner, J. Amer. Chem. Soc. **116**, 10399 (1995).
[54] J. A. Harrison, C. T. White, R. J. Colton, and D. W. Brenner, Phys. Rev. B **46**, 9700 (1992).

[55] D. VANDERBILT and S. G. LOUIE, Phys. Rev. B **30**, 6118 (1984).
[56] M. T. YIN and M. L. COHEN, Phys. Rev. Lett. **50**, 2006 (1983).
[57] A. Y. LIU, M. L. COHEN, K. C. HASS, and M. A. TAMOR, Phys. Rev. B **43**, 6742 (1991).
[58] J. R. CHELIKOWSKI and M. Y. CHOU, Phys. Rev. B **37**, 6504 (1988).
[59] J. E. HUHEEY, Inorganic Chemistry, 2nd ed., Harper & Row, New York 1978.
[60] O. SHENDEROVA, D. W. BRENNER, A. NAZAROV, A. ROMANOV, and L. YANG, Phys. Rev. B **57**, R3181 (1998).
[61] O. A. SHENDEROVA, D. W. BRENNER, and L. H. YANG, Phys. Rev. B **60**, 7043 (1999).
[62] O. A. SHENDEROVA and D. W. BRENNER, Phys. Rev. B **60**, 7053 (1999).
[63] O. A. SHENDEROVA and D. W. BRENNER, unpublished.
[64] C. T. WHITE, J. W. MINTMIRE, R. C. MOWREY, D. W. BRENNER, D. H. ROBERTSON, J. A. HARRISON, and B. I. DUNLAP, in: Buckminsterfullerenes, Eds. W. E. BILLUPS and M. CIUFOLINI, VCH Publ., New York 1993 (p. 125).
[65] D. W. BRENNER, J. D. SCHALL, J. P. MEWKILL, O. A. SHENDEROVA, and S. B. SINNOTT, J. British Interplanetary Soc. **51**, 137 (1998).
[66] D. SRIVASTAVA, D. W. BRENNER, J. D. SCHALL, K. D. AUSMAN, M. F. YU, and R. S. RUOFF, J. Phys. Chem. B **103**, 4330 (1999).
[67] J. N. GLOSLI and F. H. REE, Phys. Rev. Lett. **82**, 4659 (1999).

phys. stat. sol. (b) **217**, 41 (2000)

Subject classification: 71.15.–m; S12

A Self-Consistent Charge Density-Functional Based Tight-Binding Method for Predictive Materials Simulations in Physics, Chemistry and Biology

TH. FRAUENHEIM (a), G. SEIFERT (a), M. ELSTNER (a,c), Z. HAJNAL (a), G. JUNGNICKEL (a), D. POREZAG (b), S. SUHAI (c), and R. SCHOLZ (d)

(a) *Theoretische Physik, Fachbereich Physik, Universität Paderborn, D-33098 Paderborn, Germany*

(b) *Complex Systems Theory Branch, Naval Research Laboratory, Washington D.C. 20375-5345, USA*

(c) *German Cancer Research Center, Department of Molecular Biophysics, D-69120 Heidelberg, Germany*

(d) *Theoretische Physik III, Fachbereich Physik, Technische Universität Chemnitz-Zwickau, D-09107 Chemnitz, Germany*

(Received August 10, 1999)

We outline recent developments in quantum mechanical atomistic modelling of complex materials properties that combine the efficiency of semi-empirical quantum-chemistry and tight-binding approaches with the accuracy and transferability of more sophisticated density-functional and post-Hartree-Fock methods with the aim to perform highly predictive materials simulations of technological relevant sizes in physics, chemistry and biology. Following Harris, Foulkes and Haydock, the methods are based on an expansion of the Kohn-Sham total energy in density-functional theory (DFT) with respect to charge density fluctuations at a given reference density. While the zeroth order approach is equivalent to a common standard non-self-consistent tight-binding (TB) scheme, at second order by variationally treating the approximate Kohn-Sham energy a transparent, parameter-free, and readily calculable expression for generalized Hamiltonian matrix elements may be derived. These matrix elements are modified by a **S**elf-**C**onsistent redistribution of Mulliken **C**harges (SCC). Besides the usual "band-structure" and short-range repulsive terms the final approximate Kohn-Sham energy explicitly includes Coulomb interaction between charge fluctuations. The new SCC-scheme is shown to successfully apply to problems, where defficiencies within the non-SCC standard TB-approach become obvious. These cover defect calculations and surface studies in polar semiconductors (see M. Haugk et al. of this special issue), spectroscopic studies of organic light-emitting thin films, briefly outlined in the present article, and atomistic investigations of biomolecules (see M. Elstner et al. of this special issue).

1. Introduction

Prerequisite to an atomistic understanding of materials fabrication and response to external stimuli are a repertoire of computational methods for the description of processes that occur on many different temporal, spatial and thermodynamic scales. Further, such methods should be chemically accurate, computationally fast and applicable to an arbitrary assortment of atomic assemblies. While it is safe to say that such a code does not exist presently and that full success in such an endeavor will require another decade, significant progress in several subfields of computational materials science has been made which addresses different aspects of this problem.

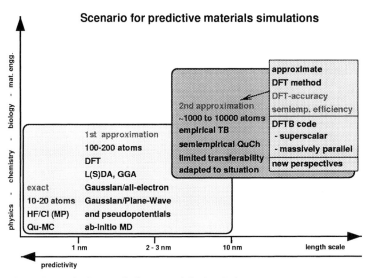

Fig. 1. Scenario for predictive materials simulations

When considering the scenario for materials simulations in cross-disciplinary fields of physics, chemistry, biology and materials engineering the highest accuracy and predictivity in chemical bonding, reaction energies, structures and physico-chemical properties can be obtained only on the level of atomistic studies based on electronic-structure-theory, see Fig. 1. Obviously, the highest accuracy is achieved by the so-called "first principle" methods, post-Hartree-Fock (Configuration Interaction – CI, Many Body Perturbation Theory – MBT) and Quantum-Monte-Carlo, including properly both, electronic exchange and correlation within a many-body treatment [1, 2]. Since such calculations are computationally very demanding, present applications can only address systems with up to 20 to 30 atoms or slightly more [3, 4].

Focusing on larger clusters, molecules or periodic structures including 100 to 200 independent atoms, density-functional theory (DFT) gives accurate solutions. In practice local-density and generalized gradient approximations (LDA/GGA) for the exchange-correlation functional within density-functional-theory (DFT) [5, 6] can be applied (see also references in M. R. Pederson et al. [7]). They use either all-electron or pseudopotential concepts, and the various implementations differ in their basis representations, including Gaussian-type orbitals [7], Slater-type orbitals [8] numerical basis sets [9], LMTOs [10], plane waves [11] or a mixed-type basis [12]. Converged basis sets result in accurate geometric structures, reaction energies and electronic and vibrational properties, but as these large bases are again a computational bottleneck, many technologically relevant problems are still out of reach. In particular for approaching the time evolution in dynamic structure formation, considerable progress has been achieved by Car and Parrinello (CP) with their ab initio molecular-dynamics method based on the DFT [13]. This approach has provoked a revolution in computational condensed matter physics and chemistry over the last 14 years. Applications of the CP method to systems of ≈ 100 atoms have demonstrated an almost unlimited transferability, see [14]. However, due to the high energy cut-off required for the plane-wave basis, these schemes become computationally very demanding for larger structures.

Therefore nowadays, the common semi-empirical methods of quantum chemistry [15] and physics [16] are becoming more and more popular, especially for nanoscale-sized materials including several hundreds or thousands of atoms, compare Fig. 1. By direct modelling of the electronic structure, albeit in a simplified parametrized way, TB (Tight-Binding) incorporates quantum effects going beyond classical empirical potentials. Therefore, they can be considered as a link between classical concepts and the full-scale modelling of the quantum nature of chemical bonding.

The modern tight-binding history starts with Slater and Koster's seminal paper [17] "Simplified LCAO-method for the Periodic Potential Problem" in 1954, and since then, this concept has addressed some important topics in computational materials science, namely the development of rapid, robust, generally transferable and accurate methods for the calculation of atomic and electronic structures, energies and forces of large molecular and complex condensed systems.

The standard TB method expands the eigenstates of a system in terms of an orthogonalized basis of atomic-like orbitals, representing the exact many-body Hamilton operator with a parametrized Hamiltonian matrix, where the matrix elements are fitted to the electronic structure of a suitable reference system.

Although the original Slater-Koster scheme was intended for investigations of the electronic band structure of periodic solids, later on the tight-binding ideas have been generalized to an atomistic total energy method. In 1979, Froyen and Harrison proposed a r^{-2} dependence of the matrix elements for problems with varying interatomic distances [18, 19], and in the same period, Chadi applied the method to surface energy minimizations of semiconductors [20]. He proposed to write the total energy as a function of all atomic coordinates,

$$E_{\text{tot}} = E_{\text{bs}} + E_{\text{rep}}, \tag{1}$$

where E_{bs} (band-structure energy) is the sum over the occupied electronic eigenstates of the TB Hamiltonian and E_{rep} stands for a short-range repulsive two-particle interaction, including the ionic repulsion and corrections due to approximations made in E_{bs}. The repulsive interactions versus distance may be determined in a parametrized functional form for reproducing cohesive energies and elastic constants (bulk-moduli) for crystalline systems.

Hence, such a TB calculation and its results clearly depend on the parametrization scheme used, and the transferability to different systems and problems is rather limited. Successful applications include band calculations in semiconductor heterostructures [21, 22], device simulations for optical properties [23], simulations of amorphous solids [24] and predictions of low-energy silicon clusters [25, 26], compare Ref. [16] for a review.

However, as the accuracy of the scheme is particularly tuned for dealing with a well-defined chemical surrounding, certain deficiencies are unavoidable for bonding geometries not covered by the parametrization. Non-orthogonality through the years has been discussed as one further step towards a better transferability [26].

In order to reduce the difficult parametrization within a multiconfigurational space and to achieve general chemical transferability at a high accuracy for opening widely spanned new application fields, more sophisticated, yet efficient, TB schemes have been developed. These methods include the TB-LMTO (linear-muffin-tin-orbitals) method of Andersen and Jepsen [27], a successful DFT parametrization of TB by Cohen et al. [28], our DF-based two-center TB approach [29] and the ab initio multicenter TB of Sankey and Niklewski [30].

Within the DF-based two-center [29] and multicenter TB approaches [30], the Hamiltonian matrix elements are explicitly calculated within a non-orthogonal basis of localized atomic orbitals. Generally, these schemes yield accurate results for a broad range of bonding situations where the superposition of overlapping atom-like electronic densities yields a good approximation for the many-atom structure.

Porezag et al. follow the track of Seifert and coworkers [31 to 33] and try to avoid any empirical parametrization by calculating the Hamiltonian and overlap matrices out of a DFT-LDA-derived local orbital basis and some integral approximations. The method includes first-principle concepts in relating the Kohn-Sham orbitals of the atomic configuration to a minimal basis of the localized atomic-like valence orbitals of all atoms, which along with the single atomic potentials are determined self-consistently within the local-density approximation (LDA). Each valence orbital is represented by a set of twelve Slater-type functions.

Making use of a simplified non-self-consistent DF scheme for the many-atom configuration the effective one-electron potential in the Kohn-Sham Hamiltonian is approximated as a sum of contracted pseudoatom potentials. These pseudoatom potentials are introduced in consistency with limiting the range of the valence electron orbital extension by an additional potential. As the result the charge densities are contracted similarly to self-consistently obtained charge densities in molecular and crystalline modifications. Consistent with this approximation one can neglect several contributions to the Hamiltonian matrix elements in the secular equation. This leaves to solve a simplified general eigenvalue problem for determining the many-atom electron eigenvalues and wave functions for clusters and solids yielding exactly the same energy expression as in common standard TB-schemes,

$$E_{\text{tot}}(\{\mathbf{R}_l\}) = E_{\text{bs}}(\{\mathbf{R}_l\}) + E_{\text{rep}}(\{\mathbf{R}_l - \mathbf{R}_k\}).$$

The first term appearing as the sum over all occupied Kohn-Sham energies represents the so-called band-structure term, and the second, as an repulsive energy, comprises the core–core repulsion between the atoms and corrections due to the Hartree and XC double-counting terms and the nonlinearity in the superposition of exchange–correlation contributions. Finally, the repulsive energy is constructed as a superposition of short-range repulsive two-particle potentials by taking the difference of the SCF-LDA cohesive energy and the corresponding TB band-structure energy versus distance for suitable reference systems.

This approach defines the density-functional tight-binding (DFTB) scheme developed in the Chemnitz/Dresden, now Paderborn group. Over the years, it has been applied successfully to various problems in systems ranging from molecules over clusters and liquids to crystalline and amorphous solids. The topics addressed include surfaces, interfaces, adsorbates and defects in covalently bonded materials, covering materials like carbon [29, 34 to 37], silicon and germanium modifications [38 to 40], boron and carbon nitrides [41 to 44], silicon carbide [45] and oxide [46], III–V semiconductors [47] and Zintl phases [48]. Provided the input charge densities have been chosen properly, the geometries, energies and forces, vibrational modes, valence electron spectra and charge densities are almost as accurate as obtained within DFT-LDA calculations.

There is a particularly close connection of our DFTB method to the local-orbital DFT-LDA scheme of Sankey and Niklewski [30]. In this approach, local atom-like basis orbitals are derived from a self-consistent LDA computation for a free pseudo-atom using Hamann-Schlüter-Chiang-type pseudopotentials. These orbitals are then strictly set to zero at and beyond a certain critical radius ("fireball" orbitals), reducing the number of matrix elements needed in an extended system. The Hamiltonian matrix elements are calculated using pseudopotentials and all two- and three-center integral contributions are taken into account. This is done efficiently by interpolating between the values of precomputed one- and two-dimensional integral tables. Finally, the exchange-correlation energy and matrix elements are evaluated in a linearized way using a "average-density" approximation.

This method has the advantage of being strictly first-principle because no fitting to any reference structures is required. However, as discussed by Seifert and co-workers [32], the inclusion of three-center integral contributions within a *non-self-consistent* DFT scheme is not expected to improve, in general, the accuracy of the method.

The theoretical justification of non-self-consistent DFT and hence TB-like schemes has been given by Foulkes and Haydock [49] who thereby generalized results obtained by Harris [50]. These authors considered an expansion of the "double-counting" terms within a Kohn-Sham total energy about the reference density used for constructing the effective electronic potential. It could be shown that the correct non-self-consistent Kohn-Sham energy for the potential used depends only to second order on the deviation of the reference density from the correct one. As a consequence, the so-called Harris or Harris-Foulkes functional obtained by neglecting these second-order terms is stationary and agrees with the ordinary Kohn-Sham functional at the correct self-consistent density, but it can be evaluated much more easily and within a non-self-consistent way using a suitable input density.

For homonuclear systems, such a reference density can be deduced by simply adding the densities obtained for some sort of free atoms (pseudo atoms). This approach works well for homopolar systems and has been used widely as the underlying idea of any standard non-self-consistent TB scheme. However, problems arise when a delicate charge balance is required for establishing chemical bonding between different types of atoms. In such cases, a self-consistent adjustment of the charge density may prove to be necessary.

However, full self-consistency would sacrifice the advantages of simple and physically descriptive TB schemes. Some authors have therefore proposed techniques achieving at least some kind of restricted self-consistency without loosing the efficiency of TB-like methods – see e.g. [51].

There are three major steps towards the inclusion of self-consistency in the charge distribution within the tight-binding framework: The Hubbard U, the Hubbard U plus self-consistent Coulomb interaction and the self-consistent *monopolar/multipolar* Charge Tight-Binding SCC-TB – for an overview see e.g. [16].

A straightforward approach consists in adjusting the occupation numbers of the atomic orbitals within the total charge density. This was applied first by Sambe and Felton [52] for computations of molecules using a Hartree-Fock-Slater method.

The same idea was utilized by Demkov and co-workers [53] for a self-consistent generalization of the Sankey-Niklewski Hamiltonian described above within the Harris-functional framework. In order to make the scheme applicable to extended systems, one has to include the long-range Coulomb contributions within the Hamiltonian matrix elements, and it was shown that a dipole expansion of these terms and Ewald summation techniques have to be used. The new charge density computed in every SCF step has to be projected back to a sum of weighted atomic densities, an approach allowing for a re-use of the precomputed integral tables. This method has been applied successfully to a number of heteronuclear systems such as several silica (SiO_2) polymorphs [53] and GaN [55].

An alternative technique to treat the long-range interactions within the same Hamiltonian was proposed earlier by Tsai and co-workers [56]. These authors transformed the non-neutral atomic charges to neutral ones by adding appropriately chosen Gaussian densities, and then evaluated the long-range contributions of these Gaussian charges by Ewald techniques.

Starting out of this background, our DFTB method [29, 32] has recently included a systematic extension of the tight-binding formalism in order to derive a generalized self-consistent charge (SCC) methodology [57]. It differs from previous approaches in that we base the modification of the TB total energy expression on a strict second-order expansion of the Kohn-Sham energy functional [6] with respect to charge density fluctuations. The new methodology ensures a proper distribution of the charge, overcomes the requirement of local charge neutrality [58] and accounts explicitly for long-range Coulomb interactions in multi-component systems.

In maintaining the simple two-center picture of the non-SCC standard DFTB, the method combines a high computational efficiency with an improved accuracy and transferability. Based on these improvements, the method becomes promising for predictive simulations of the electronic structure of various materials, including cross-disciplinary interfaces between physics, chemistry and biology. Besides the use on scalar workstations, the method has been implemented on a massive parallel system (CRAY-T3E) [59], and this parallelized version was used for studies of extended defects in GaN [60], see also Haugk et al. [61]. Furthermore, a non-orthogonal localised orbital order N-method (NOON) became available [62]. Hopefully, in the near future these schemes will allow for quantum mechanical treatments of nanoscale material sizes up to several thousands of atoms.

This paper is organized as follows. Section 2 explains the DFT foundations of our tight-binding scheme, and on this basis, the non-self-consistent DFTB model is derived (Section 3). Section 4 describes the self-consistent-charge SCC-DFTB extension, and it is shown in detail how this generalized scheme ensures a self-consistent distribution of Mulliken charges, a crucial improvement for correctly describing complex chemical bonding situations in anorganic, organic and biological systems. The possible extension of the method to spin-polarized systems is discussed in Section 5. Recent successfull results concerning defect and surface problems in polar semiconductors and very promising investigations of biomolecular systems are covered by two subsequent papers in this special issue, see Ref. [61] by M. Haugk et al. and Ref. [63] by M. Elstner et al. so that this article will focus on first applications to organic materials. After briefly discussing relevant benchmarks for organic molecular test sets in Section 6, we report in Section 7 details of the analysis of the Raman-active vibrational modes of the organic semiconductor PTCDA (3-4-9-10-perylene-tetracarboxylic-dianhydride). Finally, we summarize this paper in the last section and discuss some future perspectives and challenges in computational materials science.

2. Density-Functional Basis of TB Theory

Within density-functional theory (DFT), the total energy of a system of M electrons in the field of N nuclei at positions **R** is expressed as a unique functional of a charge density $n(\mathbf{r})$.

Using the Kohn-Sham eigenstates Ψ_i for representation of $n(\mathbf{r})$:

$$n(\mathbf{r}) = \sum_i^{occ} |\Psi_i(\mathbf{r})|^2, \quad (2)$$

the total energy may be written as:

$$E = \sum_i^{occ} \langle \Psi_i | -\frac{\Delta}{2} + V_{ext} + \int \frac{n(\mathbf{r}')}{|\mathbf{r} - \mathbf{r}'|} d^3 r' + V_{xc}[n] | \Psi_i \rangle$$
$$- \frac{1}{2} \iint \frac{n(\mathbf{r}) n(\mathbf{r}')}{|\mathbf{r} - \mathbf{r}'|} d^3 r d^3 r' - \int V_{xc}[n] n(\mathbf{r}) d^3 r + E_{xc}[n(\mathbf{r})] + \frac{1}{2} \sum_{\alpha\beta}^N \frac{Z_\alpha Z_\beta}{|\mathbf{R}_\alpha - \mathbf{R}_\beta|}. \quad (3)$$

For simplicity we use in the following the abbreviations:

$$\int d^3 r \to \int, \quad \int d^3 r' \to \int', \quad n(\mathbf{r}) \to n, \quad n(\mathbf{r}') \to n'.$$

The first term in eq. (3) is equivalent to the sum over the occupied single particle energies ϵ_i, the latter terms account for the 'double counting' contributions to the total energy (Hartree and excange–correlation (xc) contributions) and the nuclear repulsion, E_N.

Following the ideas of Foulkes and Haydock [49], we expand the total energy at a reference or input density n_0 up to second order in the density fluctuations $\delta n \equiv \delta n(\mathbf{r}) = n - n_0$:

$$E = \sum_i^{occ} \langle \Psi_i | \hat{H}^0 | \Psi_i \rangle - \frac{1}{2} \iint' \frac{n_0' n_0}{|\mathbf{r} - \mathbf{r}'|} + E_{xc}[n_0] - \int V_{xc}[n_0] n_0 + E_N$$
$$+ \frac{1}{2} \iint' \left(\frac{1}{|\mathbf{r} - \mathbf{r}'|} + \frac{\delta^2 E_{xc}}{\delta n \, \delta n'} \bigg|_{n_0} \right) \delta n \, \delta n'. \quad (4)$$

Note that the linear terms in δn cancel each other at any arbitrary input density n_0. The Hamilton operator $\hat{H}^0$ depends only on the reference density n_0:

$$\hat{H}^0 = -\frac{\Delta}{2} + V_{ext} + \int' \frac{n_0'}{|\mathbf{r} - \mathbf{r}'|} + V_{xc}[n_0]. \quad (5)$$

Further, we define a repulsive energy contribution E_{rep} as in standard tight-binding theory [49],

$$E_{rep}[n_0] = -\frac{1}{2} \iint' \frac{n_0' n_0}{|\mathbf{r} - \mathbf{r}'|} + E_{xc}[n_0] - \int V_{xc}[n_0] n_0 + E_N \quad (6)$$

by combining the nuclear repulsion and all energy contributions depending on n_0 only.

With the abbreviation E^2 for the terms second order in δn, the approximated DFT total energy contains three contributions, which are contributions to zero and second order in the density fluctuations,

$$E_{tot}[n] = \sum_i^{occ} \langle \Psi_i | \hat{H}^0 | \Psi_i \rangle + E_{rep}[n_0] + E^2[\delta n, n_0]. \quad (7)$$

The Harris functional approach [50] is recovered by neglecting $E^2[\delta n, n_0]$, which is also the basis for a non-self-consistent tight-binding sheme. Writing the initial charge density as a superposition of atomic-like neutral charge densities

$$n_0 = \sum_\alpha n_0^\alpha, \tag{8}$$

centered at the atoms α, the repulsive energy E_{rep} does not depend on the charge fluctuations and contains no long-range Coulomb interactions due to the neutrality of the atomic-like densities n_0^α. Further, E_{rep} can be approximated by a sum of two-center contributions only. Formally, it can be expanded in a cluster series [49]:

$$E_{\text{rep}}[n_0] = \sum_\alpha E_{\text{rep}}[n_0^\alpha] + \frac{1}{2} \sum_\alpha \sum_{\beta \neq \alpha} \left(E_{\text{rep}}[n_0^\alpha + n_0^\beta] - E_{\text{rep}}[n_0^\alpha] - E_{\text{rep}}[n_0^\beta] \right) + \cdots \tag{9}$$

The neglect of three-center contributions can be justified by screening arguments. Since n_0^α corresponds to the charge density of a neutral atom the three center terms in the electron–electron interactions are strongly canceled by the nucleus-nucleus interactions.

Due to this screening the two-center contributions can be assumed to be short-ranged. However, the repulsive energy as defined above does not got to zero for large interatomic distances $R_{\alpha\beta}$ but to a constant given by the atomic contributions:

$$E_{\text{rep}}[n_0] = \sum_\alpha E_{\text{rep}}[n_0^\alpha], \qquad R_{\alpha\beta} \to \infty. \tag{10}$$

Therefore, E_{rep} can be approximated as a sum of short-ranged two-center terms with respect to the energies $E_{\text{rep}}[n_0^\alpha]$ of neutral atomic fragments:

$$\tilde{E}_{\text{rep}}[n_0] \equiv E_{\text{rep}}[n_0] - \sum_\alpha E_{\text{rep}}[n_0^\alpha] = \frac{1}{2} \sum_{\alpha\beta} U[n_0^\alpha, n_0^\beta]. \tag{11}$$

For given densities n_0^α, E_{rep} could be calculated in principle. However, it is convenient to fit this expression to ab initio calculations, as will be shown below. On the basis of these considerations a DFT based tight-binding model can be formulated.

3. Zeroth Order Non-SCC Approach, Standard DFTB

In the TB approach the Kohn-Sham wave functions Ψ_i are expanded into a suitable set of localized atomic orbitals φ_ν, denoting the expansion coefficients by $c_{\nu i}$.

$$\Psi_i = \sum_\nu c_{\nu i} \varphi_\nu(\mathbf{r} - \mathbf{R}_\alpha). \tag{12}$$

This LCAO ansatz leads to a secular problem in the form:

$$\sum_\nu^M c_{\nu i}(H_{\mu\nu}^0 - \varepsilon_i S_{\mu\nu}) = 0, \quad \forall \mu, i, \tag{13}$$

The matrix elements of the Hamiltonian $H_{\mu\nu}^0$ and the overlap matrix elements $S_{\mu\nu}$ are defined as

$$H_{\mu\nu}^0 = \langle \varphi_\mu | \hat{H}_0 | \varphi_\nu \rangle; \qquad S_{\mu\nu} = \langle \varphi_\mu | \varphi_\nu \rangle; \qquad \forall \mu \in \alpha, \nu \in \beta.$$

The $H_{\mu\nu}^0$ are related to the first term in eq. (7):

$$\sum_i^{\text{occ}} \langle \Psi_i | \hat{H}^0 | \Psi_i \rangle = \sum_i^{\text{occ}} \sum_{\mu\nu} c_\mu^i c_\nu^i \langle \phi_\mu | \hat{T} + v_{\text{eff}}[n_0] | \phi_\nu \rangle \equiv \sum_i^{\text{occ}} \sum_{\mu\nu} c_\mu^i c_\nu^i H_{\mu\nu}^0 \tag{14}$$

with the effective Kohn-Sham potential

$$v_{\text{eff}}[n_0] = V_{\text{ext}} + \int' \frac{n_0'}{|r-r'|} + V_{\text{xc}}[n_0] \tag{15}$$

The potential may be decomposed into atomic-like contributions

$$v_{\text{eff}}[n_0] = V(\mathbf{r}) = \sum_\alpha V_\alpha(\mathbf{r}_\alpha) \quad \text{with} \quad \mathbf{r}_\alpha = \mathbf{r} - \mathbf{R}_\alpha \tag{16}$$

The Hamilton matrix elements $H_{\mu\nu}^0$ can be distinguished into:

(I) $\quad H_{\mu\nu} = \langle \phi_\nu | \hat{T} + V_\alpha(\mathbf{r}_\alpha) | \phi_\mu \rangle + \sum_{\alpha \neq \beta} \langle \phi_\nu | V_\beta(\mathbf{r}_\beta) | \phi_\mu \rangle \quad \mu, \nu \in \{\alpha\} \quad \text{and} \tag{17}$

(II) $\quad H_{\mu\nu} = \langle \phi_\nu | \hat{T} + V_\alpha(\mathbf{r}_\alpha) + V_\beta(\mathbf{r}_\beta) | \phi_\mu \rangle + \sum_{\gamma \neq \alpha, \gamma \neq \beta} \langle \phi_\nu | V_\gamma(\mathbf{r}_\gamma) | \phi_\mu \rangle \quad \nu \in \{\alpha\}, \mu \in \{\beta\},$

$$\tag{18}$$

where $\{\alpha\}$ denotes the atomic orbitals at the centre α. Restricting the LCAO ansatz to a valence basis only, one has to assure the orthogonality of the basis functions to the core states of the other centres. (Using atomic orbitals as basis functions, the orthogonality to the core states at the same centre is guaranteed.) Denoting $|\phi_\mu\rangle$ as an orthogonalized basis function, $|\tilde{\phi}_\mu\rangle$ is the corresponding non-orthogonalized basis function, and $|\phi_c^\beta\rangle$ the core function at the centre β, then one obtains

$$|\phi_\mu\rangle = |\tilde{\phi}_\mu\rangle - \sum_{\beta \neq \alpha} \sum_{c\beta} |\phi_c^\beta\rangle(\phi_c^\beta)|\tilde{\phi}_\mu\rangle \quad \mu, \nu \in \{\alpha\}. \tag{19}$$

Applying this orthogonalization procedure to the potential contributions of the matrix elements, the matrix elements (I) are modified as

$$\langle \phi_\nu | V_\beta(\mathbf{r}_\beta) | \phi_\mu \rangle \longrightarrow (\tilde{\phi}_\mu | [V(r_\beta) - \sum_{c\beta} |\phi_c^\beta\rangle \varepsilon_c^\beta(\phi_c^\beta|] | \tilde{\phi}_\nu) \quad \alpha \neq \beta, \tag{20}$$

ε_c^β is the energy of the core state c at centre β.

Similar expressions can be obtained for the potential contributions ($\sum_{\gamma \neq \alpha, \gamma \neq \beta} \langle \phi_\nu | V_\gamma(\mathbf{r}_\gamma) | \phi_\mu \rangle$). These expressions can be interpreted as pseudopotentials [64]. Neglecting these pseudopotential contributions, which are of course distinctly smaller than the "full" potential contributions, one obtains a representation of the matrix elements containing only two-centre contributions. The approximation $V_\alpha(\mathbf{r}_\alpha) \approx V_\alpha^0(r_\alpha)$ and the use of atomic orbitals as eigenfunctions to $\hat{h}_\alpha^0 = \hat{T} + V_\alpha^0(r_\alpha)$ allows a simple representation for the matrix elements. In addition, it was shown [32] that the neglect of the pseudopotential contributions may also be justified by screening arguments. That means, this treatment could also be viewed as an LCAO formulation of a cellular Wigner-Seitz method – see [32].

Due to the nonlinear behaviour of $V_{\text{xc}}[n]$ in principle $V_\alpha(\mathbf{r}_\alpha) + V_\beta(\mathbf{r}_\beta)$ cannot simply be written as $V_\alpha^0(r_\alpha) + V_\beta^0(r_\beta)$. Therefore, we write

$$V_\alpha(\mathbf{r}_\alpha) + V_\beta(\mathbf{r}_\beta) = V[n_\alpha^0 + n_\beta^0]. \tag{21}$$

This means, instead of superposing atomic-like potentials, as in our previous works [32, 29], we superpose the atomic densities for calculating the potential contributions to the two-centre matrix elements, which is consistent to the treatment of E_{rep}, and also could be derived by applying the cluster expansion eq. (9) to the effective potential. With this approximation, we evaluate the matrixelements $H^0_{\mu\nu}$

$$H^0_{\mu\nu} = \begin{cases} \epsilon^{\text{free atom}}_\mu, & \mu = \nu \\ \langle \phi_\mu | \hat{T} + v_{\text{eff}}[n^0_\alpha + n^0_\beta] | \phi_\nu \rangle, & \mu \in \alpha, \nu \in \beta. \end{cases} \quad (22)$$

Since indices α and β indicate the atoms on which the wave functions and potentials are centered, only two-center Hamiltonian matrix elements are treated and explicitly evaluated in combination with the two-center overlap matrix elements $S_{\mu\nu}$. As follows from eq. (22), the eigenvalues of the free atom serve as diagonal elements of the Hamiltonian, thus guaranteeing the correct limit for isolated atoms.

The basis functions ϕ_ν and the input densities n^0_α are evaluated in accord with the previous scheme [29]. We employ confined Slater-type atomic orbitals for the LCAO basis ϕ_μ. These are determined by solving a modified Schrödinger equation for a free atom within SCF-LDA/GGA:

$$\left[\hat{T} + v_{\text{eff}}[n^0_\alpha] + \left(\frac{r}{r_0} \right)^2 \right] \phi_\nu(\mathbf{r}) = \epsilon_\nu \phi_\nu(\mathbf{r}). \quad (23)$$

With this approach, we determine an optimal minimal basis set by confining the atomic orbitals. These confined basis functions have been shown to form a better basis in condensed matter applications [8]. Similar ideas of confined orbitals have been discussed by Jansen and Sankey [65]. But also for applications to molecules, atomic wavefunctions turned out to be too diffuse. For the determination of the compression r_0 a variational principle can be applied [8, 9, 66]. From this and various calculations we conclude, that the choice of $r_0 = 1.85 r_{\text{cov}}$ (r_{cov} is the covalent radius of the considered element) leads to an accurate representation of the atomic-like basis functions. The initial densities are evaluated similarly. In principle, the electronic structure of small molecules can be represented well by atomic charge densities. However, numerous self-consistent calculations on molecules and solids have shown that the electron densities in these structures can be better approximated as a superposition of confined atomic densities. Since the atoms in molecules are better represented by slightly contracted pseudo-atoms [8], we confine the atomic charge densities in space by using eq. (23). By this procedure, we resemble pseudo-atoms in the molecular environment.

With the confined ϕ_μ and n^0_α, the Hamilton and overlap matrix elements can be determined with respect to the distance of the atompairs, and are tabulated. Therefore, during a geometry optimization or MD run, no integrals have to be evaluated in this scheme.

4. Second-Order Self-Consistent Charge Extension, SCC-DFTB

The previous scheme discussed above is suitable when the electron density of the many-atom structure may be represented as a sum of atomic-like densities in good approximation. The uncertainties within the standard DFTB variant, however, increase if the chemical bonding is controlled by a delicate charge balance between different atomic constituents, especially in heteronuclear molecules and in polar semiconductors. Therefore, we have extended the approach in order to improve total energies, forces, and transferability in the presence of considerable long-range Coulomb interactions. We start from equation (4) and now explicitly consider the second-order term in the density fluctuations.

In order to include associated effects in a simple and efficient TB concept, we first decompose $\delta n(\mathbf{r})$ into atom-centered contributions which decay fast with increasing distance from the corresponding center. The second-order term then reads:

$$E^2[n, n_0] = \frac{1}{2} \sum_{\alpha\beta} \iint' \left[\frac{1}{|\mathbf{r} - \mathbf{r}'|} + \frac{\delta^2 E_{xc}[n]}{\delta n \delta n'} \right] \delta n_\alpha(\mathbf{r}) \delta n_\beta(\mathbf{r}'), \qquad (24)$$

where δn is expanded analogous to eq. 8 into atom-centered contributions. Second, the δn_α may be expanded in a series of radial and angular functions:

$$\delta n_\alpha(\mathbf{r}) = \sum_{lm} K_{ml} F_{ml}^\alpha(|\mathbf{r} - \mathbf{R}_\alpha|) Y_{lm}\left(\frac{\mathbf{r} - \mathbf{R}_\alpha}{|\mathbf{r} - \mathbf{R}_\alpha|}\right) \approx \Delta q_\alpha F_{00}^\alpha(|\mathbf{r} - \mathbf{R}_\alpha|) Y_{00}, \qquad (25)$$

where F_{ml}^α denotes the normalized radial dependence of the density fluctuation on atom α for the corresponding angular momentum. While the angular deformation of the charge density, e.g. in covalently bonded systems, is usually described very well within the non-SCC approach, charge transfers between different atoms are not properly handled in many cases. Truncating the multipole expansion (25) after the monopole term accounts for the most important contributions of this kind while avoiding a substantial increase in the numerical complexity of the scheme. Also, it should be noted that higher-order interactions decay much more rapidly with increasing interatomic distance. Finally, expression (25) preserves the total charge in the system, i.e. $\sum_\alpha \Delta q_\alpha = \int \delta n(\mathbf{r})$. Substitution of (25) into (24) yields the simple final expression for the second-order energy term:

$$E_{2\text{nd}} = \frac{1}{2} \sum_{\alpha\beta}^N \Delta q_\alpha \Delta q_\beta \gamma_{\alpha\beta}, \qquad (26)$$

where

$$\gamma_{\alpha\beta} = \iint' \Gamma[\mathbf{r}, \mathbf{r}', n_0] \frac{F_{00}^\alpha(|\mathbf{r} - \mathbf{R}_\alpha|) F_{00}^\beta(|\mathbf{r}' - \mathbf{R}_\beta|)}{4\pi} \qquad (27)$$

is introduced as shorthand. In the limit of large interatomic distances, the XC contribution vanishes within LDA and $E_{2\text{nd}}$ may be viewed as a pure Coulomb interaction between two point charges Δq_α and Δq_β. In the opposite case, where the charges are located at one and the same atom, a rigorous evaluation of $\gamma_{\alpha\alpha}$ would require the knowledge of the actual charge distribution. This could be calculated with some numerical effort by expanding the charge density into an appropriate basis set. But it would require also a proper knowledge of E_{xc}, and the well known LDA for E_{xc} could even not describe accurately the important self-interaction of the on-site charge fluctuations. In order to avoid the numerical effort associated with a basis set expansion of δn and to consider – at least approximately the self-interaction contributions – we apply a simple approximation for $\gamma_{\alpha\alpha}$, which is widely used in semi-empirical quantum chemistry methods relying on Pariser's observation [67] that $\gamma_{\alpha\alpha}$ can be approximated by the difference of the atomic ionisation potential and the electron affinity. This is related to the chemical hardness η_α [68], or the Hubbard parameter U_α: $\gamma_{\alpha\alpha} \approx I_\alpha - A_\alpha \approx 2\eta_\alpha \approx U_\alpha$. Within the monopole approximation, U_α can be calculated for any atom type within DFT as the second derivative of the total energy of atom α with respect to the atomic charge. In this approximation, the influence of the enviroment on the intra-atomic electron–electron interaction, represented by U_α, is neglected,

$$E^2[n, n_0] \approx \frac{1}{2} \frac{\partial^2 E^{\text{at}}[\varrho_\alpha^0]}{\partial^2 q_\alpha} \Delta q_\alpha^2 = \frac{1}{2} U_\alpha \Delta q_\alpha^2 \rightarrow \frac{\partial \epsilon_{\text{homo}}}{\partial n_{\text{homo}}} . \quad (28)$$

The second derivative with respect to the atomic charge may be rewritten as a second derivative with respect to the atomic occupation number. Then, due to Janak's theorem [69], the second derivative of the atomic energy can be expressed as the derivative of the energy of the highest occupied atomic orbital with respect to its occupation number.

The expression for $\gamma_{\alpha\beta}$ then only depends on the distance between the atoms α and β and on the parameters U_α and U_β.

These values are therefore neither adjustable nor empirical parameters. Indeed, the necessary corrections for a TB total energy in the presence of charge fluctuations turns out to be a typical Hubbard-type correlation in combination with a long-range interatomic Coulomb interaction. Common functional forms for $\gamma_{\alpha\beta}$ have been presented by Klopman [70], and Mataga-Nishimoto [71]. However, they are based on empirical studies and may cause severe numerical problems when applied to periodic systems since Coulomb-like behavior is only accomplished for large interatomic distances. Using these expressions for periodic systems yield ill-conditioned energies with respect to the Hubbard parameters, i.e. small changes in the Hubbard parameters may result in considerable variations of the total energy, and can therefore not be used.

In order to obtain a well defined expression useful for all scale systems and consistent with the previous approximations we make an analytical approach to obtain the functional $\gamma_{\alpha\beta}$. In accordance with the Slater-type orbitals used as a basis set to solve the Kohn-Sham equations (12), we assume an exponential decay of the normalized spherical charge densities

$$n_\alpha(r) = \frac{\tau_\alpha^3}{8\pi} e^{-\tau_\alpha |\mathbf{r} - \mathbf{R}_\alpha|} .$$

Neglecting for the moment the second order contributions of E_{xc} in (24) we obtain:

$$\gamma_{\alpha\beta} = \iint{}' \frac{1}{|r - r'|} \frac{\tau_\alpha^3}{8\pi} e^{-\tau_\alpha |\mathbf{r}' - \mathbf{R}_\alpha|} \frac{\tau_\beta^3}{8\pi} e^{-\tau_\beta |\mathbf{r} - \mathbf{R}_\beta|} .$$

Integration over r' gives:

$$\gamma_{\alpha\beta} = \int \left[\frac{1}{|\mathbf{r} - \mathbf{R}_\alpha|} - \left(\frac{\tau_\alpha}{2} + \frac{1}{|\mathbf{r} - \mathbf{R}_\alpha|} \right) e^{-\tau_\alpha |\mathbf{r} - \mathbf{R}_\alpha|} \right] \frac{\tau_\beta^3}{8\pi} e^{-\tau_\beta |\mathbf{r} - \mathbf{R}_\beta|} . \quad (29)$$

Setting $R = |R_\alpha - R_\beta|$, after some coordinate transformations one gets:

$$\gamma_{\alpha\beta} = \frac{1}{R} - \mathcal{S}(\tau_\alpha, \tau_\beta, R) . \quad (30)$$

$\mathcal{S}$ is an exponentially decaying short–range function (see [57]) with

$$\mathcal{S}(\tau_\alpha, \tau_\alpha, R) \stackrel{R \rightarrow 0}{=} \frac{5}{16} \tau_\alpha + \frac{1}{R} . \quad (31)$$

If we assume that at $R = 0$ the second order contribution can be expressed approximately via the so-called chemical hardness for a spin-unpolarized atom or Hubbard parameter U_α, we obtain

$$\frac{1}{2} \Delta q_\alpha^2 \gamma_{\alpha\alpha} = \frac{1}{2} \Delta q_\alpha^2 U_\alpha$$

and therefore from (34) for the exponents:

$$\tau_\alpha = \frac{16}{5} U_\alpha .$$

This result can be interpreted by noting that elements with a high chemical hardness tend to have localized wave functions. The chemical hardness for a spin-unpolarized atom is the derivative of the highest molecular orbital with respect to its occupation number. We calculate this chemical hardness with a fully self-consistent ab initio method and therefore include the influence of the second-order contribution of E_{xc} in $\gamma_{\alpha\beta}$ for small distances where it is important. In the limit of large interatomic distances, $\gamma_{\alpha\beta} \to 1/R$ and thus represents the Coulomb interaction between two point charges Δq_α and Δq_β. This accounts for the fact that at large interatomic distances the exchange–correlation contribution vanishes within the local density approximation. In periodic systems, the long-range part can be evaluated using the standard Ewald technique, whereas the short-range part S decays exponentially and can therefore be summed over a small number of unit cells. Hence (30) is a well defined expression for extended and periodic systems.

Finally, the DFT total energy (4) is conveniently transformed into a transparent TB form,

$$E_2^{TB} = \sum_i^{occ} \langle \Psi_i | \hat{H}_0 | \Psi_i \rangle + \frac{1}{2} \sum_{\alpha,\beta}^N \gamma_{\alpha\beta} \Delta q_\alpha \Delta q_\beta + E_{rep} , \qquad (32)$$

where $\gamma_{\alpha\beta} = \gamma_{\alpha\beta}(U_\alpha, U_\beta, |\mathbf{R}_\alpha - \mathbf{R}_\beta|)$. As discussed earlier, the contribution due to $\hat{H}_0$ depends only on n_0 and is therefore exactly the same as in the previous non-SCC studies [29].

As in the non-SCC scheme (as described above) the Kohn-Sham wave functions Ψ_i are expanded into atomic orbitals (see eq. (12)). Applying the variatonal principle to eq. 32 and employing the Mulliken charge analysis for estimating the charge fluctuations $\Delta q_\alpha = q_\alpha - q_\alpha^0$ (q_α^0 represents the number of electrons of an isolated atom),

$$q_\alpha = \frac{1}{2} \sum_i^{occ} n_i \sum_{\mu \in \alpha} \sum_\nu^N \left(c_{\mu i}^* c_{\nu i} S_{\mu\nu} + c_{\nu i}^* c_{\mu i} S_{\nu\mu} \right) , \qquad (33)$$

secular equations similar to the non-scc scheme (see eq. 13) are obtained but with modified Hamilton matrix elements:

$$\begin{aligned} H_{\mu\nu} &= \langle \varphi_\mu | \hat{H}_0 | \varphi_\nu \rangle + \frac{1}{2} S_{\mu\nu} \sum_\xi^N (\gamma_{\alpha\xi} + \gamma_{\beta\xi}) \Delta q_\xi \\ &= H_{\mu\nu}^0 + H_{\mu\nu}^1 ; \quad \forall \mu \in \alpha, \nu \in \beta . \end{aligned} \qquad (34)$$

The terms $H_{\mu\nu}^0$ and the overlap matrix elements $S_{\mu\nu}$ are identical with the corresponding matrix elements of the non-SCC scheme (see eq. (14)). Since the atomic charges depend on the one-particle wave functions Ψ_i, a self-consistent procedure is required. Since the overlap matrix elements $S_{\mu\nu}$ generally extend over a few nearest neighbour distances, they introduce multiparticle interactions. The second-order correction due to charge fluctuations is now represented by the non-diagonal Mulliken charge dependent contribution $H^1_{\mu\nu}$ to the matrix elements $H_{\mu\nu}$.

In consistency with equation (11), we determine the repulsive potential as a superposition of short-range repulsive pair interactions as a function of distance,

$$E_{\text{rep}} = \frac{1}{2} \sum_{\alpha\beta} U(\mathbf{R}_{\alpha\beta}), \qquad (35)$$

by taking the difference of the SCF-LDA/GGA cohesive energy and the corresponding SCC-DFTB electronic energy for a suitable reference structure.

Since charge transfer effects are now considered explicitly, the transferability of E_{rep} is improved compared to the non-SCC approach.

A simple analytic expression for the interatomic forces for use in MD simulations is easily derived by taking the derivative of the final TB energy (32) with respect to the nuclear coordinates,

$$\mathbf{F}_\alpha = -\sum_i^{\text{occ}} n_i \sum_{\mu\nu} c_{\mu i} c_{\nu i} \left(\frac{\partial H^0_{\mu\nu}}{\partial \mathbf{R}_\alpha} - \left(\varepsilon_i - \frac{H^1_{\mu\nu}}{S_{\mu\nu}} \right) \frac{\partial S_{\mu\nu}}{\partial \mathbf{R}_\alpha} \right) - \Delta q_\alpha \sum_\xi^N \frac{\partial \gamma_{\alpha\xi}}{\partial \mathbf{R}_\alpha} \Delta q_\xi - \frac{\partial E_{\text{rep}}}{\partial \mathbf{R}_\alpha}. \qquad (36)$$

5. Extension to Spin-Polarized Systems

In many systems, the electron density is spin-polarized, i.e. the density of spin-up and spin-down electrons is different. This is the case for most free atoms, many transition metal systems, and radicals. For this reason, a generalization of the SCC tight-binding formalism for spin-polarized systems is desirable. In order to obtain such an extension, we need to consider the dependence of the exchange–correlation energy on the spin densities. In the spin polarized case:

$$E_{\text{xc}} = E_{\text{xc}}[n(\mathbf{r}), n_s(\mathbf{r})], \qquad n(\mathbf{r}) = n_\uparrow(\mathbf{r}) + n_\downarrow(\mathbf{r}), \qquad n_s(\mathbf{r}) = n_\uparrow(\mathbf{r}) - n_\downarrow(\mathbf{r})$$

$$n_\sigma(\mathbf{r}) = \sum_i^{\text{occ}} |\Psi_{i\sigma}|^2, \qquad \Psi_{i\sigma} = \sum_\nu^N c_{\nu i\sigma} \varphi_\nu(\mathbf{r} - \mathbf{R}_\alpha), \qquad \sigma = \uparrow, \downarrow. \qquad (37)$$

Since we start from our spin-unpolarized self-consistent charge Hamiltonian, we have to expand E_{xc} at $n_s(\mathbf{r}) = 0$ which results in:

$$E_{\text{xc}} \approx E_{\text{xc}}[n_0, 0] + \int \left. \frac{\delta E_{\text{xc}}}{\delta n} \right|_{n_0,0} \delta n + \int \left. \frac{\delta E_{\text{xc}}}{\delta n_s} \right|_{n_0,0} \delta n_s + \frac{1}{2} \iint' \left. \frac{\delta^2 E_{\text{xc}}}{\delta n \, \delta n'} \right|_{n_0,0} \delta n \, \delta n'$$

$$+ \iint' \left. \frac{\delta^2 E_{\text{xc}}}{\delta n \, \delta n'_s} \right|_{n_0,0} \delta n \, \delta n'_s + \frac{1}{2} \iint' \left. \frac{\delta^2 E_{\text{xc}}}{\delta n_s \, \delta n'_s} \right|_{n_0,0} \delta n_s \, \delta n'_s \qquad (38)$$

which is correct up to second order in δn and δn_s and thus at the same level of accuracy as the SCC tight binding scheme described earlier. Analyzing eq. (38), we note that the first, second, and fourth term are not dependent on the spin density and included in the SCC-DFTB model. The third and fifth term vanish since $E_{\text{xc}}[n, n_s]$

$= E_{xc}[n, -n_s]$ for all density functionals that do not include spin–orbit coupling. The remaining contribution is the sixth term which will be called $E_{xc}^s[n, n_s]$. For local and semilocal density functionals, the expression for E_{xc}^s can be replaced by:

$$E_{xc}^s = \frac{1}{2} \int \left.\frac{\delta^2 E_{xc}}{\delta n_s^2}\right|_{n_0,0} \delta n_s^2 . \qquad (39)$$

In order to simplify eq. (39) further, we expand the spin density n_s into contributions arising from the different angular momentum states l of the atoms α in the system:

$$n_s(\mathbf{r}) = \sum_\alpha \sum_{l \in \alpha} p_{\alpha l s} f_{\alpha l}(\mathbf{r} - \mathbf{R}_\alpha) \qquad (40)$$

In the following, we assume that the contributions arising from different atoms are non-overlapping and that $f_{\alpha l}(\mathbf{r})$ is radially symmetric and universal for a specific atom type. These approximations are similar to those made in the derivation of the SCC-DFTB method. Then, E_{xc}^s can be written as:

$$\begin{aligned}
E_{xc}^s &\approx \frac{1}{2} \sum_\alpha \sum_{l \in \alpha} \sum_{l' \in \alpha} p_{\alpha l s} p_{\alpha l' s} \int f_{\alpha l}(\mathbf{r} - \mathbf{R}_\alpha) \left.\frac{\delta^2 E_{xc}}{\delta n_s^2}\right|_{n_0,0} f_{\alpha l'}(\mathbf{r} - \mathbf{R}_\alpha) \\
&\approx \frac{1}{2} \sum_\alpha \sum_{l \in \alpha} \sum_{l' \in \alpha} p_{\alpha l s} p_{\alpha l' s} W_{\alpha l l'} .
\end{aligned} \qquad (41)$$

The spin populations $p_{\alpha l s}$ can be determined from a Mulliken analysis:

$$p_{\alpha l s} = p_{\alpha l \uparrow} - p_{\alpha l \downarrow} , \qquad p_{\alpha l \sigma} = \frac{1}{2} \sum_i^{occ} n_{i\sigma} \sum_{\mu \in l,\alpha} \sum_\nu^N \left(c_{\mu i \sigma}^* c_{\nu i \sigma} S_{\mu\nu} + c_{\nu i \sigma}^* c_{\mu i \sigma} S_{\nu\mu} \right) \qquad (42)$$

and the quantities $W_{\alpha l l'}$ are determined from free-atom calculations within the density-functional used to construct the DFTB parameters. As a result, the (now spin-dependent) Hamiltonian matrix elements are defined by:

$$\begin{aligned}
H_{\mu\nu\uparrow} &= H_{\mu\nu}^0 + H_{\mu\nu}^1 + \frac{1}{2} S_{\mu\nu} \sum_{l'' \in \alpha} (W_{ll''\alpha} + W_{l'l''\alpha}) p_{\alpha l'' s} , \\
H_{\mu\nu\downarrow} &= H_{\mu\nu}^0 + H_{\mu\nu}^1 - \frac{1}{2} S_{\mu\nu} \sum_{l'' \in \alpha} (W_{ll''\alpha} + W_{l'l''\alpha}) p_{\alpha l'' s} ; \qquad \forall \mu \in \alpha, l \text{ and } \nu \in l', \alpha ,
\end{aligned} \qquad (43)$$

where $H_{\mu\nu}^0$ and $H_{\mu\nu}^1$ are defined by the SCC-DFTB defined earlier. One can clearly see that eq. (43) only modifies on-site matrix elements, i.e. the orbitals μ and ν must belong to the same atoms in order to obtain a non-vanishing correction.

6. Results

In order to validate the SCC-DFTB approach we now concentrate on presenting results of first successful applications of the SCC-DFTB scheme to a wide class of systems which are of interest in chemistry, physics and biology. Since applications to anorganic semiconductor structures and investigations of biomolecular systems are covered by two subsequent papers in this Special Issue [61, 63], we outline here only benchmarks and first applications to organic materials.

Table 1

Hydrogenation reactions (kcal/mol) for small organic molecules in comparison with DFT-LSD calculations and experiment [39]

reaction	SCC-DFTB	LSD	exp.
$CH_3CH_3 + H_2 \rightarrow 2CH_4$	20	18	19
$CH_3NH_2 + H_2 \rightarrow CH_4 + NH_3$	23	24	26
$CH_3OH + H_2 \rightarrow CH_4 + H_2O$	32	28	30
$NH_2NH_2 + H_2 \rightarrow 2NH_3$	30	43	48
$HOOH + H_2 \rightarrow 2H_2O$	101	80	86
$CH_2CH_2 + 2H_2 \rightarrow 2CH_4$	71	67	57
$CH_2NH + 2H_2 \rightarrow CH_4 + NH_3$	66	67	64
$CH_2O + 2H_2 \rightarrow CH_4 + H_2O$	65	67	59
$NHNH + 2H_2 \rightarrow 2NH_3$	56	89	68
$C_2H_2 + 3H_2 \rightarrow 2CH_4$	124	131	105
$HCN + 3H_2 \rightarrow CH_4 + NH_3$	88	102	76
$CO + 3H_2 \rightarrow CH_4 + H_2O$	83	93	63
$N_2 + 3H_2 \rightarrow 2NH_3$	37	71	37

6.1 Organic molecules

For our first benchmark, we have calculated the reaction energies of 36 processes between small closed shell molecules containing oxygen, nitrogen, carbon, and hydrogen from [72], some of them shown in Table 1. We have found a mean absolute deviation from experiment of 12.5 kcal/mol for the SCC-DFTB, compared to 11.1 kcal/mol for the DFT-LSD calculations [73]. Further, considering the optimized geometries of a 63 organic molecules test set from Ref. [74], the mean absolute deviations from experiment in the bond lenghts and bond angles are $\Delta R = 0.010$ Å and $\Delta \theta = 1.95°$ [75] respectively, compared to $\Delta R = 0.017$ Å and $\Delta \theta = 2.01°$ by using the semi-empirical AM1 method [74].

The improvement over the non-SCC treatment is impressively demonstrated for systems with a delicate counterbalance between ionic and covalent bonding contributions, as e.g. in formamide (cf. Table 2). The DFTB method overestimates the equalization of single and double bonds in the amide and carboxyl groups. This is exclusively due to too much charge flow (of nearly one electron) from carbon to oxygen, clearly indicating the need for a self-consistent charge redistribution. SCC can considerably improve vibrational frequencies of simple molecules, like for example CO_2, in which a wrong charge transfer crucially affects force constants. As in the formamid molecule a too large charge transfer from carbon to oxygen is obtained, connected with an underesti-

Table 2

Optimized geometries for three different methods

formamide	DFTB	SCC-DFTB	DFT-LSD [39]	exp. [39]
C=O	1.296	1.224	1.223	1.193
C–N	1.296	1.382	1.358	1.376
N–H	1.003	0.996	1.022	1.002
C–H	1.130	1.131	1.122	1.102
OCN	127.0	125.5	124.5	123.8

mation of the CO bond strength. The symmetric and antisymmetric stretching modes ($\sum_g$ and $\sum_u$) in CO_2 change from 1458 and 1849 cm^{-1} in DFTB to 1348 and 2305 cm^{-1} in SCC-DFTB in good agreement with the experimental values, 1333 and 2349 cm^{-1}. Frequencies have further been tested for a serious of 33 (O, N, C, and H)-containig molecules from Ref. [76] yielding 6.4% mean absolute deviation of vibrational frequencies from the experiment [75].

6.2 Large organic molecules: The example of PTCDA

During the last years, organic materials with semiconducting properties have attracted much interest due to their large technological potential [77]. Albeit commercial devices like waveguide-coupled detectors or electroluminescent diodes are becoming available, the atomistic understanding of their electronic and optical properties is still limited. The main reason is the large size of the molecules under study: Already the most simple investigations require a reliable modelling of the geometry and electronic structure, and contrary to inorganic semiconductors with unit cells containing typically only two atoms, the unit cell of organic semiconductors contains at least one or two large molecules, consisting of 20 to 50 atoms each.

Among the molecules under study, 3-4-9-10-perylene tetracarboxylic dyanhydride (PTCDA) plays a special role because high quality epitaxial films can be grown on a large variety of substrates [78 to 81]. This is directly related to the special geometry of the molecule: Isolated PTCDA is a planar molecule consisting of 38 atoms in rectangular D_{2h} symmetry, forming monoclinic crystals with two nearly coplanar molecules in the unit cell [78]. For some substrates, already the first adsorbed monolayer of PTCDA has a unit cell compatible with the monoclinic bulk [80, 81], resulting in epitaxial film growth with bulk-like properties.

The first step for modelling of such a system is an investigation of the isolated molecule. As was discussed elsewhere in more detail [82], a reliable DFTB simulation of the geometric and vibrational properties of PTCDA required a careful definition of a transferable repulsive energy for the various C–C bond lengths occuring (1.39 to 1.48 Å). Based on this systematic improvement, our relaxed SCC-DFTB geometry compares well with a fully self-consistent DFT calculation based on the NRLMOL code [83, 84, 85, 86, 87]: The r.m.s. deviation of the C–C, C–O and C=O bonds between the two approaches was < 0.5%, defining a sound basis for subsequent investigations of the vibrational properties of the molecules.

Figure 2. visualizes the HOMO and LUMO orbitals of PTCDA obtained with the DFTB method, and the deformation occurring between the relaxed electronic ground and excited states is shown in Fig. 3. This overall deformation of the molecule can be projected on the elongation patterns of the internal vibrations. Due to symmetry restrictions of the group D_{2h}, only vibrational modes of the highest symmetry A_g are elongated. Expressing the reorganization energy of each mode j as

$$\Delta E_j = \alpha_j^2 \hbar \omega_j, \qquad (44)$$

it can be shown that to leading order, the resonant Raman cross sections of these modes are proportional to [82, 88, 89]

$$\sigma_j \propto \alpha_j^2 (1 + \langle n_j \rangle)(\hbar \omega_j)^2, \qquad (45)$$

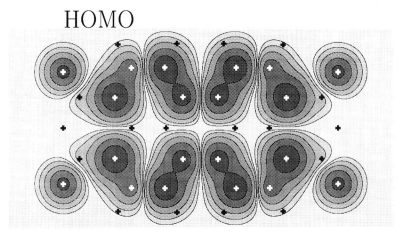

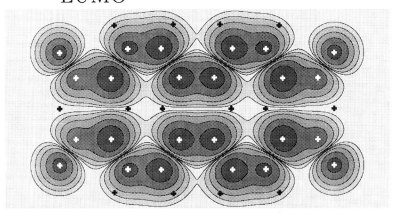

Fig. 2. Geometry of PTCDA in its electronic ground state, and charge density contours of the $A_u(xyz)$ HOMO (upper) and $B_{1g}(yz)$ LUMO (lower), at a distance of one Bohr radius above or below the plane of the molecule. The atomic positions are indicated with crosses, with the three on the l.h.s. and r.h.s. oxygen, and the four upper and four lower hydrogen, surrounding the central carbon network. As both are π-orbitals composed of atomic $2p_z$ wave functions, the hydrogen atoms do not contribute. The highest electronic charge density occuring is $|\varrho_{max}| = 10^{-1.5} e/a_B^3$, and the charge contours are shown for $|\varrho| = 10^{-2.0}$ to $10^{-5.5} e/a_B^3$ in equidistant logarithmic steps. Between consecutive lobes, both the HOMO and the LUMO wavefunctions change sign

where $\langle n_j \rangle$ is the thermal occupation of each mode. As has been shown elsewhere in more detail, the calculated Raman cross sections are in good agreement with the experimental results: The modes agree within an average absolute deviation of 22 cm^{-1} (r.m.s. deviation 28 cm^{-1}), and the cross sections of the dominating Raman modes are reproduced within a factor of about 2 [82]. Both the mode frequencies and the precision achieved for the Raman cross sections agree favourably with recent calculations of perylene [90] and PTCDA [91]. Furthermore, from a detailed comparison with the Raman cross sections at different laser energies, it was demonstrated that no significant charge transfer occurs in the region of the so-called charge-transfer exciton [54, 77, 82, 92].

PTCDA → PTCDA*

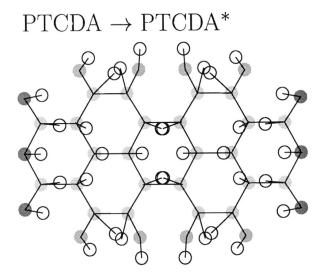

Fig. 3. Deformation patterns of PTCDA* as compared with the geometry of the molecule in the electronic ground state. Full grey circles correspond to the ground state geometry, and open circles to the geometry of the deformed PTCDA. In order ot make the geometry changes more evident, the deformations have been multiplied by a factor of 40

For single PTCDA molecules, e.g. in solution, the optical absorption spectra contain pronounced vibrational subbands. As was shown elsewhere in more detail, they can be interpreted in terms of the Franck-Condon factors of the A_g breathing modes, resulting in a Poisson distribution for the normalized transition probability of each mode j:

$$P_j(|0_g\rangle \to |n_e\rangle) = \frac{\alpha_j^{2n}}{n!} e^{-\alpha_j^2}. \quad (46)$$

The vibrational bands of the whole molecule can be obtained from a multiplication of the Poisson distributions P_j related to each breathing mode j. For the case of PTCDA dissolved in DMSO, it was shown that the deformation pattern of Fig. 3 gave the best agreement with measured absorption spectra when all reorganization energies where scaled down by a factor of 0.65, allowing for a rough estimate how the solvent dielectric function influences the dissolved molecule [82]. With this rescaling, the overall absorption lineshape obtained in the DFTB model agrees quite well with a recent calculation of the optical aborption of PTCDA [91].

7. Summary

We have presented a straightforward extension and successful implementation of the standard TB theory to operation in a self-consistent charge mode based on a second-order expansion of the Kohn-Sham total energy functional as calculated within DFT. By this we successfully address a key problem of electronic structure theory, the development of robust, accurate, rapid, and generally transferable methods for ab-initio based simulation and characterization of large scale molecular and condensed systems. The analytically derived charge dependent contribution to the total energy expression can be physically interpreted to describe the chemical hardness for vanishing distances and to represent the Coulomb interaction between point charges at large distances. Moreover, for intermediate distances the energy correction does not contain any empirical functional but is entirely consistent with approximations that frequently enter TB schemes. We show that the scheme is numerically stable for all scale systems and by describing various benchmark results clearly demonstrate the method's successful operation at sufficient accuracy on very different systems and materials. We firstly report on an extension of the method to spin-polarized systems, which shows the potential of the scheme for improving various TB applications in material science.

The method can be understood as a general SCC extension of TB theory offering the great advantage to incorporate any atom type in a straightforward manner. This will not only stimulate MD applications for large-scale semiconductor structures and biological systems, but also for other challenging types of materials.

Acknowledgements We thank Th. Köhler, M. Sternberg, U. Stephan and E. Tajkhorshid for helpful disscussions and greatfully acknowledge support from DFG, DAAD and the US-NSF.

References

[1] A. SZABO and N.S. OSTLUND, Modern Quantum Chemistry: Introduction to Advanced Electronic Structure Theory, McGraw-Hill, New York 1989.
[2] D. CEPERLY and B. ALDER, Phys. Rev. B **36**, 2092 (1987).
[3] R. AHLRICHS, S.D. ELLIOTT, and U. HUNIAR, Ber. Bunsenges. Phys. Chem. **102**, 795 (1998).
[4] W.M.C. FOULKES, M. NEKOVEE, R.L. GAUDOIN, M.L. STEDMAN, R.J. NEEDS, R.Q. HOOD, G. RAJAGOPAL, M.D. TOWLER, P.R.C. KENT, Y. LEE, W.-K. LEUNG, A.R. PORTER, and S.J. BREUER, High Performance Computing, Plenum Press, New York 1998.
[5] P. HOHENBERG and W. KOHN, Phys. Rev. **136**, B864 (1964).
[6] W. KOHN and L.J. SHAM, Phys. Rev. **140**, A1133 (1965).
[7] M.R. PEDERSON et al., phys. stat. sol. (b) **217**, 197 (2000).
[8] H. ESCHRIG, The Optimized LCAO Method and Electronic Structure of Extended Systems, Akademie-Verlag, Berlin 1988.
[9] K. KOEPERNIK and H. ESCHRIG, Phys. Rev. B **59**, 1743 (1999).
[10] O.K. ANDERSON, Phys. Rev B **12**, 3060 (1975).
H.L. SKRIVER, The LMTO Method, Springer-Verlag, Heidelberg 1984.
[11] P.R. BRIDDON, phys. stat. sol. (b) **217**, 131 (2000).
[12] J. CHELIKOWSKY et al., phys. stat. sol. (b) **217**, 173 (2000).
P. D. TEPESCH and A. A. QUONG, phys. stat. sol. (b) **217**, 377 (2000).
[13] R. CAR and M. PARRINELLO, Phys. Rev. Lett. **55**, 2471 (1985).
[14] M. PARRINELLO, Solid State Commun. **102**, 107 (1997).
[15] P. DEÁK, phys. stat. sol. (b) **217**, 9 (2000).
[16] C. M. GORINGE, D. R. BOWLER, and E. HERNANDEZ, Rep. Prog. Phys. **60**, 1447 (1997).
[17] J.C. SLATER and G.F. KOSTER, Phys. Rev. **94**, 1498 (1954).
[18] S. FROYEN and W. A. HARRISON, Phys. Rev. B **20**, 2420 (1979).
[19] W. A. HARRISON, Phys. Rev. B **34**, 2787 (1986).
[20] D. J. CHADI, Phys. Rev. Lett. **43**, 79 (1979).
[21] J.-M. JANCU, R. SCHOLZ, F. BELTRAM, and F. BASSANI, Phys. Rev. B **57**, 6493 (1998).
[22] R. SCHOLZ, J.-M. JANCU, and F. BASSANI, Mater. Res. Soc. Proc. **491**, 383 (1998).
[23] A. DI CARLO, Mater. Res. Soc. Proc. **491**, 389 (1998).
[24] C. Z. WANG and K. M. HO, Phys. Rev. Lett. **70**, 611 (1993).
[25] P. ORDEJON, D. LEBEDENKO, and M. MENON, Phys. Rev. B **50**, 5645 (1994).
[26] M. MENON and K. R. SUBBASWAMY, Phys. Rev. B **55**, 9231 (1997).
[27] O. K. ANDERSEN and O. JEPSEN, Phys. Rev. Lett. **53**, 2571 (1984).
[28] R.E. COHEN, M.J. MEHL, and D.A. PAPACONSTANTOPOULOS, Phys. Rev. B **54**, 14694 (1994).
[29] D. POREZAG, TH. FRAUENHEIM, TH. KÖHLER, G. SEIFERT, and R. KASCHNER, Phys. Rev. B **51**, 12947 (1995).
[30] O. F. SANKEY and D. J. NIKLEWSKI, Phys. Rev. B **40**, 3979 (1989).
[31] W. BIEGER, G. SEIFERT, H. ESCHRIG, and G. GROSSMANN, Z. Phys. Chemie (Leipzig) **266**, 751 (1985).

[32] G. Seifert, H. Eschrig, and W. Bieger, Z. Phys. Chemie (Leipzig) **267**, 529 (1986).
[33] P. Blaudeck, Th. Frauenheim, D. Porezag, G. Seifert, and E. Fromm, J. Phys. C**4**, 6389 (1992).
[34] Th. Frauenheim, G. Jungnickel, Th. Köhler, and U. Stephan, J. Non-Cryst. Solids **182**, 186 (1995).
Th. Frauenheim, G. Jungnickel, Th. Köhler, P. Sitch, and P. Blaudeck, Proc. 1-st Specialist Meeting of Amorphous Carbon, Eds. S. R. P. Silva, J. Robertson, W. I. Milne, and G. A. J. Amaratunga, World Scientific Publ. Co., Singapore 1997 (p. 59).
[35] Th. Köhler, M. Sternberg, D. Porezag, and Th. Frauenheim, phys. stat. sol. (a) **154**, 69 (1996).
[36] D. Porezag, G. Jungnickel, Th. Frauenheim, G. Seifert, A. Ayuela, and M.R. Pederson, Appl. Phys. A**64**, 321 (1997).
[37] G. Seifert, K. Vietze, and R. Schmidt, J. Phys. B**29**, 5183 (1996).
[38] Th. Frauenheim, F. Weich, Th. Köhler, S. Uhlmann, D. Porezag, and G. Seifert, Phys. Rev. B **52**, 11492 (1995).
[39] A. Sieck, D. Porezag, Th. Frauenheim, M. R. Pederson, and K. A. Jackson, Phys. Rev. A **56**, 4890 (1997).
[40] P. Sitch and Th. Frauenheim, J. Phys. C **8** (1996) 6873.
[41] J. Widany, Th. Frauenheim, D. Porezag, Th. Köhler and G. Seifert, Phys. Rev. B **53**, 443 (1996).
[42] Th. Frauenheim, G. Jungnickel, P. Sitch, M. Kaukonen, F. Weich, J. Widany, and D. Porezag, Diamond & Rel. Materials, **7**, 348 (1998).
[43] G. Seifert, P.W. Fowler, D. Mitchell, D. Porezag, and Th. Frauenheim, Chem. Phys. Lett. **268**, 352 (1997).
[44] F. Weich, J. Widany, Th. Frauenheim, Phys. Rev. Lett. **78**, 3326 (1997).
[45] E. Rauls, J. Elsner, R. Gutierrez, and Th. Frauenheim, Solid State Commun. **111**, 459 (1999).
[46] R. Kaschner, Th. Frauenheim, Th. Köhler, and G. Seifert, J. Comp.-Aided Mater. Design **4**, 53 (1997).
[47] J. Elsner, M. Haugk, G. Jungnickel, and Th. Frauenheim, J. Mater. Chem. **6**, 1649 (1996).
[48] G. Seifert, R. Kaschner, M. Schöne, and G. Pastore, J. Phys. C **10**, 1175 (1998).
[49] W. Foulkes and R. Haydock, Phys. Rev. B **39**, 12520 (1989).
[50] J. Harris, Phys. Rev. B **31**, 1770 (1985).
[51] A.P. Horsfield, Phys. Rev. B **56**, 6594 (1997).
[52] H. Sambe, R.H. Felton, J. Chem. Phys. **61**, 3862 (1974).
[53] A. A. Demkov, J. Ortega, O. F. Sankey, and M. P. Grumbach, Phys. Rev. B **52**, 1618 (1995).
[54] V. Bulović, and S.R. Forrest, Chem. Physics **210**, 13 (1996).
[55] P. Stumm and D. Drabold, Phys. Rev. Lett. **79**, 677 (1997).
[56] M.-H. Tsai, O. F. Sankey, and J. D. Dow, Phys. Rev. B **46**, 10464 (1992).
[57] M. Elstner, D. Porezag, G. Jungnickel, J. Elsner, M. Haugk, Th. Frauenheim, S. Suhai, and G. Seifert, Phys. Rev. B **58**, 7260 (1998).
[58] A. P. Sutton, M. W. Finnis, D. G. Pettifor, and Y. Ohata, J. Phys. C **21**, 35 (1988).
[59] M. Haugk, J. Elsner, Th. Heine, Th. Frauenheim, and G. Seifert, J. Comp. Mater. Sci. **13**, 239 (1999).
[60] J. Elsner, R. Jones, P.K. Sitch, D. Porezag, M. Elstner, Th. Frauenheim, M. Heggie, S. Öberg, and P. Briddon, Phys. Rev. Lett. **79**, 3672 (1997).
[61] M. Haugk et al., phys. stat. sol. (b) **217**, 473 (2000).
[62] M. Sternberg, G. Galli, and Th. Frauenheim, Comp. Phys. Commun. **118**, 200 (1999).
[63] M. Elstner et al., phys. stat. sol. (b) **217**, 357 (2000).
[64] J.C. Phillips and L. Kleinman, Phys. Rev. **116**, 287 (1959).
[65] R. Jansen and O. Sankey, Phys. Rev. B **36**, 6520 (1987).
[66] G. Seifert and H. Eschrig, Proc. 18th Internat. Symp. Electronic Structure of Metals and Alloys, Ed. P. Ziesche, Dresden 1988 (p. 105).
[67] R. Pariser, J. Chem. Phys. **24**, 125 (1956).
[68] R.G. Parr and R.G. Pearson, J. Amer. Chem. Soc. **105**, 1503 (1983).
[69] J.F. Janak, Phys. Rev. B **18**, 7165 (1978).
[70] G. Klopman, J. Amer. Chem. Soc. **28**, 4550 (1964).
[71] N. Mataga and K. Nishimoto, Z. phys. Chemie (Frankfurt) **13**, 140 (1957).
[72] J. Andzelm and E. Wimmer, J. Chem. Phys. **96**, 1280 (1992).
[73] M. Elstner, D. Porezag, G. Jungnickel, Th. Frauenheim, S. Suhai, and G. Seifert, Mater. Res. Soc. Symp. Proc. **491**, 131 (1998).

[74] J. S. Dewar, E. Zoebisch, E. F. Healy, J. J. P. Stewart, J. Amer. Chem. Soc. **107**, 3902 (1985).
[75] M. Elstner, D. Porezag, G. Jungnickel, Th. Frauenheim, and S. Suhai, to be published.
[76] J. A. Pople, H. B. Schlegel, R. Krishnan, D. J. Defrees, J. S. Binkley, M. S. Frisch, R. A. Whiteside R. F. Hout, and W. J. Hehre, Internat. J. Quant. Chem., Quantum Chem. Symp. **15**, 269 (1981).
[77] S.R. Forrest, Chem. Rev. **97**, 1793 (1997) and references therein.
[78] A.J. Lovinger, S.R. Forrest, M.L. Kaplan, P.H. Schmidt, and T. Venkatesan, J. Appl. Phys. **55**, 476 (1984).
[79] K. Akers, R. Aroca, A.-M. Hor, and R.O. Loutfy, J. Phys. Chem. **91**, 2954 (1987).
[80] E. Umbach, C. Seidel, J. Taborski, R. Li, and A. Soukopp, phys. stat. sol. (b) **192**, 389 (1995).
[81] T. Schmitz-Hübsch, T. Fritz, F. Sellam, R. Staub, and K. Leo, Phys. Rev. B **55**, 7972 (1997).
[82] R. Scholz, A.Yu. Kobitski, T.U. Kampen, M. Schreiber, and D.R.T. Zahn, G. Jungnickel, M. Elstner, M. Sternberg, and Th. Frauenheim, submitted to Phys. Rev. B (1999).
[83] M.R. Pederson and C.C. Lin, Phys. Rev. B **35**, 2273 (1987).
[84] M.R. Pederson and K.A. Jackson, Phys. Rev. B **41**, 7453 (1990).
[85] K.A. Jackson and M.R. Pederson, Phys. Rev. B **42**, 3276 (1990).
[86] M.R. Pederson and K.A. Jackson, Phys. Rev. B **43**,7312 (1991).
[87] D.V. Porezag and M.R. Pederson, Phys. Rev. B **54**, 7830 (1996).
[88] A.B. Myers and R.A. Mathies, in: Biological Applications of Raman Spectroscopy, Vol. 2, Ed. T.G. Spiro, Wiley, New York 1987 (p. 1).
[89] F. Markel, N.S. Ferris, I.R. Gould, and A.B. Myers, J. Amer. Chem. Soc. **114**, 6208 (1992).
[90] H. Shinohara, Y. Yamakita, and K. Ohno, J. Molec. Struct. **442**, 221 (1998).
[91] K. Gustav, M. Leonhardt, and H. Port, Monatsh. Chemie **128**, 105 (1997).
[92] S.R. Forrest, M.L. Kaplan, and P.H. Schmidt, J. Appl. Phys. **55**, 1492 (1984).

phys. stat. sol. (b) **217**, 63 (2000)

Subject classification: 71.15.–m; 71.20.Ps; 71.55.Ht; 75.25.+z; 75.50.Ee; S9.11

The Periodic Hartree-Fock Method and Its Implementation in the CRYSTAL Code

R. DOVESI (a), R. ORLANDO (a), C. ROETTI (a), C. PISANI (a), and V.R. SAUNDERS (b)

(a) *Dipartimento di Chimica IFM, Università di Torino, via Giuria 5, I-10125 Torino*

(b) *CLRC Laboratory, Daresbury, Warrington, Cheshire, WA4 4AD, U.K.*

(Received August 10, 1999)

The present chapter discusses the Hartree-Fock (HF) method for periodic systems with reference to its implementation in the CRYSTAL program. The HF theory is shortly recalled in its Closed Shell (CS), Unrestricted (UHF) and Restricted open shell (RHF) variants; its extension to periodic systems is illustrated. The general features of CRYSTAL, the periodic ab initio linear combination of atomic orbitals (LCAO) program, able to solve the CS, RHF and UHF, as well as Kohn-Sham equations, are presented. Three examples illustrate the capabilities of the CRYSTAL code and the quality of the HF results in comparison with those obtained with the Local Density Approximation using the same code and basis set: NiO in its ferro-magnetic and anti-ferromagnetic structure, trapped electron holes in doped alkaline earth oxides, and F-centres in LiF.

1. Introduction

For three decades the Hartree-Fock (HF) method has been the most popular scheme for the investigation of the electronic structure of atoms, molecules and clusters, either as such or as a well defined starting point for more sophisticated techniques. Since the beginning of the seventies, theoretical chemists have succeeded in developing powerful, general purpose, universally adopted HF-based ab initio computer programs [1]. The features of the most successful codes are essentially the same: the linearized HF-Roothaan equations are solved by using a few localized functions per atom (usually indicated as "atomic orbitals", AOs) as a basis set; in turn, the AOs are expressed as linear combinations of Gaussian type functions (GTF) with appropriate exponents and "contraction" coefficients; after obtaining the molecular orbitals (MO), eigenvectors of the Fock matrix, through a self consistent field (SCF) procedure, the correlation correction to the ground state wave function and properties, and the description of excited states are usually performed by means of configuration interaction (CI) or perturbation techniques. A posteriori, theoretical quantum chemistry appears to have followed quite a linear development from the original formulation of the HF equations by Fock, Hartree and Slater [2, 3], their linearized expression by Roothaan and Hall [4, 5], Boys' anticipation of present day techniques [6 to 8], to the full implementation of efficient computational tools. This has required a huge effort and ingenuity as concerns the definition of standard and well assessed basis sets [9, 10], the development of powerful *analytic* integration techniques [11 to 13], the efficient treatment of hundreds of thousand configurations in CI calculations [14], the new ideas for the solution of the correlation problem [15]. In parallel, a set of semi-empirical electronic structure theories have been developed which, starting from the HF equations, on the one hand drastically reduce the computational effort by disregarding or approximating most of the bielectronic integrals, on the other incorporate correlation effects by way of largely intuitive approximations. A survey of these methods can be found in Ref. [16].

Density Functional Theory [17, 18] (DFT), that had played for long time a minor role in molecular quantum chemistry, has become more and more popular in the last years, and DFT related computational schemes are now available in standard ab initio molecular codes [19].

The situation in computational solid state physics and chemistry is quite different from the one sketched above for molecular quantum chemistry.

A great variety of formal schemes, computational methods and techniques are adopted for the determination of the electronic structure of crystalline compounds; the various proposals (many of them are presented in this book) differ in many respects, including the basis set (numerical, plane waves, localized, mixed), solution techniques of the basic equations, selected Hamiltonian and all its ingredients. The relative merits and limits of the various proposals are difficult to assess at the moment, as many of the computer codes are not available to the scientific community and are often in rapid evolution. The main reason for this situation is that computational solid state physics is a relatively young science with respect to molecular quantum chemistry: ab initio methods appeared in the late seventies, and the first general, portable, publicly available code, CRYSTAL88 [20], was distributed only ten years ago; in solid state computational physics, it is just at the beginning that "filtering" process that in molecular quantum chemistry led, for example, to abandon Slater and lobe type orbitals [21], owing to the difficulties encountered in the analytic evaluation of many mono- and bielectronic integrals in the former case [22], and because of the poor numerical accuracy obtained for integrals involving high angular quantum number AOs with the latter basis type.

There are, however, two features that are common to most periodic computer programs:

a) the use of DFT based Hamiltonians, the only exception being the CRYSTAL [23, 24] code, that can solve both the HF and Kohn-Sham (KS) equations;

b) the prevalent use of numerical rather than analytical techniques.

As a matter of fact, the HF approach has never been very popular among solid state physicists, for many reasons; probably the most important two are:

– HF performs poorly for the electron gas, the simplest of all periodic systems; it has very often been assumed that this remains true (or possibly gets worse) for most properties of real solids.

– The non-local exchange term makes the HF equations more difficult to solve than the KS ones, where the exchange–correlation potential $\mu_{xc}(\mathbf{r};[\varrho])$ is simply a multiplicative operator, no matter how complicated its determination.

Early attempts at formulating and implementing HF computational schemes for periodic systems had scant success, in spite of the high quality and generality of some of them [25 to 30]. Only in recent years has it become possible to formulate a fair judgement about the usefulness of the HF approach in solid state physics, since the advent of powerful computers and of general purpose computational schemes [20, 23, 24, 31], which have allowed us to assess its performance for a variety of systems [32].

In this chapter we will recall the general features of the HF scheme, both in its formulation for closed and open shell systems, and its generalization to periodic compounds when a local basis is used. Explicitly referring to the implementation in the CRYSTAL program, we will shortly recall some of the most critical points of this implementation and show that all the relevant interactions are evaluated analytically. This requires a much bigger implementation effort, but ensures a very high numerical accuracy, that is necessary in many cases, for example, in chemical reactions [33, 34] and in the study of magnetic interactions [35 to 37], that are of the order of a fraction of a kJ/mol.

The chapter ends up with a few applications; all of them refer to cases in which paramagnetic species are involved; the UHF results will be compared with DFT data obtained with the same basis set.

2. Hartree-Fock Theory

The non-relativistic Born-Oppenheimer Hamiltonian operator for a finite cluster of N nuclei and m electrons has the form

$$\widehat{H} = -\frac{1}{2}\sum_{i=1}^{m} \nabla_i^2 - \sum_{i=1}^{m}\sum_{A=1}^{N} \frac{Z_A}{|\mathbf{r}_i - \mathbf{A}|} + \sum_{i=1}^{m}\sum_{j=1}^{i-1} |\mathbf{r}_i - \mathbf{r}_j|^{-1} + \mathcal{N}, \qquad (1)$$

where Z_j denotes the atomic number of the j-th nucleus, and

$$\mathcal{N} = \sum_{A=1}^{N}\sum_{B=1}^{A-1} \frac{Z_A Z_B}{|\mathbf{A} - \mathbf{B}|} \qquad (2)$$

is the nuclear repulsion energy. In the above equations atomic units are used. The Hamiltonian of Eq. (1) is parametrically dependent only on the nuclear coordinates, $\mathbf{A}$. A trial HF wave function is a single determinantal function constructed by antisymmetrizing a Hartree product of m single particle spin orbitals,

$$\Psi = \widehat{A}\left[\prod_{i}^{m} (\psi_i(\mathbf{r}_i)\, \chi(i))\right], \qquad (3)$$

(ψ and χ denote the orbital and spin functions, respectively; $\widehat{A}$ is the antisymmetrizing operator) and is obviously an approximate solution to Eq. (1). The HF electronic wave function is variationally *the best* among these single determinantal functions.

2.1 Restricted closed shell (CS) theory for molecular systems

In the CS [5, 6] theory, the same molecular orbital (MO) is used to define both an α- and a β-spin orbital; the wave function is then constructed from m_d ($m_d = m/2$) doubly occupied orbitals

$$\Psi_{\text{CS}} = \widehat{A}\left[\prod_{i}^{m_d} (\psi_i(\mathbf{r}_{2i-1}\alpha)\, \psi_i(\mathbf{r}_{2i}\beta))\right]. \qquad (4)$$

The expectation value of the Hamiltonian with respect to the trial HF wave function, an approximation to the total electronic energy, takes the form

$$E_{\text{CS}} = \langle \Psi_{\text{CS}} | H | \Psi_{\text{CS}} \rangle$$
$$= 2 \sum_{i}^{m_d} \int d\mathbf{r} \, \{\psi_i^*(\mathbf{r}) \, [-\nabla^2/2 + v(\mathbf{r})] \, \psi_i(\mathbf{r})\}$$
$$+ \sum_{ij}^{m_d} \int d\mathbf{r} \, d\mathbf{r}' \, [\psi_i^*(\mathbf{r}) \, \psi_i(\mathbf{r})] \, [\psi_j^*(\mathbf{r}') \, \psi_j(\mathbf{r}')]/|\mathbf{r} - \mathbf{r}'|$$
$$- \tfrac{1}{2} \sum_{ij}^{m_d} \int d\mathbf{r} \, d\mathbf{r}' \, [\psi_i^*(\mathbf{r}) \, \psi_j(\mathbf{r})] \, [\psi_j^*(\mathbf{r}') \, \psi_i(\mathbf{r}')]/|\mathbf{r} - \mathbf{r}'| \, . \tag{5}$$

The above total energy expression is to be minimized with respect to variations in the orbitals, subject to the orthonormality constraint among the orbitals,

$$\int \psi_i^*(\mathbf{r}) \, \psi_j(\mathbf{r}) \, d\mathbf{r} = \delta_{ij} \, . \tag{6}$$

The resulting HF equation takes the form ($\hat{f}$ stands for the Fock Hamiltonian)

$$\hat{f} \, \psi_i(\mathbf{r}) \equiv [-(1/2) \, \nabla^2 + v(\mathbf{r}) + \int d\mathbf{r}' \, \varrho(\mathbf{r}')/|\mathbf{r} - \mathbf{r}'|] \, \psi_i(\mathbf{r})$$
$$- \sum_{j}^{m_d} \int d\mathbf{r}' \, [\psi_j^*(\mathbf{r}') \, \psi_i(\mathbf{r}')]/|\mathbf{r} - \mathbf{r}'|] \, \psi_j(\mathbf{r})$$
$$= \varepsilon_i \psi_i(\mathbf{r}) \, , \tag{7}$$

where the electron charge density $\varrho(\mathbf{r})$ is defined in terms of the $\psi_j(\mathbf{r})$ functions as follows:

$$\varrho(\mathbf{r}) = 2 \sum_{j}^{m_d} |\psi_j(\mathbf{r})|^2 \tag{8}$$

and the factor two is due to the double occupation of the orbitals. The first three terms (kinetic, nuclear attraction, inter-electronic Coulomb repulsion) in Eq. (7) coincide with the corresponding ones which appear in the KS equations of DFT. The last term represents the "exact" non-local exchange operator, that in the KS equation is substituted by the effective exchange–correlation potential, $\mu_{\text{xc}}(\mathbf{r}; [\varrho])$. Like the KS equations, the HF equations must be solved through a Self-Consistent (SC) procedure, since both the Coulomb and the exchange operators depend on the set of functions, $\psi_i(\mathbf{r})$.

If we express the molecular orbitals, ψ_i, as linear combinations of n_b local (and real) functions φ_μ (AOs, the basis set),

$$\psi_i = \sum_{\mu}^{n_b} c_{\mu i} \varphi_\mu \, . \tag{9}$$

Eq. (7) transforms into the following matrix equation:

$$FC = SCE \, , \tag{10}$$

where F and S are the Fock and the overlap matrices, C and E the eigenvectors and eigenvalues. The expression of the various contributions to F is given below

$$F_{\mu\nu} = T_{\mu\nu} + Z_{\mu\nu} + B_{\mu\nu} + X_{\mu\nu} \, , \tag{11}$$

where

$$T_{\mu\nu} = -\tfrac{1}{2} \int \varphi_\mu(\mathbf{r}) \, \nabla_\mathbf{r}^2 \varphi_\nu(\mathbf{r}) \, d\mathbf{r} \, , \tag{12}$$

$$Z_{\mu\nu} = \sum_{A=1}^{N} \int \varphi_\mu(\mathbf{r}) \frac{Z_A}{|\mathbf{r}-\mathbf{A}|} \varphi_\nu(\mathbf{r}) \, d\mathbf{r}, \tag{13}$$

$$B_{\mu\nu} = \sum_{\lambda\varrho}^{n_b} P_{\lambda\varrho} (\mu\nu \mid \lambda\varrho), \tag{14}$$

$$X_{\mu\nu} = -\frac{1}{2} \sum_{\lambda\varrho}^{n_b} P_{\lambda\varrho} (\mu\lambda \mid \nu\varrho), \tag{15}$$

and

$$(\mu\nu \mid \lambda\varrho) = \int\int \varphi_\mu(\mathbf{r}) \varphi_\nu(\mathbf{r}) |\mathbf{r}-\mathbf{r}'|^{-1} \varphi_\lambda(\mathbf{r}') \varphi_\varrho(\mathbf{r}') \, d\mathbf{r} \, d\mathbf{r}' \tag{16}$$

$$P_{\mu\nu} = 2 \sum_i c^*_{\mu i} c_{\nu i}, \tag{17}$$

$$E_{CS} = \frac{1}{2} \sum_{\mu\nu}^{n_b} P_{\mu\nu}(F_{\mu\nu} + T_{\mu\nu} + Z_{\mu\nu}). \tag{18}$$

2.2 Unrestricted (UHF) and restricted Half-closed-shell (RHF) Hartree-Fock theories

For the description of systems containing unpaired electrons (such as atoms, molecules with an odd number of electrons, radicals) a single determinant is not, in the most general case, an appropriate wave function. In order to get the correct spin eigenfunction of these systems, it is necessary to choose a linear combination of Slater determinants (whereas, in closed shell systems, a single determinant always gives the appropriate spin eigenfunction). The UHF and RHF approximations permit, however, to describe open shell systems while maintaining the simplicity of the single determinant approximation.

In the RHF approximation [38, 39] the wave function is a single determinant constructed from m_d doubly occupied orbitals (the same MO being used to define both an α- and a β-spin orbital), and a further m_s singly occupied MOs whose spins are all aligned in parallel and which constitute the half-closed-shell. We therefore have $m = 2m_d + m_s$.

In the UHF theory [40] the spin orbitals are taken to be of the form

$$\psi(\mathbf{x}) = \psi^\alpha(\mathbf{r}) \alpha(\sigma) \quad \text{or} \quad \psi^\beta(\mathbf{r}) \beta(\sigma). \tag{19}$$

The double occupancy constraint allows the RHF approach to obtain solutions that are eigenfunctions of the spin operator, $\hat{S}^2$, whereas UHF solutions are formed by a mixture of spin states (both are eigenfunctions of $\hat{S}_z$). The greater variational freedom allows the UHF method to produce wave functions that are energetically more stable than the corresponding RHF ones; another advantage of the UHF method is that it allows solutions with locally negative spin density (i.e. anti-ferromagnetic systems), a feature that RHF solutions cannot possess.

The molecular orbital diagram for the UHF and RHF cases is shown in Fig. 1. We refer to the original papers quoted above, or to molecular quantum mechanics textbooks for the definition of the expressions equivalent to Eqs. (10) to (18) in the UHF and RHF cases.

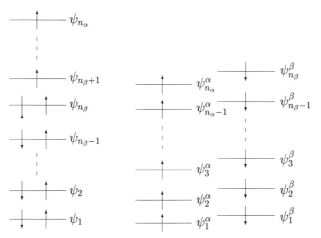

Fig. 1. Molecular orbital diagram for the Restricted Open Shell (RHF, left) and the Unrestricted (UHF, right) Hartree-Fock methods

3. Hartree-Fock Theory for Periodic Systems; the CRYSTAL Program

At first sight, the reformulation of the HF-Roothaan equations (Eqs. (10) to (18)) sketched above, to make them suitable for crystalline systems is a relatively easy task, and has been proposed more than thirty years ago by various authors [26 to 28]. Starting from a set of localized functions in the unit reference cell of the crystal, $\varphi_\mu(\mathbf{r})$ ($\mu = 1, ..., p$), it is possible to build Bloch functions (BF),

$$\phi_\mu(\mathbf{k}, \mathbf{r}) = \sum_\mathbf{G} \exp(i\mathbf{k} \cdot \mathbf{G})\, \varphi_\mu(\mathbf{r} - \mathbf{G}), \tag{20}$$

where the sum is extended to all $\mathbf{G}$ vectors of the direct lattice, while $\mathbf{k}$ is a general vector in the first Brillouin zone (BZ). The resulting set of equations is in one-to-one correspondence with Eqs. (10) to (18).

3.1 The formal equations for closed shell systems

In the basis of the Bloch functions, the Fock matrix becomes

$$F_{\mu\nu}(\mathbf{k}) = \sum_\mathbf{G} e^{i\mathbf{k}\mathbf{G}}\, F_{\mu\nu}(\mathbf{G}), \tag{21}$$

where $F_{\mu\nu}(\mathbf{G})$ is the matrix element of the Fock operator between the μ-th AO located in the *zero* cell and the ν-th AO located in the $\mathbf{G}$ cell. The row index can be limited to the *zero* cell for translational symmetry. Matrices represented in the Bloch basis (or in "$\mathbf{k}$ space") take a block diagonal form, as Bloch functions are bases for the Irreducible Representations (IR) of the Translation Group (TG); each block has the dimensions of the AO basis in the unit cell, n_b,

$$F(\mathbf{k})\, C(\mathbf{k}) = S(\mathbf{k})\, C(\mathbf{k})\, E(\mathbf{k}). \tag{22}$$

In principle the above equation should be solved for all the (infinite) $\mathbf{k}$ points of the Brillouin zone; fortunately, only a finite and usually small subset of these blocks, corresponding to a suitable sampling of $\mathbf{k}$ points, needs to be diagonalized, because interpolation techniques can be used for eigenvalues and eigenvectors throughout the first Brillouin zone.

$$F_{\mu\nu}(\mathbf{G}) = \mathbf{T}_{\mu\nu}(\mathbf{G}) + \mathbf{Z}_{\mu\nu}(\mathbf{G}) + \mathbf{B}_{\mu\nu}(\mathbf{G}) + \mathbf{X}_{\mu\nu}(\mathbf{G}) , \qquad (23)$$

$$T_{\mu\nu}(\mathbf{G}) = \int \varphi_\mu(\mathbf{r}) \, \nabla_\mathbf{r}^2 \varphi_\nu(\mathbf{r} - \mathbf{G}) \, d\mathbf{r} , \qquad (24)$$

$$Z_{\mu\nu}(\mathbf{G}) = \sum_{A=1}^{N} \sum_{\mathbf{M}} \int \varphi_\mu(\mathbf{r}) \, \frac{Z_A}{|\mathbf{r} - \mathbf{A} - \mathbf{M}|} \, \varphi_\nu(\mathbf{r} - \mathbf{G}) \, d\mathbf{r} , \qquad (25)$$

$$B_{\mu\nu}(\mathbf{G}) = \sum_{\lambda\varrho}^{n_b} \sum_{\mathbf{G}'} \mathbf{P}_{\lambda\varrho}(\mathbf{G}') \sum_{\mathbf{M}} (\mu \mathbf{0} \nu \mathbf{G} \mid \lambda \mathbf{M} \varrho (\mathbf{M} + \mathbf{G}')) , \qquad (26)$$

$$X_{\mu\nu}(\mathbf{G}) = -\tfrac{1}{2} \sum_{\lambda\varrho}^{n_b} \sum_{\mathbf{G}'} P_{\lambda\varrho}(\mathbf{G}') \sum_{\mathbf{M}} (\mu \mathbf{0} \lambda \mathbf{M} \mid \nu \mathbf{G} \varrho (\mathbf{M} + \mathbf{G}')) , \qquad (27)$$

$$P_{\lambda\varrho}(\mathbf{G}') = 2 \int d\mathbf{k} \, e^{i\mathbf{k}\mathbf{G}'} \sum_{j}^{n_b} c_{\lambda j}^*(\mathbf{k}) \, c_{\varrho j}(\mathbf{k}) \, \theta(\epsilon_F - \epsilon_j(\mathbf{k})) , \qquad (28)$$

$$E_{CS} = \tfrac{1}{2} \sum_{\mu\nu}^{n_b} \sum_{\mathbf{G}} P_{\mu\nu}(\mathbf{G}) \, (\mathbf{F}_{\mu\nu}(\mathbf{G}) + \mathbf{T}_{\mu\nu}(\mathbf{G}) + \mathbf{Z}_{\mu\nu}(\mathbf{G})) , \qquad (29)$$

where ϵ_F is the Fermi energy, the integration in Eq. (28) extends to the first Brillouin zone, and $c_j(\mathbf{k})$ and $\epsilon_j(\mathbf{k})$ are eigenvectors and eigenvalues of the $F(\mathbf{k})$ matrix.

The above equations, as such, are useless, because the $\mathbf{G}$, $\mathbf{G}'$ and $\mathbf{M}$ summations extend to the infinite set of translation vectors; a strategy must then be specified for the treatment of the infinite Coulomb and exchange series, as well as for the substitution of the integral that appears in Eq. (28) with a weighted sum extended to a finite set of $\mathbf{k}$ points. An accurate and efficient solution to these problems has been implemented in the CRYSTAL code [23, 24, 31, 32]. Some of the features of CRYSTAL will be summarized in the next section. Before moving to these technical aspects, we want however to spend a few more words on the periodic UHF method, as all the applications discussed in the last section concern paramagnetic species.

3.2 The UHF equations

In the UHF case, two sets of matrix equations must be solved self-consistently, for α and β electrons,

$$F^\alpha(\mathbf{k}) \, C^\alpha(\mathbf{k}) = S(\mathbf{k}) \, C^\alpha(\mathbf{k}) \, E^\alpha(\mathbf{k}) , \qquad (30)$$

$$F^\beta(\mathbf{k}) \, C^\beta(\mathbf{k}) = S(\mathbf{k}) \, C^\beta(\mathbf{k}) \, E^\beta(\mathbf{k}) , \qquad (31)$$

where the $F^\alpha(\mathbf{k})$, $F^\beta(\mathbf{k})$ and $S(\mathbf{k})$ matrices are obtained by Fourier transform from the corresponding "direct space" equivalent quantities as in Eq. (21).

We now define the *total density* (P) and *spin density* (P^{spin}) matrices,

$$P(\mathbf{G}) = P^\alpha(\mathbf{G}) + P^\beta(\mathbf{G}) , \qquad (32)$$

$$P^{\mathrm{spin}}(\mathbf{G}) = P^\alpha(\mathbf{G}) - P^\beta(\mathbf{G}) , \qquad (33)$$

where $P^\alpha(\mathbf{G})$ and $P^\beta(\mathbf{G})$ are obtained as in Eq. (28) by using the α and β eigenvectors obtained from Eq. (30) and (31), respectively.

The $F^\alpha(\mathbf{G})$ and $F^\beta(\mathbf{G})$ matrices are defined as follows:

$$F^\alpha_{\mu\nu}(\mathbf{G}) = F_{\mu\nu}(\mathbf{G}) - X^{\rm spin}_{\mu\nu}(\mathbf{G}), \qquad (34)$$

$$F^\beta_{\mu\nu}(\mathbf{G}) = F_{\mu\nu}(\mathbf{G}) + X^{\rm spin}_{\mu\nu}(\mathbf{G}). \qquad (35)$$

The $F(\mathbf{G})$ matrix is defined as in Eq. (23), where the total density matrix $P(\mathbf{G})$ defined in Eq. (32) is used in the Coulomb and exchange terms; $X^{\rm spin}_{\mu\nu}(\mathbf{G})$ is defined as $X_{\mu\nu}(\mathbf{G})$ in Eq. (27), where however the spin density matrix $P^{\rm spin}(\mathbf{G})$ is used instead of the total density matrix $P(\mathbf{G})$.

4. The Implementation of the Periodic HF Equations in the CRYSTAL Program

CRYSTAL is an ab initio periodic computer program implemented by the Group of Theoretical Chemistry of the University of Torino (Italy) in collaboration with the Computational Group at the Daresbury Laboratory (UK). The first version was distributed in 1988 [20], and for about ten years it has been the only public general-purpose ab initio periodic computer program; three improved versions of CRYSTAL were subsequently released; [23, 41, 42] (the last one is in distribution since the beginning of 1999). More than 200 copies of both the '92 and '95 versions have been distributed over the world.

4.1 The general features

We summarize here the main features of the CRYSTAL98 program:

1. It is a "direct space" program, in the sense that all the relevant quantities (mono- and bi-electronic integrals, overlap and Fock matrices, see Eqs. (23) to (29)), are computed in the configuration space. Just before the diagonalization step the Fock matrix is

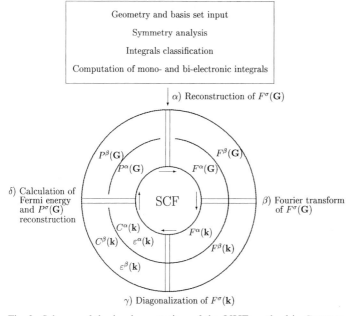

Fig. 2. Scheme of the implementation of the UHF method in CRYSTAL

Fourier transformed to reciprocal space (Bloch functions basis; see Eq. (21)), then the eigenvalues and eigenvectors of the $F(\mathbf{k})$ matrices are combined to generate the "direct space" density matrix for the next SCF cycle (Eq. (28)).

2. In its most recent version, CRYSTAL can solve the HF as well as the KS equations; as regards the latter, the most popular *local* and *non local* functionals are available, as well as *hybrid* schemes, such as the so called B3-LYP, which combines the HF exchange term with the Becke [43] and Lee-Yang-Parr [44] functionals according to the formula proposed by Becke [45]. As regards the former, the CS, RHF and UHF options are available. Schemes are also available, that permit to correct the HF total energy by estimating the correlation energy *a posteriori*, integrating a correlation-only functional of the HF charge density. This latter scheme has been shown to provide accurate binding energies for a large family of compounds [32].

The general structures of the program is illustrated in Fig. 2, for the case of the UHF path. The scheme shows that the most relevant quantities (Fock and density matrices, eigenvalues and eigenvectors) are duplicated, so that it is possible to differentiate α from β orbitals. The two sections of the code, corresponding to α and β electrons, are independent until the Fermi energy calculation and Fock matrix reconstruction. It is then possible to force the system into a state with a particular S_z value, by imposing the desired $m_\alpha - m_\beta$ value when the crystalline monoelectronic energy levels are populated at each cycle of the SC step.

Many steps of the calculation are common to the HF and DFT options (for example the treatment of the Coulomb series, which are evaluated analytically, as discussed below); the main difference concerns, obviously, the exchange (and correlation) contribution to the Hamiltonian matrix and total energy: in the HF case the exchange bielectronic integrals are evaluated analytically, and the exchange series is truncated after a certain number of terms, as discussed below. In the DFT calculations [46], the exchange–correlation potential is expanded in an auxiliary basis set of GTF, with even tempered exponents. At each SCF iteration the auxiliary basis set is fitted to the actual analytic form of the exchange–correlation potential, which changes with the evolving charge density. For numerical integration, the atomic partition method proposed by Becke [47] has been adopted, combined with the Gauss-Legendre (radial) and Lebedev (angular) quadratures.

3. It can treat systems periodic in 0 (molecules), 1 (polymers), 2 (slabs) and 3 (crystals) directions with similar accuracy; this permits the evaluation of energy differences such as bulk-minus-molecule (lattice energy of a molecular crystal), bulk-minus-slab (surface energy), bulk-minus-chain (inter-chain interactions) with high accuracy, as well as energy differences between crystals with different cell size, shape, and number of atoms.

4. All the crystalline space, layer and rod groups are available, simply by specifying the label of the group according to the international crystallographic notations.

5. Several geometrical options are available, which permit an easy manipulation of the cell (creating defects, distorting the cell, building supercells, cutting slabs from the bulk, extracting molecules from the bulk, and so on).

4.2 The treatment of the infinite Coulomb and exchange series

As anticipated, the problem with the formal periodic equations (22) to (29), is that they contain three infinite summations ($\mathbf{G}$, $\mathbf{G}'$ and $\mathbf{M}$), that extend to all lattice vectors. We refer to previous papers for an exhaustive presentation of the strategy adopted in CRYSTAL for the treatment of the Coulomb [31, 48] and exchange [24] series. Here we summarize the main ideas:

– *The general strategy is to evaluate the integrals analytically, whenever possible.* The core of the integral package in the CRYSTAL code derives from the GAUSSIAN70 [1] and ATMOL [49] integral packages; both have been heavily modified and generalized for many reasons, and new parts have been added. As regards the latter point, the following new kinds of integrals are required for the construction of the Fock operator:

a) multipolar integrals of a product of two GTFs;

b) bipolar expansions of two interacting charge distributions represented by a product of two AOs each;

c) interaction of a charge distribution $\varrho_{\mu\nu}$ with the infinite array of point charges and higher multipoles of a charge distribution. These integrals are evaluated analytically using Ewald type techniques through a combination of recursion relations involving Hermite polynomials and spherical harmonics, so that all the integrals can be obtained from the generalized error functions, as for molecular integrals [31, 48].

Specific routines are available for periodicity in one, two and three dimensions (see Refs. [31, 50]).

– *The overlap between two GTFs decays exponentially with distance.* This property can be exploited for the truncation of the **G** and **G**′ sums in the Coulomb term (see Eq. (26) and (29)) and for two of the exchange summations (Eq. (28)). In a similar way the overlap, kinetic, nuclear attraction integrals can be reduced to the consideration of a finite number of neighbours of the reference cell. The screening criteria are applied both to the contracted AOs, and to the single Gaussians of a contraction. An efficient sorting of the **G** vectors and evaluation of the overlap between shell couples allow a rapid selection of the integrals to be computed.

– *The density matrix elements decay with distance between the involved functions.* In a localized basis set, the elements of the density matrix of an insulator decay exponentially with the distance between the two centres, the larger the gap the faster the decay. Density matrix in direct space decays to zero with distance for conductors, too, but much more slowly [24, 51].

This behaviour is exploited for an efficient truncation of the exchange series in the Fock matrix (Eq. (27)) and total energy (Eq. (29)) expressions. The two above equations show that the **M** summation is limited by the exponential decay of the product $\varphi_\mu \varphi_\lambda$ and for a similar reason the $|\mathbf{G} - \mathbf{M} - \mathbf{G}'|$ distance cannot be too large. These two conditions imply that the **G** and **G**′ vectors cannot be very different, although their moduli could be large. However, the exponential decay of the density matrix permits us to disregard integrals involving **G** and **G**′ vectors with large moduli. The exchange series are then truncated according to these criteria. Long range effects are not taken into account, as they are negligible. Exchange bielectronic integrals are evaluated as such or through a *bipolar expansion* (see below), when the two charge distributions do not overlap.

– *The multipolar expansion and the Ewald method for the Coulomb series.* Let us define a partition of the cell charge distribution, for example in terms of Mulliken atomic charges ϱ_A,

$$\varrho(\mathbf{r}) = \sum_A \sum_{\mathbf{M}} \varrho_A(\mathbf{r} - \mathbf{A} - \mathbf{M}), \qquad (36)$$

The Periodic Hartree-Fock Method

where

$$\varrho_A(\mathbf{r} - \mathbf{A}) = \sum_{\lambda \in A} \sum_{\varrho} \sum_{\mathbf{G}'} P_{\lambda\varrho}(\mathbf{G}') \, \varphi_\lambda(\mathbf{r}) \, \varphi_\varrho(\mathbf{r} - \mathbf{G}') - Z_A \, \delta(\mathbf{r} - \mathbf{A}) \quad (37)$$

is translationally invariant and the last term is the nuclear charge contribution. The sum of the electron–electron and electron–nuclei contributions to the Fock matrix, $R_{\mu\nu}(\mathbf{G}) = B_{\mu\nu}(\mathbf{G}) + Z_{\mu\nu}(\mathbf{G})$ (Eq. (25) and (26)), can then be written as

$$R_{\mu\nu}(\mathbf{G}) = \sum_A \sum_{\mathbf{M}} \iint \varrho_{\mu\nu}(\mathbf{r}_1 - \mathbf{G}) \, |\mathbf{r}_1 - \mathbf{r}_2|^{-1} \, \varrho_A(\mathbf{r}_2 - \mathbf{A} - \mathbf{M}) \, d\mathbf{r}_1 \, d\mathbf{r}_2 \quad (38)$$

with

$$\varrho_{\mu\nu}(\mathbf{r}_1 - \mathbf{G}) = \varphi_\mu(\mathbf{r}_1) \, \varphi_\nu(\mathbf{r}_1 - \mathbf{G}). \quad (39)$$

If the charge distributions ϱ_A and $\varrho_{\mu\nu}$ do not overlap for any $\mathbf{M}$, Eq. (38) can be written as

$$R_{\mu\nu}(\mathbf{G}) = \sum_A \sum_{\ell m} \left(\sum_{\mathbf{M}} \gamma_{\ell m}(A) \, \Phi_{\mu\nu\ell m}(\mathbf{G}, \mathbf{A} + \mathbf{M}) \right), \quad (40)$$

so that $R_{\mu\nu}$ is expressed in terms of the multipole moments of ϱ_A ($N_{\ell m}$ is a normalization factor),

$$\gamma_{\ell m}(A) = \int \varrho_A(\mathbf{r}_2 - \mathbf{A}) \, N_{\ell m} X_{\ell m}(\mathbf{r}_2 - \mathbf{A}) \, d\mathbf{r}_2 \quad (41)$$

and of the field integrals

$$\Phi_{\mu\nu\ell m}(\mathbf{G}, \mathbf{A} + \mathbf{M}) = \int \varphi_\mu(\mathbf{r}_1) \, \varphi_\nu(\mathbf{r}_1 - \mathbf{G}) \, X_{\ell m}(\mathbf{r}_1 - \mathbf{A} - \mathbf{M}) \, |\mathbf{r}_1 - \mathbf{A} - \mathbf{M}|^{-2\ell - 1} \, d\mathbf{r}_1. \quad (42)$$

where $X_{\ell m}$ is a real solid spherical harmonic.

The infinite $\mathbf{M}$ summation can be performed by using the Ewald technique, which reduces it to a finite number of new integrals, that can be evaluated analytically [31]. The main advantage of Eq. (38), with respect to Eq. (25) and (26), is that a large number of terms (typically, more than 1000 for a packed system like MgO with a double-ζ basis set) are summed up (Eq. (37)) and treated at the same time with a corresponding saving factor in computational effort. If $\varrho_{\mu\nu}$ and ϱ_A do overlap for some $\mathbf{M}$ vectors close to the origin (let us call this set $\mathbf{M}_B$), the same Ewald technique is used. A correction is however applied, consisting in subtracting by direct summation the contribution from the $\mathbf{M}_B$ set in Eq. (40), and adding the corresponding contribution from Eq. (25) and (26). The resulting finite (possibly small) number of bielectronic Coulomb integrals and nuclear attraction integrals is then evaluated with standard techniques; it is, however, to note that a non negligible fraction of integrals corresponding to the $\mathbf{M}_B$ set can be approximated by a *bipolar expansion* of both charge distributions, because it may happen that, although ϱ_A is penetrating $\varrho_{\mu\nu}$, some of the contributions to ϱ_A are not, and can then be approximated (see Ref. [24], pp. 47 to 52).

– *Point symmetry.* A crucial point for the speed and accuracy of the calculation is the complete exploitation of symmetry. Point symmetry is exploited at various levels in CRYSTAL, both in the direct (AO) and reciprocal space (Bloch functions) representations.

As concerns the former aspect, two points are worth noting:

– The contributions to the Fock matrix $F_{\mu\nu}(\mathbf{G})$ are evaluated only for the set of indices $\mu\nu\mathbf{G}$ belonging to the *irreducible set*, i.e. to the minimal set necessary for the generation of the complete matrix through the use of the point symmetry operations. The saving factor in CPU time and space requirements can be as large as h, the number of the symmetry operations in the point group. We remind here that *translational symmetry* has already been used for limiting the first index μ to the *zero* cell.

— Disk space requirements for the storage of the bielectronic integrals can be reduced, and the Fock matrix reconstruction at each step of the SCF stage can be accelerated by storing symmetrized sums of bielectronic integrals [24, 52].

Point symmetry is also exploited in those parts of the program which work in the Bloch function representation.

— In principle the Fock matrix must be diagonalized at all the **k** points of the first BZ. The eigenvalues are then used for the determination of the Fermi energy, and the eigenvectors for building the density matrix of the next cycle. It is, however, easily shown that the Fock matrix can be diagonalized only in the so-called *irreducible part* of the BZ (IBZ); the full set of eigenvectors can then be obtained by rotation of the eigenvectors of the IBZ.

— The set of operators that do not move a **k** point of the IBZ form the so-called little co-group [53] of the **k** point, and can be used to block-diagonalize the Fock matrix $F(\mathbf{k})$, by building symmetry adapted crystalline orbitals [54, 55]. This point is of minor importance when small unit cell systems are considered, as most of the **k** point have no symmetry at all; however, when large unit cell, highly symmetric systems are considered, the exploitation of this particular aspect of symmetry becomes essential, as the order of the little co-group of the few **k** points to be considered is quite high [56].

5. Examples of Applications

Applications of the HF-Crystalline Orbitals SCF method as implemented in the CRYSTAL program in many different areas have been performed by the present authors since 1980, when the first studies devoted to diamond and graphite were published [57, 58]; a complete list of our publications can be found at the web site *http://www.ch.unito.it/ifm/teorica/crystal.html*. Here, we concentrate on three applications involving uncoupled electrons, namely a transition metal oxide (NiO), a trapped hole in CaO bulk where a Li is substituted for a Ca atom, an F-centre (an electron trapped in an anion vacancy) in LiF. In all cases the UHF option has been used; for comparison, the calculations have been repeated at the spin polarized LDA (LSDA) level.

5.1 The electronic and magnetic properties of transition metal compounds: the case of NiO

NiO is a prototype compound in the family of transition metal oxides, and has been the subject of extensive theoretical investigations [59 to 64]. From an experimental point of view, these materials are classified as antiferromagnetic (AFM) ionic insulators. The band structures, obtained at the UHF and LSDA level (we refer to previous papers for the basis set description and other technical details [63, 64]), are reported in Fig. 3; the figure includes the valence bands, and three conducting bands: two of them are the empty (at the HF level) e_g levels, the third is an anti-bonding oxygen p band. The two band structures differ in many respects:

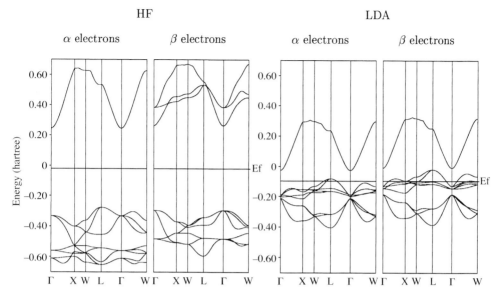

Fig. 3. HF and LDA NiO band structure; ferromagnetic solution

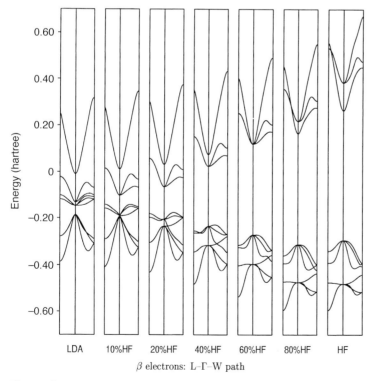

Fig. 4. NiO valence band structure obtained with Hamiltonians containing various mixing of the LDA and HF exchange potential. The three highest bands are empty, and correspond to the e_g empty states and to an antibonding p state. The LDA solution is metallic

(i) the HF solution presents a large gap, whereas the LDA solution is metallic; this behaviour is a direct consequence of the local character of the exchange term, as demonstrated by Fig. 4, where the results of various "hybrid" solutions, obtained by mixing in various percentages the HF and LDA exchange operator in the Hamiltonian, are reported.

(ii) The highest part of the valence band structure is mainly contributed by p oxygen states in the HF solution, whereas d states are predominant in the LDA solution.

As regards bulk properties, such as the lattice parameter, they are in line with previous experience: UHF overestimates (4.26 Å), whereas LSDA underestimates (4.10 Å) the experimental result (4.17 Å).

The electronic structure can be described in terms of Mulliken charges, as shown in Table 1. It turns out that the UHF solution is very ionic, with spin moments and net charges very close to the formal ones (+2 for Ni, eight electrons in the d shell), and a small and negative Ni-O bond population, indicating that the interaction between the two ions is essentially electrostatic, with a short range repulsion contribution. The ferromagnetic (FM) and AFM solutions are very similar, apart from the small spin polarization on the oxygen atom in the latter case; this is in qualitative agreement with the very small energy differences between the two solutions, that must imply very small differences in the wave function. The LSDA situation is quite different: the net charges are about 0.5 electrons smaller, and when a solution corresponding to two uncoupled electrons is imposed ($S_z = 1$), the spin polarization involves also the oxygen atom for about 0.5 electrons, whereas in the UHF case the oxygen spin polarization was much smaller (0.08 electrons). The LSDA FM and AFM solutions are quite different. The bond population, though small, is positive.

The very different spin polarization of the oxygen obtained with the two Hamiltonians in the FM state is also evident from the spin density map (Fig. 5, left), whereas it disappears, for symmetry reasons, in the AFM solution (Fig. 5, right). As the superexchange mechanism is mainly related to this oxygen polarization (see, for example, the discussion in Ref. [37]) it is clear why the LSDA Hamiltonian produces much higher energy differences between the FM and AFM states than UHF. The behaviour

Table 1

Net charges (q), spin moments (μ) and bond populations (q_{AB}) in the ferromagnetic and anti-ferromagnetic states of NiO evaluated according to a Mulliken partition. In the second row the number of electrons in the Ni d shell is reported. Units: electrons

		HF		LDA	
		FM	AF	FM	AF
q	Ni	+1.87	+1.86	+1.55	+1.50
	d_{Ni}	8.098	8.102	8.387	8.443
	O	−1.87	−1.86	−1.55	−1.50
μ	Ni	1.923	±1.913	1.556	±1.185
	d_{Ni}	1.919	±1.912	1.550	±1.185
	O	0.077	0	0.432	0
q_{AB}	Ni-O	−0.032	−0.032	+0.010	+0.018

of this quantity (ΔE) as a function of the lattice parameter in the UHF case is given in Fig. 6: it shows the rapid fall-down to zero (that is: the FM and AFM states have the same energy) for large Ni–O distance as a consequence of the exponential reduction of the short range inter-ionic repulsion; this behaviour has been documented experimentally in the case of the KMF$_3$ perovskites (M = Mn,Ni), in quite good agreement with the UHF results [37]. The LSDA ΔE value are much larger and show a different dependence on the Ni–O distance: it is ten times bigger at 3.80 Å, about

Fig. 5. HF and LDA spin density maps in the FM and AFM states; iso-density lines are drawn at intervals of 0.01 (Bohr)$^{-3}$; continuous, dashed and dash-dotted lines correspond to positive, negative and zero values, respectively

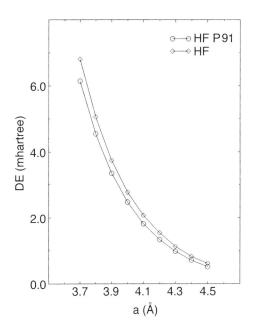

Fig. 6. Energy difference between the ferromagnetic and anti-ferromagnetic solutions for NiO as a function of lattice parameter. The two curves refer to HF energies (diamonds) and energies corrected with a correlation-only density functional (circles)

20 times bigger at the experimental geometry, 25 times bigger at 4.60 Å. In the cases where a direct comparison with experimental results is possible, as for the above mentioned KMF_3 perovskites, the UHF super-exchange constant extracted from ΔE turns out to be about one third of the corresponding experimental quantity [37].

5.2 Trapped electron holes in alkali doped alkaline earth oxides

Paramagnetic centres are formed when alkaline earth oxides, doped with alkali metal ions, are exposed to ionizing radiation at low temperature [65, 66]. EPR spectra give evidence of the formation of alkali metal trapped electron holes in oxides such as MgO, CaO and SrO [65 to 68], that are usually denoted as $[M]^0$. The interest in these materials is connected with two distinct applications: 1) their possible use as insulators in high radiation environments, as they can suppress radiation damage; 2) their catalytic activity in the formation of higher order hydrocarbons from methane, for possible exploitation in industrial processes [69, 70].

Electron Nuclear Double Resonance (ENDOR) spectroscopy has been applied to $[Li]^0$- and $[Na]^0$-doped isostructural oxides MgO, CaO and SrO for the accurate measurement of their hyperfine structure, which has provided useful information about the spin density distribution and geometry of the defects [71, 72].

The electronic structure and properties of trapped electron hole centres can also be investigated theoretically with the UHF method. If a defect is localized and its interactions with the rest of the crystal are short-ranged, it can formally be represented by a periodic model, such as that adopted by the CRYSTAL program, provided that a suitable supercell can be found [73 to 75], i.e. a multiple of the unit cell, that is large enough to make the mutual interaction among neighbouring defects negligible, but still in the range of a computationally accessible problem. As far as the condition of non-interacting defects is satisfied, the method is fairly accurate and size consistent: the numerical accuracy of the implemented algorithms is such that the calculated energy per supercell of a crystal is precisely proportional to the supercell volume [74]. An analysis of the dependence of the formation energy (E_f) of $[Li]^0$ centres in MgO, for example, on the supercell size (see Table 1 in Ref. [76]) shows that big supercells may be necessary to compute E_f very accurately, but a supercell of 32 atoms (S_{32}) allows the calculation of the defect formation energy within a few percent error and to account for the most important relaxation effects.

The UHF calculation provides very similar electronic structures for all these systems. The unpaired electron, that is formed after the substitution of an alkali metal ion for a cation in alkaline earth oxides, could in principle be found in two different electronic configurations:

– fractions of the unpaired electron almost symmetrically delocalized on the oxygen ions nearest neighbours of M^+;

– the M^+ ion coupled to one of the oxygen ions, forming a well localized electron hole.

UHF calculations lead to the localized electron hole solution unequivocally, in agreement with the experimental indications, and the localized configuration is preferred to the delocalized one by about 300 kJ/mol.

As examples, we refer to bulk $[Li]^0$ and $[Na]^0$ centres in CaO, calculated with the S_{64} supercell at the HF equilibrium geometry for CaO (lattice parameter: 4.83 Å; expt.: 4.79 Å). We refer to Refs. [76, 77] for technical details of the calculation. The band structure is particularly useful for the characterization of the defect. Figure 7 shows the valence and conduction bands of $[Li]^0$ in CaO, that are essentially due to the p-type oxygen orbitals. The levels corresponding to the p electrons of O^- are split from the valence band originated by the p atomic orbitals of the O^{2-} ions: the populated levels, i.e. the doubly degenerate p_x–p_y and the α-p_z levels (z along the ionic pair axis) move to lower energy values (α-p_z is stabilized for as much as -608 kJ/mol from the bottom of the O^{2-} p valence band), and a β-p_z level appears in the large band gap, due to the electron hole. This picture is supported by the atomic net charges of M^+ and O^- (O_1) reported in Table 2 and obtained from Mulliken population analysis, which are very close to $\pm 1\ |e|$. Also the spin moment of O_1 is very close to one, indicating that the unpaired electron is almost completely localized and only very small fractions of spin density are found on the neighbouring ions. A more detailed description of electron

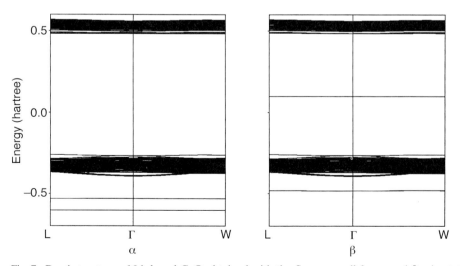

Fig. 7. Band structure of Li-doped CaO obtained with the S_{64} supercell for α- and β-spin states

Table 2

Net atomic charges (q) and spin moments (μ) evaluated according to a Mulliken partition of the periodic densities. Units: electrons. Labels of the atoms as in Fig. 8

	M		O_1		O_2	
	q	μ	q	μ	q	μ
$[Li]^0$ in CaO	0.980	0.000	−1.047	0.985	−1.932	0.002
$[Na]^0$ in CaO	0.998	−0.002	−1.053	0.986	−1.928	0.003

charge and spin density is given in Fig. 8 for a $[Li]^0$ centre in CaO. The unpaired electron is clearly localized on O_1 with some residual spin density spread in the region around the defect, oscillating from positive to negative values.

It results from energy minimization that the M^+ and O^- monovalent ions move in opposite directions and increase their mutual distance, R (Table 3), as they are less attracted by each other than by their neighbouring bivalent O^{2-} and Ca^{2+} ions. This is in reasonable agreement with the experimental indications, despite the rough approximations in the model used to derive R from primary experimental data [66, 71, 72]. The theoretical method allows us to distinguish the individual displacements of M and O_1 from their lattice positions: the M^+ ions are between 3 and 4 times more mobile than O^-. This increase in the $M-O_1$ distance with respect to the perfect lattice interionic distance (4.83 Å) is remarkable when M = Li. The higher mobility of the small Li^+ ion allows a stronger electrostatic interaction with the surrounding ions and corresponds to a higher relaxation energy, which is nearly three times as large as for Na. The displacements of the neighbouring oxygen ions from their lattice positions are not as large as for M, but the corresponding relaxation energies are far from negligible, especially for M = Na (31 kJ/mol), whereas it is less important for Li (5 kJ/mol).

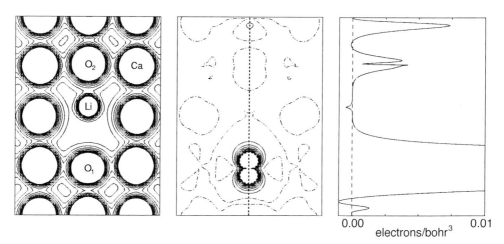

Fig. 8. Total electron charge (left) and spin (middle) density maps in the (100) plane through the Li^+-O^- pair. On the right, the spin density function along the vertical path dashed in the spin map. The separation between contiguous iso-density curves in the maps corresponds to 0.01 e/Bohr^3 in the total charge density and 0.005 e/Bohr^3 in the spin map. Symbols as in Fig. 5

Table 3
Relaxation of M and O_1, and M–O_1 distances (R) in Å. The experimental values (R^{exp}) reported were obtained with different interpreting models

	Δz_M	Δz_O	R	R^{exp}
CaO:[Li]^0	0.44	−0.11	2.91	2.86–3.05[a]), 2.83 ± 0.19[b])
CaO:[Na]^0	0.25	−0.08	2.74	2.23 to 2.46[c])

[a]) Ref. [72]; [b]) Ref. [66]; [c]) Ref. [71].

A straight comparison between calculated and experimental data is possible for the parameters that determine the hyperfine structure of the spectra [72]: the hyperfine coupling constants a and b and the nuclear quadrupole coupling constant P. These constants appear in the formulation of the spin Hamiltonian in terms of which the ENDOR spectra are interpreted.

The isotropic (a) and anisotropic (b) hyperfine coupling constants (in MHz) are expressed for each nucleus N as [78]

$$a_N = \frac{2\mu_0}{3h} g\beta_e g_N \beta_n \varrho^{\alpha-\beta}(0), \tag{43}$$

$$b_N = \frac{2\mu_0}{3h} g\beta_e g_N \beta_e \left[T_{11} - \frac{1}{2}(T_{22} + T_{33}) \right], \tag{44}$$

where the spin density $\varrho^{\alpha-\beta}$ at N, the elements of the hyperfine coupling tensor **T** and the electron g factor are the only terms which depend on the electronic structure of the system (in our calculation we use the free electron g to approximate the true g factor, which is usually an acceptable approximation for our purpose). The other multiplicative factors in Eq. (43) and (44) are all tabulated constants [78, 79] (h, the Planck constant, β_e and β_n, the electronic and nuclear magnetons, μ_0, the permeability of the vacuum and g_N, the nuclear g factor). **T** is a tensor of rank two, that is obtained as the quadrupole moment of the spin density at N. Its generic element has the form:

$$T_{ij}^N = \sum_{\mu\nu} \sum_{G} P_{\mu\nu G}^{spin} \int \varphi_\mu(\mathbf{r}) \left(\frac{r^2 \delta_{ij} - 3 r_i r_j}{r^5} \right) \varphi_\nu(\mathbf{r} - \mathbf{G}) \, d\mathbf{r}, \tag{45}$$

where the origin of the reference system is at nucleus N and r_i denotes the i-th component of r. T_{ii} in equation (44) are the elements of **T** in diagonal form, following the convention that T_{11} is the maximum module component.

The other important parameters in the spectrum are the nuclear quadrupole coupling constants P_N, that measure the coupling of the electric field gradient (q_N^{efg}) with the nuclear quadrupole moment (Q_N) at N,

$$P_N = \frac{3}{h} \frac{e^2 q_N^{efg} Q_N}{4I(2I-1)}, \tag{46}$$

where I is the total nuclear spin quantum number, e the electron charge magnitude, Q_N is measured experimentally and q_N^{efg} is obtained from a tensor of the same form as **T** where the spin density matrix in (45) is replaced by the total density matrix.

The calculated and experimental values of a, b and P for the coupling of the unpaired electron to the M nucleus in M-doped CaO are compared in Table 4. The best

Table 4

Spin Hamiltonian hyperfine isotropic (a) and anisotropic (b) coupling constants and nuclear quadrupole coupling constant (P) at nucleus M in doped CaO. Comparison between calculated (HF) and experimental [72] values. Units: MHz

M	a calc.	a exp.	b calc.	b exp.	P calc.	P exp.
Li	−0.819	−2.472	1.468	1.317	0.001	0.009
Na	−2.538	−9.145	1.558	1.877	0.449	0.446

agreement between the calculated and experimental results is obtained for b, with a percentage error within 15% of the experimental value. The same error range for b has been found also for MgO and SrO [77].

The calculated value of P_N is surprisingly good for CaO when N = Na, though it must be remarked that the variability of the values of Q reported by different authors is rather large and the calculation of P is obviously affected by the same uncertainty (data in Table 4 were obtained for $Q_{Li} = -0.03$ and $Q_{Na} = 0.14$ barn, as reported in the International Tables of Physics and Chemistry [79], Vol. 9, p. 31). In the Li case, both experimental and calculated P values are negligible.

As regards a, an accurate estimation is a more difficult task, since it requires a very precise determination of spin density at nuclei where it can be nearly zero. This is particularly true in this case, where the unpaired electron is very well localized on O$^-$ and very low spin densities are found anywhere else in the crystal (see the spin moment values in Table 2. Nevertheless, it has already been emphasized that Fig. 8 reveals a well defined pattern for the residual spin density, with an alternation of positive and negative zones, that is not random; it is rather a consequence of spin polarization [80]: the electrons in the defect region with the same spin (α) as the unpaired electron, are stabilized by exchange and super-exchange interactions and, for this reason, they interact with it in a slightly more favourable way. Therefore, not only the absolute value of a is important, but also its sign, as it specifies if α- or β-spin electrons prevail at the nuclei. In Fig. 8 (right) the profile of spin density in CaO along the Li–O$^-$ axis is shown in detail. What determines the value of a_{Li} is the depth of the small negative peak (about 5×10^{-4} $|e|$/Bohr^3) at Li, i.e. almost three orders of magnitude less than spin density at O$_1$ (O$^-$) and one order of magnitude less than that at O$_2$ (incidentally, the twin peak at O$_2$ implies p$_z$ character, like for O$_1$). The agreement of the calculated a_N with the experimental data obtained from ENDOR spectra (Table 4) is only semi-quantitative, though the sign and the order of magnitude are correct, and the calculated a_N values scale approximately with the same trend as the experimental data, as can also be seen for MgO and SrO in Ref. [77]. From the computational point of view, particular caution is required to the level of convergence of the SCF cycle which is needed for a spin density of the order of 10^{-4} $|e|$/Bohr^3 to be determined with the necessary precision. Convergence of the order of 1×10^{-5} Hartree on total energy, which is adequate for the calculation of most observables, may be insufficient to evaluate a correctly, as is shown in Table 5 for Li-doped CaO. Attempts at improving the basis sets (either the valence or the core functions) and altering the geometry of the defect, as well as better approximating the Coulomb and exchange series or using a denser net in reciprocal

Table 5
Spin Hamiltonian hyperfine isotropic (*a*) and anisotropic (*b*) coupling constants and nuclear quadrupole coupling constant (*P*) at Li in doped CaO as a function of the level of convergence of the SCF cycle, when the precision in the calculation of total energy is 10^{-x} Hartree. Units: MHz

x	a_{Li}	b_{Li}	P_{Li}
5	−0.952	1.468	0.001
6	−0.920	1.468	0.001
7	−0.854	1.468	0.001
8	−0.819	1.468	0.001
9	−0.809	1.468	0.001
10	−0.807	1.468	0.001
11	−0.805	1.468	0.001
12	−0.805	1.468	0.001

space, do not lead to any significant improvement in the evaluation of a_N and the poor agreement between calculation and experiment is likely to be considered as a limit of the method: the spin densities involved are so low that the lack of electron correlation and the spin contamination intrinsic in the UHF approximation (the UHF wave function is an eigenfunction of the $\hat{S}_z$ operator and not of the total spin momentum $\hat{S}^2$) may be important.

In order to have some indications about the importance of electron correlation in the determination of *a*, a comparison with results obtained with DFT based Hamiltonians would have been useful. However, we did not succeed in localizing the electron hole onto one oxygen ion with DFT methods. In particular, the LSDA unpaired electron is nearly completely delocalized over the six oxygens around the defect, in contrast with the experimental evidence. This result seems to confirm the tendency of LSDA, already observed for NiO (Fig. 5), to delocalize unpaired electrons in tightly bound states onto neighbouring atoms. In the case of NiO, however, stoichiometry limits charge transfer from each Ni to a single O ion (the overall delocalization is about 25%), whereas, in the present case, six O ions are available and delocalization can proceed further.

5.3 F-centres in lithium fluoride

The capability of the method to reproduce the experimental values of *b* and *P* with a quite satisfactory accuracy and, at the same time, the difficulty in computing *a* as accurately as *b* and *P* have stimulated the investigation of other paramagnetic species, which are characterized by a stronger Fermi contact, such as F-centres in alkali halides, and for which very accurate ENDOR measurements [81] are available in the literature. Alkali halides are similar to alkali earth oxides as concerns crystal symmetry (same space group) and ionicity, and the F-centres offer the advantage of a simpler electronic structure than the electron holes in doped oxides.

As an example, we consider F-centres in lithium fluoride. The model is a 64 atom supercell crystal at the experimental lattice parameter, from which an F atom has been removed. The F-centre consists of an unpaired electron localized inside the vacancy, as can be seen in Fig. 9 and from the Mulliken population analysis (Table 6), but the picture differs from that of the trapped holes in alkaline earth oxides in some respects: the electron wave function in the vacancy has spherical symmetry and is much more diffuse; relaxation of the nuclei surrounding the defect is negligible in this case (the Li ions around the vacancy are displaced by only 0.03 Å far from it, with an energy gain of just 1.7 kJ/mol). The unpaired electron in the vacancy is obviously more weakly bound than the electron of O^- in the trapped hole centre. A bound α-spin level appears in the valence–conduction band gap in this case, whereas the unpaired electron state of $[M]^0$ centres was stabilized below the O^{2-} p valence band. As in the previous examples, we compared HF with LDA results, using Dirac-Slater exchange potential [82] and the Perdew-Zunger parametrization of the Ceperley-Alder free electron gas correlation [83]. A discrete bound (α) and the corresponding unbound (β) levels appear in the gap also in the LSDA band structure, which is qualitatively similar, but differ from UHF quantitatively: the band gap is 936 kJ/mol (UHF: 2130 kJ/mol) and the energy difference between the defect α and β levels is 156 kJ/mol (UHF: 981 kJ/mol). This great difference in stability of the F-centre corresponds indeed to different electron charge and spin density distributions (Fig. 9). Mulliken analysis (Table 6) assigns almost exactly one electron to the UHF vacancy, while there are only 0.887 electrons in the LSDA vacancy. At the atoms around the defect, the spin moments calculated with both methods (Table 6) are nearly three orders of magnitude smaller than in the vacancy. Nevertheless, the spin density profiles in Fig. 9 show that the two sharp symmetric side peaks, which correspond to the Li nuclei, are about as high as the large middle peak centred at the vacancy. Since the heights of the Li peaks measure the spin density at those nuclei and the Fermi contact interaction at Li depends on it, the corre-

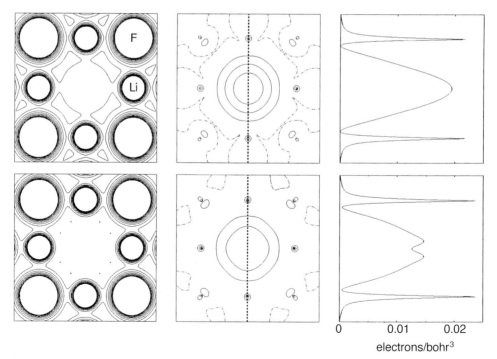

Fig. 9. HF (top) and LDA (bottom) total electron charge (left) and spin (middle) density maps in the (100) plane through the vacancy. On the right, the spin density function along the vertical path dashed in the spin map. Symbols and units as in Fig. 8

Table 6

Net atomic charges (q) and spin moments (μ) evaluated with the HF and LDA approximations according to a Mulliken partition of the periodic densities. Units: electrons. Labels of the atoms as in Fig. 9

	F-centre		Li		F	
	q	μ	q	μ	q	μ
HF	−1.002	1.085	0.978	0.003	−0.975	−0.009
LDA	−0.887	0.951	0.986	0.007	−0.961	0.001

Table 7

Spin Hamiltonian hyperfine isotropic (a) and anisotropic (b) coupling constants at first (Li) and second (F) nearest neighbours of the F-centre in LiF, as calculated in the HF and LDA approximations and measured experimentally with ENDOR spectroscopy [81]. Units: MHz

	Li		F	
	a	b	a	b
HF	38.571	3.244	76.375	11.257
LDA	41.811	3.150	105.56	13.720
expt.	39.0	3.2	105.9	14.9

sponding a constant value is large and positive. The same happens for the F ions around the vacancy. Table 7 shows that, in this case, where we deal with spin densities at the nuclei of the order of $10^{-2}\,|e|/\text{Bohr}^{-3}$ (two orders of magnitudes larger than in the trapped hole case) the agreement between the calculated and experimental values is satisfactory not only for b, but also for a, especially when the first nearest neighbours (Li) are involved. For farther neighbours of the vacancy the UHF error in a and b tends to increase (see Table 7 for the F ions), whereas the LSDA approximation seems to perform better. As a matter of fact, the tendency of LSDA to delocalize unpaired electrons, which prevents from describing well localized states (3d electrons in NiO and 2p electrons in $[M]^0$) correctly, is particularly suitable in the present case, where the unpaired electron occupies a very diffuse 1s state and leads to results in better agreement with experiment than UHF.

6. Conclusions

In this chapter we have recalled the fundamentals of the HF method and discussed how it can be used for the study of periodic structures. For the latter purpose we have made reference to a specific implementation, that embodied in the CRYSTAL public code. While this is not the most general possibility (in particular, the choice of using localized Gaussian functions to describe the crystalline orbitals pervades the very structure of the code and dictates many technical solutions), we have been able in this way to show concretely some of the subtleties which are needed to realize from the basic equations a powerful, efficient and useful tool.

Our second main objective has been to demonstrate through examples the usefulness of the HF method in solid state applications. In this sense we have been necessarily confronted with DFT, the alternative approach which represents nowadays the favourite choice for the overwhelming majority of workers in this field of studies. In spite of all its merits, DFT is known to meet with serious difficulties in the description of systems with well localized spin densities which cannot be described through statistical local averages of exchange interactions. The examples here reported (which have been partly elaborated specially for the present work) are all concerned with this kind of problems. In our opinion, they prove convincingly the merits of the HF technique if not else as an auxiliary technique to DFT calculations.

There is, however, a more fundamental issue which justifies the effort towards the continuous improvement of HF-based techniques, and this comes from molecular quantum mechanics. At variance with DFT, standard quantum-chemical schemes can provide (in principle, at least) results of any required accuracy concerning ground and excited states of many-electron systems. In this respect, programs like CRYSTAL provide a unique opportunity to implement and test post-HF schemes for crystalline systems.

Acknowledgements The present work is part of a project coordinated by A. Zecchina and co-financed by the Italian MURST (Cofin98, Area 03). Financial support from the Italian C.N.R. is acknowledged.

References

[1] D.J. HEHRE, W.A. LATHAN, M.D. NEWTON, R. DITCHFIELD, and J.A. POPLE, GAUSSIAN70, Program number 236, QCPE, Indiana University, Bloomington (Indiana) 1970.
[2] V. FOCK, Z. Phys. **61**, 126 (1930).
[3] J.C. SLATER, Phys. Rev. **35**, 210 (1930).
[4] C.C.J. ROOTHAAN, Rev. Mod. Phys. **23**, 69 (1951).
[5] G.G. HALL, Proc. Roy. Soc. **A205**, 541 (1951).
[6] S.F. BOYS, Proc. Roy. Soc. **A200**, 542 (1950).
[7] S.F. BOYS, Proc. Roy. Soc. **A201**, 125 (1950).
[8] S.F. BOYS, G.B. COOK, C.M. REEVES, and I. SHAVITT, Nature (London) **178**, 1207 (1956).
[9] E. CLEMENTI and R. ROETTI, Atomic Data Nucl. Data Tables **14**, 177 (1974).
[10] S. HUZINAGA, Gaussian Basis Sets for Molecular Calculations, Elsevier Publ. Co., Amsterdam 1984.
[11] V. R. SAUNDERS, in: Computational Techniques in Quantum Chemistry and Molecular Physics, Eds. G.H.F. DIERCKSEN, B.T. SUTCLIFFE, and A. VEILLARD, Reidel, Dordrecht 1975 (p. 347).
[12] L.E. MCMURCHIE and E.R. DAVIDSON, J. Comput. Phys. **26**, 218 (1978).
[13] V. R. SAUNDERS, in: Methods in Computational Molecular Physics, Eds. G.H.F. DIERCKSEN and S. WILSON, Reidel, Dordrecht 1983 (p. 1).
[14] V.R. SAUNDERS and J.H. VANLENTHE, Mol. Phys. **48**, 923 (1983).
[15] M.F. GUEST and S. WILSON, Chem. Phys. Lett. **75**, 66 (1980).
[16] P. DEAK, in: Properties of Crystalline Silicon, Ed. R. HULL, EMIS Datareview Series, Vol. 20., INSPEC, London 1999 (p. 245).
[17] P. HOHENBERG and W. KOHN, Phys. Rev. **136**, B864 (1964).
[18] W. KOHN and L.J. SHAM, Phys. Rev. **140**, A1133 (1965).
[19] M.J. FRISCH, G.W. TRUCKS, H.B. SCHLEGEL, P.M.W. GILL, B.G. JOHNSON, M.A. ROBB, J.R. CHEESEMAN, T.A. KEITH, G.A. PETERSON, J.A. MONTGOMERY, K. RAGHAVACHARI, M.A. AL-LAHAM, V.G. ZAKRZEWSKI, J.V. ORTIZ, J.B. FORESMAN, J. CIOLOWSKI, B.B. STEFANOV, A. NANAYAKKARA, M. CHALLACOMBE, C.Y. PENG, P.Y. AYALA, W. CHEN, M.W. WONG, J.L. ANDRES, E.S. REPLOGE, R. GOMPERTS, R.L. MARTIN, D.J. FOX, J.S. BINKLEY, D.J. DEFREES, J. BAKER, J.P. STEWART, M. HEAD-GORDON, C. GONZALEZ, and J.A. POPLE, GAUSSIAN94, revision C.3, Gaussian, Inc., Pittsburg (P.A.) 1995.
[20] R. DOVESI, C. PISANI, C. ROETTI, M. CAUSÀ, and V.R. SAUNDERS, CRYSTAL 88 Program number 577, QCPE, Indiana University, Bloomington (Indiana) 1989.
[21] H. PREUSS, Z. Naturf. **11a**, 823 (1956).

[22] C.A. WEATHERFORD and H.W. JONES (Eds.), ETO Multicenter Molecular Integrals, Reidel, Dordrecht 1982.
[23] V. R. SAUNDERS, R. DOVESI, C. ROETTI, M. CAUSÀ, N.M. HARRISON, R. ORLANDO, and C.M. ZICOVICH-WILSON, CRYSTAL98, User's Manual, Università di Torino, Torino 1998.
[24] C. PISANI, R. DOVESI, and C. ROETTI, Hartree-Fock ab initio Treatment of Crystalline Systems, Lecture Notes Chem., Vol. 48, Springer-Verlag, Berlin 1988.
[25] R. MCWEENY, Proc. Phys. Soc. (London) **74**, 385 (1959).
[26] J.M. ANDRÉ, L. GOUVERNEUR, and G. LEROY, Internat. J. Quantum Chem. **1**, 451 (1967).
J.L. BRÉDAS, J.M. ANDRÉ, J.G. FRIPIAT, and J. DELHALLE, Gazz. Chim. Ital. **108**, 307 (1978).
[27] G. DEL RE, J. LADIK, and G. BICZÓ, Phys. Rev. **155**, 997 (1967).
[28] A. KARPFEN, Internat. J. Quantum Chem. **19**, 1207 (1981).
[29] A.B. KUNZ, Phys. Rev. B **6**, 606 (1972).
[30] R.N. EUWEMA, D.L. WILHITE, and G.T. SURRATT, Phys. Rev. B **7**, 818 (1973).
[31] V.R. SAUNDERS, C. FREYRIA-FAVA, R. DOVESI, L. SALASCO, and C. ROETTI, Mol. Phys. **77**, 629 (1992).
[32] R. DOVESI, in: Quantum-Mechanical ab initio Calculation of the Properties of Crystalline Materials, Ed. C. PISANI, Lecture Notes Chem., Vol. 67, Springer-Verlag, Berlin 1996 (p. 179).
[33] M. CATTI, G. VALERIO, R. DOVESI, and M. CAUSÀ, Phys. Rev. B **49**, 14179 (1994).
[34] M. CATTI, F. FREYRIA FAVA, C. ZICOVICH, and R. DOVESI, Phys. Chem. Minerals **26**, 389 (1999).
[35] M.D. TOWLER, R. DOVESI, and V.R. SAUNDERS, Phys. Rev. B **52**, 10150 (1995).
[36] M. CATTI and G. SANDRONE, Faraday Discuss. **106**, 189 (1997).
[37] R. DOVESI, F. FREYRIA FAVA, C. ROETTI, and V.R. SAUNDERS, Faraday Discuss. **106**, 173 (1997).
[38] R. MCWEENY, Proc. Roy. Soc. (London) **A241**, 239 (1957).
[39] I.H. HILLIER and V.R. SAUNDERS, Internat. J. Quantum Chem. **4**, 503 (1970).
[40] J.A POPLE and R.K. NESBET, J. Chem. Phys. **22**, 571 (1954).
[41] R. DOVESI, V. R. SAUNDERS, and C. ROETTI, CRYSTAL92, User's Manual, Università di Torino, Torino 1993.
[42] R. DOVESI, V. R. SAUNDERS, C. ROETTI, M. CAUSÀ, N.M. HARRISON, R. ORLANDO, and E. APRÀ, CRYSTAL95, User's Manual, Università di Torino, Torino 1996.
[43] A.D. BECKE, Phys. Rev. A **38**, 3098 (1988).
[44] C. LEE, W. YANG, and R.G. PARR, Phys. Rev. B **37**, 785 (1988).
[45] A.D. BECKE, J. Chem. Phys. **98**, 5648 (1993).
[46] M.D. TOWLER, A. ZUPAN, and M. CAUSÀ, Comp. Phys. Commun. **98**, 181 (1996).
[47] A.D. BECKE, J. Chem. Phys. **88**, 2547 (1988).
[48] R. DOVESI, C. PISANI, C. ROETTI, and V.R. SAUNDERS, Phys. Rev. B **28**, 5781 (1983).
[49] D. MONCRIEFF and V.R. SAUNDERS, Doc NAT648, University of Manchester Regional Computing Center, 1986.
[50] V.R. SAUNDERS, C. FREYRIA-FAVA, R. DOVESI, and C. ROETTI, Comp. Phys. Commun. **84**, 156 (1993).
[51] M. CAUSÀ, R. DOVESI, R. ORLANDO, C. PISANI, and V.R. SAUNDERS, J. Phys. Chem. **92**, 909 (1988).
[52] R. DOVESI, Internat. J. Quantum Chem. **29**, 1755 (1986).
[53] M. LAX, Symmetry Principles in Solid State and Molecular Physics, John Wiley & Sons, New York 1974.
[54] C.M. ZICOVICH-WILSON and R. DOVESI, Internat. J. Quantum Chem. **67**, 299 (1998).
[55] C.M. ZICOVICH-WILSON and R. DOVESI, Internat. J. Quantum Chem. **67**, 311 (1998).
[56] C.M. ZICOVICH-WILSON and R. DOVESI, Chem. Phys. Lett. **277**, 227 (1997).
[57] R. DOVESI, C. PISANI, F. RICCA, and C. ROETTI, Phys. Rev. B **22**, 5963 (1980).
[58] R. DOVESI, C. PISANI, and C. ROETTI, Internat. J. Quantum Chem. **17**, 517 (1980).
[59] T. OGUCHI, K. TERAKURA, and A.R. WILLIAMS, Phys. Rev. B **28**, 6443 (1983).
[60] K. TERAKURA, T. OGUCHI, A.R. WILLIAMS, and J. KÜBLER, Phys. Rev. B **30**, 4734 (1984).
[61] A. SVANE and O. GUNNARSON, Phys. Rev. Lett. **65**, 1148 (1990).
[62] S. MASSIDA, M. POSTERNAK, and A. BALDERESCHI, Phys. Rev. B **46**, 11705 (1992).
[63] W.C. MACKRODT, N.M. HARRISON, V.R. SAUNDERS, N.L.ALLAN, M.D. TOWLER, E. APRÀ, and R. DOVESI, Phil. Mag. **68**, 653 (1993).
[64] M.D. TOWLER, N.L. ALLAN, N.M. HARRISON, V.R. SAUNDERS, W.C. MACKRODT, and E. APRÀ, Phys. Rev. B **50**, 5041 (1994).
[65] G. RIUS, R. COX, R. PICARD, and C. SANTIER, C.R. Acad. Sci. (France) **271**, 824 (1970).

[66] O.F. SCHIRMER, J. Phys. Chem. Solids **32**, 499 (1971).
[67] H.T. TOHVER, B. HENDERSON, Y. CHEN, and M.M. ABRAHAM, Phys. Rev. B **5**, 3276 (1972).
[68] M.M. ABRAHAM, Y. CHEN, J.L. KOLOPUS, and H.T. TOHVER, Phys. Rev. B **5**, 4945 (1972).
[69] D.J. DRISCOLL and J.H. LUNSFORD, J. Phys. Chem. **89**, 4415 (1985).
[70] T. ITO, J. WANG, C.H. LIN, and J.H. LUNSFORD, J. Amer. Chem. Soc. **107**, 5062 (1985).
[71] G. RIUS and A. HERVÉ, Solid State Commun. **15**, 399 (1974).
[72] M.M. ABRAHAM, W.P. UNRUH, and Y. CHEN, Phys. Rev. B **10**, 3540 (1974).
[73] C. FREYRIA-FAVA, R. DOVESI, V.R. SAUNDERS, M. LESLIE, and C. ROETTI, J. Phys.: Condensed Matter **5**, 4793 (1993).
[74] R. ORLANDO, P. AZAVANT, N.M. HARRISON, and V.R. SAUNDERS, J. Phys.: Condensed Matter **6**, 8573 (1994).
[75] R. ORLANDO, P. AZAVANT, M.D. TOWLER, R. DOVESI, and C. ROETTI, J. Phys.: Condensed Matter **8**, 1123 (1996).
[76] A. LICHANOT, C. LARRIEU, R. ORLANDO, and R. DOVESI, J. Phys. Chem. Solids **59**, 7 (1998).
[77] A. LICHANOT, C. LARRIEU, C. ZICOVICH-WILSON, C. ROETTI, R. ORLANDO, and R. DOVESI, J. Phys. Chem. Solids **59**, 1119 (1998); *Erratum:* J. Phys. Chem. Solids **60**, 855 (1999).
[78] J.A. WEIL, J.R. BOLTON, and E. WERTZ, Electron Paramagnetic Resonance – Elementary Theory and Practical Applications, John Wiley & Sons, New York 1994.
[79] D.R. LIDE (Ed.), Handbook of Chemistry and Physics, 72nd ed., CRC Press, Boca Raton 1991/1992.
[80] O.F. SCHIRMER, J. Phys. C **6**, 300 (1973).
[81] B. HENDERSON, Defects in Crystalline Solids, Edward Arnold, Ltd., London 1972.
[82] P.A.M. DIRAC, Proc. Cambridge Phil. Soc. **26**, 376 (1930).
[83] J.P. PERDEW and A. ZUNGER, Phys. Rev. B **23**, 5048 (1981).

phys. stat. sol. (b) **217**, 89 (2000)

Subject classification: 71.15.–m; 71.20.Mq; S5.11

An Introduction to the Third-Generation LMTO Method

R. W. Tank and C. Arcangeli

Max-Planck-Institut für Festkörperforschung, Heisenbergstr. 1, D-70569 Stuttgart, Germany

(Received August 10, 1999)

The LMTO method for *ab inito* electronic structure calculations has been used for many years. During this time it has been evolving and has now reached what is termed its *third generation*. In this paper we give an introduction to this latest formalism showing how the LMTO basis set is constructed and used to solve Schrödinger's equation. In addition we discuss the topics of downfolding and evaluation of the total energy within LMTO.

1. Overview

The LMTO method [1] has been used for many years as an *ab initio* method for electronic structure calculations. It is related to both KKR methods [2, 3], which also use muffin tin orbitals and the LAPW method [4, 5], in which the orbitals are augmented by plane waves. What sets it apart and characterises it, is its basis set, the Linear Muffin Tin Orbitals (LMTOs), which are both minimal and localised in real space.

The small basis produces a small Hamiltonian matrix, which makes computation more efficient. In addition the minimal quality makes it an intelligible set in the sense that it is very easy in a given system to extract information about which orbitals are playing the major role in determining the system properties. For example this has been exploited in recent LMTO based studies of high T_c compounds [6 to 9] where detailed information about the orbital nature of bands near the Fermi surface has been obtained. It also lends itself naturally to the use of both ELF and COHP techniques [10, 11] to study in detail the chemical bonding of a compound. In a similar fashion GW calculations [12, 13] can be done that are based upon an underlying LMTO calculation. The localised quality of the LMTOs means they can be used to study non-periodic solids. Calculations can be performed in real space in conjunction with recursion [14, 15] or Green function [16] techniques allowing studies of metallic glasses [17], amorphous solids [18], and impurities [19]. In addition to the study of the interior of solids, LMTO can be used for surfaces [20, 21] and polymers [22, 23].

Being an all electron method, heavy atoms can be incorporated without the difficulties that may occur in a pseudopotential calculation and with only a minimal increase to the size of the basis set.

Over the years there have been offshoots such as the development of full potential LMTO methods [24 to 29] which give highly accurate total energies, as well as the development of the PAW scheme [30] and a linear-response LMTO scheme [31] that gives excellent phonon spectra.

The LMTO method itself has been evolving in the intervening time. The second generation LMTO [32, 33] commonly known as tight-binding or TB-LMTO introduced screening to make the LMTOs as localised as possible, and has been very successful. The formalism has now entered a third generation [9, 34, 35]. The region of space between the spheres is now treated on the same footing as the region inside the spheres. This has resulted in numerous improvements, for example the eigenvalues are now correct to third order in energy around the linearisation energy. At the same time the formalism is now more complete and opens up a way forward to higher order calculations such as quadratic QMTO or cubic CMTO.

In this paper we shall present an introduction to the formalism and underlying concepts of the third generation LMTO method. It is intended to be useful for those who have no knowledge of LMTO or only knowledge of the formalism of previous generations. The outline of the paper is as follows. In Section 2 we shall review the use of the muffin tin form for the Kohn-Sham potential. Section 3 will provide a one-dimensional example that introduces the concepts used later in the paper. In Section 4 we shall define the *kinked partial waves* which will be used to construct an exact solution of the Schrödinger equation for the muffin tin potential. In Section 5 we show how to linearise and create the LMTO basis set. The last two sections will then cover the topics of downfolding and evaluation of the total energy.

1.1 Starting point; density functional theory

The task we are interested in is to study the electronic structure of solid materials. This means that we wish to treat quantum mechanically a system of many electrons moving in the potential of a set of nuclei with fixed position. The electrons are interacting, and this fact vastly complicates the problem. What saves us is density functional theory, which allows us to reduce the problem to an *equivalent* one in which the electrons are non-interacting. This equivalent problem is then easy to solve using LMTO. For an excellent review of density functional theory the reader is referred to [36]. Here we merely repeat the relevant results for us.

Hohenberg and Kohn [37] showed that the total energy of an interacting system is a functional of the charge density, and furthermore is minimised by the ground state charge density, $E[n] \geq E_{gs} = E[n_{gs}]$. Our task is then to find this ground state charge density, n_{gs} from which we can obtain the total energy and other ground state properties. The second important result for us is that of Kohn and Sham [38], who derived an exact equivalent single-particle (i.e. non-interacting) description of the many-particle interacting system. They rewrote the expression for the total energy as

$$E[n] = T_0[n] + \int d\mathbf{r} \left[V_{\text{ext}}(\mathbf{r}) + \tfrac{1}{2} \Phi(\mathbf{r}) \right] n + E_{\text{xc}}[n],$$

where $T_0[n]$ would be the kinetic energy if there were no electron–electron interactions, Φ is the electrostatic potential of the electrons and the exchange correlation energy E_{xc} is implicitly defined by this equation. Then by minimising E while constraining the particle number to be constant one obtains

$$\frac{\delta T_0}{\delta n} + \underbrace{V_{\text{ext}} + \Phi + \frac{\delta E_{\text{xc}}}{\delta n}}_{V'} = \mu, \qquad (1)$$

which is formally identical to the non-interacting problem

$$\frac{\delta T_0}{\delta n} + V' = \mu.$$

Here V' can be thought of an effective local potential seen by the electrons. Hence if we know V' we need only to solve the one-electron Schrödinger equation

$$[-\nabla^2 + V'(\mathbf{r})]\psi_i(\mathbf{r}) = \varepsilon_i\psi_i(\mathbf{r}) \quad (2)$$

and then the ground state charge density is given by

$$n_{gs}(\mathbf{r}) = \sum_i^N |\psi_i(\mathbf{r})|^2, \quad (3)$$

where the sum is over the lowest states up to the required particle number. The LMTO method is a general way to solve this SE for any given system to obtain the single particle eigenfunctions ψ and the Kohn-Sham energies ε.

The biggest difficulty now is that we do not know in advance V'. From its definition in (1) it is clear that V' itself depends upon the charge density and so (2), (3) need to be solved self-consistently. Moreover the exact functional form of the exchange correlation potential $\delta E_{xc}/\delta n = V_{xc}$ is unknown. We need some approximation for V_{xc}. Many are known and used, the most common being the local density approximation or LDA [36]. We shall not discuss the relative advantages of different approximations to the exchange correlation potential in this paper. In all our examples we shall use the LDA approximation.

Given a choice for V_{xc}, the LMTO method provides an efficient scheme to find the self-consistent solution of (2) and hence find n_{gs} from which all other properties follow.

2. Dividing up Space and the Muffin Tin Approximation

Rather than try to solve Schrödinger's Equation (SE) for the whole of space in one go, we choose to divide up space into smaller regions for which we individually solve SE and then attach together the spatially separate solutions. This approach has two main advantages. Firstly we can use the same division of space to simplify our potential, via the muffin tin approximation. Secondly this approach is then straightforward to linearise. In order to find a suitable way to divide space we first note that free atoms are spherical. Furthermore, when atoms are collated[1] to make a solid they retain their sphericity to a certain degree. Inspired by this we choose to divide up space using spheres, placing one sphere at each atomic site $\mathbf{R}$, with radius $s_\mathbf{R}$. In open systems, such as the diamond structure, there can be quite large voids or empty spaces between the atoms. In such cases we usually include additional spheres centred at some chosen points in this empty space rather than on nuclei of real atoms. These additional spheres are called *empty spheres*.

We shall now make an assumption about the functional form of the Kohn-Sham potential V'. We shall assume that it is *spherical inside each chosen sphere* and *constant between the spheres*. More specifically,

$$V'(\mathbf{r}) = \sum_\mathbf{R} [V_\mathbf{R}(|\mathbf{r}-\mathbf{R}|)\Theta(|\mathbf{r}-\mathbf{R}|-s_\mathbf{R})] + V_{mtz}\Theta_I(\mathbf{r}) \quad (4)$$

with

$$\Theta_I(\mathbf{r}) = \left[1 - \sum_\mathbf{R} \Theta(|\mathbf{r}-\mathbf{R}|-s_\mathbf{R})\right]. \quad (5)$$

---

[1] collate: collect and arrange systematically

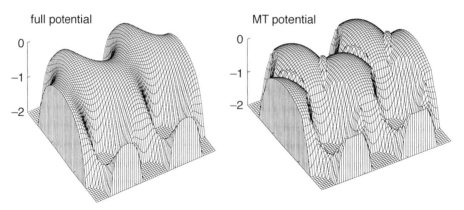

Fig. 1. Full (left) and MT (right) potential for Si in the (110) plane

It is often useful to rewrite this as

$$V(\mathbf{r}) = \sum_{\mathbf{R}} [v_{\mathbf{R}} \Theta(|\mathbf{r} - \mathbf{R}| - s_{\mathbf{R}})] + V_{\text{mtz}} \quad (6)$$

with

$$v_{\mathbf{R}} = V_{\mathbf{R}} - V_{\text{mtz}}. \quad (7)$$

With our choice of Θ_l (5) we can allow the spheres to overlap. This is useful because overlapping spheres tend to give a better representation of the full potential. The potential in the overlap region between two spheres 1 and 2 is

$$V(\mathbf{r}) = v_1 + v_2 + V_{\text{mtz}} = V_1 + V_2 - V_{\text{mtz}}.$$

The *total* potential inside each sphere will now no longer be spherical, as it contains intrusions from neighbouring spheres that overlap with the central sphere. From now on we shall refer to the spheres of radius s that divide up space as *potential spheres*.

This muffin tin approximation for the potential (4) greatly simplifies the solution of SE and makes LMTO efficient compared to full potential calculations where no such assumptions are made for the shape of V'. Experience has shown it to be a very good approximation for a vast number of systems.

It is natural to ask how closely this muffin tin potential reproduces the full potential. In Fig. 1 we compare the full and MT potentials for silicon in the (110) plane. Two things are worth noting. Firstly, the presence of empty spheres. These provide maxima or a repulsive hump in the potential at the interstitial sites. Secondly, the spheres are overlapping. This leads to significant non-spherical components in the total potential around each Si atom. It is quite clear that the potential in the bond direction is lower than in the back bond direction. In Fig. 2 we show these same two potentials now compared along some specific lines. It is notable how many features of the full potential are reproduced by the muffin tin form, showing that it is not a so drastic approximation as it may seem at first sight.

3. A One-Dimensional Example

For a potential of the muffin tin form, with non-overlapping spheres, is is possible to write down the exact solution of SE. To do this we need only to solve SE for each

An Introduction to the Third-Generation LMTO Method

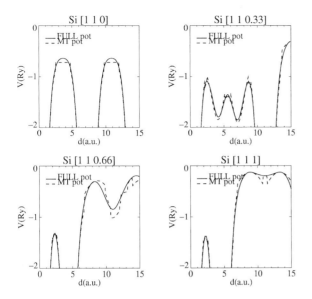

Fig. 2. Full (full line) and MT (dashed line) potential for Si along various directions

sphere individually and for the interstitial region where the potential is flat. These "parts" of the solution must then be matched continuously and differentially at every sphere boundary. The construction of the individual parts will be covered in the next section and is in places rather involved. Therefore to illustrate the basic principles, and to introduce notation that will be used later we shall here consider the exact solution of SE for an extremely simple system, namely the one dimensional square well.

Such a well is illustrated in Fig. 3. This single square well is an analogue of a single muffin tin sphere. We have three regions of space, I ($x < -a$), II ($-a \leq x \leq a$), and III ($x > a$) for which we shall solve SE individually and then match the parts. We shall look for the even solution. It is trivial to construct the solution ϕ in each region separately,

$$\phi_I = e^{\sqrt{-E}x} e^{\sqrt{-E}a},$$
$$\phi_{II} = \cos\left(\sqrt{(E-V_0)}\,x\right) \left[\cos\left(\sqrt{(E-V_0)}\,a\right)\right]^{-1},$$
$$\phi_{III} = e^{-\sqrt{-E}x} e^{\sqrt{-E}a}.$$

These parts of the solution (or *partial waves*) have been normalised so that they take the value 1 on the boundary of the well. This trivially satisfies the requirement that the

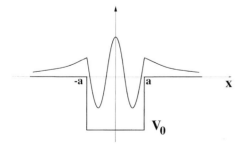

Fig. 3. Kinked partial wave for a single square well

total solution be continuous at the boundary. There is, however, still a discontinuity in the slope at $x = \pm a$. We refer to this as a *kink* and it has the value

$$K(E) = -\sqrt{(E - V_0)} \tan\left(\sqrt{(E - V_0)}\, a\right) + \sqrt{-E}.$$

The combination $\phi_I + \phi_{II} + \phi_{III}$ we will then call a *kinked partial wave*. This kinked partial wave is now almost the required eigenfunction. It is not because it is kinked. To find the eigenfunction we must look for a specific energy E for which the kink of the kinked partial wave will vanish. This energy is then the eigenenergy and the kinked partial wave will become the eigenfunction. That is, the exact solution is given by the energy E for which the *kink cancellation condition* $K(E) = 0$ is satisfied.

The case of two square wells is shown in Fig. 4. In this case we first construct 4 kinked partial waves Φ_1 to Φ_4. Two are centred on each well, one even in character and one odd in character. In the region between the wells we have some extra freedom, and use this to make the choice that each kinked partial wave should vanish at the boundary of the other well. For instance,

$$\Phi_1(x) = \left(\frac{e^{-k(x-b)} - e^{k(x-b)}}{e^{-k(a-b)} - e^{k(a-b)}}\right); \quad a < x < b.$$

The exact solution of this twin well system will be a superposition of the Φ_i,

$$\Psi = \sum_i \Phi_i(E)\, v_i.$$

This must be done in such a way that all the kinks cancel so that the result is smooth. The eigenvalue E and the eigenvector v_i are again given by the *kink cancellation condition*, which is now a set of linear equations,

$$\sum_j K_{ij}(E)\, v_j = 0.$$

Here K_{ij} is the odd or even kink at site i of Φ_j.

The case of a muffin tin potential in three dimensions is solved in the same way. For each individual sphere we construct a kinked partial wave that is continuous but kinked at the sphere boundaries and vanishing on all spheres except its own. The analogue of "even" and "odd" character becomes angular momentum character in the 3D case. The exact solution of SE for the muffin tin potential will be the superposition of kinked

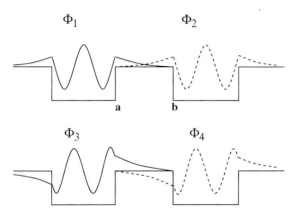

Fig. 4. Kinked partial wave for a double square well

4. Kinked Partial Waves

4.1 Partial waves and the solution of the radial Schrödinger equation

We now consider the solution of the Schrödinger equation inside each individual sphere. For heavier atoms it is important to take into account relativistic effects, as the potential is large close to the nucleus. Therefore in practice one usually solves the Dirac equation with the so-called *scalar relativistic approximation*. The potential inside the sphere is spherical, and for this the solution of the Dirac equation is well known. Rather than explicitly deriving this solution we shall again quote the important results [39 to 41]. The form of the solution is

$$\phi(\mathbf{r}, \mathbf{s}) = \begin{bmatrix} g(r)\,\chi(\hat{\mathbf{r}}, \mathbf{s}) \\ f(r)\,\tilde{\chi}(\hat{\mathbf{r}}, \mathbf{s}) \end{bmatrix}. \qquad (8)$$

where $\mathbf{s}$ represents the spin components. The radial functions f and g satisfy the coupled differential equations

$$cf' = c\left(\frac{\mathcal{K}-1}{r}\right)f - (E-V)\,g\,,$$

$$cg' = -c\left(\frac{\mathcal{K}-1}{r}\right)g + (E-V+2mc^2)\,f\,.$$

Here "′" denotes radial differentiation. The operator $\mathcal{K} = -(\sigma \cdot \mathbf{l} - 1)$ contains the product of the spin and orbital angular momentum operators. It has possible eigenvalues $\mathcal{K} = l$ (when the total angular momentum $j = l - 1/2$) and $\mathcal{K} = -l + 1$ (when $j = l + 1/2$). Eliminating f from the two equations, and using the identity $\mathcal{K}(\mathcal{K}+1) = l(l+1)$ gives the following equation for g:

$$-\frac{1}{2M}\left[g'' + \frac{2}{r}g' - \frac{l(l+1)}{r^2}g\right] - \frac{V'g'}{4M^2c^2} + Vg - \left[\frac{\mathcal{K}+1}{r}\frac{V'g}{4M^2c^2}\right] = Eg\,. \qquad (9)$$

Here $M = m + (E-V)/(2c^2)$. The last term is the only one to contain spin and can thus be thought of as a spin-orbit term. In the scalar relativistic approximation we drop this term. The result is thus

$$\phi(\mathbf{r}, \mathbf{s}) = \begin{bmatrix} g_l\,\mathrm{Y}_{lm}\chi(\mathbf{s}) \\ \dfrac{1}{2Mc}\left(g_l' - \dfrac{1}{r}g_l\sigma \cdot \mathbf{L}\right)\mathrm{Y}_{lm}\tilde{\chi}(\mathbf{s}) \end{bmatrix}. \qquad (10)$$

Upon dropping the spin orbit term we regain the angular momentum l as a good quantum number. The resulting radial differential equation is easily solved numerically. The solution ϕ we shall call a partial wave.

Note that for each partial wave (10) there are two radial components, an upper and lower one in the vector. The lower component is very small except close to the nucleus, and would vanish if we neglected relativistic effects. In Fig. 5 we show the s, p and d partial waves for both silicon and osmium. On each graph we have plotted both the upper and lower components. This shows clearly how the lower components are truly

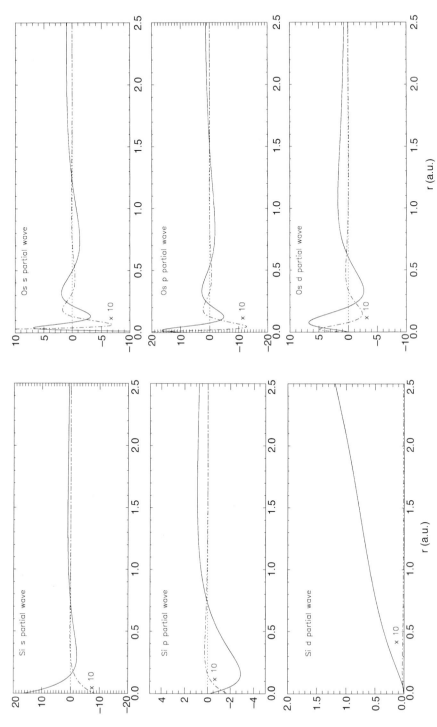

Fig. 5. Upper and lower components of the s, p and d partial waves for Si and Os. The lower components are shown multiplied by a factor of 10

small except close the the nucleus, and secondly that they are bigger for the heavier atom Os.

The charge density of a given partial wave inside each potential sphere is given by the sum of the square of the two radial components,

$$n(r) = g_l^2 + \frac{1}{(2Mc)^2}\left[g_l'^2 + \frac{l(l+1)}{r^2}g_l^2\right]$$

$$= g_l^2\left[1 + \frac{1}{(2Mc)^2}\frac{l(l+1)}{r^2}\right] + \frac{1}{(2Mc)^2}g_l'^2.$$

At the sphere boundary $r = s$ the second component has effectively vanished and hence $M(s) = m = 1/2$. This leads to a particularly simple expression for the integral of the square of the partial wave over the sphere volume (i.e its total charge). Consider the integral

$$\langle\phi(E_2)|\,H\,|\phi(E_1)\rangle = E_1\langle\phi(E_2)|\phi(E_1)\rangle$$
$$+ s^2 g(E_2, s)\,g'(E_1, s) - \left[\frac{1}{2Mc^2}V'(s)\right]s^2 g(E_2, s)g(E_1, s).$$

This is Hermitian and so $\langle\phi(E_2)|\,H\,|\phi(E_1)\rangle - \langle\phi(E_1)|\,H\,|\phi(E_2)\rangle = 0$,

$$0 = (E_1 - E_2)\langle\phi(E_2)\,|\,\phi(E_1)\rangle + s^2[g(E_2, s)\,g'(E_1, s) - g(E_1, s)\,g'(E_2, s)].$$

Rearranging this gives

$$\langle\phi(E_2)\,|\,\phi(E_1)\rangle = -\frac{s^2[g(E_2, s)g'(E_1, s) - g(E_1, s)\,g'(E_2, s)]}{(E_1 - E_2)}$$
$$- sg(E_2, s)\,g(E_1, s)\left[\frac{D(E_1) - D(E_2)}{(E_1 - E_2)}\right].$$

Here we have introduced the logarithmic derivative $D = rg'(r)/g(r)$. Taking the limit $E_1 \to E_2$ gives

$$\langle\phi\,|\,\phi\rangle = -sg^2\dot{D}\,. \tag{11}$$

The integral is thus proportional to the energy derivative of the logarithmic derivative of g. Although this expression only involves the slope of the upper component it should be stressed that it is the integral of $n(r)$ which is the sum of the squares of both the upper and the lower components. In other words the relativistic effects do alter the slope of the partial wave at the sphere boundary.

4.2 Screened spherical waves and the solution of Schrödinger's equation in the interstitial

In the previous section we introduced the partial waves which solve SE inside the spheres. We need to match them to the solution in the interstitial. As the potential in the interstitial is constant and equal to V_{mtz} the SE reduces to the wave equation

$$\nabla^2\psi = -(E - V_{\text{mtz}})\,\psi = -\kappa^2\psi\,,$$

of which many solutions are well known. Our choice of solution is driven by two factors. Firstly we wish to make the procedure of matching to the partial waves as easy as

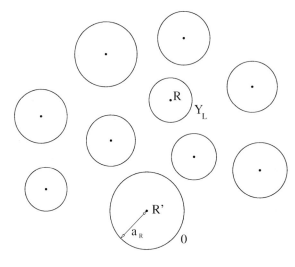

Fig. 6. Screening spheres and boundary conditions for the screened spherical wave ψ_{RL}

possible. We would like one single interstitial function attached to each partial wave. Secondly we would like this function to be well localised. This is partly so that we can use them for real space calculations and partly because when we come to linearise the problem we shall obtain a wider energy window. Our choice of functions are the screened spherical waves (SSW) which solve SE in the interstitial with specific boundary conditions. In order to specify these conditions we first place a *screening sphere* of radius $a_\mathbf{R}$ at each site $\mathbf{R}$. These screening spheres (or a spheres) are defined to be non-overlapping. The boundary conditions that we chose are then imposed at these spheres. For an SSW $\psi_{\mathbf{R}L}(\mathbf{r})$[2]) centred at site $\mathbf{R}$ and of character L we require that

$$\psi_{\mathbf{R}L}(|\mathbf{r} - \mathbf{R}'| = a_{\mathbf{R}'}) = Y_L \quad \text{if} \quad \mathbf{R}' = \mathbf{R},$$
$$\psi_{\mathbf{R}L}(|\mathbf{r} - \mathbf{R}'| = a_{\mathbf{R}'}) = 0 \quad \text{if} \quad \mathbf{R}' \neq \mathbf{R}. \tag{12}$$

That is, an SSW $\psi_{\mathbf{R}L}(\mathbf{r})$ should vanish on all screening spheres except for the screening sphere at its own site $\mathbf{R}$ on which it equals the spherical harmonic Y_L. The screening spheres and this boundary condition are shown schematically in Fig. 6. The a spheres cannot overlap because then there would be points on the sphere at $\mathbf{R}$ where we require $\psi_{\mathbf{R}L}$ to vanish and to be non-zero simultaneously. By forcing the SSW to vanish on neighbouring spheres we localise it for both positive and negative energies. In addition the SSWs will be shaped according to the structure being studied, via the positioning of the screening spheres. This feature will make our basis set minimal. The matching condition to the partial waves is made easy by the fact that we have required the SSW to equal a spherical harmonic on a particular

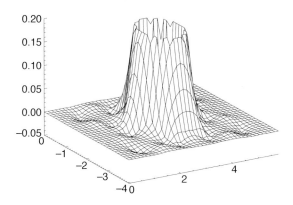

Fig. 7. Screened spherical wave in the (100) plane for an f.c.c. structure

---

[2]) Notation: $L \equiv \{l, m\}$

chosen sphere. It is important to realise that although each SSW is labelled with an angular momentum L, the SSWs do not have a pure angular momentum character (except of course on the screening sphere a_R). Fig. 7 shows an SSW in the [110] plane of an f.c.c. structure. We have plotted an $l = 0$ SSW, and one can clearly see how it becomes spherical towards the central sphere. However, as one moves towards the neighbouring spheres it becomes non-spherical and at the same time rapidly drops off in value.

4.2.1 Screening and screened structure constants

The boundary conditions used to define the SSWs have an electrostatic analogy. The screening spheres can be thought of as earthed metallic spheres and the electrostatic potential between the spheres the analogy of our SSW. We chose one sphere at $\mathbf{R}$ on which is affixed a static multipole charge. The electrostatic potential must vanish on all the earthed metallic spheres except for the one on which we glued the multipole. There it should equal a spherical harmonic. This analogy illustrates the concept of screening in LMTO [32, 33]. The multipole is screened by the metal spheres that surround it. Image charges are induced on these spheres and the total electrostatic potential is that of the original multipole plus the electrostatic potential of the image charges. The total potential is, like the SSW, short ranged.

The screening procedure used to construct the SSWs follows this electrostatic analogy, and is shown schematically in Fig. 8. In essence we shall start from an *unscreened* spherical wave, $n_l(\kappa r) Y_{lm}(\hat{\mathbf{r}})$ where n_l is a Neumann function[3]) centred on one site $\mathbf{R}$ and *screen* it by adding additional spherical waves of differing l and placed at differing sites. We do this in such a way that the boundary conditions (12) are satisfied.

In Fig. 8a we show a single spherical wave centred at site $\mathbf{R}$. Any spherical wave $n_{\mathbf{R}l} Y_{\mathbf{R}L}$ can be expanded about another site $\mathbf{R}'$ in Bessel functions,

$$n_{\mathbf{R}l} Y_{\mathbf{R}L} = n_{\mathbf{R}l} Y_{\mathbf{R}L} \delta_{\mathbf{R}',\mathbf{R}} - \sum_{L'} j_{\mathbf{R}'l'} Y_{\mathbf{R}'L'} S^0_{\mathbf{R}'L',\mathbf{R}L}. \tag{13}$$

This is an example of a so-called "one-center expansion", in the sense that we have made an expansion of a given function in terms of other functions all of which are centred at *the same (i.e. one)* site. In Fig. 8a the dashed line indicates a region of space around another site where we might well choose to use this expansion. Such an expansion can be done for any given function, as it is no more than an expansion in spherical harmonics on a series of shells with radius r about that chosen site. Equation (13) is exact as long as the summation is taken to infinity. The S^0 are known as the *unscreened structure constants* (or bare structure constants) and have a well known analytical form [42, 43],

$$S^0_{\mathbf{R}'L',\mathbf{R}L} = -4\pi \sum_{l''} \frac{2(\kappa\omega)^{l+l'-l''} (2l''-1)!!}{(2l'-1)!! (2l-1)!!}$$
$$\times C_{LL'L''}(-1)^{-l} Y_{L''}(\widehat{\mathbf{R}-\mathbf{R}'}) n_l(|\mathbf{R}-\mathbf{R}'|). \tag{14}$$

---

[3]) The Neumann function n and Bessel function j solve the radial part of the wave equation
$$r^2 \frac{d^2 f}{dr^2} + 2r \frac{df}{dr} + \left[\kappa^2 r^2 - l(l+1)\right] f = 0.$$
The solutions for f that are regular at the origin are the j_l and those that are irregular are the n_l.

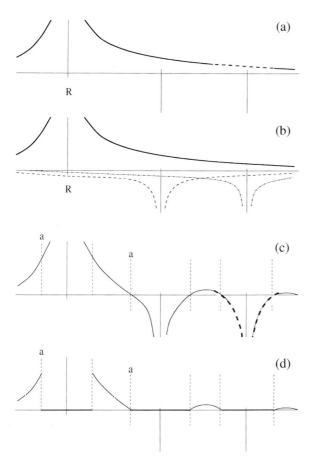

Fig. 8. Pictorial representation of screening and truncation, as explained in the text

Note that we have chosen a slightly different normalisation for j and n compared to the "textbook" definitions [44], they have been rescaled by the following factors:

$$n_l \Rightarrow -n_l \frac{(\kappa\omega)^{(l+1)}}{(2l-1)!!}, \qquad j_l \Rightarrow j_l \frac{(2l-1)!!}{2(\kappa\omega)^l}.$$

Here ω is an arbitrary length scale. Some people find ω a strange choice of symbol for a length scale, but the reason is historical, as it is often set equal to the Wigner-Seitz sphere radius. These factors are chosen to give a simple limit for $\kappa = 0$ (zero energy). In this limit $n_l = (\omega/r)^{(l+1)}$ and $j_l = 1/[2(2l+1)] \, (r/\omega)^l$.

In Fig. 8b we show the first stage of the screening process. Here we have added further spherical waves at all sites. These additional spherical waves can be thought of as screening multipoles. The screened spherical wave is then the sum of the initial spherical wave at **R** plus all of these screening multipoles,

$$\psi_{\mathbf{R}L}(\mathbf{r}) = \sum_{\mathbf{R}'L'} n_{\mathbf{R}'l'} Y_{L'} M_{\mathbf{R}'L', \mathbf{R}L} \qquad (15)$$

The matrix M still needs to be found, and is specified via the boundary conditions for ψ. Equation (15) is an example of a so-called multicenter expansion, the function $\psi(\mathbf{r})$ being expanded in functions that are centred on *many (i.e. a multitude of)* different sites.

The result of this superposition is shown in Fig. 8c. One sees that the SSW now has divergences at all sites. The vertical dotted lines indicate the screening spheres, on which we require the SSW to vanish. Consider now the one-center expansion of ψ about another site. Again the heavy dashed line in Fig. 8c shows a region where we might choose to make such an expansion. In analogy to (13) we choose to write it in the form

$$\psi_{RL} = n^a_{\mathbf{R}l}\, Y_{\mathbf{R}L}\, \delta_{\mathbf{R}',\mathbf{R}} - \sum_{L'} j^a_{\mathbf{R}'l'}\, Y_{\mathbf{R}'L'}\, S^a_{\mathbf{R}'L',\mathbf{R}L}, \qquad (16)$$

where we must now specify the radial functions n^a and j^a for each site. As ψ solves the wave equation, so must n^a_l and j^a_l. Hence they can only be linear combinations of Bessel and Neumann functions,

$$n^a_{\mathbf{R}l} = t_1 n_l + t_2 j_l,$$
$$j^a_{\mathbf{R}l} = t_3 n_l + t_4 j_l. \qquad (17)$$

In addition, from the boundary conditions on the SSW ψ we have $n^a_{\mathbf{R}l}(a_\mathbf{R}) = 1$ and $j^a_{\mathbf{R}l}(a_\mathbf{R}) = 0$. Finally in order to completely specify $t_1 \ldots t_4$ we make the following choice for their radial slope on the a spheres:

$$n^{a\prime}_{\mathbf{R}l}(a_\mathbf{R}) = 0 \quad \text{and} \quad j^{a\prime}_{\mathbf{R}l}(a_\mathbf{R}) = -1/a_\mathbf{R}. \qquad (18)$$

With these we can find $t_1 \ldots t_4$ at all sites and for all l,

$$\begin{pmatrix} t_1 & t_2 \\ t_3 & t_4 \end{pmatrix}_{\mathbf{R}l} = \frac{2a_\mathbf{R}^2}{\omega} \begin{pmatrix} j'_{\mathbf{R}l}(a_\mathbf{R}) & -n'_{\mathbf{R}l}(a_\mathbf{R}) \\ j_{\mathbf{R}l}(a_\mathbf{R})/a_\mathbf{R} & -n_{\mathbf{R}l}(a_\mathbf{R})/a_\mathbf{R} \end{pmatrix}. \qquad (19)$$

Here we have used the fact that $r^2(n_{\mathbf{R}l}(r)\, j'_{\mathbf{R}l}(r) - n'_{\mathbf{R}l}(r)\, j_{\mathbf{R}l}(r)) = \omega/2$. Now it is straightforward to find both M and S^a.

From (13) and (15) we form an expansion of $\psi_{RL}(\mathbf{r})$ about a chosen site R'',

$$\psi_{RL} = \sum_{L''} Y_{L''} \left[\sum_{\mathbf{R}'L'} [n_{\mathbf{R}''l''}\delta_{\mathbf{R}'',\mathbf{R}'}\delta_{L'',L'} - j_{\mathbf{R}''l''} S^0_{\mathbf{R}''L'',\mathbf{R}'L'}] M_{\mathbf{R}'L',\mathbf{R}L} \right].$$

We can get a similar expression from (16) and (17),

$$\psi_{RL} = \sum_{L''} Y_{L''} [n_{\mathbf{R}''l''}(t_{1,\mathbf{R}''L''}\delta_{\mathbf{R}'',\mathbf{R}'}\delta_{L'',L'} - t_{3,\mathbf{R}''L''} S^a_{\mathbf{R}''L'',\mathbf{R}'L'})$$
$$+ j_{\mathbf{R}''l''}(t_{2,\mathbf{R}''L''}\delta_{\mathbf{R}'',\mathbf{R}'}\delta_{L'',L'} - t_{4,\mathbf{R}''L''} S^a_{\mathbf{R}''L'',\mathbf{R}'L'})]$$

Comparing coefficients of n and j gives

$$M_{\mathbf{R}''L'',\mathbf{R}L} = (t_{1,\mathbf{R}L}\delta_{\mathbf{R}''L'',\mathbf{R}L} - t_{3,\mathbf{R}''L''} S^a_{\mathbf{R}''L'',\mathbf{R}L}),$$

$$\sum_{\mathbf{R}'L'} S^0_{\mathbf{R}''L'',\mathbf{R}'L'} M_{\mathbf{R}'L',\mathbf{R}L} = -(t_{2,\mathbf{R}L}\delta_{\mathbf{R}''L'',\mathbf{R}L} - t_{4,\mathbf{R}''L''} S^a_{\mathbf{R}''L'',\mathbf{R}L}).$$

For simplicity in the last step we shall now drop the **R**L subscripts and then matrix multiplication shall be implied. Eliminating M gives

$$S^a = \frac{t_1}{t_3} + \frac{1}{t_3}\left(-\frac{t_4}{t_3} - S^0\right)^{-1}\frac{1}{t_3}[t_1 t_4 - t_2 t_3]. \qquad (20)$$

These are known as the *screened structure constants*, and together with n^a and j^a, defined in (17), they fully specify each SSW.

To summarise this section, the SSWs are solutions of SE for the flat interstitial potential with certain boundary conditions imposed on the screening spheres. Each individual SSW can be expressed in a multi-center expansion as a sum of spherical waves of differing l character on differing sites. The coefficients of the expansion are given by $M = t_1 - t_3 S^a$. Note that the matrix M has an inverse. This means that the set of SSWs $\{\psi_{\mathbf{R}L}\}$ and the set of spherical waves $\{n_l(r_{\mathbf{R}}) Y_L(\widehat{(\mathbf{r} - \mathbf{R})})\}$ are equivalent basis sets. Any function $f(\mathbf{r})$ that solves the wave equation with energy E can be expanded in SSWs of the same energy. The expansion coefficients are trivial, the coefficient of $\psi_{\mathbf{R}L}$ will just be the value of the L component of the spherical harmonic expansion of f on the sphere $a_{\mathbf{R}}$,

$$f(\mathbf{r}) = \sum_{\mathbf{R}L} \psi_{\mathbf{R}L}(\mathbf{r}) f_L(a_{\mathbf{R}}). \tag{21}$$

Furthermore any two sets of SSWs, each with different a sphere radii, will be fully equivalent. The transformation between the two sets can be obtained from (21).

4.2.2 "Finite l" screening and the choice of screening spheres

In practice we do not enforce the SSW to vanish *exactly* on each screening sphere $a_{\mathbf{R}}$. Rather we require the components of the spherical harmonic expansion of $\psi(\mathbf{r})$ on the $a_{\mathbf{R}}$ spheres to vanish up to some $l = l_{\max}$. This means that we only allow screening multipoles with angular character $l \leq l_{\max}$. By doing this the matrices in (20) become finite sized. In terms of the one-center expansion,

$$\psi_{\mathbf{R}L} = n^a_{\mathbf{R}l} Y_{\mathbf{R}L} \delta_{\mathbf{R}',\mathbf{R}} - \sum_{L'} j^a_{\mathbf{R}'l'} Y_{\mathbf{R}'L'} S^a_{\mathbf{R}'L',\mathbf{R}L},$$

as j^a cannot contain any Neumann function part for $l > l_{\max}$, we must now impose that j^a_l should equal the Bessel function j_l for $l > l_{\max}$. The cutoff $l_{\max}$ can depend on site $\mathbf{R}$ and is chosen to equal the maximum l for which we explicitly solve the radial Dirac equation on that site. This is because each individual SSW is created to match to one particular partial wave.

Choice of screening radii: It is possible to let the screening radii depend not just on site $\mathbf{R}$ but also on angular momentum l, $a_{\mathbf{R}} \Rightarrow a_{\mathbf{R}l}$. In terms of the boundary conditions, this means we now impose that the $l = 0$ component of the SSW to vanish on the sphere with radius $a_{\mathbf{R}0}$ while the $l = 1$ components vanish on a sphere with different radius $a_{\mathbf{R}1}$ and so on. The values for $t_1 \ldots t_4$ are changed accordingly but the analysis of the previous section proceeds in the same way and (20) is unchanged.

As all sets of SSWs at the same energy are equivalent, we are in some sense free to make whatever choice we like for $a_{\mathbf{R}}$ without changing the answer. However, the choice does become important when we come to linearisation. Then we shall make a set of LMTOs from a set of SSWs. The LMTOs will change according to which set of $a_{\mathbf{R}}$ spheres we started from. Usually we chose the screening spheres to give the shortest range for the SSW. This will give us the largest energy window over which the LMTO basis set is good. In practice, for a wide range of structures setting the $a_{\mathbf{R}}$ to be 90% of touching sphere radii produces good localisation.

The use of angular momentum dependent screening spheres gives additional freedom to shape the LMTOs. This can be useful in certain cases to optimise a basis set for a particular compound.

4.2.3 Truncation

The SSWs as defined above exist in all space. As we wish to use them only in the interstitial region, where they solve SE, we must truncate them. To do this we cut out inside each $a_{\mathbf{R}'}$ sphere the *low* (i.e. $l \leq l_{\max}$) angular momentum components of the SSW ψ. In terms of the one-center expansion (16) we set

$$n^a_{\mathbf{R}'l}(r) = 0, \qquad j^a_{\mathbf{R}'l}(r) = 0 \quad \text{for} \quad r < a_{\mathbf{R}'}, \qquad l \leq l_{\max}. \tag{22}$$

This means that j^a now has a kink at $r = a$ and n^a has a discontinuity at $r = a$ (but no change in slope). This is seen clearly from (18).

The result is shown schematically in Fig. 8d. Each SSW $\psi_{\mathbf{R}L}$ now has a kink at $a_{\mathbf{R}'}$; $\forall \mathbf{R}'$ and a discontinuity at $a_{\mathbf{R}}$. The slope of the L' projection of ψ on the sphere $a_{\mathbf{R}'}{}^+$ is just

$$\frac{1}{a_{\mathbf{R}'}} S^a_{\mathbf{R}'L', \mathbf{R}L}. \tag{23}$$

For this reason the structure constant S^a matrix is also called the *slope matrix*. As the slope of the SSW just inside the $a_{\mathbf{R}}$ sphere must vanish then (23) will also be the kink of the SSW at that sphere.

The SSWs still exist in all space, but now inside the screening spheres they only have angular momentum components for $l > l_{\max}$. (These are not shown in Fig. 8d.) These high angular momentum components are referred to as high partial waves. Their radial dependence is just that of a Bessel function j_l. Note that in the radial Dirac equation, as l increases the centrifugal term will start to dominate the potential term. In the limit of large l the Dirac equation thus reduces to the wave equation, and its solutions will tend towards Bessel functions.

4.2.4 Integral of a product of SSWs

In the same way as we reduced the integral of a product of partial waves to a surface term at $r = s$ we can reduce the integral of a product of SSWs to surface terms on the screening spheres. From

$$\langle \psi_{\mathbf{R}L}(E_1) | H - E | \psi_{\mathbf{R}'L'} \rangle = \langle \psi_{\mathbf{R}L}(E_1) | -\nabla^2 + (V_{\text{mtz}} - E) | \psi_{\mathbf{R}'L'}(E_2) \rangle$$
$$= \langle \psi_{\mathbf{R}L}(E_1) | -\nabla^2 | \psi_{\mathbf{R}'L'}(E_2) \rangle + (V_{\text{mtz}} - E_2) \langle \psi_{\mathbf{R}L}(E_1) | \psi_{\mathbf{R}'L'}(E_2) \rangle$$

look at the difference $\langle \psi_{\mathbf{R}L}(E_1) | H | \psi_{\mathbf{R}'L'}(E_2) \rangle - \langle \psi_{\mathbf{R}'L'}(E_2) | H | \psi_{\mathbf{R}L}(E_1) \rangle \equiv 0$,

$$0 = \langle \psi_{\mathbf{R}L}(E_1) | -\nabla^2 | \psi_{\mathbf{R}'L'}(E_2) \rangle - \langle \psi_{\mathbf{R}'L'}(E_2) | -\nabla^2 | \psi_{\mathbf{R}L}(E_1) \rangle$$
$$+ (E_1 - E_2) \langle \psi_{\mathbf{R}L}(E_1) | \psi_{\mathbf{R}'L'}(E_2) \rangle.$$

Applying Green's theorem to the first two terms we obtain surface integrals over the screening spheres,

$$(E_1 - E_2) \langle \psi_{\mathbf{R}L}(E_1) | \psi_{\mathbf{R}'L'}(E_2) \rangle$$
$$= -\sum_{\mathbf{R}} \int_{a_{\mathbf{R}}} (\psi_{\mathbf{R}L}(E_1) \nabla \psi_{\mathbf{R}'L'}(E_2) - \psi_{\mathbf{R}'L'}(E_2) \nabla \psi_{\mathbf{R}L}(E_1)) \, d\sigma$$
$$= -a_{\mathbf{R}L} S^a_{\mathbf{R}L, \mathbf{R}'L'}(E_2) + a_{\mathbf{R}'L'} S^a_{\mathbf{R}'L', \mathbf{R}L}(E_1).$$

And so
$$\langle \psi_{\mathbf{R}L}(E_1) | \psi_{\mathbf{R}'L'}(E_2)\rangle = \frac{a_{\mathbf{R}'L'}S^a_{\mathbf{R}'L',\mathbf{R}L}(E_1) - a_{\mathbf{R}L}S^a_{\mathbf{R}L,\mathbf{R}'L'}(E_2)}{(E_1 - E_2)}.$$

Taking the limit $E_1 \to E_2$ we obtain
$$\langle \psi_{\mathbf{R}L} | \psi_{\mathbf{R}'L'}\rangle = a_{\mathbf{R}L}\dot{S}^a_{\mathbf{R}L,\mathbf{R}'L'}. \qquad (24)$$

This particularly simple expression is the result of choosing the SSW to vanish on all but one screening sphere. When the screening spheres are l dependent the derivation can be generalised easily, and the final result (24) is unchanged.

4.3 Matching the partial wave and SSW: the kinked partial wave

We now have the solution of SE for the potential both inside the spheres and in the interstitial. The next step is to match the two parts continuously to make a kinked partial wave. This should be done at $r = s$, the potential sphere radius. However, we have defined the SSWs via the screening spheres which have a smaller radii, and hence we cannot immediately join the two. This problem is illustrated in the top part of Fig. 9. The figure shows a Si p partial wave ϕ and SSW ψ plotted along the [111] line joining two nearest neighbours. (Note that the high components of ψ are just visible inside the a sphere.) It is natural to conclude from this picture that we should have set the screening spheres equal to the potential spheres, $a_{\mathbf{R}} = s_{\mathbf{R}}$. However, if the potential spheres overlap this would cause problems. The boundary conditions on the SSWs would then be extreme. Indeed they would be impossible to satisfy in the limit $l_{\max} \to \infty$, which is why we imposed that the screening spheres should not overlap in the first place. If we did set $a_{\mathbf{R}} = s_{\mathbf{R}}$, the resulting SSWs would be poorly localised. It would no longer be possible to use a real space cluster calculation to evaluate the screened structure constants and the LMTOs would have a small energy window.

Conversely if we tried to use smaller potential spheres, then the muffin tin potential would no longer be a good approximation to the full potential.

To solve this dilemma we keep our two sets of spheres at differing radii and introduce a third function, the *back-extrapolated partial wave*, $\varphi_l(r)$. This radial function exists only in the range $a < r < s$ and is defined to solve SE for the flat interstitial potential,
$$(-\nabla^2 + (V_{\text{mtz}} - E))\,\varphi_l(r)\,Y_L = 0$$
with the boundary condition that it matches the partial wave at $r = s$ both in value and slope.

We then normalise both the partial wave and the back-extrapolated partial wave so that $\varphi_l(a_{\mathbf{R}}) = 1$. By doing this the back-extrapolated wave will automatically match the SSW $\psi_{\mathbf{R}}$ continuously at the screening sphere. This is shown in the bottom part of Fig. 9. The φ can be seen to smoothly match the ϕ at the potential sphere boundary.

The **kinked partial wave (KPW)** is then made from the combination of the partial wave, back-extrapolated partial wave and the screened spherical wave,
$$\Phi_{\mathbf{R}L} = (\phi_l - \varphi_l)\,Y_L + \psi_{\mathbf{R}L}. \qquad (25)$$

It is shown by the dotted line in the bottom part of Fig. 9. The KPW is continuous, but has a kink (i.e. a discontinuous slope) at every $a_{\mathbf{R}}$ sphere.

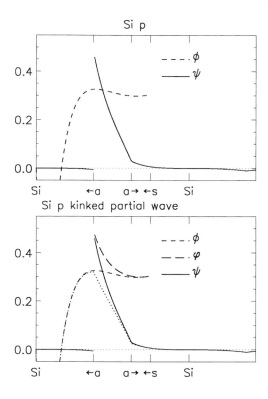

Fig. 9. Top: Si p partial wave (dashed line) and SSW (full line) along the [111] direction. Bottom: the Si p KPW (dotted line) and its three component functions along the [111] direction

The reader might have noticed an apparent mismatch of the φ and ψ at the screening sphere. This is an "optical illusion" due to the fact that we have finite l screening. The $l = 1$ component of ψ at $a_\mathbf{R}$ *does match* φ. What we see in addition in the figure are the higher components of ψ (those with $l \leq 4$) which exist both for $r < a$ and $r \geq a$ and are continuous across the screening sphere. What matters is that the total jump in ψ at $a_\mathbf{R}$ does match the total jump in φ. This makes the KPW continuous.

4.3.1 The kink matrix

We shall define the kink to be the change of slope of a KPW upon entering a screening sphere. The kink matrix K is then defined such that $K_{\mathbf{R}'L',\mathbf{R}L} = a^2 \times$ (kink at $\mathbf{R}'$ in the L' projection of $\Phi_{\mathbf{R},L}$). From (16) we obtain for $\mathbf{R}' \neq \mathbf{R}$ or $L' \neq L$

$$K_{\mathbf{R}'L',\mathbf{R}L} = -a_{\mathbf{R}'} S^a_{\mathbf{R}'L',\mathbf{R}L}$$

and for $\mathbf{R}'L' = \mathbf{R}L$

$$K_{\mathbf{R}L,\mathbf{R}L} = a_\mathbf{R}^2 \frac{\partial \varphi_{\mathbf{R}l}}{\partial r} - a_\mathbf{R} S^a_{\mathbf{R}L,\mathbf{R}L} = a_\mathbf{R} D^a - a_\mathbf{R} S^a_{\mathbf{R}L,\mathbf{R}L}.$$

Here we have introduced the notation D^a for the logarithmic derivative of the back-extrapolated partial wave at $r = a_\mathbf{R}$. Hence the kink matrix can be written in the simple form

$$K^a(E) = a(D^a - S^a). \tag{26}$$

The kink is illustrated in detail in Fig. 10. Two kinks are visible, one on each of the two screening spheres. The kink at the left sphere is shown enlarged in the inset. Although on one level the kink matrix stores no more than the values of the discontinuities in slope of the KPW, it actually has all the information we need to know about the KPWs. The LMTO overlap and Hamiltonian matrices will be shown to be made up from solely the kink matrix and its energy derivatives.

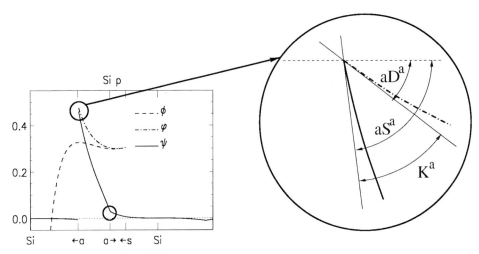

Fig. 10. Kinks for the Si p KPW

4.4 The kink cancellation condition

We are now in a position to construct the exact solution of SE for the muffin tin potential. The KPWs contain the partial waves ϕ which are the correct solutions of SE inside the potential spheres. They also contain the SSWs which solve SE for the muffin tin potential in the interstitial. If it were not for the fact that it was kinked, a KPW would be an eigenfunction of the SE for the whole muffin tin potential. The exact eigenfunction Ψ with energy E can be written as a superposition of KPWs, as long as this is done such that the kinks of all the individual KPWs cancel each other. That is

$$\Psi(E) = \sum_{\mathbf{R}L} \Phi_{\mathbf{R}L}(E)\, v_{\mathbf{R}L},$$

where both E and $\mathbf{v}$ are given by the kink cancellation condition,

$$\sum_{\mathbf{R}'L'} K(E)_{\mathbf{R}L,\mathbf{R}'L'}\, v(E)_{\mathbf{R}',L'};\quad \forall \mathbf{R},L.$$

Which is much more pretty when written in matrix notation,

$$K(E) \cdot \mathbf{v} = 0. \tag{27}$$

This equation, which ensures that Ψ is smooth[4]), is also known as the screened KKR equation. It is the solution of this equation that gives us the Kohn-Sham eigenvalues and eigenfunctions.

When (27) is satisfied, then inside the potential spheres ($r < s$) only the partial waves ϕ_l remain whereas in the interstitial only the SSWs remain (which is the required solution). To show this examine the one-center expansion of Φ inside the R potential sphere. This can be obtained from their definition (25) and the one-center expansion of the screened spherical waves

$$\Phi_{\mathbf{R}L} = \left(\phi_{\mathbf{R}l} - \varphi_{\mathbf{R}l} + n^a_{\mathbf{R}l}\right) Y_{\mathbf{R}L}\, \delta_{\mathbf{R}',\mathbf{R}} - \sum_{L'} j^a_{\mathbf{R}'l'}\, Y_{\mathbf{R}'L'}\, S^a_{\mathbf{R}'L',\mathbf{R}L}.$$

---

[4]) In the LMTO literature the word "smooth" is used to mean that the function in question is both continuous and differentiable. It does not, however, imply that the function does not oscillate.

Now $(-\varphi_{\mathbf{R}l} + n^a_{\mathbf{R}l})$ will vanish at the $a_{\mathbf{R}}$ sphere with slope $-1/a_{\mathbf{R}} D^a_{\mathbf{R}L}$. Also, both φ and ψ are solutions of the wave equation. These two facts mean that

$$(-\varphi_{\mathbf{R}l} + n^a_{\mathbf{R}l}) = j^a_{\mathbf{R}l} D^a_{\mathbf{R}L}.$$

Thus the one-center expansion of the KPW in the region $r < s$ can be written

$$\Phi_{\mathbf{R}L} = \phi_{\mathbf{R}l} Y_{\mathbf{R}L} \delta_{\mathbf{R}',\mathbf{R}} + j^a_{\mathbf{R}l} Y_{\mathbf{R}L} D^a_{\mathbf{R}L} \delta_{\mathbf{R}',\mathbf{R}} - \sum_{L'} j^a_{\mathbf{R}'l'} Y_{\mathbf{R}'L'} S^a_{\mathbf{R}'L',\mathbf{R}L},$$

$$\Phi_{\mathbf{R}L} = \phi_{\mathbf{R}l} Y_{\mathbf{R}L} \delta_{\mathbf{R}',\mathbf{R}} + \sum_{L'} j^a_{\mathbf{R}'l'} Y_{\mathbf{R}'L'} \frac{1}{a_{\mathbf{R}'}} K^a_{\mathbf{R}'L',\mathbf{R}L}. \tag{28}$$

From this we can obtain the one-center expansion of the exact eigenfunction as

$$\Psi = \sum_L \phi_{\mathbf{R}l} Y_{\mathbf{R}L} v_{\mathbf{R}L} + \sum_L \left[j^a_{\mathbf{R}l} \underbrace{\left[\sum_{\mathbf{R}'L'} K^a_{\mathbf{R}L,\mathbf{R}'L'} v_{\mathbf{R}'L'} \right]}_{=0} \right] Y_{\mathbf{R}L}.$$

Inside the potential spheres ($r < s$) the last term vanishes and the wavefunction is a superposition of only the partial waves ϕ, as it should be. The kink cancellation condition is illustrated in Fig. 11, where one can see that the resulting eigenfunction follows the partial wave ϕ all the way up to the potential sphere s where it joins smoothly onto the superposition of SSWs ψ.

The back extrapolated partial waves, which were introduced as an aid to construction for the KPWs will drop out at kink cancellation.

It should be noted that the kink cancellation condition will strictly give the exact solution of SE only if the potential spheres do not overlap. When they do overlap then an error will be introduced. However, for linear overlaps less than about 15% this error can be neglected. It is possible to correct for this overlap error, and with this overlap correction one can use potential spheres with much larger overlaps (~40%) [45].

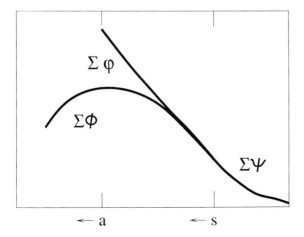

Fig. 11. Pictorial representation of kink cancellation

5. Linearisation

In the previous section we showed that the eigenfunctions of the muffin tin potential are given as superpositions of the KPWs with coefficients determined by the kink cancellation condition. For an eigenfunction with energy E the KPWs need to be constructed at this energy E. Hence the KPW set changes for each individual eigenfunction. In this sense the KPWs can be thought of as an energy-dependent basis set. In this section we shall derive from the KPWs an *energy-independent* basis set, that is a fixed, minimal basis set that can be used equally well to expand all of the eigenfunctions. This basis set will be the Linear Muffin Tin Orbitals (LMTOs). It is possible to construct such a set because we are usually only interested in eigenfunctions with energies within a certain range (e.g valence band plus a few conduction bands) about some central energy E_v. Given such a basis set of LMTOs one can then go on to construct Hamiltonian and overlap matrices for the muffin tin potential. The eigenfunctions can then be obtained from a linear diagonalisation of the Hamiltonian, which is computationally much more efficient than solving the kink cancellation condition.

In Fig. 12 we compare the p KPWs for Si in the diamond structure for two energies; the bottom and top (E_f) of the valence band. One can see that there is a rather small variation in $\Phi(E)$ over this energy range. One of the reasons for this small change is that the KPWs we constructed are well localised. To understand this it is useful to con-

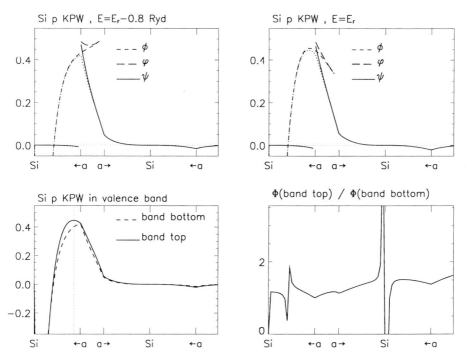

Fig. 12. Si p KPW and its three component functions along the [111] direction calculated for an energy equal to the valence band bottom (top left) and to the valence band top (top right). Bottom left: comparison of the two KPWs shown above. Bottom right: ratio of the two KPWs shown above

An Introduction to the Third-Generation LMTO Method

sider the following. Suppose we want to solve the radial SE, starting at $r = 0$ with fixed value and slope and integrating outwards. If we compare two solutions with slightly different energies, then we will find that close to $r = 0$ they will be similar, but as one goes further and further out their differences become larger. Therefore, when we restricted the partial waves to inside the potential spheres and pinned each SSW to be localised close to its own site, we at the same time reduced its variation with energy. This is how the division of space aids linearisation.

The small energy dependence of the KPWs allows us to approximate them with a Taylor series in energy about E_ν. Inserting this into the expansion of a given eigenfunction gives

$$\Psi = \sum \Phi(E) \, v(E)$$
$$= \sum [\Phi(E_\nu) + \dot{\Phi}(E_\nu) (E - E_\nu)] \, v(E) + O(E - E_\nu)^2. \qquad (29)$$

The term $O(E - E_\nu)^2$ will be small for energies in our range of interest and we can neglect it. Therefore if we wish to construct an LMTO basis set we should use $\Phi(E_\nu)$ and its energy derivative $\dot{\Phi}(E_\nu)$ both evaluated at the fixed energy E_ν.

5.1 Construction of the LMTO basis set

Starting from (29) we drop the second order terms and postulate replacing $(E - E_\nu)$ by a *matrix* h^1. This gives

$$\Psi = \sum \underbrace{[\Phi(E_\nu) + \dot{\Phi}(E_\nu) \, h^1]}_{\Downarrow \text{ LMTO}} \cdot v(E).$$

The term in square brackets now has no E dependence whatsoever and hence could become our LMTO. To do this we must ensure that this expansion in LMTOs will still give the eigenfunction Ψ correct to first order. For this to be true we require

$$h^1 \cdot v(E) = (E - E_\nu) \, v(E) + O(E - E_\nu)^2.$$

That is h^1 should be a first order Hamiltonian. To find this matrix we again start from kink cancellation condition, but now expand the kink matrix in a Taylor series,

$$K(E) \cdot v(E) = 0,$$
$$[K(E_\nu) + \dot{K}(E_\nu) (E - E_\nu) + O(E - E_\nu)^2] \cdot v(E) = 0,$$

rearranging gives

$$-\dot{K}^{-1}(E_\nu) \, K(E_\nu) \cdot v(E) = (E - E_\nu) \, v(E) + O(E - E_\nu)^2.$$

So the required first order Hamiltonian is $h^1 = -\dot{K}^{-1}(E_\nu) \, K(E_\nu)$ and the LMTO is given by

$$\chi = \Phi(E_\nu) - \dot{\Phi}(E_\nu) \, \dot{K}^{-1}(E_\nu) \, K(E_\nu). \qquad (30)$$

The set of LMTOs $\{\chi_{\mathbf{R}L}\}$ are then an energy independent basis set that will reproduce the eigenfunctions of the muffin tin potential correct to first order in $(E - E_\nu)$. Note that this basis set is very small. If the eigenfunction has no f character then we only need 9 LMTOs per site. This minimal quality of the LMTOs makes them both an attractive and efficient basis set. It arises because they are constructed from the KPWs

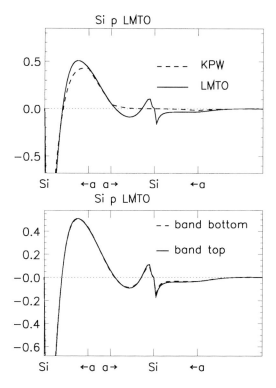

Fig. 13. Top: Si p KPW and LMTO along the [111] direction. Bottom: comparison of two Si p LMTO calculated at two different energies

which are themselves shaped according to the structure and composition of each individual compound. If f character is important for a particular site the number of LMTOs centred on that site rises to 16.

The Si p LMTO is shown in the top part of Fig. 13 where it is compared to the KPW at the same energy. Note that the LMTO is differentiable everywhere[5]), the total kink being $K - \dot{K}\dot{K}^{-1}K = 0$. In this sense the LMTO is smooth. In general the set of LMTOs have a longer range than the set of KPWs from which they are constructed. They are nevertheless still localised.

5.2 *Alternative construction of the LMTO basis set*

There are many ways to construct the LMTOs. Although they all give the same answer, the different methods illustrate different aspects of the LMTO basis set. We shall therefore present a second, complementary construction. This time we shall start by *assuming* a form for an *energy-dependent* muffin tin orbital,

$$\chi(E) = \Phi(E) - \dot{\Phi}(E_\nu) N K(E),$$

where N is at the moment an unspecified matrix. Notice that because of the factor of $K(E)$ in the $\dot{\Phi}$ term, the energy dependent $\{\chi(E)\}$ set can be used to expand the exact eigenfunctions of the muffin tin potential,

$$\begin{aligned}\sum \chi(E)\,v(E) &= \sum [\Phi(E) - \dot{\Phi}(E_\nu) N K(E)]\,v(E) \\ &= \sum \Phi(E)\,v(E) - \sum \dot{\Phi}(E_\nu) N K(E)\,v(E) \\ &= \sum \Phi(E)\,v(E)\,.\end{aligned}$$

This is true because for the exact eigenfunctions that kink cancellation condition $K(E) \cdot v(E) = 0$ is satisfied. If we now Taylor expand each muffin tin orbital $\chi(E)$ we get

$$\begin{aligned}\Psi &= \sum \chi(E)\,v(E) \\ &= \sum [\chi(E_\nu) + \dot{\chi}(E_\nu)(E - E_\nu)]\,v(E) + O(E - E_\nu)^2\,.\end{aligned}$$

---

[5]) except of course at the nuclei

An Introduction to the Third-Generation LMTO Method

Once again we can drop the second order terms. Also, as before we would like to make the energy dependence of the square bracket zero, so that that bracket becomes the LMTO. Hence we require N to be such that $\dot{\chi}(E_\nu) = 0$,

$$\dot{\chi}(E_\nu) = 0 = \dot{\Phi}(E_\nu) - \dot{\Phi}(E_\nu) N \dot{K}(E_\nu)$$
$$= \dot{\Phi}(E_\nu) \left[1 - N\dot{K}(E_\nu)\right].$$

Thus $N = \dot{K}^{-1}$. We then recover the same expression for the LMTO found in the previous section by evaluating the $\chi(E)$ at $E = E_\nu$, giving

$$\chi(E_\nu) = \chi = \Phi - \dot{\Phi}\dot{K}^{-1}K.$$

The second derivation shows that two LMTOs χ_1 and χ_2 created at two energies E_1, E_2 should show less difference than the KPWs $\Phi(E_1)$ and $\Phi(E_2)$. This is indeed true and is illustrated in the bottom part of Fig. 13 where we now compare two Si p LMTOs, one constructed with energy at the top of the valence band, and the other at the bottom. The differences are small, being second order in energy.

The LMTOs in the region $a < r < s$: By differentiating (28) with respect to energy one can obtain the one-center expansion for $\dot{\Phi}$. The one-center expansion of an LMTO for $r < s$ can then be shown to be

$$\chi_{\mathbf{R}L} = \phi_{\mathbf{R}L}\delta_{\mathbf{R}',\mathbf{R}} - \sum_{L'} \dot{\phi}_{\mathbf{R}'L'}(\dot{K}^{-1}K)_{\mathbf{R}'L',\mathbf{R}L} - \sum_{L'} \dot{j}^a_{\mathbf{R}'l} \frac{1}{a_{\mathbf{R}'}} (K\dot{K}^{-1}K)_{\mathbf{R}'L',\mathbf{R}L}. \quad (31)$$

There is therefore a partial cancellation between the SSWs and the back-extrapolated partial waves. What remains is a term containing the energy derivative of j^a, namely $\dot{j}^a$. It is easy to show that $\dot{j}^a \sim (r-a)^3$, and thus this term is small.

5.3 Hamiltonian and overlap matrices for the LMTO basis set

Given the LMTO basis set and the muffin tin potential, we can form both the overlap (O) and Hamiltonian matrices (H). For this we require the Hamiltonian matrix elements between the KPWs and their energy derivatives. Unfortunately these are complicated by the fact that the three parts of each KPW (ϕ, φ and ψ) coexist between the screening and potential spheres. However, we know that at solution the SSW ψ and back-extrapolated partial wave φ will cancel, leaving just ϕ. Therefore when evaluating matrix elements we shall be able to drop cross terms between the three parts (e.g. $\langle\phi\,|\,\psi\rangle$). This dramatically simplifies the expressions for H and O.

5.3.1 The overlap matrix between KPWs

The true overlap operator is $\hat{1}$ and this has the following matrix elements between two KPWs,

$$\langle\Phi|\,\hat{1}\,|\Phi\rangle = \langle\phi - \varphi + \psi\,|\,\phi - \varphi + \psi\rangle$$
$$= \langle\phi\,|\,\phi\rangle - \langle\varphi\,|\,\varphi\rangle + \langle\psi\,|\,\psi\rangle + \langle\psi - \varphi\,|\,\phi - \varphi\rangle + \langle\phi - \varphi\,|\,\psi - \varphi\rangle.$$

In the last two terms we can use the one-center expansion (28) to expand $(\psi - \varphi)$ in the s sphere of the $(\phi - \varphi)$. This gives

$$\langle\Phi|\,\hat{1}\,|\Phi\rangle = \underbrace{\langle\phi\,|\,\phi\rangle - \langle\varphi\,|\,\varphi\rangle + \langle\psi\,|\,\psi\rangle}_{\langle\Phi|\,\hat{O}\,|\Phi\rangle} + \underbrace{K\langle j^a/a\,|\,\phi - \varphi\rangle + \langle\phi - \varphi\,|\,j^a/a\rangle K}_{\langle\Phi|\,\Delta\hat{O}\,|\Phi\rangle}.$$

Here we have conceptually split the overlap operator into two parts, $\hat{O}$ and $\Delta\hat{O} = \hat{1} - \hat{O}$. The $\Delta\hat{O}$ part contains all of the cross terms between the three parts of the KPW. Now the diagonal matrix element of $\Delta\hat{O}$ with the exact eigenfunction Ψ vanishes due to the kink cancellation condition,

$$\langle \Psi | \Delta\hat{O} | \Psi \rangle = \mathbf{v} \cdot \langle \Phi | \Delta\hat{O} | \Phi \rangle \cdot \mathbf{v}$$
$$= \mathbf{v} \cdot K \langle j^a/a | \phi - \varphi \rangle \cdot \mathbf{v} + \mathbf{v} \cdot \langle \psi - \varphi | j^a/a \rangle K \cdot \mathbf{v}$$
$$= 0.$$

This means we can approximate the overlap matrix between KPWs by only the $\langle \Phi | \hat{O} | \Phi \rangle$ term, and still *retain the correct normalisation* for all the eigenfunctions. This gives

$$\langle \Phi | \Phi \rangle \approx \langle \Phi | \hat{O} | \Phi \rangle = \langle \phi | \phi \rangle - \langle \varphi | \varphi \rangle + \langle \psi | \psi \rangle. \tag{32}$$

This is called the *backwards–forwards* definition of the overlap. Note that it is not guaranteed to be positive. In practice it is positive as long as the distance between $a_\mathbf{R}$ and $s_\mathbf{R}$ does not become unusually large. In previous sections we have reduced the individual parts of (32) to surface terms on the a and s spheres. The $\langle \phi | \phi \rangle$ term was given in (11). The $\langle \varphi | \varphi \rangle$ term can be derived in exactly the same way as for $\langle \phi | \phi \rangle$ (remembering that for φ effectively $V(r) = V_{\text{mtz}}$ and $c = \infty$) and gives a surface term at both boundaries, the a sphere and the s sphere. The $\langle \psi | \psi \rangle$ was given in (24). This gives

$$\langle \Phi | \Phi \rangle = \underbrace{-s\phi^2(s) \dot{D}(\phi(s)) + (s\varphi^2(s) \dot{D}(\varphi(s)))}_{\text{cancel}} - a\varphi(a) \dot{D}(\varphi(a))) + a\dot{S}^a ,$$
$$\langle \Phi | \Phi \rangle = -\dot{K} . \tag{33}$$

In addition we will require the overlap matrix elements involving the energy derivative $\dot{\Phi}$. These can be obtained from going back to the finite difference formula $\langle \Phi(E_1) | \Phi(E_2) \rangle = -(K(E_1) - K(E_2))/(E_1 - E_2)$ differentiating with respect to either E_1 or E_2 and then taking the limit $E_1 \to E_2$,

$$\langle \Phi | \Phi \rangle = -\dot{K}, \qquad \langle \Phi | \dot{\Phi} \rangle = -\tfrac{1}{2}\ddot{K},$$
$$\langle \dot{\Phi} | \Phi \rangle = -\tfrac{1}{2}\ddot{K}, \qquad \langle \dot{\Phi} | \dot{\Phi} \rangle = -\tfrac{1}{6} \ldots K. \tag{34}$$

Hence all the required overlap integrals can be reduced to expressions involving only energy derivatives of the kink matrix.

5.3.2 Hamiltonian matrix elements between KPWs

The Hamiltonian operator for the muffin tin potential is

$$\hat{H}_{\text{MT}} = -\nabla^2 + \sum_\mathbf{R} v_\mathbf{R} + V_{\text{mtz}}.$$

Acting with this on the KPW $\Phi = \phi - \varphi + \psi$ gives

$$\hat{H}_{\text{MT}} | \Phi \rangle = E | \Phi \rangle + \sum_{\mathbf{R}L} \frac{Y}{a_\mathbf{R}^2} \delta(r_\mathbf{R} - a_\mathbf{R}) K + \sum_\mathbf{R} v_\mathbf{R}(-|\varphi\rangle + |\psi\rangle).$$

An Introduction to the Third-Generation LMTO Method

The second term on the right hand side arises due to the kink on all screening spheres. The last term arises because both the SSW and the back-extrapolated partial wave (which both solve SE for the flat potential V_{mtz}) extend inside the potential spheres, where the total potential is $V_{\text{mtz}} + v_\mathbf{R}$. In the last term we again use (28) to expand $(-\varphi + \psi)$ about site $\mathbf{R}$,

$$\hat{H}_{\text{MT}}|\Phi\rangle = \underbrace{E|\Phi\rangle + \sum_{\mathbf{R}L} \frac{Y}{a_\mathbf{R}^2}\delta(r_\mathbf{R} - a_\mathbf{R})K}_{\hat{H}|\Phi\rangle} + \underbrace{\sum_{\mathbf{R},L} v_\mathbf{R} j^a \frac{Y}{a}K}_{\Delta\hat{H}|\Phi\rangle}.$$

Here we have conceptually split the Hamiltonian operator into two parts, $\hat{H}$ and $\Delta\hat{H}$. The last term contains a factor of the kink matrix K. This means that the action of $\Delta\hat{H}$ on any exact eigenfunction Ψ will give zero, $\Delta\hat{H}|\Psi\rangle = 0$. Hence eigenfunctions of $\hat{H}_{\text{MT}}$ are also eigenfunctions of $\hat{H}$. This means that we are justified in dropping $\Delta\hat{H}$ and approximating the matrix elements of $\hat{H}_{\text{MT}}$ by those of $\hat{H}$.

We require in addition matrix elements involving the energy derivative $\dot{\Phi}$. The action of $\hat{H}$ on $\dot{\Phi}$ gives

$$\hat{H}|\dot{\Phi}\rangle = E|\dot{\Phi}\rangle + |\Phi\rangle + \sum_{\mathbf{R}L} \frac{Y}{a_\mathbf{R}^2}\delta(r_\mathbf{R} - a_\mathbf{R})\dot{K},$$

and all the required matrix elements are the following:

$$\langle\Phi|\hat{H}|\Phi\rangle = E\langle\Phi|\Phi\rangle + K,$$
$$\langle\Phi|\hat{H}|\dot{\Phi}\rangle = E\langle\Phi|\dot{\Phi}\rangle + \langle\Phi|\Phi\rangle + \dot{K},$$
$$\langle\dot{\Phi}|\hat{H}|\Phi\rangle = E\langle\dot{\Phi}|\Phi\rangle,$$
$$\langle\dot{\Phi}|\hat{H}|\dot{\Phi}\rangle = E\langle\dot{\Phi}|\dot{\Phi}\rangle + \langle\dot{\Phi}|\Phi\rangle. \tag{35}$$

5.4 The linear eigenvalue problem

From the definition of the LMTO (30) and the matrix elements of the KPWs (34), (35) the Hamiltonian and overlap matrices for the LMTOs can easily be shown to be

$$H - E_\nu O = K - \tfrac{1}{2}\dot{K}^{-1}\ddot{K}\dot{K}^{-1}K, \tag{36}$$

$$O = -\dot{K} + \tfrac{1}{2}[K\dot{K}^{-1}\ddot{K} + \ddot{K}\dot{K}^{-1}K] - \tfrac{1}{6}K\dot{K}^{-1}\dddot{K}\dot{K}^{-1}K, \tag{37}$$

Given these one can then solve the linear eigenvalue problem $(H - EO)\cdot\mathbf{v} = 0$ to obtain the eigenvectors $\mathbf{v}$ and eigenvalues E. For a given k point all eigenvalues are obtained simultaneously. This is much more efficient than solving the kink cancellation condition directly, which in essence has to be done for each state individually. Furthermore in the LMTO basis set the Hamiltonian is much smaller than for non-minimal sets such as plane waves. Correspondingly the time taken to diagonalise the matrix, which has N^3 scaling is much reduced.

In Fig. 14 we show the silicon band structure as calculated from this eigenvalue equation compared to that obtained from solving directly the kink cancellation condition (i.e. the exact solution). The LMTOs were all constructed at an energy E_ν placed at the band center. The two band structures are practically indistinguishable, showing that the

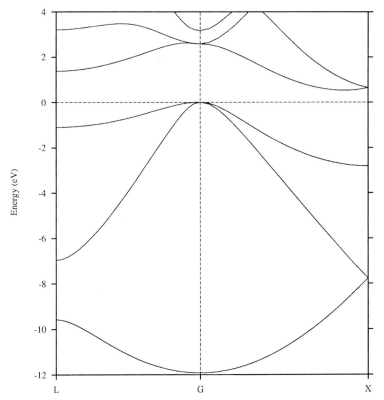

Fig. 14. Si band structure calculated solving the kink cancellation condition (full line) and the linear eigenvalue equation (dashed line)

energy window over which the linearisation procedure works is rather wide. The eigenfunctions from the linear eigenvalue equation will be correct to first order in $(E - E_\nu)$. Correspondingly the eigenvalues are correct to order $(E - E_\nu)^3$. Thus in those cases where one is just interested in the valence band and the lower lying conduction states, the linearisation procedure gives us a large increase in computational efficiency with only a minimal decrease in accuracy.

Triple-valuedness: If one examines the LMTO eigenfunctions they will show only a partial cancellation of the φ and ψ parts in the region between the screening sphere a and potential sphere s. This is known as *triple-valuedness*. The L component of the remainder inside the $s_\mathbf{R}$ sphere from this partial cancellation is

$$\sum_{\mathbf{R}'L'} j^a_{\mathbf{R}l} \frac{1}{a_\mathbf{R}} Y_L \big(K\dot{K}^{-1}K\big)_{\mathbf{R}L,\mathbf{R}'L'} \cdot v_{\mathbf{R}'l'}.$$

This error is small, as can be seen from the following:

$$j^a K \dot{K}^{-1} K \cdot v \sim j^a (E - E_\nu) K \cdot v$$
$$\sim j^a (E - E_\nu)^2$$
$$\sim (r - a)^3 (E - E_\nu)^2.$$

Here we have used the fact that $\dot{K}^{-1}K$ is a first order Hamiltonian. The $(E-E_\nu)^2$ factor clearly shows that this error is a *linearisation error*, and is therefore acceptable.

One last point remains to be addressed. This is the question of normalisation. One might wonder if an energy dependent renormalisation for the KPWs could change the energy window for linearisation. Consider renormalising by a factor $N(E)$, $\Phi(E) \Rightarrow \Phi(E) N(E) = \bar{\Phi}(E)$. The LMTO gets modified $\chi \Rightarrow \bar{\chi}$ as follows

$$\chi = \Phi - \dot{\Phi}\dot{K}^{-1}K$$
$$= (\Phi N) N^{-1} - [(\dot{\Phi}N) N^{-1} + (\Phi N)(\dot{N}^{-1})]\dot{K}^{-1}K,$$
$$\chi = [\bar{\Phi} - \dot{\bar{\Phi}}\dot{\bar{K}}^{-1}\bar{K}], \qquad [N^{-1} - (\dot{N}^{-1})\dot{K}^{-1}K],$$
$$\chi = \bar{\chi}[N^{-1} - (\dot{N}^{-1})\dot{K}^{-1}K].$$

Therefore the LMTO basis set simply gets transformed by a matrix. This linear transformation leaves the eigenvalues unchanged, and hence the energy window is unchanged. We are free to choose any normalisation for the KPWs that we wish.

6. Downfolding

Downfolding [46] is a simple concept that often leads to much confusion. It is, however, a very useful and powerful procedure. Therefore in this section we shall explain in detail the downfolding transformation and its underlying concepts.

The LMTO basis set we have derived will have one element for every orbital on each site, for example 9 per site for an spd basis. No attempt has been made so far to distinguish the relative importance of these orbitals to the chemical binding of the system. In the silicon examples shown so far we have used an spd basis set. For silicon the bonding is sp^3 character, and therefore the d orbitals play a minor (but non-zero) role. It would be nice to have an LMTO basis set that has only s and p LMTOs in it. It would be wrong to simply neglect the d orbitals, instead they must be *downfolded* into the remaining s and p LMTOs. Downfolding is therefore a way to reduce the size of an LMTO basis set. The character of the orbitals "removed" is somehow included into the modified orbitals that are "retained".

Apart from the obvious advantage of reducing the size of the basis set there are circumstances when it is a necessity. This is when ghost bands occur. These are a feature of any linearisation method and can arise if we have an LMTO in our basis set for an orbital that plays no major role in any of the bands within the linearisation window. There may then be one band corresponding to this orbital lying above the window and one lying below it. The linear eigenvalue equation will then not know which band it should pick, and as a result of this schizophrenia it may produce a very obvious and spurious band lying in the middle. This is a ghost band. The solution to this problem is simple, the offending orbital should be downfolded.

There is in principle no restriction to the number of orbitals that one can downfold at the same time. Suppose for example that one wishes to study in detail the bands that lie close to the Fermi surface. There may be only handful of orbitals (e.g 4) that play a major role in the states close to E_f. We can downfold all the other orbitals and obtain a tiny 4×4 Hamiltonian that accurately describes our chosen bands [9]. The hopping parameters can then be read off and possibly used in a further study such as a many body calculation.

6.1 The downfolded KPWs

The downfolding transformation is performed at the level of the KPWs, *before* linearisation. At this level everything is still energy dependent. In Section 4 we saw that we can freely transform between different KPW sets. Downfolding of the KPWs is therefore an exact procedure. In this section we shall first give a formal definition of the downfolding transition via the Löwdin procedure [47] and then give a physical interpretation.

We shall divide our set of KPWs into two sets, those which we wish to retain (called the *lower* set) and those which we wish to remove (called the *intermediate* set), $\{\Phi\} = \{\Phi_\mathcal{L}\} + \{\Phi_\mathcal{I}\}$. We then make a similar division of the kink cancellation condition into lower and upper parts to give a block matrix equation

$$\begin{pmatrix} K_{\mathcal{LL}} & K_{\mathcal{LI}} \\ K_{\mathcal{IL}} & K_{\mathcal{II}} \end{pmatrix} \begin{pmatrix} \mathbf{v}_\mathcal{L} \\ \mathbf{v}_\mathcal{I} \end{pmatrix} = \begin{pmatrix} 0 \\ 0 \end{pmatrix}. \tag{38}$$

We then follow the Löwdin downfolding procedure and eliminate $\mathbf{v}_\mathcal{I}$ to give

$$\mathbf{v}_\mathcal{I} = -K_{\mathcal{II}}^{-1} K_{\mathcal{IL}} \mathbf{v}_\mathcal{L} \tag{39}$$

and

$$\underbrace{\left[K_{\mathcal{LL}} - K_{\mathcal{LI}} K_{\mathcal{II}}^{-1} K_{\mathcal{IL}} \right]}_{K_{\mathcal{LL}}^D} \mathbf{v}_\mathcal{L} = 0. \tag{40}$$

Here $K_{\mathcal{LL}}^D$ is the downfolded kink matrix. It is smaller than the original kink matrix. Equation (40) is the downfolded kink cancellation condition. Note that as this procedure is exact both (40) and (38) have the same solutions. To obtain the corresponding downfolded KPWs we take the expansion of any solution $\Psi(E)$ in the original set $\{\Phi(E)\}$ and again eliminate the $\mathbf{v}_\mathcal{I}$.

$$\begin{aligned} \Psi(E) &= \sum \Phi(E) \mathbf{v}(E) \\ &= \sum_\mathcal{L} \Phi_\mathcal{L}(E) \mathbf{v}_\mathcal{L}(E) + \sum_\mathcal{I} \Phi_\mathcal{I}(E) \mathbf{v}_\mathcal{I}(E) \\ &= \sum_\mathcal{L} \underbrace{\left[\Phi_\mathcal{L}(E) - \Phi_\mathcal{I}(E) K_{\mathcal{II}}^{-1} K_{\mathcal{IL}} \right]}_{\Phi_\mathcal{L}^D(E)} \mathbf{v}_\mathcal{L}(E). \end{aligned} \tag{41}$$

In (97) the summation runs only over low terms, and the function in the square bracket is the downfolded KPW, $\Phi_\mathcal{L}^D(E)$. Note that matrix multiplication is implied within this square bracket. This set of downfolded KPWs is smaller than the original one, but fully equivalent. Each individual $\Phi_\mathcal{L}^D(E)$ is made up from a single original low KPW $\Phi(E)_\mathcal{L}$ which is modified by adding to it a particular combination of the original intermediate KPWs.

In the undownfolded set, each $\Phi_{\mathbf{R}L}$ contained only a single partial wave, ϕ_l, inside the potential sphere on its own site $\mathbf{R}$. In the downfolded set, by contrast, each KPW will contain additional partial waves ϕ_l in every potential sphere. This is how the "removed" orbitals ϕ get folded into the ones that we retain.

To make this physical picture more concrete we will consider the example of silicon. In Fig. 15 we show a Si d KPW, which we wish to remove form our set so that we have only Si s and p downfolded KPWs. Fig. 16 shows pictorially how we do this. Here we

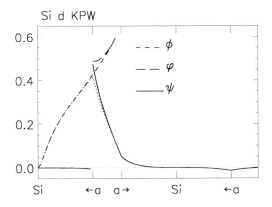

Fig. 15. Si d KPW and its three component functions along the [111] direction

show an undownfolded Si p KPW. This contains a partial wave ϕ_p in the potential sphere on its own site. On the neighbouring Si site the potential sphere is currently empty. Into this empty space we wish to insert the Si d partial waves ϕ_d. But to do this we "must make the hole fit". We must change the SSW part of the Si p KPW, altering its boundary conditions so that now it will *smoothly* match the ϕ_d partial wave on the neighbouring site at $r = s$. If we do this we can then truncate the SSW inside $r = s$ and insert the partial wave ϕ_d. The resulting downfolded KPW is shown in Fig. 17, where one can clearly see the ϕ_d orbital sitting on the neighbouring Si site. Note that in doing this the downfolded KPW no longer has any kink in its $l = 2$ component on any site. Therfore its kink matrix will reduce in size. The downfolding transition can thus be thought of as one in which the boundary conditions for the SSWs are altered. For the intermediate channels we now require j^a not to vanish on the a sphere but instead to match the partial wave ϕ_l at $r = s$ smoothly. This means we require $j_l^a = \varphi_l$, the back extrapolated partial wave. Hence the one-center expansion for a downfolded SSW is just

$$\psi_{\mathbf{R}L}^D = n_{\mathbf{R}l}^a \, Y_{\mathbf{R}L} \, \delta_{\mathbf{R}',\mathbf{R}} - \sum_{L'}^{\text{low+high}} j_{\mathbf{R}'l'}^a \, Y_{\mathbf{R}'L'} \, S_{\mathbf{R}'L',\mathbf{R}L}^D - \sum_{L'}^{\text{int.}} \varphi_{\mathbf{R}'l'} \, Y_{\mathbf{R}'L'} \, S_{\mathbf{R}'L',\mathbf{R}L}^D. \quad (42)$$

The first sum is over L' that exist in the low and high sets, and the second over those in the intermediate set. This downfolded SSW can be expanded in the original set of SSWs. The expansion coefficients will just be the values of the spherical harmonic projections of ψ^D on the screening spheres. Remembering that the back-extrapolated par-

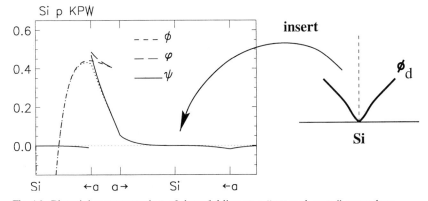

Fig. 16. Pictorial representation of downfolding as a "cut and paste" procedure

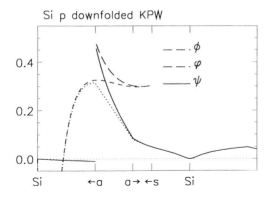

Fig. 17. Si p downfolded KPW and its three component functions along the [111] direction

tial waves equal 1 on the a spheres it is easy to show that

$$\psi^D_{\mathbf{R}L} = \psi_{\mathbf{R}L} + \sum_{\mathbf{R}'L'}^{\text{int.}} \psi_{\mathbf{R}'L'} S^D_{\mathbf{R}'L',\mathbf{R}L}. \tag{43}$$

From (42), (43) we can find the relationship between the corresponding structure constants,

$$S^D_{\mathcal{LL}} = S^a_{\mathcal{LL}} + S^a_{\mathcal{LI}} S^D_{\mathcal{IL}},$$
$$D^a_I S^D_{\mathcal{IL}} = S^a_{\mathcal{IL}} + S^a_{\mathcal{II}} S^D_{\mathcal{IL}}.$$

Once again we have used block matrix notation. Eliminating $S^D_{\mathcal{IL}}$ we get the transformation between the structure constants,

$$S^D_{\mathcal{LL}} = S^a_{\mathcal{LL}} + S^a_{\mathcal{LI}} \left(D^a_I - S^a_{\mathcal{II}} \right)^{-1} S^a_{\mathcal{IL}}. \tag{44}$$

Multiplying by $-a$ and adding $D^a_\mathcal{L}$ we recover the Löwdin expression for the downfolded kinked partial wave K^D (40). This shows that this pictorial "cut and paste" procedure is truly equivalent to the Löwdin downfolding procedure.

6.2 The downfolded LMTOs

The downfolded LMTOs are obtained from the downfolded KPWs by the linearisation procedure described in Section 5. The downfolded LMTO is therefore

$$\chi^D = \Phi^D(E_\nu) - \dot{\Phi}^D(E_\nu) \dot{K}^{D^{-1}}(E_\nu) K^D(E_\nu). \tag{45}$$

From these the Hamiltonian and overlap matrices can be constructed. The expressions (36), (37) are unchanged, we merely replace K by K^D.

It is not possible to expand the downfolded LMTOs in the original set χ. This means that, unlike the KPWs, the two sets of LMTOs are not equivalent. They will both still give the eigenfunction correct to first order in $(E - E_\nu)$ but they will have different linearisation errors. From Fig. 17 is is clear that the downfolded KPWs have a longer range. This means that the downfolded LMTOs will have a smaller linearisation window. The more orbitals that are downfolded, the smaller the range of energy around E_ν over which the bands are reproduced correctly. Another way to see this is that we have fixed the relative ratios of the orbitals retained with those removed, and have thus

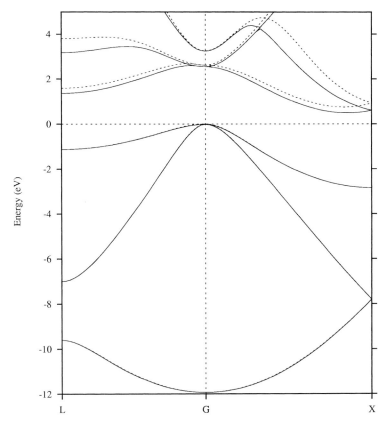

Fig. 18. Band structure for Si calculated using an spd basis set (full line) and an sp-d downfolded basis set (dashed line)

slightly reduced the flexibility if the basis set. In Fig. 18 we compare the bandstructures obtained form an Si spd LMTO basis to that of a Si sp basis with the d orbitals downfolded. The valence band is still reproduced correctly, but now linearisation errors are apparent in the conduction band.

7. The Charge Density

In the previous sections we showed how we can solve Schrödinger's equation for the muffin tin potential using the LMTO basis set. Given the eigenfunctions Ψ_n one can then construct the charge density $n(\mathbf{r})$. The density operator is

$$\hat{\varrho} = \sum_n |\Psi_n\rangle \langle \Psi_n|,$$

where the sum is taken over the occupied states. Each eigenfunction can be expanded in the LMTO basis set $\Psi_n = \sum_i \chi_i v_{\{n,i\}}$ where we have used i as shorthand for $\mathbf{R}L$. This gives

$$\hat{\varrho} = \sum_{ij} |\chi_i\rangle \underbrace{\left[\sum_n v_{\{n,i\}}\, v_{\{n,j\}}\right]}_{D^0_{ij}} \langle \chi_j|. \qquad (46)$$

Here D^0 is the LMTO density matrix. Each LMTO χ_i can in turn be expanded in the KPWs, $\chi_i = \Phi_i - \dot{\Phi}_j(\dot{K}^{-1}K)_{ji}$. Inserting this into (46) we can express the density operator in terms of the KPWs,

$$\hat{\varrho} = \sum_{ij} |\Phi_i\rangle D^0_{ij} \langle \Phi_j| + \sum_{ij} |\dot{\Phi}_i\rangle D^1_{ij} \langle \Phi_j|$$
$$+ \sum_{ij} |\Phi_i\rangle D^1_{ji} \langle \dot{\Phi}_j| + \sum_{ij} |\dot{\Phi}_i\rangle D^2_{ij} \langle \dot{\Phi}_j|. \qquad (47)$$

Here $D^1 = -\dot{K}^{-1}KD^0$, and $D^2 = -D^1 K \dot{K}^{-1}$. The matrices D^0, D^1, D^2 are the three KPW density matrices. The charge density is then $n(\mathbf{r}) = \langle \mathbf{r}| \hat{\varrho} |\mathbf{r}\rangle$,

$$n(\mathbf{r}) = \sum_{ij} [\Phi_i(\mathbf{r}) \Phi_j(\mathbf{r}) D^0_{ij} + 2\dot{\Phi}_i(\mathbf{r}) \Phi_j(\mathbf{r}) D^1_{ij} + \dot{\Phi}_i(\mathbf{r}) \dot{\Phi}_j(\mathbf{r}) D^2_{ij}]. \qquad (48)$$

As the KPWs are localised this can be evaluated in real space using a cluster method. We take only those KPWs Φ and their energy derivatives $\dot{\Phi}$ that lie within a certain distance from $\mathbf{r}$ and evaluate (48) directly.

When forming the product of two KPWs we must do this in a way that is consistent with the forwards–backwards definition of the overlap matrix. That is we must take

$$\Phi_{\mathbf{R}L}(\mathbf{r}) \Phi_{\mathbf{R}'L'}(\mathbf{r}) = (\phi_l \phi_{l'} - \varphi_l \varphi_{l'}) Y_L Y_{L'} \delta_{\mathbf{R},\mathbf{R}'} + \psi_{\mathbf{R}L} \psi_{\mathbf{R}'L'}.$$

In this way we ensure that the charge density has the correct total charge. Note that the forwards–backwards definition leads to the charge density being made up of three distinct components,

$$n(\mathbf{r}) = n_\phi(\mathbf{r}) - n_\varphi(\mathbf{r}) + n_\psi(\mathbf{r}),$$

arising from the partial waves, back-extrapolated partial waves and the screened spherical waves, respectively.

7.1 One-center expansion of the charge density

We now wish to calculate the coefficients of the spherical harmonic expansion of the charge density about a given site $\mathbf{R}$,

$$n(\mathbf{r}) = \sum_L n_{\mathbf{R}L}(r_\mathbf{R}) Y_L. \qquad (49)$$

There will again be three contributions to $n_{\mathbf{R}L}(r_\mathbf{R})$. Let us consider the contributions from the partial waves first. The forwards–backwards definition does not allow cross terms between partial waves that are on different sites, even if the potential spheres overlap. Therefore we can write

$$n_\phi = \sum_{R'} n_{\phi,\mathbf{R}'}. \qquad (50)$$

For $R = R'$ the one-center expansion of $n_{\phi,\mathbf{R}'}$ is simply given by

$$n_{\phi,\mathbf{R}} = \sum_{L'',L'} [\phi_{l''}\phi_{l'} D^0_{\mathbf{R}L'',\mathbf{R}L'} + 2\dot{\phi}_{l''}\phi_{l'} D^1_{\mathbf{R}L'',\mathbf{R}L'} + \dot{\phi}_{l''}\dot{\phi}_{l'} D^2_{\mathbf{R}L'',\mathbf{R}L'}] Y_{L''} Y_{L'}$$
$$= \sum_L \left[\sum_{L'',L'} [\phi_{l''}\phi_{l'} D^0_{\mathbf{R}L'',\mathbf{R}L'} + 2\dot{\phi}_{l''}\phi_{l'} D^1_{\mathbf{R}L'',\mathbf{R}L'} + \dot{\phi}_{l''}\dot{\phi}_{l'} D^2_{\mathbf{R}L'',\mathbf{R}L'}] C_{L,L''L'} \right] Y_L. \qquad (51)$$

Here we have introduced the Gaunt coefficients $C_{L,L''L'}$. The large term in square brackets will be the contribution of $n_{\phi,\mathbf{R}}$ to the one-center expansion of $n(\mathbf{r})$ about $\mathbf{R}$. The contribution from $n_{\phi,\mathbf{R}'}$ for sites $\mathbf{R}'$ that are neighbours of $\mathbf{R}$ is slightly more difficult. It can be thought of as an overlap contribution because the sphere at $\mathbf{R}'$ overlaps the region of space around $\mathbf{R}$. To evaluate it one first makes the expansion of $n_{\phi,\mathbf{R}'}$ about $\mathbf{R}'$ using (51), and then re-expands about $\mathbf{R}$ using an extension of a procedure originally developed by Löwdin [48, 49]. This overlap correction is quick to evaluate.

The contribution from n_ψ is easier. Each individual SSW can be expanded about $\mathbf{R}$ using (16) and then the product made. All products of spherical harmonics can then be reduced using the Gaunt coefficients to a sum over single spherical harmonics, $Y_{L''}Y_{L'} = \sum_L C_{L,L''L'} Y_L$. This procedure is fast as long as one first contracts the structure constants with the density matrices and then forms the product. Note that in the expansion of an SSW about a site $\mathbf{R}$ one should include high terms.

If the region around $\mathbf{R}$ for which we construct the one-center expansion extends into screening spheres on neighbouring sites then there will be an additional overlap contribution due to the truncation of the SSW. This is, however, easily dealt with using the Löwdin re-expansion procedure [48, 49].

In this way the expansion of $n(\mathbf{r})$ about any site $\mathbf{R}$ can be built up. Although seemingly complicated, this can in practice be done very fast. The one-center expansion can really only be used close to $\mathbf{R}$. As one moves further out, then the convergence of (50) in L becomes poor, in general even before the potential sphere is reached.

7.2 The spheridised charge density and the self-consistent loop

As mentioned in Section 1 the SE needs to be solved self-consistently. This is achieved by a loop cycle. Given an input charge density one constructs the muffin tin potential, the SE is then solved to give an output charge density. This is then mixed with the input and the whole process repeated until input and output are the same.

There is more than one way to construct a muffin tin potential from a given density [45]. For the sake of simplicity we do this by first spheridising the charge density. We replace the charge density inside each potential sphere by its spherical average,

$$n(\mathbf{r}) \Rightarrow n_{\text{ASA}}(\mathbf{r}) = \sum_{\mathbf{R}} \tilde{n}_{\mathbf{R}0} \frac{1}{\sqrt{4\pi}} \Theta(|\mathbf{r}-\mathbf{R}|-s_{\mathbf{R}}). \qquad (52)$$

Here n_{ASA} is known as the ASA charge density. In order to keep the correct total charge n_{ASA} needs to be renormalised, and this is indicated by the tilde over the $n_{\mathbf{R}0}$. This only works if the volume of the potential spheres equals the total volume of the system. In practice n_{ASA} is accumulated directly from the wavefunctions, and the contribution to n_{ASA} from each eigenfunction is renormalised individually.

Given a charge of the form (52) it is trivial to construct a potential that is automatically in the muffin tin form. For each $\tilde{n}_{\mathbf{R}0}$ one can solve Poisson's equation numerically to obtain the electrostatic potential. The long range part of the electrostatic potential is then taken care of by a Madelung summation, in which the charge of each sphere can be treated as point-like. The result of the Madelung sum will simply shift the individual potentials inside each sphere relative to each other. The exchange-correlation potential is trivially given by

$$v_{\text{xc}}(\mathbf{r}) = \sum_{\mathbf{R}} v_{\text{xc}} \left[\tilde{n}_{\mathbf{R}0} \frac{1}{\sqrt{4\pi}} \right] \Theta(|\mathbf{r}-\mathbf{R}|-s_{\mathbf{R}})$$

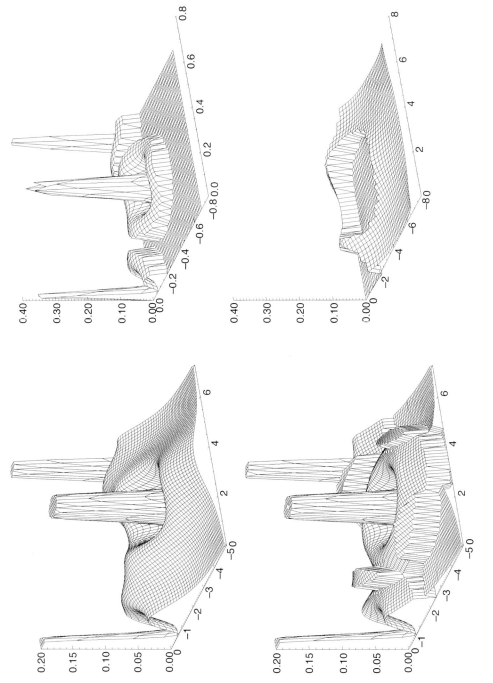

Fig. 20. Intra-sphere (top) and interstitial (bottom) regions for the description of the full charge density

Fig. 19. Full (top) and ASA (bottom) charge densities for Si in the (110) plane

and can be added to the electrostatic potential. Finally the muffin tin zero is given as the average of the potential at all $s_\mathbf{R}$, weighted by $(s_\mathbf{R})^2$.

Although the spheridisation procedure makes the self-consistency loop fast, there is a loss of information. In Fig. 19 we compare the full charge density $n(\mathbf{r})$ with $n_\text{ASA}(\mathbf{r})$. The difference is clearly visible. If one evaluates the total energy functional $E[n]$ using n_ASA then although the result is good enough for a test for self-consistency, it is not good enough to exact structural energy differences nor elastic constants. Such quantities depend upon the non-spherical parts of the charge. To calculate the total energy correctly we must use the full charge density $n(\mathbf{r})$. To do this we will be required to expand $n(\mathbf{r})$ in a set of functions that allows us to solve Poisson's equation. A common choice is to expand in plane waves, but we shall expand in screened spherical waves.

7.3 The full charge expansion

Once again we start by dividing up space. At each atomic site $\mathbf{R}$ we place a *second* screening sphere $a_\mathbf{R}$. We do not place them on empty sphere sites. These screening spheres will form the boundaries for a second set of SSWs with which we shall expand the charge. In addition they divide up space into intra-sphere ($|\mathbf{r}| < a_\mathbf{R}$) and interstitial ($|\mathbf{r}| > a_\mathbf{R}; \forall \mathbf{R}$) regions. These two regions are illustrated schematically in Fig. 20.

In the intra-sphere regions we use the one-center expansion for the charge density (49). As the screening spheres have a small radius this expansion is well converged for $L = 3$. The spherical part of the charge density, $n_{\mathbf{R}0}$ is known accurately beyond the screening spheres. As this spherical part is the most important we would like to use this extra information that we have. Therfore we let the $l = 0$ part of the one-center expansion extend further out to the potential sphere. To this $l = 0$ component we then attach a *back-extrapolated charge density* $n^0(r)$. This is defined in analogy to the back-extrapolated partial wave. It exists only in the range $a < r < s$ and is defined to solve the wave equation for the same energy as the SSWs. Furthermore it matches the $n_{\mathbf{R}0}(r)$ at $r = s$ both in value and slope. However we do not renormalise it to equal 1 at $r = a$. The intra-sphere part of the charge density expansion is thus

$$n(\mathbf{r}) = \sum_{\mathbf{R}L} n_{\mathbf{R}L}(r_\mathbf{R})\, Y_L \Theta(r_\mathbf{R} - a_\mathbf{R}) \\ + \sum_\mathbf{R} (n_{\mathbf{R}0}(r_\mathbf{R}) - n_\mathbf{R}^0(r_\mathbf{R}))\, Y_0 \Theta(r_\mathbf{R} - s_\mathbf{R})\, \Theta(a_\mathbf{R} - r_\mathbf{R}). \qquad (53)$$

In the interstitial region we expand the charge density in the SSWs ψ and their first two energy derivatives. As well as having different boundary conditions to the SSWs used to construct the KPWs, these will have a different energy. Experience has shown that a negative energy of between -1 Ry and -2 Ry works best. Together with (53) our chosen expansion for the full charge density is

$$n(\mathbf{r}) = \sum_{\mathbf{R}L} n_{\mathbf{R}L}(r_\mathbf{R})\, Y_L \Theta(r_\mathbf{R} - a_\mathbf{R}) \\ + \sum_\mathbf{R} (n_{\mathbf{R}0}(r_\mathbf{R}) - n_\mathbf{R}^0(r_\mathbf{R}))\, Y_0 \Theta(r_\mathbf{R} - s_\mathbf{R})\, \Theta(a_\mathbf{R} - r_\mathbf{R}) \\ + \sum_{\mathbf{R}L} [\psi_{\mathbf{R}L} A_{\mathbf{R}L} + \dot\psi_{\mathbf{R}L} B_{\mathbf{R}L} + \ddot\psi_{\mathbf{R}L} C_{\mathbf{R}L}]. \qquad (54)$$

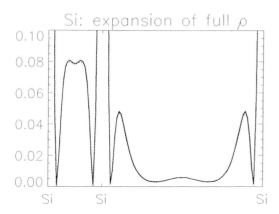

Fig. 21. Full charge density (full line) as compared to the expansion (dotted line) as explained in the text for Si along the [111] direction

The use of SSWs to expand the charge density has two main advantages. Firstly they form a minimal set and secondly as they are localised this expansion can be done in real space. To find the coefficients A, B, and C a multi step process is used.

First we require the expansion to be continuous at the screening spheres. As both ψ and $\dot{\psi}$ vanish on all screening spheres this continuity condition gives

$$A_{\mathbf{R}L} = \begin{cases} n^0_{\mathbf{R}}(a) & \text{if } L = 0 \\ n_{\mathbf{R}_L}(a) & \text{if } L > 0 \end{cases}.$$

Secondly we require the expansion to be differentiable at the screening spheres. From (23) we have that the slope of $\psi_{\mathbf{R}L}$ on the screening spheres is $(1/a)S_{\mathbf{R}'L',\mathbf{R}L}$. The slope of the energy derivative $\dot{\psi}_{\mathbf{R}L}$ is $(1/a)\dot{S}_{\mathbf{R}'L',\mathbf{R}L}$. From this it is easy to show that

$$B_{\mathbf{R}L} = \sum_{\mathbf{R}'L'}\left[\dot{S}^{-1}_{\mathbf{R}L,\mathbf{R}'L'}a_{\mathbf{R}'}\frac{\partial n_{\mathbf{R}'L'}}{\partial r}(a_{\mathbf{R}'}) - (\dot{S}^{-1}S)_{\mathbf{R}L,\mathbf{R}'L'}n_{\mathbf{R}'L'}(a_{\mathbf{R}'})\right],$$

where in *this expression* one should replace $n_{\mathbf{R}'0}$ by $n^0_{\mathbf{R}'}$.

This continuous and differentiable expansion is a remarkable good fit to the full charge density, and requires only a final correction. This is done by adjusting both B and C using a least squares fit in which we sample the full charge density $n(\mathbf{r})$ at points in the interstitial. As the correction required is small these points are sparsely placed. We choose them using a Sobol sequence [50, 51]. The least squares fit is constrained so that the expansion has the correct total charge.

In Fig. 21 we compare this expansion with the full charge density for the case of silicon. The [111] line is shown, and one can see that the differences between the two are very small. As silicon in the diamond structure is an open system the fraction of space occupied by the region where we expand in SSWs is large, in this case 77%.

7.4 Evaluating the total energy

The total energy functional is a sum of three terms, a kinetic term $T_0[n]$, and exchange-correlation term $E_{xc}[n]$ and an electrostatic or Hartree term $E_{es}[n]$.

The kinetic term is

$$T_0[n] = \sum_n E_n - \int n(\mathbf{r})\, V(\mathbf{r})\, d^3\mathbf{r},$$

where the sum of the eigenvalues runs over the occupied states. Inserting into this the form for the muffin tin potential (6) gives

$$T_0[n] = \sum_n E_n + \sum_{\mathbf{R}} \left[\sqrt{4\pi} \int n_{\mathbf{R}0}(r) \, v_{\mathbf{R}}(r) \, r^2 \, dr\right] - V_{\text{mtz}} Q \, .$$

Here Q is the total charge of the system. The radial integrals are easy to evaluate, and the eigenvalues are known, so T_0 is easily found. The sum of the eigenvalues is usually carried out using a tetrahedron method.

The exchange-correlation term in the LDA approximation is

$$E_{\text{xc}}[n] = \int n(\mathbf{r}) \, \epsilon_{\text{xc}}(\mathbf{r}) \, d^3 \mathbf{r} \, ,$$

where ϵ_{xc} is the exchange-correlation energy density per electron. The full charge density is expanded in the form (54). We make a similar expansion for ϵ_{xc}. For this we require the one-center expansion of ϵ_{xc}, to be used in the intra-sphere region. This can be obtained from the one-center expansion of the charge by evaluating $\epsilon_{\text{xc}}(\mathbf{r})$ at points on successive shells around $\mathbf{R}$ and directly projecting out the components. Thus ϵ_{xc} is expanded as

$$\begin{aligned}\epsilon_{\text{xc}}(\mathbf{r}) &= \sum_{\mathbf{R}L} \epsilon_{\text{xc},\mathbf{R}L}(r_{\mathbf{R}}) \, Y_L \Theta(r_{\mathbf{R}} - a_{\mathbf{R}}) \\ &+ \sum_{\mathbf{R}} \left(\epsilon_{\text{xc},\mathbf{R}0}(r_{\mathbf{R}}) - \epsilon_{\text{xc},\mathbf{R}}^0(r_{\mathbf{R}})\right) Y_0 \Theta(r_{\mathbf{R}} - s_{\mathbf{R}}) \, \Theta(a_{\mathbf{R}} - r_{\mathbf{R}}) \\ &+ \sum_{\mathbf{R}L} \left[\psi_{\mathbf{R}L} A_{\mathbf{R}L}^{\text{xc}} + \dot{\psi}_{\mathbf{R}L} B_{\mathbf{R}L}^{\text{xc}} + \ddot{\psi}_{\mathbf{R}L} C_{\mathbf{R}L}^{\text{xc}}\right]. \end{aligned} \quad (55)$$

Where the A^{xc}, B^{xc} and C^{xc} are found in the same way as for the charge density. The integral of the product of $\epsilon_{\text{xc}}(\mathbf{r})$ and $n(\mathbf{r})$ is now easy to do. Cross terms in the product that involve the intra-sphere parts $\epsilon_{\text{xc},\mathbf{R}L}$ and $n_{\mathbf{R}L}$ can be reduced to radial integrals. Care must be taken when the potential spheres overlap. The resulting overlap corrections are small but not insignificant. They can be evaluated simply using the extended Löwdin re-projection [48, 49] procedure described before. The remaining cross terms will involve products of two SSWs or their derivatives. Integrals of these products have been calculated earlier (24),

$$\langle \psi \mid \psi \rangle = a \dot{S}^a \, , \qquad \langle \psi \mid \dot{\psi} \rangle = a \tfrac{1}{2} \ddot{S}^a \, ,$$

$$\langle \psi \mid \ddot{\psi} \rangle = a \tfrac{1}{3} \dddot{S}^a \, , \qquad \langle \dot{\psi} \mid \dot{\psi} \rangle = a \tfrac{1}{6} \dddot{S}^a \, ,$$

$$r\langle \dot{\psi} \mid \ddot{\psi} \rangle = a \tfrac{1}{12} \dddot{S}^a \, , \qquad \langle \ddot{\psi} \mid \ddot{\psi} \rangle = a \tfrac{1}{30} \dddot{S}^a \, ,$$

and all reduce to energy derivatives of the structure constants. The integrals of these cross terms can then be summed to obtain the total exchange-correlation energy.

The electrostatic contribution to the total energy is

$$E_{\text{es}}[n] = \frac{1}{2} \int n(\mathbf{r}) \, v_{\text{es}}(\mathbf{r}) \, d^3 \mathbf{r} + \int n(\mathbf{r}) V_{\text{nuc}}(\mathbf{r}) \, d^3 \mathbf{r} + \sum_{\mathbf{R},\mathbf{R}'} \frac{Z_{\mathbf{R}} Z_{\mathbf{R}'}}{|\mathbf{R} - \mathbf{R}'|} \, . \quad (56)$$

Here v_{es} is the electrostatic potential of the electronic charge density $n(\mathbf{r})$ and is given by the solution of Poisson's equation

$$\nabla^2 v_{\text{es}}(\mathbf{r}) = -8\pi n(\mathbf{r}) \, . \quad (57)$$

With $n(\mathbf{r})$ expanded in the form (54) it is easy to solve this equation. For the intrasphere parts (53) the Poisson equation can be separated in spherical polar coordinates, and the resulting radial differential equation solved by standard techniques. The resulting potential can be written in the form

$$v_{\text{intra}}(\mathbf{r}) = \sum_{\mathbf{R},L\neq 0} v_{\mathbf{R}L}(r_{\mathbf{R}})\, Y_L \Theta(r_{\mathbf{R}} - a_{\mathbf{R}}) + \sum_{\mathbf{R}} v_{\mathbf{R}}(r_{\mathbf{R}})\, Y_0 \Theta(r_{\mathbf{R}} - s_{\mathbf{R}}). \tag{58}$$

The interstitial parts are expanded in SSWs (54). Each ψ solves the wave equation with energy E. That is

$$\nabla^2 \psi = -E\psi. \tag{59}$$

Differentiating with respect to energy we obtain

$$\nabla^2 \dot{\psi} = -E\dot{\psi} - \psi, \tag{60}$$

$$\nabla^2 \ddot{\psi} = -E\ddot{\psi} - 2\dot{\psi}. \tag{61}$$

The solution of the Poisson equation for the interstitial parts of the charge is then

$$v_{\text{inter}}(\mathbf{r}) = 8\pi \sum_{\mathbf{R}L} \Bigg[\left(\frac{\psi_{\mathbf{R}L} - \psi_{\mathbf{R}L}(0)}{E}\right) A_{\mathbf{R}L}$$

$$+ \left(\dot{\psi}_{\mathbf{R}L} - \left(\frac{\psi_{\mathbf{R}L} - \psi_{\mathbf{R}L}(0)}{E}\right)\right) \frac{1}{E} B_{\mathbf{R}L}$$

$$+ \left[\ddot{\psi}_{\mathbf{R}L} - 2\left(\dot{\psi}_{\mathbf{R}L} - \left(\frac{\psi_{\mathbf{R}L} - \psi_{\mathbf{R}L}(0)}{E}\right)\right)\frac{1}{E}\right] \frac{1}{E} C_{\mathbf{R}L} \Bigg]. \tag{62}$$

Here we have introduced $\psi_{\mathbf{R}L}(0)$ which is an SSW defined with the same screening spheres but with energy $E = 0$. Equation (62) can be verified by direct application of (57) and using (59), (60), (61).

To (58) and (62) we must add a "Laplace" part to the solution, v_{lap}, where $\nabla^2 v_{\text{lap}} = 0$. This Laplace part is chosen so that the total potential satisfies the boundary conditions of being everywhere finite, continuous and differentiable. Inside the screening spheres v_{lap} can therefore be at most some linear combination of zero energy Bessel functions $j_l(E=0)$. In the interstitial the zero energy SSWs $\psi(0)$ form a complete set in which we can expand any v_{lap}. Hence the Laplace potential can always be written in the form

$$v_{\text{lap}}(\mathbf{r}) = \sum_{\mathbf{R}L} [j_l(E=0, r)\, Y_L \alpha_{\mathbf{R}L} \Theta(r_{\mathbf{R}} - a_{\mathbf{R}}) + \psi_{\mathbf{R}L}(E=0, \mathbf{r})\, \beta_{\mathbf{R}L}], \tag{63}$$

where as stated above α and β are such that the total potential is smooth. We shall not include a long derivation of α and β here as it is tedious but straightforward. As the nuclear potential solves $\nabla^2 V_{\text{nuc}} = 0$ it can easily be incorporated into v_{lap}. The total electrostatic potential is then

$$v_{\text{es}}(\mathbf{r}) = v_{\text{intra}}(\mathbf{r}) + v_{\text{inter}}(\mathbf{r}) + v_{\text{lap}}(\mathbf{r}).$$

The integral of the product of $v_{\text{es}}(\mathbf{r})\, n(\mathbf{r})$ can be performed in the same way as the integral for the exchange-correlation energy. Once again all intra-sphere terms reduce to radial integrals and the interstitial terms reduce to derivatives of the structure constants.

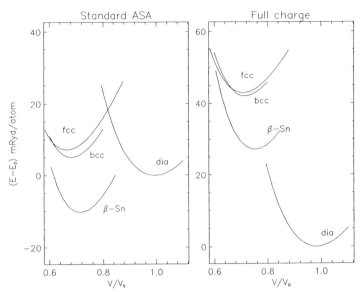

Fig. 22. Calculated energy versus volume curves for different crystalline phases in Si. Left: results for n_{ASA}. Right: full charge results

In this manner the total energy can be evaluated for the full charge density $n(\mathbf{r})$. In Fig. 22 we show the total energy versus volume curves for silicon in a number of different structures. In these calculations the self-consistency cycle was carried out as described above, with the muffin tin potential being calculated from n_{ASA}. Once a self-consistent potential had been obtained we then expanded the full charge density $n(\mathbf{r})$ as described here and evaluated the total energy $E[n]$. For comparison we show the total energy evaluated using n_{ASA} instead of the full charge. While the full charge expansion correctly predicts the diamond structure to be the most stable, $E[n_{\mathrm{ASA}}]$ predicts the β-tin structure to be the most stable one. This is the origin of the commonly held belief that LMTO does not give good total energies. LMTO does give good total energies, as long as they are evaluated will the full charge $n(\mathbf{r})$.

In Table 1 we show elastic constants calculated for silicon in the diamond structure. Here we compare the values calculated from n_{ASA} and those from the full charge $n(\mathbf{r})$ to both experimental values and those from a full potential calculation [52, 53]. The $E[n_{\mathrm{ASA}}]$ produces reasonable results for both the equilibrium lattice constant a_0 and bulk modulus B. However it fails for the elastic constants $C_{11} - C_{12}$ and C_{44}. These involve deformations that lower the symmetry and hence have a large dependence on the non-spherical parts of the charge density. The C_{44} also has an internal relaxation, measured by the parameter ζ. The energy minimum should lie between $\zeta = 0$ and $\zeta = 1$. However the ASA energy shows no minimum, just a steady decrease from $\zeta = 1$ to $\zeta = -1$. In contrast the full charge expansion gives both a minimum and a reasonable value for C_{44}.

Table 2 shows elastic constants for GaAs. Again the full charge values reproduce the experimental values well. It is worth noting that although evaluating $E[n]$ is more involved than $E[n_{\mathrm{ASA}}]$ it only needs to be done once, at the end of the calculation once self-consistency has been achieved.

Table 1

Lattice parameter and elastic constants for Si in the diamond structure

	Si			
	standard ASA	full charge	experiment	FP-LMTO
a_0 (a.u.)	10.245	10.198	10.263	10.223
B (Mbar)	0.92	0.97	0.992	0.99
$C_{11} - C_{12}$ (Mbar)	2.14	1.08	1.025	1.02
C_{44} (Mbar)	–	0.84	0.801	0.83
ζ	–	0.40	0.54	0.51

Table 2

Lattice parameter and elastic constants for GaAs

	GaAs		
	standard ASA	full charge	experiment
a_0 (a.u.)	10.632	10.636	10.676
B (Mbar)	0.73	0.74	0.784
$C_{11} - C_{12}$ (Mbar)	1.52	0.60	0.652
C_{44} (Mbar)	–	0.61	0.600
ζ	–	0.57	0.48

8. Concluding Remarks

In this paper we have presented the basic principles of the third-generation LMTO method. The philosophy is one in which great emphasis is placed on the construction of the basis set, so that it is not only minimal, but also the best possible basis set for its size. Each LMTO in itself contains useful information about the system under study.

It is perhaps useful at this point to reflect on the improvements over second-generation LMTO. Now that the interstitial is treated properly the results are far more accurate. Linearisation works better with bigger energy windows. This in turn allows us to downfold as many orbitals as we wish. The formalism itself is simpler and now internally complete. Those who know TB-LMTO will have noted the absence here of any mention of "combined corrections" or "downfolding terms" in the Hamiltonian. Everything is now wrapped up properly inside the kink matrix.

The prospects for further development look good. Already here we have shown how it is possible to extract good total energies from an LMTO-ASA calculation. The formalism no longer requires the potential spheres to be non-overlapping, and work is underway to use this to our advantage to get rid of the empty spheres [45], by making the atomic spheres fatter with bigger overlaps. The self-consistency loop will of course have to be modified, but the full charge expansion shows the way forward.

References

[1] O. K. ANDERSEN, Phys. Rev. B **12**, 3060 (1975).
[2] J. KORRIGA, Physica **13**, 392 (1947).
[3] W. KOHN and N. ROSTOKER, Phys. Rev. **94**, 111 (1954).
[4] D. D. KOELLING and G. O. ARBMAN, J. Phys. F (Metal Phys.) **5**, 2041 (1975).
[5] D. J. SINGH, Planewaves, Pseudopotentials and the LAPW Method, Kluwer Academic Publ., Dordrecht 1994.
[6] O. K. ANDERSEN, O. JEPSEN, A. I. LIECHTENSTEIN, and I. I. MAZIN, Phys. Rev. B **49**, 4145 (1994).
[7] O. K. ANDERSEN, A. I. LIECHTENSTEIN, O. JEPSEN, and F. PAULSEN, J. Phys. Chem. Solids **56**, 1573 (1995).
[8] O. JEPSEN, O. K. ANDERSEN, I. DASGUPTA, and S. Y. SAVRASOV, J. Phys. Chem. Solids **59**, 1718 (1998).
[9] O. K. ANDERSEN, C. ARCANGELI, R. W. TANK, T. SAHA-DASGUPTA, G. KRIER, O. JEPSEN, and I. DASGUPTA, in: Tight-Binding Approach to Computational Material Science, Eds. L. COLOMBO, A. GONIS, and P. TURCHI, Mater. Res. Soc. Symp. Proc. **491**, 3 (1998).
[10] D. JOHRENDT, C. FELSER, O. JEPSEN, O. K. ANDERSEN, A. MEWIS, and J. ROUXEL, J. Solid State Chem. **130**, 254 (1997).
[11] F. BOUCHER and R. ROUSSEAU, Inorg. Chem. **37**, 2351 (1998).
[12] F. ARYASETIAWAN and O. GUNNARSSON, Phys. Rev. Lett. **74**, 3221 (1995).
[13] F. ARYASETIAWAN and O. GUNNARSSON, Phys. Rev. B **49**, 7219 (1994).
[14] P. R. PEDUTO, S. FROTA-PESSOA, and M. S. METHFESSEL, Phys. Rev. B **44**, 13283 (1991).
[15] P. VARGAS, in: Lectures on Methods of Electronic Structure Calculations, Eds. V. KUMAR, O. K. ANDERSEN, and A. MOOKERJEE, World Scientific Publ. Co., Singapore 1994 (p. 147).
[16] O. GUNNARSSON, O. JEPSEN, and O. K. ANDERSEN, Phys. Rev. B **27**, 7144 (1983).
[17] S. K. BOSE, S. S. JASWAL, O. K. ANDERSEN, and J. HAFNER, Phys. Rev. B **37**, 9955 (1988).
[18] H. J. NOWAK, O. K. ANDERSEN, T. FUJIWARA, O. JEPSEN, and P. VARGAS, Phys. Rev. B **44**, 3577 (1991).
[19] S. FROTA-PESSOA, Phys. Rev. B **46**, 14570 (1992).
[20] H. L. SKRIVER and N. M. ROSENGAARD, Phys. Rev. B **43**, 9538 (1991).
[21] H. L. SKRIVER and N. M. ROSENGAARD, Phys. Rev. B **46**, 7157 (1992).
[22] M. SPRINGBORG and O. K. ANDERSEN, J. Chem. Phys. **87**, 7125 (1987).
[23] M. SPRINGBORG, Internat. Rev. Phys. Chem. **12**, 241 (1993).
[24] M. METHFESSEL, Phys. Rev. B **38**, 1537 (1988).
[25] M. METHFESSEL, C. O. RODRIGUEZ, and O. K. ANDERSEN, Phys. Rev. B **40**, 2009 (1989).
[26] S. Y. SAVRASOV and D. Y. SAVRASOV, Phys. Rev. B **46**, 12181 (1992).
[27] K. H. WEYRICHS, Phys. Rev. B **37**, 10269 (1988).
[28] J. H. WILLS, D. PRICE, and B. COOPER, Phys. Rev. B **39**, 4945 (1989); **46**, 11368 (1992).
[29] E. BOTT, M. METHFESSEL, W. KRABS, and P. C. SCHMIDT, J. Math. Phys. **39**, 3393 (1998).
[30] P. BLOCHL, Phys. Rev. B **50**, 17953 (1994).
[31] S. Y. SAVRASOV, Phys. Rev. B **54**, 16470 (1996).
[32] O. K. ANDERSEN and O. JEPSEN, Phys. Rev. Lett. **53**, 2571 (1984).
[33] O. K. ANDERSEN, O. JEPSEN, and M. SOB, in: Lecture Notes in Physics: Electronic Band Structure and Its Applications, Ed. M. YUSSOUFF, Springer-Verlag, Berlin 1987.
[34] O. K. ANDERSEN, O. JEPSEN, and G. KRIER, in: Lectures Notes on Methods of Electronic Structure Calculations, Eds. V. KUMAR, O. K. ANDERSEN, and A. MOOKERJEE, World Scientific Publ. Co., Singapore 1994 (p. 63).
[35] O. K. ANDERSEN, T. SAHA-DASGUPTA, R. W. TANK, C. ARCANGELI, O. JEPSEN, and G. KRIER, Ab Initio Calculation of the Physical Properties of Solids, Ed. H. DREYSSE, Lecture Notes in Physics, Vol. 535, Springer-Verlag, Berlin 1999.
[36] R. O. JONES and O. GUNNARSSON, Rev. Mod. Phys. **61** 689, (1989).
[37] P. HOHENBERG and W. KOHN, Phys. Rev. **136**, B864 (1964).
[38] W. KOHN and L. J. SHAM, Phys. Rev. **140**, A1133 (1965).
[39] M. E. ROSE, Relativistic Electron Theory, Wiley, New York 1961.
[40] P. WEINBERGER, Electron Scattering Theory for Ordered and Disordered Matter, Chap. 3, Clarendon Press, Oxford 1990.
[41] D. D. KOELLING and B. N. HARMON, J. Phys. C (Solid State Phys.) **10**, 3107 (1977).

[42] O. K. Andersen, in: The Electronic Structure of Complex Systems, Eds. P. Phariseau and W. M. Temmerman, Plenum Publ. Corp., New York 1984.
[43] W. Hobson, Theory of Spherical and Ellipsoidal Harmonics, Chelsea, New York 1955.
[44] For example, see: G. B. Arfken and H. J. Weber, Mathematical Methods for Physicists, 4th edn., Academic Press, London/New York 1996.
[45] C. Arcangeli, R. W. Tank, and O. K. Andersen, unpublished.
[46] W. R. L. Lambrecht and O. K. Andersen, Phys. Rev. B **34**, 2439 (1986).
[47] P. O. Löwdin, J. Chem. Phys. **19**, 1396 (1951).
[48] P. O. Löwdin, Adv. Phys. **5**, 1 (1956).
[49] C. Arcangeli and O. K. Andersen, unpublished.
[50] I. M. Sobol, USSR Comp. Math. and Math. Phys. **7**, 86 (1967).
[51] P. Bratley and B. L. Fox, ACM Trans. Math. Software, **14**, 88 (1988).
[52] O. K. Andersen, M. Methfessel, C. O. Rodriguez, P. Blöchl, and H. M. Polatoglou, in: Atomistic Simulation of Materials, Eds. V. Vitek and D. J. Srolovitz, Plenum Publ. Corp., New York 1989 (p. 1).
[53] O. H. Nielsen and R. M. Martin, Phys. Rev. B **32**, 3792 (1985).

phys. stat. sol. (b) **217**, 131 (2000)

Subject classification: 71.15.Mb; 71.15.Pd

LDA Calculations Using a Basis of Gaussian Orbitals

P.R. BRIDDON (a) and R. JONES (b)

(a) Department of Physics, University of Newcastle, Newcastle upon Tyne, NE1 7RU, UK

(b) Department of Physics, University of Exeter, Exeter, Devon, EX4 4QL, UK

(Received August 10, 1999)

In this paper we will consider some of the methods involved in carrying out density functional calculations within the framework of localised basis sets, specifically those of Gaussian type orbitals. Particular emphasis is placed on the methods used in the AIMPRO (Ab Initio Modelling PROgram) code, but mention is made of a number of developments and strategies used in other programs.

1. Introduction

We begin by reviewing the Kohn-Sham equations of density functional theory and pseudopotentials which are fundamental to this work. From this starting point, we briefly consider the relative merits of localised orbitals and plane waves, two of the most common basis set expansions used to solve the Kohn-Sham equations. We then look briefly at the setting up and quality of some standard Gaussian basis sets and then consider in some detail how the total energy and equilibrium structure of a cluster of atoms may be evaluated within this formalism. To finish, a sample of applications that have been studied successfully using the AIMPRO code is reviewed.

In this paper, *atomic units* will be used unless otherwise indicated. In these units, $\hbar = m_e = e = 4\pi\epsilon_0 = 1$. The unit of length is the Bohr radius, $a_0 \approx 0.529$ Å, and the unit of energy the Hartree or 27.2116 eV.

2. Fundamental Equations

2.1 The Kohn-Sham equations of density functional theory

According to the Hohenberg-Kohn theorem [1], the total energy of an electron gas in the presence of a background potential $V^{\text{ext}}(\mathbf{r})$ may be written as a functional of the charge density. In the Kohn-Sham approach [2] this is written as

$$E[n] = T_s[n] + \int V^{\text{ext}}(\mathbf{r})\, n(\mathbf{r})\, d\mathbf{r} + \frac{1}{2} \int \frac{n(\mathbf{r})n(\mathbf{r}')}{|\mathbf{r} - \mathbf{r}'|}\, d\mathbf{r}\, d\mathbf{r}' + E_{\text{xc}}[n], \quad (1)$$

where $T_s[n]$ is the kinetic energy of a non-interacting system of electrons that has the same charge density as the real system,

$$n(\mathbf{r}) = \sum_\lambda f_\lambda |\psi_\lambda(\mathbf{r})|^2, \quad (2)$$

where $\psi_\lambda(\mathbf{r})$ are the single particle orthonormal orbitals in this system and the sum is over all occupied electron states. f_λ can be considered as the occupancy of state λ, usually in insulators either 0 or 1. With the introduction of these states, the kinetic energy of this non-interacting system is readily evaluated as

$$T_s[n] = -\tfrac{1}{2} \sum_\lambda f_\lambda \int \psi_\lambda^*(\mathbf{r}) \nabla^2 \psi_\lambda(\mathbf{r}) \, d\mathbf{r}. \tag{3}$$

Some approximation is used for what is usually termed the exchange-correlation energy, $E_{xc}[n]$ in Eq. (1).

According to the second theorem of Hohenberg-Kohn [1] the true charge density of the system is that $n(\mathbf{r})$ which minimises this functional subject to it having the correct normalisation. In the Kohn-Sham approach, this minimisation is carried out by varying $\psi_\lambda(\mathbf{r})$ and by introducing Lagrange multipliers ϵ_λ to handle the constraints $\int |\psi_\lambda(\mathbf{r})|^2 \, d\mathbf{r} = 1$. The resulting equations are called the Kohn-Sham equations

$$-\tfrac{1}{2} \nabla^2 \psi_\lambda(\mathbf{r}) + V(\mathbf{r}) \psi_\lambda(\mathbf{r}) = \epsilon_\lambda \psi_\lambda(\mathbf{r}), \tag{4}$$

where the potential $V(\mathbf{r})$ introduced is given by

$$V(\mathbf{r}) = V^{\text{ext}}(\mathbf{r}) + \int \frac{n(\mathbf{r}') \, d\mathbf{r}'}{|\mathbf{r} - \mathbf{r}'|} + \frac{\delta E_{xc}}{\delta n(\mathbf{r})}. \tag{5}$$

Eqs. (4) and (5) must be solved self-consistently together with (2). The resulting orbitals $\psi_\lambda(\mathbf{r})$ are called the Kohn-Sham orbitals. Once the density has been determined, the total energy is found relatively straightforwardly using Eq. (1).

One important generalisation of this formalism has been made to the case where the external field includes a spin-dependent potential (as is the case with a magnetic field). This makes a fundamental difference in that the potential felt by spin-up and spin-down electrons will no longer be equal. The consequence of this is that the energy is now a unique functional of the two variables $n_\uparrow(\mathbf{r})$ and $n_\downarrow(\mathbf{r})$ the spin-up and spin-down densities which together determine the external potential [3, 4]. This is sometimes referred to as spin density functional theory. The main advantage of this formalism has turned out not to be that magnetic fields can be described, but rather that when coupled with a local spin density approximation (see Section 2.2) the description of open-shell atoms, molecules and more complex systems such as paramagnetic defects in semiconductors is vastly improved. If we restrict ourselves to this scenario, then the above equations are easily generalised to

$$-\tfrac{1}{2} \nabla \psi_{\lambda s}(\mathbf{r}) + V_s(\mathbf{r}) \psi_{\lambda s}(\mathbf{r}) = \epsilon_{\lambda s} \psi_{\lambda s}(\mathbf{r}), \tag{6}$$

$$V_s(\mathbf{r}) = V^{\text{ext}}(\mathbf{r}) + \int \frac{n(\mathbf{r}') \, d\mathbf{r}'}{|\mathbf{r} - \mathbf{r}'|} + \frac{\delta E_{xc}[n_\downarrow(\mathbf{r}), n_\uparrow(\mathbf{r})]}{\delta n_s(\mathbf{r})}, \tag{7}$$

$$n_s(\mathbf{r}) = \sum_\lambda f_{\lambda s} |\psi_{\lambda s}(\mathbf{r})|^2, \tag{8}$$

$$n(\mathbf{r}) = \sum_s n_s(\mathbf{r}), \tag{9}$$

where the subscript $s = \pm 1$ labels the spin (either up or down). Spin-orbit coupling terms can also be included within this framework [5] although we will not consider this further here.

One common ingredient of modern calculational procedures is that of fractional occupancies in which the energy functional $E[n]$ is replaced by a free energy $F[n] = E[n] - TS$ and the occupancies f_λ are fractional. This will be discussed in more detail later. A full discussion on the foundations of density functional theory can be found in, for example, [6, 7]. A much briefer introduction focusing more on theory relevant to materials simulation can be found in [8].

2.2 Approximations for E_{xc}

The most common approximation for E_{xc} is the local density approximation (LDA). This treats the inhomogeneous electron case as uniform locally,

$$E_{xc} = \int n(\mathbf{r})\, \epsilon_{xc}(n(\mathbf{r}))\, d\mathbf{r},$$

where $\epsilon_{xc}(n)$ is the exchange-correlation energy per electron for a uniform electron gas of density n. An improved description is that of the local spin-density approximation (LSDA) [3] in which $\epsilon_{xc}(n(\mathbf{r}))$ is replaced by the result for the polarised uniform electron gas, $\epsilon_{xc}(n_\uparrow, n_\downarrow)$.

Typically we split this term into exchange and correlation effects $E_{xc}[n_\uparrow, n_\downarrow] = E_x[n_\uparrow, n_\downarrow] + E_c[n_\uparrow, n_\downarrow]$. The exchange part is easily evaluated via an analytic Hartree-Fock treatment of the uniform electron gas:

$$E_x[n_\uparrow, n_\downarrow] = -\frac{3}{2}\left(\frac{3}{4\pi}\right)^{1/3}[n_\uparrow^{4/3} + n_\downarrow^{4/3}], \tag{10}$$

and the correlation part is determined by a mixture of analytic treatments and Monte Carlo simulation. Usually these numerical results for E_c are fitted to a simple parameterised form, one of the most popular early parametrisations being that of Perdew and Zunger (PZ) [9] who fitted the non-polarised result $\epsilon_c(n/2, n/2)$ and the completely spin-polarised result $\epsilon_c(n, 0)$. The result for a partially polarised gas is obtained by a weighting procedure [3]:

$$\epsilon_c(n_\uparrow, n_\downarrow) = \epsilon_c(n/2, n/2) + f(\xi)\left[\epsilon_c(n, 0) - \epsilon_c(n/2, n/2)\right], \tag{11}$$

where

$$\xi = [n_\uparrow - n_\downarrow]/n \quad \text{and} \quad f(\xi) = \frac{(1+\xi)^{4/3} + (1-\xi)^{4/3} - 2}{2^{4/3} - 2}.$$

An alternative parametrisation has been given by Vosko-Wilk-Nusair (VWN) [10]. More recently, a new form has been suggested by Perdew and Wang [11] with a number of improvements over the previous work. This may be the most accurate representation available at present, but for most computational purposes the PZ, VWN and PW parametrisations give very similar results.

These formulae have been used widely over two decades to model an enormous variety of systems ranging over atoms, molecules, clusters and solids and have made a significant contribution to a number of areas of physics, chemistry and materials science. The accuracy has been remarkable, and more than expected from such a simple approach.

The most obvious way to improve on the LDA is to develop a function $\epsilon_{xc}(n, |\nabla n|)$ which would take into account the inhomogeneity of the gas. This has to be done in a way that satisfies certain sum rules [12, 13], co-ordinate scaling laws [14] and limits [15] The resulting approximations, known as generalised gradient approximations (GGAs) are now coming into widespread use.

An early functional referred to as Becke-Perdew (BP) consisted of the Perdew formula for correlation [13] combined with an extremely accurate but empirical formula for exchange developed by Becke [16, 17]. A more widely used functional was developed by Perdew and Wang in 1991 [18] and is referred to as PW91 in the literature. Applications of this show improvements over the LDA, particularly for atoms and molecules and alkali metals [19]. This has now been superseded by a new functional, PBE96 [20 to 22]. Other functionals involve the mixing in a component of Hartree-Fock exchange.

2.3 Pseudopotentials

The use of pseudopotentials has proved an extremely important step in using *ab initio* methods to model large systems. The basic idea is to replace the nuclear potential $-Z/r$ with a more complex potential felt by the valence electrons and which removes the need to consider the core electrons. An excellent review of this technique has been compiled by Pickett [23] and modern developments are discussed elsewhere in this volume [24].

Here we will just note the form of the pseudopotentials as this is important for the discussion of their manipulation with Gaussian orbitals. They are *nonlocal* in the sense that they are angular momentum dependent:

$$\hat{V}^{ps}(r) = \sum_{lm} |lm\rangle V_l^{ps}(r) \langle lm|.$$

It is possible to split the potential into a local and nonlocal part by choosing any l component of the pseudopotential as a local term, $V^{loc}(r)$, and subtracting this from each term in the sum over l,

$$\hat{V}^{ps}(r) = V^{loc}(r) + \sum_{lm} |lm\rangle V_l^{nl}(r) \langle lm|, \tag{12}$$

where $V^{loc}(r) = V_L^{ps}(r)$ and $V_l^{nl}(r) = V_l^{ps}(r) - V_L^{ps}(r)$. In this case the sum over l can be truncated to $l = 2$ for many elements with $l = 3$ being necessary for the $5d$ transition metals. This choice will make all of the non-local terms $V^{nl}(r)$ short ranged (in fact they will be zero outside the core radius).

An alternative choice that is often made for $V^{loc}(r)$ is a function with the asymptotic form at large r of $-Z_{ion}/r$ as this choice will also make all of the non-local terms $V^{nl}(r)$ short ranged (this time very small and decaying rapidly outside the core radius).

A number of different methods exist for constructing potentials of the general form above [25 to 28]. The generation and properties of these is considered elsewhere in this volume [24], and in this paper, only the methods for dealing with these using Gaussian orbitals will be discussed in a later section. More recently, a new type of pseudopotential has appeared [29, 30], being designed for use with plane wave calculations. This is of a different form from those above and will not be discussed further here. Also omitted from the discussion here is the Kleinman-Bylander form of the pseudopotential [31, 32]. This is extremely important in plane wave calculations but less so in local orbital approaches.

3. Adoption of a Basis Set

The most common methods for solving the Kohn-Sham equations proceed by expanding the Kohn-Sham spin orbitals in a basis set. In passing, we note that other approaches include discretisation of the equations and the apparatus associated with

3.1 Reduction to matrix eigenvalue problem

Denoting the basis functions as $\phi_i(\mathbf{r})$, we write

$$\psi_{\lambda s}(\mathbf{r}) = \sum_{i=1}^{N} c_i^{\lambda s} \phi_i(\mathbf{r}), \qquad (13)$$

where the accuracy of this approach is limited by the error made in this expansion. As in the variational principle of elementary quantum mechanics, this error is determined by the number, N, of functions used and the suitability of the choice of the functions $\phi_i(\mathbf{r})$. In this paper the index λ is used to denote the spatial part of an electron state, s to denote a spin state (i.e. up or down) and $i, j, \ldots$ to label basis functions. In terms of these basis functions, the normalisation of a Kohn-Sham orbital is expressed as

$$\int |\psi_{\lambda s}(\mathbf{r})|^2 \, d\mathbf{r} = \sum_{i,j} c_i^{\lambda s} c_j^{\lambda s} S_{ij} = 1,$$

where we have introduced the overlap matrix

$$S_{ij} = \int \phi_i(\mathbf{r}) \phi_j(\mathbf{r}) \, d\mathbf{r}. \qquad (14)$$

Note that we have not written in complex conjugate symbols into these equations. This is because the basis functions are real, and all matrix elements formed from these will be real as well and so the coefficients will also be real. Complex conjugate symbols will therefore be dropped from now on unless they are specifically needed.

The spin densities are given by

$$n_s(\mathbf{r}) = \sum_{\lambda} f_{\lambda s} |\psi_{\lambda s}(\mathbf{r})|^2 \qquad (15)$$

$$= \sum_{ij} b_{ij}^s \phi_i(\mathbf{r}) \phi_j(\mathbf{r}), \qquad (16)$$

where the spin-density matrices b_{ij}^s in this localised orbital representation are given by

$$b_{ij}^s = \sum_{\lambda} f_{\lambda s} c_i^{\lambda s} c_j^{\lambda s} \qquad (17)$$

and the charge density matrix b_{ij} is

$$b_{ij} = \sum_{s} b_{ij}^s.$$

In this basis representation, our fundamental quantities are the b_{ij}^s and, according to the Hohenberg-Kohn theorem [1], the energy is uniquely determined by these. The Kohn-Sham equations in this representation are found by minimising this with respect to the quantities $c_i^{\lambda s}$,

$$\frac{\partial}{\partial c_i^{\lambda s}} \left[E[b_{ij}^{\uparrow}, b_{ij}^{\downarrow}] - \sum_{\lambda', s'} \epsilon_{\lambda' s'} \left(\sum_{ij} S_{ij} c_i^{\lambda' s'} c_j^{\lambda' s'} - 1 \right) \right] = 0. \qquad (18)$$

The differentiation is readily accomplished using

$$\frac{\partial E}{\partial c_i^{\lambda s}} = \sum_{jk} \frac{\partial E}{\partial b_{jk}^s} \frac{\partial b_{jk}^s}{\partial c_i^{\lambda s}} = \sum_j H_{ij}^s c_j^{\lambda s}, \qquad (19)$$

where we have introduced the two Hamiltonian matrices (one for each spin)

$$H_{ij}^s = \frac{\partial E}{\partial b_{ij}^s}. \qquad (20)$$

This results in the matrix equations

$$\sum_j H_{ij}^s c_j^{\lambda s} = c_{\lambda s} \sum_j S_{ij} c_j^{\lambda s}. \qquad (21)$$

This is a well-known problem in linear algebra, the generalised real-symmetric eigenvalue problem. There are a number of ways of tackling this with both standard "out of the box" approaches and others which have been developed, specifically with this application in mind being commonly used. This part of the problem will be discussed later.

3.2 Choice of basis set

The two most common choices of basis function are plane waves and Gaussian type orbitals.

The usual choice has been plane waves as these fit naturally with periodic boundary conditions, which are generally used by condensed matter theorists. In this case, the Kohn-Sham orbitals are expanded in terms of plane waves (we drop any notation associated with **k** points as this is not relevant here):

$$\psi_{\lambda s}(\mathbf{r}) = \sum_{\mathbf{G}} c_{\lambda s}(\mathbf{G}) \exp[i\mathbf{G} \cdot \mathbf{r}]. \qquad (22)$$

This expansion has the advantage that it is very transparent (there are no hidden parameters), it is very stable (plane waves with different **G** are of course orthogonal, so no instability can creep into the calculation through near-linear dependencies in the basis set), it is clear how to improve the expansion (usually, all plane waves with $G^2/2 < E_{\text{cut}}$ are included – to improve this we just increase E_{cut}), it is not biased (charge can move without restriction to any point of the unit cell) and finally it is very simple to program – if a "difficult" element requiring more basis functions is modelled, no further coding is needed, just a higher E_{cut} (and correspondingly more CPU time).

There are also some disadvantages with plane wave expansions. The principal of these is that an extremely large number of functions need to be used. For example, with an "easy" element such as silicon, a minimum of 100 planes waves per atom need to be used. Some elements are particularly difficult – for example first period elements such as carbon and oxygen or the $3d$ transition metals. This means that the time and memory requirements of such a code will be considerable. Furthermore, if a unit cell contains 100 silicon atoms and just one oxygen atom, the presence of one difficult atom (as is often present in a defect) controls the size of the basis and results in an unnecessarily flexible descriptive power having to be employed at all points in the unit cell. One way of avoiding this is to use a technique pioneered by Gygi [35] and applied to solids by Hamann [36]. The alternative is to use localised orbitals:

$$\psi_\lambda(\mathbf{r}) = \sum_i c_i^\lambda \phi_i(\mathbf{r}). \qquad (23)$$

A common choice for the functions ϕ_i is that of Cartesian Gaussian functions which consist of Gaussians multiplied by polynomials of the position vector,

$$\phi_i(\mathbf{r}) = (x - R_{ix})^{n_1} (y - R_{iy})^{n_2} (z - R_{iz})^{n_3} e^{-\alpha_i(\mathbf{r} - \mathbf{R}_i)^2}, \tag{24}$$

where n_1, n_2 and n_3 are integers. In later sections of this paper where we are primarily concerned with the prefactor and its manipulation, we will denote this function by

$$\phi_i(\mathbf{r}) = |i; n_1, n_2, n_3\rangle = |i; \mathbf{n}\rangle \tag{25}$$

to make the nature of the polynomial explicit. Clearly linear combinations of these Cartesian Gaussian functions can be chosen to form functions that transform like spherical harmonics under rotation,

$$r^l Y_{lm} e^{-\alpha r^2} = \sum_{\substack{n_1, n_2, n_3 \\ n_1 + n_2 + n_3 = l}} A_l[m, n_1, n_2, n_3] x^{n_1} y^{n_2} z^{n_3} e^{-\alpha r^2},$$

explaining the labelling of functions as s, p- or d-type. If n_1, n_2 and n_3 are all zero the function corresponds to an s-orbital of spherical symmetry. Orbitals of p-symmetry correspond to one of these integers being unity and the others zero, whereas five d-like and one s-like orbital can be generated if $\sum_i n_i = 2$.

This expansion has the advantage that it is very efficient (using techniques discussed in Section 3.3.2 results can be obtained with only four orbitals per atom, and quite well converged results with 10 to 20 functions), efficiently applicable to all elements of the periodic table (the first-period elements such as carbon, nitrogen and oxygen can be treated just as easily as silicon, gallium, arsenic etc. and calculations are not dependent on using pseudopotentials), it is flexible (if we have one difficult atom, additional orbitals can be placed on just that atom so the overall speed of the calculation is not significantly affected). For example when modelling an element of a transition element as an impurity in silicon, higher angular momentum functions need to be placed on that atom. However, the rest of the system can be treated with the standard basis set.

Disadvantages include the fact that the functions can become over-complete (numerical noise can enter a calculation if two functions with similar exponents are placed on the same atom), that they are difficult to program (especially if high angular momentum functions are needed), that it is difficult to test or to demonstrate absolute convergence (many things can be changed – the number of functions, the exponents, the location of the function centres).

One final advantage of localised orbitals is that the Hamiltonian matrix becomes sparse as the system size increases, and this is one feature that is important for the development of linear scaling methodologies which may eventually replace todays conventional methods. This will not addressed further here, but is considered elsewhere in this volume [37].

3.3 Types of Gaussian basis set

Two types of parameters are required to specify a Gaussian basis for an atom – the coefficients of the functions and their exponents. The optimal coefficients for an atom are readily determined by solving Eq. (21) in the way discussed in the next section. However, the exponents α_i should also be varied to minimise the energy and this is very much more difficult and time consuming as the energy depends on them in a non-linear way. This is generally only done for atoms.

There are a number of ways of exporting the exponents and coefficients determined for atoms and using them for molecular calculations and as a result many types of basis exist, differing in details which vary in importance. As a result, this has proved to be one of the most fertile grounds for the development of acronyms. Instead of going through these one by one, a few general observations will be made showing the development of some of the more common basis sets. In this section few references to the original papers will be given as a full list would just be too long. More comprehensive information is given in reviews and reference works [38 to 40]. A convenient source of the exponents and coefficients for some common basis sets is the *Extensible Computational Chemistry Environment Basis Set Database*, developed and distributed by the Molecular Science Computing Facility, Environmental and Molecular Sciences Laboratory which is part of the Pacific Northwest Laboratory funded by the U.S. Department of Energy. This is best viewed from the WWW interface (links can be found from *www.emsl.pnl.gov*).

Firstly, it is clear that the demands placed on a basis set are very different, depending on whether an all-electron or pseudopotential calculation is being attempted. The all-electron calculation is significantly more demanding as tightly bound, oscillating core states must be described and, most seriously, these have cusps at the nucleus of each atom which are easily described by simple exponential functions $\exp[-\alpha|\mathbf{r} - \mathbf{R}|]$ but not so easily described by the Gaussian function $\exp[-\alpha|\mathbf{r} - \mathbf{R}|^2]$ which is analytic at $\mathbf{r} = \mathbf{R}$. The difficulty is greatest with the most tightly bound core states (which dominate the total energy of the system) and these need many Gaussian orbitals to describe them accurately, far more than would be necessary to describe the chemically important valence states.

A second shortcoming of Gaussian functions arises from the fact that the wavefunctions should decay exponentially at large distance from the nucleus, but Gaussian orbitals decay much faster. This is a problem in pseudopotential calculations as well as all-electron calculations (but not in bulk condensed matter simulation, where there is no vacuum). Because of these considerations, it is clear that a considerable number of Gaussian functions is required to form a reasonable basis, often more than it is practical to use in a large-scale calculation.

In this section, three issues will be addressed. Firstly, there is the optimisation of the exponents of the primitive Gaussian functions, secondly there is the contraction scheme to be used to attempt to make the basis set more compact and thirdly there is the incorporation of polarisation functions. The basis sets discussed are generally early ones, chosen mainly to illustrate the construction process, rather than as recommendations for use.

3.3.1 Exponent optimisation

The optimisation of the exponents of the Gaussian functions (often referred to as primitives) in an atomic calculation is an extremely demanding task. This is because the energy surface that must be searched has regions which are very flat and corrugated and there are many (quite different) sets of exponents that give very similar energies. If one really wishes to get the global minimum, then considerable care must be taken.

There have been numerous publications of these exponents (for very early and more recent examples see [41] and [42]. Furthermore, it is important that in the minimisation process, no two exponents should come close together as that will make the underlying evaluation of the energy unstable.

These problems can always be overcome with sufficient care and ingenuity, but the difficulties led to the development of the concept of even tempered basis sets in which the exponents were given in terms of a smaller number of variables, for example $\alpha_i = ab^{i-1}$ [43] in which a is the smallest exponent and b a multiplicative factor that gets us from one to the next. This prevents exponents crossing and it is much easier to find the minimum of a function of two variables. An extension of this idea [44] led to the concept of well-tempered functions which approached the quality of independently optimised sets. This approach is often used today if an intermediate fit has to be performed and evolve as part of a simulation with no human intervention.

The result of this optimisation is a set of exponents that is suitable for use as a basis in a molecular or solid-state environment. In a pseudopotential calculation this is indeed a good starting point. In an all electron calculation, many functions may be needed to treat the atom well so that this basis would make larger calculations too demanding. The solution to this is through contraction strategies.

3.3.2 Contraction

One of the earliest basis sets is labelled STO–3G [45]. In this Slater type orbitals which are good approximations to the atomic states are fitted to linear combinations of Gaussians. For example if we consider the state $\psi_{nlm}(\mathbf{r}) = r^l Y_{lm}(\theta, \phi) R_{nl}(r)$, then we write

$$R_{nl} = \sum_i c_i^{nl} \exp\left[-\alpha_i^{nl} r^2\right].$$

The simplest approach would then be to use as basis functions for a solid state or cluster simulation the functions

$$\phi_{nlm}(\mathbf{r}) = r^l Y_{lm}(\theta, \phi) \sum_i c_i^{nl} \exp\left[-\alpha_i^{nl} r^2\right]$$

or more precisely the Cartesian form of this in which we use $x^{n_1} y^{n_2} z^{n_3}$ as the prefactor instead of $r^l Y_{lm}(\theta, \phi)$. The important point here is that in the molecular calculation, only the coefficients of $\phi_{nlm}(\mathbf{r})$ are allowed to vary – the contraction coefficients remain fixed at their atomic values c_i^{nl}. This basis is termed a *minimal basis* as the atomic orbitals have no freedom to change shape during bonding – the molecular wavefunctions produced will only be appropriate superpositions of these. In the STO–3G basis, three Gaussians are used to represent each atomic state. In fact this is not really enough to represent the $r = 0$ cusps at all well, and further sets such as STO–6G attempt to remedy this. Also note that the exponents are the same for s and p states with the same principal quantum numbers (although the contraction coefficients are of course different).

The most serious problem remaining with STO–NG is that there is still no freedom for the valence states to change shape to describe bonding in molecules or solids. This is overcome at the next level of sophistication – "double-ζ" basis sets in which the valence orbitals are split into two components, the coefficients of which are allowed to vary in the molecule or solid (but the contraction coefficients determining the weights of primitive Gaussians that make up that component remain fixed as in the atom).

A typical acronym describing a double-ζ basis set would be 6–31G [46] in which six Gaussians are contracted together to describe the core states, and the valence states are described by four Gaussians divided into two groups, the three with the largest exponents being contracted together to form one function and the Gaussian with smallest exponent varying independently. Thus a silicon atom is described by 13 functions – the 1s, 2s and 2p states are each expanded in terms of 6 primitives and each of the 3s and 3p states being described by two functions, the first of which is made up of a combination of three Gaussians, the second being a single Gaussian). The contraction is described as (16s, 10p)$\to$ [4s, 3p] which shows the starting number of primitive Gaussians and the final number of contracted sets. The atomic energy is reproduced by these states to an accuracy of order 10^{-2} Rydbergs. The exponents used vary dramatically – for example, in silicon they range from 16115.9 to 0.0778369.

A further step forward takes us to triple-zeta bases such as 6-311G [47]. This basis has increased flexibility in the valence region relative to the 6-31G basis because it uses three functions to represent each valence atomic orbital.

The basis sets described so far have *segmented contractions*, or in other words a given primitive may appear in only a single contraction. Another approach is the method of *general contraction* [48]. In this a given primitive may contribute to every contracted function. The contraction process is thus simply a transformation with rank reduction. Flexibility (mainly in the valence shell) can be provided by adding to these states, the primitive Gaussians with the smallest few exponents as independent functions.

Other basis sets that are commonly used include Dunning's segmented contraction of the Huzinaga primitives into double and triple zeta bases; the Dunning-Hay split valence and double zeta sets; Huzinaga's MINI, MIDI and MAXI sets and sets with acronyms such as cc-pVDZ. (cc stands for "correlation consistent", a term associated with the Hartree Fock methodology that lies behind the generation of the basis).

3.3.3 Polarisation functions

One problem with the basis sets described to date is that they are optimised for atoms and not for solids or molecules. For example, the description of a hydrogen atom is not improved by adding p-type functions, but the addition of such functions to a hydrogen atom in a molecule or solid will make a big difference. Similarly, the addition of d-type functions will improve the description of atoms such as carbon or silicon in a molecular or solid state environment. Such functions are referred to as polarisation functions as they would allow polarisation (distortion) of atomic states in an external field (as could be provided indirectly by the neighbouring atoms in a molecule or solid). Suggested values for the exponents of these functions are contained as additions to 6-31G in basis sets such as 6-31G* (which adds d functions to all elements apart from H and He) and 6-31G** (which also adds p-type polarisation states to H and He atoms) or 6-31+G* which also adds additional diffuse s- and p-type functions. Similar extensions exist for triple-zeta basis sets, so for example we have 6-311G* and so on. Large modern basis sets have many polarisation functions often of very high angular momenta. For example, for carbon, the triple-zeta "correlation consistent" basis cc-pVTZ has the contraction (10s, 5p, 2d, 1f) $\to$ [4s, 3p, 2d, 1f] and the (extremely large) cc-pV6Z basis set has (16s, 10p, 5d, 4f, 3g, 2h, 1i) $\to$ [7s, 6p, 5d, 4f, 3g, 2h, 1i].

Another method of incorporating polarisation functions is to place basis functions at the centres of bonds between two atoms. This can be an effective way to increase the quality of a basis set without the need to incorporate higher momenta functions. The disadvantage of this method of improving a basis is that there has to be a subjective judgment of "what constitutes a bond" – this is not always clear, especially when considering diffusion mechanisms where bonds are being broken and reformed near the saddle point of the diffusion path.

The choice of polarisation exponent can be made either on the basis of atomic considerations or on the basis of a molecular or condensed matter simulation (for example, a new exponent could be introduced and chosen to minimise the energy of a specimen bulk material or solid). Suitable values are obtainable from the chemical literature.

3.3.4 Basis set superposition error

The basis set superposition error (BSSE) is important when calculating quantities such as dissociation energies. For example, if we consider the binding energy curve of a diatomic molecule, the problem arises as the basis set on one atom will improve the description of orbitals on the other atom. The most common method for estimating the size of this is the counterpoise method [49] in which the energy of each fragment of the dissociating molecule is calculated using the full molecular basis. This gives an *upper bound* for the error. The problem is most evident when the atomic bases used are not well balanced – a good treatment of the valence states has been combined with a poor treatment of the core. Apart from dissociation and similar situations where bonds are being broken or formed, most molecular properties do not appear to be too significantly affected by the BSSE.

3.3.5 Pseudopotential basis sets

Most of the comments made so far pertain to all electron calculations. Standard basis sets are not so readily available for pseudopotential calculations as they would differ depending on the construction of the pseudopotential. However, the situation is much easier in this case – sensible answers (although, not of course the best) for most elements can be obtained using any four or five approximately evenly tempered exponents ranging from 0.1 to 5. In AIMPRO, the smallest basis used for an element like C or Si consist of using four exponents on each atom giving around 16 functions. These are not usually contracted together, but the coefficients of each function are allowed to vary independently. The exponents are chosen by optimising either the energy of a pseudoatom or the bulk solid. These are augmented with functions placed at the centre of bonds – typically one or two exponents are used for these giving four to eight basis functions per bond. In tests on bulk diamond, it was found that a basis of 28 functions per atom plus bond centres gives a convergence similar to a 125 Rydberg cut-off in a plane wave calculation (1800 plane waves). Basis sets have been developed either by optimising the energy of isolated atoms or, in the case of carbon or silicon, by optimising the energy of a unit cell of 2 atoms.

4. Evaluation of the Total Energy and Hamiltonian

In this section, we will consider the evaluation of the total energy and the Hamiltonian matrix for a cluster of atoms within the framework of an expansion of the Kohn-Sham orbitals in terms of Gaussian orbitals.

As described in the introduction, the total energy is written as

$$E[n] = T_s[n] + \int V^{\text{ext}}(\mathbf{r}) \, n(\mathbf{r}) \, d\mathbf{r} + \frac{1}{2} \int \frac{n(\mathbf{r})n(\mathbf{r}')}{|\mathbf{r} - \mathbf{r}'|} \, d\mathbf{r} \, d\mathbf{r}' + E_{\text{xc}}[n_\uparrow, n_\downarrow] + \sum_{a<b} \frac{Z_a Z_b}{|\mathbf{R}_a - \mathbf{R}_b|}, \quad (26)$$

where the symbols have the meanings introduced above. The charge density is constructed in terms of localised orbitals using Eqs. (16) and (17). The final term describes the coulombic repulsion between either the nuclei (in an all electron calculation) or between the ion-cores (in a pseudopotential calculation).

We will now consider in some detail the evaluation of the terms in the total energy as given by Eq. (26) and the corresponding contribution to the Hamiltonian matrix as defined by Eq. (20). The final term is of course trivial to evaluate and indeed does not contribute to the Hamiltonian matrix.

It should be noted that the evaluation of the above expression is essentially a problem of numerical integration – quantities of the form $\int f(\mathbf{r}) \, n(\mathbf{r}) \, d\mathbf{r}$ need to be found where $f(\mathbf{r})$ could be the external potential, the Hartree potential or a complicated function of $n(\mathbf{r})$ in the exchange-correlation term. Two schools of thought have developed, one being that because of the complexity of the formula for exchange-correlation, numerical integration cannot be avoided if an exact calculation is to be done and therefore that this technique might as well be embraced fully in all parts of the calculation. A second school of thought is to do things analytically wherever possible. The latter approach is used in the AIMPRO code and most space will be given to discussing this. The philosophy of AIMPRO is to produce a formalism that can *routinely* treat systems of 200 atoms (in other words these should not be considered "large" and should be treatable in reasonable time with moderate resources). This has led to techniques being chosen which optimise speed so as to achieve this objective. For example, intermediate fits are used to evaluate the Hartree energy rather than methods based on four-centre integrals and an approximate treatment of exchange-correlation is preferred to the numerical integration approach that would yield the exact answer albeit more slowly.

This section will however commence by looking at a couple of strategies adopted for numerical integration before looking in more detail at the analytic methods.

4.1 Numerical integration

Performing integrals of the form $\int f(\mathbf{r}) \, n(\mathbf{r}) \, d\mathbf{r}$ is obviously most straightforward in pseudopotential calculations in which the charge density varies relatively slowly in all regions and permits use of a uniform or near uniform mesh. This is essentially the approach in the plane wave method for a periodic system (the Gaussian quadrature mesh for a periodic function consists of exactly equally spaced points). This contrasts with all electron treatments in which the rapid variation of the density near nuclei places much greater demands on the choice of grid.

A commonly used method for performing integrals of the above form over a cluster is due to Becke [50, 51] in which the density is split up into a set of (overlapping) atom centred distributions

$$n_a(\mathbf{r}) = P_a(\mathbf{r}) \, n(\mathbf{r}), \qquad (27)$$

where $P_a(\mathbf{r})$ is some projector function which is normalised $\sum_a P_a(\mathbf{r}) = 1$ and a labels an atom. $P_a(\mathbf{r})$ is chosen to approach unity when $\mathbf{r}$ is close to atom a and to decay smoothly to zero further away. Any desired integral can then be divided into atom centred terms using

$$\int f(\mathbf{r})\, d\mathbf{r} = \sum_a \int P_a(\mathbf{r}) f(\mathbf{r})\, d\mathbf{r}.$$

Now that we have reduced the problem to atom centred integrations, it is more straightforward to introduce suitable meshes for integration. Lebedev grids [52] are used for the angular part of the integral and some appropriate radial grid is used which takes into account the rapid variation near $r = 0$. This approach is attractive as only the one-dimensional mesh needs to take care of the rapid variation near nuclei.

A different approach to integration is proposed by Pederson and Jackson [53] in which a variational mesh is proposed. Space is divided into parallelepipeds where any parallelepiped which contains an atom is forced to be a cube. An atom-centred sphere is placed inside each of these cubes so that a numerical integral over the system being modelled may overlap three types of region – the interstitial parallelepipeds in which the charge density is slowly varying, the atom-centred spheres in which the density is approximately spherical and rapidly varying in the radial direction and what Pederson termed "excluded cubical volumes", the parts of the cubes which are outside the spheres. The integration method proceeds by making a continuous transformation on a one-dimensional Gaussian quadrature mesh according to the formula $x' = -\log x/\gamma$ where γ is an arbitrary parameter which is then varied to optimise the performance of the mesh for a wide variety of integrands. It was shown that the error decreases exponentially with the number of points employed. Force algorithms were implemented to refine the meshes. This approach has been proved to be one of the bedrocks upon which the NRLMOL code has developed. Many aspects of NRLMOL depend on robust and accurate numerical integration [54, 55]. The penalty that must be paid though is that the accuracy is bought at the price of speed.

4.2 Evaluation of the overlap matrix, S_{ij} and integral screening

We begin by considering the overlap matrix defined by Eq. (14) as this is the simplest term. If we initially restrict ourselves to s-type Gaussian functions, then the product of two Gaussian functions can be seen to be also a Gaussian function, centred on some point between the original two functions:

$$\begin{aligned}\phi_i(\mathbf{r})\, \phi_j(\mathbf{r}) &= \exp[-\alpha_i(\mathbf{r} - \mathbf{R}_i)^2 - \alpha_j(\mathbf{r} - \mathbf{R}_j)^2] \\ &= \exp\left[-\frac{\alpha_i \alpha_j}{\alpha_i + \alpha_j}(\mathbf{R}_i - \mathbf{R}_j)^2\right] \exp-\bar{\alpha}_{ij}(\mathbf{r} - \overline{\mathbf{R}}_{ij})^2],\end{aligned} \qquad (28)$$

where

$$\bar{\alpha}_{ij} = \alpha_i + \alpha_j \quad \text{and} \quad \overline{\mathbf{R}}_{ij} = \frac{\alpha_i \mathbf{R}_i + \alpha_j \mathbf{R}_j}{\alpha_i + \alpha_j}. \qquad (29)$$

Use of this important relation makes the integral trivial on the change of variable $\mathbf{r}' = \mathbf{r} - \overline{\mathbf{R}}_{ij}$ with the result

$$S_{ij}^{ss} = \left(\frac{\pi}{\alpha_i + \alpha_j}\right)^{3/2} \exp\left[-\frac{\alpha_i \alpha_j}{\alpha_i + \alpha_j}(\mathbf{R}_i - \mathbf{R}_j)^2\right].$$

The superscripts indicate that both functions are of s-type symmetry.

Now we will consider the result if the functions have higher momenta. The the function $\phi_i(\mathbf{r})$ is of p-symmetry, we may use the result

$$(\mathbf{r} - \mathbf{R}_i)_\mu \exp[-\alpha_i(\mathbf{r} - \mathbf{R}_i)^2] = \frac{1}{2\alpha_i} \frac{\partial}{\partial \mathbf{R}_{i\mu}} \exp[-\alpha_i(\mathbf{r} - \mathbf{R}_i)^2], \tag{30}$$

where μ represents x, y or z. Differentiating the above expression gives

$$S_{ij}^{ps} = [\bar{\mathbf{R}}_{ij} - \mathbf{R}_i]_\mu S_{ij}^{ss},$$
$$S_{ij}^{sp} = [\bar{\mathbf{R}}_{ij} - \mathbf{R}_j]_\mu S_{ij}^{ss}.$$

It is possible to find more of these by continuing the differentiation process using a generalisation of Eq. (30),

$$\frac{\partial}{\partial R_{ix}} |i; n_x, n_y, n_z\rangle = 2\alpha_i |i; n_x + 1, n_y, n_z\rangle + n_x |i; n_x - 1, n_y, n_z\rangle, \tag{31}$$

where we have used the notation in Eq. (25). This however rapidly becomes cumbersome and results for higher angular momentum basis functions are best evaluated by devising a recurrence relation that enables a difficult term such as S_{ij}^{dp} to be written town in terms of simpler results such as S_{ij}^{pp}, S_{ij}^{ds} and S_{ij}^{sp}. Such recurrence relations are easily found for simple integrals like S_{ij}, and a collection is given by Obara and Saika [56]. Using these formulae, the overlap matrix is easily built up.

Clearly, if $S_{ij} \ll 1$, then the product of the two functions is negligible and any other more complex term such as $\int \phi_i(\mathbf{r}) V(\mathbf{r}) \phi_j(\mathbf{r}) \, d\mathbf{r}$ is bounded by $S_{ij} V_{\max}$, where $V_{\max}$ is the maximum value attained by $|V(\mathbf{r})|$. This is central to the writing of efficient code and enables many (usually the vast majority) of time-consuming integrals to be set to zero without being evaluated. This approach is often referred to in the quantum chemistry literature as *integral screening*.

Indeed, for a given function ϕ_i there will only be a fixed number of functions ϕ_j such that S_{ij} is non-negligible, as the centre $\mathbf{R}_j$ must lie within a fixed distance of $\mathbf{R}_i$. This number will not depend on the number of atoms being modelled, so we can conclude that there will be of order N non-zero elements in the overlap matrix and in the Hamiltonian, where N is the size of these matrices. The coefficient of N can however be very large though as the smallest exponent α_i will be of order 0.1 a.u. This means, that all functions within a distance of around 15 Å will need to be considered before they can be neglected and a few hundred atoms may lie within this volume.

4.3 Evaluation of the kinetic energy term, $T_s[n]$

The kinetic energy term is given by

$$T_s[n] = -\tfrac{1}{2} \sum_s \int \psi_{\lambda s}(\mathbf{r}) \nabla^2 \psi_{\lambda s}(\mathbf{r}) \, d\mathbf{r} = \sum_{ij} b_{ij} T_{ij},$$

where the matrix elements of the kinetic energy operator in the localised basis are

$$T_{ij} = -\tfrac{1}{2} \int \phi_i(\mathbf{r}) \nabla^2 \phi_j(\mathbf{r}) \, d\mathbf{r}. \tag{32}$$

The integral is readily evaluated if the functions are of s-type symmetry and again recurrence relations are given by Obara and Saika [56] for higher angular momenta basis functions.

4.4 Matrix elements of the external potential

This contribution to the total energy may be written as

$$V[n] = \int n(\mathbf{r}) V^{\text{ext}}(\mathbf{r}) \, d\mathbf{r} = \sum_{ij} b_{ij} V_{ij}, \qquad (33)$$

where the matrix elements V_{ij} are given by

$$V_{ij} = \int \phi_i(\mathbf{r}) \phi_j(\mathbf{r}) V^{\text{ext}}(\mathbf{r}) \, d\mathbf{r}.$$

The evaluation of this depends on the details of the external potential. If an all-electron treatment is being attempted, then V^{ext} consists of the Coulombic attractions of the nuclei present:

$$V^{\text{ext}}(\mathbf{r}) = -\sum_a \frac{Z_a}{|\mathbf{r} - \mathbf{R}_a|}.$$

The integration may be done analytically and the result is a standard one [56].

If a pseudopotential is being used, then the integral must be done either numerically (if the pseudopotential is specified in this way), or if the pseudopotential is parameterised in terms of analytic functions, the integral may be done analytically. In either case, we write

$$V^{\text{ext}}(\mathbf{r}) = \sum_a \hat{V}^{\text{ps}}(\mathbf{r} - \mathbf{R}_a),$$

where the sum index a runs over atoms in the system and $\mathbf{R}_a$ is the position of atom a. We will only consider pseudopotentials of the general form given in Eq. (12).

In the remainder of this section it is assumed that we have shifted the origin of coordinates to put the atom whose pseudopotential is being considered at the origin. $\mathbf{R}_i$ therefore represents $\mathbf{R}_i - \mathbf{R}_a$ in this section and similarly $\mathbf{R}_j$ represents $\mathbf{R}_j - \mathbf{R}_a$. The potential will therefore be referred to as $V(r)$ rather than $V(\mathbf{r} - \mathbf{R}_a)$.

4.4.1 Matrix elements of numerical pseudopotentials

If the pseudopotential is given in numerical form, then the matrix elements must be found at least in part numerically. The local term can be treated by splitting off the asymptotic dependence, for example using

$$V^{\text{loc}}(r) = \frac{Z_{\text{ion}} \text{erf}(\alpha r)}{r} + \tilde{V}^{\text{loc}}(r),$$

where α is chosen to ensure that $\text{erf}(\alpha R_c) \approx 1$, where R_c is the cut-off radius. This ensures that the remainder term $\tilde{V}^{\text{loc}}$ is now short ranged, being small for $r > R_c$ and decreasing approximately exponentially after this.

The matrix elements of the analytic part of this potential can be evaluated by recognising that the error function arises as the potential of a Gaussian distribution of charge:

$$\frac{\text{erf}(\alpha r)}{r} = \left(\frac{\alpha}{\pi}\right)^{3/2} \int \frac{\exp[-\alpha r'^2] \, d\mathbf{r}'}{|\mathbf{r} - \mathbf{r}'|}. \qquad (34)$$

The matrix elements of this can then be evaluated as a three-centre integral

$$V_{ij}^{\text{loc}-1} = -Z\left(\frac{\alpha}{\pi}\right)^{3/2} \int \frac{\phi_i(\mathbf{r}) \phi_j(\mathbf{r}) \exp[-\alpha r'^2]}{|\mathbf{r} - \mathbf{r}'|} \, d\mathbf{r} \, d\mathbf{r}' \qquad (35)$$

with this again being a standard result. An expression for this can be found in Ref. [56] by adapting the result for a four-centre integral presented there.

We are then left with the numerical integration:

$$\begin{aligned}
V_{ij}^{\text{loc}-2} &= \int \tilde{V}^{\text{loc}}(r)\,\phi_i(\mathbf{r})\,\phi_j(\mathbf{r})\,\mathrm{d}\mathbf{r} \\
&= \exp\left[-\frac{\alpha_i\alpha_j}{\alpha_i+\alpha_j}(\mathbf{R}_i-\mathbf{R}_j)^2\right] \int \tilde{V}^{\text{loc}}(r)\exp\left[-(\alpha_i+\alpha_j)(\mathbf{r}-\overline{\mathbf{R}})^2\right]\mathrm{d}\mathbf{r} \\
&= \frac{\pi}{(\alpha_i+\alpha_j)R}\exp\left[-\frac{\alpha_i\alpha_j}{\alpha_i+\alpha_j}(\mathbf{R}_i-\mathbf{R}_j)^2\right] \\
&\quad \times \int r\tilde{V}^{\text{loc}}(r)\left[\mathrm{e}^{-\alpha_{ij}(r-R)^2}-\mathrm{e}^{-\alpha_{ij}(r+R)^2}\right]\mathrm{d}r,
\end{aligned}$$

where $\alpha_{ij} = \alpha_i + \alpha_j$, $R = |\mathbf{R}|$ and we have used the results of Section 4.2 to take the second step and the third step results after evaluating the angular integrals. The remaining integrand is short ranged and the lower limit can be reduced to just above the cut-off radius, R_c, used to construct the pseudopotential. This integral is evaluated numerically. It can be clearly seen that many of these integrals will be zero. If S_{ij} is zero the first exponential factor kills the matrix element, and if $\alpha_{ij}(R-R_c)^2 > 30$ say, then the integral is vanishingly small. The result for higher angular momenta functions is obtained by differentiation and again a simple recurrence relation is set up. This is most easily obtained by introducing the modified spherical Bessel function $\mathrm{i}_0(x) = \sinh x/x$ and rewriting the integral as

$$V_{ij}^{\text{loc}-2} = 4\pi \exp[-\alpha_i R_i^2 - \alpha_j R_j^2] \int r^2 \tilde{V}^{\text{loc}}(r)\,\mathrm{i}_0[2\alpha_{ij}rR]\exp\left[-\alpha_{ij}r^2\right]\mathrm{d}r$$

and exploiting the recurrence relationship for the derivative of i_0.

The evaluation of the matrix elements of the non-local part of Eq. (12) is less often discussed, a consequence of the prevalence of all-electron calculations in quantum chemistry. We need to evaluate the integral

$$V_{ij}^{\text{nl}} = \sum_{lm} \int r^2 F_i^{lm}(r)\,F_j^{lm*}(r)\,V_l^{\text{nl}}(r)\,\mathrm{d}r, \tag{36}$$

where we have defined the quantity $F_i^{lm}(r) = \int Y_{lm}(\theta,\varphi)\,\phi_i(\mathbf{r})\sin\theta\,\mathrm{d}\theta\,\mathrm{d}\varphi$. This is readily found analytically by first considering the case where ϕ_i is an s-type orbital. We expand ϕ_i using the Bauer series expansion:

$$\begin{aligned}
\mathrm{e}^{-\alpha(\mathbf{r}-\mathbf{R})^2} &= \mathrm{e}^{-\alpha(r^2+R^2)}\,\mathrm{e}^{2\alpha\mathbf{r}\cdot\mathbf{R}} \\
&= 4\pi\,\mathrm{e}^{-\alpha(r^2+R^2)}\sum_{l,m}\mathrm{i}_l(2\alpha_i R_i r)\,Y_{lm}^*(\hat{\mathbf{r}})\,Y_{lm}(\hat{\mathbf{R}}_i),
\end{aligned}$$

where $R = |\mathbf{R}|$, $\mathrm{i}_l(z)$ is the modified spherical Bessel function and it is understood that the arguments of the spherical harmonics are unit vectors. On inserting this into our expression for F_l^{lm} the angular integrals are trivially done and we obtain

$$F_i^{lm}(r) = 4\pi \mathrm{i}_l(2\alpha_i r R_i)\,Y_{lm}(\hat{\mathbf{R}}_i)\,\mathrm{e}^{-\alpha(r^2+R^2)}.$$

We can now easily work out analytic expressions for $F_i^{lm}(r)$ if $\phi_i(\mathbf{r})$ is a p or d-type orbital by differentiation using Eq. (31) together with the well known recurrence relations for derivatives of i_l and Y_{lm}. Finally, we are left with the one-dimensional integral (36), where all quantities are now known. This is well behaved as we can write

$$i_l(2\alpha_i r R_i)\, e^{-\alpha(r^2+R_i^2)} = f_l(2\alpha_i r R_i)\, e^{-\alpha(r-R_i)^2},$$

where $f_l(x)$ is a well-behaved function which for small x varies as x^l and for large x decays as $1/x$. The most straightforward approach is that of Gaussian quadrature. As the non-local potential is zero outside a core radius, R_c, the upper limit can be reduced to give an integral over a finite range. This step takes only a very small percentage of the overall time of the calculation.

4.4.2 Matrix elements of analytic pseudopotentials

In this section we will briefly consider the case of the Bachelet, Hamann and Schlüter potentials [27] which are fitted to analytic functions of the form

$$V^{\text{loc}}(r) = -\frac{Z}{r} \sum_{p=1}^{2} c_p \operatorname{erf}(\beta_p r), \quad \text{where} \quad c_1 + c_2 = 1, \tag{37}$$

$$\hat{V}_l^{\text{nl}}(r) = \sum_{p=1}^{3} (A_p^l + A_{p+3}^l r^2) \exp[-\beta_p^l r^2]. \tag{38}$$

The matrix elements of the local part of this potential can be evaluated by recognising that the error function arises as the potential of a Gaussian distribution of charge (see Eq. (34)) and transforming the integral to a three-centre integral

$$V_{ij}^{\text{loc}} = -Z_{\text{ion}} \sum_{p=1}^{2} c_p \left(\frac{\beta_p}{\pi}\right)^{3/2} \int \frac{\phi_i(\mathbf{r})\, \phi_j(\mathbf{r})\, \exp[-\beta_p r'^2]}{|\mathbf{r}-\mathbf{r}'|}\, d\mathbf{r}\, d\mathbf{r}' \tag{39}$$

with this again being a standard result. An expression for this can be found in Ref. [56] by adapting the result for a four centre integral presented there, letting the exponent of one function tend to zero.

The evaluation of the matrix elements of the non-local part is probably best done in the numerical way described above, but if wished, an analytic result can be obtained starting by substituting our expression for F_i^{lm} into Eq. (36):

$$V_{ij}^{\text{nl}} = 4\pi \sum_l (2l+1)\, e^{-(\alpha_i R_i^2 + \alpha_j R_j^2)} P_l(\mu) \sum_{p=1}^{3} \int_0^\infty r^2 (A_p^l + A_{p+3}^l r^2)$$

$$\times e^{-(\alpha_i+\alpha_j+\beta_p^l)r^2} i_l(2\alpha_i R_i r)\, i_l(2\alpha_j R_j r)\, dr$$

$$= 4\pi \sum_l (2l+1)\, e^{-(\alpha_i R_i^2 + \alpha_j R_j^2)} P_l(\mu) \sum_{p=1}^{3} \left(A_p^l - A_{p+3}^l \frac{\partial}{\partial \beta_p^l}\right)$$

$$\times \int_0^\infty r^2 e^{-\alpha r^2} i_l(2\alpha_i R_i r)\, i_l(2\alpha_j R_j r)\, dr,$$

where $\mu = \mathbf{R}_i \cdot \mathbf{R}_j / (R_i R_j)$ and $\alpha = \alpha_i + \alpha_j + \beta_p^l$.

Somewhat surprisingly, the integral can be done analytically and the derivative evaluated using the recurrence relation $i'_l(z) = i_{l-1}(z) - (l+1)\,i_l(z)/z$ giving a final result:

$$V_{ij}^{nl} = (2l+1)\left(\frac{\pi}{\alpha}\right)^{3/2} \exp\left[\frac{\alpha_i^2 R_i^2 + \alpha_j^2 R_j^2}{\alpha} - \alpha_i R_i^2 - \alpha_j R_j^2\right] P_l(\mu)$$

$$\times \sum_p \left\{\left[A_p^l + A_{p+3}^l\left(\frac{1-2l}{2\alpha} + \frac{\alpha_i^2 R_i^2 + \alpha_j^2 R_j^2}{\alpha^2}\right)\right] i_l(z) + \frac{zA_{p+3}}{\alpha}\,i_{l-1}(z)\right\},$$

where $z = 2\alpha_i\alpha_j R_i R_j / \alpha$.

The expression for the matrix elements of this potential in p- or d-type basis functions can be found by differentiation or by establishing recurrence relations. In the AIMPRO code, the latter approach has been adopted which enables arbitrarily high angular momenta functions to be used.

4.5 Evaluation of the Hartree energy and potential

The Hartree energy presents special problems with the adoption of a localised basis set. In terms of our basis set expansion, the Hartree energy and matrix elements of the Hartree potential are given by

$$E^H = \frac{1}{2}\sum_{ijkl} b_{ij}b_{kl}\int \frac{\phi_i(\mathbf{r})\,\phi_j(\mathbf{r})\,\phi_k(\mathbf{r'})\,\phi_l(\mathbf{r'})}{|\mathbf{r}-\mathbf{r'}|}\,d\mathbf{r}\,d\mathbf{r'}, \tag{40}$$

$$V_{ij}^H = \sum_{kl} b_{kl}\int \frac{\phi_i(\mathbf{r})\,\phi_j(\mathbf{r})\,\phi_k(\mathbf{r'})\,\phi_l(\mathbf{r'})}{|\mathbf{r}-\mathbf{r'}|}\,d\mathbf{r}\,d\mathbf{r'}. \tag{41}$$

The evaluation of the four centre integral can be done analytically, although the process is expensive in terms of computer time. Numerous strategies exist to facilitate the evaluation of integrals involving p, d and functions of higher angular momenta. An enormous literature exists [56 to 62] testifying to the efforts of workers to evaluate these quantities. A review of work up to 1994 is given by Gill [63].

In this paper, we will look at schemes that have evolved to avoid the calculation of so many integrals.

4.5.1 Hierarchical multipole methods

Clearly, if $S_{ij} \ll 1$ or $S_{kl} \ll 1$, then the four-centre integrals I_{ijkl} and I_{klij} can be neglected for all k and l. As we have found that there are only $O(N)$ nonzero elements of the overlap matrix there will only be $O(N^2)$ four-centre integrals that are non-negligible. The coefficient can be very large, of order 10^4 to 10^6. This is still a considerable challenge and large numbers of atoms would need to be considered before this begins to be dominated by the $O(N^3)$ work involved in diagonalisation. Until recently this was the bottleneck that limited the size of systems treatable by standard quantum chemistry packages.

Calculation of the four centre integrals is obviously a non-optimal strategy as $O(N^2)$ of them are non-zero whereas after contraction with b_{ij}, only $O(N)$ non-negligible quantities remain. An obvious aim would be to attempt to find the V_{ij}^H matrix directly without needing to find the intermediate quantities. One method of doing this is the so-called *J matrix engine* [64] in which the matrix elements of the Hartree potential (called the *J* matrix in quantum chemical literature) are found directly by summing the density matrix into the underlying Gaussian integration formulas.

More ambitious approaches have appeared, based on hierarchical methods such as tree codes and the fast multipole method. The continuous fast multipole method [65] is an extension of fast summation methods used for point particles. The important idea is to split the system into a hierarchy of cells. The density distribution in each cell is expressed as a multipole expansion and the far-distance limit of this can be used to compute the interaction between well separated components. Various applications of these and related ideas have been proposed [65 to 68] and eventually for large enough systems these methods should approach linear scaling.

4.5.2 Solutions of Poisson's equation

A completely different philosophy is to solve Poisson's equation to find the potential and then either multiply by the charge density and integrate numerically to find the Hartree energy or multiply by pairs of Gaussians and integrate to evaluate the matrix elements.

One implementation of this [69] first uses the Becke projector method to divide the charge density into atom-centred pieces $n_a(\mathbf{r})$ as described above (see Eq. (27)). These density components are then decomposed into different angular momentum components,

$$n_a(\mathbf{r}) = \sum_{lm} n_{lma}(r) Y_{lm}(\Omega_a).$$

The Hartree potential is then determined from a solution of the one-dimensional Poisson equation,

$$\left[\frac{d^2}{dr^2} - \frac{l(l+1)}{r^2}\right] U_{lma}(r) = -4\pi r n_{lma}(r)$$

and then recombined with the angular dependence to give the potential in the molecule as

$$V(\mathbf{r}) = \sum_a \sum_{lm} \frac{U_{lma}(|\mathbf{r} - \mathbf{R}_a|) Y_{lm}(\Omega_a)}{|\mathbf{r} - \mathbf{R}_a|}.$$

One important issue with this approach is the cut-off imposed on the sum over *l*. Chen et al. stopped at $l = 3$ and proposed a method for dealing with the residual charge.

4.5.3 Use of an intermediate fit

The important idea here is to introduce a second or *intermediate* fit to the charge density,

$$\tilde{n}(\mathbf{r}) = \sum_k c_k g_k(\mathbf{r}),$$

where the functions g_k are once again commonly chosen to be Gaussians. The important consideration is now how to choose the coefficients c_k. The most obvious choice is to use a least squares fitting procedure, in which a quantity such as

$$\int [n(\mathbf{r}) - \tilde{n}(\mathbf{r})]^2 \, d\mathbf{r}$$

is minimised. This however does not give the most accurate result for the total energy. A better approach which shows a much faster and more uniform convergence is given by minimising the error in the Hartree energy. The exact Hartree energy is

$$E_H[n] = \frac{1}{2} \int \frac{n(\mathbf{r}_1)\,n(\mathbf{r}_2)}{|\mathbf{r}_1 - \mathbf{r}_2|}\,d\mathbf{r}_1\,d\mathbf{r}_2,$$

and is written as

$$E_H = \tilde{E}_H + \Delta E_H,$$

where $\tilde{E}_H$ is an approximate value, and ΔE_H is the error in this approximation. Our choice follows the work of Dunlap et al. [70] and Jones and Sayyash [71]. In this approach, the approximation is

$$\tilde{E}_H = \int \frac{n(\mathbf{r}_1)\,\tilde{n}(\mathbf{r}_2)}{|\mathbf{r}_1 - \mathbf{r}_2|}\,d\mathbf{r}_1\,d\mathbf{r}_2 - \frac{1}{2} \int \frac{\tilde{n}(\mathbf{r}_1)\,\tilde{n}(\mathbf{r}_2)}{|\mathbf{r}_1 - \mathbf{r}_2|}\,d\mathbf{r}_1\,d\mathbf{r}_2$$

and the error is

$$\Delta E_H = \frac{1}{2} \int \frac{[n(\mathbf{r}_1) - \tilde{n}(\mathbf{r}_1)][n(\mathbf{r}_2) - \tilde{n}(\mathbf{r}_2)]}{|\mathbf{r}_1 - \mathbf{r}_2|}\,d\mathbf{r}_1\,d\mathbf{r}_2. \tag{42}$$

Several important points should be noted here. First it is clear that the error is positive (it is only zero if $\tilde{n} \equiv n$). This means that our approximation is always *smaller* than the true value, so that as we improve the quality of the fit, the Hartree energy systematically *increases* towards the correct value. A second point is that the error ΔE is second order with respect to $n - \tilde{n}$. This means that quite good energies can be obtained with even quite poor $\tilde{n}$. Finally, it should be noted that as the Hartree energy is positive, underestimating it will result in the total energy being lower (i.e. under normal circumstances, more negative) than it should be, so that in this approximation an energy lower than the true, exact energy is obtained. This is not a problem as by systematically improving the quality of the intermediate fit, the effect on the energy can be closely monitored.

Differentiating Eq. (42) with respect to c_k to determine the minimum gives

$$G_{kl}c_l = t^H_{ijk}b_{ij}. \tag{43}$$

Here,

$$t^H_{ijk} = \int \frac{\phi_i(\mathbf{r}_1)\,\phi_j(\mathbf{r}_1)\,g_k(\mathbf{r}_2)}{|\mathbf{r}_1 - \mathbf{r}_2|}\,d\mathbf{r}_1\,d\mathbf{r}_2,$$

$$G_{kl} = \int \frac{g_k(\mathbf{r}_1)\,g_l(\mathbf{r}_2)}{|\mathbf{r}_1 - \mathbf{r}_2|}\,d\mathbf{r}_1\,d\mathbf{r}_2.$$

The Eqs. (43) are readily solved, as this is a standard linear equation problem. In addition the matrix G_{kl} is symmetric and positive definite making the solution process more stable. However, this step is potentially a weak link in the overall formalism as the matrix G becomes ill-conditioned as more intermediate fitting functions are introduced with the result that the solution vector c_k will contain noise. This has not generally been found to be a problem, over many years, but a very small number of cases have been found when care has had to be taken.

LDA Calculations Using a Basis of Gaussian Orbitals

In terms of these symbols, our approximation $\tilde{E}^H$ to the Hartree energy is given by

$$E^H = c_k b_{ij} t^H_{ijk} - \tfrac{1}{2} c_k c_l G_{kl} = \tfrac{1}{2} c_k c_l G_{kl}, \tag{44}$$

where the last simplification is obtained using Eq. (43).

We now consider the choice of the fitting functions $g_k(\mathbf{r})$. The most obvious choice consists of Gaussian functions $e^{-b_k(\mathbf{r}-\mathbf{R}_k)^2}$ defined by a site $\mathbf{R}_k$ and an exponent b_k. The sites need not correspond to the location of atoms but can include, for example, bond centres. All the integrals can be computed analytically which leads to considerable time saving. However, for clusters of less than about 100 atoms, it is the evaluation of the t^H_{ijk} which is often the most time consuming procedure. The number of fitting functions g_k is usually roughly proportional to the number of basis functions ϕ_i, i.e. N, and hence there are $O(N^3)$ integrals of the type t^H_{ijk} of which $O(N^2)$ (with a large coefficient) are not negligible. These cannot be stored in main memory and must either be evaluated once and stored on disk or, for todays very fast processors, be repeatedly evaluated during each of the self-consistent cycles. There is then an advantage in choosing a set of g_k which leads to a simple analytical form for t^H_{ijk} and which minimises the number that are non-negligible.

This can be done when g_k is divided into two sets. The first set has g_k defined by

$$g_k = \left[1 - \frac{2b_k}{3}(\mathbf{r}-\mathbf{R}_k)^2\right] e^{-b_k(\mathbf{r}-\mathbf{R}_k)^2}. \tag{45}$$

These functions give a potential of Gaussian form,

$$\int \frac{g_k(\mathbf{r}_1)}{|\mathbf{r}-\mathbf{r}_1|} d\mathbf{r}_1 = \frac{3b_k}{2\pi} e^{-b_k(\mathbf{r}-\mathbf{R}_k)^2}, \tag{46}$$

and thus integrals in t^H_{ijk} involve a product of three Gaussian functions and can be evaluated very quickly. Also, many of these are now zero, as all three Gaussians must overlap to give a nonzero value (indeed only $O(N)$ are now non-negligible). However, in order to get the short-ranged Gaussian potential and to avoid the long-ranged Coulomb potential, the integral of g_k must vanish. It is readily verified that this indeed happens for the functions in Eq. (45). It is therefore necessary to add to this set additional functions which do carry a net charge. We have done this using functions which are purely Gaussian and whose integrals do contribute to the total number of electrons. It is however sometimes possible to choose the coefficients of these as fixed quantities related to the anticipated total charge on the atom or ion.

For example, we could fit the charge on an atom by

$$\tilde{n}(\mathbf{r}) = A e^{-\alpha(\mathbf{r}-\mathbf{R}_k)^2} + \sum_k c_k \left[1 - \frac{2b_k}{3}(\mathbf{r}-\mathbf{R}_k)^2\right] e^{-b_k(\mathbf{r}-\mathbf{R}_k)^2}.$$

The coefficients A and c_k must then be varied in the approach towards self-consistent. The exponent α is a fixed constant, chosen to make the first function have a range similar to the size of the atom in question. The first term carries all the charge, and the second term sum over k redistributes this radially to the required shape.

A more ambitious procedure in terms of the speed of the formalism would be to replace the first term by

$$Z \left(\frac{\alpha}{\pi}\right)^{3/2} e^{-\alpha(\mathbf{r}-\mathbf{R}_k)^2},$$

where Z is the ionic charge or nuclear charge of the atom in question, depending on whether we are carrying out a pseudopotential or all-electron calculation. The coefficient of the first term is now *fixed*, so that the potential arising from it does not change during the approach to self-consistency. It can therefore be calculated at the same time as the pseudopotential so that only the (much simpler) three-centre overlap integrals need be recalculated as the approach towards self-consistency proceeds. A second advantage of this approach is that the sum of the local part of the pseudopotential (long-ranged with a $-Z/r$ tail) and the potential from this fixed fitting function (also long ranged, this time with a tail $+Z/r$) is *short-ranged* (in fact decaying as a Gaussian) making this term much faster to evaluate. This of course makes the task harder for the functions g_k in Eq. (45), especially if the material modelled is ionic or partly ionic. It should also be noted that this approach doesn't work if the cluster being modelled is charged.

A second even more ambitious possibility replaces the first term by a pair of Gaussians where the two values of α are the same as for example the exponents in the local part of the pseudopotential. In this case, the two terms cancel exactly and no three-centre Coulombic interaction integrals (i.e. with denominator $|\mathbf{r} - \mathbf{r}'|$) need be evaluated at all. Needless to say, this method produces the fastest code, but we have not followed this approach as the exponents in the local part of the pseudopotential are much larger than is appropriate for a description of the spatial extent of the valence charge density of an atom with the result that a much more challenging radial redistribution of charge is left for the second term to achieve.

These strategies illustrate one attractive feature of the intermediate fitting approach. Not only is it significantly faster than the direct method, but accuracy can be traded for speed. The method can be made extremely accurate if large numbers of functions are used including fitting functions of high angular momentum. However, an order of magnitude gain in speed can be obtained by simplifying the fitting basis *and the calculation retains its internal consistency* so that the results remain sensible. This is not true with other methods which can fail completely if the accuracy is not sufficient.

We finish this section by giving the expression for the matrix elements of the Hartree potential. These are given by

$$V_{ij}^{\mathrm{H}} = \frac{\partial \tilde{E}}{\partial b_{ij}}$$

$$= \frac{\partial \tilde{E}}{\partial c_k} \frac{\partial c_k}{\partial b_{ij}}$$

$$= [G_{kl} c_l] [(G)^{-1}_{km} t_{ijm}^{\mathrm{H}}]$$

$$= c_k t_{ijk}^{\mathrm{H}}$$

completing our discussion of this term in the Hamiltonian.

4.6 Evaluation of the exchange correlation energy and potential

The evaluation of the exchange-correlation energy also presents a challenge for localised orbital methodologies, as it is not possible to evaluate this term in a straightforward analytic way. The obvious approach is to do the integral numerically, and the apparatus necessary was described earlier in Section 4.1. Here, an extremely fast approximate analytic treatment is presented.

4.6.1 Use of intermediate fit

In this approach, the exact expression for this energy,

$$E_{xc} = \int n(\mathbf{r}) \, \epsilon_{xc}(n_\uparrow, n_\downarrow) \, d\mathbf{r},$$

is replaced by an approximate one $\tilde{E}_{xc}$ involving approximate spin densities, $\tilde{n}_s$

$$\tilde{E}_{xc} = \int \tilde{n} \epsilon_{xc}(\tilde{n}_\uparrow, \tilde{n}_\downarrow) \, d\mathbf{r}. \tag{47}$$

Clearly the error we make is negligible if $\tilde{n}_s$ is close to n_s. The first step then is to fit n_s to a set of functions. It is possible to choose the same $g_k(\mathbf{r})$ as used in the construction of $\tilde{E}_H$. However, the least squares procedure used there minimizes the electrostatic energy associated with the error in the charge density, i.e. $n - \tilde{n}$, and it does not mean that at each value of $\mathbf{r}$, $n(\mathbf{r})$ and $\tilde{n}(\mathbf{r})$ are as close as possible. Moreover, the choice of g_k was selected to reflect the difficulty of working out the integrals t_{ijk}^H. Hence, in dealing with $\tilde{E}_{xc}$ it is better to use a sum of simple Gaussian functions h_k so that

$$\tilde{n}_s(\mathbf{r}) = \sum_k d_{k,s} h_k(\mathbf{r}), \tag{48}$$

where $d_{k,s}$ is found from minimizing

$$\int [n_s(\mathbf{r}) - \tilde{n}_s(\mathbf{r})]^2 \, d\mathbf{r}.$$

Differentiating this with respect to the coefficients $d_{k,s}$ leads to the equations

$$H_{kl} d_{l,s} = t_{ijk}^E b_{ij}^s, \tag{49}$$

where we have introduced

$$H_{kl} = \int h_k(\mathbf{r}) h_l(\mathbf{r}) \, d\mathbf{r}, \tag{50}$$

$$t_{ijk}^E = \int \phi_i(\mathbf{r}) \phi_j(\mathbf{r}) h_k(\mathbf{r}) \, d\mathbf{r}. \tag{51}$$

We note that the integrals are the same for each spin-index s and that in view of Eq. (46) t_{ijk}^E are simply proportional to t_{ijk}^H if g_k is chosen as in Eq. (45) above:

$$t_{ijk}^E = \frac{2\pi}{3b_k} t_{ijk}^H,$$

where the repeated index k is not summed over. This saves a considerable amount of computer time.

4.6.2 Non-polarised case – approximate analytic treatment

We consider first the non-polarized or spin-averaged case where we can dispense with the spin label and write

$$\tilde{E}_{xc} = \sum_k d_k \int h_k(\mathbf{r}) \epsilon_{xc}(\tilde{n}) \, d\mathbf{r}. \tag{52}$$

If it were desired to perform a fast numerical treatment then this integral represents a good starting point. The charge density $\tilde{n}$ is rapidly found and in a large system, the convergence of the integral is controlled in a very clear way by the Gaussian decay of $h_k(\mathbf{r})$. Lebedev integration can be used to perform the angular integral and a form of

Gaussian quadrature to do the radial integral. We have used this mainly in small molecule runs and program tests.

If h_k is chosen to be a positive definite localized function such as a Gaussian, then each integral is proportional to the average value of the exchange-correlation density under h_k,

$$\langle \epsilon_{xc}(\tilde{n}) \rangle_k .$$

We next note that $\epsilon_{xc}(n)$ varies slowly with n and hence we expect

$$\langle \epsilon_{xc}(\tilde{n}) \rangle_k \approx \epsilon_{xc}(\langle \tilde{n} \rangle_k) .$$

This approximation is tantamount to replacing the exact exchange-correlation density at **r** by its homogeneous electron gas value for the average density

$$m_{1k} = \langle \tilde{n} \rangle_k = \frac{\int h_k(\mathbf{r}) \, \tilde{n}(\mathbf{r}) \, d\mathbf{r}}{\int h_k(\mathbf{r}) \, d\mathbf{r}} = \frac{H_{kl} d_l}{I_k} , \qquad (53)$$

where I_k is simply the normalisation integral of h_k. This is not accurate enough for a real calculation but forms a useful starting approximation. We can improve on this approximation as follows. According to Taylor's theorem

$$\epsilon_{xc}(n) = \epsilon_{xc}(m_{1k}) + (n - m_{1k}) \, \epsilon'_{xc}(m_{1k}) + \frac{(n - m_{1k})^2}{2!} \, \epsilon''_{xc}(\bar{n}) ,$$

where $\bar{n}$ is an unknown density between n and m_{1k} which varies with n. On substituting this into Eq. (52), the second term vanishes on account of the definition of m_{1k} (Eq. (53)) and we find that

$$E_{xc} = \sum_k d_k \int h_k(\mathbf{r}) \, \epsilon_{xc}(\mathbf{r}) \, d\mathbf{r}$$

$$= \sum_k d_k \left[\epsilon_{xc}(m_{1k}) \, I_k + \frac{1}{2!} \int (n_s(\mathbf{r}) - m_{1k})^2 \, h_k(\mathbf{r}) \, \epsilon''_{xc}[\bar{n}(\mathbf{r})] \, d\mathbf{r} \right]$$

$$= \sum_k d_k [\epsilon_{xc}(m_{1k}) \, I_k + (\text{const}) \, I_k] ,$$

where the last step follows from the weighted integral mean value theorem and the constant will be positive as $\epsilon''_{xc} > 0$. The two terms have opposite signs, so that the (smaller) second term partially cancels the first. We write this as

$$E_{xc} = \sum_k d_k \epsilon_{xc}(m_{1k}) \, e^{f_k} I_k , \qquad (54)$$

where $f_k < 0$ remains to be found.

So far the analysis has been exact. We will determine the small correction f_k using an approximation, noting that the exchange-correlation density can be fitted to a reasonable accuracy with $\epsilon_{xc}(n) = An^s$ with $s = 0.30917$. Our strategy will be to initially regard s as a variable, to find $f_k(s)$ analytically for the simple cases of $s = 0, 1$ and 2 and then use interpolation to find $f_k(0.30917)$. Clearly for $s = 0$ or 1, $f(s)$ is 0, while for $s = 2$, $f(s)$ must be a positive quantity. Since we are interested in values of s around 0.3, we can approximate $f(s)$ by $f_k(s) = s(s-1) f_k(2)/2$, where

$$f_k(2) = \ln \left[\frac{m_{2k}}{(m_{1k})^2} \right]$$

and m_{2k} is the second moment of $\tilde{n}$, i.e.

$$m_{2k} = \langle \tilde{n}^2 \rangle_k = \frac{\sum_{lm} d_l d_m u_{klm}}{I_k}, \tag{55}$$

where $u_{klm} = \int h_k h_l h_m \, d\mathbf{r}$. These integrals can all be evaluated analytically. This completes the evaluation of the exchange-correlation energy.

4.6.3 Spin-polarised case – approximate analytic treatment

This theory has been extended to the spin-polarized case [72]. We follow the treatment of the previous section by writing the energy per unit volume of a uniform polarised gas as

$$E^{xc}(n_\uparrow, n_\downarrow) = \sum_{i,s} A_i \int n_s^{p_i+1} n_{-s}^{q_i},$$

where A_i, p_i and q_i are all given in Table 1. The error in this is less than 0.001 for $n_\uparrow, n_\downarrow < 1$.

In an inhomogeneous environment, we have

$$E^{xc}(n_\uparrow, n_\downarrow) = \sum_{i,s} A_i \int n(\mathbf{r})_s^{p_i+1} n(\mathbf{r})_{-s}^{q_i} \, d\mathbf{r},$$

and on replacing n_s on right hand side by $\tilde{n}_s$ the expression becomes

$$\tilde{E}_{xc} = \sum_{ks} d_{k,s} \epsilon_{k,s}, \tag{56}$$

where

$$\epsilon_{k,s} = \sum_i A_i I_k \langle \tilde{n}_s^{p_i} \tilde{n}_{1-s}^{q_i} \rangle_k.$$

Once again, proceeding in the spirit as for the spin-averaged case, we define the quantity f by

$$\langle \tilde{n}_s^p \tilde{n}_{-s}^q \rangle_k = \langle \tilde{n}_s \rangle_k^p \langle \tilde{n}_{-s} \rangle_k^q \, e^{f(p,q)},$$

$$f(p, q) = \ln \left(\frac{\langle \tilde{n}_s^p \tilde{n}_{-s}^q \rangle_k}{\langle \tilde{n}_s \rangle_k^p \langle \tilde{n}_{-s} \rangle_k^q} \right).$$

We now approximate f by the formula

$$f(p, q) = \tfrac{1}{2} p(p - 1) f(2, 0) + \tfrac{1}{2} q(q - 1) f(0, 2) + pq f(1, 1),$$

which interpolates f between the known integer values. In this way the spin-polarized exchange-correlation energy is evaluated.

Table 1
Parameterization of the spin-polarised exchange-correlation energy

i	A_i	p_i	q_i
1	−0.9305	0.3333	0
2	−0.0361	0	0
3	0.2327	0.4830	1
4	−0.2324	0	1

4.6.4 Matrix elements of exchange-correlation potential

We now turn to the evaluation of the matrix elements of the exchange-correlation potential that are also required in this formalism. The matrix required is given by

$$V_{ij}^{xc} = \frac{\partial E_{xc}[\tilde{n}]}{\partial b_{ij}} = \frac{\partial E_{xc}}{\partial d_k}\frac{\partial d_k}{\partial b_{ij}} = p_k t_{ijk}^E, \qquad (57)$$

where we have defined $p_k = H^{-1} q_l$ and $q_k = \partial E_{xc}/\partial d_k$. This derivative is evaluated in a straightforward albeit rather tedious way as

$$q_k = \epsilon_{xc}(m_{1k})\, e^{f_k} I_k \qquad (58)$$

$$+ \sum_l d_l \epsilon'_{xc}(m_{1l})\, e^{f_l} H_{kl} \qquad (59)$$

$$+ s(s-1) \sum_l d_l \epsilon_{xc}(m_{1l})\, e^{f_l} \left[\frac{d_m u_{klm}}{m_{2l}} - \frac{H_{kl}}{m_{1l}}\right], \qquad (60)$$

which is easily evaluated in $O(N)$ operations as

$$q_k = \epsilon_{xc}(m_{1k})\, e^{f_k} I_k + q_l^A H_{kl} + q_l^B d_m u_{klm},$$

where

$$q_l^A = d_l\, e^{f_l}\left[\epsilon'_{xc}(m_{1l}) - \frac{s(s-1)\,\epsilon_{xc}(m_{1l})}{m_{1l}}\right] \quad \text{and} \quad q_l^B = \frac{s(s-1)\,d_l\epsilon_{xc}(m_{1l})\, e^{f_l}}{m_{2l}}.$$

The expression in the spin-polarised case is similarly determined [73].

4.7 Summary

In this section, a description has been given, indicating how, given an arbitrary charge density matrix b_{ij} in a Gaussian orbital representation, one can calculate the total energy and the Hamiltonian matrix within the framework of Gaussian orbitals. The next important task is to determine the charge density matrix itself. This will prove to be an iterative procedure and approaches to it will be outlined in the following section.

5. Self-Consistency: Traditional Approaches

This part of the paper is slightly less dependent on the precise details of the choice of basis functions than the previous section. Here we will be concerned with the determination of the charge density matrix b_{ij}. This is an iterative procedure as traditional calculations have proceeded on the following lines:

1. A starting charge density matrix is obtained. This can be derived from a series of atomic calculations. If the charge density matrices are known for all the atoms making up the cluster, then an approximate cluster charge density matrix will consist of a superposition of these down the diagonal of the cluster density matrix, producing a block diagonal form. This will be referred to as the *input* charge density matrix. If an intermediate fit is being used, then the atomic charge densities can be superposed instead.

2. This matrix has been used to set up the Hamiltonian matrix using the formalism detailed above.

3. The generalised eigenvalue problem (Eq. (21)) is then solved in some way (traditionally by calling a standard LAPACK routine or equivalent) giving the wavefunction

expansion coefficients, c_i^λ. These are then used to construct an *output* charge density matrix or the spin density matrices using Eq. (17). This will not be the same as the input charge density (unless for some reason the atoms genuinely do not interact!).

4. From a consideration of the input and output charge density matrices, a new input charge density for the next iteration is produced. The Hamiltonian is set up using this and is diagonalised again, producing a new output density.

5. This procedure is continued until some stopping criterion is satisfied.

It should be pointed out that many calculations performed today do not follow precisely this approach, with numerous methods being adopted to avoid the workload associated with the solution of the eigenvalue problem. These are absolutely essential when a basis of plane waves is being used as in that case, the basis set is so large that the direct diagonalisation is just not possible on more than approximately 50 atoms. In contrast, the Gaussian basis set is much more compact, and many calculations are still performed in this way as diagonalisation is not necessarily the time dominant part of the code. Alternative strategies will be outlined later in Section 7.

Also it should be noted in passing that although the above discussion has considered iterating the charge density matrix b_{ij} as this is the quantity that is used to calculate the total energy, we could equally well iterate the charge density itself as this is all that is required to set up the Hamiltonian matrix. In calculations employing an intermediate fit to the charge density, the coefficients of the intermediate fit are therefore commonly used instead of the b_{ij} to control the iteration towards self consistency. In this case one iteration consists of taking a set of input coefficients c_k^{in} using the notation in the previous section, setting up and diagonalising the Hamiltonian and calculating the output density matrix b_{ij}. An output intermediate fit c_k^{out} can then be obtained via Eq. (43) and a comparison between this and c_k^{in} used to identify a new input c_k. In this section, this notation will be used to describe the iteration strategies. However, the equations will work if all the c_ks are replaced with b_{ij}s (or even, if the charge density has been discretised on a grid, with the list of tabulated points of the charge density).

In this section, we will briefly look at the two questions – how to produce a new input charge density and when to stop iterating.

5.1 Charge density mixing

The most obvious method of selecting a new input charge density is to take a linear combination of the previous input and the corresponding output charge density:

$$c'_{in} = (1 - w) c_{in} + w c_{out}, \qquad (61)$$

where w is the mixing parameter and $w = 0$ corresponds to no change being made since the previous iteration and $w = 1$ corresponds to the output density from one iteration being used as the input density for the next. This second limit is analogous to simple fixed point iteration strategies but rarely works in practice as large oscillations in density result. Smaller values of w lead to more stable convergence, larger values to faster but less reliable progress towards a self-consistent solution.

A number of strategies exist for choice of w:

1. A simple but non-optimal approach uses a fixed value, say $w = 0.3$. This method has the problem that the value chosen has to be small enough to be stable at the early stages of self-consistency and then is invariably too small to be effective at the later stages.

2. Another obvious approach starts from noting that this is a standard problem in numerical analysis, that of the determination of the minimum of a function of a single variable. By defining a function, $E_{\text{TOT}}[w]$, we can invoke any procedure to find the optimum w. A suitable method is that of Brent [74]. Indeed, other standard methods of numerical analysis can be utilised here [75, 76].

3. A slightly less standard method is adopted in the AIMPRO code. In this we first note that the output charge density is a (non-linear) function of the input charge density:

$$c_{\text{out}} = \hat{L} c_{\text{in}},$$

where we have written c to stand for the vector c_k, then if we use a mixing procedure to determine the input charge density c'_{in} for the next iteration, then the output density produced by this will be

$$\begin{aligned} c'_{\text{out}} = \hat{L} c'_{\text{in}} &= \hat{L}\left[c'_{\text{in}} + w(c'_{\text{out}} - c'_{\text{in}})\right] \\ &= \hat{L} c'_{\text{in}} + w \hat{L}'(c'_{\text{out}} - c'_{\text{in}}) \\ &= c_{\text{out}} + w \hat{L}'(c'_{\text{out}} - c'_{\text{in}}), \end{aligned}$$

where the second step is possible if w is small enough to allow linearisation. The condition for self-consistency is that the input and output charge densities are equal or, equivalently $c'_{\text{out}} - c'_{\text{in}} = 0$. Using the above results in this gives

$$(1 - w)(c_{\text{in}} - c_{\text{out}}) - w \hat{L}'(c_{\text{out}} - c_{\text{in}}) = 0.$$

The final term can be evaluated using a small value of w, say $\bar{w}$ and determining the output density $\bar{c}_{\text{out}}$ for this explicitly so that

$$\bar{c}_{\text{out}} = c_{\text{in}} + \bar{w} \hat{L}'(c_{\text{out}} - c_{\text{in}}).$$

Solving this for $\hat{L}'(c_{\text{out}} - c_{\text{in}})$ and substituting gives

$$(1 - w)(c_{\text{in}} - c_{\text{out}}) - w(\bar{c}_{\text{out}} - c_{\text{in}})/\bar{w} = 0. \tag{62}$$

Denoting this quantity by e_k (where we restore the vector notation), we require that this be minimised with respect to w. In AIMPRO, the Coulomb energy associated with this difference in charge density is minimised, i.e. the quantity

$$\tfrac{1}{2} e_k e_l G_{kl}.$$

It is possible to generalize this procedure so that the predicted charge density is built up from several previous iterates i.e. we write the input charge density in the $(n+1)$st iteration as

$$c_{\text{in}}^{(n+1)} = \sum_{l=0}^{2} \left[c_{\text{in}}^{n-l} + w_l(c_{\text{out}}^{n-l} - c_{\text{in}}^{(n-l)}) \right].$$

5.2 Problems with achieving self-consistency

In practice, the self-consistency cycle converges quickly, taking between four to ten iterations with the difference in the input and output Hartree energies typically becoming less than 10^{-5} a.u. Convergence is particularly rapid when there is a gap between the highest filled and lowest empty level but problems can arise when this gap is very small or vanishes. When simulating defects in materials, this often happens due to an attempted crossing of an occupied and unoccupied energy level whereupon the output charge density changes discontinuously as a function of the mixing parameter.

Level crossing problems can be overcome by resorting to fractional occupancy of the Kohn-Sham levels, giving the level $\epsilon_{\lambda s}$ an occupation of $f_{\lambda s}$. If this is done it is important to replace the energy by a free energy $F[n] = E[n] - TS[n]$, where the entropy term must be added to make sure that the energy is variational or forces with respect to these occupancies will appear when the structure is optimised [77, 78].

The most physical form of occupancy is that given by the Fermi-Dirac distribution [77]

$$f_{\lambda s} = \frac{1}{e^{(\epsilon_{\lambda s} - \mu)/k_B T} + 1},$$

where the chemical potential μ is found by demanding that $\sum_{\lambda s} f_{\lambda s} = M$, the total number of electrons being simulated, and the entropy that must be added is given by

$$-TS = \sum_{\lambda s} [f_{\lambda s} \ln f_{\lambda s} + (1 - f_{\lambda s}) \ln (1 - f_{\lambda s})].$$

In practice, $k_B T$ is taken to be about 0.04 eV. Often, where two energy levels separated by of order 0.1eV that cross in the approach to self-consistency, this will remove the discontinuous change, but when self-consistency is achieved, provided the final splitting is more than 0.04 eV one state is found to be fully occupied and the other empty. If this does not happen, the temperature can gradually be reduced to zero after self consistency has been attained at a higher temperature. In this sense, we are using variable filling purely as a computational tool and are not attempting to simulate materials at finite temperatures.

Other broadening schemes are possible with Gaussian broadening being popular [79]. In this the occupancy is given by

$$f_\lambda = \frac{1}{2} \left[1 - \mathrm{erf} \left(\frac{\epsilon_\lambda - \mu}{k_B T} \right) \right].$$

In this case a different "entropy" term must be added [80].

An alternative approach to the level crossing problem can be used when the levels concerned have different symmetry properties. This frequently turns out to be the case. For example, many defects undergo small symmetry distortions due to the Jahn-Teller effect. For example, a substutional negatively charged nickel atom moves off-centre in silicon due to a Jahn-Teller effect [81]. In the process, as the symmetry drops from T_d to C_{2v} three degenerate states of t_2 symmetry split into states of a_1, b_1 and b_2 symmetry. The symmetry splitting is often small and there is a danger that if smearing is used $k_B T$ will be greater than the final splitting. In this case, the filling of Kohn-Sham levels can be restricted to 0 or 1, but the states filled are not necessarily those of lowest symmetry, but rather those chosen by the user. Clearly, if this approach is taken, then it is important to make sure that choice of occupancies really is the one that gives the lowest energy.

5.3 Stopping criteria

A number of different criteria have been used for stopping the iterations. A common measure of convergence is the change in total energy since the previous iteration, with values in the range 10^{-4} to 10^{-6} Hartrees typically being used. The more exacting end of this range will guarantee accurate forces which can be used to calculate second derivatives which will also be reasonable. However, care should always be taken when using this measure as the total energy converges fastest of all the quantities of interest. Properties that are calculated from the charge density (e.g. dipole moments, hyperfine couplings) are not produced as accurately as they are not variationally protected in the way the total energy is. Another approach monitors a quantity such as $n_{\rm in}(\mathbf{r}) - n_{\rm out}(\mathbf{r})$ and looks at some functional of this — the AIMPRO code follows the electrostatic energy associated with this difference in charge density.

6. Determination of Forces and Structural Optimisation

It is well known that density functional theory is used to predict the ground state energy of an interacting electron gas, in most cases that of a molecule or a solid. The equilibrium microscopic structure of such a system is of fundamental importance to our understanding. This is determined by the solution of the equations

$$F_{I\alpha} = -\frac{\partial E_{\rm TOT}}{\partial R_{I\alpha}} = 0,$$

where $F_{I\alpha}$ is the force on atom I in the direction α, where $\alpha = 1, 2, 3$ represents x, y or z. In this section we will discuss the evaluation of these forces and their use in optimising structures.

6.1 The Hellmann-Feynman theorem

Fundamental to the determination of forces is the Hellmann-Feynman theorem [82, 83]. For our purposes this implies that in a calculation in which the Hamiltonian depends on a parameter R, then if we have

$$E = \langle \psi | H(R) | \psi \rangle$$

then the derivative is given by

$$\frac{\partial E}{\partial R} = \langle \psi | \frac{\partial H(R)}{\partial R} | \psi \rangle.$$

This expression is routinely used in calculations involving plane waves to find the forces, but it cannot be used if Gaussian orbitals are used as the basis functions as these are in general centred on atoms and this creates a hidden dependence, giving rise to additional terms known as Pulay corrections. The complexity of these makes the determination of forces more difficult than in plane-wave calculations. It is not possible to neglect these terms and this gives rise to the need to differentiate all the expressions and matrix elements detailed in Section 4.

6.2 Determination of forces

We begin with the expression for the total energy using our localised orbital methodology. In this section, it will be assumed that an intermediate fit is being used to evaluate

LDA Calculations Using a Basis of Gaussian Orbitals

the Hartree energy and the approximate analytic treatment of exchange-correlation.

$$E[n] = T_s[n] + \int V^{\text{ext}}(\mathbf{r})\, n(\mathbf{r})\, d\mathbf{r} + \frac{1}{2} \int \frac{n(\mathbf{r})n(\mathbf{r}')}{|\mathbf{r} - \mathbf{r}'|}\, d\mathbf{r}\, d\mathbf{r}' + E_{\text{xc}}[n] + E_{i-i}$$

$$= T_{ij} b_{ij} + V^{\text{ext}}_{ij} b_{ij} + \frac{1}{2} c_k c_l G_{kl} + d_k \epsilon_{\text{xc}}(m_{1k})\, e^{f_k} I_k + \frac{1}{2} \sum_{a \neq b} \frac{Z_a Z_b}{|\mathbf{R}_a - \mathbf{R}_b|}.$$

The quantities involved in this equation are first those representing the charge density (the matrix b_{ij} and the intermediate fits c_k and d_k) and matrix elements that depend directly on atom positions, the matrices T_{ij}, t^{H}_{ijk}, G_{kl} and so on. All of these terms must be differentiated. We begin with

$$\frac{\partial E}{\partial R} = \frac{\partial E}{\partial b_{ij}} \frac{\partial b_{ij}}{\partial R} + \left[\frac{\partial T_{ij}}{\partial R} + \frac{\partial V^{\text{ext}}_{ij}}{\partial R}\right] b_{ij} + \frac{\partial c_k}{\partial R} c_l G_{kl} + \frac{1}{2} c_k c_l \frac{\partial G_{kl}}{\partial R}$$

$$+ \frac{\partial E_{\text{xc}}}{\partial d_k} \frac{\partial d_k}{\partial R} + \frac{\partial E_{\text{xc}}}{\partial H_{kl}} \frac{\partial H_{kl}}{\partial R} + \frac{\partial E_{\text{xc}}}{\partial u_{klm}} \frac{\partial u_{klm}}{\partial R} + \frac{\partial E_{i-i}}{\partial R},$$

where the symbol R refers to the position vector of an atom, $R_{I\alpha}$, but we will suppress the unnecessary indices I and α in this derivation. We may use Eq. (43) to evaluate the derivative of c_k:

$$\frac{\partial c_k}{\partial R} = \frac{\partial G^{-1}_{kl}}{\partial R} b_{ij} t^{\text{H}}_{ijl} + G^{-1}_{kl} \frac{\partial b_{ij}}{\partial R} t^{\text{H}}_{ijl} + G^{-1}_{kl} b_{ij} \frac{\partial t^{\text{H}}_{ijl}}{\partial R}.$$

We use these in the above equation together with $(G^{-1})' = -G^{-1} G' G^{-1}$, and Eq. (43) to write the derivative of the Hartree energy as

$$c_k t^{\text{H}}_{ijk} \frac{\partial b_{ij}}{\partial R} + c_k b_{ij} \frac{\partial t^{\text{H}}_{ijk}}{\partial R} - \frac{1}{2} c_k c_l \frac{\partial G_{kl}}{\partial R}.$$

A similar procedure with the term involving the derivative of d_k gives

$$p_k t^{E}_{ijk} \frac{\partial b_{ij}}{\partial R} + p_k b_{ij} \frac{\partial t^{E}_{ijk}}{\partial R} - p_k d_l \frac{\partial H_{kl}}{\partial R}.$$

First, gathering together all the terms involving derivatives of b_{ij} we have

$$\left[T_{ij} + V^{\text{ext}}_{ij} + c_k t^{\text{H}}_{ijk} + p_k t^{E}_{ijk}\right] \frac{\partial b_{ij}}{\partial R} = H_{ij} \frac{\partial b_{ij}}{\partial R},$$

where the term in brackets has been recognised as the Hamiltonian matrix H_{ij}. This term therefore becomes

$$H_{ij} \frac{\partial b_{ij}}{\partial R} = H_{ij} \frac{\partial (c^{\lambda}_i c^{\lambda}_j)}{\partial R} = 2 \frac{\partial c^{\lambda}_i}{\partial R} H_{ij} c^{\lambda}_j = 2 \frac{\partial c^{\lambda}_i}{\partial R} \epsilon^{\lambda} S_{ij} c^{\lambda}_j = \epsilon_{\lambda} S_{ij} \frac{\partial (c^{\lambda}_i c^{\lambda}_j)}{\partial R},$$

where we have used the matrix Eq. (21) in the last step. We now use the chain rule to write

$$\epsilon_{\lambda} S_{ij} \frac{\partial (c^{\lambda}_i c^{\lambda}_j)}{\partial R} = \epsilon^{\lambda} \frac{\partial (S_{ij} c^{\lambda}_i c^{\lambda}_j)}{\partial R} - \epsilon_{\lambda} c^{\lambda}_i c^{\lambda}_j \frac{\partial S_{ij}}{\partial R} = -\left[\epsilon_{\lambda} c^{\lambda}_i c^{\lambda}_j\right] \frac{\partial S_{ij}}{\partial R}$$

as $S_{ij}c_i^\lambda c_j^\lambda = 1$ because the wavefunctions are normalised. This short derivation shows that the derivative of the charge density is not explicitly required to determine the force. This final term involving the derivative of the overlap matrix is a contribution to the Pulay force. This result may be considered to be a restatement of the content of the Hellmann-Feynman theorem in the case where the wavefunction has been expanded in terms of an incomplete basis which depends on the parameter being differentiated.

Our final expression for the negative of the force is then

$$\frac{\partial E}{\partial R} = \left[\frac{\partial T_{ij}}{\partial R} + \frac{\partial V_{ij}^{\text{ext}}}{\partial R} + c_k\frac{\partial t_{ijk}^H}{\partial R} + p_k\frac{\partial t_{ijk}^E}{\partial R}\right]b_{ij} - \left[\epsilon_\lambda c_i^\lambda c_j^\lambda\right]\frac{\partial S_{ij}}{\partial R}$$
$$- \frac{1}{2}c_k c_l \frac{\partial G_{kl}}{\partial R} - p_k d_l \frac{\partial H_{kl}}{\partial R} + q_k^A d_l \frac{\partial H_{kl}}{\partial R} + q_k^B d_l d_m \frac{\partial u_{klm}}{\partial R} + \frac{\partial E_{i-i}}{\partial R}.$$

The expression is somewhat complex, but is in fact simple in the following way. Looking first at the derivatives of a two centre term, we first find the vector of quantities

$$\frac{\partial T_{ij}}{\partial R}b_{ij} = b_{ij}\frac{\partial T_{ij}}{\partial \mathbf{R}_i}\frac{\partial \mathbf{R}_i}{\partial R} + b_{ij}\frac{\partial T_{ij}}{\partial \mathbf{R}_j}\frac{\partial \mathbf{R}_j}{\partial R} = 2b_{ij}\frac{\partial T_{ij}}{\partial \mathbf{R}_i}\frac{\partial \mathbf{R}_i}{\partial R},$$

where the two terms are seen to be equal by first interchanging the dummy indices i and j and then using the fact that T and b are symmetric matrices. The vector of quantities

$$v_i = 2\sum_j b_{ij}\frac{\partial T_{ij}}{\partial \mathbf{R}_i}$$

is first found for all the basis functions that are centred on atoms whose forces are needed. This is then trivially contracted with the final term $\partial \mathbf{R}_i/\partial R$ (which contains either zeros, ones or — in the case of bond-centred basis functions — halves) to give derivatives with respect to atoms. A typical three-centre term involves t_{ijk}^H which is symmetric in i and j, and the derivative with respect to $\mathbf{R}_k$ is given by

$$\frac{\partial t_{ijk}^H}{\partial \mathbf{R}_k} = -\frac{\partial t_{ijk}^H}{\partial \mathbf{R}_i} - \frac{\partial t_{ijk}^H}{\partial \mathbf{R}_j}$$

on account of the translational invariance of t_{ijk}. We therefore have

$$b_{ij}c_k\frac{\partial t_{ijk}^H}{\partial R} = 2b_{ij}c_k\frac{\partial t_{ijk}^H}{\partial \mathbf{R}_i}\left[\frac{\partial \mathbf{R}_i}{\partial R} - \frac{\partial \mathbf{R}_k}{\partial R}\right]$$

which is found by first evaluating the vector quantities

$$v_i = \sum_{jk} 2b_{ij}c_k\frac{\partial t_{ijk}^H}{\partial \mathbf{R}_i} \quad \text{and} \quad w_k = \sum_{ij} 2b_{ij}c_k\frac{\partial t_{ijk}^H}{\partial \mathbf{R}_i}$$

and then performing the contractions as before.

One final point should be made regarding the calculation. It is important that the c_k used should be those determined from b_{ij} by Eq. (43). In a practical implementation it is likely that b_{ij} will be the output density matrix, but the copies of vectors such as q_k to hand will have been calculated from the input charge density which are used to set up the Hamiltonian. To get the best convergence care is needed here as at self-consistency the difference between input and output charge densities may be small in the sense that the total energy is only marginally affected, but the forces will depend more sensitively on this [84]. Vectors such as q_k should first be recalculated using the output density before calculating forces. This is discussed in terms of a different methodology in Ref. [54].

The evaluation of the forces takes a proportionately longer time in Gaussian orbital calculations as compared with plane wave calculations in which the Hellmann-Feynman theorem can be applied directly. This part of the calculation is however still far faster than the self-consistent cycle as the determination of the forces essentially scales linearly with system size whereas diagonalisation is involved in the main part of the calculation.

6.3 Optimisation of the structure

Having obtained the forces, it now remains only to move the atoms so as minimise the total energy. Methods for achieving this are numerous and are not directly concerned with the nature of the underlying basis set, so only a brief account will be given here. This is a standard problem of numerical analysis and a simple introduction is given in Ref. [85]. A common approach and one of the simplest is to use the conjugate gradient method. This requires knowledge of the function to be minimised and its first derivatives (i.e. the forces we have been discussing). If we denote the force on atom a in direction m in the nth iteration of structural optimisation by f_{am}^n, then the method proceeds by first calculating a conjugate direction d_{im}^n using

$$d_{am}^n = f_{am}^n - x d_{am}^{n-1},$$

where

$$x = \frac{\sum_{am} f_{am}^n (f_{am}^n - f_{am}^{n-1})}{\sum_{am} (f_{nm}^{n-1})^2}$$

and in the first iteration, x is set to zero. The atoms are moved simultaneously according to

$$R'_{am} = R_{am} + w d_{am}^n,$$

where w is chosen so as to minimise the energy. This one dimensional optimisation is readily accomplished. In the AIMPRO code cubic interpolation is used taking the energy at the start of the iteration, the gradient of this along the search direction, and these same two quantities after the atoms have moved a fixed distance along this direction. The projected gradient is easily found using the expression

$$f_{am} - \frac{\sum_{bl} f_{bl} d_{bl}}{\sum_{bl} d_{bl}^2} d_{am}$$

and the fixed distance used is chosen so that the atom moving most in this iteration moves twice as far as the atom which moved most in the previous iteration (subject to various cut-offs and limits to ensure numerical stability). A preset value is used in the first iteration.

The number of iterations required depends on a number of factors, one of which is the co-ordination of the atoms – atoms with low co-ordination (e.g. at the surface of a cluster) can keep wobbling for many iterations before settling down. This can be cured by transforming to internal co-ordinates to perform the The forces are most readily found in Cartesian co-ordinates as described above, but they are transformed and fed to the optimisation algorithm in terms of internal co-ordinates. The displacements of the atoms are then made according to the same internal co-ordinate system. There has been some argument as to the optimum set of co-ordinates, but this is a little beyond the scope of this review.

7. The Generalised Eigenvalue Problem

It has been explained above how the adoption of a basis produces a generalised eigenvalue problem,

$$Hc = \epsilon Sc, \tag{63}$$

where H and S are matrices and c represents an eigenvector with eigenvalue ϵ.

The conventional way of solving this is to first reduce it to a standard eigenvalue problem. This is done by carrying out the Cholesky factorisation of S. That is we find a lower triangular matrix L such that $S = LL^T$, where the superscript T indicates the matrix transpose. This is possible as S is symmetric and positive definite. The generalised problem can then be reduced by substituting this into Eq. (100) and rearranging to get

$$H'c' = \epsilon c',$$

where

$$H' = (L^{-1}) H (L^{-1})^T \quad \text{and} \quad c' = L^T c.$$

All the operations here are standard ones of linear algebra and "off the peg" high performance software exists to implement this (e.g. LAPACK for serial implementations, ScaLAPACK for parallel). The floating point operation count involved in this is however considerable – $O(3N^3)$, where N is the size of the Hamiltonian.

One disadvantage with standard approaches is that as the system being modelled increases in size, the matrices H and S become increasingly sparse, but this property is ignored in the above approach. Indeed, unless extreme care is taken H' will be a dense matrix. Sparse linear algebra packages are available, but are not in general designed for this particular problem (they normally assume a far smaller percentage of eigenvectors are needed). An implementation of such an algorithm is in the ARPACK package.

Other approaches have combined the solution of the eigenvalue problem with the approach to self-consistency and have proposed combined algorithms. This was especially important for plane wave approaches as so many such basis functions are required per atom, that the Hamiltonian becomes so large that it cannot be stored, never mind diagonalised. Alternative approaches developed in the 1980s and were initially used exclusively by plane wave calculations. A brief account of them will be given here as they are also used to a lesser extent in Gaussian orbital calculations.

7.1 Car-Parrinello molecular dynamics

The Car-Parrinello methodology has become a widespread method of solving the Kohn-Sham equations. Molecular dynamics has been a popular way of modelling systems using empirical potentials for many years. The principle has been to form an interatomic potential that returns the total energy as a function of the atom positions $V(\{\mathbf{R}_i\})$. Motion of the atoms is governed by equations of motion

$$M_i \ddot{\mathbf{R}}_i = -\frac{\partial V}{\partial \mathbf{R}_i}$$

which are integrated and properties of the system evaluated. For example the temperature of the system can be increased, leading to melting, the liquid studied, the system could then be rapidly cooled and the defects frozen in studied and so on.

Of course, this calculation is not quantum-mechanical. However, Car and Parrinello [86] have put forward a formalism which enables the function $V(\{\mathbf{R}_i\})$ to be treated quantum mechanically within the LDA. The Lagrangian is of the form

$$L = \tfrac{1}{2}\sum_i M_i \dot{R}_i^2 + \tfrac{1}{2}\mu \sum_\lambda \int |\dot{\psi}_\lambda|^2 \, d\mathbf{r} - E[\{\mathbf{R}_i\},\{\psi_\lambda\}],$$

where E is the Kohn-Sham energy functional, μ is an arbitrary fictitious mass the ψ_λ are subject to the holonomic constraints of orthonormality. The resulting equations of motion are integrated, and if the kinetic energy of the nuclei and fictitious kinetic energy of the electronic degrees of freedom are gradually reduced (either by directly scaling the velocities, by including a dissipative term or by attaching a Nosé-Hoover thermostat) the equilibrium state of minimum energy is attained. Advantages of this include the fact that the Hamiltonian does not need to be stored, it is not explicitly diagonalised, reducing the workload from $O(N^3)$ to $O(NM^2)$, where N is the number of basis functions and M the number of occupied levels. Further discussion of practical details (how orthonormality of the orbitals is maintained and how the equations of motion are best integrated) is given by Payne et al. [87].

7.2 Direct minimisation of the Kohn-Sham functional

An alternative strategy is to regard the Kohn-Sham energy as a function of all the wavefunction coefficients and to use a standard technique from numerical analysis to carry out the minimisation process. One significant advantage of this is that we can guarantee that the energy always decreases from iteration to iteration, removing instabilities which it has been claimed [87] makes the Car-Parrinello molecular dynamics method unsuitable for application to large systems [87]. The conjugate gradient method achieves this minimisation by making a series of one-dimensional line minimisations. Again, a significant advantage is that the Hamiltonian doesn't need to be stored, and that only the action of the Hamiltonian on the wavefunctions is required. The details of the procedure applied with a plane wave basis is given in [87] and in localised orbitals in [69]. Key issues are how the bands are updated, how orthogonality constraints are imposed and what form of preconditioning is applied.

8. Applications

In this section we will take a brief look at a few of the successful applications of the AIMPRO code. Many applications of the formalism have been made to molecular systems such as fullerenes [88]. Here however, we shall focus on some of the applications to point and line defects in bulk solids.

8.1 Point defects in diamond

Nitrogen is one of the most important impurities in diamond occurring in concentrations as large as 10^{20} cm^{-3}. It readily complexes with itself and with other impurities and intrinsic defects and the resulting complexes are often important optical centres. In type Ib or synthetic diamonds, N_s is present as an isolated defect, but in annealed synthetic diamonds, the nitrogen aggregates to give complexes with more than one N atom. These complexes are also found in the great majority of natural (type Ia) diamonds. It is a long-standing problem to elucidate the final fate of N aggregation in diamond when it is annealed for long periods. The ab initio calculations have helped to clarify the properties of many of these nitrogen complexes.

The substitutional defect was investigated by Briddon et al. [89] and led to an explanation of the anomalous vibrational mode associated with the defect. This local mode at 1344 cm^{-1} was observed [90] and its intensity correlated with the EPR signal due to N_s suggesting that it is associated with the vibrations of N_s. Surprisingly, however, the mode did not shift with ^{15}N doping. The cluster calculation revealed that for the neutral substitutional defect, not only N was displaced from a lattice site along [111] by 0.2 Å, but also the neighbouring C atom was displaced along [$\bar{1}\bar{1}\bar{1}$] by the same amount thus leading to back C–C bonds about 5% shorter than the normal C–C bonds. This was independently found by plane wave pseudopotential calculations [91]. The vibrational modes of the defect were found by a Green function method. This gave three bands centred at 1320, 1122 and 1032 cm^{-1} in good agreement with observed ones at 1344, 1130 and 1080 cm^{-1} [90]. The highest mode was localized on the unique C atom and its C neighbours and does not shift with a change in the N isotope. This explains the anomalous mode. The N related modes fell below the Raman frequency.

The N_s defect is not stable during prolonged annealing at high temperatures and aggregates firstly into A centres, which are believed to be N_s dimers [92]. The calculated vibrational modes of the defects [93] are in good agreement with observation giving further support to the assignments. Furthermore, they give a clue as to why N atoms should aggregate. The highest filled level in the A centre is around mid-gap which is considerably lower than that of the N_s donor. Thus the driving force for aggregation is the lowering of the one-electron energy. It is an insight such as this which makes the theory so useful.

Vacancies and interstitials were investigated by Breuer and Briddon [94] who confirmed the importance of many-body effects and the need to determine the energies of different multiplets. The theory found that V$^-$ was a spin 3/2 defect, in agreement with experiment, and the calculated optical transition energy agreed well with the observed value. In many cases, vacancies will complex with impurities and a recent study of N–V and Si–V centres [95] concluded that they possess very different structures. Whereas the N–V defect has C_{3v} symmetry, Si–V possesses D_{3d} symmetry where the Si atom sits mid-way between two adjacent vacancies. This finding explains the surprising optical properties of the defect.

8.2 Point defects in silicon

A great deal of effort has been devoted to understanding the properties of the light impurities: H, B, C, O and N in silicon, and especially their vibrational frequencies as local mode spectroscopy has been such a valuable experimental tool. Calculations have

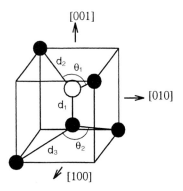

Fig. 1. Structure of the interstitial C atom in silicon

been made for the various substitutional and interstitial defects with substantial agreement obtained with experimental results for the local vibrational modes in each case.

One example is the C_i–O_i defect which turned out to have very surprising structure. The C interstitial on its own takes the form of a [100] oriented split-interstitial as shown in Fig. 1 [96]. Two of the C–Si bonds along [011] are 1.8 Å long and pull in the Si neighbours lying there. This leaves the Si–Si bonds along this direction extended and hence are favorable sites for attack by oxygen. The calculation [97] shows that O does not lie at a bond centre site within these dilated bonds but rather moves towards the Si radical shown in Fig. 1. The reason for this is the electronegativity of C exceeds Si rendering the Si radical positively charged. This in turn attracts the O atom so that its becomes over-coordinated leading to rather long Si–O bonds. The three Si–O bonds are by no means equal in strength. A consequence is that the O-related vibrational mode lies well below that of interstitial and even substitutional oxygen. The same process occurs for interstitial N. But now the state arising from the dangling bond on the Si_3 atom is occupied. This has led to a remarkable finding [98]: the O atom, being negatively charged, squeezes itself into dilated Si–Si bonds adjacent to N and pushes up the donor level due to Si_3. For N_i–O_2, the level is displaced almost to the conduction band. This defect might explain the occurrence of shallow thermal donors which arise when Czochralski Si, containing N, is annealed to 650 °C [99]. This 'wonderbra' mechanism of deep to shallow level conversion is not unique to N – shallow donor level also arises when a C–H unit replaces N.

8.3 Point defects in compound semiconductors

The AIMPRO cluster calculations were the first to describe and detail the structure and modes of H passivated Si donors and Be acceptors in GaAs [100]. Subsequent studies of trigonal C–H complexes in GaAs have been particularly fruitful. The local modes of the defect [101] exhibit several unusual properties. The E^--mode, which involves a movement of H perpendicular to the C_3 axis, and out of phase with C, was placed around 715 cm^{-1}. The C-related A_1 and E^+ modes, which involve motion of H in phase with C in respective directions parallel and perpendicular to the C–H bond, were calculated to lie at 413 and 380 cm^{-1} respectively. Infra-red spectroscopy on GaAs containing high concentrations of C and H grown by molecular beam epitaxy and chemical vapor deposition methods located modes at 453 (X) and 563 cm^{-1} (Y) [102]. Both were subsequently shown to be due to the C–H defect as they exhibited shifts with C and H isotopes. A Raman scattering experiment assigned the 453 cm^{-1} mode to C–A_1. Y is now believed to be the E^+ mode [103]. The E^- mode was not observed in these early experiments. However, in deuterated samples, the E^- mode was detected at 637 cm^{-1}.

This must imply that the unobserved E⁻ mode in the H samples lies above 637 cm⁻¹. The failure of the early infra-red experiments to locate the H–E⁻ mode was explained by the ab initio cluster theory as the consequence of a small transition dipole moment. This mode has since been detected at 739 cm⁻¹ by Raman scattering experiments.

The C–H defect is unusual in possessing a resonant electron trap which has profound consequences for the dissociation of the defect. The calculated activation energy for dissociation is about 1 eV lower in the presence of minority carriers which can be trapped in the resonant level [104]. This calculation anticipated experimental results confirming this reduction in the activation energy.

8.4 Line defects

In addition to the work carried out on point defects there has been a considerable attempt to understand the structure and kinetics of dislocations in group IV and III-V semiconductors. In these materials dislocations are dissociated into partials separated by a stacking fault. Commonly occurring partials are 90° and 30° ones. AIMPRO calculations were the first ab initio simulations to reveal that 90° partial dislocations in Si [105] and GaAs [106] are reconstructed as shown in Fig. 2. The reconstruction leads to electrical inactivity of the line and is to be contrasted with earlier models of deep states arising from a line of dangling bonds. Intriguingly, impurities like P and N have a pronounced effect on the reconstruction in Si and actually break it [107]. This effect might explain the very strong locking effect of these impurities – especially N – which has important technological implications.

An important question concerns the mobility of dislocations as this controls their rate of growth and ultimately their density in the crystal. This is especially important as dislocations bind point defects like vacancies and interstitials as well as impurities, all of which possess deep gap levels which can greatly affect the electronic and optical properties of the material. Now, it is believed that dislocations propagate by creating double kinks as shown in Fig. 2 which then expand under the influence of stress leading to motion of the dislocation. The energetics of this process can be followed by embedding the dislocation in a cluster. The kink formation energy was found [106] to be a very small value in these materials: about 0.1 eV, whereas the activation energy necessary to

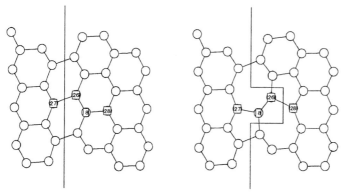

Fig. 2. The reconstructed 90° dislocation and double kink in silicon. Vertical axis is [01$\bar{1}$], horizontal axis is [$\bar{2}$11]

break the reconstructed bonds was considerable. The total activation energy for dislocation motion was found to be 1.9 eV in Si, in excellent agreement with the observed 2.1 eV [108].

Acknowledgements The many developments and applications of the formalism presented here have been carried out by a great number of collaborators. Among these, Malcolm Heggie, Sven Öberg, S. J. Breuer and Grenville Lister are especially to be thanked.

References

[1] P. Hohenberg and W. Kohn, Phys. Rev. **136**, B864 (1964).
[2] W. Kohn and L. J. Sham, Phys. Rev. **140**, A1133 (1965).
[3] U. von Barth and L. Hedin, J. Phys. C **5**, 1629 (1972).
[4] M. N. Pant and A. K. Rajagopal, Solid State Commun. **10**, 1157, (1972).
[5] O. Gunnarsson and B. I. Lundqvist, Phys. Rev. B **13**, 4274 (1976).
[6] R. G. Parr and W. Yang, Density Functional Theory of Atoms and Molecules, Oxford University Press, New York 1989.
[7] R. M. Dreizler and E. K. U. Gross, Density Functional Theory, Springer-Verlag, Berlin 1990.
[8] P. R. Briddon, in: Properties of Crystalline Silicon, EMIS Datareview Series No. 20, Ed. R. Hull, INSPEC, London 1999.
[9] J. P. Perdew and A. Zunger, Phys. Rev. B **23**, 5048 (1981).
[10] S. H. Vosko, L. Wilk, and M. Nusair, Canad. J. Phys. **58**, 1200 (1980).
[11] J. P. Perdew and Y. Wang, Phys. Rev. B **45**, 13244 (1992).
[12] J. P. Perdew and Y. Wang, Phys. Rev. B **33**, 8800 (1986).
[13] J. P. Perdew, Phys. Rev. B **33**, 8822 (1986).
[14] M. Levy, in: Density Functional Theory, Ed. E. K. U. Gross and R. M. Dreizler, Plenum Press, New York 1995.
[15] M. Levy and J. P. Perdew, Phys. Rev. B, **48**, 11638 (1993).
[16] A. D. Becke, J. Chem. Phys. **85**, 7184 (1986).
[17] A. D. Becke, Phys. Rev. A **38**, 3098 (1988).
[18] J. P. Perdew, in: Electronic Structure of Solids '91, Ed. P. Ziesche and H. Eschrig, Akademie-Verlag, Berlin 1991.
[19] J. P. Perdew, J. A. Chevary, S. H. Vosko, K. A. Jackson, M. R. Pederson, D. J. Singh, and C. Fiolhais, Phys. Rev. B **46**, 6671 (1992).
[20] J. P. Perdew, K. Burke, and M. Ernzerhof, Phys. Rev. Lett. **77**, 3865 (1996).
[21] J. P. Perdew, K. Burke, and Y. Wang, Phys. Rev. B **54**, 16533 (1996).
[22] D. C. Patton, D. V. Porezag, and M.R.Pederson, Phys. Rev. B **55**, 7454 (1997).
[23] W. E. Pickett, Comput. Phys. Rep. **9**, 115, (1989).
[24] D. V. Porezag, M. R. Pederson, and A. Y. Liu, phys. stat. sol. (b) **217**, 219 (2000).
[25] G. P. Kerker, J. Phys. C **13**, L189 (1980).
[26] D. R. Hamann, M. Schlüter, and C. Chiang, Phys. Rev. Lett. **48**, 1425 (1982).
[27] G. B. Bachelet, D. R. Hamann, and M. Schlüter, Phys. Rev. B **26**, 4199 (1982).
[28] N. Troullier and J. L. Martins, Phys. Rev. B **43**, 1993 (1991).
[29] D. Vanderbilt, Phys. Rev. B **41**, 7892 (1990).
[30] K. Laasonen, R. Car, C. Lee, and D. Vanderbilt, Phys. Rev. B **43**, 6796 (1991).
[31] L. Kleinman and D. M. Bylander, Phys. Rev. Lett. **48**, 1425 (1982).
[32] D. M. Bylander and L. Kleinman, Phys. Rev. B **41**, 907 (1990).
[33] J. R. Chelikowsky, X. D. Jing, K. Wu, and Y. Saad, Phys. Rev. B **53**, 12071 (1996).
[34] E. L. Briggs, D. J. Sullivan, and J. Bernholc, Phys. Rev. B **54**, 14362 (1996).
[35] F. Gygi, Phys. Rev. B **48**, 11692 (1993).
[36] D. R. Hamann, Phys. Rev. B **51**, 9508 (1995).
[37] G. Galli, phys. stat. sol. (b) **217**, 231 (2000).
[38] S. Huzinaga, Gaussian Basis Sets for Molecular Calculations, Elsevier Publ. Co., Amsterdam 1984.

[39] R. POIRER, R. KARI and I. G. CZISMADIA, Handbook of Gaussian Basis Sets, Elsevier Publ. Co., Amsterdam 1985.
[40] E. R. DAVISON and D. FELER, Chem. Rev. **86**, 681 (1986).
[41] S. HUZINAGA, J. Chem. Phys. **42**, 1293 (1965).
[42] H. PARTRIDGE, J. Chem. Phys. **90**, 1043 (1988).
[43] R. D. BARDO and K. RUEDENBERG, J. Chem. Phys. **60**, 918 (1974).
[44] S. HUZINAGA, M. KLOBUKOWSKI, and H. TATEWAKI, Canad. J. Chem. **63**, 1812 (1985).
[45] W. J. HEHRE, R. F. STEWART, and J. A. POPLE, J. Chem. Phys. **51**, 2657 (1969).
[46] W. J. HEHRE, R. DITCHFIELD, and J. A. POPLE, J. Chem. Phys. **56**, 2257 (1972).
[47] R. KRISHNAN, J. S. BINKLEY, R. SEEGER, and J. A. POPLE, J. Chem. Phys. **72**, 650 (1980).
[48] R. C. RAFENETTI, J. Chem. Phys. **58**, 4452 (1973).
[49] S. F. BOYS and F. BERNADI, Mol. Phys. **19**, 553 (1970).
[50] A. D. BECKE, J. Chem. Phys. **88**, 2547 (1988).
[51] A. D. BECKE and R. M. DICKSON, J. Chem. Phys. **89**, 2993 (1988).
[52] V. I. LEBEDEV, Sib. Mat. Zh. **18**, 132 (1977).
[53] M. R. PEDERSON and K. A. JACKSON, Phys. Rev. B **41**, 7453 (1990).
[54] K. JACKSON and M. R. PEDERSON, Phys. Rev. B **42**, 3276 (1990).
[55] A. BRILEY, M. R. PEDERSON, K. A. JACKSON, D. C. PATTON, and D. V. POREZAG, Phys. Rev. B **58**, 1786 (1998).
[56] S. OBARA and A. SAIKA, J. Chem. Phys. **84**, 3963 (1986).
[57] J. A. POPLE and W. J. HEHRE, J. Comput. Phys. **27**, 161 (1978).
[58] H. F. KING and M. DEPUIS, J. Comput. Phys. **21**, 144 (1976).
[59] L. E. MCMURPHIE and E. R. DAVIDSON, J. Comput. Phys. **26**, 218 (1978).
[60] J. RYS, M. DUPUIS, and H. F. KING, J. Comput. Chem. **4**, 154 (1983).
[61] M. HEAD-GORDON and J. A. POPLE, J. Chem. Phys. **89**, 5777 (1988).
[62] R. LINDH, U. RYU and B. LIU, J. Chem. Phys. **95**, 5889 (1991).
[63] P. M. W. GILL, Adv. Quantum Chem. **25**, 141 (1994).
[64] C. A. WHITE and M. HEAD-GORDON, J. Chem. Phys. **104**, 2620 (1996).
[65] C. A. WHITE, B. G. JOHNSON, P. M. W. GILL, and M. HEAD-GORDON, Chem. Phys. Lett. **230**, 8 (1994).
[66] M. CHALLACOMBE, E. SCHWEGLER, and J. ALMLÖF, J. Chem. Phys. **104**, 4685 (1996).
[67] M. CHALLACOMBE and E. SCHWEGLER, J. Chem. Phys. **106**, 5526 (1997).
[68] J. M. PÉREZ-JORDÁ and W. YANG, J. Chem. Phys. **107**, 1218 (1997).
[69] X. J. CHEN, J. M. LANGLOIS, and W. A. GODDARD, Phys. Rev. B **52**, 2348 (1995).
[70] B. I. DUNLAP, W. J. D. CONNOLLY, and J. R. SABIN, J. Chem. Phys. **71**, 4993 (1979).
[71] R. JONES and A. SAYYASH, J. Phys. C **19**, L653 (1986).
[72] G. M. LISTER and R. JONES, unpublished (1988).
[73] R. JONES and P. R. BRIDDON, The ab initio Cluster Method and the Dynamics of Defects in Semiconductors, in: Identification of Defects in Semiconductors, Vol. 51, Ed. M. STAVOLA, Semiconductors and Semimetals, Treatise, Eds. R. K. WILLARDSON, A. C. BEER, and E. R. WEBER, Academic Press, New York 1998 (p. 287).
[74] Algorithms for Minimisation without Derivatives, Prentice Hall, Englewood Cliffs (NJ) 1973.
[75] C. G. BROYDEN, Math. Comput. **19**, 577 (1965).
[76] D. D. JOHNSON, Phys. Rev. B **38**, 12807 (1988).
[77] M. WEINERT and J. W. DAVENPORT, Phys. Rev. B **45**, 13709 (1992).
[78] M. SPRINGBORG. R. C. ALBERS, and K. SCHMIDT, Phys. Rev. B **57**, 1427 (1998).
[79] C. L. FU and K. M. HO, Phys. Rev. B **28**, 5480 (1983).
[80] C. ELSÄSSER, M. FÄHNLE, C. T. CHAN, and K. M. HO, Phys. Rev. B **49**, 13975 (1994).
[81] R. JONES, S. ÖBERG, J. GOSS, P. R. BRIDDON, and A. RESENDE, Phys. Rev. Lett. **75**, 2734 (1995).
[82] H. HELLMANN, Einführung in die Quantenchemie, Franz Deuticke, Leipzig 1937.
[83] R. P. FEYNMAN, Phys. Rev. **56**, 340 (1939).
[84] P. R. BRIDDON, PhD Thesis, Exeter 1990.
[85] W. H. PRESS, B. P. FLANNERY, S. A. TEUKOLSKY, and W. T. VETTERLING, Numerical Recipes, Cambridge University Press, Cambridge 1987.
[86] R. CAR and M. PARRINELLO, Phys. Rev. Lett. **55**, 2471 (1985).
[87] M. C. PAYNE, M. P. TETER, D. C. ALLAN, T. A. ARIAS, and J. D. JOANNOPOULOS, Rev. Mod. Phys. **64**, 1045 (1992).

[88] B. R. Eggen, M. I. Heggie, G. Jungnickel, C. D. Latham, R. Jones, and P. R. Briddon, Science **272**, 87 (1996).
[89] P. R. Briddon, M. I. Heggie, and R. Jones, Mater. Sci. Forum **83-7**, 457 (1991).
[90] A. T. Collins and G. S. Woods, Phil. Mag. B **46**, 77 (1982).
[91] S. A. Kajihara, A. Antonelli, J. Bernholc, and R. Car, Phys. Rev. Lett. **66**, 2010 (1991).
[92] G. Davies, J. Phys. C **9**, L537 (1976).
[93] R. Jones, P. R. Briddon, and S. Öberg, Phil. Mag. Lett. **66**, 67 (1992).
[94] S. J. Breuer and P. R. Briddon, Phys. Rev. B **51**, 6984 (1995).
[95] J. P. Goss, R. Jones, S. J. Breuer, P. R. Briddon, and S. Öberg, Phys. Rev. Lett. **77**, 3041 (1996).
[96] J. F. Zheng, M. Stavola, and G. D. Watkins, in: The Physics of Semiconductors, Ed. D. J. Lockwood, World Scientific Publ. Co., Singapore 1984 (p. 2363).
[97] R. Jones and S. Öberg, Phys. Rev. Lett. **68**, 86 (1992).
[98] C. P. Ewels, R. Jones, S. Öberg, J. Miro, and P. Deák, Phys. Rev. Lett. **77**, 865 (1996).
[99] M. Suezawa, K. Sumino, H. Harada, and T. Abe, Jpn. J. Appl. Phys. **25**, L859 (1986).
[100] P. R. Briddon and R. Jones, Phys. Rev. Lett. 64, 2535 (1990).
[101] R. Jones and S. Öberg, Phys. Rev. B **44**, 3673 (1991).
[102] K. Woodhouse, R. C. Newman, T. J. deLyon, J. M. Woodall, G. J. Scilla, and F. Cardone, Semicond. Sci. and Technol. **6**, 330 (1991).
[103] B. R. Davidson, R. C. Newman, T. J. Bullough, and T. B. Joyce, Phys. Rev. B **48**, 17106 (1993).
[104] S. J. Breuer, R. Jones, P. R. Briddon, and S. Öberg, Phys. Rev. B **53**, 16289 (1996).
[105] M. I. Heggie, R. Jones, and A. Umerski, Inst. Phys. Conf. Series **117**, 125 (1991).
[106] S. Öberg, P. K. Sitch, R. Jones, and M. I. Heggie, Phys. Rev. B **51**, 13138 (1995).
[107] M. I. Heggie, R. Jones, and A. Umerski, phys. stat. sol. (a) **138**, 383 (1993).
[108] I. Yonenaga and K. Sumino, J. Appl. Phys. **65**, 85 (1989).

phys. stat. sol. (b) **217**, 173 (2000)

Subject classification: 61.46.+w; 71.15.Mb; S5.11; S5.12

Electronic Structure Methods for Predicting the Properties of Materials: Grids in Space

J. R. CHELIKOWSKY (a), Y. SAAD (b), S. ÖĞÜT (a), I. VASILIEV (a), and A. STATHOPOULOS (c)

(a) *Department of Chemical Engineering and Materials Science, Institute of Technology, University of Minnesota, Mineapolis, Minnesota 55455, USA*

(b) *Department of Computer Science Minnesota Supercomputing Institute, University of Minnesota, Minneapolis, Minnesota 55455, USA*

(c) *Department of Computer Science, College of William and Mary, Williamsburg, VA 23187, USA*

(Received August 10, 1999)

If the electronic structure of a given material is known, then many physical and chemical properties can be accurately determined without resorting to experiment. However, determining the electronic structure of a realistic material is a difficult numerical problem. The chief obstacle faced by computational materials and computer scientists is obtaining a highly accurate solution to a complex eigenvalue problem. We illustrate a new numerical method for calculating the electronic structure of materials. The method is based on discretizing the *pseudopotential density functional method* (PDFM) in real space. The eigenvalue problem within this method can involve large, sparse matrices with up to thousands of eigenvalues required. An efficient and accurate solution depends increasingly on complex data structures that reduce memory and time requirements, and on parallel computing. This approach has many advantages over traditional plane wave solutions, e.g., no fast Fast Fourier Transforms (FFTs) are needed and, consequently, the method is easy to implement on parallel platforms. We demonstrate this approach for localized systems such as atomic clusters.

1. Introduction: The Electronic Structure Problem

A fundamental problem in condensed matter physics is the prediction of the electronic structure of complex systems such as amorphous solids and glasses or small atomic clusters. Many materials properties can be predicted if an accurate solution of the electronic structure for the system of interest exists. For example, the structural properties of a material can be determined if the total electronic energy of the system is known as a function of atomic positions. Likewise, response functions such as optical and dielectric constants can be determined if the electronic wave functions are known. Beyond the scientific merit of verifying experimental results and establishing the validity of new scientific concepts, these electronic structure calculations also facilitate testing of hypothetical materials without laboratory experiments.

There are numerous approaches to the electronic structure problem [1]. These approaches range from simple empirical methods where experiment is used to fix adjustable parameters to first principles methods where no experimental data are needed. Here we focus on a first principles approach. It is important to recognize the advantages of first principles methods. Such methods avoid *ad hoc* constructs and the prejudice of "preconceived" ideas relative to the nature of the chemical bonds in condensed matter.

The construction of efficient first principles methods is among the most challenging tasks in computational materials science today. The heart of the computation problem is to obtain highly accurate values for the total electronic energy of matter from the solution of a large, Hermitian eigenvalue problem. Part of the challenge stems from the fact that the number of eigenvalues and eigenvectors (i.e., eigenpairs) required can be very large, say in the order of thousands. This number is proportional to the number of atoms in the system which can be in the thousands (or more) for realistic models.

The electronic structure of matter is described by a many body wavefunction Ψ which obeys the Schrödinger equation

$$\mathcal{H}\Psi = \mathcal{E}\Psi,$$

where $\mathcal{H}$ is the Hamiltonian operator for the system and $\mathcal{E}$ is the total energy. This expression can be simplified through several approximations. These approximations are all based on the removal of degrees of freedom. For example, the Born-Oppenheimer approximation separates the nuclear degrees freedom and the electronic degrees of freedom. Within this approximation, the nuclear coordinates are treated as classical objects. Another simplification is the utilization of density functional theory [2 to 4] to map the many body problem on to a one-electron problem. These two approximations yield the following:

$$\left[\frac{-\hbar^2 \nabla^2}{2m} + V_{\text{tot}}[\varrho(\mathbf{r}), \mathbf{r}]\right] \psi_i(\mathbf{r}) = E_i \psi_i(\mathbf{r}), \tag{1}$$

where $\hbar$ is Planck's constant, m is the electron mass, V_{tot} is the total potential at some point $\mathbf{r}$ in the system, and $\varrho(\mathbf{r})$ is the charge density at that point. The potential depends explicitly on the charge density, which in turn depends on the wavefunctions ψ_i as follows:

$$\varrho(\mathbf{r}) = -e \sum_i |\psi_i(\mathbf{r})|^2, \tag{2}$$

where the sum is over occupied states. The electronic structure problem can be viewed as a nonlinear eigenvalue problem because of the nonlinear dependence of the operator on the left-hand side on the eigenfunctions.

Within the *local density approximation* theory [2], the potential V_{tot} may be written as a sum of three distinct terms, specifically,

$$V_{\text{tot}}(\mathbf{r}) = V_{\text{ion}}(\mathbf{r}) + V_{\text{H}}(\mathbf{r}) + V_{\text{xc}}[\mathbf{r}, \varrho(\mathbf{r})], \tag{3}$$

where V_{ion} is the unscreened potential. In the case of an atom, it would correspond to the bare nuclear potential. V_{H} is the Hartree potential, and V_{xc} is the exchange–correlation potential. Once the charge density $\varrho(r)$ is known, the Hartree potential is obtained by solving the Poisson equation

$$\nabla^2 V_{\text{H}} = -4\pi e \varrho(r). \tag{4}$$

The exchange–correlation potential depends on the charge density at the point of interest. Both potentials V_{H} and V_{xc} have a local character. The density functional approximation reduces the number of degrees to those of a "one-electron" problem.

Within the local density approximation, the total potential and the wave functions are interdependent through the charge density. Equations (1), (3), and (4) constitute a set of nonlinear equations. These are typically solved by the construction of a self-consistent field (SCF). The procedure is usually initiated by superposing atomic charge densities to obtain an approximate charge density for the system of interest. From this density, the "input" Hartree and exchange–correlation potentials are formed. One solves a *Kohn-Sham* eigenvalue problem [2],

$$\left[\frac{-\hbar^2 \nabla^2}{2m} + V_{\text{ion}}^{\text{p}}(\mathbf{r}, \varrho(\mathbf{r})) + V_H(\mathbf{r}) + V_{\text{xc}}(\mathbf{r}) \right] \psi_i(r) = E_i \psi_i(r), \tag{5}$$

for the eigenvalues and eigenvectors using the input potentials. With the eigenvalues and eigenvectors determined, we can obtain an "output" charge density. Using the "input" and "output" charge densities new V_H and V_{xc} potentials can be obtained. If the input and output charge densities are identical, then a self-consistent field is obtained. Since the superposition of atomic charge densities is not identical to the charge density in condensed matter phases, the input and output densities are significantly different. The input and output densities are mixed and a new density is formed and input into a new SCF cycle. The resulting V_H and V_{xc} are inserted into Eq. (5) and new eigenvalues and eigenvectors are obtained. This process is repeated until the difference between input and output potentials is below some specified tolerance.

The total electronic structure of the material, $E_{\text{tot}}^{\text{el}}$, can be written as

$$E_{\text{tot}}^{\text{el}} = \sum_i E_i - \tfrac{1}{2} \int d^3 r \, V_H(\mathbf{r}) \, \varrho(\mathbf{r}) + \tfrac{1}{2} \int d^3 r \, [E_{\text{xc}}[\varrho(\mathbf{r})] - V_{\text{xc}}[\varrho(\mathbf{r})]] \, \varrho(\mathbf{r}). \tag{6}$$

The total energy of the system is given by

$$E\{\mathbf{R}\} = E_{\text{tot}}^{\text{el}} + E_{\text{ion-ion}}\{\mathbf{R}\}. \tag{7}$$

The second term represents the ion–ion interaction, i.e., the Coulombic interaction between the ion cores whose positions are given by $\{\mathbf{R}\}$. $E_{\text{xc}}[\varrho(\mathbf{r})]$ is the exchange–correlation energy density [2]. If we are given $E\{\mathbf{R}\}$, any property related to the structure of matter can be calculated, at least in principle.

2. Solving the Eigenvalue Problem

A major difficulty in solving the eigenvalue problem in Eq. (5) are the length and energy scales involved. The inner (core) electrons are highly localized and tightly bound compared to the outer (valence electrons). A simple basis function approach is frequently ineffectual. For example, a plane wave basis might require 10^5 to 10^6 waves to represent converged wavefunctions for a core electron whereas only 10^2 waves are required for a valence electron. The pseudopotential overcomes this problem by removing the core states from the problem and replacing the all electron potential by one that replicates only the chemically active, valence electron states [5]. By construction, the pseudopotential reproduces the valence state properties such as the eigenvalue spectrum and the charge density outside the ion core. The unscreened pseudopotential, $V_{\text{ion}}^{\text{p}}(\mathbf{r})$ replaces $V_{\text{ion}}(\mathbf{r})$ in Eq. (5).

Since the pseudopotential is weak, simple basis sets such as a plane wave basis are extremely effective. For example, in the case of crystalline silicon only 50 to 100 plane waves need to be used. The resulting matrix representation of the Schrödinger operator is dense on the Fourier (plane wave) space, but it is not formed explicitly, Instead, matrix–vector product operations are performed with the help of Fast Fourier Transforms (FFT). This approach is akin to spectral techniques used in solving certain types of partial differential equations. The plane wave method uses a basis of the form

$$\psi_{\mathbf{k}}(\mathbf{r}) = \sum_{\mathbf{G}} a(\mathbf{k}, \mathbf{G}) \exp\left(i(\mathbf{k} + \mathbf{G}) \cdot \mathbf{r}\right), \tag{8}$$

where $\mathbf{k}$ is the wave vector, $\mathbf{G}$ is a reciprocal lattice vector and $a(\mathbf{k}, \mathbf{G})$ represent the coefficients of the basis. In a plane wave basis, the Laplacian term of the Hamiltonian is represented by a diagonal matrix. The potential term V_{tot}^p gives rise to a dense matrix. In practice, these matrices are never formed explicitly, since with appropriate use of FFT we can easily operate with this matrix by going back and forth between real space and Fourier space. Indeed, in real space it is trivial to operate with the potential term which is represented by a diagonal matrix, and in Fourier space it is trivial to operate with the Laplacian term which is also represented by a diagonal matrix. The use of plane wave bases also leads to natural preconditioning techniques which are obtained by simply employing a matrix obtained from a smaller plane wave basis, neglecting the effect of high frequency terms on the potential. For periodic systems, where $\mathbf{k}$ is a good quantum number, the plane wave basis coupled to pseudopotentials is quite effective. However, for non-periodic systems such as clusters, liquids or glasses, the plane wave basis must be combined with a *supercell method* [5]. The supercell repeats the localized configuration to impose periodicity to the system. There is also again a parallel to be made with spectral methods which are quite effective for simple periodic geometries, but lose their superiority when more generality is required. In addition to these difficulties the two FFTs performed at each iteration can be costly, requiring $n \log n$ operations, where n is the number of plane waves, versus $O(N)$ for real space methods where N is the number of grid points. Usually, the matrix size $N \times N$ is larger than $n \times n$ but only within a constant factor. This is exacerbated in high performance environments where FFTs require an excessive amount of communication and are particularly difficult to implement efficiently.

Another popular basis employed with pseudopotentials include Gaussian orbitals, see e.g. [6]. Gaussian bases have the advantage of yielding analytical matrix elements provided the potentials are also expanded in Gaussians. However, the implementation of a Gaussian basis is not as straightforward as with plane waves. For example, numerous indices must be employed to label the state, the atomic site, and the Gaussian orbitals employed. On the positive side, a Gaussian basis yields much smaller matrices and requires less memory than plane wave methods. For this resason Gaussians are especially useful for describing transition metal systems.

An alternative approach is to avoid the use of a basis. For example, one can use a real space method that avoids the use of plane waves and FFTs altogether. This approach has become popular and different versions of this general approach been implemented by several groups. Here we illustrate a particular version of this approach called the *Finite-Difference Pseudopotential Method* (FDPM) [7].

A real space approach overcomes some of the complications involved with non-periodic systems, and although the resulting matrices can be larger than with plane waves, they are sparse and the methods are easier to parallelize. Even on sequential machines, we find that real space methods can be an order of magnitude faster than the traditional approach.

Our *real space* algorithms avoid the use of FFTs by performing all calculations in real physical space instead of Fourier space. A benefit of avoiding FFTs is that the new approaches have very few global communications. In fact, the only global operation remaining in *real space* approaches is that of the inner products. These inner products are required when forming the orthogonal basis used in the generalized Davidson procedure as discussed below.

Our approach utilizes finite difference discretization on a real space grid. A key aspect to the success of the finite difference method is the availability of *higher order finite difference expansions* for the kinetic energy operator, i.e., expansions of the Laplacian [10]. Higher order finite difference methods significantly improve convergence of the eigenvalue problem when compared with standard finite difference methods. If one imposes a simple, uniform grid on our system where the points are described in a finite domain by (x_i, y_j, z_k), we approximate $\partial^2 \psi / \partial x^2$ at (x_i, y_j, z_k) by

$$\frac{\partial^2 \psi}{\partial x^2} = \sum_{n=-M}^{M} C_n \psi(x_i + nh, y_j, z_k) + O(h^{2M+2}), \qquad (9)$$

where h is the grid spacing and M is a positive integer. This approximation is accurate to $O(h^{2M+2})$ upon the assumption that ψ can be approximated accurately by a power series in h. Algorithms are available to compute the coefficients C_n for arbitrary order in h [10].

With the kinetic energy operator expanded as in Eq. (9), one can set up a one-electron Schrödinger equation over a grid. One may assume a uniform grid, but this is not a necessary requirement. $\psi(x_i, y_j, z_k)$ is computed on the grid by solving the eigenvalue problem

$$-\frac{\hbar^2}{2m} \left[\sum_{n_1=-M}^{M} C_{n_1} \psi_n(x_i + n_1 h, y_j, z_k) + \sum_{n_2=-M}^{M} C_{n_2} \psi_n(x_i, y_j + n_2 h, z_k) \right.$$
$$\left. + \sum_{n_3=-M}^{M} C_{n_3} \psi_n(x_i, y_j, z_k + n_3 h) \right] + [V_{\text{ion}}(x_i, y_j, z_k)$$
$$+ V_H(x_i, y_j, z_k) + V_{xc}(x_i, y_j, z_k)] \psi_n(x_i, y_j, z_k) = E_n \psi_n(x_i, y_j, z_k). \qquad (10)$$

If we have L grid points, the size of the full matrix resulting from the above problem is $L \times L$.

A complicating issue in setting up an algorithm is the ionic pseudopotential term. This term is easy to cast in Fourier space, but it may also be expressed in real space. The interactions between valence electrons and pseudo-ionic cores may be separated into a local potential and a Kleinman and Bylander [8] form of a nonlocal pseudopotential in *real space* [9],

$$V_{\text{ion}}(\mathbf{r}) \psi_n(\mathbf{r}) = \sum_{a} V_{\text{loc}}(|\mathbf{r}_a|) \psi_n(\mathbf{r}) + \sum_{a,n,lm} G^a_{n,lm} u_{lm}(\mathbf{r}_a) \Delta V_l(r_a), \qquad (11)$$

$$K^a_{n,lm} = \frac{1}{\langle \Delta V^a_{lm} \rangle} \int u_{lm}(\mathbf{r}_a) \Delta V_l(r_a) \psi_n(\mathbf{r}) \, d^3 r \qquad (12)$$

and $\langle \Delta V^a_{lm} \rangle$ is the normalization factor,

$$\langle \Delta V^a_{lm} \rangle = \int u_{lm}(\mathbf{r}_a) \Delta V_l(r_a) u_{lm}(\mathbf{r}_a) \, d^3 r, \qquad (13)$$

where $\mathbf{r}_a = \mathbf{r} - \mathbf{R}_a$, and u_{lm} are the atomic pseudopotential wave functions of angular momentum quantum numbers (l, m) from which the l-dependent ionic pseudopotential, $V_l(r)$, is generated. $\Delta V_l(r) = V_l(r) - V_{\text{loc}}(r)$ is the difference between the l component of the ionic pseudopotential and the local ionic potential.

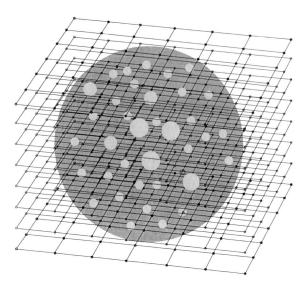

Fig. 1. Uniform grid illustrating a typical configuration for examining the electronic structure of a localized system. The gray sphere represents the domain where the wave functions are allowed to be nonzero. The light spheres within the domain are atoms

The grid we use is based on points uniformly spaced in a three-dimensional cube as shown in Fig. 1, with each grid point corresponding to a row in the matrix. However, many points in the cube are far from any atoms in the system and their negligible charge may then be replaced by zero. Special data structures may be used to discard these points and keep only those having a nonzero charge. The size of the Hamiltonian matrix is usually reduced by a factor of two to three with this strategy, which is quite important considering the large number of eigenvectors which must be saved. Further, since the Laplacian can be represented by a simple stencil, and since all local potentials sum up to a simple diagonal matrix, the Hamiltonian needs not be stored. Handling the ionic pseudopotential is complex as it consists of a local and a nonlocal term (Eqs. (11) and (12)). In the discrete form, the nonlocal term becomes a sum over all atoms a, and quantum numbers (l,m) of rank-one updates

$$V_{\text{ion}} = \sum_a V_{\text{loc},a} + \sum_{a,l,m} c_{a,l,m} U_{a,l,m} U_{a,l,m}^{\text{T}}, \tag{14}$$

where $U_{a,l,m}$ are sparse vectors which are only nonzero in a localized region around each atom, $c_{a,l,m}$ are normalization coefficients.

There are several difficulties with the eigenproblems generated in this application in addition to the size of the matrices. First, the number of required eigenvectors is proportional to the atoms in the system, and can grow up to thousands. Besides storage, maintaining the orthogonality of these vectors can be a formidable task. Second, the relative separation of the eigenvalues becomes increasingly poor as the matrix size increases and this has an adverse effect on the rate of convergence of the eigenvalue solvers. Preconditioning techniques attempt to alleviate this problem.

On the positive side, the matrix needs not be stored as was mentioned earlier and this reduces storage requirement. In addition, good initial eigenvector estimates are available at each iteration from the previous SCF loop. An iterative method should be able to use this information.

In this work, we developed a code based on the generalized Davidson [11] method, in which the preconditioner is not restricted to be a diagonal matrix as in the Davidson method. The code addresses the problems mentioned above by using implicit deflation (locking), a windowing approach to gradually compute all the required eigenpairs, and special targeting and reorthogonalization schemes. A more detailed description can be found in [12].

The preconditioning technique we used in our approach is based on a filtering idea and the fact that the Laplacian is an elliptic operator [13]. The eigenvectors corresponding to the few lowest eigenvalues of ∇^2 are smooth functions and so are the corresponding wavefunctions. When an approximate eigenvector is known at the points of the grid, a smoother eigenvector can be obtained by averaging the value at every point with the values of its neighboring points. Assuming a Cartesian (x, y, z) coordinate system, the low frequency filter acting on the value at the point (i, j, k), which represents one element of the eigenvector, is described by

$$\psi_{i,j,k} := \frac{\psi_{i-1,j,k} + \psi_{i,j-1,k} + \psi_{i,j,k-1} + \psi_{i+1,j,k} + \psi_{i,j+1,k} + \psi_{i,j,k+1}}{12} + \frac{\psi_{i,j,k}}{2}. \quad (15)$$

It is worth mentioning that other preconditioners that have been tried have resulted in mixed success. The use of shift-and-invert [14] involves solving linear systems with $\mathbf{A} - \sigma\mathbf{I}$, where $\mathbf{A}$ is the original matrix and the shift σ is close to the desired eigenvalue. These methods would be prohibitively expensive in our situation, given the size of the matrix and the number of times that $\mathbf{A} - \sigma\mathbf{I}$ must be factored. Alternatives based on an approximate factorization such as ILUT [15] are ineffective beyond the first few eigenvalues. Methods based on approximate inverse techniques have been somewhat more successful, performing better than filtering at additional preprocessing and storage cost. Preconditioning 'interior' eigenvalues, i.e., eigenvalues located well inside the interval containing the spectrum, is still a very hard problem. Current solutions only attempt to dampen the effect of eigenvalues which are far away from the ones being computed. This is in effect what is achieved by filtering and sparse approximate inverse preconditioning. These techniques do not reduce the number of steps required for convergence in the same way that shift-and-invert techniques do. However, filtering techniques are inexpensive to apply and result in fairly substantial savings in iterations.

3. Parallel Implementation

For distributed memory parallel computers, the SPMD (Single Program Multiple Data) model has emerged as the most popular programming paradigm. In our implementation of the SCF procedure we have followed a hybrid of the SPMD and the master-worker paradigm. The master performs most of the preprocessing, computing of scalar values, and processing of the new potential at each SCF iteration. The master is also responsible for applying the mixing scheme on the potentials. The workers solve the eigenvalues and eigenvectors, update the charge density, and solve the Poisson equation for the Hartree potential in an SPMD fashion.

There are several reasons dictating the master-worker choice. First, there are some inherently sequential parts in the code which require large memory but short execution time. It is also common that one of the nodes in a parallel environment is equipped with larger memory than the others. Second, the code calls several library routines which have been written by various research groups over a long period of time. Despite their importance, these routines take only a few seconds to execute. Parallelizing them all would require an inordinate amount of effort with doubtful results as to the achievable gains. Third, this paradigm allows incremental parallelization of the code, implementing first the most time consuming procedures, such as the eigensolver, then gradually adding parallelism to other parts. Correctness of the code is also easier to maintain by this strategy. Finally, the resulting code is portable to other parallel platforms without requiring large amounts of memory for all the worker processors.

The primary sources of parallelism intrinsically available in the application are: 1. the multitude of required eigenvectors, and 2. parallelism from spatial decomposition. Assigning each processor the task of calculating all the eigenpairs in a segment of the spectrum would provide excellent coarse grain parallelism and parallel efficiency. For each eigenpair, one could use inverse iteration with some iterative method. However, the linear systems to be solved are highly indefinite and iterative methods for the inverse iteration converge extremely slowly. As was mentioned earlier shift-and-invert is impractical for the large matrices at hand. An alternative is to use a polynomial preconditioning approach. A polynomial p can be found such that the dominant eigenvalues of $p(\mathbf{A})$ are the transforms by p of the eigenvalues in the desired subinterval. Then these dominant eigenvalues and associated eigenvectors can be computed and the corresponding eigenvalues of the original matrix can then be evaluated. A major advantage with this approach is that global orthogonality does not need to be maintained since the eigenvectors of a Hermitian matrix are orthogonal if they are computed accurately enough; only eigenvectors associated with eigenvalues in a given subinterval must be orthogonalized during the computation. This is a workable approach but the book-keeping required in order to ensure that no eigenvalues are missed and that they are all represented only once may be quite cumbersome. In addition high degree polynomials may be needed that reduce the gains from parallelism.

Instead of this 'spectral decoupling' idea, we have adopted a domain decomposition approach based on partitioning the physical space. The problem is mapped onto the processors in a data parallel way because of the fine granularity parallelism present in the matrix–vector multiplication and orthogonalization operations. The rows of the Hamiltonian (and therefore the rows of the eigenvectors and potential vectors) are assigned to processors according to a partitioning of the physical domain. The subdomains can be chosen naturally as sub-cubes or slabs of the cube, but since the zero-charge areas can be arbitrarily distributed in the domain, a general partitioning is more appropriate. This is illustrated in Fig. 2. We have designed the mapping routines to be independent of the partitioner, requiring only a function $P(i, j, k)$ which returns the number of the processor where point (i, j, k) resides. This facilitates the use of many publicly available partitioning tools. We have tested two ways of partitioning. The first is a greedy approach that optimizes load balancing by ordering the points and assigning the same number of points to each processor, but it ignores the amount of communication which is induced. The second approach uses the popular partitioning package METIS [16] which seeks to optimize both load balancing and the communication volume between processors.

Since the matrix is not actually stored, an explicit reordering can be considered so that the rows on a processor are numbered consecutively. Under this conceptually easier scheme, only a list of pointers is needed that denote where the rows of each processor start. The nonlocal part of the matrix, which is a sum of rank-one updates, is mapped in a similar way. For each atom and for each pair of quantum numbers, the

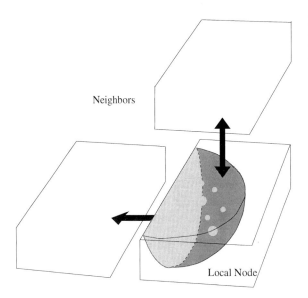

Fig. 2. An example of a possible decomposition. The subdomains illustrated are assigned to a particular processor. Although the subdomains are shown as cubic, they can be chosen to be an arbitrary configuration. See Fig. 1

sparse vector $U_{a,l,m}$ in Eq. (14) is partitioned according to the rows it contributes to. Even though the number of nonzero elements of the U vectors is small, their partitioning is fairly well balanced if the matrix partitioning is well balanced. With this mapping, the large storage requirements of the program are distributed.

The tools that we developed for mapping, setting up the data structures and performing the communication, are independent of the nature of our problem and can be embedded in other applications for unstructured stencil computations, which use any of the general data structures described in the following sections.

In the Davidson algorithm, the basis vectors and long work arrays, follow the same distribution as the eigenvectors. Thus, all vector updates (saxpy operations) can be performed in parallel, and all reduction operations (e.g., sdot operations) require a global reduction (e.g., global sum) of the partial results on each processor.

The matrix–vector multiplication is performed in three steps. First, the contributions of the diagonals (potentials and the Laplacian diagonal) is computed in parallel on all processors. Second, the contribution of the Laplacian is considered on the rows of each processor. As in the sequential code, this is performed by using the stencil information. In the parallel implementation communication is necessary, since some of the neighboring points of the local subdomain may reside on different processors. For this reason, each processor maintains the following data structure, which maps the local grid points to the local rows, and appends the needed interface points from other processors at the end of the local row list,

$$\text{index}(i,j,k) = \begin{cases} \text{row number in local ordering}, & \text{if } (i,j,k) \text{ is on local processor}, \\ \text{index below local ordering}, & \text{if } (i,j,k) \text{ is a needed interface row}, \\ \text{special index}, & \text{if } (i,j,k) \text{ is not considered (zero charge)}. \end{cases}$$
(16)

The workers build this and other supporting data structures during the setup phase, by locating which of their rows are needed in the stencils of other processors. In the second step of the matrix–vector multiplication, this interface information is exchanged among nearest neighbors and the stencil multiplication can proceed in parallel. In the third step, each of the rank-one updates of the nonlocal components is computed as a sparse, distributed dot product. All local dot products are first computed before a global sum of their values takes place. The solution of Eq. (4) for the Hartree potential with the Conjugate Gradient method and the preconditioning operation also require the stencil, and therefore, they have the same communication pattern as the second step of the matrix–vector multiplication.

Orthogonalization is an expensive phase, and as the number of required eigenvectors increases, it is bound to dominate the cost. Reorthogonalization is performed every time a vector norm reduces significantly after orthogonalization. Although reorthogonalization recovers the numerical accuracy lost in the Gram-Schmidt procedure, its nature is sequential and induces several synchronization points. In the current application, global sums of the dot products are delayed so that only one synchronization is needed. In addition, by performing the reorthogonalization test through easily obtained estimates of the vector norms, we introduce only two synchronization points in the procedure.

To demonstrate the scalability of the code, we examined a large quantum dot involving 191 silicon atoms and 148 hydrogen atoms. The matrix size involved 83200 grid points, i.e., in principle the Hamiltonian matrix contains 83200×83200 entries. For the electronic structure calculation, 560 eigenvalues were obtained. The overall scalability is illustrated in Fig. 3. Clusters or quantum dots present a difficult problem as the environment for each atom can be very different, e.g., a surface atom has far fewer neighbors than does an interior atom. Owing to this issue, our scale up efficiency of $\approx 80\%$ is quite good.

We note that there are several groups utilizing this real space approach, examples of similar approaches can be found in [17]. There are some notable differences between the current approaches and these approaches. A non-uniform grid is often incorporated. Non-uniform grids can be used to accommodate systems with highly heterogeneous environments. As an example, consider a system with two atomic constituents: one with highly localized wave functions and another with delocalized wave functions. For a uniform grid, the spacing will be fixed by the highly localized species and, consequently, will be "over converged" for the delocalized species. Non-uniform grids can be adapted so that regions with rapidly fluctuating wave functions are represented by a

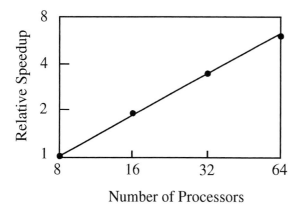

Fig. 3. Speedup efficiency for a large silicon cluster on a massively parallel platform

fine grid and regions with slowly fluctuating wavefunctions are represented by a coarse grid. This advantage can be considerable for some systems, but the real space approach loses its ease of implementation in this case. For example, the Hamiltonian matrix loses its highly structured form and expressions for the interatomic forces become quite complex. Also, implementing and optimizing a uniform grid, especially in systems where the atoms are allowed to move, can significantly increase the computational load. To date, no molecular dynamics simulations have been performed with non-uniform grids for this reason.

An alternative approach to the finite difference code is to use a *finite elements* method. The finite element approach shares some of the advantages associated with methods based on non-uniform grids. Finite elements can be adapted to enhance convergence over specific regimes in real space. In addition, finite element approaches are variational, since they correspond to a basis oriented approach. (In contrast, finite difference methods are not variational with the grid spacing and the total energy can converge from above or below.) However, finite element methods also share the disadvantages of non-uniform grids, i.e., they are difficult to implement, and more computationally intensive.

4. Properties of Confined Systems: Clusters

The electronic and structural properties of atomic clusters stand as one of the outstanding problems in materials science. Clusters possess properties that are characteristic of neither the atomic nor solid state. For example, the energy levels in atoms may be discrete and well separated in energy relative to kT. In contrast, solids have continuum of states (energy bands). Clusters may reside between these limits, i.e., the energy levels may be discrete, but with a separation much less than kT.

Real space methods are ideally suited for investigating these systems. In contrast to plane wave methods, real space methods can examine non-periodic systems without introducing artifacts such as supercells. Also, one can easily examine charged clusters. In supercell configurations, unless a compensating background charge is added, the Coulomb energy diverges for charged clusters. A closely related issue concerns electronic excitations. In periodic systems, it is nontrivial to consider localized excitation, e.g., exciting an atom in one cell, excites all atoms. Density functional formalisms avoid these issues by considering localized or non-periodic systems.

4.1 Structure

Perhaps the most fundamental issue in dealing with clusters is the structure. Before any accurate theoretical calculations can be performed for a cluster, the atomic geometry must be known. However, determining the atomic structure of clusters can be a formidable exercise. Serious problems arise from the existence of multiple local minima in the potential energy surface of these systems. This is especially true for some clusters such as those involving semiconducting species. In these clusters, strong many body forces can exist.

A convenient method to determine the structure of small clusters is *simulated annealing*. Within this technique, atoms are randomly placed within a large cell and allowed to interact at a high (usually fictive) temperature. The atoms will sample a large number of configurations. As the system is cooled, the number of high energy configurations sampled is restricted. If the anneal is done slowly enough, the procedure should quench out structural candidates for the ground state structures.

Langevin molecular dynamics appears well suited for such simulated anneals. In Langevin dynamics, the ionic positions $\mathbf{R}_j$ evolve according to

$$M_j\ddot{\mathbf{R}}_j = \mathbf{F}(\{\mathbf{R}_j\}) - \gamma M_j\dot{\mathbf{R}}_j + \mathbf{G}_j, \tag{17}$$

where $\mathbf{F}(\{\mathbf{R}_j\})$ is the interatomic force on the j-th particle, and $\{M_j\}$ are the ionic masses. The last two terms on the right-hand side of Eq. (17) are the dissipation and fluctuation forces, respectively. The dissipative forces are defined by the friction coefficient γ. The fluctuation forces are defined by random Gaussian variables $\{\mathbf{G}_i\}$ with a white noise spectrum,

$$\langle G_i^\alpha(t) \rangle = 0 \quad \text{and} \quad \langle G_i^\alpha(t)\, G_j^\alpha(t') \rangle = 2\gamma M_i k_B T \delta_{ij}\, \delta(t-t'). \tag{18}$$

The angular brackets denote ensemble or time averages, and α stands for the Cartesian component. The coefficient of T on the right-hand side of Eq. (18) insures that the fluctuation–dissipation theorem is obeyed, i.e., the work done on the system is dissipated by the viscous medium [18, 19]. The interatomic forces can be obtained from the Hellmann-Feynman theorem using the pseudopotential wavefunctions.

Our simulations can be contrasted with other techniques such as the Car-Parrinello method. We do not employ fictitious electron dynamics; at each time step the system is quenched to the Born-Oppenheimer surface. Our approach requires a full self-consistent treatment of the electronic structure problem; however, because the interatomic forces are true, quantum forces of the resulting molecular dynamics simulation can be performed with much larger time steps. Typically, it is possible to use steps an order of magnitude larger than in the Car-Parrinello method.

To illustrate the procedure, we consider a germanium cluster of seven atoms. With respect to the technical details for this example, the initial temperature of the simulation was taken to be 2800 K; the final temperature was taken to be 300 K. The annealing schedule lowered the temperature 500 K each 50 time steps. The time step was taken to be 7 fs. The friction coefficient in the Langevin equation was taken to be 6×10^{-4} a.u. After the clusters reached a temperature of 300 K, they were quenched to 0 K. The ground state structure was found through a direct minimization by a steepest descent procedure.

Choosing an initial atomic configuration for the simulation takes some care. If the atoms are too far apart, they will exhibit Brownian motion and may not form a stable cluster as the simulation proceeds. If the atoms are too close together, they may form a metastable cluster from which the ground state may be kinetically inaccessible even at the initial high temperature. Often the initial cluster is formed by a random placement of the atoms with a constraint that any given atom must reside within 1.05 and 1.3 times the dimer bond length of at least one atom. The cluster in question is placed in a spherical domain. Outside of this domain, the wavefunction is required to vanish. The radius of the sphere is such that the outmost atom is at least 6 a.u. from the boundary. Initially, the grid spacing was 0.8 a.u. For the final quench to a ground state structure, the grid spacing was reduced to 0.5 a.u. As a rough estimate, one can compare this grid spacing with a plane wave cut-off of $(\pi/h)^2$ or about 40 Ry for $h = 0.5$ a.u.

In Fig. 4, we illustrate the simulated anneal for this Ge_7 cluster. While the initial cluster contains several of bonds, the structure is still somewhat removed from the

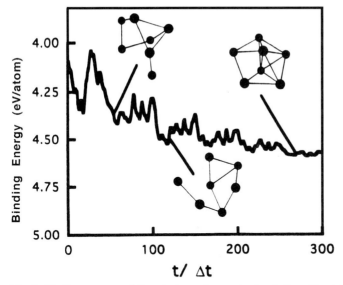

Fig. 4. Binding energy of Ge$_7$ during a Langevin simulation. The initial temperature is 2800 K; the final temperature is 300 K. Bonds are drawn for interatomic distances of less than 2.5 Å. The time step is 7 fs

ground state. After ≈200 time steps, the ground state structure is essentially formed. The ground state of Ge$_7$ is a bicapped pentagon, as is the corresponding structure for the Si$_7$ cluster. The binding energy shown is relative to the isolated Ge atom. We have not included gradient corrections, or spin polarization [20] in our work. Therefore, the values indicated are likely to overestimate the binding energies by about 20% or so.

In Fig. 5, we present the ground state structures for Ge$_n$ for $n \leq 10$. The structures for Ge$_n$ are very similar to Si$_n$. The primary difference resides in the bond lengths. The Si bond length in the crystal is 2.35 Å, whereas in Ge the bond length is 2.44 Å. This difference is reflected in the bond lengths for the corresponding clusters. Ge$_n$ bond lengths are typically a few percent larger than the corresponding Si$_n$ clusters.

It should be emphasized that this annealing simulation is an optimization procedure. As such, other optimization procedures may be used to extract the minimum energy structures. Recently, a genetic algorithm has been used to examine carbon clusters [21]. In this algorithm, an initial set of clusters is "mated" with the lowest energy offspring "surviving". By examining several thousand generations, it is possible to extract a reasonable structure for the ground state. The genetic algorithm has some advantages over a simulated anneal, especially for clusters which contain more than ≈20 atoms. One of these advantages is that kinetic barriers are more easily overcome. However, the implementation of the genetic algorithm is more involved than an annealing simulation, e.g., in some cases "mutations," or *ad hoc* structural rearrangements, must be introduced to obtain the correct ground state.

4.2 Photoemission spectra

A very useful probe of condensed matter involves the photoemission process. Incident photons are used to eject electrons from a solid. If the energy and spatial distributions of the electrons are known, then information can be obtained about the electronic structure of the materials of interest. For crystalline matter, the photoemission spectra can be related to the electronic density of states. For confined systems, the interpretation is not as straightforward. One of the earliest experiments performed to examine

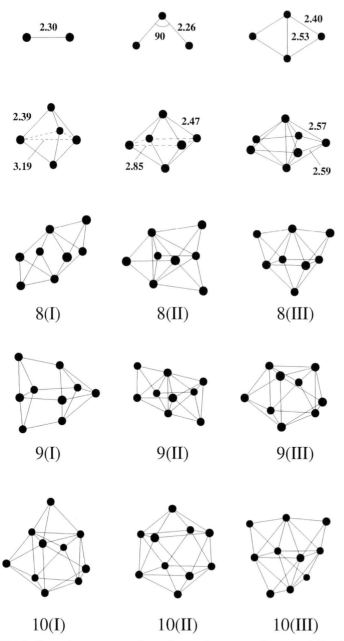

Fig. 5. Ground state geometries and some low-energy isomers of Ge_n ($n \leq 10$) clusters. Interatomic distances (in Å) are given for clusters with $n \leq 7$. For $n > 8$, the lowest energy isomer is given by (I)

the electronic structures of small semiconductor clusters examined negatively charged Si_n and Ge_n ($n \leq 12$) clusters [22]. The photoemission spectra obtained in this work were used to gauge the energy gap between the highest occupied state and the lowest unoccupied state. Large gaps were assigned to the "magic number" clusters, while other clusters appeared to have vanishing gaps. Unfortunately, the first theoretical estimates [23] for these gaps showed substantial disagreements with the measured values. It was proposed by [22], that sophisticated calculations including transition cross sections and final states were necessary to identify the cluster geometry from the photoemission data. The data were first interpreted in terms of the gaps obtained for *neutral* clusters; it was later demonstrated that *atomic relaxations* within the *charged* cluster are important in analyzing the photoemission data [24]. In particular, atomic relaxations as a result of charging may change dramatically the electronic spectra of certain clusters. These charge induced changes in the gap were found to yield very good agreement with the experiment.

The photoemission spectrum of Ge_{10}^- illustrates some of the key issues. Unlike Si_{10}^-, the experimental spectrum for Ge_{10}^- does not exhibit a gap. Cheshnovsky et al. interpreted this to mean that Ge_{10}^- does not exist in the same structure as Si_{10}^-. This is a strange result. Si and Ge are chemically similar and the calculated structures for both *neutral* structures are similar. The lowest energy structure for both ten atom clusters is the tetracapped trigonal prism (labeled by (I) in Fig. 5). The photoemission spectra for these clusters can be simulated by using Langevin dynamics. Within the Langevin framework, the clusters are immersed in a fictive heat bath, and as such, subjected to stochastic forces. If one maintains the temperature of the heat bath and averages over the eigenvalue spectra, a density of states for the cluster can be obtained. The heat bath resembles a buffer gas as in the experimental set-up, but the time intervals for collisions are not similar to the true collision processes in the atomic beam. The simulated photoemission spectrum for Si_{10}^- is in very good agreement with the experimental results, reproducing both the threshold peak and other features in the spectrum. If a simulation is repeated for Ge_{10}^- using the tetracapped trigonal prism structure, the resulting photoemission spectrum is *not* in good agreement with experiment. Moreover, the calculated electron affinity is 2.0 eV in contrast to the experimental value of 2.6 eV. However, there is no reason to believe that the tetracapped trigonal prism structure is correct for Ge_{10} when charged. In fact, we find that the bicapped antiprism structure is

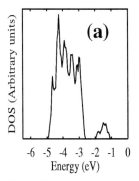 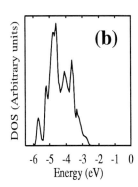 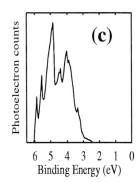

Fig. 6. Calculated density of states for Ge_{10}^- a) for the tetracapped trigonal prism structure, b) for the bicapped antiprism structure. c) Experimental photoemission spectra from Ref. [5]

lower in energy for Ge_{10}^-. The resulting spectra using both structures (I and II in Fig. 5) are presented in Fig. 6, and compared to the photoemission experiment. The calculated spectrum using the bicapped antiprism structure is in very good agreement with the photoemission. The presence of a gap is indicated by a small peak removed from the density of states (Fig. 6a). This feature is absent in the bicapped antiprism structure (Fig. 6b) and is consistent with experiment. For Ge_{10}, charging the structure reverses the relative stability of the two structures. This accounts for the major differences between the photoemission spectra.

4.3 Vibrational modes

Experiments on the vibrational spectra of clusters can provide us with very important information about their physical properties. Recently, Raman experiments have been performed on clusters which have been deposited on inert substrates [25]. Since different structural configurations of a given cluster can possess different vibrational spectra, it is possible to compare the vibrational modes calculated for a particular structure with experiment. If the agreement between experiment and theory is good, this is a necessary condition for the validity of the theoretically predicted structure.

There are two common approaches for determining the vibrational spectra of clusters. One approach is to calculate the dynamical matrix for the ground state structure of the cluster,

$$M_{i\alpha, j\alpha} = \frac{1}{m} \frac{\partial^2 E}{\partial R_i^\alpha \partial R_j^\alpha} = -\frac{1}{m} \frac{\partial F_i^\alpha}{\partial R_j^\alpha}, \qquad (19)$$

where m is the mass of the atom, E is the total energy of the system, F_i^α is the force on atom i in the direction α, R_i^α is the α component of coordinate for atom i. One can calculate the dynamical matrix elements by calculating the first-order derivative of force versus atom displacement numerically. From the eigenvalues and eigenmodes of the dynamical matrix, one can obtain the vibrational frequencies and modes for the cluster of interest [26].

The other approach to determine the vibrational modes is to perform a molecular dynamics simulation. The cluster in question is excited by small random displacements. By recording the kinetic (or binding) energy of the cluster as a function of the simulation time, it is possible to extract the power spectrum of the cluster and determine the vibrational modes. This approach has an advantage for large clusters in that one never has to do a mode analysis explicitly. Another advantage is that anharmonic modes and mode coupling can be examined. It has the disadvantage in that the simulation must be performed over a long time to extract all the modes.

Table 1

Calculated and experimental vibrational frequencies in a Si_4 cluster. See Fig. 7 for an illustration of the normal modes. The frequencies are given in cm^{-1}

	B_{3u}	B_{2u}	A_g	B_{3g}	A_g	B_{1u}
experiment [25]			345		470	
dynamical matrix (this work)	160	280	340	460	480	500
MD simulation (this work)	150	250	340	440	490	500
HF [28]	117	305	357	465	489	529
LCAO [27]	55	248	348	436	464	495

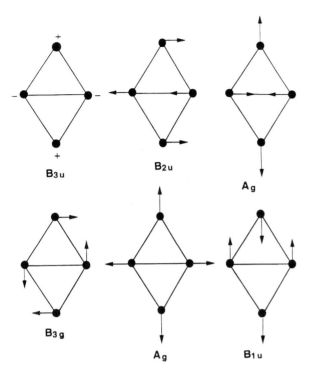

Fig. 7. Normal modes for a Si$_4$ cluster. The + and − signs indicate motion in and out of the plane, respectively

As a specific example, consider the vibrational modes for a small silicon cluster: Si$_4$. The starting geometry was taken to be a planar structure for this cluster as established from a higher order finite difference calculation [26].

It is straightforward to determine the dynamical matrix and eigenmodes for this cluster. In Fig. 7, the fundamental vibrational modes are illustrated. In Table 1, the frequency of these modes are presented. One can also determine the modes via a simulation. To initiate the simulation, one can perform a Langevin simulation [24] with a fixed temperature at 300 K. After a few dozen time steps, the Langevin simulation is turned

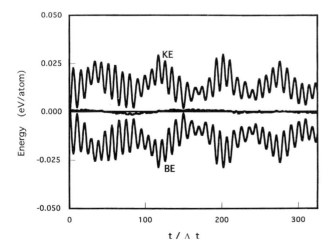

Fig. 8. Simulation for a Si$_4$ cluster. The kinetic energy (KE) and binding energy (BE) are shown as a function of simulation time. The total energy (KE+BE) is also shown with the zero of energy taken as the average of the total energy. The time step, Δt, is 7.4 fs

off, and the simulation proceeds following Newtonian dynamics with "quantum" forces. This procedure allows a stochastic element to be introduced and establish initial conditions for the simulation without bias toward a particular mode. For this example, the time step in the MD simulation was taken to be 3.7 fs, or approximately 150 a.u. The simulation was allowed to proceed for 1000 time steps or roughly 4 ps. The variation of the kinetic and binding energies is given in Fig. 8 as a function of the simulation time. Although some fluctuations of the total energy occur, these fluctuations are relatively small, i.e., less than ≈ 1 meV, and there is no noticeable drift of the total energy. Such fluctuations arise, in part, because of discretization errors. As the grid size is reduced, such errors are minimized [26]. Similar errors can occur in plane wave descriptions using supercells, i.e., the artificial periodicity of the supercell can introduce erroneous forces on the cluster. By taking the power spectrum of either the KE or BE over this simulation time, the vibrational modes can be determined. These modes can be identified with the observed peaks in the power spectrum as illustrated in Fig. 9.

A comparison of the calculated vibrational modes from the MD simulation and from a dynamical matrix calculation are listed in Table 1. Overall, the agreement between the two simulations and the dynamical matrix analysis is quite satisfactory. In particular, the softest mode, i.e., the B_{3u} mode, and the splitting between the (A_g, B_{1u}) modes are well replicated in the power spectrum. The splitting of the (A_g, B_{1u}) modes is less than 10 cm^{-1}, or about 1 meV, which is probably at the resolution limit of any *ab initio* method.

The theoretical values are also compared to experiment. The predicted frequencies for the two A_g modes are surprisingly close to Raman experiments on silicon clusters [25]. The other allowed Raman line of mode B_{3g} is expected to have a lower intensity and has not been observed experimentally.

The theoretical modes using the formalism outlined here are in good accord (except the lowest mode) with other theoretical calculations given in Table 1 (LCAO calculation [27] and a Hartree-Fock (HF) calculation [28]). The calculated frequency of the lowest mode, i.e., the B_{3u} mode, is problematic. The general agreement of the B_{3u} mode as calculated by the simulation and from the dynamical matrix is reassuring.

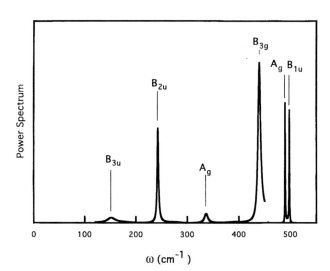

Fig. 9. Power spectrum of the vibrational modes of the Si$_4$ cluster. The simulation time was taken to be 4 ps. The intensity of the B_{3g} and (A_g,B_{1u}) peaks has been scaled by 10^{-2}

Moreover, the real space calculations agree with the HF value to within ≈20 to 30 cm^{-1}. On the other hand, the LCAO method yields a value which is 50 to 70% smaller than either the real space or HF calculations. The origin of this difference is not apparent. For a poorly converged basis, vibrational frequencies are often overestimated as opposed to the LCAO result which underestimates the value, at least when compared to other theoretical techniques. Setting aside the issue of the B$_{3u}$ mode, the agreement between the measured Raman modes and theory for Si$_4$ suggests that Raman spectroscopy can provide a key test for the structures predicted by theory.

4.4 Polarizabilities

Recently, polarizability measurements [29] have been performed for small semiconductor clusters. These measurements allow us to compare our computed values with experiment.

The polarizability tensor α_{ij} is defined as the second derivative of the energy with respect to electric field components. For a noninteracting quantum mechanical system, the expression for the polarizability can be easily obtained by using second-order perturbation theory where the external electric field $\mathcal{E}$ is treated as a weak perturbation.

Within the density functional theory, since the total energy is not the sum of individual eigenvalues, the calculation of polarizability becomes a nontrivial task. One approach is to use density functional perturbation theory which has been developed recently in Green's function and variational formulations [30, 31].

Another approach, which is very convenient for handling the problem for *confined* systems, like clusters, is to solve the full problem exactly within the one-electron approximation. In this approach, the external ionic potential $V_{\text{ion}}(\mathbf{r})$ experienced by the electrons is modified to have an additional term given by $-e\mathcal{E} \cdot \mathbf{r}$. The Kohn-Sham equations are solved with the full external potential $V_{\text{ion}}(\mathbf{r}) - e\mathcal{E} \cdot \mathbf{r}$. For quantities like polarizability, which are derivatives of the total energy, one can compute the energy at a few field values, and differentiate numerically. Real space methods are very suitable for such calculations on confined systems, since the position operator $\mathbf{r}$ is not ill-defined, as is the case for supercell geometries in plane wave calculations.

In Table 2, we present some recent calculations for the polarizability of small Si and Ge clusters. (This procedure has recently been extended to heteropolar clusters such as

Table 2
Static dipole moments and average polarizabilities of small silicon and germanium clusters

silicon			germanium		
cluster	$\|\mu\|$ (D)	$\langle\alpha\rangle$ (Å^3/atom)	cluster	$\|\mu\|$ (D)	$\langle\alpha\rangle$ (Å^3/atom)
Si$_2$	0	6.29	Ge$_2$	0	6.67
Si$_3$	0.33	5.22	Ge$_3$	0.43	5.89
Si$_4$	0	5.07	Ge$_4$	0	5.45
Si$_5$	0	4.81	Ge$_5$	0	5.15
Si$_6$ (I)	0	4.46	Ge$_6$ (I)	0	4.87
Si$_6$ (II)	0.19	4.48	Ge$_6$ (II)	0.14	4.88
Si$_7$	0	4.37	Ge$_7$	0	4.70

Ga$_m$As$_n$, see [32].) It is interesting to note that some of these clusters have permanent dipoles. For example, Si$_6$ and Ge$_6$ both have nearly degenerate isomers. One of these isomers possesses a permanent dipole, the other does not. Hence, in principle, one might be able to separate the one isomer from the other via an inhomogeneous electric field.

4.5 Optical spectra

While the theoretical background for calculating ground state properties of many-electron systems is now well established, excited state properties such as optical spectra present a challenge for computational methods. Recently developed linear response theory within the time-dependent density-functional formalism provides a new tool for calculating excited states properties [33]. This method, known as the time-dependent LDA (TDLDA), allows one to compute the true excitation energies from the conventional, time independent Kohn-Sham transition energies and wavefunctions.

Within the TDLDA, the electronic transition energies Ω_n are obtained from the solution of the following eigenvalue problem [33]:

$$\left[\omega_{ij\sigma}^2 \delta_{ik}\delta_{jl}\delta_{\sigma\tau} + 2\sqrt{f_{ij\sigma}\omega_{ij\sigma}}\, K_{ij\sigma,kl\tau}\, \sqrt{f_{kl\tau}\omega_{kl\tau}} \right] \mathbf{F}_n = \Omega_n^2 \mathbf{F}_n, \tag{20}$$

where $\omega_{ij\sigma} = \epsilon_{j\sigma} - \epsilon_{i\sigma}$ are the Kohn-Sham transition energies, $f_{ij\sigma} = n_{i\sigma} - n_{j\sigma}$ are the differences between the occupation numbers of the i-th and j-th states, the eigenvectors $\mathbf{F}_n$ are related to the transition oscillator strengths, and $K_{ij\sigma,kl\tau}$ is a coupling matrix given by

$$K_{ij\sigma,kl\tau} = \int\int \phi_{i\sigma}^*(\mathbf{r})\, \phi_{j\sigma}(\mathbf{r}) \left(\frac{1}{|\mathbf{r}-\mathbf{r}'|} + \frac{\partial v_\sigma^{xc}(\mathbf{r})}{\partial \varrho_\tau(\mathbf{r}')} \right) \phi_{k\tau}(\mathbf{r}')\, \phi_{l\tau}^*(\mathbf{r}')\, \mathrm{d}\mathbf{r}\, \mathrm{d}\mathbf{r}', \tag{21}$$

where i, j, σ are the occupied state, unoccupied state, and spin indices, respectively, $\phi(\mathbf{r})$ are the Kohn-Sham wavefunctions, and $v^{xc}(\mathbf{r})$ is the LDA exchange–correlation potential.

The TDLDA formalism is easy to implement in real space within the higher-order finite difference pseudopotential method [7]. The real-space pseudopotential code represents a natural choice for implementing TDLDA due to the real-space formulation of the general theory. With other methods, such as the plane wave approach, TDLDA calculations typically require an intermediate real-space basis. After the original plane wave calculation has been completed, all functions are transferred into that basis, and the TDLDA response is computed in real space [34]. The additional basis complicates calculations and introduces an extra error. The real-space approach simplifies implementation and allows us to perform the complete TDLDA response calculation in a single step.

We illustrate the TDLDA technique by calculating the absorption spectra of a sodium cluster. We chose sodium clusters as well-studied objects, for which accurate experimental measurements of the absorption spectra are available [35]. The ground-state structures of the clusters were determined by simulated annealing [24]. In all cases the obtained cluster geometries agreed well with the structures reported in other works [36]. Since the wavefunctions for the unoccupied electron states are very sensitive to the boundary conditions, TDLDA calculations need to be performed within a relatively large boundary domain. For sodium clusters we used a spherical domain with a radius

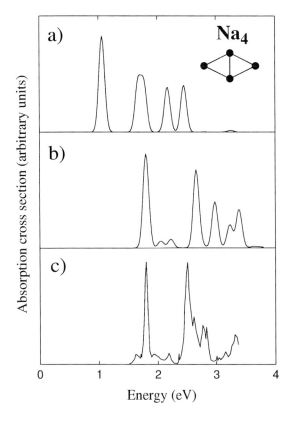

Fig. 10. The calculated and experimental absorption spectrum for Na_4. a) Local density approximation to the spectrum using Kohn-Sham eigenvalues. b) TDLDA calculation. Technical details of the calculation can be found in [39]. c) Experiment from [35]

of 25 a.u. and a grid spacing of 0.9 a.u. We carefully tested convergence of the calculated excitation energies with respect to these parameters.

The calculated absorption spectrum for Na_4 is shown in Fig. 10 along with experiment. In addition, we illustrate the spectrum generated by considering transitions between the LDA eigenvalues. The agreement between TDLDA and experiment is remarkable, especially when contrasted with the LDA spectrum. TDLDA correctly reproduces the experimental spectral shape, and the calculated peak positions agree with experiment within 0.1 to 0.2 eV. The comparison with other theoretical work demonstrates that our TDLDA absorption spectrum is as accurate as the available CI spectra [37]. Furthermore, the TDLDA spectrum for the Na_4 cluster seems to be in better agreement with experiment than the GW absorption spectrum calculated in Ref. [38].

5. Conclusions

We have presented in this paper a real space method for describing the structural and electronic properties of materials and, in particular, confined systems. Real space methods offer a powerful approach to these systems. A few of the advantages of real space methods over "traditional" plane wave methods to the electronic structure problem are as follows: Real space methods are far easier to implement than plane wave codes with no loss of accuracy. This is especially true for parallel implementations where real space methods appear to be roughly an order of magnitude faster than comparable implementations with plane wave methods. They do not require the use of supercells for localized systems. No cell–cell interactions are present. Charged systems can be handled directly without artificial compensating backgrounds. Replication of vacuum is natural and minimized compared to extended basis sets. No Fast Fourier Transforms are required and, consequently, global communications are minimized.

We have illustrated how this method can be applied to confined media. Specifically, we used real space methods to calculate the photoemission spectra, Raman or vibrational spectra, polarizabilities and optical absorption spectra of clusters. By making comparisons with available experimental data, we have confirmed the accuracy and utility of real space methods.

While we focused in this review on small clusters, it is possible to apply these techniques to quite large systems. For example, quantum dots with over 800 atoms have been examined with real space methods [40]. With increasingly efficient computer platforms and with new advances in algorithm developments, it is likely that larger systems will become routine in the near future.

References

[1] J.R. Chelikowsky and S.G. Louie (Eds.), Quantum Theory of Real Materials, Kluwer Academic Publ., Dordrecht 1996, and references therein.
[2] W. Kohn and L. Sham, Phys. Rev. **140**, A1133 (1965).
[3] P. Hohenberg and W. Kohn, Phys. Rev. **136**, B864 (1964).
[4] S. Lundqvist and N.H. March, Theory of the Inhomogeneous Electron Gas, Plenum Press, New York 1983, and refences therein.
[5] J.R. Chelikowsky and M.L. Cohen, Ab initio Pseudopotentials for Semiconductors, Handbook on Semiconductors, Vol. 1, Ed. P. Landsberg, Elsevier Publ. Co., Amsterdam 1992 (p. 59).
[6] A. Briley, M.R. Pederson, K.A. Jackson, D.C. Patton, and D.V. Porezag, Phys. Rev. B **58**, 1786 (1997).
K.A. Jackson, M.R. Pederson, D.V. Porezag, Z. Hajnal, and Th. Fraunheim, Phys. Rev. B **55**, 2549 (1997).
J.R. Chelikowsky and S.G. Louie, Phys. Rev. B **29**, 3470 (1984).
R.W. Jansen and O.F. Sankey, Phys. Rev. B **36**, 6520 (1987), and references therein.
[7] J.R. Chelikowsky, N. Troullier, and Y. Saad, Phys. Rev. Lett. **72**, 1240 (1994).
J.R. Chelikowsky, N. Troullier, K. Wu, and Y. Saad, Phys. Rev. B **50**, 11355 (1994).
J.R. Chelikowsky, N. Troullier, X. Jing, D. Dean, N. Binggeli, K. Wu, and Y. Saad, Comput. Phys. Commun. **85**, 325 (1995).
J.R. Chelikowsky, N. Troullier, K. Wu, and Y. Saad, Algorithms for Predicting Properties of Real Materials on High Performance Computers, Proc. Toward Teraflop Computing Conf. and New Grand Challenge Applications, Baton Rouge (LA) 1994, Eds. R.K. Kalia and P. Vashista, Nova, New York 1995 (p. 13).
J.R. Chelikowsky, X. Jing, K. Wu, and Y. Saad, Phys. Rev. B **53**, 12071 (1996).
[8] L. Kleinman and D.M. Bylander, Phys. Rev. Lett. **48**, 1425 (1982).
[9] N. Troullier and J.L. Martins, Phys. Rev. B **43**, 1993 (1991).
[10] B. Fornberg and D.M. Sloan, Acta Numerica 94, Ed. A. Iserles, Cambridge University Press, 1994.
[11] R.B. Morgan and D.S. Scott, SIAM J. Sci. Stat. Comput. **7**, 817 (1986).
[12] Y. Saad, A. Stathopoulos, J.R. Chelikowsky, K. Wu, and S. Öğüt, BIT **36**, 563 (1996).
[13] C.H. Tong, T.F. Chan, and C.C. J. Kuo, SIAM J. Sci. Stat. Comput. **13**, 227 (1992).
[14] B.N. Parlett, The Symmetric Eigenvalue Problem, Prentice Hall, Englewood Cliffs 1980.
[15] Y. Saad, Iterative Methods for Sparse Linear Systems, PWS Publ. Co., Boston 1996.
[16] G. Karypis and V. Kumar, Parallel Multilevel Graph Partitioning, Proc. 10th Internat. Parallel Processing Symp., Bloomington (Illinois) 1996 (p. 314).
[17] E.L. Briggs, D.J. Sullivan, and J. Bernholc, Phys. Rev. B **52**, 5471 (1995).
F. Gygi and G. Galli, Phys. Rev. B **52**, 2229 (1995).
G. Zumbac, N. Modine, and E. Kaxiras, Solid State Commun. **99**, 57 (1996).
T. Hoshi and T. Fujiwra, J. Phys. Soc. Jpn. **66**, 3710 (1997), and references therein.
[18] R. Kubo, Rep. Progr. Theor. Phys. **29**, 255 (1966).
[19] H. Risken, The Fokker-Planck Equation, Springer-Verlag, Berlin 1984.
[20] F.W. Kutzler and G.S. Painter, Phys. Rev. B **45**, 3236 (1992).
[21] D. Deaven and K.M. Ho, Phys. Rev. Lett. **75**, 288 (1995).

[22] O. CHESHNOVSKY, S.H. YANG, C.L. PETTIETT, M.J. CRAYCRAFT, Y. LIU, and R.E. SMALLEY, Chem. Phys. Lett. **138**, 119 (1987).
[23] D. TOMANEK and M. SCHLÜTER, Phys. Rev. Lett. **56**, 1055 (1986).
[24] N. BINGGELI and J.R. CHELIKOWSKY, Phys. Rev. B **50**, 11764 (1994).
[25] E.C. HONEA, A. OGURA, C.A. MURRAY, K. RAGHAVACHARI, O. SPRENGER, M.F. JARROLD, and W.L. BROWN, Nature **366**, 42 (1993).
[26] X. JING, N. TROULLIER, J.R. CHELIKOWSKY, K. WU, and Y. SAAD, Solid State Commun. **96**, 231 (1995).
[27] R. FOURNIER, S.B. SINNOTT, and A.E. DEPRISTO, J. Chem. Phys. **97**, 4149 (1992).
[28] C. ROHLFING and K. RAGHAVACHARI, J. Chem. Phys. **96**, 2114 (1992).
[29] R. SCHÄFER, S. SCHLECT, J. WOENCKHAUS, and J.A. BECKER, Phys. Rev. Lett. **76**, 471 (1996).
[30] S. BARONI, P. GIANOZZI, and A. TESTA, Phys. Rev. Lett. **58**, 1861 (1987).
[31] X. GONZE, D.C. ALLAN, and M.P. TETER, Phys. Rev. Lett. **68**, 3603 (1992).
[32] I. VASILIEV, S. ÖĞÜT, and J.R. CHELIKOWSKY, Phys. Rev. Lett. **78**, 4805 (1997).
[33] M. E. CASIDA, in: Recent Advances in Density-Functional Methods, Ed. D.P. CHONG, World Scientific Publ. Co., Singapore 1995, Part I, Chap. 5; in: Recent Developments and Applications of Modern Density Functional Theory, Ed. J.M. SEMINARIO, Elsevier Publ. Co., Amsterdam 1996.
[34] X. BLASE, A. RUBIO, S.G. LOUIE, and M.L. COHEN, Phys. Rev. B **52**, R2225 (1995).
[35] C.R.C. WANG, S. POLLACK, D. CAMERON, and M.M. KAPPES, Chem. Phys. Lett. **93**, 3787 (1990).
[36] I. MOULLET, J.L. MARTINS, F. REUSE, and J. BUTTET, Phys. Rev. B **42**, 11598 (1990). J.L. MARTINS, J. BUTTET, and R. CAR, Phys. Rev. B **31**, 1804 (1985).
[37] V. BONACIC-KOUTECKY, P. FANTUCCI, and J. KOUTECKY, J. Chem. Phys. **93**, 3802 (1990); Chem. Phys. Lett. **166**, 32 (1990).
[38] G. ONIDA, L. REINING, R.W. GODBY, R. DEL SOLE, and W. ANDREONI, Phys. Rev. Lett. **75**, 181 (1995).
[39] I. VASILIEV, S. ÖĞÜT, and J.R. CHELIKOWSKY, Phys. Rev. Lett. **82**, 1919 (1999).
[40] S. ÖĞÜT, J.R. Chelikowsky, and S.G. LOUIE, Phys. Rev. Lett. **79**, 1770 (1997).

phys. stat. sol. (b) **217**, 197 (2000)

Subject classification: 71.15.Mb; 71.15.Pd

Strategies for Massively Parallel Local-Orbital-Based Electronic Structure Methods

M. R. Pederson, D. V. Porezag, J. Kortus, and D. C. Patton

Center for Computational Materials Science, Naval Research Laboratory, Washington DC 20375, USA

(Received August 10, 1999)

We discuss several aspects related to massively parallel electronic structure calculations using the gaussian-orbital based Naval Research Laboratory Molecular Orbital Library (NRLMOL). While much of the discussion is specific to gaussian-orbital methods, we show that all of the computationally intensive problems encountered in this code are special cases of a general class of problems which allow for the generation of parallel code that is automatically dynamically load balanced. We refer to the algorithms for parallelizing such problems as "honey-bee algorithms" because they are analogous to nature's way of generating honey. With the use of such algorithms, BEOWULF clusters of personal computers are roughly equivalent to higher performance systems on a per processor basis. Further, we show that these algorithms are compatible with more complicated parallel programming architectures that are reasonable to anticipate. After specifically discussing several parallel algorithms, we discuss applications of this program to magnetic molecules.

1. Introduction

In contrast to the period 1985 to 1994 when the speed of computers was primarily increased by vector machines, during the last five years enhancements in computational speed have been largely driven by the development of massively parallel computational platforms. Further, if current trends continue during the beginning of the next millennium many computational problems will be most efficiently addressed by performing calculations on relatively inexpensive clusters of "personal computers" (PCs). In this paper we present a discussion of the practical problems that we have encountered during the development of massively parallel methodologies for gaussian-orbital based electronic structure calculations. While one generality is gleaned from our experiences, the discussion here is centered on our experiences with parallelization of the Naval Research Laboratory Molecular Orbital Library (NRLMOL). Most of the primary points related to the numerical algorithms are discussed in Refs. [1 to 9] and will not be repeated here. Instead, we will identify three types of algorithms which encompass all of the algorithms used in NRLMOL and show that they are amenable to massive parallelization. Further, these three types of algorithms each fulfill a single set of simple requirements which allow for the facile generation of parallel software that is dynamically load balanced. The primary goal of this chapter is not to provide a comprehensive review of either the present state of parallel computing or of gaussian-orbital methodologies. Instead we show by example how a serial code can be migrated onto massively parallel architectures, and we illustrate how local-orbital methods can be used in conjunction with parallel computer architectures to perform electronic structure calculations on large scale molecules.

The chapter is organized as follows. In Section 2, we discuss general issues related to parallel programming. These issues range from identifying a general class of problems for which parallelization is easily achieved to practical issues related to use of existing parallel programming tools and hardware. In Section 3, we discuss gaussian-orbital based methods from a general point of view and also discuss the algorithms used in NRLMOL. We discuss methods for parallelization of these computer codes. Since the goal of this volume is to highlight the usefulness of computational methodologies for investigating real materials, we will intersperse several timings on real systems during the course of discussion. We briefly discuss some of the practical points associated with the construction of a massively parallel BEOWULF cluster and provide a list of references for users who may be interested in constructing their own BEOWULF cluster. In Section 4, we present applications of this methodology to the problem of magnetic molecules. In addition to a discussion of the technological and scientific interest to these molecules, we include actual timings for the calculations on our pentium BEOWULF cluster.

2. General Approach to Parallel Processing with Dynamical Load Balancing

Present day parallel platforms can be divided into two main categories. The first category consists of an array of many identical processors. Such computers are generally shared by many users which necessitates the use of "batch jobs" or queueing systems that decide, with some limited user input, when and on which processors each job is to be run. On such platforms the queues are usually set up in a way that allow a user to have unfettered access to a given number of such processors for a certain period of time. In other words, during this period of time the user can expect that any given processor will only be busy when the user (or the user's code) requests that the processor perform a task. Further, since these processors are identical, the user can count on each processor requiring the same amount of time for similar tasks. Examples of such systems include SGI Origin 2000's, IBM SP2's and SP3's, CRAY T3E's and BEOWULF-type clusters of identical pentiums. An important figure of merit for these types of platforms is the communication time between the processors. We will refer to these platforms as homogeneous parallel platforms. The second category of parallel platforms, referred to as heterogeneous parallel platforms, consists of a collection of nonidentical processors. Processors may be nonidentical for a variety of reasons, which for the purposes of discussion here are indistinguishable. One example would be a BEOWULF-type cluster which consists of processors that are intrinsically different. An example of this would be several personal computers (possibly of different speed) and several IBM or SUN workstations. Another example would be a collection of intrinsically identical processors which are not entirely allocated to a single user at a time. While the processors are intrinsically identical, at any given time the actual speed of a particular node can be decreased by any factor if other jobs start running on that node. A heterogeneous network is often constructed from a collection of desktop workstations with the goal of allowing users at desk A to have access to user B's computer when user B is not accessing his or her computer.

In addition to the classification of homogeneous and heterogeneous parallel architectures there is another widely used classification regarding the access of memory. Shared memory systems such as dual- or quad- processor symmetric multiprocessor machines with the Intel XEON contain several processors which have simultaneous access to the same memory. This allows each and every processor to have fast access to any and all data without explicit communication between the processors. This advantage used by so-called threaded programming techniques [10] makes the hardware much more expensive and limits the scalability to tens of processors. Problems which may be easier to parallelize on these machines include matrix algebra, such as diagonalization, matrix multiplies and finite element methods. Distributed memory systems, such as the IBM SP2 or a cluster of workstations, consist of separate nodes connected by fast network connections which are cheaper and show scalability of up to thousands of processors primarily because the single node can be built from standard components. The lower hardware price has to be compensated by more complicated programming techniques, because data stored on or generated by a given processor can only be addressed by another processor by direct communication between the processors.

As an introduction to the goal of load balancing on both hetero- and homo-geneous parallel platforms, let us consider what can occur on a simple two-processor heterogeneous machine. Suppose user A starts one task on processor 1 and a second task on processor 2. Further suppose that if each processor was allocated entirely to user A that each task would require the same amount of time to complete. This would mean that user A would complete his or her two tasks in half the time if both processors are used. Now suppose that user B, who "owns" processor 2, decides to start four jobs on processor 2. In doing so, the task that user A started on processor 2 will require five times as much wall clock time to complete as the task that user A started on processor 1. So instead of realizing a factor of two speedup, user A is now faced with a factor of 2.5 slowdown. A similar situation would occur on a system containing two very fast nodes and one very slow node. If user A asked a queueing system to start his job on two nodes and the queueing system chose one of the very slow nodes, user A would again be faced with a net slowdown rather than a net speedup. However, since user A has no idea when both fast nodes may be simultaneously free, telling the queueing system to wait until both of the fast nodes are free is not necessarily a good strategy either. The latter problem also would arise on a homogeneous parallel platform. The point here is that on a heterogeneous network it is simply impossible to predict the actual speed of any given processor since the load on that processor changes due to the needs of other users. Further on all networks it is impossible for a user to reliably guess when a given percentage of the entire machine will be available. The situation is significantly more complicated when a heterogeneous network expands to include more processors and more users. A common albeit imperfect solution to this problem is to develop queueing systems which allocate a portion of the parallel platform to a single user for a given amount of time.

Our goal then is to determine programming techniques which achieve the theoretical speedup that should be possible on both hetero- and homo-geneous parallel platforms. Further, since it is impossible to predict the actual speed of a given processor on a heterogeneous platform, we want to use programming techniques which will allow the program to dynamically adjust to the variable speeds of each computer node. In other words we wish to develop programming techniques which are automatically dynamically load balanced on either heterogeneous or homogeneous parallel platforms. While such a goal may be either difficult or even impossible for some problems a class of problems exists for which this goal is quite easy to achieve.

In nature an illustrative example of this class is the problem of honey bees making honey. For this problem there is generally a field of flowers which greatly outnumber the number of worker bees in the beehive. In the beehive is a single Queen (or master). The master is not interested in requiring each worker in the hive to supply the hive with the same amount of nectar. Instead, the master's goal is to obtain as much nectar as possible with the workers available. To do this the master directs each worker to fly to a flower in the field, extract nectar from the flower and return to the beehive with the nectar. Every time a worker returns to the beehive they drop off their nectar and are then told to return to the field and obtain some more pollen. The amount of time required to direct a worker to leave the hive is small compared to the amount of time required for the worker to find a flower and extract the nectar. Further, the workers that fly the fastest or are significantly more efficient at finding flowers with bountiful supplies of nectar end up supplying the hive with a disproportionate fraction of the nectar. As long as the number of flowers greatly surpasses the number of workers, this is the best way for the Queen bee to manage the affairs of the beehive.

Pictured in Fig. 1 is a simplified schematic diagram of a parallel architecture and the method we use for massive parallelization. With the exception that we sometimes require the master to perform subtasks, this method for massive parallelization is identical to nature's method for producing honey. Therefore in the remainder of the paper we refer to these algorithms as "honey-bee" algorithms. Sufficient conditions for problems which allow for facile dynamical load balancing are as follows: First, the communication time per subtask between the master and worker processors must be small compared to the computation time. This condition is rather obvious since it is clear that if communication is the bottleneck, it does not matter how long the computations take. This condition applies to both dynamic and static load balancing. Second, it must be possible to break the total task into a number of subtasks that is at least an order of magnitude larger than the total number of processors that one plans to use during the

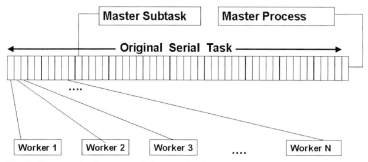

Fig. 1. Schematic diagram of the method for parallelization used for converting the serial version of NRLMOL to a massively parallel version. A time-intensive serial task is divided into a number of subtasks that is large compared to the total number of processors. While we require that the number of subtasks is large compared to the total number of processors we do not require that the intrinsic computational complexity be exactly the same for each subtask. It is only necessary to require that the intrinsic complexity of any given subtask is significantly smaller than the intrinsic complexity of the original serial task. The subtasks are sequentially sent to each worker processor until all the subtasks have been completed. Because this approach to parallel programming is essentially identical to the algorithm that bees use for making honey we refer to this programming method as "honey-bee" algorithms

run. If the second condition holds, one simply has a master processor which sends new subtasks to the worker processors whenever a worker completes a subtask. Since there are many subtasks, a fast worker is automatically sent more subtasks then a slow worker. However, since the amount of time for each subtask is still small compared to the total amount of time for all of the subtasks, we remove the possibility of having many workers sitting idle for a long period of time while a slow processor completes its work.

Before continuing, it is worthwhile to compare and contrast the two parallel tools we have used during the course of conversion. The present version of NRLMOL has employed the Message Passing Interface software (MPI) which appears to have become the de facto standard for all parallel applications. The earliest version employed the Parallel Virtual Machine (PVM) software which is no longer widely used. While both programming tools allow for parallelization within the honey-bee algorithm, each has some advantages. A primary and overwhelming advantage of the MPI software is that it is widely used and has significantly more technical support. Further the MPI programming environment is more structured and seems to be more conducive for use with programs that are themselves evolving. However, one aspect of PVM which is especially advantageous to the efficient use of sub-state-of-the-art machines is that the worker processes can have intrinsically smaller memory requirements. With the advent of allocatable memory in C and Fortran 90 software, this feature is less important, but it did prove to be useful in developing early parallel versions of NRLMOL for use on heterogeneous platforms which included aging machines.

Finally, it is worth pointing out that the general structure of parallel programming which is discussed here is general enough to be used with the next generation of parallel programming tools which are often referred to as multiple threading approaches [10]. As discussed above, present queueing systems on homogeneous parallel platforms generally allocate a fixed number of processors to a given user for the entire duration of a job. To more efficiently allocate the resources of a parallel platform it would be better if the total number of processors could dynamically change during a run. For example, suppose several or many jobs are initially using 100 percent of the processors on a parallel architecture and a large job finishes freeing up a portion of the processors. If at this time there are no additional requests for new jobs it would be most efficient if one or more of the current jobs could "hire" the idle workers immediately. Within the task partitioning scheme discussed above, this can be accomplished quite easily since the master node assigns a very large number of small tasks sequentially. If instead the master had broken a task into exactly M subtasks (M the number of currently available nodes), it would not be possible to use any new nodes until all M subtasks are finished. Similar increase in efficiency of resource allocation could be achieved if the master job could release or "fire" some of its workers during periods of increased overall demands on the system. Since most jobs will always have certain tasks that are intrinsically serial in nature the capability to hire and fire workers during a given job would lead to maximal efficiency of a multiuser parallel platform. While each user would immediately see the advantage to dynamically hiring additional worker processors, the advantage of firing worker processors would only be observed if each user was charged in some way for the use of the machine on an integrated per processor basis. If the charge per processor was further allowed to fluctuate inversely with the demands for system resources, such programming strategies would maximize the utilization of the system. Such a "capitalist queueing system" would indeed be the best approach to dynamical parallel resource allocation since it would allow computationally intensive jobs to use all resources during overall periods of low system demands but would not prevent less intensive jobs from receiving significant priorities during periods of high system demand.

We now turn to the problem of the implementation of gaussian-orbital based methods for electronic structure calculations and show how many problems fit into this general class of problems.

3. Gaussian-Orbital Based Methods for Electronic Structure Calculations

Within the density-functional theory (DFT) [11] and the Born-Oppenheimer approximation, the energy for a nanoscale system consisting of nuclei and electrons is written according to

$$E_t = \sum_{i\sigma} \langle \psi_{i\sigma} | -\frac{1}{2} \nabla^2 + V_{\text{nuc}} | \psi_{i\sigma} \rangle + \frac{1}{2} \int d^3r \int d^3r' \frac{\varrho(\mathbf{r}) \varrho(\mathbf{r}')}{|\mathbf{r} - \mathbf{r}'|} + \int d^3r g\left[\varrho(\mathbf{r}), \frac{d\varrho(\mathbf{r})}{dx}, \ldots\right] + \frac{1}{2} \sum_{\mu \neq \nu} \frac{Z_\mu Z_\nu}{|\mathbf{R}_\mu - \mathbf{R}_\nu|}. \tag{1}$$

The only unknown quantity in the above expression is the exchange-correlation energy which is represented by the third term of the above equation. Approximations for this term have been developed by Perdew and coworkers [12 to 14] and by Becke [15] and Lee et al. [16]. The best energy functionals and potentials currently depend on the spin densities, their gradients and their second derivatives. Once an approximation to the energy functional is determined, an application of the variational principle tells us that it is necessary to self-consistently solve a Schrödinger-like equation of the form

$$\langle \delta \psi_{i\sigma} | -\tfrac{1}{2} \nabla^2 + V_{\text{scf}}(\mathbf{r}) - \epsilon_i | \psi_{i\sigma} \rangle = 0. \tag{2}$$

In order to make further progress researchers generally expand the Kohn-Sham orbitals in terms of a product of spatial basis functions ($f_i(\mathbf{r})$) and spinors (χ_σ) according to

$$\psi_{is}(\mathbf{r}) = \sum_{j\sigma} C_{j\sigma}^{is} f_j(\mathbf{r}) \chi_\sigma. \tag{3}$$

Once this ansatz is introduced, one varies the expansion coefficients ($C_{j\sigma}^{is}$) rather than the value of each wavefunction at each point in space which leads to a secular equation. One aspect of the approximations to the density functional theory is that some of the integrals required to solve the above equation cannot be done in closed form. Therefore, some numerical work is required to solve the above equations. Many algorithms for the solution of the above equations exist, and the algorithm that one uses depends rather strongly on the form of the spatial basis functions that are used. In our work, we have used gaussian orbitals for the solution of Schrödinger's equation. There are several reasons that gaussian orbitals have been traditionally used for basis sets. Originally, it was realized that integrals of gaussian orbitals have simple analytic forms which aid in construction of the secular equations and calculations of total energies and forces. As system sizes have increased, two other aspects of gaussian orbitals have been determined to be useful. First, because the gaussian orbitals are localized in space, as the system size increases the overlap matrix becomes sparse which leads to an $O(N)$

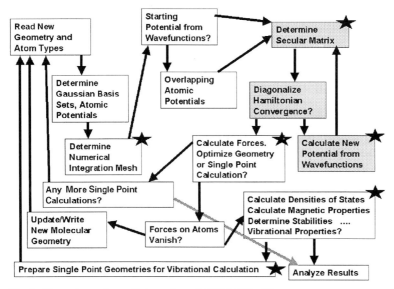

Fig. 2. Flow chart of parallel version of NRLMOL. The gray area represents the iterative part of the self-consistency cycle which is the computationally intensive part of the problem. The stars on the boxes represent the tasks which are massively parallelized and fit into the general class of problem discussed herein

storage rather than $O(N^2)$. Second, as will be discussed extensively below, the computational problems one faces within a gaussian-orbital framework are easily parallelized. An extensive literature on gaussian-orbital based methods exists and the interested readers are referred to Refs. [1 to 9, 17 to 23] for information on all the variations of gaussian-type methods. In Fig. 2, we present a simplified flow chart that describes the standard tasks which are used in NRLMOL to self-consistently solve Schroedingers equations.

As shown in the flow chart, it is first necessary to determine the locations and charges of the atoms. Once these are determined, the program either generates a basis set or uses a previously generated basis set [9]. Secondly, the self-consistent potential is numerically determined for each isolated atom and a least-square representation of the self-consistent atomic potentials is generated. These potentials are expanded as a sum of bare spherical gaussians or as a sum of gaussian-screened $1/r$ potentials. Given the basis sets and the gaussian-representation of the atomic potentials, it is possible to obtain very good insight into the class of multicenter integrands that need to be integrated, and this information is used to generate a numerical variational integration mesh [2] that allows us to precisely determine integrals required for calculation of secular matrices, total energies and derivatives according to

$$I = \int d^3r \, Q(\mathbf{r}) = \sum_i Q(\mathbf{r_i}) \, V_i, \qquad (4)$$

where V_i is the volume associated with point $\mathbf{r}_i$. Once the variational mesh is determined the calculation starts. For a new problem, a guess at the Hamiltonian matrix or self-consistent wavefunctions is not available so we rely on the least-square fit representation of overlapping atomic potentials to determine a starting Hamiltonian. Once the wavefunctions are determined, it is necessary to calculate the potential due to these wavefunctions. Within the density-functional theory the coulomb potential due to the electrons and the nuclei as well as the exchange-correlation energy density and potentials are required. For the most complicated approximation to the density-functional theory, the latter terms require the evaluation of spin densities and the first and second derivatives of the spin densities. Given the potential it is necessary to determine a secular equation which is then used to determine new wavefunctions. The equations are then solved self-consistently by iterating until the total energy is converged to a μHartree. It is important to point out here that the number of iterations required to reach self-consistency can be significantly decreased by using complicated mixing of input and output potentials which were originally advanced by Broyden and others. We use the Broyden algorithm of Johnson in our calculations [24]. Once self-consistency is achieved the forces acting on each atom are determined from the Hellmann-Feynman-Pulay theorem [25]. For gaussian-orbital methods, determination of the Hellmann-Feynman force is relatively inexpensive but the Pulay force is computationally intensive and requires parallelization. Given the forces on all the atoms a conjugate-gradient method, or other force-based algorithms, can be used to determine a new set of atomic coordinates. Once a new set of atomic coordinates is determined we find that the wavefunction expansion coefficients provide the best starting point for a calculation on this geometry. Eventually the geometrical optimization leads to the situation where the forces on all atoms vanish.

Once an equilibrium geometry and Kohn-Sham wavefunctions are determined there are many physical observables which one might be interested in calculating. These include local, total and joint electronic densities of states, polarizabilities, vibrational frequencies, infrared and Raman spectra, magnetic moments, charge states, magnetic anisotropy barriers, and potential and density contour plots. In the last subsection of this section we discuss how these quantities may be calculated on parallel architectures.

3.1 Parallel construction of numerical grids

Most of the details related to the construction of the variational mesh have been reported in Ref. [2] and will not be repeated here. Here we simply note that the algorithm relies on placing all atoms in a box which is large compared to the extent of the most delocalized wavefunction. The integrals required for solution of the Kohn-Sham equations have the following characteristics. First, they have singularities and/or strong peaks at each and every atomic position. Secondly there are many length scales in the problem. A well defined algorithm for slicing the box into a large collection of small parallelpipeds was developed and the resulting boxes are characterized by two types. The first type of parallelpiped is a perfect cube which encloses a single atom at the center of the cube. The second type of parallelpiped is not a perfect cube and does not enclose any atoms. The first type of parallelpiped can be further decomposed into two regions which consist of a sphere enclosing the central atom and the region outside the sphere and inside the cube. In Ref. [2] three different numerical quadrature methods were developed which allow for the determination of the numerical integrals within these regions. In Table 1, we present the total number of boxes as a function of the number of atoms for several different molecules. What is clear from this table is that the number of regions is roughly proportional to the number of atoms and very rapidly becomes larger than the number of processors available on most parallel platforms. While the method for determining the integration mesh is quite complex, the information that is required for each region scales as $O(N)$ for a molecule with less than 100 to 500 atoms. For a molecule with $N < 100$ to 500 atoms, this data set consists of approximately $63N$ numbers and can be broadcast to each processor prior to construction of the mesh.

The only additional information required for construction of a mesh within each region is the information about the boundaries and position of each region. This requires that the master process sends at most 24 numbers to the masters for each subtask. Even for a small molecule the amount of time spent in a subregion is large compared to the amount of time required to send 24 numbers from a master to a worker process. Therefore, parallelization of this part of the code is easily achieved. For very large molecules ($N > 500$ or so) the data set required for the construction of submesh would consist of approximately $63M$ numbers (with M the number of atoms within 5 to 10 Bohr of the region). This point is made only to note that, rather than $O(N^2)$, the actual scaling for mesh construction is $O(NM)$ with M a characteristic number of neighbors for each region.

While most of the information associated with mesh construction is contained in Ref. [2], there have been a few changes in the atom-partioning scheme which have lead to a variational mesh which attains the same desired precision with fewer mesh points. For the purpose of completeness, we include these changes here to show why it is now necessary to include atomic potentials for mesh construction. In the course of the mesh construction, one needs to decide where to put the plane which separates the two atoms. The plane should be positioned so that areas with large potential are contained in the atomic cubes which can be integrated more efficiently with high accuracy than the interstitial regions. This way, the error of the numerical integration is minimized for the interstitial regions. Assume the cut will be performed in the yz-plane and the x-coordinates of the two atoms are x_1 and $x_2 > x_1$. The x-coordinate of the cut x_c must fulfill $x_1 < x_c < x_2$. If x_c is too close to x_1 (or x_2), the potential on the surface of the atomic cube corresponding to atom 1 (or 2) will be very low. This is unfavorable since it requires high accuracy of the interstitial mesh near that atom. The optimal value for x_c should lead to a situation where the potential on the surface of the atomic cubes is

Table 1

Number of subregions ($N_{regions}$) required for construction of variational mesh as a function of molecular size (N_{atoms}). The total number of subregions is large compared to the number of processors on most parallel platforms even for small molecules. Also included is the average number of (Mega) floating point operations per subregion. An item of importance is that the average number of floating point operations per region depends primarily on local connectivity and atom type but is roughly independent of the number of atoms

molecule	N_{atoms}	$N_{regions}$	$N_{regions}/N_{atoms}$	N_{op}
CH_4	5	50	10	117
C_6H_6	12	186	15	53
$SnGe_5S_7H_9$	22	376	17	58
C_{60}	60	1128	19	71

approximately the same for atoms 1 and 2. If we assume that the potential in the vicinity of an atom is roughly given by the free-atom potential, then the condition which needs to be fulfilled by x_c is $V_1(x_c - x_1) = V_2(x_2 - x_c)$. We have found that this condition is a good approximation for the best value of x_c.

3.2 The Coulomb potential, spin densities and their derivatives

Given a set of Kohn-Sham wavefunctions which are expanded in terms of basis functions the density may be determined according to

$$\varrho(\mathbf{r}) = \sum_{i\sigma} |\psi_{i\sigma}(\mathbf{r})|^2 = \sum_{mn} \gamma_{mn} f_m(\mathbf{r}) f_n(\mathbf{r}). \tag{5}$$

While general basis sets lead to an $O(N^2)$ pairs of basis-function products, localized basis sets decrease this number to $O(NM)$ with M a characteristic number of neighbors. As shown in Table 2 the number of points at which the density must be calculated is also linear in the number of atoms. However since each product of localized basis functions is also localized the determination of the density is intrinsically an $O(N)$ problem. Further simplification occurs when the gaussian orbitals are used. Since the product of a pair of gaussian functions can always be expressed as a single gaussian function, it follows that the total and spin densities can always be analytically expressed as an $O(N)$ sum given by

$$\varrho(\mathbf{r}) = \sum_{\mathbf{A}_i, l, m} P(\mathbf{r} - \mathbf{A}_i) \exp\left(-\beta_i (\mathbf{r} - \mathbf{A}_i)^2\right), \tag{6}$$

where $P(\mathbf{r} - \mathbf{A}_i)$ is a polynomial function centered at $\mathbf{A}_i$. Once the spin densities are analytically decomposed according to the above expression, it is possible to determine the gradients and second derivatives as well as the Coulomb potential of each of these charge distributions analytically in the near-field region or by using multipole methods in the far-field regions. In Table 2, we give the total number of gaussian charge distributions as a function of the number of atoms. Again, this number increases rapidly as a function of the number of atoms and very quickly becomes large compared to the total number of processors one can expect to have access to. As such parallelization, with automatic dynamical load balancing, of Poisson's equation is accomplished by sending each processor a set of charge distributions.

Table 2
Number of mesh points determined from the variational mesh program as a function of molecular size (N_{atoms}). Even for small molecules the total number of mesh points is large compared to the number of processors on most parallel platforms. Also included is the number of charge distributions determined by decomposing the density into pairs of basis functions. Again, the total number of charge distributions, while linear in the number of atoms, is large compared to the number of processors on a parallel platform

molecule	N_{atoms}	N_{pts}/N_{atoms}	N_{ops}	N_{chgs}
CH_4	5	10230	4003	598
C_6H_6	12	7335	34002	4047
$SnGe_5S_7H_9$	22	4047	149790	18796
C_{60}	60	7717	4335096	102672
$Fe_8O_2(OH)_{12}(C_6N_3H_{15})_6H_8$	186	7769		

3.3 Construction of Hamiltonian and related problems

For the calculation of the Hamiltonian matrix, Pulay corrections, joint densities of states, magnetic anisotropies and many other quantities one generally is interested in performing calculations of the form

$$I = \int d^3r \, f_\alpha(\mathbf{r}) \, \Theta g_\beta(\mathbf{r}) = \sum_i V_i f_\alpha(\mathbf{r}_i) \, (\Theta g_\beta(\mathbf{r}_i)). \tag{7}$$

For construction of the secular equation, (f_α, g_β) would be a pair of basis functions and Θ would be the Hamiltonian operator. For calculation of the joint density of states Θ would be the dipole operator and (f_α, g_β) would be a pair occupied and unoccupied Kohn-Sham orbitals. As discussed below, for the calculation of magnetic anisotropy barriers (f_α, g_β) would be derivatives of occupied and unoccupied Kohn-Sham orbitals and Θ would be the total Coulomb potential. For the calculation of Pulay corrections, $(\Theta = H - \epsilon_\alpha)$ and f_α would be the explicit derivative of α-th occupied Kohn-Sham wavefunction with respect to an atomic position while g_β would be the occupied Kohn-Sham wavefunction.

Since there are roughly (5000 to 10000) N mesh points for an N-atom molecule, one method for parallelization is to break the mesh up into many smaller sets of mesh points and assign each worker the task of performing the sum over the smaller set of mesh points,

$$I = \sum_{j, i_j} V_{j, i_j} f_\alpha(\mathbf{r}_{j, i_j}) \, (\Theta g_\beta(\mathbf{r}_{j, i_j})). \tag{8}$$

This is essentially the strategy that we have used to parallelize the calculation of the joint densities of states, Pulay corrections and the magnetic anisotropy energies. It is possible to make use of sparsity by ordering the mesh points in subsets that correspond to small regions in space.

An alternative more efficient use of sparsity is to note that with the use of localized basis functions it is always possible to formulate any desired integral as a sum over pairs of atoms. As molecules get very large the basis functions on a given atom are only nonzero at meshpoints that lie within a radius of 10 to 15 atomic units. This is illustrated in Fig. 3 for a fullerene tubule. In the two panels we show the effective scaling and the derivative of the computational time with respect to molecular size changes as a function of tubule size. For small tubules, the *measured* scaling is observed to be N^3 as expected. However, for tubules in the range of 150 to 200 atoms the effective scaling approached $O(N)$ because the functions are localized. Alternatively this is illustrated in the lower panel by showing that the derivative of the total computational time approaches a constant for large tubules. In Fig. 3, we include a circle which represents the length scale that corresponds to the longest-range gaussian function on each atom in the tubule. It is when the system is larg compared to this length scale that the $O(N)$ behavior sets in. These calculations are performed by looping over all pairs of atoms in the system. We start with the first atom and determine all the (L) mesh points (typically those within 10 to 15 Bohr) at which the three longest-range gaussian functions on this atom are nonzero. The values of these functions are temporarily saved. Then as we loop over pairs of atoms $(i, j > i)$ it is only necessary to recalculate the short range functions associated with atom i and all the functions associated with atom j on this smaller mesh of points. Since most functions on both atom i and atom j vanish

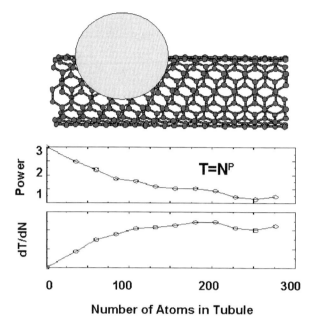

Fig. 3. Time (arbitrary units) required to construct the Hamiltonian matrix for an N-atom fullerene tubule as a function of the number of atoms. For tubules larger than 150 to 200 atoms, the time required for the calculation becomes linear in the number of atoms since basis functions on one atom only overlap with those that are within 10 to 15 atomic units. The circle represents the size of the longest-range gaussian that would be found on each atom in the tubule

outside a range of 1 to 4 Bohr most of the required work is associated with the initial determination of the long-range functions on atom i. This is especially true as the separation between atom i and atom j increases. While this method is the most efficient serial method, making full use of this algorithm requires an additional layer of communication between workers. As we loop over pairs of atoms $(i, j > i)$, worker m may be the first worker to determine the submesh for atom i. When another worker (n) is assigned a pair of atoms that includes atom i, it is possible for the master to further inform worker n that worker m has already, or is in the process of, determining the mesh for atom i. Then worker n must either wait for worker m to send this information or decide to recalculate the information itself. Either way some loss in efficiency would result. For the purpose of this paper we only wish to illustrate that use of sparsity and parallelization can at times lead to additional complications but that these complications can indeed be overcome in practice.

3.4 Vibrational modes

In other chapters of this issue and elsewhere, there have been discussions on algorithms for the calculation of vibrational modes and Raman and infrared intensities within mean-field methods. Here, we only point out that for a molecule containing N atoms, these algorithms require $(3N + 12)$ SCF calculations to determine all of the vibrational frequencies and the IR and Raman intensities. Each of these calculations is independent of all the other calculations so the parallel calculation of vibrational frequencies can be accomplished in parallel by sequentially sending the $(3N + 12)$ tasks to a number of master processors. To stay within the class of honey-bee problems, the total number of master processors should not exceed $(3N + 12)/10$. Each master processor could further enlist worker processors to perform the SCF calculations. The multi-tiered approach to parallelization is the best approach for this problem since it further inhibits limitations due to Amdahl's law. We now discuss the performance of these algorithms on a cluster of pentium workstations.

3.5 Commercial supercomputers versus clusters of PC class workstations

The early years of high-performance parallel computing were dominated by commercial supercomputers such as the IBM SP2, CRAY T3E, and SGI Origin 2000. However, as the PC chip technology has become more and more advanced, the CPUs of PC class systems now show a speed which is comparable to those used in the commercial supercomputers. For example, a Pentium III working at 550 MHz is only about 20% slower than the CPU on a current SGI Origin 2000 system. Further, since the PC server and workstation market is very competitive, PC technology comes at a much more favorable price/performance ratio. For this reason, there has been a lot of work on building parallel machines comprised of off-the-shelf PCs during the last five years. These are known as BEOWULF [26] projects, named after the first developments at the Center of Excellence in Space Data and Information Sciences which is part of the NASA Goddard Space Flight Center.

The main idea is to use inexpensive PC hardware which runs a free UNIX-like operating system such as Linux, connect the individual computers over a fast network, and use specific software for the communication between the different hosts. As mentioned earlier, MPI has become a standard for parallel computing. There are at least two free MPI implementations available that run under the Linux OS: MPICH [28] and LAM MPI [27]. Furthermore, the user-friendliness of a BEOWULF cluster may be improved by using queueing systems to schedule and run jobs. Presently, most systems use the free PBS [29] software.

Setting up a BEOWULF-type cluster of PCs requires a good knowledge of the hardware, operating system, and software that is to be used. Usually, a compromise must be found on how to split the available funds between network equipment (Fast-Ethernet switches/hubs or even Gigabit-Ethernet) and the actual PCs. The network bandwidth and latency time is usually the weak point of a BEOWULF system when compared to the more expensive commercial supercomputers. However, for many applications (including NRLMOL) it is possible to define worker subtasks that require much more CPU time than communication. For a detailed description on how to build and maintain a BEOWULF system see Ref. [30].

3.6 Use of parallel architectures

Pictured in Fig. 4 is a BEOWULF cluster of pentium workstations that was constructed by two of the authors (Porezag and Kortus). The cluster consists of 37 dual-processor 450 MHZ pentiums with 512 MBytes of memory per processor, 11 single-processor 500 MHz pentiums with 1 GByte of memory per processor and 56 single-processor 500 MHz pentiums with 384 MBytes of memory. The processors are connected by several high speed switches and MPI software is used for communication between nodes. For applications where the "honey-bee" approach to parallel processing is possible such BEOWULF clusters are significantly more cost-effective than supercomputers. For example, for a $Cu_{32}Xe$ cluster we have performed electronic structure calculations using 32 processors on a CRAY T3E and on our BEOWULF cluster. We find that our BEOWULF cluster is *faster* than the CRAY T3E. Similarly for a 40-atom cluster containing

Fig. 4. Picture of BEOWULF cluster used for timings in this paper

Al and N atoms, we have compared the speed of our BEOWULF cluster to a state-of-the-art SGI ORIGIN 2000. Using 32 processors on each machine we find that the SGI ORIGIN 2000 is only 30% faster than the BEOWULF cluster.

Pictured in Fig. 5 is the 100-atom $Mn_{12}O_{12}(COOH)_{16}(H_2O)_4$ molecule which is currently receiving a significant amount of scientific attention because it exhibits resonant tunneling of spins and retains a stable magnetic moment at relatively high temperatures. We have performed calculations on this molecule and the results are discussed in Refs. [31 to 33]. Here we concentrate on the timings for these calculations as a function of the number of processors.

In Table 3 we show the total wall clock time to perform the two computationally intensive SCF tasks which are (i) the construction of the density and Coulomb potential

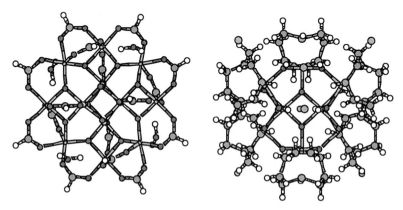

Fig. 5. Two of the more complicated molecules that have been addressed by the parallel NRLMOL code. The $Fe_8O_2(OH)_{12}(C_6N_3H_{15})_6H_8$ molecule (right) is ferrimagnetically ordered with two minority spin Fe atoms and six majority spin Fe atoms. While the net spin of this system is the same as for the $Mn_{12}O_{12}(COOR)_{16}(H_2O)_4$ molecule (left) our calculations show that the moments localized on the transition metal atoms are quite different

Table 3

Timings (seconds) and relative speeds as a function of the number of processors on a BEOWULF cluster of pentiums. Included here is the wall clock time for the calculation of the Hamiltonian matrix (H) and the determination of the density and Coulomb potential (P) and the total time (T). The observed speed ups (R_A) and actual speed ups (R_T) are also included

number of processors	H s	P s	T s	R_A	R_T
8	1077	7404	8481	1.0	1.0
32	339	1673	2012	4.2	4.0
64	184	864	1048	8.1	8.0
131	94	565	659	12.9	16.4

and (ii) the construction of the Hamiltonian matrix. In addition, the relative speed of the calculation is included along with the theoretical speedup. For the 32- and 64-processor calculations we actually do better than the theoretical speedups. This is primarily due to the fact that for the smaller calculations some of the subtasks are performed by the master. While this is clearly the correct thing to do for two to four processors, the better than theoretical speedup observed suggests that we already are introducing some system latency when we allow the master to perform subtasks on 8-processor runs. For the 131-processor run, the actual speedup relative to the 8-processor run is only 12.2 compared to the theoretical value of 16.4. There are several reasons that this occurs. First, because of the current way we are taking advantage of sparse matrices in construction of the secular equation we lose some efficiency in parallelization for molecules approaching the size of 100 to 200 atoms. For the 131-atom cluster we find speedups of 11.5 and 13.1 for construction of the secular equation and solution of Poisson's equation, respectively. For the most intensive step, the efficiency degrades to only 82% in the 70- to 131-processor range. The decrease in efficiency is probably caused by a combination of the small amount of scalar time that the master processor uses in constructing the charge distributions as well as an increase in communication time due to more network traffic within the cluster. While we are confident that it is in principle possible to determine how to increase the efficiency back toward 95 to 98% percent, Amdahl's law now dictates that other parts of the problem need to be made more efficient and this point is now discussed.

One of the attractive features of gaussian-orbital basis sets is that they are very compact and diagonalization of the secular equation is significantly less time-consuming than what one expects from more delocalized basis functions such as plane-waves. Nevertheless regardless of basis-set choice the size of the Hamiltonian matrix increases linearly with the number of atoms which implies that the amount of time required to diagonalize the Hamiltonian matrix increases as N^3. For the calculations on this molecule the time required to diagonalize the Hamiltonian matrix was 647 seconds which is now comparable to the amount of time required for the rest of the iterative procedure. Unfortunately, we are unaware of highly efficient algorithms for the massively parallel diagonalization of general eigenvalue problems. Further, since the ratio of computation to communication appears to be roughly unity for such problems it does not appear that future algorithms will fit nicely into the honey bee strategy discussed in this work.

Fifteen years ago, the plane-wave community was confronted by this problem and Car and Parrinello suggested methodologies for simultaneously treating the nuclear and electronic degrees of freedom and for dealing with the so-called N^3 bottleneck in plane-wave-based electronic-struture applications [34]. Within this volume Bernholc et al., Galli, and others have discussed how work along these lines have evolved since this time. Further Chelikowsky et al. have demonstrated by example that efficient scalable algorithms for *sparse* matrices exist (see Fig. 3 of that work). Algorithms similar in spirit to those proposed by Car and Parinello will now need to be used to deal with the N^3 bottleneck in local-orbital based methods before spending any additional serious effort at parallelizing the parts of the codes that do not scale as N^3. In addition to simultaneous variation of electronic and nuclear degrees of freedom, for gaussian-orbital methods, exploitation of sparsity for *both* basis functions and localized orthogonal Kohn-Sham orbitals will be useful. Possible ways of achieving this within an all-electron gaussian-based methodology have been discussed in Refs. [1, 6] and we hope to implement some reasonable combination of these methodologies, sparse matrix methodologies and scalable methodologies in the near future.

4. Applications to Magnetic Molecules

Several applications of this massively parallel version of NRLMOL are already contained within this volume. This includes the pseudopotential-based investigations of Porezag et al., a discussion of vibrational spectra due to Jackson et al. and a discussion of the photoelectron spectra of negatively charged halide-deficient alkali-halide clusters due to Ashman et al. Here we would like to review a new method for the study of magnetic molecules and present some new applications. The full details for the theory in this work can be found in a recent series of papers due to Pederson and Khanna [31 to 33]. Here we offer a short overview and then turn to questions about shape anisotropies in several different putative transition metal structures.

The magnetic molecules of current interest are typically composed of approximately 8 to 12 transition metal atoms which are locked at their lattice sites by a carefully arranged host consisting of organic molecules and ligands [35 to 45]. Without these ligands the transition metal lattice is found to be unstable [31, 32]. Two examples of these molecules have molecular formulae of $Fe_8O_2(OH)_{12}(C_6N_3H_{15})_6H_8$ (with H a halide ion) and $Mn_{12}O_{12}(COOH)_{16}(H_2O)_4$, respectively. Our calculations on the latter molecule have been discussed in detail in Refs. [31 to 33]. These molecules are of interest from both a fundamental and technical point of view. The fundamental interest in these molecules is that they have large magnetomolecular anisotropy energies which means their moments are thermally stable at relatively high temperatures (45 to 65 K). Further it is possible to observe resonant tunneling of spins when magnetic fields of certain magnitudes are applied. For the original discussion of magnetic anisotropy energies in crystals see Ref. [46]. An early discussion for the calculation of magnetic anisotropy energies within electronic-structure methods can be found in Ref. [47]. The potential for technological interest lies in the fact that they could be useful for molecular-scale magnetic storage if the anisotropy barriers could be increased by a factor of ten or so. In Fig. 5 we present pictures of the Mn- and Fe-based magnetic molecules. These molecules are characterized by low symmetry, ferrimagnetic ordering and large overall moments. Both molecules have net moments of $10\mu_B$.

As discussed in Ref. [33] both the resonant tunneling magnetic-field strengths and the magnetic anisotropy barrier are primarily determined by the second-order spin–orbit coupling interaction. Classically, the spin–orbit coupling interaction is caused by an electron moving with velocity **v** through the electric field (**E**) due to the other electrons and the nuclei. Since the electron has a spin, the interaction energy is given by

$$U(\mathbf{r}, \mathbf{p}, \mathbf{S}) = -\frac{1}{2c^2} \mathbf{S} \cdot \mathbf{p} \times \nabla \Phi(\mathbf{r}). \tag{9}$$

For further discusion see Refs. [48] and [49]. In the above, $\mathbf{E} = -\nabla \Phi(\mathbf{r})$ with Φ the Coulomb potential, and the electronic velocity (**v**) has been replaced by the momentum operator (**p**). The factor of two in the denominator, also derivable from the Dirac equation, is due to the Thomas precession. Given a spherically symmetric potential $\Phi(r)$ and some simple algebraic reductions the above expression is usually rewritten according to

$$U(r, \mathbf{L}, \mathbf{S}) = \frac{1}{2c^2} \mathbf{S} \cdot \mathbf{L} \frac{1}{r} \frac{d\Phi(r)}{dr}. \tag{10}$$

While the above equation is exact for spherical systems, an attempt to approximate the spin–orbit coupling in multicenter systems as a superposition of such terms on a lattice could omit nonspherical corrections that may be especially important for anisotropy energies.

Instead of using the traditional $L \cdot S$ for the spin–orbit coupling we return to more fundamental (**r**, **p**) representation and note that since all basis-set oriented mean-field methods expand the single-electron wavefunctions according to Eq. (3), it is only necessary to determine matrix elements of the form

$$U_{j\sigma, k\sigma'} = \langle f_j \chi_\sigma | U(\mathbf{r}, \mathbf{p}, \mathbf{S}) | f_k \chi_{\sigma'} \rangle = \sum_x \frac{1}{i} \langle f_j | V_x | f_k \rangle \langle \chi_\sigma | S_x | \chi_{\sigma'} \rangle \tag{11}$$

with the operator V_x defined according to

$$\langle f_i | V_x | f_j \rangle = \frac{1}{2c^2} \langle f_i | \left(\frac{d\Phi}{dy} \frac{d}{dz} - \frac{d\Phi}{dz} \frac{d}{dy} \right) | f_j \rangle. \tag{12}$$

In Ref. [33] it is shown that an integration by parts reduces the above matrix element to the simplified form

$$\langle f_i | V_x | f_j \rangle = \frac{1}{2c^2} \left(\left\langle \frac{df_i}{dz} \middle| \Phi \middle| \frac{df_j}{dy} \right\rangle - \left\langle \frac{df_i}{dy} \middle| \Phi \middle| \frac{df_j}{dz} \right\rangle \right), \tag{13}$$

with matrix elements for V_y and V_z determined by cyclic permutations of the coordinate labels in the above equation. As explained in Ref. [33] the above equation is valid for both periodic systems and finite systems. This representation for the spin–orbit coupling matrix offers several advantages over the more usual $L \cdot S$ representation. First, it does not require the determination of the electric field and depends only on the ability to accurately determine the Coulomb potential and the gradient of each basis function in the problem. This representation for the spin–orbit coupling matrix is especially ideal for basis functions constructed from gaussian-type orbitals, slater-type functions and plane waves. For numerical basis functions such as those used in other chapters of this volume it should still be useful since it is generally necessary to determine the gradient of a numerical function for determination of the kinetic energy matrix.

Upon inclusion of spin–orbit coupling, the first-order change in the total energy vanishes under most conditions and the second-order energy of a collection of Kohn-Sham spin–orbitals ($\psi_{i\sigma} = \phi_{i\sigma}(\mathbf{r}) |\chi_\sigma\rangle$) is perturbed according to the following approximation:

$$\Delta = \sum_{\sigma\sigma'} \sum_{xy} M_{xy}^{\sigma\sigma'} S_x^{\sigma\sigma'} S_y^{\sigma'\sigma},$$

$$M_{xy}^{\sigma\sigma'} = M_{yx}^{\sigma\sigma'*} = -\sum_{ij} \frac{\langle \phi_{i\sigma}| V_x |\phi_{j\sigma'}\rangle \langle \phi_{j\sigma'}| V_y |\phi_{i\sigma}\rangle}{\epsilon_{i\sigma} - \epsilon_{j\sigma'}},$$

$$S_x^{\sigma\sigma'} = \langle \chi_\sigma| S_x |\chi'_\sigma\rangle. \quad (14)$$

The $\phi_{i\sigma}$ and $\phi_{j\sigma'}$ are occupied and unoccupied Kohn-Sham orbitals, respectively. The operators **V** and **S** operate on the spatial and spin degrees of freedom respectively and transform like pseudovectors under the symmetry operations. Now, taking a general unitary transformation on a fixed set of spinors,

$$|\chi_1\rangle = \cos\theta |\uparrow\rangle + e^{i\beta}\sin\theta |\downarrow\rangle,$$
$$|\chi_2\rangle = -e^{-i\beta}\sin\theta |\uparrow\rangle + \cos\theta |\downarrow\rangle, \quad (15)$$

and assuming uniaxial symmetry for the molecule we find that the second-order spin–orbit coupling energy as a function of spin projection reduces to

$$\Delta = (M_{xx}^{11} + M_{xx}^{22} + M_{zz}^{12} + M_{zz}^{21})\frac{\sin^2(2\theta)}{4}$$
$$+ (M_{zz}^{11} + M_{zz}^{22} + M_{xx}^{12} + M_{xx}^{21})\frac{\cos^2(2\theta)}{4} + (M_{xx}^{12} + M_{xx}^{21})\frac{1}{4}. \quad (16)$$

For the uniaxial systems discussed here, we have used the fact that all off-diagonal cartesian matrix elements vanish and that $M_{xx}^{\sigma\sigma'} = M_{yy}^{\sigma\sigma'}$. Because of the symmetry there is also no dependence on the phase β. Further since $\langle S_z \rangle = \Delta N \cos(2\theta)/2$ the equation can be simply written according to

$$\Delta = A + \frac{\gamma}{2}\langle S_z \rangle^2 \quad (17)$$

with $A = (M_{xx}^{11} + M_{xx}^{22} + M_{zz}^{12} + M_{zz}^{21} + M_{xx}^{12} + M_{xx}^{21})/4$ and

$$\gamma = \frac{2}{\Delta N^2}(M_{zz}^{11} + M_{zz}^{22} + M_{xx}^{12} + M_{xx}^{21} - M_{xx}^{11} - M_{xx}^{22} - M_{zz}^{12} - M_{zz}^{21}). \quad (18)$$

In the above equation, A is the average second-order contribution due to the spin–orbit interaction. If we assert further that we may identify $\langle S_z \rangle$ with M, we can determine that the energy spacing between different M states is given by $\gamma/2\,(M_1^2 - M_2^2)$. Regardless of whether we rely on the identity $M = \langle S_z \rangle$, the classical energy difference is given by $\gamma/2 \times S^2$. If γ is positive the minimum energy corresponds to $\langle S_z \rangle = 0$ and the maximum energy corresponds to $\langle S_z \rangle = \pm\Delta N/2$. If γ is negative the maximum energy corresponds to $\langle S_z \rangle = 0$ and the minimum energy corresponds to $\langle S_z \rangle = \pm\Delta N/2$.

From the equation for the anisotropy barrier we see that there are several competing factors which are responsible for the formation of the anisotropy barrier. One figure of merit is the size of the energy denominator. A small energy denominator will typically make the magnitude of the M-matrices larger. It is also evident that the size of the M-matrices requires some degree of spatial overlap between a pair of occupied and

Table 4
Magnetic anisotropy energy barrier or valley (U) for several magnetically stable spin states of Mn_6 bicapped squares. Included in the table are the energy (E), the geometrical parameters (A, B), the HOMO-LUMO gaps for each pair of spins, and the net moment (S). Antiferromagnetically ordered clusters are designated with a superscript

	E (eV)	A (Å)	B (Å)	energy gap (eV) for spin pairing				S (μ_B)	U (K)
				1-1	2-2	1-2	2-1		
Mn_6^{AF}	1.27	1.48	2.03	0.50	0.73	0.50	0.73	4	−3.0
Mn_6	1.09	1.90	1.40	0.27	0.60	0.46	0.41	14	−7.7
Mn_6^{AF}	0.19	2.02	1.40	0.54	0.30	0.30	0.54	12	3.0
Mn_6	0.00	1.87	1.87	1.87	0.16	0.89	1.14	26	0.0

unoccupied electrons. This implies that pairs of occupied and unoccupied states which share the same principal quantum numbers and total angular momentum allow for large numerators in Eq. (14).

In Ref. [33] the above method has been applied to the uniaxial Mn-acetate molecule. The predicted value of the second-order anisotropy barrier was 56.7 K which is in excellent agreement with the second order experimental barrier of 56.6 + K. Here we present new results on two types of uniaxial molecules to begin to understand what effects are important for enhancing the anisotropy energy in such molecules. We have performed calculations on several different uniaxial Mn_6 clusters as a function of magnetic order and net moment. For each magnetic structure we have fully relaxed the geometry of the clusters. The geometry of each cluster is a bicapped square with atoms at $(A, 0, 0)$, $(−A, 0, 0)$, $(0, A, 0)$, $(0, −A, 0)$, $(0, 0, B)$ and $(0, 0, −B)$. Within uniaxial symmetry is possible to have either ferromagnetic ordering or antiferromagnetic ordering. The results for four low lying structures are presented in Table 4. As discussed in Ref. [50] the lowest lying geometry is found to be a perfect ferromagnetic octahedron ($A = B$) with a net moment of $26\,\mu_B$. The second lowest geometry is significantly distorted from an ideal octahedron, has a net spin of 12, and exhibits antiferromagnetic ordering with in-plane majority and out-of-plane minority spin Mn atoms. The third lowest geometry is also a distorted octahedron but is primarily ferromagnetic with a net spin of $14\,\mu_B$. Analysis of Eq. (18) and the results of these calculations show that a large tunneling barrier is not necessarily correlated with a large net moment. Indeed the highest moment cluster, which collapsed to cubic symmetry, has no $1/4c^4$ contributions to the anisotropy energy as shown numerically in Table 4 and analytically in Eq. (18). Also included in Table 4 are the HOMO-LUMO gaps for each spin (here "1" refers to majority spin and "2" refers to minority spin).

We now turn to an application of this method to several different metal-oxide M_6O_6 clusters where M represents a transition metal atom. For M = Mn, clusters of this stoichiometry have been observed by Ziemann and Castleman [51] in the gas phase. They have suggested that this cluster as well as larger isostoichometric clusters are characterized by layers of hexagonal $(MnO)_3$ rings. These structures exhibit uniaxial symmetry. As shown in Table 5, for M = V, Mn, Fe and Co we have optimized the electronic structure and geometry of the two-layer hexagonal geometry and have found that the magnetic ground state leads to a system with a nonvanishing energy gap between the

Table 5

Magnetic anisotropy energy barrier (B) for the highest spin state of four different transition metal oxide clusters M_6O_6 with the hexagonal tower topology suggested by Ziemann and Castleman [51]. A superscript "R" is placed on the gap that contributed the largest contribution to the anisotropy energy

cluster	energy gap (eV) for spin pairing				S	B
	1-1	2-2	1-2	2-1	(μ_B)	(K)
V	0.27	4.68	0.11	4.95	18	3.86
Mn	0.78	0.52	0.24^R	3.4	30	3.02
Fe	1.98^R	0.60	0.38	2.2	24	1.90
Co	2.28	0.24^R	0.24	2.28	18	25

highest occupied molecular orbital and the lowest unoccupied molecular orbital. The total moments for these towers are found to be largest for Mn ($S = 30\mu_B$) and smallest for the V and Co towers ($S = 18\mu_B$). Again we observe that a large anisotropy barrier is not necessarily correlated with a large magnetic moment. For example V and Mn which have similar anisotropy barriers have very different moments and the tower with the largest anisotropy barrier (containing Co) corresponds to the smallest barrier. Also included in the table are energy gaps as a function of spin pairing. From Eq. (14), we see that the anisotropy barrier is determined by interactions between pairs of occupied and unoccupied electrons and that in addition to coupling between electrons with parallel spins there is a coupling between electrons with antiparallel spins. In some cases it was possible to attribute most of the anisotropy energy to a coupling between specific spin–spin correlations. For example, the Mn tower which is composed of Mn atoms with a 3d^5 majority valence the largest contribution to the anisotropy barrier is due to coupling between the majority spin occupied states and the minority spin unoccupied states. This is not surprising since in addition to the fact that this spin pair leads to the smallest energy gap, there will also be good spatial overlap between the occupied majority and unoccupied minority Mn d states. In contrast, for Co it is a coupling between the minority spin occupied and minority spin unoccupied electrons that is primarily responsible for the formation of the barrier despite the fact that the minority–minority and majority–minority gaps are the same. This is partially due to the fact that the fully occupied majority d shell will be more tightly bound than the partially occupied minority d shell. However, the unoccupied minority d electrons will also be delocalized which implies that the numerators of Eq. (14), which depend on overlap between occupied and unoccupied electrons, will be larger for the minority–minority case than for the majority–minority case. The results of these simple applications show that we are now at a point where we can determine how differences in bonding and chemistry can lead to quantitative differences in magnetomolecular anisotropy. Computer aided design and optimization of molecular scale magnets represent one of many new technological areas that will be impacted by density-functional-based methods in the next ten years. Further, the relative simplicity related to construction of cost effective scalable BEOWULF clusters should further enhance the role of Computational Materials Science in the design and optimization of new materials.

Acknowledgement This work was supported in part by ONR (Grant No. N0001400WX2011 and N0001498WX20709).

References

[1] M. R. PEDERSON and C. C. LIN, Phys. Rev. B **35**, 2273 (1987).
[2] M. R. PEDERSON and K. A. JACKSON, Phys. Rev. B **41**, 7453 (1990).
[3] K. A. JACKSON and M. R. PEDERSON, Phys. Rev. B **42**, 3276 (1990).
[4] M. R. PEDERSON and K. A. JACKSON, Phys. Rev. B **43**, 7312 (1991).
[5] A. A. QUONG, M. R. PEDERSON, and J. L. FELDMAN, Solid State Commun. **87**, 535 (1993).
[6] M. R. PEDERSON, J. Q. BROUGHTON and B. M. KLEIN, Phys. Rev. B **38**, 3825 (1988).
[7] D. V. POREZAG and M. R. PEDERSON, Phys. Rev. B **54**, 7830 (1996).
[8] A. BRILEY, M. R. PEDERSON, K. A. JACKSON, D. C. PATTON, and D. V. POREZAG, Phys. Rev. B **58**, 1786 (1998).
[9] D. POREZAG and M. R. PEDERSON, Phys. Rev. A **60**, 9566 (1999).
[10] D. R. BUTENHOF, Programming with Posix Threads, Addison-Wesley, Reading (Mass.) 1998.
[11] P. HOHENBERG and W. KOHN, Phys. Rev. **136**, B864 (1964).
W. KOHN and L. J. SHAM, Phys. Rev. **140**, A1133 (1965).
[12] J. P. PERDEW and A. ZUNGER, Phys. Rev. B **23**, 5048 (1981).
[13] J. P. PERDEW, J. A. CHEVARY, S. H. VOSKO, K. A. JACKSON, M. R. PEDERSON, D. SINGH, and C. FIOLHAIS, Phys. Rev. B **45**, 6671 (1992).
[14] J. P. PERDEW, K. BURKE, and M. ERNZERHOF, Phys. Rev. Lett. **77**, 3865 (1996).
[15] A. D. BECKE, Phys. Rev. A **38**, 3098 (1988); J. Chem. Phys. **98**, 13 (1993).
[16] C. LEE, W. YANG, and R. G. PARR, Phys. Rev. B **37**, 785 (1988).
[17] E. E. LAFON and C. C. LIN, Phys. Rev. **152**, 579 (1966).
[18] G. S. PAYNTER and F. W. AVERILL, Phys. Rev. B **26**, 1781 (1982).
[19] B. I. DUNLAP, J. W. CONNOLLY, and J. R. SABIN, J. Chem. Phys. **71**, 3396 (1979).
[20] J. A. POPLE et al., GAUSSIAN94, Gaussian Inc., Pittsburgh (PA) 1994.
[21] J. ANDZELM, E. RADZIO, and D. R. SALAHUB, J. Comput. Chem. **6**, 520 (1985).
[22] R. JONES and A. SAYYASH, J. Phys. C **19**, L653 (1986).
[23] W. J. HEHRE, L. RADOM, P. V. R. SCHLEYER, and J. A. POPLE, Ab-Initio Molecular Orbital Theory, John Wiley and Sons, New York 1986.
[24] D. D. JOHNSON, Phys. Rev. B **38**, 12807 (1998), and references therein.
[25] H. HELLMANN, Einführung in die Quantentheorie, Deuticke, Leipzig 1937.
R. P. FEYNMAN, Phys. Rev. **56**, 340 (1939).
P. PULAY, Mol. Phys. **17**, 197 (1969).
[26] http://www.beowulf.org
[27] http://www.mpi.nd.edu/lam/
[28] http://www-unix.mcs.anl.gov/mpi/mpich/
[29] http://pbs.mrj.com/
[30] T. L. STERLING, J. SALMON, D. J. BECKER, and D. SAVARESE, How to Build a BEOWULF, MIT Press, Cambridge (MA) 1998.
[31] M. R. PEDERSON and S. N. KHANNA, Phys. Rev. B **59**, R691 (1999).
[32] M. R. PEDERSON and S. N. KHANNA, Chem. Phys. Lett. **303**, 373 (1999).
[33] M. R. PEDERSON and S. N. KHANNA, Phys. Rev. B. **60**, 9572 (1999).
[34] R. CAR and M. PARRINELLO, Phys. Rev. Lett. **55**, 2471 (1985).
[35] T. LIS, Acta Cryst. Soc. B **36**, 2042 (1980).
[36] J. R. FRIEDMAN, M. P. SARACHIK, J. TEJADA, and R. ZIOLO, Phys. Rev. Lett. **76**, 3830 (1996).
[37] L. THOMAS, F. LIONTI, R. BALLOU, D. GATTESCHI, R. SESSOLI, and B. BARBARA, Nature **383**, 145 (1996).
[38] A. CANESCHI, D. GATTESCHI, and R. SESSOLI, J. Amer. Chem. Soc. **113**, 5873 (1991).
[39] R. SESSOLI, H.-L. TSAI, A. R. SCHAKE, S. WANG, J. B. VINCENT, K. FOLTING, D. GATTESCHI, G. CHRISTOU, and D. N. HENDRICKSON, J. Amer. Chem. Soc. **115**, 1804 (1993).
[40] R. SESSOLI, D. GATTESCHI, A. CANESCHI, and M. A. NOVAK, Nature **365**, 141 (1993).
[41] J. HERNANDEZ, X. ZHANG, F. LOUIS, J. BARTOLOME, J. TEJADA, and R. ZIOLO, Europhys. Lett. **35**, 301 (1996).

[42] J. Villain, F. Hartman-Boutron, R. Sessoli, and A. Rettori, Europhys. Lett. **27**, 159 (1994).
[43] F. Fominaya, J. Villain, P. Gandit, J. Chaussy, and A. Caneschi, Phys. Rev. Lett. **79**, 1126 (1997).
[44] A. L. Barra, D. Gatteschi, and R. Sessoli, Phys. Rev. B **56**, 8192 (1997).
[45] A. Fort, A. Rettori, J. Villain, D. Gatteschi, and R. Sessoli, Phys. Rev. Lett. **80**, 612 (1998).
[46] J. Van Vleck, Phys. Rev. B **52**, 1178 (1937).
[47] H. J. F. Jansen, Phys. Rev. B **38**, 8022 (1988).
[48] S. Gasiorowicz, Quantum Physics, John Wiley and Sons, New York 1974 (p. 272).
[49] C. Kittel, Introduction to Solid State Physics, John Wiley and Sons, New York 1963 (p. 181).
[50] M. R. Pederson, F. Reuse, and S. N. Khanna, Phys. Rev. B, 5632 (1998).
[51] P. J. Ziemann and A. W. Castleman, Jr., Phys. Rev. B **46**, 13480 (1992).

Subject classification: 71.15.Hx; 71.15.Mb

The Accuracy of the Pseudopotential Approximation within Density-Functional Theory

D. Porezag (a, b), M. R. Pederson (a), and A. Y. Liu (b)

(a) *Center for Computational Materials Science, Naval Research Laboratory, Washington, DC 20375, USA*

(b) *Department of Physics, Georgetown University, Washington, DC 20057, USA*

(Received August 10, 1999)

We have investigated the accuracy of pseudopotential (PSP) density-functional calculations with respect to the corresponding all-electron (AE) results for a variety of atoms and small molecules. It is found that most of the deviations between ab-initio PSP and AE calculations are due to the linearization of the exchange-correlation functional within the PSP approach. This problem can be eliminated by applying nonlinear core corrections (NLCC). We find that a correct description of spin-polarized states requires the NLCC, even for first-row atoms. This is essential for simulations of magnetic systems and reaction processes which involve radicals. The NLCC is also essential for a realistic description of elements with more long-range core states such as alkali atoms. A further improvement of pseudopotential accuracy may be achieved by explicitly including semi-core states in the calculation.

1. Introduction

Recent implementations of electronic structure methodologies can be divided into two groups. On one side, there are all-electron (AE) approaches which consider both core and valence electrons explicitly in the calculation. Since core electrons are strongly localized, these methods need to use localized basis functions such as augmented plane waves (APW) [1], linearized muffin-tin orbitals (LMTO) [2], Slater-type orbitals (STOs), or gaussian orbitals [3 to 5]. Alternatively, one can employ pseudopotentials (PSPs) which effectively project out the core states from the problem while retaining the physical properties of the valence region. Extensive discussions and reviews of this approach have appeared in Refs. [6 to 10]. Many pseudopotential applications use a plane-wave basis set [11 to 13] but a variety of local-orbital based implementations exist as well [14 to 21]. Currently, a lot of work is also aimed at mesh-based approaches [22 to 27]. Nevertheless, many PSP applications use plane-waves. The size of the basis needed for a calculation of this type depends strongly on the shape (hardness) of the atomic pseudopotentials. Several approaches have been proposed to construct soft pseudopotentials which minimize the numerical costs while retaining the accuracy of the approach [28 to 31]. In this work, we use a modification of the scheme proposed by Troullier and Martins [31]. It will be discussed in Section 2.

Although the aspect of PSP softness is vital for practical applications, the more important problem is pseudopotential transferability, i.e. the ability of the PSP to deal with electronic configurations that differ from the one used for its construction. By definition, all PSPs that are created from first principles must reproduce the all-electron eigenvalues of the electronic reference state. In order to avoid an explicit spin dependence of the PSP, the spin-averaged ground state of the corresponding atom is usually chosen. However, the actual ground states of most atoms are spin-polarized and the PSP must be transferable in order to describe these states accurately. This may not be the case for a PSP with a poor transferability. Similar problems arise when atoms are placed in a molecular environment and donate or accept electrons. It is therefore important to systematically study the results of PSP calculations and compare them with the corresponding all-electron data. This way, systematic PSP deficiencies can be found, understood, and corrected.

Our paper is organized as follows. In Section 2, we will describe the different levels of pseudopotential approximations that may be used in practical calculations and how to implement them. Section 3 focuses on discussing the accuracy of these approximations for calculations on atoms and small molecules. The results are summarized in Section 4.

2. Method

If the density-functional formalism is used in the usual Kohn-Sham approach [32, 33], the total energy may be written as:

$$E = T_0[\varrho(\mathbf{r})] + \int d\mathbf{r}\, \varrho(\mathbf{r})\, V_{\text{ext}}(\mathbf{r}) + E_H[\varrho(\mathbf{r})] + E_{\text{xc}}[\varrho(\mathbf{r})], \tag{1}$$

where

$$E_H[\varrho(\mathbf{r})] = \frac{1}{2} \int d\mathbf{r}\, \varrho(\mathbf{r})\, V_H(\mathbf{r}), \qquad V_H(\mathbf{r}) = \int d\mathbf{r}'\, \frac{\varrho(\mathbf{r}')}{|\mathbf{r} - \mathbf{r}'|},$$

$$\varrho(\mathbf{r}) = \sum_i^{\text{occ}} |\Psi_i(\mathbf{r})|^2, \qquad T_0[\varrho(\mathbf{r})] = -\sum_i^{\text{occ}} \int d\mathbf{r}\, \Psi_i^*(\mathbf{r})\, \frac{\nabla^2}{2}\, \Psi_i(\mathbf{r}). \tag{2}$$

Note that atomic units for energy (1 a.u. = 27.211 eV) and length (1 a_B = 0.529 Å) have been used in the above equations. $V_{\text{ext}}(\mathbf{r})$ is the external potential, for an atomic calculation $V_{\text{ext}}(\mathbf{r}) = -Z/|r|$. $T_0[\varrho(\mathbf{r})]$ is the kinetic energy of an interaction-free electron gas which may be calculated easily using the Kohn-Sham orbitals $\Psi_i(\mathbf{r})$. $E_{\text{xc}}[\varrho(\mathbf{r})]$ is the exchange-correlation energy which is a nonlinear functional of the density $\varrho(\mathbf{r})$. The exact expression for E_{xc} is unknown but there are reasonable approximations to it. In this paper, we have used both local-density and the Perdew-Burke-Ernzerhof generalized-gradient approximations [34]. A generalized total energy expression for spin-polarized systems may be obtained by allowing different Kohn-Sham orbitals for up and down electrons and summing over both spin systems.

Minimization of the expression (1) leads to the Kohn-Sham equations

$$\left[-\frac{\nabla^2}{2} + V_{\text{ext}}(\mathbf{r}) + V_H[\varrho(\mathbf{r})] + V_{\text{xc}}[\varrho(\mathbf{r})] \right] \Psi_i(\mathbf{r}) = \varepsilon_i\, \Psi_i(\mathbf{r}),$$

$$V_{\text{xc}}[\varrho(\mathbf{r})] = \frac{\delta E_{\text{xc}}[\varrho(\mathbf{r})]}{\delta \varrho(\mathbf{r})}, \qquad V_{\text{eff}}(\mathbf{r}) = V_{\text{ext}}(\mathbf{r}) + V_H[\varrho(\mathbf{r})] + V_{\text{xc}}[\varrho(\mathbf{r})]. \tag{3}$$

In an all-electron calculation, both core and valence states are contained in the set of Kohn-Sham orbitals Ψ_i used in the calculation. Since core orbitals usually change very little if the atom is put in a different environment, it is a good approximation to assume that they are fixed for a certain atom type and that their effect may be simulated by a modified potential (pseudopotential). Experience has shown that a reliable pseudopotential should fulfill the following conditions:

1. The (node-less) lowest-energy eigenfunction (pseudo-wavefunction) of the pseudopotential has the same eigenvalue as the all-electron valence wave function.

2. The pseudo-wavefunction is identical to the all-electron wave function outside a chosen cut-off radius r_{cut} (norm conservation). There are ways to relax the second condition, for a discussion of these methods, see e.g. Ref. [30]. Since the all-electron valence functions Ψ_l for different angular momenta l (s, p, ...) are different, the corresponding pseudopotentials V_l^{ps} will be l-dependent. Therefore, they are called semi-local.

There is a number of schemes leading to pseudopotentials which obey the two conditions stated above. Here, we will focus on the approach suggested by Troullier and Martins [31]. Within this method, the pseudo-wavefunction Ψ_l^{ps} is written as

$$\Psi_l^{ps}(r) = \begin{cases} \exp\left(\sum_{n=0}^{m+2} c_{nl} r^{2n}\right) & \text{for } r \leq r_{cut}, \\ \Psi_l(r) & \text{for } r \geq r_{cut}. \end{cases} \quad (4)$$

The c_{nl} are determined from the following conditions:

$$\frac{d^n}{dr^n} \Psi_l^{ps}(r_{cut}) = \frac{d^n}{dr^n} \Psi_l(r_{cut}) \quad \forall n = 0, m \quad \text{(match first } m \text{ derivatives)},$$

$$\int_0^{r_{cut}} r^2 ((\Psi_l^{ps}(r))^2 - (\Psi_l(r))^2) = 0 \quad \text{(norm conservation)},$$

$$\frac{d^2}{dr^2} V_l^{ps/scr}(0) = 0 \quad \text{(smoothness condition)}, \quad (5)$$

where $V_l^{ps/scr}(r)$ is the screened pseudopotential replacing the effective all-electron potential in Eq. (3) and related to $\Psi_l^{ps}(r)$ by

$$V_l^{ps/scr}(r) = \varepsilon_l + \frac{\nabla^2 \Psi_l^{ps}(r)}{2\Psi_l^{ps}(r)} \quad \text{(inverted Kohn-Sham equation)}. \quad (6)$$

In the original Troullier-Martins approach, only the first four derivatives of $\Psi_l^{ps}(r)$ were matched at $r = r_{cut}$. In order to improve the PSP smoothness in this area, we match the first six derivatives. As a result, the eigenvalue of the all-electron valence wave function is usually reproduced by the pseudo-wavefunction within 1 meV. As an example, Fig. 1 shows pseudo and all-electron valence wave functions for the sulphur atom.

As has been noted earlier, the screened PSP $V_l^{ps/scr}(r)$ in a pseudopotential calculation is equivalent to the effective potential $V_{eff}(\mathbf{r})$ in an all-electron calculation,

$$\left[-\frac{\nabla^2}{2} + V_l^{ps/scr}(r)\right] \Psi_l^{ps}(r) = \varepsilon_l \Psi_l^{ps}(r). \quad (7)$$

However, what is needed for applications of the formalism to arbitrary systems is the unscreened PSP $V_l^{ps/unscr}(r)$ which is equivalent to the external potential in an all-electron calculation. In the traditional PSP approach, the pseudo-density $\varrho^{ps}(r)$ takes the

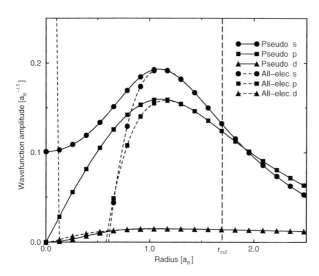

Fig. 1. Pseudo and all-electron valence wave functions for sulphur

place of the total density $\varrho(r)$ in an all-electron calculation. Therefore, the PSP Kohn-Sham equation reads

$$\left[-\frac{\nabla^2}{2} + V_l^{\mathrm{ps/unscr}}(r) + V_{\mathrm{H}}[\varrho^{\mathrm{ps}}(r)] + V_{\mathrm{xc}}[\varrho^{\mathrm{ps}}(r)]\right] \Psi_l^{\mathrm{ps}}(r) = \varepsilon_l \Psi_l^{\mathrm{ps}}(r). \tag{8}$$

and hence

$$V_l^{\mathrm{ps/unscr}}(r) = V_l^{\mathrm{ps/scr}}(r) - V_{\mathrm{H}}[\varrho^{\mathrm{ps}}(r)] - V_{\mathrm{xc}}[\varrho^{\mathrm{ps}}(r)]. \tag{9}$$

However, this approach neglects that in contrast to the Coulomb potential V_{H} which is linear in the charge density, E_{xc} and V_{xc} are nonlinear functionals. Consequently, core and valence contributions to the exchange-correlation energy cannot be separated in a simple manner. For this reason, Louie et al. [35] introduced a scheme which takes the nonlinear nature of the E_{xc} functional into account by explicitly adding a core charge density, i.e. replacing $\varrho^{\mathrm{ps}}(r)$ by $\varrho^{\mathrm{ps}}(r) + \varrho^c(r)$. Hence, the unscreening procedure is modified

$$V_l^{\mathrm{ps/unscr}}(r) = V_l^{\mathrm{ps/scr}}(r) - V_{\mathrm{H}}[\varrho^{\mathrm{ps}}(r)] - V_{\mathrm{xc}}[\varrho^{\mathrm{ps}}(r) + \varrho^c(r)]. \tag{10}$$

This correction to the standard PSP approach is commonly known as nonlinear core correction (NLCC). If ϱ^c is chosen as $\varrho^c(r) = \varrho(r) - \varrho^{\mathrm{ps}}(r) = \varrho_{\mathrm{full}}^c(r)$ (full NLCC), the corresponding PSP should give the same results as an all-electron calculation, provided that core relaxation effects are negligible. However, such an approach is not feasible for plane-wave calculations because of the strongly localized nature of the true core charge density. For this reason, Louie et al. [35] suggested a partial NLCC characterized by a pseudo core density that is equivalent to the true core charge density for a radius $r \geq r_{\mathrm{core}}$ and smooth for smaller radii. Unfortunately, the original ansatz for $\varrho^c(r)$ of Ref. [35] does not lead to a smooth transition at $r = r_{\mathrm{core}}$. Therefore, we have developed a different procedure to determine this quantity. We set

$$\varrho^c(r) = \begin{cases} \exp\left(\sum_{n=0}^{m+1} d_n r^{2n}\right) & \text{for } r \leq r_{\mathrm{core}}, \\ \varrho_{\mathrm{full}}^c(r) & \text{for } r \geq r_{\mathrm{core}}, \end{cases} \tag{11}$$

and the d_n are determined based on the following conditions:

$$\frac{d^n}{dr^n} \varrho^c(r_{\text{core}}) = \frac{d^n}{dr^n} \varrho^c_{\text{full}}(r_{\text{core}}) \qquad \forall n = 0, m \quad \text{(match first } m \text{ derivatives)},$$

$$\int_0^{r_{\text{core}}} \left(\frac{d^2}{dr^2} \varrho^c(r)\right)^2 \to \min \qquad \text{(smoothness condition)}, \qquad (12)$$

Figure 2 illustrates the PSP unscreening without the NLCC and with a partial NLCC for the sulphur atom. Note that the unscreened potentials depend on the magnitude of the NLCC.

After the unscreened PSP has been defined, it can be used in electronic structure calculations. All results presented here have been obtained using the gaussian-orbital based density functional code NRLMOL developed by Pederson, Porezag, and Jackson [4, 5, 20]. This method is based on an accurate numerical integration scheme. Besides the expansion of the wave functions into gaussian orbitals, no additional approximations are made. The program allows for mixed all-electron and pseudopotential descriptions within the same calculation, i.e. some atoms may be treated within an all-electron

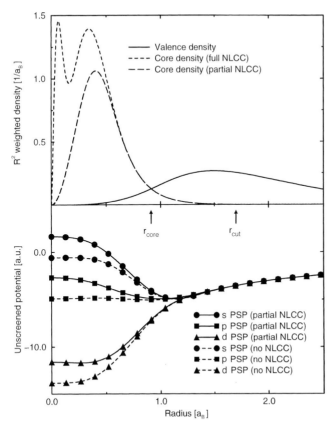

Fig. 2. Upper part: valence density and core densities with full and partial NLCC for sulphur. Lower part: semilocal PSPs without and with partial NLCC for sulphur (a.u. atomic units)

scheme while others may be described with PSPs. We have employed very large basis sets and checked that all results are converged with respect to basis size. The generalized-gradient (GGA-PBE) and local-density (LDA) approximations [34] have been used to describe exchange-correlation effects.

3. Results and Discussion

Before presenting our results for molecules, it is worthwhile to look at the calculated properties of free atoms. In fact, while the PSP construction is usually performed based on spin-unpolarized electronic states, the actual atomic ground state is often characterized by a non-vanishing total spin. It is therefore important to compare the relative energies of different electronic configurations and compare them to the corresponding all-electron atoms. One of the most important figures is the energy difference between the spin-averaged electronic state used in the PSP construction and the (frequently spin-polarized) atomic ground state. This quantity will be referred to as the atomic spin polarization energy (SPE) in the following. Table 1 summarizes the cut-off and NLCC core radii used to construct the PSPs for the elements discussed here. The spin-unpolarized electronic ground state of the corresponding atom was used as the PSP construction reference state.

3.1 Atoms

Figure 3 shows the calculated differences between PSP and all-electron SPEs with and without the NLCC for the first-row atoms B through Ne. These atoms are characterized by a single, strongly localized 1s core shell. Therefore, it is usually assumed that they can be treated without the NLCC. However, since the valence states in these systems are also fairly compact, there is still a significant overlap between valence and core charge densities. As a result, the PSP based spin polarization energies show significant errors as compared to the corresponding all-electron values if nonlinear core corrections are ignored. The deviations are largest for N which has a fully polarized 2p shell and a total SPE of 3.02 eV within LDA and 3.12 eV within GGA-PBE. If the NLCC is not applied, the SPE errors amount to about 10% of the total SPE. The deviations can be as large as 0.3 eV (for N and O within GGA-PBE). Errors of this magnitude will lead to discrepancies of approximately 0.6 eV if binding energies for homo-nuclear dimers are calculated. For example, the AE GGA-PBE dissociation energy for O_2 is 6.2 eV whereas it is determined to be 5.7 eV in the PSP approach without the NLCC.

Table 1

Cut-off and core radii [in a_B] used for the pseudopotential construction. Mn^c corresponds to a conservative Mn pseudopotential which includes the 3s and 3p semi-core states in the valence basis

	H	Li	B	C	N	O	F	S	Mn	Mn^c
s	0.80	2.65	1.40	1.14	0.96	0.83	0.73	1.53	2.53	0.81
p	0.80	1.62	1.40	1.14	0.96	0.83	0.73	1.71	2.75	0.81
d	0.80	1.62	1.40	1.14	0.96	0.83	0.73	1.71	0.81	0.81
NLCC	0	1.62	0.72	0.55	0.44	0.36	0.31	0.92	0.43	0.43

The Accuracy of the Pseudopotential Approximation

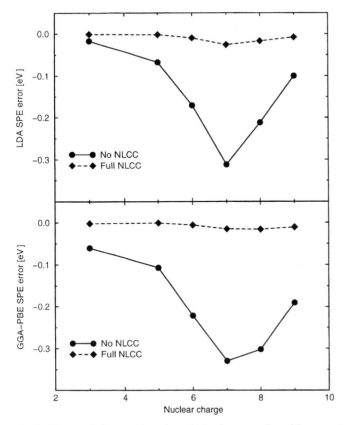

Fig. 3. Errors of the atomic spin-polarization energies with respect to all-electron values as calculated with full NLCC and without NLCC

Including the NLCC greatly improves the performance of the PSP approach, leading to a maximum deviation of only 0.02 eV (for the SPE of N). Hence, it can be concluded that SPEs of first-row atoms are very well described if a full NLCC is applied. The question which remains to be answered is how the SPE error depends on the choice of the core radius if only a partial NLCC is applied. Table 2 shows results for a GGA-PBE calculation on oxygen (the behavior is very similar for the other first-row atoms as well).

For core radii of about $0.4a_B$ or smaller, the spin polarization energies determined with a partial NLCC do not differ from the full NLCC results.

Table 2

Atomic GGA-PBE spin polarization energies [in eV] for oxygen as calculated with all-electron and PSP methods using NLCC and different core radii r_{core} (first line, given in a_B). $r_{core} = \infty$ corresponds to a neglect of NLCC

	AE	0	0.36	0.75	1.10	∞
SPE	−1.86	−1.88	−1.88	−1.95	−2.07	−2.16
error	—	−0.02	−0.02	−0.09	−0.21	−0.30

Table 3
Atomic LDA spin polarization energies [in eV] for Mn as calculated with all-electron and PSP methods using different core radii r_{core} (first line, given in a_B). $r_{core} = \infty$ corresponds to a neglect of NLCC

AE	0	0.43	0.65	0.80	2.50	∞
−5.28	−5.31	−5.31	−5.40	−5.57	−7.58	−12.37

The deficiencies of the PSP approach without the NLCC become even more apparent for transition metal atoms. These systems have strongly localized d electrons. In fact, the d charge density in the 3d transition metals overlaps substantially with the 3s and 3p core density. In the following we consider the Mn atom which has a fully spin-polarized 3d shell. Table 3 shows results for the LDA spin polarization energies as calculated within the all-electron and PSP approaches using different core radii for the NLCC. It is apparent that the PSP method without the NLCC gives completely unreasonable results. Further, the core radius must be chosen fairly small ($0.43 a_B$) in order to achieve converged results. However, if the NLCC is applied correctly, it leads to a very good agreement with the all-electron calculation.

Another interesting and important property is the s–d transfer energy (i.e. the total energy differences between the atomic $s^2 d^n$ and $s^1 d^{n+1}$ electronic configurations). We have calculated this quantity and give the results in Table 4. The all-electron and converged PSP + NLCC s–d transfer energies are found at 1.04 and 1.20 eV, respectively. Note that the error increases significantly if r_{core} is chosen too large. Further, although the NLCC leads to a much improved description of the s–d electron transfer, there is still a difference of 0.16 eV between PSP and all-electron results which needs further discussion. The calculations presented so far were based on PSPs which assigned the 3s and 3p states to the core. However, when we examine the spatial range of these states, we find that they strongly overlap with the spatial range of the the 3d electrons. Transferring an electron from the rather diffuse 4s state to the localized 3d state will increase the Coulomb repulsion in this area and hence lead to a relaxation of the 3s and 3p states. Assigning these states to the core makes such a relaxation impossible. For this reason, we have constructed a 'conservative' pseudopotential for Mn which includes the 3s and 3p states in the valence basis. As can be seen from Table 4, the conservative approach leads to results which are in excellent agreement with the all-electron data. Therefore, it can be concluded that the PSP approach can describe transition metal atoms with a similar accuracy as other atom types if both nonlinear core–valence interactions and semi-core relaxation effects are included in the method.

Table 4
Atomic LDA s–d transfer energies [in eV] for Mn as calculated with all-electron and PSP methods using different core radii r_{core} (first line, given in a_B). $r_{core} = \infty$ corresponds to a neglect of NLCC. The Ar core line refers to a calculation where the 3sp electrons are treated as core electrons

	AE	0	0.43	0.65	0.80	2.50	∞
Ar core	1.04	1.20	1.20	1.24	1.33	2.26	4.22
Ne core	1.04	1.06	1.06				1.35

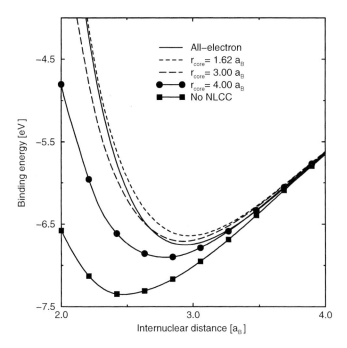

Fig. 4. LDA cohesive energy for the LiF molecule as a function of internuclear distance. The PSP+NLCC results for $r_{core} = 1.62 a_B$ are identical to the full NLCC

3.2 LiF diatomic molecule

The preceding section was aimed at discussing energy differences induced by simply changing the occupation numbers of the atomic one-electron orbitals. However, in many practical applications, the most important change in the atomic electron density occurs when the atom is incorporated into a solid, cluster, or molecule. The largest relative changes will occur for atoms with a very diffuse valence density if they are paired with atoms characterized by compact localized states. Prominent examples for such systems are alkali-halogenide compounds such as LiF. In order to demonstrate the problems which occur for this combination of elements, we focus on the LiF diatomic molecule. Figure 4 shows the cohesive energy of this system as a function of interatomic distance as calculated within LDA. The PSP calculation without the NLCC overestimates the binding energy by about 1 eV and underestimates the bond length by more than $0.5 a_B$. Applying a full NLCC corrects this deficiency and leads to a good agreement with the all-electron result. Partial and full NLCC results agree for $r_{core} \leq 1.62 a_B$. The remaining difference between PSP + NLCC and AE data can be attributed to the neglect of the 1s core relaxation. Note that in order to rule out additional sources of error, the F atom has also been treated in an all-electron fashion here.

3.3 Structural and vibrational properties of small molecules

The preceding sections have established that it is often necessary to apply the NLCC in order to obtain reliable results. While there are also many cases where the NLCC has negligibly small effects, we advocate to always use this correction since it usually does not increase the numerical effort of a calculation very much, especially not in calculations with gaussian-orbital basis sets. Consequently, the results presented in the follow-

Table 5

Bond lengths [in a_B] and angles for some molecules as calculated with the all-electron (AE) and PSP + NLCC approaches. The GGA-PBE exchange-correlatioin functional has been used

		AE	PSP			AE	PSP
H_2	r(HH)	1.418	1.418	LiH	r(LiH)	3.039	3.045
Li_2	r(LiLi)	5.200	5.200	LiF	r(LiF)	2.981	2.974
N_2	r(NN)	2.085	2.083	H_2O	r(OH)	1.833	1.832
O_2	r(OO)	2.309	2.310		$\angle$(HOH)	104.1°	104.1°
F_2	r(FF)	2.679	2.684	CH_4	r(CH)	2.071	2.071
CO	r(CO)	2.148	2.146	SO_2	r(SO)	2.774	2.774
HF	r(HF)	1.760	1.759		$\angle$(OSO)	118.9°	118.9°

ing are based on calculations with pseudopotentials that employ the NLCC and use the cut-off and core radii summarized in Table 1. For a more detailed discussion on the proper choice of the NLCC core radii please refer to Ref. [36].

Table 5 lists the calculated bond lengths and angles for twelve molecules as calculated with the all-electron and pseudopotential approaches and the GGA-PBE exchange-correlation functional. PSP and AE results are in excellent agreement with a root-mean-square (RMS) error of only $0.003a_B$ for the bond lengths. The largest errors occur for LiH and LiF which can be attributed to the neglect of the Li core relaxation effects. Table 6 contains information about calculated atomization energies. Again, AE and PSP data are in very good agreement, the deviations do not exceed 0.04 eV per atom. The current PSP results show much smaller errors than the ones published in an earlier paper [20] which were based on BHS pseudopotentials without nonlinear core corrections. This is especially true for the molecules containing Li where the previous PSP calculations gave binding energies with errors of several eV as compared to the corresponding all-electron values. This illustrates that the NLCC is especially important for alkali atoms. Note that all atomization energies are consistently underestimated. This is most likely due to the neglect of core relaxation effects for the molecular species.

Table 7 summarizes the vibrational properties of the molecules investigated here. Frequencies, IR intensities, and Raman scattering activities have been determined with a method published in Ref. [37]. The PSP based vibrational frequencies are in excellent agreement with the AE calculations. The two methodologies usually differ by only a few wavenumbers, or at most 1%. However, the behavior of the IR intensities and Ra-

Table 6

Atomization energies [in eV] for some molecules as calculated with the all-electron (AE) and PSP + NLCC approaches. The GGA-PBE exchange-correlation functional has been used. Zero-point motion has been included in the values presented here

	AE	PSP	Δ		AE	PSP	Δ
H_2	4.27	4.27	0	HF	5.93	5.91	−0.02
Li_2	0.84	0.82	−0.02	LiH	2.22	2.18	−0.04
N_2	10.35	10.28	−0.07	LiF	5.96	5.89	−0.07
O_2	6.10	6.05	−0.05	H_2O	9.61	9.57	−0.04
F_2	2.23	2.17	−0.06	CH_4	17.03	17.01	−0.02
CO	11.50	11.46	−0.04	SO_2	11.52	11.47	−0.05

Table 7

Harmonic vibrational frequencies [in cm^{-1}], IR intensities [in (D/Å)^2 amu^{-1}] and Raman activities [in Å^{4}amu^{-1}] for some molecules as calculated with the all-electron (AE) and PSP + NLCC approaches. The GGA-PBE exchange-correlation functional has been used

		AE			PSP		
		freq.	IR	Ram.	freq.	IR	Ram.
H_2	Σ_g	4311	–	165	4312	–	167
Li_2	Σ_g	327	–	1700	324	–	2060
N_2	Σ_g	2345	–	26.1	2344	–	24.3
O_2	Σ_g	1547	–	13.8	1543	–	14.4
F_2	Σ_g	994	–	8.08	984	–	8.08
CO	Σ	2123	1.54	19.4	2125	1.54	18.4
HF	Σ	3959	2.18	42.6	3972	2.14	41.4
LiH	Σ	1373	3.27	851	1360	2.94	924
LiF	Σ	891	3.33	10.1	888	3.13	8.97
H_2O	A_1	1594	1.61	0.77	1597	1.60	0.78
		3695	0.05	107	3702	0.04	106
	B_2	3800	1.21	26.2	3805	1.23	26.3
CH_4	T_2	1282	0.91	0.27	1284	0.91	0.32
		3079	1.32	142	3078	1.24	141
	E	1506	–	4.67	1509	–	4.69
	A_1	2965	–	235	2960	–	235
SO_2	A_1	478	0.53	2.29	479	0.53	2.35
		1087	0.52	35.0	1085	0.53	33.8
	B_2	1268	4.36	8.10	1268	4.33	8.57

man activities (which depends strongly on subtle changes of the molecular charge distribution as a function of the nuclear coordinates) is different. While many IR and Raman strengths agree very well (within 5%) there are also cases with significantly larger deviations. For example, the spectral intensities of all molecules containing Li show errors of about 10%, indicating that the effect of the Li core electrons cannot be completely neglected. However, even the largest errors do not exceed the typical differences between theory and experiment which were found to be 10 to 30% in Ref. [37].

4. Summary and Conclusions

We have investigated the accuracy of the pseudopotential approximation as compared to the all-electron approach for a variety of small molecules using norm-conserving Troullier-Martins PSPs and a gaussian-orbital based numerical scheme. It is found that nonlinear core corrections are often necessary to obtain reliable results, therefore, we suggest to always use them in applications of the method to larger systems. We have shown that the PSP approach with NLCC predicts equilibrium structures, energies, and vibrational frequencies with a very high accuracy. Results for IR and Raman intensities show larger errors but are still sufficiently accurate for most practical applications. Remaining deviations between the PSP and AE methods can be explained by core relaxation effects. Consequently, the performance of the PSP approximation can be substantially improved by using conservative pseudopotentials, i.e. by including semi-core states explicitly in the calculation. In summary, it can be concluded that the combined PSP + NLCC approach is very well suited for the class of systems investigated here.

Acknowledgements DVP greatfully acknowledges support from the Alexander-von-Humboldt foundation of Germany. This work was supported in part by the ONR Molecular Design Institute N0001498WX20709 and by National Science Foundation Grant No. DMR9627778.

References

[1] E. WIMMER, H. KRAKAUER, M. WEINERT, and A. J. FREEMAN, Phys. Rev. B **24**, 864 (1981).
[2] O. K. ANDERSEN, Phys. Rev. B **12**, 3060 (1975).
[3] Gaussian 94, M. J. FRISCH, G. W. TRUCKS, H. B. SCHLEGEL, P. M. W. GILL, B. G. JOHNSON, M. A. ROBB, J. R. CHEESEMAN, T. KEITH, G. A. PETERSSON, J. A. MONTGOMERY, K. RAGHAVACHARI, M. A. AL-LAHAM, V. G. ZAKRZEWSKI, J. V. ORTIZ, J. B. FORESMAN, J. CIOSLOWSKI, B. B. STEFANOV, A. NANAYAKKARA, M. CHALLACOMBE, C. Y. PENG, P. Y. AYALA, W. CHEN, M. W. WONG, J. L. ANDRES, E. S. REPLOGLE, R. GOMPERTS, R. L. MARTIN, D. J. FOX, J. S. BINKLEY, D. J. DEFREES, J. BAKER, J. P. STEWART, M. HEAD-GORDON, C. GONZALEZ, and J. A. POPLE, Gaussian Inc., Pittsburgh PA, 1994.
[4] M. R. PEDERSON and K. A. JACKSON, Phys. Rev. B **41**, 7453 (1990).
[5] K. A. JACKSON and M. R. PEDERSON, Phys. Rev. B **42**, 3276 (1990).
[6] M. L. COHEN and V. HEINE, Solid State Phys. **24**, 37 (1970).
[7] W. E. PICKETT, Comp. Phys. Rep. **9**, 116 (1989).
[8] J. R. CHELIKOWSKY and M. L. COHEN, in: Handbook on Semiconductors, Vol. 1, Ed. P. T. LANDSBERG, Elsevier Publ. Co., Amsterdam 1992 (p. 59).
[9] M. C. PAYNE, M. P. TETER, D. C. ALLAN, T. A. ARIAS, and J. D. JOANNOPOULOS, Rev. Mod. Phys. **64**, 1045 (1992).
[10] S. GOEDECKER and K. MASCHKE, Phys. Rev. B **45**, 88 (1992).
[11] R. CAR and M. PARRINELLO, Phys. Rev. Lett. **55**, 2471 (1985).
[12] R. M. WENTZCOVICH and J. L. MARTINS, Solid State Commun. **78**, 831 (1991).
[13] M. BOCKSTEDTE, A. KLEY, J. NEUGEBAUER, and M. SCHEFFLER, Comput. Phys. Commun. **107**, 187 (1997).
[14] J. BERNHOLC and S. T. PANTELIDES, Phys. Rev. B **18**, 1780 (1978).
[15] D. VANDERBILT and S. G. LOUIE, Phys. Rev. B **30**, 6118 (1984).
[16] J. R. CHELIKOWSKY and S. G. LOUIE, Phys. Rev. B **29**, 3470 (1984).
[17] P. J. FEIBELMAN, Phys. Rev. B **35**, 2626 (1987).
[18] O. F. SANKEY, D. A. DRABOLD, and G. B. ADAMS, Bull. Amer. Phys. Soc. **36**, 924 (1991).
[19] R. JONES and A. SAYYASH, J. Phys. C **19**, L653 (1986).
[20] A. BRILEY, M. R. PEDERSON, K. A. JACKSON, D. C. PATTON, and D. V. POREZAG, Phys. Rev. B **58**, 1786 (1998).
[21] D. SANCHEZ PORTAL, P. ORDEJON, E. ARTACHO, and J. M. SOLER, Internat. J. Quantum Chem. **65**, 453 (1997).
[22] F. GYGI, Europhys. Lett. **19**, 617 (1992).
[23] D. R. HAMANN, Phys. Rev. B **51**, 7337 (1995).
[24] A. DEVENYI, K. CHO, T. A. ARIAS, and J. D. JOANNOPOULOS, Phys. Rev. B **49**, 13373 (1994).
[25] N. A. MODINE, G. ZUMBACH, and E. KAXIRAS, Phys. Rev. B **55**, 10289 (1997).
[26] J. R. CHELIKOWSKY, N. TROULLIER, and K. WU, Phys. Rev. B **16**, 11355 (1994).
[27] E. L. BRIGGS, D. J. SULLIVAN, and J. BERNHOLC, Phys. Rev B **52**, R5471 (1995).
[28] D. R. HAMANN, M. SCHLÜTER, and C. CHIANG, Phys. Rev. Lett. **43**, 1494 (1979).
[29] A. M. RAPPE, K. M. RABE, E. KAXIRAS, and J. D. JOANNOPOULOS, Phys. Rev. B **41**, 1227 (1990).
[30] D. VANDERBILT, Phys. Rev. B **41**, R7892 (1990).
[31] N. TROULLIER and J. L. MARTINS, Phys. Rev. B **43**, 1993 (1991).
[32] P. HOHENBERG and W. KOHN, Phys. Rev. **136**, B864 (1964).
[33] W. KOHN and L. J. SHAM, Phys. Rev. **140**, A1133 (1965).
[34] J. P. PERDEW, K. BURKE, and M. ERNZERHOF, Phys. Rev. Lett. **77**, 3865 (1996).
[35] S. G. LOUIE, S. FROYEN, and M. L. COHEN, Phys. Rev. B **26**, 1738 (1982).
[36] D. POREZAG, M. R. PEDERSON, and A. Y. LIU, Phys. Rev. B **60**, 14132 (1999).
[37] D. POREZAG and M. R. PEDERSON, Phys. Rev. B **54**, 7830 (1996).

phys. stat. sol. (b) **217**, 231 (2000)

Subject classification: 71.15.Fv; 71.15.Mb

Large-Scale Electronic Structure Calculations Using Linear Scaling Methods

G. Galli

Lawrence Livermore National Laboratory, P.O. Box 808, Livermore, CA 94551, USA

(Received August 10, 1999)

We describe linear scaling methods for electronic structure calculations and quantum molecular dynamics simulations, and discuss the basic differences and similarities between the various frameworks proposed in the literature.

1. Introduction

Quantum simulations are aimed at modeling materials as well as physical and chemical processes at the microscopic level, by solving numerically the equations governing the atomic motion. In order to obtain an accurate microscopic description of most materials properties, the interaction between atoms must be described using the laws of quantum mechanics. Interatomic forces can be computed by solving the Schrödinger equation for electrons, thus determining the electronic ground state at given positions of the nuclei. In many cases of interest the nuclei can be considered as classical objects. Once atomic trajectories are determined, using, e.g., molecular dynamics, a variety of materials properties can be calculated.

The computer time required by a quantum simulation, and ultimately its feasibility, are mainly determined by the time necessary to solve the Schrödinger equation for the electrons. In the last two decades, many theoretical investigations of materials have adopted mean field theories to solve the Schrödinger equation for the electrons, in particular density functional theory (DFT) [1], within the local density approximation (LDA) [2], and tight-binding (TB) one-particle models [4]. While DFT within the LDA is a first-principles approach requiring no input from experiment, most TB Hamiltonians are semiempirical.

Standard approaches [4] to the solution of the Schrödinger-like equations derived within LDA as well as within a TB framework require a workload proportional to the cube of the number of atoms involved in the simulation (i.e. a number of operations of order N^3 ($O(N^3)$): Doubling the size of the system amounts to multiplying the computing time by eight. This unfavorable scaling poses limitations to the kind of problems which can be tackled with quantum simulations. The size of systems that can be treated with first-principles DFT approaches is limited to a few hundred atoms, even when using the most powerful modern computers. The size of systems that can be studied in conventional TB-MD simulations is bigger, i.e. limited to about a thousand atoms, since TB models rely on much simpler Hamiltonians than ab-initio LDA calculations, and on smaller basis sets.

Recently new methods for solving the electronic Schrödinger equation have been developed, which imply a workload growing linearly with the system size [5], a recent review can be found in [6]. These approaches, called *linear scaling methods* (or $O(N)$ methods), are aimed at simulating systems much larger than previously accessible, thus widening the range of problems that can be addressed. At present, linear scaling methods using tight-binding Hamiltonians allow one to perform simulations involving up to thousands of atoms on small workstations, and up to ten thousand atoms for tens of picoseconds when using supercomputers. This has made it possible to study problems such as large organic molecules in water [27], thin film growth [37], for a review, see [38], extended defects [39] and dislocations [40] in semiconductors. Although the implementation of first-principles linear scaling methods is less advanced than that of semi-empirical methods, promising results [31 to 33] have already been published.

In the recent literature, different $O(N)$ methods have been proposed, with some approaches being based on an orbital formulation of the electronic properties, [7, 9 to 13, 18, 19, 21, 22, 25, 32, 33] and others based on the calculation of the Green function [8, 17], or the density matrix [14, 15, 24 to 28, 34] or the density [29, 30]. A key point of $O(N)$ methods is the evaluation of total energy and forces without computing the eigenvalues and eigenstates of an effective single particle Hamiltonian. In orbital and density matrix $O(N)$ approaches, this is accomplished by dividing the full system into subsystems and then defining electronic degrees of freedom — either orbitals or density matrices — which are localized in the subsystems. Schrödinger-like equations are then solved for these localized degrees of freedom. The subsystems are overlapping portions of the full system, often called localization [9] or support regions [25]. The important point is that the extension of a localization region (LR) depends on the physical and chemical properties of the system but not on its volume. The size of a LR is the parameter controlling the accuracy of the calculation.

The reason why linear scaling methods can exist, and thus the reason why the extension of a localization region can depend only on the physical properties of a given system and not on its volume, has been discussed by Kohn [26] in terms of *nearsightedness* of the equilibrium system. This is a physical principle widely applicable to *many* quantum mechanical particles moving in an external potential $v(r)$ (i.e. many electrons in the field of ions or ionic cores). Loosely, the *nearsightedness* of equilibrium systems can be expressed as follows, for systems without long-range electric fields: a change Δv in the external potential, no matter how large, has a small effect on a static property F, if Δv is limited to a region "distant from F", i.e. if $\Delta v(r')$ is such that $|r' - r| \gg \lambda$, with r being the center of mass of the coordinates which F depends upon, and λ being a typical de Broglie wavelength occurring in the ground state wavefunction of the system. In other words, F does not "see" Δv if r' is "far" from its center of mass r. As stated in Ref. [26], the *nearsightedness* of equilibrium systems does require the presence of *many* particles (not necessarily interacting) and it is the consequence of wave-mechanical destructive interference. We note that the principle is not universally valid; amongst notable exceptions quoted in Ref. [26] are systems with translationally invariant long-range order, like a Wigner crystal in a torus.

$O(N)$ approaches based on the direct calculation of the one-particle Green function [8, 17] are different in spirit from orbital or density matrix based frameworks, although they rely upon the same principle of *nearsightedness* of equilibrium systems. Charge and energy densities are obtained directly from the Green function by using the recursion method [3], without solving any Schrödinger-like equation. The accuracy of the calculation is determined by the truncation of the continuous fraction used to express the elements of the Green function, i.e. by the choice of the so-called terminator for the continuous fraction. Green function formulations [41] have been the basis for the development of bond order potentials, reviewed in a recent book by Pettifor [42].

In the following we will review implementations of $O(N)$ methods based upon orbital and density matrix formulations. In the condensed matter physics community a popular approach is based on energy functional minimizations with respect to either localized functions or localized density matrices. In the chemistry community, the divide-and-conquer strategy introduced by Yang [7] for Kohn-Sham Hamiltonians and then generalized to Hartree-Fock and semiempirical calculations [27, 28] is probably the most popular $O(N)$ method. In next sections we describe in some detail the approaches used in both communities. In particular, in Section 2 we briefly recall the basic equations of conventional orbital and density matrix formulations. In Section 3 we present $O(N)$ orbital and density matrix approaches, and in Section 4 we describe the divide-and-conquer method. Section 5 concludes the paper.

2. Energy Functionals and Their Minimization

In the Kohn-Sham formulation of density functional theory [2], the problem of N interacting electrons in an external field (e.g. the field of ions) is mapped onto the problem of an effective system of non-interacting particles, where the effective Hamiltonian $\hat{H}_{\text{eff}}(\mathbf{r}) = -\hbar/2m \, \nabla^2 + V_{\text{eff}}(\mathbf{r})$ is

$$\hat{H}_{\text{eff}}(\mathbf{r}) = -\frac{\hbar}{2m} \nabla^2 + V_{\text{ion}}(\mathbf{r}) + V_{\text{H}}(\mathbf{r}) + V_{\text{xc}}(n(\mathbf{r})). \tag{1}$$

Here $V_{\text{ion}}(\mathbf{r})$ is the Coulomb potential of the nuclei, $V_{\text{H}}(\mathbf{r}) = e^2 \int d\mathbf{r}' \, n(\mathbf{r})/|\mathbf{r} - \mathbf{r}'|$ is the Hartree potential and $V_{\text{xc}}(n(\mathbf{r}))$ is the exchange and correlation potential, which in the LDA depends only on the value of the charge density $n(\mathbf{r})$.

Within an *orbital representation*, the charge density of the system is expressed in terms of single particle orbitals, $n(\mathbf{r}) = \sum_i^{N/2} f_i \psi_i(\mathbf{r}) \, \psi_i(\mathbf{r})$, and the total electronic energy of the system is given by

$$\begin{aligned} E &= 2 \sum_i^{N/2} \langle \psi_i | \hat{H}_{\text{eff}} | \psi_i \rangle - E_{\text{DC}}[n(\mathbf{r})] \\ &= 2 \, \text{Tr}[\mathbf{H}_{\text{eff}}] - E_{\text{DC}}[n(\mathbf{r})], \end{aligned} \tag{2}$$

where

$$E_{\text{DC}}[n(\mathbf{r})] = \int d\mathbf{r} \, n(\mathbf{r}) \left[V_{\text{xc}}(n(\mathbf{r})) - \epsilon_{\text{xc}}(n(\mathbf{r})) + \tfrac{1}{2} V_{\text{H}}(\mathbf{r}) \right]. \tag{3}$$

Here $\epsilon_{\text{xc}}(n(\mathbf{r}))$ is the exchange-correlation energy density. In Eq. (2), $\text{Tr}[\mathbf{H}_{\text{eff}}]$ is the trace of the $(N/2 \times N/2)$ matrix $\mathbf{H}_{\text{eff}}$, with $(H_{\text{eff}})_{ij} = \langle \psi_i | \hat{H}_{\text{eff}} | \psi_j \rangle$. This term is often called the band structure energy term contribution (E_{BS}) to the total electronic energy. For simplicity, I have assumed that the occupation numbers f_i are all equal to 2 and that $\{\psi\}_i$ (and in general electronic orbitals) are real functions.

Within a *density matrix formulation*, the charge density of the system is given in terms of the density matrix operator $\hat{n}(\mathbf{r}, \mathbf{r}')$ in the coordinate representation, and the total energy is written as

$$E = -\frac{\hbar}{2m} \int d\mathbf{r} \, \nabla_r [\nabla_{r'} n(\mathbf{r},\mathbf{r}')]_{\mathbf{r}'=\mathbf{r}} + \int d\mathbf{r} \, d\mathbf{r}' \, V_{\text{eff}}(\mathbf{r},\mathbf{r}') \, n(\mathbf{r},\mathbf{r}') - E_{\text{DC}}[n(\mathbf{r},\mathbf{r})]$$
$$= \text{Tr}[\hat{n}\hat{H}_{\text{eff}}] - E_{\text{DC}}[n(\mathbf{r},\mathbf{r})] \, . \tag{4}$$

Here $\text{Tr}[\hat{n}\hat{H}_{\text{eff}}] = E_{\text{BS}}$ is the trace of the products of the two operators $\hat{n}$ and $\hat{H}_{\text{eff}}$, which in coordinate representation is given by $\int H_{\text{eff}}(\mathbf{r},\mathbf{r}') \, n(\mathbf{r}',\mathbf{r}) \, d\mathbf{r} \, d\mathbf{r}'$.

Within DFT, the ground state energy E_0 is determined by minimizing the functional E with respect to either the single-particle wavefunctions $\{\psi_i\}$ (Eq. (2)) or the density matrix $n(\mathbf{r},\mathbf{r}')$ (Eq. (4)), using the variational principle (Hohenberg-Kohn theorem [1]). In conventional orbital formulations, the minimization is carried out with respect to the set of functions satisfying the *orthonormality condition* $\langle \psi_i | \psi_j \rangle = \delta_{ij}$; in iterative minimizations this condition is explicitly enforced at each step [4]. In general, the $\{\psi_{0i}\}$ minimizing $E(E[\{\psi_{0i}\}] = E_0)$ is a set of extended orbitals related to the eigenfunctions of the single particle Hamiltonian H_{eff} by a unitary transformation. Within density matrix formulations, the minimization of E is carried out with respect to the set of *idempotent density matrices* $n(\mathbf{r},\mathbf{r}')$, subject to the condition $\int n(\mathbf{r},\mathbf{r}) \, d\mathbf{r} = N$.

Within an empirical tight-binding model, the total electronic energy is written as

$$E = E_{\text{BS}} + \sum_{LL'} V_{\text{R}}(|\mathbf{R}_L - \mathbf{R}_{L'}|) \, , \tag{5}$$

where V_{R} is a repulsive two-body potential between atoms defined by the position coordinates $\{\mathbf{R}\}_L$. In this case $E_{\text{BS}} = 2 \sum_i^{N/2} \langle \psi_i | \hat{H}_{\text{eff}} | \psi_i \rangle$, where here the effective Hamiltonian $\hat{H}_{\text{eff}}$ is a TB Hamiltonian which in most approximations does not depend on the charge density of the system.

Whether performed by direct diagonalization of $\mathbf{H}_{\text{eff}}$ or using iterative techniques, the minimization of the functional E (both within LDA and TB frameworks) is in general of $O(N^3)$. This means that the minimization of E requires a number of operations scaling as the third power of the number of electrons (and thus of atoms) in the system. Direct diagonalizations require a number of operations proportional to $\mathcal{M}^3$, with $\mathcal{M}$ being the number of basis functions used to expand the N single particle electronic states. Iterative minimizations [4, 43] are mostly used when $\mathcal{M}$ is much larger than N (e.g. when the electronic orbitals are expanded in plane waves or on a real space grid) and they are usually performed in two steps: first the gradient of the total energy $\delta E / \delta \psi_i = -\hat{H}_{\text{eff}} \psi_i$ is evaluated and then new single-particle wavefunctions are predicted; secondly, the predicted single particle states are orthonormalized. If the $\{\psi_i\}$'s are expanded, e.g. in plane waves, the first step of the minimization can be made in $O(N \log N \mathcal{M}) \simeq O(N\mathcal{M})$ operations. The second step is instead of $O(N^2 \mathcal{M})$. For systems larger than few hundred atoms, the orthogonalization step is the most time consuming and the workload of an iterative minimization becomes of $O(N^3)$. Indeed for extended states, $\mathcal{M}$ grows linearly with the volume of the system, i.e. $\mathcal{M} \propto N$.

3. Linear Scaling Methods

In order to minimize the functional E using $O(N)$ operations, instead of $O(N^3)$ operations, two basic concepts have been introduced, both within density matrix (DM) and orbital based formulations. (i) In DM approaches, the idempotency condition on the density matrix is not enforced, but a weaker condition is required when minimizing E; similarly, in orbital based approaches the orthonormality condition of the single particle electronic orbitals is never explicitly enforced. Weakening either the idempotency or the orthonormality condition leads to the definition of an energy functional of n or ψ, respectively, which is different from the energy functional minimized in conventional approaches, but which has the same absolute minimum. (ii) In DM frameworks, this energy functional is minimized with respect to spatially localized density matrices; in orbital based approaches, the functional is minimized with respect to spatially localized single particle orbitals, which we call localized functions (LF).

I now turn to the discussion of issues (i) and (ii) in detail, for both density matrix and orbital based formulations. I will then discuss how $O(N)$ scaling is achieved, when minimizing the functional E, in both approaches.

3.1 Density matrix formulation

3.1.1 Trial density matrix and purification transformations

An $O(N)$ density matrix approach was first introduced by Li, Nunes and Vanderbilt (LNV) [14] and independently by Daw [15]. Starting from a *trial density matrix* $\varrho(\mathbf{r},\mathbf{r}')$ which in general is not idempotent, a class of density matrices $\tilde{\varrho}$ with eigenvalues lying in the interval [0,1] was constructed. We call this condition on the eigenvalues a *weak* idempotency condition. Then the functional

$$E[\tilde{\varrho}] = \mathrm{Tr}[\hat{\tilde{\varrho}}(\hat{H}_{\mathrm{eff}} - \mu)] - E_{\mathrm{DC}}[\tilde{\varrho}(\mathbf{r},\mathbf{r})] \tag{6}$$

is minimized with respect to $\tilde{\varrho}$. Here μ is a Lagrange multiplier which is interpreted as an electronic chemical potential, introduced to set the correct total electronic charge at N. For non self-consistent Hamiltonians one can prove [14, 15, 19] that $\min_{\tilde{\varrho}} E[\tilde{\varrho}] = \min_n E[n] = E_0$.

The weak idempotency condition can be imposed e.g. by taking $\tilde{\varrho} = I - (I - \varrho)^p$, where p is an even integer [19]. This choice of the polynomial relating ϱ to $\tilde{\varrho}$ (i.e. of the *purification transformation*) defines a class of positive definite $\tilde{\varrho}$'s, with eigenvalues less than unity. Another meaningful choice is the so-called McWeeny transformation [14, 44], $\tilde{\varrho} = 3\varrho^2 - 2\varrho^3$, provided the range of $\tilde{\varrho}$ lies in the interval $[-0.5, 1.5]$. Note that for idempotent ϱ matrices one has $\tilde{\varrho} = \varrho$. Recently, modified versions of the LNV and McWeeny purification transformations have been proposed in the literature [34, 35], in an attempt to improve the computational efficiency of the minimization techniques used to find the ground state energy.

In TB calculations, the trial density matrix can be expressed as a linear combination of TB basis functions and the coefficients of the expansion can be considered as variational parameters. This is the approach chosen in Refs. [14, 15].

In LDA calculations it is convenient that both ϱ and $\tilde{\varrho}$ be representable in the form

$$\varrho(\mathbf{r},\mathbf{r}') = \sum_{ij}^{M} L_{ij} \phi_i(\mathbf{r}) \phi_j(\mathbf{r}') \tag{7}$$

and

$$\tilde{\varrho}(\mathbf{r},\mathbf{r}') = \sum_{ij}^{M} K_{ij} \phi_i(\mathbf{r}) \phi_j(\mathbf{r}'), \tag{8}$$

where ϕ are called support functions and in general are overlapping orbitals. M is equal to or larger than the number of occupied single particle electronic states $N/2$. This is the framework introduced in Ref. [25]. Eqs. (7) to (8) constitute the basic link between DM and orbital based formulations.

3.1.2 Trial density matrix and penalty functionals

Another approach to energy functional minimizations without enforcing the idempotency condition has been introduced by Kohn [26]. Instead of purification transformations to satisfy a weak idempotency condition, a penalty functional for violating the idempotency condition was introduced, which was then added to the total energy functional. In particular, the functional

$$\mathcal{Q}[\tilde{\varrho}] = E[\tilde{\varrho}] + \alpha P[\varrho]$$
$$= \text{Tr}[\hat{\tilde{\varrho}}(\hat{H}_{\text{eff}} - \mu)] - E_{\text{DC}}[\tilde{\varrho}(\mathbf{r},\mathbf{r})] + \alpha P[\varrho] \qquad (9)$$

is varied with respect to the non-negative definite matrices $\tilde{\varrho} = \varrho^2$, where α is a constant and P is a penalty functional defined in terms of the trial density matrix ϱ

$$P[\varrho] = [\int d\mathbf{r}\, [\tilde{\varrho}(1-\varrho)^2]_{\mathbf{r}'=\mathbf{r}}]^{\frac{1}{2}}. \qquad (10)$$

In the non-interacting case, Kohn proved that for a given μ and for α larger than a critical value α_c, the minimization of $Q[\tilde{\varrho}]$ *without imposing the idempotency constraint* leads to the correct ground state E_0 and then to an idempotent density matrix.

The penalty function introduced in Ref. [26] is difficult to use in practice. The branch point introduced by the presence of the square root in Eq. (10) means that the gradient of the functional Q is undefined at the ground state. Therefore the functional Q cannot be minimized using standard techniques such as conjugate gradients algorithms. Haynes and Payne [33] recently introduced a generalized functional with a penalty function which enforces idempotency of the energy matrix only approximately, but which permits the use of efficient minimization methods. Calculations obtained for the Si crystal [33] with the Haynes and Payne functional indicate that the approach can yield reasonably accurate results.

3.1.3 Localized density matrices

In general, the minimization of the functionals E (Eq. (6)) or $\mathcal{Q}$ (Eq. (9)) is of $O(N^3)$, as it is the minimization of the energy functional of Eq. (4). However, having weakened the idempotency condition, the functionals (6) and (9) can now be minimized in $O(N)$ operations, *if* the minimization is restricted to *localized density matrices*. Restricting the minimization to *localized density matrices* is often referred to as "applying localization constraints". It is important to stress that, applying straightforwardly *localization constraints* to the minimization of the functional in Eq. (4) does *not* change the scaling of the minimization from $O(N^3)$ to $O(N)$. Instead, applying *localization constraints* to the minimization of the functionals in Eq. (6) or Eq. (9) *does* lead to a change in scaling, from $O(N^3)$ to $O(N)$, because the idempotency condition on the DM has been weakened.

Restricting the minimization to *localized density matrices* can be accomplished, e.g., by searching over $\tilde{\varrho}$ such that

$$\tilde{\varrho}(\mathbf{r},\mathbf{r}') = 0, \qquad |\mathbf{r}-\mathbf{r}'| > R_c. \qquad (11)$$

Here R_c is a cut-off radius depending upon the physical and chemical properties of a given system but not on its size. If $\tilde{\varrho}$ is represented in terms of orbitals (Eq. (8)), its localization can be imposed by requiring that the support functions be non zero only within chosen regions, called localization (or support) regions, and that the coefficients K_{ij} vanish if the separation of the localization regions of ϕ_i and ϕ_j exceeds a chosen cut-off [25].

We note that the existence, for a large class of systems, of a finite R_c which does *not* depend upon the entire volume of the system stems from the *nearsightedness* of equilibrium systems containing *many* quantum mechanical particles, as discussed in the Introduction. The *nearsightedness* of equilibrium systems implies that the one-electron DM is short ranged for a large class of physical systems, unlike KS orbitals, i.e. eigenstates of the KS Hamiltonian, which are in general extended fields.

An $O(N)$ method based on the finite temperature density matrix, or Fermi matrix was proposed by Goedecker and Colombo [20] for TB Hamiltonians, and is reviewed in Ref. [6]. Within this approach $E_{BS} = \text{Tr}[\hat{H}_{\text{eff}} \hat{F}]$, where $F = f_\beta(\mu - H_{\text{eff}})$ and f_β is the Fermi-Dirac distribution function ($\beta = k_B T$, where k_B is the Boltzmann constant and T the temperature). The function f_β is then approximated by a given polynomial p_f of degree n_f in the interval spanned by the lowest ($e_{\min}$) and the highest ($e_{\max}$) eigenvalues of the Hamiltonian. Once p_f is known, each column of the matrix F is computed by applying p_f to the TB basis set functions. This calculation, which in principle requires $O(N^2)$ operations, can be performed in $O(N)$ operations by forcing the Fermi matrix to be localized in a given volume of real space, i.e. by enforcing condition (11) for the Fermi matrix. In general $n_f \simeq (e_{\max} - e_{\min})/k_B T$ and therefore the efficiency of the method decreases as the temperature decreases; however, for an insulator with electronic gap e_{gap} one can find a polynomial of degree $(e_{\max} - e_{\min})/e_{\text{gap}}$ which accurately approximates f_β in the valence band region even at low temperatures.

Orbital-free density functionals have been recently proposed also by Madden and coworkers [29] and by Carter and coworkers [30]. These are based on an expression of the energy solely in terms of the charge density of the system. The kinetic energy functional is written in terms of the electron density via the response functionals of the non-interacting electron gas. This approach leads naturally to a $O(N)$ method for electronic structure calculations and MD simulations. The proposed orbital-free representations are particularly advantageous for metals, while for insulating phases the agreement with conventional Kohn-Sham calculations is significantly poorer.

3.2 Localised function formulation

3.2.1 Energy functionals with implicit orthonormalization constraints

A basic ingredient of orbital based $O(N)$ methods is the definition of energy functionals with *implicit* orthonormalization constraints, as opposed to conventional energy functionals which are minimized by *explicitly* enforcing orbital orthonormalization. Such functionals were first introduced by Mauri et al. [11] and independently by Ordejón et al. [12], and then generalized by Hierse and Stechel [18] and by Kim et al. [22].

The definition of functionals with implicit orthonormalization constraints proceeds as follows. First a Löwdin transformation [45] is applied to the $\{\psi_i\}$: $\psi_i = \sum_j S_{ij}^{-1/2} \Phi_j$, where $S_{ij} = \langle \Phi_i | \Phi_j \rangle$. In terms of the $\{\Phi_i\}$, the total energy and charge density are given by

$$E[\Phi] = 2\sum_{ij}^{N/2} S_{ij}^{-1}\langle\Phi_i|\hat{H}_{\text{eff}}|\Phi_j\rangle - E_{\text{DC}}[n]$$
$$= 2\text{Tr}[\mathbf{S}^{-1}\mathbf{H}_{\text{eff}}] - E_{\text{DC}}[n] \qquad (12)$$

and

$$n(r) = 2\sum_{ij}^{N/2} S_{ij}^{-1}\Phi_i\Phi_j. \qquad (13)$$

A new functional, different from that of Eq. (12), is defined by replacing the inverse of the overlap matrix $\mathbf{S}^{-1}$ by a truncated Taylor expansion [19]

$$\mathbf{Q} = \sum_{n=0}^{\mathcal{N}} (\mathbf{I}-\mathbf{S})^n \qquad (14)$$

up to an odd order $\mathcal{N}$; here it is assumed that $\mathbf{S} < 2\mathbf{I}$. The new energy functional is

$$E[\Phi] = 2\sum_{ij}^{N/2} Q_{ij}\langle\Phi_i|\hat{H}_{\text{eff}} - \eta|\Phi_j\rangle - E_{\text{DC}}[\tilde{\varrho}] + \eta N$$
$$= 2\,\text{Tr}[\mathbf{Q}\mathbf{H}_{\text{eff}}] + 2\eta\,\text{Tr}[\mathbf{I}-\mathbf{QS}] - E_{\text{DC}}[\tilde{\varrho}] \qquad (15)$$

with

$$\tilde{\varrho}(\mathbf{r},\mathbf{r}) = 2\sum_{ij}^{N/2} Q_{ij}\Phi_i(\mathbf{r})\,\Phi_j(\mathbf{r}). \qquad (16)$$

Although $E[S^{-1},\Phi]$ and $E[Q,\Phi]$ are different functionals, there are instances under which their absolute minimum coincides. In particular, for the non-interacting case, if the constant η is such that, given a basis set, the operator $(\hat{H}_{\text{eff}} - \eta)$ is negative definite,

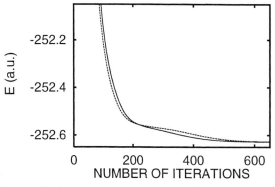

Fig. 1. Total energy (E) as a function of the number of iterations for a steepest descent minimization of 64 Si atoms in the diamond structure, described within LDA with a plane wave basis set. The solid and dotted lines correspond to the minimization of $E[Q]$ (see Eq. (17)) and the functional of Eq. (2) (see text), respectively. The matrix $\mathbf{Q}$ (see Eq. (14)) was defined with $\mathcal{N} = 1$. We used kinetic energy cut-offs of 12 and 36 Ry for the wavefunctions and charge density, respectively, and we set η at 3 Ry above the top of the valence band. Each run was started from the same set of random Fourier coefficients. From Ref. [19]

then the two following statements can be proved [19]: (i) the absolute minimum of $E[Q, \Phi]$ with respect to Φ is E_0, and (ii) the minimization of $E[\Phi]$ yields orthonormal orbitals, i.e. at the minimum $S_{ij} = \delta_{ij}$. Note that the orthonormality condition is never explicitly enforced and it is attained only when the functional E reaches its absolute minimum. Here **S** and **Q** are real symmetric matrices, since we assumed that the electronic wavefunctions are real fields. In Fig. 1 we compare the minimization of $E[Q, \Phi]$ to that of the conventional energy functional of Eq. (2), for a Si sample of 64 atoms. The calculations are carried out within the local density approximation of DFT, using a plane wave basis sets. It is seen that at the end of the iterative procedures the two functionals attain the same minimum value. Correspondingly, the integral of the charge density reaches the value N, which is the total number of electrons in the system (see Fig. 2).

In most practical applications, including the one shown in Figs. 1 and 2, **Q** has been defined by taking $\mathcal{N} = 1$ in Eq. (14), i.e. $\mathbf{Q} = 2\mathbf{I} - \mathbf{S}$. The energy and charge density are then given by

$$E[\Phi] = 2 \sum_{ij}^{N/2} (2\delta_{ij} - S_{ij}) \langle \Phi_i | \hat{H}_{\text{eff}} - \eta | \Phi_j \rangle - E_{\text{DC}}[\tilde{\varrho}] + \eta N \qquad (17)$$

and

$$\tilde{\varrho}(\mathbf{r}, \mathbf{r}) = 2 \sum_{ij}^{N/2} (2\delta_{ij} - S_{ij}) \Phi_i(\mathbf{r}) \Phi_j(\mathbf{r}). \qquad (18)$$

In terms of density matrices we have: $\tilde{\varrho}(\mathbf{r}, \mathbf{r}') = 2\varrho(\mathbf{r}, \mathbf{r}') - \varrho^2(\mathbf{r}, \mathbf{r}')$, where $\varrho(\mathbf{r}, \mathbf{r}') = \sum_i^{N/2} \Phi_i(\mathbf{r}) \Phi_i(\mathbf{r}')$ is a trial density matrix and $(2\varrho - \varrho^2)$ is the purification transformation from the trial density matrix to the set of matrices on which the variation of the energy functional is performed.

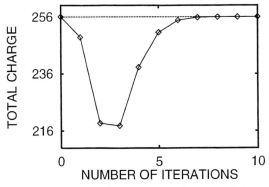

Fig. 2. Total electronic charge as a function of the number of iterations for the energy minimizations reported in Fig. 1. The total number of electrons in the system is 256. For $\mathcal{N} = \infty$, $\Delta N = N - \int d\mathbf{r}\, \tilde{\varrho}(\mathbf{r}) = N - \text{Tr}(\mathbf{QS})$ is given by $\Delta N = \text{Tr}((\mathbf{I} - \mathbf{S})^2)$. This is a positive quantity which goes to zero as the orbitals become orthonormal. The picture shows that the difference ΔN between the total number of electrons and the integrated charge reaches a value very close to zero ($\simeq 10^{-6}$) after 10 iterations, showing that the single particle wavefunctions are orthonormal already well before reaching the minimum of the total electronic energy. From Ref. [19]

The functionals (15) and (17) can be generalized so as to depend on an arbitrary number M of orbitals, in general larger than the number of occupied states $N/2$, as suggested by Kim et al. [22]. The functionals can be further generalized by expanding the so called trial states Φ_i over the set of M orbitals, as first proposed in Ref. [18]: $\Phi_i = \sum_k^M b_{ik}\phi_k$. If we set $\mathbf{b}^2 = \mathbf{L}$, then Eq. (17) and Eq. (18) become

$$E[\phi] = 2 \sum_{ij}^M (2\mathsf{L}_{ij} - (LS'L)_{ij}) \langle \phi_i | \hat{H}_{\text{eff}} - \eta | \phi_j \rangle - E_{\text{DC}}[\tilde{\varrho}] + \eta N$$
$$= 2 \operatorname{Tr}[(2\mathbf{L} - \mathbf{LS'L}) \mathbf{H}_{\text{eff}}] + 2\eta \operatorname{Tr}[\mathbf{I} - (2\mathbf{LS'} - \mathbf{LS'LS'})] - E_{\text{DC}}[\tilde{\varrho}] \qquad (19)$$

and

$$\tilde{\varrho}(\mathbf{r},\mathbf{r}) = 2 \sum_{ij}^M (2\mathsf{L}_{ij} - (LS'L)_{ij}) \, \phi_j(\mathbf{r}) \, \phi_i(\mathbf{r}), \qquad (20)$$

where $S'_{ij} = \langle \phi_i | \phi_j \rangle$. Here η must be used as a chemical potential to fix the integral of the charge density at the total number of electrons ($\eta = \mu$). In order to improve the variational freedom in the minimization procedure, Hierse and Stechel [18] suggested to vary E also with respect to the $\mathbf{L}$ matrix and proved that the minimum of E with respect to $\mathbf{L}$ is unique. In the formulation of Ref. [18] M is taken equal to $N/2$. The minimization of the functional E of Eq. (19) with respect to both $\{\phi_i\}$ and $\mathbf{L}$ yields overlapping orbitals. This is true also for the variation of E with $\mathbf{L} = \mathbf{I}$, when $M > N/2$ [22].

In the language of DM formulations, here the trial density matrix and the density matrix are given by Eqs. (7) and (8), respectively, with $K_{ij} = 2\mathsf{L}_{ij} - (LS'L)_{ij}$. If the McWeeny purification $\tilde{\varrho} = 3\varrho^2 - 2\varrho^3$ is chosen, as in Ref. [25], then $\mathbf{K} = 3\mathbf{LS'L} - 2\mathbf{LS'LS'L}$. We note that the $\mathbf{L}$ matrix plays the role of an inverse overlap matrix between the functions defining the trial density matrix, e.g. the TB basis functions chosen to represent the trial density in the formulation of Ref. [46]. For $\mathbf{L} = \mathbf{S}^{-1}$, one has $\mathbf{K} = \mathbf{S}^{-1}$, irrespective of the chosen purification, and the energy functional of Eq. (12) is recovered.

An orbital approach similar in spirit to the one introduced by Kohn [26] within a DM formulation was proposed by Wang and Teter [10]. The minimization of the electronic energy functional is performed by relaxing the orthogonality condition and by using a penalty function $P[\phi_i]$ to mimic orthonormality constraints. The penalty function is written as $P[\phi_i] = \sum_{ij}^{N/2} |\langle \phi_j | \phi_i \rangle|^2$ and the total energy to be minimized by variation is $\mathcal{Q} = 2 \sum_i^{N/2} \langle \phi_i | \hat{H}_{\text{eff}} | \phi_i \rangle + \lambda P[\phi_i]$, where the term E_{DC} is not included since the method was originally proposed for TB Hamiltonians. Here λ is a constant which in practical calculations with localization constraints turns out to be larger than but of the same order as the Hamiltonian matrix elements.

3.2.2 Localised functions

In general, the minimization of the functional $E[Q]$ takes of the order N^3 operations. However, because the minimization of $E[Q]$, contrary to the minimization of E in Eq. (2), does not require any explicit orthonormalization of the orbitals, it may be done in $O(N)$ operations, *if* it is carried out with respect to *localized functions* [9, 10] (ϕ^{loc}).

The minimization of $E[Q,\phi]$ with respect to functions ϕ^{loc} is accomplished, for example, by searching only over those ϕ_i which have zero components outside a region of real space defined by an appropriate radius R_c. Similar to localized density matrices, restricting the search to localized ϕ's, when minimizing E, is often referred to as "applying localization constraints", during the minimization procedure. When varying E also with respect to the **L** matrix, the additional constraint that K_{ij} vanishes if the separation of the localization regions of ϕ_i^{loc} and ϕ_j^{loc} exceeds a chosen cut-off is enforced [25]. We note that localization constraints would not introduce any approximation if the resulting localized functions could be obtained by a unitary transformation of the occupied eigenstates. Therefore the use of localized orbitals ϕ^{loc} is well justified for, e.g., periodic insulators, for which exponentially localized Wannier functions can be constructed by a unitary transformation of occupied Bloch states [47, 48].

The localized orbitals minimizing $E[\phi]$ can be regarded as generalized Wannier functions. They are nearly orthogonal functions in the formulation of Refs. [11, 12], whereas they are overlapping orbitals in the generalized formulations of Refs. [18, 22]. The use of maximally localized overlapping wave functions for a generic system was first introduced by Anderson [49], as a generalization of Wannier functions for periodic systems. A way of obtaining orthogonal generalized Wannier functions by using a Löwdin-like orthogonalization was proposed in recent years by Kohn [13]. Real space Wannier-like formulations have been also used to describe the response of an insulator to an external electric field [50].

When using localized functions, the number (m) of basis orbitals or real space points needed to express the electronic degrees of freedom becomes independent of the system size, being determined only by the extension of the localization (support) region. Therefore the evaluation of the total energy, and of the gradients of the total energy with respect to electronic ($\{\phi_i^{\text{loc}}\}$ and **L**) and ionic ($\{\mathbf{R}_I\}$) degrees of freedom, amounts to computing products of sparse matrices and thus becomes of $O(N)$. Indeed all double sums entering, e.g. the energy expression of Eq. (19), run over orbitals defined in neighboring localization regions. In particular, in iterative minimizations the evaluation of the energy gradient, which in conventional calculation requires $O(\mathcal{M}N)$ operations, becomes of $mO(N)$. Furthermore the orthogonalization step is not needed since one minimizes a functional with implicit orthogonalization constraints and therefore the whole calculation is of $O(N)$. The $O(N)$ scaling was verified in several electronic structure and molecular dynamics calculations using both TB and LDA Hamiltonians [19, 25, 58].

3.3 Accuracy of localization constraints

Similar to DM calculations, the accuracy of LF calculations is controlled by the parameter R_c. In general $\min_{\phi^{\text{loc}}} E[\phi^{\text{loc}}] = E_0^{\text{loc}} > \min_\phi E[\phi] = E_0$, but

$$\lim_{R_c \to \infty} (\min_{\phi^{\text{loc}}} E[\phi^{\text{loc}}]) = \min_\phi E[\phi]. \tag{21}$$

By choosing appropriately the size of the localization regions, the error $\Delta E_{\text{loc}} = (E_0^{\text{loc}} - E_0)$ can be made acceptably small, as it has been shown for a variety of systems in practical calculations [19, 21, 22, 51]. Two examples using a TB Hamiltonian for an insulating and a metallic carbon system are displayed in Fig. 3, and one example for bulk silicon using a LDA Hamiltonian is shown in Fig. 4. The density matrix of systems with a gap decays exponentially with separation whereas that of metals decays with an inverse

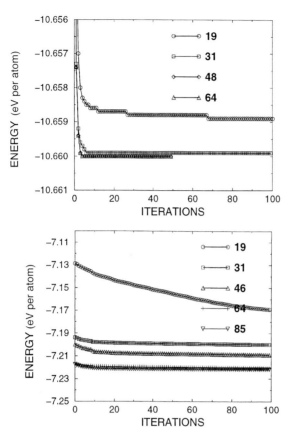

Fig. 3. Convergence of the total energy as a function of the localization region size (determined by the radius R_c, see text) for π- and σ-2D graphite for a supercell containing 576 carbon atoms. In π-2D graphite, each carbon atom has 4 valence electrons and the system is metallic. σ-2D graphite is a model insulating system, where each carbon atom has artificially three valence electrons per atom. The calculations have been performed using a TB Hamiltonian [51]. The upper panel shows the convergence from a random start for the σ-2D graphite for localization regions containing between 19 and 64 atoms. (The value of the total energy obtained by direct diagonalization of the Hamiltonian is 10.66 eV.) The lower panel shows the convergence for π graphite for localization regions containing between 19 and 85 atoms starting from an initial state that has been converged for 100 steps with a localization region containing 19 atoms. (The value of the total energy obtained by direct diagonalization of the Hamiltonian is 7.28 eV.) From Ref. [66]

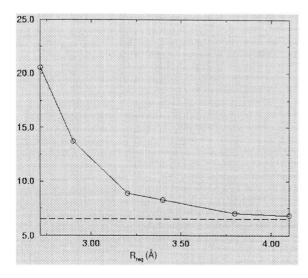

Fig. 4. Convergence of the total energy as a function of the localization region size (determined by the radius R_{reg}, called R_c in the text) for a Si crystal (circles joined by solid line). Calculations were performed in Ref. [61], within DFT, in the LDA, using the code CONQUEST. Dashed lines shows the ground state energy computed with a plane wave code using extended states. From Ref. [61]

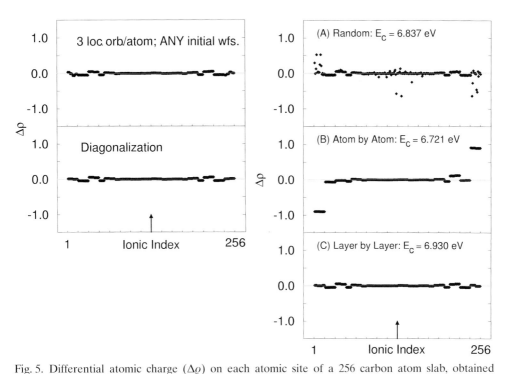

Fig. 5. Differential atomic charge ($\Delta\varrho$) on each atomic site of a 256 carbon atom slab, obtained using a TB Hamiltonian [51]. The slab, consisting of 16 layers, represents bulk diamond terminated by a C(111)-2×1 Pandey reconstructed surface on each side. The ionic index indicates individual atomic sites belonging to the slab, which are ordered layer by layer, starting from the uppermost surface. The arrow indicates the slab center. $\Delta\varrho_K = \varrho_K - \varrho^0$, where $\varrho_K = 2\sum_{ij=1}^{M}\sum_{l}\langle\phi_i|a_{Kl}\rangle$ $\times (2\delta_{ij} - S_{ij})\langle a_{Kl}|\phi_j\rangle$, $\varrho^0 = 4$; K is the atomic site and α represents atomic basis functions. E_c is the cohesive energy of the slab. On the right side, in panels A, B and C we show the results of calculations performed with two localized orbitals per atomic site, i.e. $M = N/2$ in Eq. (17) and with three different wave function inputs. *Random* input: the wave function expansion coefficients on each site of a localization region (LR) are random numbers, and orbitals belonging to the same LR are orthonormalized at the beginning of the calculations. *Atom by atom* input: each orbital has a nonzero coefficient only on the atomic site to which it is associated, and for each atomic site this coefficient is chosen to be the same. *Layer by layer* input: each orbital has a nonzero coefficient only on the atomic site to which it is associated, and the value of this coefficient is chosen to be the same for each equivalent atom in a layer. In the case of *atom by atom* and *layer by layer* inputs, the initial wave functions are an orthonormal set. On the left side, in the upper panel we report the results of a calculation carried out with three orbitals per atomic site, i.e. $M = 3N/4 > N/2$ in Eq. (19), and with a totally random input for the initial wave functions. In this case $E_c = -6.978$ eV/atom. In the lower panel, we show results obtained by direct diagonalization ($E_c = -7.04$ eV/atom). Contrary to the calculation started from a totally random input and performed with $M = N/2$ (see right side), the calculation with $M = 3N/4$ gives a ground state charge density very close to that obtained by diagonalization. From Ref. [22]

power of the separation. This was rigorously proven for one-dimensional systems [47] and suggests that the error ΔE_{loc} goes to zero exponentially for insulators and algebraically for gapless systems, as the radius R_c is increased. In Fig. 3 it is seen that for a given size of the localization regions (i.e. value of R_c) the error ΔE_{loc} is much smaller for the insulating carbon model system than for the metallic system, as expected.

The minimization of the total energy functional with respect to *extended* states can be easily performed so as to lead directly to the ground state energy E_0, without traps at local minima or metastable configurations [4]. On the contrary, the minimization of E (Eq. (17)) with respect to localized functions can lead to a variety of minima, if a number of orbitals equal to the number of occupied states is used. An example was given in Ref. [22] and it is displayed in Fig. 5. This figure shows that in some cases (e.g., reconstructed surfaces) information about the bonding properties of the system has to be included in the input wavefunctions, in order to attain the minimum representing the ground state, when minimizing the functional of Eq. (17). This implies a knowledge of the system that may be available only in particular cases. The inclusion of extra orbitals ($M > N/2$) and of the chemical potential variable, as in the formulations of Refs. [22, 25], overcomes the multiple minima problem, and makes it possible to perform $O(N)$ calculations for an arbitrary system, with totally unknown bonding properties. In the generalized schemes of Refs. [22, 25], the appropriate filling of the orbitals is determined by the *global* variable μ. This allows for long-range charge transfers in the minimization process, irrespective of the extent of the localization in the wavefunctions.

The generalized formulations have also other advantages with respect to the original one. While retaining the same computational cost, they allow one to decrease the error in the variational estimate of E_0, for a given size of the localization region, and to improve the energy conservation during a molecular dynamics run [22].

3.4 Electronic structure calculations and quantum molecular dynamics simulations using orbital and density matrix O(N) methods

Using the energy functionals described in the previous sections and localized functions or density matrices, quantum molecular dynamics simulations can be performed, with the computational cost of each step requiring $O(N)$ operations. If we consider, e.g., LFs and use the Helmann-Feynman theorem, the forces acting on a given atom I can be obtained by computing $\mathbf{F}_I = -\nabla_I E[\{\phi^{\text{loc}}\}, \{\mathbf{R}_I\}]$; here $(\mathbf{R}_I)$ denotes ionic positions and $\{\phi^{\text{loc}}\}$ are the localized functions minimizing E, which are determined at each ionic move.

So far, most large scale applications using $O(N)$ methods have been carried out using TB Hamiltonians [37, 52 to 60]. Molecular dynamics simulations involving about one thousand atoms can routinely be performed on workstations, while the efficient implementations of the $O(N)$ TB algorithm on a massively parallel computer [57, 58] allow one to simulate 5000 to 10000 atoms for tens of picoseconds [37]. An example showing a snapshot from a TB quantum simulation of small fullerenes deposited on a carbon surface is reported in Fig. 6 [37].

LDA applications are less advanced than TB calculations. Ordejón et al. [32] have recently presented an LCAO implementation of first-principles LDA-MD simulations and applied it to a variety of systems, using the code SIESTA. A review of these applications is given in another paper of this book. Hernandez et al. [25] have carried out a real space implementation of the LDA $O(N)$ method for total energy calculations. Later they [31] have developed a mixed scheme involving both finite differences in real space and finite elements (in terms of "blip-functions"), proposing as well a preconditioning procedure [61] specific to linear scaling methods for DFT-LDA Hamiltonians.

Large-Scale Electronic Structure Calculations

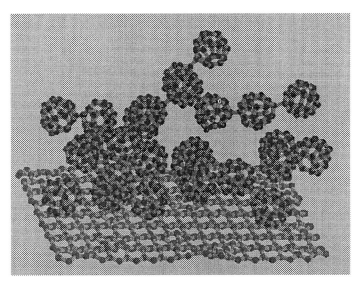

Fig. 6. Section of the system investigated in a $O(N)$-TB quantum simulation, mimicking C_{28}'s deposition on a diamond surface [37]. The full system consisted of 4472 atoms and was monitored for about 70 ps. The picture shows a side view of the top layer of the diamond substrate: only the fullerenes which are *not* bonded to the surface are shown. The tendency of the fullerenes to form polymer type structures is clearly visible. From Ref. [37]. We note that low frequency modes associated with the oscillations of these polymer-like structures introduce very slow time scales in the problem, which would make the equilibration of a film containing more that $\simeq 50$ fullerenes extremely long and eventually unaffordable with present computer capabilities, even when using linear scaling methods. This highlights the importance of simulating 'long' time scales when simulating 'large' systems

The method has been used for total energy minimizations and not yet for MD simulations. Very recently Fattebert and Bernholc [62] have presented a real space implementation of an orbital based $O(N)$ method [9], using multigrid algorithms to minimize the energy functional. This implementation, as well as those of Ref. [31] and of the code SIESTA, have been carried out within a pseudopotential approach. Regarding all-electron methods, Wang et al. [23] proposed an all-electron real space multiple scattering approach and applied it to the study of Cu–Zn metal alloys. But for the implementation of Wang et al. [23], the calculation of the Madelung potential has been done straightforwardly, without implementing the fast multiple expansion [63] to achieve $O(N)$ scaling. Indeed, this part of the calculation becomes dominant for a number of atoms much larger than tractable at present.

4. Divide-and-Conquer Approaches

A divide-and-conquer strategy to the solution of KS equations, leading to a $O(N)$ scaling of the calculations, was proposed by Yang [7]. Similarly to $O(N)$ density matrix and localized function methods, this approach is based on the division of a given system into overlapping subsystems. Here the key idea is to divide the electron density into contributions from subsystems using partition functions, and then determine each contribution using localized basis sets. We note that the use of localized *basis sets* (not to be confused with the LF of the previous section) is a distinctive feature of divide-and-conquer approaches; in LF and DM methods, localized basis sets may be used but are not a basic ingredient of the formulation.

Consider the charge density written in the coordinate representation and a partition $1 = \sum_\alpha p^\alpha(\mathbf{r})$, where p^α is a positive function which is large in the subsystem α and small away from it. The partition function p^α plays the same role as the filtering function and the bucket potential introduced in Ref. [9]

$$n(\mathbf{r}) = 2 \sum_\alpha p^\alpha \langle \mathbf{r}| f_\beta(\mu - \hat{H}_{\text{eff}}) |\mathbf{r}\rangle. \tag{22}$$

Here we have introduced the finite temperature representation of the charge density, by using the Fermi function f_β instead of the Heaviside function which one has at $T = 0$ ($\beta \to \infty$). Equation (22) is exact. An approximation is introduced by replacing $\hat{H}_{\text{eff}}$ with $\hat{H}^\alpha = \hat{P}^\alpha \hat{H}_{\text{eff}} \hat{P}^\alpha$, where $\hat{P}^\alpha = \sum_{ij} |\phi_j^\alpha\rangle s_{ij}^{-1} \langle \phi_i^\alpha|$ is a projection operator onto the subsystem α. The set $\{\phi_j^\alpha(\mathbf{r})\}$ is a set of nonorthogonal basis functions *localized* in the subsystem α and $\mathbf{s}$ is the overlap matrix between these basis functions. The total electron density is then approximated by

$$\tilde{n}(\mathbf{r}) \simeq 2 \sum_\alpha p^\alpha(\mathbf{r}) \sum_i f_\beta(\mu - \epsilon_i^\alpha) |\psi_i^\alpha(\mathbf{r})|^2, \tag{23}$$

where ϵ_i^α and ψ_i^α are the eigenvalues and eigenfunctions of H^α, respectively. The value of the chemical potential μ is determined by the normalization constraint $N = \int d\mathbf{r}\, \tilde{n}(\mathbf{r})$. We note that a finite β is essential to guarantee a unique solution for μ, at a given N.

The approximate charge density $\tilde{n}(\mathbf{r})$ can now be used to evaluate the total energy (Eq. (2)), where the eigenvalue sum $\sum_i \epsilon_i = 2 \sum_i^{N/2} \langle \psi_i| \hat{H}_{\text{eff}} |\psi_i\rangle$ is approximated in a similar fashion

$$\sum_i \epsilon_i = 2 \int d\mathbf{r} \langle \mathbf{r}| \hat{H}_{\text{eff}} f_\beta(\mu - \hat{H}_{\text{eff}}) |\mathbf{r}\rangle$$
$$\simeq 2 \sum_\alpha \sum_i f_\beta(\mu - \epsilon_i^\alpha) \langle \psi_i^\alpha| p^\alpha |\psi_i^\alpha\rangle. \tag{24}$$

The eigenvalues (ϵ_i) and eigenvectors (ψ_i) of the Hamiltonian of the full system are never computed. Similarly to LF and DM $O(N)$ formulations, the coupling between subsystems (or localization regions) occurs through the value of the chemical potential and the local potential. At variance with LF and DM $O(N)$ approaches, the divide-and-conquer method guarantees neither upper nor lower bound to the exact Kohn-Sham energy.

Recently a density matrix formulation of the DC approach was proposed, in which the partition of the system into subsystems is achieved in the space of atomic orbitals [24, 27]. The drawback of the division in the atomic orbital space is that the orbitals localization in the physical space depends on the basis set and decreases as more diffuse atomic functions are used. A major advantage of the density matrix formulation with atomic orbital partitions is that it can easily be generalized to include Hartree-Fock and semiempirical Hamiltonians. This has been done by several groups [27, 28] and the method is now being used to study large molecules and in general non-periodic systems.

The $O(N)$ scaling of the method is achieved when the calculation of the matrix elements of the matrices H^α and the evaluation of the electrostatic potential are carried out in $O(N)$ operations. These objectives are accomplished by a truncation of the basis functions in the physical space and by the use of fast multiple methods, respectively.

As a final remark, we note that in the chemical literature the partition of a system into building blocks and the development of localized orbital theories date back to the work of Adams [49a] and Gilbert [49b]. The relationship between Adams' theory and Anderson's [49] pseudopotential theory of maximally localized functions is nicely discussed in Ref. [65].

5. Conclusions

The linear scaling methods described here are promising frameworks for the study of broader classes of problems than affordable with conventional techniques. The implementation of $O(N)$ methods for tight-binding and semiempirical Hamiltonians has already given remarkable results, including the study of growth processes [37] on a surface and large organic molecules in water [27].

One of the major technical problems of $O(N)$ calculations based on localized functions or density matrices is the large number of iterations needed to converge the electronic structure at each molecular dynamics step. This is sensibly larger than in calculations using extended states [22]. A possible solution to the problem could come from the use of finite temperature calculations [66] or specific preconditioning techniques [61]. In this respect multigrid methods [62] adopted in LDA applications seem to be very promising. A more fundamental problem regards the computation of spectral properties, since eigenstates and eigenfunctions are never computed when using $O(N)$ methods. This might be accomplished by combining the present $O(N)$ methods with inverse iteration approaches [67] or with methods similar to those recently proposed for the calculation of the density of states [16].

As a final remark, I would like to note that as the size of the system which can be simulated increases, the time scales involved in the problem can change considerably and become unaffordable using conventional techniques. Therefore the development of approaches extending the application of quantum MD to larger systems must be accompanied by the developments of methods capable of handling slower time scales, in order to build efficient simulation tools.

Acknowledgement This work performed by the Lawrence Livermore National Laboratory under the auspices of the U.S. Department of Energy, Office of Basic Energy Sciences, Division of Materials Science, Contract No. W–7405–ENG–48.

References

[1] P. HOHENBERG and W. KOHN, Phys. Rev. B8 **136**, 64 (1964).
[2] W. KOHN and L. J. SHAM, Phys. Rev. A1 **140**, 133 (1965).
[3] For a review see, e.g., R. HAYDOCK, in: Solid State Physics, Vol. 35, Academic Press, New York 1980 (p. 215),
V. HEINE, in: Solid State Physics, Vol. 35, Academic Press, New York 1980 (p. 1).
[4] For a review see, e.g., M. C. PAYNE, M. P. TETER, D. C. ALLAN, T. A. ARIAS, and J. D. JOANNOPOULOS, Rev. Mod. Phys. **64**, 1045 (1993).
G. GALLI and A. PASQUARELLO, in: Computer Simulation in Chemical Physics, Eds. M.P. ALLEN and D.J. TILDESLEY, Kluwer Academic Publ., Dordrecht 1993 (p. 261).
[5] A recent review can be found in G. GALLI, Current Opinion in Solid State and Materials Science **1**, 864 (1996).

[6] S. Goedecker, Rev. Mod. Phys. (1999), to appear.
[7] W. Yang, Phys. Rev. Lett. **66**, 1438 (1991).
[8] S. Baroni and P. Giannozzi, Europhys. Lett. **17**, 547 (1991).
[9] G. Galli and M. Parrinello, Phys. Rev. Lett. **69**, 3547 (1992).
[10] W.-L. Wang and M. Teter, Phys. Rev. B **46**, 12798 (1992).
[11] F. Mauri, G. Galli, and R. Car, Phys. Rev. B **47**, 9973 (1993).
[12] P. Ordejón, D. Drabold, M. Grunbach, and R. Martin, Phys. Rev. B **48**, 14646 (1993).
[13] W. Kohn, Chem. Phys. Lett. **208**, 167 (1993).
[14] X.-P. Li, R. Nunes, and D. Vanderbilt, Phys. Rev. B **47**, 10891 (1993).
[15] M. S. Daw, Phys. Rev. B **47**, 10895 (1993).
[16] A. D. Drabold and O. Sankey, Phys. Rev. Lett. **70**, 3631 (1993).
[17] M. Aoki, Phys. Rev. Lett. **71**, 3842 (1993).
[18] W. Hierse and E.B. Stechel, Phys. Rev. B **50**, 17811 (1994).
[19] F. Mauri and G. Galli, Phys. Rev. B **50**, 4316 (1994).
[20] S. Goedecker and L. Colombo, Phys. Rev. Lett. **73**, 122 (1994).
[21] P. Ordejón, D. Drabold, R. Martin, and M. Grunbach, Phys. Rev. B **51**, 1456 (1995).
[22] J. Kim, F. Mauri, and G. Galli, Phys. Rev. B **52**, 1640 (1995).
[23] Y. Wang, G. M. Stocks, W.A. Shelton, D. M. C. Nicholson, Z. Szotek, and W. M. Temmerman, Phys. Rev. Lett. **75**, 2867 (1995).
[24] W. Yang and T.S. Lee, J. Chem. Phys. **103**, 5674 (1995).
[25] E. Hernandez and M.J. Gillan, Phys. Rev. B **51**, 10157 (1995).
   E. Hernandez, M.J. Gillan, and C.M. Goringe, Phys. Rev. B **53**, 7147 (1996).
[26] W. Kohn, Phys. Rev. Lett. **76**, 3168 (1996).
[27] T.S. Lee, D.M. York, and W. Yang, J. Chem. Phys. **105**, 2744 (1996).
[28] S. L. Dixon and K. M. Jr. Merz, J. Chem. Phys. **104**, 6643 (1996).
[29] M. Foley and P. A. Madden, Phys. Rev. B **53**, 10589 (1996), and references therein.
[30] Y. A. Wang, N. Govind, and E. Carter, Phys. Rev. B **58**, 13465 (1998).
[31] E. Hernandez, M. J. Gillan, and C. M. Goringe, Phys. Rev. B **55**, 13485 (1997).
[32] P. Ordejón, E. Artacho, and J. M. Soler, Phys. Rev. B **53**, R10441 (1996).
[33] P. D. Haynes and M. C. Payne, Phys. Rev. B **59**, 12173 (1999).
[34] M. Challacombe, J. Chem. Phys. **110**, 2332 (1999).
[35] D. Bowler and M. J. Gillan, Comput. Phys. Commun.; in press.
[36] R. Baer and M. Head-Gordon, Phys. Rev. Lett. **79**, 3962 (1997).
[37] A. Canning, G. Galli, and J. Kim, Phys. Rev. Lett. **78**, 4442 (1997).
   J. Kim, G. Galli, J. Wilkins, and A. Canning, J. Chem. Phys. **108**, 2631 (1998).
[38] G. Galli, Comp. Mater. Sci. **12**, 242 (1998).
[39] J. Kim, J. W. Wilkins, F. S. Khan, and A. Canning, Phys. Rev. B **55**, 16186 (19997).
[40] J. Bennetto, R. W. Nunes, and D. Vanderbilt, Phys. Rev. Lett. **79**, 245 (1997).
[41] A. P. Horsfield, D. R. Bowler, C. M. Gorringe, D. G. Pettifor, and M. Aoki, Mater. Res. Soc. Symp. Proc. **491**, 417 (1998).
[42] D. G. Pettifor, Clarendon Press, Oxford 1995.
[43] R. Car and M. Parrinello, Phys. Rev. Lett. **55**, 2471 (1985).
[44] R. McWeeny, Rev. Mod. Phys. **32**, 335 (1960).
[45] P. Löwdin, J. Chem. Phys. **18**, 365 (1950).
[46] R.W. Nunes and D. Vanderbilt, Phys. Rev. B **50**, 17611 (1994).
[47] W. Kohn, Phys. Rev. **115**, 809 (1959).
[48] W. Kohn, Phys. Rev. B **7**, 4388 (1973).
[49] P. W. Anderson, Phys. Rev. Lett. **21**, 13 (1968).
   D. Bullet, in: Solid State Physics, Vol. 35, Academic Press, New York 1980 (p. 173).
[50] R. W. Nunes and D. Vanderbilt, Phys. Rev. Lett. **73**, 712 (1994).
   A. Dal Corso and F. Mauri, Phys. Rev. B **50**, 5756 (1994).
[51] S.-Y. Qiu, C.Z. Wang, and C.T. Chan, J. Phys.: Condensed Matter **6**, 9153 (1994).
[52] G. Galli and F. Mauri, Phys. Rev. Lett. **73**, 3471 (1994).
[53] P. Ordejon, D. A. Drabold, R. M. Martin, and S. Ito, Phys. Rev. Lett. **75**, 1324 (1995).
[54] S. Ito, P. Ordejon, and R. M. Martin, Phys. Rev. B **53**, 2132 (1996).
[55] R. W. Nunes, J. Bennetto, and D. Vanderbilt, Phys. Rev. Lett. **77**, 1516 (1996).
[56] J. Kim, J. W. Wilkins, and F. S. Khan, Phys. Rev. B **55**, 16186 (1997).

[57] S. ITO, P. ORDEJÓN, and R. M. MARTIN, Comput. Phys. Commun. **88**, 173 (1995).
[58] A. CANNING, G. GALLI, F. MAURI, A. DE VITA, and R. CAR, Comput. Phys. Commun. **94**, 89 (1996).
[59] M. STERNBERG, G. GALLI, and T. FRAUENHEIM, Comput. Phys. Commun. **118**, 200 (1999).
[60] R. HAERLE, G. GALLI, and A. BALDERESCHI, Appl. Phys. Lett. (1999), in press.
[61] D. BOWLER and M. J. GILLAN, Comput. Phys. Commun. **112**, 103 (1998).
[62] J.-L. FATTEBERT and J. BERNHOLC, Bull. Amer. Phys. Soc. **44**, 1341 (1999).
J.-L. FATTEBERT, private communication.
[63] M. C. STRAIN, G. E. SCUSERIA, and M. J. FRISCH, Science **271**, 51 (1996).
[64] W. H. ADAMS, J. Chem. Phys. **34**, 89 (1961); **37**, 2009 (1962).
R. PAUNCZ and M. COHEN, Adv. Atom. Mol. Phys. **7**, 97 (1971).
[65] J. D. WEEKS, P. W. ANDERSON, and A. G. H. DAVIDSON, J. Chem. Phys. **58**, 1388 (1973).
[66] G. GALLI, J. KIM, A. CANNING, and R. HAERLE, see [41] (p. 425).
[67] K. A. MÄDER, L.-W. WANG, and A. ZUNGER, Phys. Rev. Lett. **74**, 2555 (1995), and references therein.

phys. stat. sol. (b) **217**, 251 (2000)

Subject classification: 62.20.–x; 71.15.–m; S5.11; S10.1

Concurrent Coupling of Length Scales in Solid State Systems

R. E. RUDD (a) and J. Q. BROUGHTON (b)

(a) *Department of Materials, University of Oxford, Oxford OX1 3PH, UK, and SFA, Inc., 1401 McCormick Drive, Largo, MD 20774, USA*
e-mail: robert.rudd@materials.oxford.ac.uk

(b) *Complex Systems Theory Branch, Naval Research Lab., Washington D.C. 20375, USA, and Yale School of Management, New Haven, CT 06520, USA*

(Received August 10, 1999)

A strategic objective of computational materials physics is the *accurate* description of specific materials on length scales spanning the electronic to the macroscopic. We describe progress towards this goal by reviewing a seamless coupling of quantum to statistical to continuum mechanics, involving two models, implemented via parallel algorithms on supercomputers, for unifying finite elements (FE), molecular dynamics (MD) and semi-empirical tight-binding (TB). The first approach, FE/MD/TB Coupling of Length Scales (FE/MD/TB CLS), consists of a hybrid model in which simulations of the three scales are run concurrently with the minimal coupling that guarantees physical consistency. The second approach, Coarse-Grained Molecular Dynamics (CGMD), introduces an effective model, a scale-dependent generalization of finite elements which passes smoothly into molecular dynamics as the mesh is reduced to atomic spacing. These methodologies are illustrated and validated using the examples of crack propagation in silicon and the dynamics of micro-resonators. We also briefly review a number of other approaches to multiscale modeling.

1. Introduction

The burgeoning field of multiscale modeling offers great promise to computational science as we prepare to enter the third millennium. It employs models at different physical scales in combination to build a comprehensive description of systems that could not be modeled otherwise. The challenge is that these systems have phenomena at one scale that require a very accurate and computationally expensive description, and phenomena at another scale for which a coarser description is satisfactory and in fact necessary to avoid prohibitively large computations. These may be length, time or energy scales, and in general there may be a hierarchy of many such scales. Traditionally models have been developed to describe each scale individually, but this is not appropriate for systems with multiple scales. No one of these models alone would suffice to describe the entire multiscale system, but it may be possible to combine the models of different scales, effectively concentrating the computational power where it is needed most. The quantities of interest could then be computed at the level of accuracy of the most intensive model, but at a manageable fraction of the computational cost. This is what is meant by a multiscale model: a model that takes advantage of the multiple scales present in many systems in order to balance accuracy and efficiency.

Consider the case of crack propagation (see Fig. 1) [1]. This is one of the systems which we consider in detail in this Review, and it is a particularly clear example of a

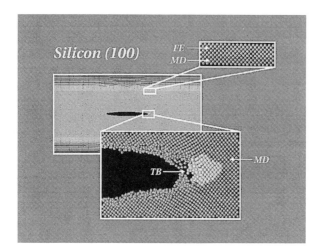

Fig. 1. The crack system as simulated in Ref. [1], with expanded views of the atomistic/continuum interface in the periphery and the quantum/classical interface at the crack tip. Note the decomposition of the system into five regions of differing dynamics: those regions governed by the three conventional Hamiltonians, finite elements (FE), molecular dynamics (MD), tight binding (TB), and the two handshaking regions, FE-MD and MD-TB

multiscale system. In order for the crack to grow, chemical bonds between the atoms at the crack tip must be broken. These bonds are formed by electrons concentrated in orbitals between the neighboring atoms. The bond breaking is characterized by a marked deformation of the electron density at the Angstrom scale (10^{-10} m). Surrounding the crack is a region in which the crystal lattice is significantly deformed, even to the extent that the structure becomes amorphous in regions around the crack tip. Some bonds are significantly distorted, but not to the point of breaking. Deformations of the lattice take place over a longer distance, and the region is dominated by atomistic physics at the nanometer scale (10^{-9} m). Far from the crack, the lattice is less deformed, but the strain fields persist for great distances and the system is governed by micron scale (10^{-6} m) continuum mechanics. These are the three length scales relevant to the silicon crack shown in Fig. 1. Additional scales may be relevant to cracks propagating in other materials or even in other directions in silicon. For example, a mesoscale may result from microstructure in the material due to existing grain boundaries or due to plastic deformations spawning dislocations about the crack.

Individually each of the scales in the crack propagation system is described well by an existing computational model. The Angstrom scale physics at the crack tip is described by quantum mechanical models of the electronic bonding between atoms. The nanometer scale physics is captured in atomistic models of the lattice deformation in which the individual atoms interact via classical empirical potentials. The micron scale physics is described by continuum mechanics such as elastic theory, in which the material is treated as a continuous medium and calculations are done in a discretization such as finite element analysis.

The challenge is that the various scales depend on each other. In the case of the crack the scales are linked fairly intimately. The strength of the bonds that break at the crack tip depends on how the surrounding material is deformed, and the deformation of the material depends in turn on how the crack propagates and, in particular, on where and when the bonds break. At larger scales, the propagation of the crack is driven by relaxation of the long-range strain fields. The strain energy is dissipated through local, dynamical processes at the crack tip including heating, plastic deformations and emission of elastic waves. Each of these effects propagates away from the crack affecting the surrounding material. Since the elastic waves propagate at the speed of sound, it takes less than a nanosecond for material a micron from the crack to be affected. Any successful multiscale model must faithfully reproduce the intertwined nature of the three length scales.

In fact this is the key issue in multiscale modeling: how the different scales are linked to each other. The degree to which the scales are coupled varies tremendously from system to system, and consequently the implementation of the coupling of scales varies tremendously from technique to technique. In some systems the scales may be treated sequentially in separate calculations; in some, the scales must be treated concurrently, but enough separation exists that a hybrid technique with minimal coupling is ideal; in some, an effective scaling model is required with a smooth transition across the scales; and in others, the scales are so strongly coupled that multiscale modeling as yet offers no advantage, and the finest description (the smallest length scale, the shortest time scale or the highest energy scale) must be used throughout. In this Review we focus on multiscale simulations of dynamical systems at finite temperature using a concurrent hybrid model [2] and a monolithic effective scaling model [3, 4] with applications in fracture [1, 5] and Micro-Electro-Mechanical Systems (MEMS) [5 to 9]. First we consider examples of all four classes in a general overview of multiscale modeling.

1.1 Linking scales in serial

Science abounds with examples of multiscale systems in which the scales are only weakly coupled. Were this not so, we would have made little progress in the theoretical sciences, and the fact that it is so leads in part to what Wigner has called the "unreasonable effectiveness" of mathematics in describing Nature [10]. Simple models often work better than we have any right to expect. The universe has a panoply of interconnected scales, and yet many physical phenomena are well-described by single-scale models where the physics of all the other scales is condensed into a few parameters: e.g. the thermal conductivity of silicon can be understood completely without recourse to the details of nuclear physics — only the mass of the nucleus is relevant.

In systems in which the scales are weakly coupled, it is natural to develop models that treat the scales in a serial or sequential fashion. Parameters are calculated from a computation at a small scale and fed into a computation at a larger scale. Here we are using the term "small scale" to denote a short length or time scale or a high energy scale; i.e. the scales that are more fundamental in the sense that the large-scale physics is completely determined in principle by the small scale along with a set of boundary conditions. So the parameters of the large-scale model may be determined by a small-scale calculation.

An excellent example is the landmark work of Clementi and coworkers [11]. They studied the behavior of water from the Angstrom scale of molecules through the kilometer scale of tidal basins, the calculations at each scale parameterizing the next. The starting point was quantum-mechanical simulations of a water molecule. This computed the parameters needed to study a cluster of water molecules. A database of computations of the quantum mechanical interaction of several water molecules enabled them to parameterize an empirical potential for use in molecular dynamics simulation. These atomistic simulations in turn evaluated the viscosity of water using atomic autocorrelation functions. Finally, the computed density and viscosity were employed in a computational fluid dynamics calculation to model tidal circulation in Buzzard's Bay.

Another good example is in atmospheric and environmental science [12] where chemists use methods of high computational complexity to evaluate reaction barriers of simple chemical reactions which are then used in large rate equation coupled to spatial grid codes to determine and predict chemical meteorology. Note that this example spans time and length scales that are almost completely decoupled: the microscopic reaction times and molecular rearrangements are separated by many orders of magnitude from the macroscopic time and length scales. Recently, this kind of transition state theory has been generalized to handle situations in which the set of possible molecular configurations is enormous. Then it is not possible to compute all of the transitions ahead of time, so the microscopic model must compute a sequence of transitions, each one affecting the transitions to follow. This dependence on the history of the system effectively links the microscopic and macroscopic scales more strongly, sparking the development of new methodologies for rare event acceleration such as hyperdynamics [13].

The serial approach is suited to systems in which variations or structures occur at well-separated scales. Large-scale variations decouple from the small-scale system, or at least they appear homogeneous and quasi-static from the small-scale point of view. There are many other examples in the literature, including some nice applications in condensed matter physics. In all of these schemes an appropriate computational methodology is used at each scale, whether it be the accuracy of quantum mechanics at the shortest scales, or rate theory and/or fluid dynamics at longest scales.

1.2 Linking scales in parallel: hybridization

Recently there has been an increased effort to couple scales in parallel, driven by the need to describe systems in which the scales are not so weakly coupled. Here it is not possible to treat the scales sequentially. The crack propagation is an example of such an inherently multiscale system. The behavior at each scale depends strongly on the others, and a formulation that ensures consistency among the scales is essential. Information must be allowed to pass back and forth between the scales, so a parallel, concurrent simulation is a necessity.

The field of concurrent multiscale simulation is still very new and only a few models have been developed to date. The most successful of these have made use of the inherent inhomogeneity of many multiscale systems in order to decompose the system into domains characterized by different scales. Phenomena such as fracture have small-scale structures embedded in large-scale surroundings. These are naturally described by combining models in parallel at large and small scales. The domains abut each other, so boundary conditions must be imposed that guarantee consistency across the interface. In general the boundary conditions are chosen to enforce a natural correspondence between the degrees of freedom at the two scales while respecting all of the relevant dynamics and conservation laws. Furthermore, the thickness of the handshaking region is taken to be as small as possible, keeping the number of linking parameters to a minimum. In this sense, the hybrid approach invokes a minimal coupling between the domains.

In the case of crack propagation an Angstrom-scale domain surrounds the crack tip, a nanometer-scale domain encompasses the material surrounding the crack apart from the tip region, and a micron-scale domain covers the periphery. In the simulations of Abraham and coworkers [1, 5] the domain about the crack tip is described by a quantum-mechanical electronic structure model; the nanoscale domain, by an atomistic model; the microscale, by a continuum model. These three scales must be coupled in a way that is not only consistent with the static elastic response of the material, but also with its elastic dynamics (phonons) since elastic waves are emitted by the crack tip, and with its thermodynamics since the crack behavior is temperature-dependent. This puts a heavy demand on the quality of the interface. The details of the coupling are described in Section 2, but briefly the handshaking comes from enforcing continuity of the displacement field across the atomistic/continuum interface and from consistency of the bonding forces at the surface of the quantum-mechanical region. The continuity at the atomistic/continuum interface is implemented by reducing the finite element mesh to coincide with the atomic lattice and introducing a mean-force coupling. The consistency of the bonding forces at the electronic/atomistic interface is implemented through a bond termination prescription. We refer to this hybrid model as FE/MD/TB Coupling of Length Scales (FE/MD/TB CLS), because it implements a coupling of finite elements (FE) to molecular dynamics (MD) to tight-binding electronic structure (TB).

Another application demanding a concurrent multiscale description is the simulation of the dynamical behavior of sub-micron MEMS, as demonstrated in the simulations of Rudd and Broughton and coworkers [5 to 9]. Here again the FE/MD/TB CLS model has been used. Consider the micro-resonator shown in Fig. 2. This device consists of a long, thin bar that has been etched from a single crystal of silicon. The bar forms a high quality oscillator which is excited in the flexural mode (like a violin string) with a resonant frequency in the gigahertz regime. The bar is only 0.2 µm in length and about 100 atoms wide. This is the scale at which atomistic effects become important. Conventional MEMS simulations using continuum elastic theory and FE cannot reliably predict the behavior of these small features.

The failings of continuum elastic theory are evident in many ways for MEMS. Sub-micron devices are so miniscule that materials defects and surface effects can have a large impact on their performance. Atomistic surface processes which would be negligible in large devices are a major source of dissipation in sub-micron devices, leading to a degradation in the quality factor (Q) of the resonator (although Q is still relatively high

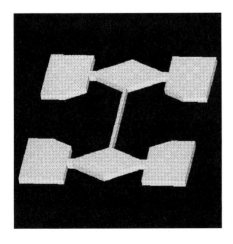

Fig. 2. The silicon micro-resonator, as modeled. The long thin bar in the center is free to oscillate above the substrate. Its oscillations are out of the plane of the device

for a single-crystal device). These effects vary with temperature and the amplitude of oscillation. Additionally, the motion of the device can couple to the dynamics of microstructure such as dislocations in the region of attachment to the substrate. These atomistic, nanoscale effects are beyond the scope of continuum elastic theory.

In certain MEMS devices, just as in fracture, Angstrom scale physics can play a decisive role as well. The nanotribology of micro-motors and micro-gears grinding against each other is determined by incessant bond breaking and reconstruction within the zone of contact. This process is governed by the electronic physics at the Angstrom scale [6]. Another example is dissipation in micro-resonator systems which can arise due to strain-induced impurity migration, a process where the hopping rate is determined by barrier heights set by Angstrom scale physics but influenced by the nanoscale configuration of the atoms [7].

The behavior of these MEMS devices is not determined just by the atomistic and electronic physics within the smallest features of the device, however; the details of the attachment to the substrate are also crucial. In the case of the oscillator, long-range strain fields extend out from the bar into the substrate. The geometry of this peripheral region has a large impact on the resonant frequency and other properties of the oscillator. This physics is well described by micron-scale continuum elastic theory.

Just as in the case of fracture, the three scales are intertwined, and the overall behavior of sub-micron MEMS devices is not just due to the physics at individual scales, but the way in which the scales interact. The single most important question about the resonators is how the energy in micron-scale modes is dissipated due to nanometer and Angstrom scale processes. The crucial question about the next-generation micro-gears is how micron-scale strain induces bonds to break at the Angstrom scale within the zone of contact.

The use of multiscale modeling is absolutely necessary for these systems. While it may be possible to use pure atomistics to model the central, active region of a device approaching a micron in size, it is beyond the capabilities of even the largest supercomputers to model the entire region of interest, including the attachment to the substrate, of a micron-scale device. The region of attachment is important, and it may involve volumes of many cubic microns and billions of atoms. So many atoms cannot be handled by empirical potentials, let alone tight-binding or other quantum mechanical models. Thus, many sub-micron MEMS are too small for finite elements and too large for atomistics, but they are well suited for multiscale modeling.

Another class of examples comes from fluid dynamics, where microscopic flows interact with macroscopic flows. Many such systems have been identified as important applications of multiscale techniques. In MEMS, an important application is flow from a fluid reservoir described by a Navier-Stokes continuum model into a microchannel with high Knudsen number; i.e. flow in the regime where the mean free path is comparable to or exceeds the dimensions of the channel. Here the continuum fluid couples to an atomistic flow [14]. The atomistic flow has been modeled with Direct Simulation Monte Carlo (DSMC) or molecular dynamics, depending on the density of the fluid. Another potential application is a microscopic description of boundary layer flow such as a detailed picture of how the no-slip boundary condition arises from molecules in the fluid interacting with the wall [15, 16]. Also simulations have been done on microscopic fluid–wall interactions in the motion of fluid interfaces in sheared liquids. The atomistic regions need not be at boundaries: there are many examples of eddies, shocks and other dynamical structures that have features down to the atomistic scale [17]. In these cases, traditional adaptive remeshing techniques would refine the mesh to the atomic scale and beyond in an effort to control the errors due to discretization in regions of strong gradients. But this neglects the fact that the continuum model itself is unreliable at these scales and a new atomistic model must be incorporated.

In the broadest terms, the challenge of coupling continuum to atomistic models of fluids arises from the use of different frames of reference in the two cases. Atomistics employs a Lagrangian picture in which particles move through the region of interest. Continuum models use an Eulerian picture in which velocity, pressure and temperature fields are associated with a fixed point in space. At the interface between a continuum model and an atomistic model, the atomistic model must generate a current consistent with the continuum velocity field and *in equilibrium*. The coupling cannot introduce any unphysical correlations. It appears that this problem has now been solved for dilute gases described by Navier-Stokes coupled to DSMC [18]. In this case, the molecules scatter off each other via contact interactions – the interaction potential has zero range – and the spatial correlations are not so important. This has allowed Alder and coworkers [18] to implement the coupling using a current generation scheme based on the Chapman-Enskog velocity distribution. It is interesting to note that while this methodology couples finite elements to atomistics, it is in a sense the very antithesis of the coupling between finite elements and atomistics in the crack and MEMS simulations, where spatio-temporal correlations are paramount and diffusion is negligible.

Another multiscale methodology for crack propagation has been propounded recently by Rafii-Tabar and coworkers [19]. It is an iterated sequential approach coupling continuum to nanoscale through a mesoscale introduced for computational purposes. The coupling in this case is non-dynamical and athermal. Continuum calculations based on finite elements at the large and intermediate scales are used to compute the boundary forces on the atomistic region. Then a zero temperature atomistic simulation computes the propagation of the crack tip within this region over short time intervals. Apart from the large-scale forces, the boundary atoms are taken to have no dynamics and they are thermally insulating. With these boundary conditions, the atomistic simulation is used to calculate an effective stochastic diffusion coefficient for the crack tip. The average velocity of the crack tip is taken from a separate atomistic calculation. This information is then fed back into the large-scale continuum in order to compute the crack propagation at much larger length and time scales. Here finite elements is used, augmented with something resembling Langevin dynamics for the crack tip, in order to simulate the postulated Ornstein-Uhlenbeck process of crack propagation. This multiscale approach to crack propagation is clearly very different from the concurrent simulations with dynamical coupling used by Abraham et al. Thus far, the application of the Rafii-Tabar method has been limited to two-dimensional simulations of crack propagation in a silver plate, where the atomistic region consists of roughly 5000 atoms, but the initial results are encouraging.

An important realization about all of the hybrid models is that once the issues of interfacing between the different regions have been solved, the overall algorithm becomes computationally very efficient. The reason is that one is using the right tool for the right part of the system, that tool having been optimized, historically, to solve a particular problem. Perhaps the most important result from this general approach is that for many purposes the handshaking can be made seamless so that the algorithm is not only efficient but also very *accurate*. "Accurate" implies that the dynamics of the phenomenon under study are indistinguishable whether they be determined from a length-scale-coupled system or from one of the same size comprising the finest description *only* (e.g. TB atoms). The many-body behavior of specific materials really can be addressed on structurally important length scales.

1.3 Linking scales in parallel: derived scaling

Hybrid models have the advantage that the constituent models have been optimized and validated for each particular scale; on the other hand, the constituent models have not been designed to couple to other scales. In principle it would be advantageous to have a model that smoothly interpolates between the scales. A single monolithic model would describe the entire system, with a parameter such as a mesh spacing that determines the computational power directed to each part. Such a model allows more control of the errors introduced in the coupling of scales, and it can offer more physical insight into the behavior of the system.

In order to construct a monolithic model, an underlying fine-scale model is envisaged for the entire system and degrees of freedom are removed explicitly at scales up to the scale of interest. This results in what is known as an effective model for that particular scale. For example, in the periphery of the crack simulation we would like to have a model with the accuracy of an atomistic simulation, but an actual atomistic simulation would be too expensive because it would carry a great deal of irrelevant information. So an effective model is constructed by introducing an irregular mesh in the periphery á la finite elements. The mesh carries a displacement field, and perhaps some other fields, at its nodes. In the effective model the equations of motion for the nodal fields are not derived from continuum models, but from the underlying atomistic model. The nodal fields represent the average properties of the underlying atoms, and the equations of motion are constructed to describe the mean behavior. Many degrees of freedom are eliminated in the process. The manner in which this is done will be discussed below for two such models: the Quasicontinuum Technique and Coarse-Grained Molecular Dynamics.

The idea that degrees of freedom can be removed scale by scale in order to derive the physical scale dependence of the system has come out of the renormalization group community. In some statistical models it is found that the Hamiltonian retains its form as degrees of freedom are removed – only the parameters in the Hamiltonian change. An equation can be written to describe how the parameters flow as the scale is changed. This is the renormalization group equation [20], and its solution determines the effective model that describes the system at any given scale. This approach has had its greatest successes in analytic models of critical phenomena, such as Ising spin systems, where the scaling of the Hamiltonian is particularly simple.

More recently the idea has been applied to the mechanics of solids. The resulting effective models would be intractable analytically, but as multiscale models they are well suited for computer simulation. The first significant use of an effective model to couple atomistics and finite elements was the Quasicontinuum Technique devised by Tadmor et al. [21], and subsequently developed by Phillips and coworkers at Brown [22]. In their formulation, the finite element mesh tessellates the entire system, and the model allows it to be refined down to atomic dimensions where needed. Unlike conventional finite elements, the energy of each cell is computed from an underlying atomistic Hamiltonian. In particular, an atomistic calculation is performed for a representative portion of each cell in order to compute the energy as a function of the FE degrees of freedom. The degrees of freedom in the FE region consist of the deformation gradient defined at the nodes of the mesh. Within each cell, the atomistic energy is computed for a single "representative" atom embedded in a crystal deformed according to the nodal deformation gradient. A weighted sum of the representative energies determines the energy of the system as a whole, and the equilibrium configuration of the system is then determined through a zero temperature relaxation technique. While the computational cost for each cell is high compared to conventional finite elements, it does not grow with the size of the cell. Thus, the number of degrees of freedom is dramatically reduced from the underlying atomistics to the effective finite element model. This approach also recovers the correct atomistic forces when the mesh is collapsed to the atomic spacing, so regions of conventional atomistics are allowed. The model is well-suited to adaptive mesh refinement, which has been used to enable the atomistic region to track defects as they propagate.

In the quasicontinuum technique the coupling between the atomistics and the generalized finite elements is almost completely natural and seamless, but there is a source of error due to the lack of continuity between neighboring cells. Near the interface where cell sizes are less than the range of the atomistic potential, it is necessary to know the position of atoms within neighboring FE cells in order to compute the atomistic energies. These positions are determined in the quasicontinuum technique using the nodal deformation gradient: the atomic configuration is not fully relaxed. This leads to forces that are anomalous and somewhat unphysical in the interface region, but it does not seem to introduce any major pathologies.

To date, the quasicontinuum technique has been applied to defective systems (such as cracks, dislocations and interfaces) in both two and three dimensions, where the defects are treated in an atomistic region embedded in a continuum. This approach has the advantage that it avoids the usual assumptions inherent in continuum models, such as the rather ad-hoc criterion typically used for failure in a given region of space. It also allows for slip within the FE cells. Recently, the quasicontinuum technique has also been used by another group to study nanoindentation [23].

While the quasicontinuum technique is a zero temperature approach, another effective theory has been developed for dynamical and finite temperature systems: Coarse-Grained Molecular Dynamics (CGMD) [3, 4]. The coupling between atomistics and finite elements in the hybrid FE/MD/TB model works very well for many applications. As described above, hybrid simulations of crack propagation have shown both the strain fields and the elastic waves that emanate from a crack tip pass fairly smoothly from the atomistic region into the finite element region [1]. In particular, there is little coherent backscatter of elastic shock waves from the FE/MD interface, a problem that has plagued pure atomistic simulations of cracks [24, 25]. It is clear, however, that finite elements does not connect perfectly smoothly with atomistics in the limit that the element size becomes atomic scale. Finite element analysis assumes that the energy density is spread smoothly throughout each element, but at the atomic scale, the potential energy is localized to the covalent bonds of silicon and the kinetic energy is localized largely to the nuclei. This small atomic scale mismatch can cause problems in some types of simulation.

CGMD has been developed as a substitute for finite elements which does connect seamlessly to molecular dynamics in the atomic limit. It also reproduces the results of finite elements (with slight improvements) in the limit of large element size. The fact that it is an effective model, derived from an underlying atomistic Hamiltonian, guarantees that the interpolation between these two limits is smooth and physically meaningful.

CGMD has been constructed to provide a consistent treatment of the short wavelength modes which are present in the underlying atomistics but are missing from the coarse finite element mesh. These modes can participate in the dynamics and the thermodynamics of the system. In CGMD the short wavelength modes missing from the mesh are taken to be in thermal equilibrium, and their average contribution is included in the dynamics of the system. In many situations, the short wavelength part of the spectrum is relatively unpopulated, and the missing modes are irrelevant to the behavior of the system. But this is not true for sufficiently small systems, or when there is a strong source of high frequency elastic waves (such as a propagating crack or when two micro-gears grind against each other). In these cases CGMD offers an important improvement over finite elements. In addition to offering properly scaled elastic forces and well-behaved thermodynamics, CGMD also models the elastic wave spectrum more accurately than conventional finite elements, and it allows the system to respond properly when long wavelength modes are driven out of equilibrium. Furthermore, CGMD includes scale-dependent nonlinear effects that are compatible with the atomistics; i.e. it effectively provides nonlinear constitutive equations that are derived from the atomistics without any free parameters. CGMD is described in detail in Section 3.

1.4 Linking scales in unison: unresolvable scales

Finally, for completeness we mention that there are important multiscale systems in which the scales are so strongly linked that the multiscale techniques developed so far offer no real advantage. Turbulence is a prime example. Energy is input by stirring at a large scale, and it cascades down through the scales until it is dissipated by viscous forces at a microscopic scale [26]. This qualitative picture is clear, but to date no unified multiscale model has been developed. The scales are too strongly coupled.

1.5 Overview

In this introduction, we have identified several approaches to multiscale modeling. All have been developed to exploit the existence of multiple scales in order to increase computational efficiency. In many cases this allows systems to be simulated which would have been too complex or too large for single-scale models. It should be pointed out that there are reasons to use multiscale models even when the system could be treated with a large single-scale calculation, and economy of computational resources is not a concern. There is the question of the value of the information that is produced by the simulation. If the finest description is used throughout a multiscale system, a great deal of extraneous data is produced. These data must be filtered in order to extract the physically significant results. The size of the data sets can be overwhelming, and a multiscale model can be an effective way to facilitate the reduction and analysis of the data. Thus, multiscale modeling is not just an algorithmic motif, but also a means to gain physical insight.

In Section 2 we consider in some detail the hybrid FE/MD/TB CLS methodology for the crack and MEMS simulations. In Section 3 we describe the basic implementation of Coarse-Grained Molecular Dynamics. Section 4 describes the issues involved with implementation of these codes on supercomputers. In Section 5 we present an overview of the results from applying coupling of length scale to different systems, and in Section 6 we give some perspective. Our discussion of the hybrid FE/MD/TB CLS methodology borrows heavily from the work of Abraham et al. [1, 2]. We describe the developments necessary to apply the methodology to the fully three-dimensional simulation of silicon and quartz oscillators.

2. Methodology: FE/MD/TB CLS

In this section we examine in detail the first of two multiscale models designed to simulate the dynamical evolution of solid state systems at finite temperature with a coupling of electronic to atomistic to continuum physics. This is the hybrid FE/MD/TB CLS model where a dynamically coupled simulation has been used to study fracture [1, 5] and MEMS [5 to 9].

The hybrid model combines FE, MD and TB models with the minimal coupling that is consistent with the dynamics of the system and the corresponding conservation laws. The ensuing discussion centers on Fig. 1. We envisage a system in which the region of Angstrom-scale physics comprises an extremely small subset of all the atoms in the system – a few hundred out of millions of atoms. This is the region of rupture and dynamical reconstruction of broken bonds, and it is this region which dictates the kinetics of the crack propagation [1] and the nanotribology of MEMS gears in contact [6], for example. It is important to describe the energetics of this part of the system very accurately. Since it is a region where bonds are breaking, it requires a quantum-mechanical description. The bonds are distorted beyond the regime of reliability for empirical interatomic potentials. We have chosen to describe this part of the system with a tight-binding Hamiltonian, since this is one of the fastest quantum-mechanical models which contains the basic physics of electronic bonding and from which it is simple to extract forces.

The computational cost of tight binding, while less than other quantum-mechanical models, is high enough that it cannot be used to describe the entire multiscale system. A model of greater computational efficiency needs to be used in the surrounding region, but one which still captures the relevant physics of lattice deformations, elastic waves and thermal fluctuations. Molecular dynamics is well-suited to this purpose. Thus the TB region is embedded in an MD region where the atoms interact with one another via an empirical potential well-parametrized for the system of interest. The system on this length scale is less perturbed from equilibrium than in the bond-ruptured region, and the empirical potential is suitable.

In addition to providing the correct dynamical and thermal environment in which to embed the TB region, the MD region may also contain a great deal of nanoscale physics that is important for the behavior of the system. In fact, many systems have negligible coupling to the Angstrom-scale physics, and MD forms the finest level of description. The MEMS resonator is just such a system, where the phenomena of interest are predominantly nanoscale surface effects. Even in cases like crack propagation where the vital processes at the crack tip are at the Angstrom scale, the MD region may also contain essential physics. The surrounding material may include defects but ones for which the primary dynamics are no longer important. The surface of a crack entrained behind a propagating crack tip is an example.

Lastly, although MD system sizes currently can run to hundreds of millions of atoms [27 to 29], when considering the mesoscopic and macroscopic scales even these formidable calculations are unable to represent properly the rest of the environment of the dynamical system. Here, the precise statistical mechanics is less important than representing the long-range strain fields and allowing the free passage of (usually long wavelength) energy into or out of the system. Long-range strain fields are important in many systems, such as nanoindentation and delamination of ceramic/insulator interfaces in addition to fracture and the dynamics of MEMS. In this domain the basic tool of engineers, finite elements, is appropriate to solve for the strain field of the system. The MD region should be embedded in a continuum-mechanical model in the far field. Here, atoms are displaced only slightly from equilibrium and elastic theory should work very well. Indeed, FE is the method of choice for this region.

The FE algorithm offers better computational efficiency than the MD. It deals only with the minimal degrees of freedom necessary to describe the correct physics. Consider, for a moment, large-scale simulations of crack propagation, where graphical representations focus mainly upon the high-energy parts of the system: the crack faces and emitted dislocations [27, 28, 30]. In the far-field regions, little of interest is happening. Many computer cycles are spent computing the motion of atoms that do little apart from vibrating around lattice sites. In this region a mean-field approach is all that is required. On the other hand, it is not appropriate to dispense with this region entirely, as evidenced by purely atomistic simulations of crack propagation. These large-scale MD simulations show that elastic shock waves radiate from the crack tip [24, 25], propagate to the edge of the simulation, bounce off the edges of the computational cell and reflect back towards the center where they contaminate the very phenomenon under study. A FE region surrounding the MD region allows these elastic waves to propagate harmlessly into the continuum (the continuum region can be made large). Note that although Fig. 1 shows FE regions at the top and bottom ends of the pseudo-one-dimensional topology (actually, a three-dimensional (3D) simulation has been performed as demonstration) there is nothing to stop the use of FE regions on all sides surrounding the central MD region. The decomposition of the system depends upon the physics under study.

This approach should not be viewed as a universal solution. It works well on a subset, albeit an important subset, of systems. These are inhomogeneous multiscale systems. In these systems, either because of their inherent structure (as in the case of MEMS) or because of the focus on a particular phenomenon within the system (as in the case of the crack), a small region plays a crucial role in the dynamics of the larger system. In fracture the region about the crack tip dictates the kinetics of the system, and it is surrounded by material only slightly perturbed from equilibrium. The MD region is chosen to be of sufficient size to allow all defects to form and/or propagate. The trajectory of the region of breaking bonds is tracked via a dynamical relocatable TB region. But once the defects reach the FE/MD interface, the algorithm, as propounded below, must terminate.

A second restriction, upon which we have not yet touched, is that of timescale. As we illustrate below, the advent of parallel architecture machines allows access to system sizes approaching the macroscale, but as yet the total elapsed time of the simulation is, and therefore those phenomena which may be accessed are, restricted to approximately one nanosecond in any simulation with an uninterrupted atomistic time step. Time must flow sequentially, and mere access to more processors does not solve, in any universal way, the issue of timescales for *dynamic* simulation. Progress in algorithms which address the timescale issue would revolutionize materials simulation.

We turn now to a detailed description of the coupling of length scales in the hybrid FE/MD/TB model, originally proposed by Abraham et al. [1, 2]. The unifying theme is that a Hamiltonian, H_{tot}, is defined throughout the entire system. It is a function of the atomic positions, $\mathbf{r}$, and their velocities, $\dot{\mathbf{r}}$, in the TB and MD regions, and the displacements, $\mathbf{u}$, and their time rates of change, $\dot{\mathbf{u}}$, in the FE regions. (The conjugate momenta are simply related to the velocities.) Equations of motion for all the relevant degrees of freedom are obtained by taking the appropriate derivatives of this Hamiltonian in a standard Euler-Lagrange procedure. The time evolution of all the variables then marches forward in lock-step using the same integrator. Thus the entire time history of the system may be obtained numerically given an appropriate set of initial conditions.

Within this framework, it is the Hamiltonian that is partitioned into FE, MD, TB and handshaking contributions during the domain decomposition. The degrees of freedom, the atomic and nodal displacements, do not necessarily fall into a unique domain, but their interactions do. Conceptually, H_{tot} may be written

$$H_{tot} = H_{FE}(\mathbf{u}_a, \dot{\mathbf{u}}_a) + H_{FE/MD}(\mathbf{r}_j, \dot{\mathbf{r}}_j, \mathbf{u}_a, \dot{\mathbf{u}}_a) \\
+ H_{MD}(\mathbf{r}_j, \dot{\mathbf{r}}_j) + H_{MD/TB}(\mathbf{r}_j, \dot{\mathbf{r}}_j) \\
+ H_{TB}(\mathbf{r}_j, \dot{\mathbf{r}}_j). \tag{1}$$

This equation should be read as implying that there are three separate Hamiltonians, one for each sub-system, as well as Hamiltonians for the handshaking regions. "MD/TB" and "FE/MD" imply such handshaking regions. Following a trajectory dictated by this Hamiltonian will result in a conserved total energy. This is an important feature of this computational approach since it ensures numerical stability. Each of these Hamiltonians is described in some detail below. Also, since this approach unifies models which traditionally have been used by different disciplines, we briefly review the basics of MD, TB and FE before we present the details of the implementation.

2.1 Molecular dynamics

Molecular dynamics computes the classical trajectories of atoms by integrating Newton's third law, $F = ma$, for the system. The forces are computed from a potential energy tuned to the scale of interest. In the MD region the interatomic force law follows from an empirical potential; in the TB region it is extracted from a quantum-mechanical Hamiltonian which solves the mean-field equations for the valence electrons in the system. We have simulated silicon and quartz microsystems, using the empirical potentials by Stillinger and Weber [31] and Vashishta [32], respectively. They involve both two-body and three-body interatomic terms:

$$V = \sum_{i<j} V^{(2)}(r_{ij}) + \sum_{i,(j<k)} V^{(3)}(\mathbf{r}_{ij}, \mathbf{r}_{ik}). \tag{2}$$

The sums run over atomic indices i, j and k, and the sum in the three-body term is written such that i is the apex of the triplet. The exact forms of these interactions, V, are given in Refs. [31] and [32].

Forces, **F**, are computed from the empirical potential by taking derivatives analytically with respect to the atomic coordinates. A trajectory through phase space for the atoms is computed via Newton's law integrated using the standard velocity-Verlet algorithm. This integrator is energy conserving (symplectic), and it is easily augmented to handle accelerated dynamics through multiple time steps [33]. The motion is calculated via iteration of the algorithm

$$\dot{\mathbf{r}}_i(t + \tfrac{1}{2}\Delta t) = \dot{\mathbf{r}}_i(t) + \frac{\Delta t}{2m}\,\mathbf{F}_i(t), \tag{3}$$

$$\mathbf{r}_i(t + \Delta t) = \mathbf{r}_i(t) + \Delta t\,\dot{\mathbf{r}}_i(t + \tfrac{1}{2}\Delta t), \tag{4}$$

$$\mathbf{F}_i(t + \Delta t) = \frac{\partial V}{\partial \mathbf{r}_i(t + \Delta t)}, \tag{5}$$

$$\dot{\mathbf{r}}_i(t + \Delta t) = \dot{\mathbf{r}}_i(t + \tfrac{1}{2}\Delta t) + \frac{\Delta t}{2m}\,\mathbf{F}_i(t + \Delta t). \tag{6}$$

Each of the steps is performed sequentially for every atom, i, in the system before continuing to the next step. After exiting the last step, the simulation time is incremented by Δt for the next iteration. For both silicon and quartz, we have used a time step of 5×10^{-16} s. The mass of the silicon atom, m, is 4.6639×10^{-26} kg; the mass of the oxygen is 2.6567×10^{-26} kg. Evaluation of the SW energy and forces is coded to take advantage of Verlet lists of atomic neighbors, so that the time for the computation scales with the number of atoms in the system, i.e. it is order-N. The two- and three-body terms truncate smoothly to zero just before the second neighbor distance in zero-pressure diamond cubic structure silicon.

2.2 Tight binding

Tight binding is a semi-empirical electronic structure methodology which relies first on an ansatz for the total energy of the system, and second on a parameterization of the integrals which arise in a mean-field treatment of electronic bonding. The total energy of the system is written as:

$$V_{\text{TB}} = \sum_{n=1}^{N_{\text{occ}}} \varepsilon_n + \sum_{i<j} V^{\text{rep}}(r_{ij}). \tag{7}$$

The sum includes all of the occupied states, N_{occ}, up to the Fermi level. The form of the total energy has been given sound justification by Foulkes and Haydock [34]. Its form is similar to the full density functional expression for the total energy, but the double-counting of the Coulomb and exchange-correlation terms inherent in the eigenvalue sum is approximately offset by the repulsive interatomic potential, V^{rep}, summed over all pairs of atoms. The eigenvalues, ε_n, corresponding to the energy of the single-particle states of a first-principles Hartree-Fock or density-functional calculation, are obtained from a non-orthogonal one-electron Hamiltonian

$$[\mathbf{H}]\boldsymbol{\Psi}_n = \varepsilon_n [\mathbf{S}] \boldsymbol{\Psi}_n, \tag{8}$$

$$\boldsymbol{\Psi}_n = \sum_{i\alpha} c_{i\alpha}^n \phi_{i\alpha}. \tag{9}$$

In Eq. (9) the single-particle wavefunctions, Ψ_n, are expanded as a linear combination of atomic basis functions, $\phi_{i\alpha}$. The index n labels the orbital number, while i and j label the atomic sites and α and β label the basis functions (in the minimal basis of silicon, these represent s, p_x, p_y and p_z atomic orbitals). The matrix elements within the Hamiltonian [**H**] and overlap [**S**] matrices are obtained by fitting the equivalent integrals within an extensive database of first-principles calculations to a particular parametric form:

$$H_{i\alpha j\beta} \equiv \langle \phi_{i\alpha} | \hat{H} | \phi_{j\beta} \rangle = h_{\alpha\beta}(\mathbf{r}_{ij}),$$
$$S_{i\alpha j\beta} \equiv \langle \phi_{i\alpha} | \phi_{j\beta} \rangle = s_{\alpha\beta}(\mathbf{r}_{ij}). \qquad (10)$$

The size of the [**H**] and [**S**] matrices in the sp basis is $4N \times 4N$. Their matrix elements are parametrized in the two-center approximation by pairwise functions, h and s, as indicated in Eq. (10). These functions smoothly truncate to zero near 5 Å, between the third and fourth neighbor distances in silicon. Note that since all of the integrals are represented by pair functions (V^{rep}, h and s), the exact form of the basis functions is not required. The functions V^{rep}, h and s are obtained by fitting to a database involving the experimental indirect band gap of the diamond cubic structure and the total energies of crystalline and defective diamond cubic and β-tin silicon at different densities. The parameters for this fit are given by Bernstein and Kaxiras [35]. For most purposes, this non-self-consistent TB Hamiltonian describes bulk, amorphous and surfaces properties of silicon very well.

Solving for the coefficients $c_{i\alpha}^n$ is a generalized eigenvalue problem: given a set of atomic coordinates, the coefficients are found by diagonalization. Single-electron states are occupied up to the Fermi-level, with a small amount of Fermi-level broadening. Forces are then computed from the derivative of the TB energy with respect to displacement of the nuclei,

$$\mathbf{F}_i^{\text{TB}} = -\left[\sum_{n=1}^{N_{\text{occ}}} \sum_{\alpha} c_{i\alpha}^n \sum_{j\beta} c_{j\beta}^n \left[\frac{\partial H_{i\alpha j\beta}}{\partial \mathbf{r}_i} - \varepsilon_n \frac{\partial S_{i\alpha j\beta}}{\partial \mathbf{r}_i} \right] \right] - \sum_{j \neq i} \frac{\partial V^{\text{rep}}(r_{ij})}{\partial \mathbf{r}_i}. \qquad (11)$$

Derivatives of the coefficients with respect to atomic positions do not occur according to the Hellman-Feynman theorem. Such contributions vanish identically by orthonormality. Knowing the forces, atomic coordinates may be advanced through time using exactly the same algorithm as that used for the SW system (cf. Eq. (3)). The same time step may be used since Einstein frequencies in both cases are nearly identical; in fact SW parameters could be adjusted to ensure exact equality.

The above describes an $\mathcal{O}(N^3)$ algorithm: brute-force diagonalization scales as the cube of the number of electrons, and it parallelizes poorly. The scaling may be improved somewhat by implementing a fictitious Lagrangian [36], but this is $\mathcal{O}(N^2)$ at best. $\mathcal{O}(N)$ schemes for electronic structure have been discussed extensively in the literature [37 to 40], but the tight binding has not been implemented in $\mathcal{O}(N)$ thus far in the hybrid algorithm following Abraham et al. [1]. There are several reasons they have advocated this approach. First, the $\mathcal{O}(N)$ methods often have problems with situations in which states wander across the Fermi level, something that is expected to happen as dangling bonds are created at the crack tip. And second, the cross-over of improved efficiency from the $\mathcal{O}(N^3)$ to the $\mathcal{O}(N)$ schemes with the accuracy required for silicon occurs at system sizes above several hundred atoms [41]. Diagonalization wins for fewer atoms. The stipulation of one second of wall-clock time per time step places the choice in the $\mathcal{O}(N^3)$ system size regime, unless a very large number of processors is devoted to the TB region. As $\mathcal{O}(N)$ schemes improve, the optimal choice may change. For the present, then, direct diagonalization has been used since it is robust and efficient in the systems of interest.

2.3 Finite elements

Finite elements is a method of computing an approximate solution of continuum partial differential equations (PDEs) by discretizing space on an irregular mesh and solving numerically the set of coupled ordinary differential equations that result. In the crack and MEMS systems, the FE equations of motion involve forces resulting from derivatives of a discretization of the elastic energy in continuum mechanics. The energy has been taken from the theory of linear elasticity, so the input parameters are simply the elastic moduli and the density of the material. These parameters are in agreement with the values computed in the MD and TB regions so that the simulation is consistent across the scales. Linear elastic theory suffices as long as the FE region is restricted to the far field where the strain gradients are very small.

To be precise, the total elastic energy in the absence of tractions and body forces within the continuum model is given by

$$H_{\text{FE}} = V_{\text{FE}} + K_{\text{FE}}, \tag{12}$$

$$V_{\text{FE}} = \frac{1}{2} \int d\Omega \sum_{\mu,\nu,\lambda,\sigma=1}^{3} \epsilon_{\mu\nu}(\mathbf{r}) \, C_{\mu\nu\lambda\sigma} \, \epsilon_{\lambda\sigma}(\mathbf{r}),$$

$$K_{\text{FE}} = \frac{1}{2} \int d\Omega \, \varrho(\mathbf{r}) \, |\dot{\mathbf{u}}(\mathbf{r})|^2.$$

Here V_{FE} is the Hookian potential energy term, which is quadratic in the symmetric strain tensor, ϵ, contracted with the elastic constant tensor, C. The Greek indices denote Cartesian directions. The kinetic energy, K_{FE}, involves the time rate of change of the displacement field, $\dot{\mathbf{u}}$, and the mass density, ϱ. The strains are related to the displacements according to

$$\epsilon_{\mu\nu} = \frac{\partial u_\mu}{\partial r_\nu} + \frac{\partial u_\nu}{\partial r_\mu}. \tag{13}$$

These are fields defined throughout space in the continuum theory, so the total energy of the system is an integral of these quantities over the volume of the sample, Ω.

The FE algorithm involves partitioning the system into cells (also known as elements). These cells are often hexahedra (such as cubes) or tetrahedra in three dimensions, and rectangles or triangles in two dimensions [42]. Tetrahedra have proven to be more useful than hexahedra for coupling of length scales in solids because the FE mesh must collapse down to the crystal lattice at the atomic scale. For any crystal lattice is is possible to find a set of tetrahedra that tessellate the entire volume of the system such that the lattice points form the apices of the tetrahedra without the need for additional nodes. This is not the case with cubes and other parallelepipeds. Tetrahedra have the added advantage that linear interpolation may be used (see below). An extensive literature exists on FE meshes including automatic mesh generation and adaptive remeshing [43], although little work has been done on remeshing techniques that are optimized for, or commensurate with, the underlying crystallography [4, 22]. We will describe the meshes used in the crack and MEMS simulations in the next subsection.

In finite elements the values of the displacements and their time derivatives are defined at the nodes of the mesh (the apices of these cells). In order to make contact with the continuum theory, interpolation functions (shape functions) are used to determine the values of these fields throughout each cell. We have used linear interpolation in which each basis function has the value of one at a single node, decreases linearly to zero at the nearest neighboring nodes, and is zero everywhere else. The result is that the displacement fields are represented in piecewise smooth manner. Linear interpolation is a common choice in tetrahedral elements for many FE applications based on second-order PDEs. In coupling of length scales, linear interpolation is one of the few schemes that guarantees a natural one-to-one correspondence between atoms and nodes in the atomic limit. Other possible choices for shape functions are discussed in Ref. [4].

Use of the interpolation functions reduces Eq. (12) to the approximate form

$$H_{FE} = \frac{1}{2} \sum_{m=1}^{N_{cell}} \sum_{p,q=1}^{p_{max}} \left[u_p^m K_{pq}^m u_q^m + \dot{u}_p^m M_{pq}^m \dot{u}_q^m \right], \quad (14)$$

where $[\mathbf{K}]$ and $[\mathbf{M}]$ are local stiffness and mass matrices, respectively. Thus, the energy and ultimately the forces are computed cell by cell. The cell index is denoted by m. The total number of FE cells is N_{cell}.

The exact form for $[\mathbf{K}]$ depends on the dimensionality of the system. In the example of crack propagation, the system is in plane-strain and a two-dimensional (2D) far-field region suffices. In this case triangular cells have been used [2]. In the example of MEMS, the far-field region is roughly in plane-stress, but in order to capture all of the relevant physics we have used a full 3D implementation with tetrahedra. In 2D the indices p and q run over the $p_{max} = 6$ degrees of freedom associated with the 2D displacements at the three apices of a triangular cell; in 3D, they run over the $p_{max} = 12$ degrees of freedom of 3D displacements at the four apices of a tetrahedron.

The total FE energy is given as a sum of products of local matrices. The stiffness matrix is of the form

$$[\mathbf{K}_m] = V_m [\mathbf{B}_m^T][\mathbf{C}][\mathbf{B}_m], \quad (15)$$

where the prefactor V_m is the volume of the m-th cell, $[\mathbf{C}]$ is the elastic constant tensor and $[\mathbf{B}_m]$ is the matrix of coordinate differences of the apices of the m-th cell. The values of the elements of $[\mathbf{C}]$ depend upon the orientation of, and the constraints on, the system; i.e. in the 2D cases the matrix must be dimensionally reduced in a manner consistent with the plane-strain or plane-stress constraint. The elements of $[\mathbf{C}]$ are functions of the three fundamental elastic constants of silicon, namely C_{11}, C_{12} and C_{44}. $[\mathbf{K}]$ is time independent, since the mesh and hence V_m and $[\mathbf{B}_m]$ are not changing. The B matrix in two dimensions is given by

$$[\mathbf{B}_m^T] = \frac{1}{2A_m} \begin{pmatrix} b_1^m & 0 & a_1^m \\ 0 & a_1^m & b_1^m \\ b_2^m & 0 & a_2^m \\ 0 & a_2^m & b_2^m \\ b_3^m & 0 & a_3^m \\ 0 & a_3^m & b_3^m \end{pmatrix}, \quad (16)$$

where A_m is the area of the m-th triangle, and a_l^m and b_l^m are the coordinate differences

$$a_l^m = x_{l+2}^m - x_{l+1}^m,$$
$$b_l^m = y_{l+1}^m - y_{l+2}^m. \tag{17}$$

Here l denotes the cyclic apex index, running from 1 to 3, and x and y denote the 2D FE mesh coordinates. In 2D plane-strain, the [**C**] matrix is given by

$$[\mathbf{C}] = \begin{pmatrix} C_{11} & C_{12} & 0 \\ C_{12} & C_{11} & 0 \\ 0 & 0 & C_{44} \end{pmatrix}. \tag{18}$$

Displacements, $\mathbf{u}_a$, are defined with respect to these coordinates. In 3D the stiffness matrix is defined in a similar manner [44].

The reduced elastic constant matrix (25) for this geometry was obtained by averaging the results of Balamane et al. [45] and Ray [46] for the zero-temperature C_{11}, C_{12} and C_{44} elastic constants of SW silicon:

$$C_{11} = 1.578 \times 10^6 \text{ Mbar},$$
$$C_{12} = 0.7930 \times 10^6 \text{ Mbar}, \tag{19}$$
$$C_{44} = 0.6365 \times 10^6 \text{ Mbar}.$$

These values of the elastic constants assure good agreement with the MD and TB regions.

The mass matrix, [**M**], is handled rather differently. In principle the kinetic energy density varies across any given cell. However, it is necessary to reduce the FE mesh in the FE/MD handshaking region to coincide with the perfect atomic lattice. In this limit the kinetic energy is localized to the nodes since that is where the atoms are situated. Thus for the FE region, we have employed the "lumped mass" approximation [47], a choice that has been validated through elastic wave dispersion and scattering comparisons [3]. This is a diagonal approximation of the actual distributed mass matrix which would have both diagonal terms and couplings between nearest neighbor nodes resulting from apportioning the kinetic energy according to the FE shape functions. The lumped mass matrix does reduce to the MD masses in the atomic limit. One third of the mass in each 2D triangular cell, and one fourth of the mass in each 3D tetrahedron, is apportioned to each apex in silicon. The kinetic energy is thus given by

$$K_{\text{FE}} = \sum_{a=1}^{N_{\text{mesh}}} \tfrac{1}{2} M_a |\dot{\mathbf{u}}_a|^2, \tag{20}$$

$$M_a = \begin{cases} \varrho \sum_{m=1}^{N_{\text{cell}}} \sum_{l=1}^{3} \delta_{am_l} V_m/3 & \text{in 2D}, \\ \varrho \sum_{m=1}^{N_{\text{cell}}} \sum_{l=1}^{4} \delta_{am_l} V_m/4 & \text{in 3D}, \end{cases} \tag{21}$$

where a labels the FE mesh points, of which there are N_{mesh} total, and m_l labels the mesh point index at each of the apices of cell m. The velocities $\dot{\mathbf{u}}_a$ in Eq. (20) are vectors of length two in 2D and three in 3D, since they relate to a node on the global mesh. The density was taken to be $\varrho = 2.3304 \times 10^3$ kg/m^3, in order to agree with a lattice constant of 5.43 Å.

In quartz, SiO$_2$, the mass matrix must be treated differently in order to get agreement with the atomic limit in which there are two different masses. The density-based formula (21) is used for large cells, in particular for any FE cell that contains over 100 atoms. For smaller cells the following CGMD-inspired formula [3] is used for polyatomic crystals:

$$M_a = \sum_j N_a(\mathbf{r}_{j0}) m_j, \qquad (22)$$

where $N_a(\mathbf{r})$ is the shape function associated with node a, $\mathbf{r}_{j0}$ is the perfect lattice position of atom j and m_j is the mass of atom j. This formula does recover the MD masses in the atomic limit, and it connects smoothly with the density-based lumped mass approximation used for large cells.

Forces for the displacement degrees of freedom in Eq. (14) are obtained by taking spatial derivatives. Displacements and their temporal rate of change may be obtained as a function of time, for given boundary conditions, using the same velocity-Verlet update algorithm, Eq. (3), and same time step as that used for the MD and TB. As Broughton and coworkers [2] first discovered, the use of identical symplectic time integrators and time steps is a critical issue in order to achieve numerical stability when length scales are coupled dynamically.

The force due to the m-th cell is given by

$$\mathbf{F}_{\text{FE}}^m = [\mathbf{K}_m] \mathbf{u}^m, \qquad (23)$$

where in keeping with Eq. (14), the vectors in Eq. (23) are of length six or twelve in 2D or 3D, respectively. The total force associated with a mesh point is then the sum of the contributions from each of the cells with apices in common with that point. Finally

$$\mathbf{F}_{\text{FE}}^a = M_a \ddot{\mathbf{u}}_a. \qquad (24)$$

Note that two FE nodes are directly coupled only when they are on a common element.

2.4 Handshaking FE/MD

The two principal issues for handshaking between FE and MD regions are the matching of the fields and the definition of the forces. The matching determines how the displacement field on the FE mesh coincides with the positions of the MD atoms. The forces are determined by the form of the handshaking Hamiltonian, and are chosen to guarantee consistent dynamics and thermodynamics.

The correspondence between the FE nodal variables and the MD atomic coordinates is a generalization of an idea due to Kohlhoff and coworkers [48]. An imaginary surface is placed at the interface between the FE and MD regions. On both sides of this fiducial surface within a fixed distance equal to the range of the MD interatomic potential, FE mesh points are placed at the equilibrium position of the MD atoms. In our silicon systems these are the ideal lattice sites. We make the assumption that there is no diffusion at the FE/MD interface, so that atoms remain near the corresponding mesh points on either side of this interface. In principle, the distinction between the two is academic: atomic motion may be viewed as displacement around a lattice (mesh) site, and the displacement field may be viewed as motion of an atom away from its perfect site. In practice, the FE forces in linear elastic theory are not correct for large strains, and diffusion is unavoidably quenched. So as formulated, diffusion must be negligible at the interface.

This methodology would work for amorphous systems as well; all that is required is a one-to-one mapping of a mesh point to an atom site. Moving away from the handshaking region into the FE region, the mesh spacing may be made larger. This is the principal reason that the FE algorithm is computationally efficient. The largest spacing depends upon the physics to be captured; for example the largest spacing determines the shortest wavelength phonons that propagate unimpeded though the FE region.

The study of crack propagation in silicon has focused on brittle fracture, and the simulations of Abraham et al. [1] have been set up accordingly. Specifically the rectilinear system has been oriented such that it has (100) faces on all sides. The FE region is represented as a 2D system which treats the third dimension in plane strain; Eq. (15) has the parameter, $L = V_m/A_m$, to represent the thickness of the sample, so that the 2D FE region is consistent with the 3D MD and TB regions. Then the projection of a diamond cubic lattice onto a (100) plane is required to generate the 2D mesh. Figure 3 shows how this is done, as well as how the projected atoms are triangulated to get the FE mesh. At the edges of the computational cell, the triangulation is able to wrap around, and periodic boundary conditions are used for FE as well as MD. A periodic triangulation has exactly twice as many cells as mesh points. Away from the handshaking region and into the FE region, the mesh expanded along the dimension perpendicular to the FE/MD interface, while keeping the mesh spacing constant in the second. The expansion was taken to be of the hyperbolic tangent form. Thus near the handshaking region there is no expansion of the atomic mesh, and far away from the handshaking region the spacing asymptotes to a constant multiple of the atomic lattice parameter. In the crack propagation example below, this multiple was ten. The transition region spanned a couple of hundred Angstroms.

Turning now to the form of the handshaking Hamiltonian, where this approach differs from the prior work of Kohlhoff et al. [48] is in the dynamics of the handshaking region. Broughton and coworkers [2] have found that it is very important for the *dynamic* coupling of length scales to define a single conservative Hamiltonian for the

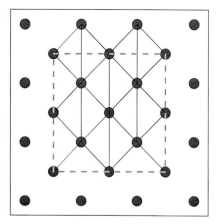

Fig. 3. Triangulation of 2D unit cell for diamond cubic lattice. The unit cell is denoted by dashed lines. Triangles are marked with solid lines. Note the solid lines on left and right but not top and bottom of unit cell boundary, with 8 mesh points and 16 triangles per unit cell

entire system including the handshaking region. Then the system may be stepped through time using uniform symplectic time evolution of the atomic and displacement trajectories across the handshaking region, as described above.

The fiducial interface is used to determine which interactions enter the FE/MD handshaking Hamiltonian. The discussion is best understood with reference to Fig. 4. In conceptualizing this Hamiltonian, two different materials are envisioned sitting on either side of the interface: in one case it is FE silicon and in the other it is SW silicon. The handshaking interactions at the interface to first order can be approximated by a mean of the two descriptions. All FE triangles which cross the interface contribute half their weight to the Hamiltonian. Any triangle which is fully in the MD region contributes zero weight. Similarly, any SW interaction (two-body or three-body) which crosses the interface contributes half its usual weight. Any SW interaction between mesh points, all of which are fully on the FE side of the interface, contributes zero weight. The SW energy formulation which concentrates upon atomic coordinates, $\mathbf{r}_j$, and the FE energy formulation which concentrates upon displacements, $\mathbf{u}_a$, can be used throughout the handshaking region because the FE and MD degrees of freedom overlap, and because of the equivalence of atoms and mesh points. The one-to-one mapping of atoms to nodes is not required at distances greater than twice the SW pair cut-off away from the interface in the FE region. This is the distance of greatest three-body range. Figure 4 indicates diagrammatically those interactions which contribute to the handshaking Hamiltonian. More precisely, the handshaking potential is given by

$$V_{\text{FE/MD}} = \frac{1}{4} \sum_{m \in \tau_c} \sum_{p,q=1}^{p_{\max}} u_p^m K_{pq}^m u_q^m$$
$$+ \frac{1}{2} \left[\sum_{(i<j) \in \beta_{2c}} v^{(2)}(r_{ij}) + \sum_{(i,(j<k)) \in \beta_{3c}} v^{(3)}(\mathbf{r}_{(ij),(ik)}) \right], \quad (25)$$

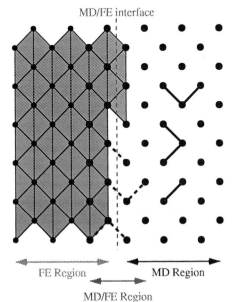

Fig. 4. Illustration of FE/MD handshaking couplings for the crack simulation, with only a few representative MD examples shown. The FE/MD fiducial interface is the dashed vertical line. FE cells contributing to the overall Hamiltonian with full weight have dark shading; those contributing with half weight have light shading. Two- and three-body SW interactions which cross fiducial interface also carry half weight and are shown with dotted lines. Continuous lines represent full-weight SW interactions. The FE/MD handshaking for MEMS differs somewhat (see text)

where τ_c is the set of FE triangles that cross the FE/MD interface, and β_{2c} and β_{3c} are the sets of two-body and three-body bonds, respectively, that cross. Indeed, $V_{FE/MD}$ is only defined for interactions which cross the boundary. Otherwise the other terms in Eq. (1) define the forces at mesh points and atoms. The formalism in Eq. (25) is meant to imply that *any* one atom of the triplet in the three-body terms can be on an opposite side of the interface to the other two.

In the present implementation of the FE/MD/TB hybrid algorithm, the MD and TB regions are 3D for the crack, but the FE is 2D (with the third dimension treated in mean-field); for the MEMS the FE is 3D as well. The coupling between 3D MD and 2D FE is done in the following way. In $V_{FE/MD}$, x and y displacements of atoms on the MD side of the FE/MD boundary that contribute to the elastic energy are obtained by averaging over all equivalent atoms in the depth, z. Conversely, the SW energy contribution to the handshaking Hamiltonian on the FE side of the interface is determined for atoms situated at ideal lattice sites in the third dimension with duplicate x and y displacements given by the FE coordinates. In the case of MEMS where the FE is 3D, no averaging is required. The overall Hamiltonian is conservative in both cases.

In making the FE/MD interface seamless, two other issues confront the definition of energy. They both involve reference state; one is potential energy, the other is thermal energy. The SW potential is referenced to infinitely separated atoms. The FE potential is referenced to a zero-temperature ($T = 0$) unstrained lattice. For the purposes of graphical analysis, therefore, a constant offset energy which does not affect the dynamics is added to each FE mesh point. The $T = 0$ energy density for SW silicon at zero pressure is -4.33444 eV/atom. The offset energy is computed for every FE point using an equation entirely analogous to that used to compute mass in the lumped mass approximation except that instead of a mass density, the SW zero-point energy density is used (see Eq. (20)). As before, this scheme ensures the correct limiting behavior as the mesh spacing is reduced to atomic dimensions. For atoms in the handshaking region, for systems with unusual orientation where the offset is non-trivial to estimate atom by atom, a $T = 0$ calculation with zero strain for the coupled FE/MD system can be performed. The offset may thereby be calculated to maintain the energy/atom constant through the interface. This is easily achieved by virtue of periodicity and symmetry.

Turning now to the thermal energy, the work of Rudd and Broughton [3] has shown the relationship between the square of the time rate of change of the displacements in the FE region and the temperature. Effectively the FE algorithm involves an average over the atomic degrees of freedom that are missing from the mesh. Thus, to bring the atomic and continuum thermal energies onto an equivalent footing, the total FE thermal energy must again be offset. These corrected energies are denoted by a prime:

$$K'_{FE} = \frac{3}{2}\Delta N k_B T + K_{FE} + \frac{\delta}{2}N_{mesh} k_B T,$$

$$V'_{FE} = \frac{3}{2}\Delta N k_B T + V_{FE} + \frac{\delta}{2}N_{mesh} k_B T, \qquad (26)$$

where $\Delta N = N_{atom} - N_{mesh}$ and δ is the codimension of the FE region (i.e., δ is 0 or 1 for 3D or 2D FE, respectively). N_{atom} is the number of atoms contained within an equivalent 3D volume, and k_B is the Boltzmann constant. Equipartition has been invoked, and fluctuations about the average energy are neglected in this expression (see below). It is further assumed that the background temperature is constant during the simulation. The first term therefore accounts for the missing degrees of atomic freedom while the last term augments the 2D FE plane-strain simulation for the missing third dimension in its degrees of freedom. As before, these offsets do not affect the dynamics of the system and the thermal corrections can be apportioned to each mesh point as described above for the zero temperature FE potential energy. For finite temperature simulations, the $\dot{\mathbf{u}}_a$ degrees of freedom are thermalized to a Maxwellian distribution. Also, the appropriate elastic constants for that temperature should be used in the FE equations of motion so as to make the MD and FE regions seamless and compatible. Further, since this methodology requires a continuation of ideal lattice sites into the FE/MD handshaking region in order to determine mesh coordinates, the appropriate lattice parameter for given temperature should be used.

We conclude this subsection with a discussion of the issue of dissipation in the FE region. In a real physical system of silicon, the long wavelength modes that are represented on the FE mesh couple through anharmonic interactions to the short wavelength degrees of freedom that are missing from the FE model. These couplings allow the system to come to thermal equilibrium, something that is not possible in a FE model based on linear elastic theory. It is possible to improve the model, however, by treating the short wavelength degrees of freedom as a Brownian heat bath weakly coupled to the FE modes. This coupling causes fluctuations in the FE displacements, and it leads to dissipation. These effects promote thermalization of the system, since the bath is taken to have the temperature at which the simulation is being performed. This coupling to the missing degrees of freedom has been discussed in many guises in the literature, and it has been formulated precisely within CGMD [4]. In this generalization of FE, the force used in the third step of the velocity-Verlet algorithm (see Eq. (5)) now includes random and dissipative terms,

$$\mathbf{F}_{FE}^a = \frac{\partial V_{FE}}{\partial \mathbf{u}_a} + \varrho^G(T, y_a) - \xi(y_a) M_a \dot{\mathbf{u}}_a, \qquad (27)$$

where y_a is the coordinate of the mesh point in the direction perpendicular to the crack (or the direction of the longest dimension of the MEMS resonator). ϱ^G is a Gaussian random variable and ξ is a friction coefficient, which are related by the fluctuation dissipation theorem. Specifically the variance, σ^G, of the Gaussian is given by

$$\sigma^G = \sqrt{\frac{2\xi M_a k_B T}{\Delta t}}. \qquad (28)$$

In order to perturb the dynamics of the active zone (i.e. MD and TB) minimally, ξ was made a function of the (time invariant) FE mesh y-coordinate. ξ was linearly ramped from zero in the handshaking region to finite value (0.1 for the crack propagation study) at the extremal outer edge of the FE regions.

There are two significant differences in the way we have implemented the FE/MD coupling for MEMS. First, we have used 3D finite elements in order to model the full geometry of the region in which the MEMS device is attached to the substrate. In this case the nodes are in a one-to-one correspondence with the atoms in the handshaking region, as shown in Fig. 5. We have used tetrahedral cells, with a tessellation based on the triangulation described above, but extended into the third dimension. Each triangle vertex corresponds to one atom per unit cell, so the triangulation extends naturally to a tessellation by triangular prisms. An edge is introduced across the short diagonal of

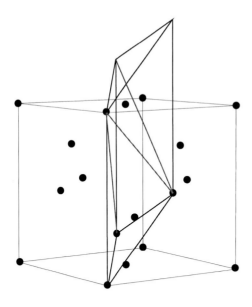

Fig. 5. Tessellation of the 3D diamond cubic unit cell with tetrahedra based on an extension of the 2D triangulation in Fig. 3

each of the side parallelograms on the prisms, giving three tetrahedra per prism. These tetrahedra then tessellate the diamond cubic lattice, and a generalization of the 2D procedure described above is used to extend the mesh away from the FE/MD interface.

Second, we have defined the FE/MD handshaking Hamiltonian slightly differently. The FE contribution is identical, but an additional MD contribution is included. The fact that the strength of the FE interactions is reduced for an entire cell at a time (as it must be to maintain translational invariance) means that the interaction between some pairs of nodes is reduced even when the edge that connects them does not cross the fiducial interface. For example, in Fig. 4 it is clear that for every triangular cell involved in handshaking, two edges cross the interface but the third does not. Nevertheless, the interaction between the nodes on the third side are reduced. In the MEMS simulations, we have included rescaled two-body MD bonds to offset this effect. Suppose both nodes in the pair $\{i,j\}$ are apices of $n_{i,j}^{\text{cell}}$ different FE cells, of which $n_{i,j}^{\text{hs}}$ are at half strength due to participation in handshaking, $n_{i,j}^{\text{zs}}$ are at zero strength because they are in the MD region and $n_{i,j}^{\text{fs}}$ are at full strength because they are in the FE region. Then an MD two-body contribution is added between the atoms i and j, weighted by a factor $c_{i,j}$ to make up for the missing FE interactions. In particular, the handshaking interactions (25) are modified to be

$$V_{\text{FE/MD}} = \sum_{m \in \tau_c} \sum_{p,q=1}^{p_{\max}} \tfrac{1}{4} u_p^m K_{pq}^m u_q^m + \sum_{i<j} c_{i,j} v^{(2)}(r_{ij}) + \sum_{(i,(j<k)) \in \beta_{3c}} \tfrac{1}{2} v^{(3)}(\mathbf{r}_{(ij),(ik)}), \quad (29)$$

where the weighting factor is given by

$$c_{i,j} = \left(n_{i,j}^{\text{zs}} + \tfrac{1}{2} n_{i,j}^{\text{hs}}\right) / n_{i,j}^{\text{cell}}. \quad (30)$$

This redefinition of the handshaking Hamiltonian leads to a small improvement in two dimensions, and a somewhat larger improvement in three dimensions, as tested by elastic wave scattering tests.

2.5 Handshaking MD/TB

The basic issues of matching the degrees of freedom and defining consistent forces must be addressed at the MD/TB interface as well. The degrees of freedom consist of atomic coordinates in both regions, so the matching is trivial and the basic issue is the definition of the forces. A satisfactory definition of the forces is non-trivial, due to differences in the way the forces are calculated in the two regions. Unlike the well-defined bonds of the MD region, in the TB region, the energy is computed for the entire cluster simultaneously, and it is difficult to apportion and localize energy to specific bonds in a computationally efficient way. The total energy is a property of the entire system. Attempts to average the MD and TB bond contributions, such as was used for the FE/MD interface, run into problems of orthonormality such that the Hellman-Feynman theorem breaks down and Eq. (11) is no longer sufficient: derivatives of the electronic coefficients with respect to the atomic coordinates are required and these must be computed numerically. In order to avoid this computational overhead, a different approach has been taken.

The MD/TB handshaking takes place conceptually across an interface *consisting* of atoms. This is in marked contrast to FE/MD handshaking where an interface *between* planes of atoms was defined. Whereas specific bonds were taken to have the mean force in the FE/MD case, in the MD/TB case the properties of the interfacial atoms are modified to ensure consistent forces throughout the TB cluster. At the surface of this cluster the Hamiltonian must be modified in order to take account of the bonds that extend from the TB region into the MD region (cf. Fig. 6). One reason the coupling of length scales has been implemented for a semiconductor like silicon is that covalent bonds are local objects. Dangling bonds at the surface of the cluster may be tied off with univalent atoms. These handshaking atoms have been termed "silogens" in Ref. [1] because they contain a single s orbital like hydrogen, but with the parameters modified to mimic the silicon surrounding the cluster. The parameters are chosen not only to give the correct Si–Si bond length and binding energy, but also to ensure that the forces within the cluster remain physical as the interior bonds are deformed. In particular, the [**H**] and [**S**] matrix elements and the repulsive pair potential V^{rep} that couple these atoms to the silicon atoms within the interior of the TB region were chosen (a) to maintain electro-neutrality (as measured by Mulliken charges) on both the silogens and silicons, (b) to locate the silogen potential energy minimum at the Si–Si distance, *not*

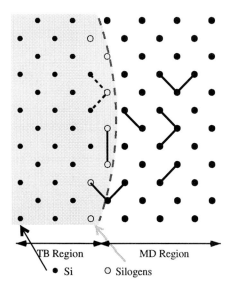

Fig. 6. Illustration of MD/TB handshaking for the crack simulation. The outer surface of the TB region is terminated with monovalent silogens (see text) constrained to sit at MD silicon sites (light dots). The TB Hamiltonian is diagonalized throughout the shaded region, with Si–Si matrix elements employed except at the silogens where Si–silogen matrix elements are used. SW interactions contributing to handshaking Hamiltonian are designated by full lines. The dashed lines represent a non-contributing SW three-body term. Only some representative SW examples are shown

the Si–H distance, (c) to provide a bond energy equal to a single Si–Si bond, (d) to provide a longitudinal force constant equal to that of silicon and (e) to minimize the error in the interior forces in a database of configurations about a defect. At the perimeter of the TB region, the silogens are placed at the silicon sites of the MD region, with more than one silogen at a given site if necessary. (At a Si(100) surface, terminating dangling bonds with silogens necessitates placing two at each empty silicon site.) Thus there are no matrix elements nor V^{rep} terms which couple *any* of the silogens to one another. Operationally, a cylinder is drawn around an inner set of atoms: these are designated TB silicons. Then any atom outside this cylinder, but within range of an inner atom, is designated as a silogen. Following Broughton et al. [2], the range criterion used was the mean of the first (r_0) and second neighbor ($\sqrt{(8/3)}r_0$) distances of the equilibrium silicon lattice.

Within the TB region both Si–Si and Si–silogen matrix elements are used, as shown in Fig. 6. Parameters of the former are given by Bernstein and Kaxiras [35]. The latter, using the same formalism, are:

$$\epsilon_s = -7.661518 \text{ eV},$$
$$V_{ss\sigma} = -1.6967418 \text{ eV},$$
$$V_{sp\sigma} = 3.8704886 \text{ eV},$$
$$S_2(r_{ij}) = \frac{S_{ss\sigma}(r_{ij}) - \sqrt{3}S_{sp\sigma}(r_{ij})}{2}. \tag{31}$$

Bernstein and Kaxiras [35] provide a prescription for relating the S overlap and the V terms above. The TB [**H**] and [**S**] generalized eigenvalue problem is solved for the entire silicon plus silogen system. Forces are extracted as defined by Eq. (11).

All that remains is to specify which SW two- and three-body terms are required to couple the silogen atoms to the MD region. Examples are shown in Fig. 6. Since silogens are not coupled to one another in the TB region, SW terms which account for them are required. All SW pair terms between a silogen and either a silicon atom in the MD region or another silogen are included. All SW triplets which include at least one MD silicon to one silogen–silicon pair are also included. The forces which arise on the silogen–silicons from these terms are added to the forces arising from the TB Hamiltonian on these atoms. Thus, the terms in the unified Hamiltonian (1) should be reinterpreted. The term $H_{\text{MD/TB}}$ involves only SW interactions crossing the boundary and the term H_{TB} involves a TB calculation for the combined silicon plus silogen system.

In the absence of a dynamic allocation of the TB region to those parts of the system where bonds are breaking during a simulation, the above prescription produces a conservative Hamiltonian. Unfortunately, for many systems (such as the crack), the hundred or so atoms, whose forces may currently be updated using a non-orthogonal TB Hamiltonian in one second of wall clock time, do not comprise as large a region as one would like. In particular, in order to ensure the fidelity of forces at the center of the cluster (e.g. at the crack tip), the radius of the cluster should be at least twice the range of the TB density matrix. Clusters of this size require very expensive computations. Part of the problem may be solved using periodic boundary conditions. Our example of crack propagation used a slab two unit cells deep; thus the TB region is a cylinder. There may be other systems, such as a void within the bulk of a material, where periodicity is not appropriate and a spherical region must be used. Another part of the problem might be ameliorated by using more than one processor per TB region to perform the diagonalization, but unfortunately such algorithms presently are not efficient on coarse-grained scalable architecture computers.

Instead, Abraham et al. [1] have introduced a "clover leaf" of TB regions to represent the region of breaking bonds. Fig. 7 gives an illustration of four overlapping TB regions. In the crack propagation simulation, described below, eight overlapping regions were used. Each of these regions is diagonalized separately on a separate processor. The silogens are positioned as shown in Fig. 6. After forces on each atom are obtained for each TB region separately, the force to be used in the velocity-Verlet (see Eq. (3)) update is obtained via an average over the different regions: where there is no overlap of TB regions, use the same prescription as for a single TB region; where only silogens overlap, use the mean value; where a silogen of one TB region overlaps silicons of others, use the mean silicon value. The number of atoms that are propagated using TB forces is therefore less than the total number within all the "clover leaf" regions. These rules are intuitive, and although it is not now possible to write an overall conservative Hamiltonian for the entire system, the atomic trajectories have proven to be well behaved, as measured by the lack of anomalies in local kinetic energies. For a system without periodicity, such as the decoration of TB regions around a void, the "clover leaf" metaphor can easily be generalized to something akin to a "raspberry" of overlapping spheres.

Finally in this subsection, the allocation algorithm is described: that is the algorithm whereby the TB "clover leaf" is made to track a region of breaking bonds. The energy and force algorithm as *implemented* in the MD and TB regions proceeds by calculating the SW energy for *all* atoms in the MD processors. TB processors calculate not only TB energies and forces but also those SW forces which must be subtracted from those double counted in the MD processors. The result is that SW energies, by suitable apportioning of two-body and three-body terms, are available for all atoms. These are then used to discriminate different regions. The apex of a crack is found, for example, by locating the atom, with a potential energy greater (more positive) than 60% of the bulk cohesive potential energy, furthest from the center of the system (see Fig. 1). The cen-

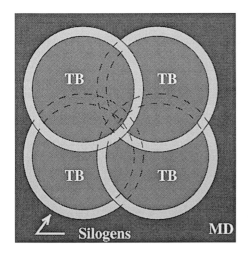

Fig. 7. Illustration of the "clover-leaf" of overlapping TB regions embedded in MD region used in the crack simulation. The crack simulation employed eight overlapping TB regions

tral TB region of the "clover leaf" is then placed at that atom. Such placement should not be performed at every time step. Abraham et al. [1] have repositioned the TB region every ten steps.

2.6 Seamless FE/MD/TB

The foregoing discussion indicates how the simulation can be made seamless. The TB region, since it is the region described at the most microscopic level, should determine the elastic constants and the atomic force fields used elsewhere in the system. Thus, firstly, a pure TB simulation is performed for a small number of atoms representing the bulk system (at given temperature and pressure). By appropriate deformation of the computational cell, the elastic constants are extracted. By movement of one atom within the cell, a local "Einstein oscillator" force constant can be found. The SW parameters for silicon may then be adjusted to reproduce the same quantities. The elastic constants from the TB region are also used for the [**K**] stiffness matrix of the FE region. Lastly, the parameters used for the Si–silogen matrix elements are adjusted so that displacement of a silogen–silicon in the coupled system gives rise to the same "Einstein oscillator" force constant of the pure bulk system.

In the crack propagation example the elastic constants of the SW and FE region are made identical, but the SW and TB elastic constants are slightly different. Nevertheless, the results indicate that even here, the objective of seamlessness is close to reality.

3. Methodology: CGMD

The second model we describe in detail is coarse-grained molecular dynamics (CGMD) in which the FE/MD coupling is based on a derivation of the physical scaling properties of the system [3, 4]. CGMD should be viewed as a replacement for finite elements in the FE/MD/TB methodology, to be used in cases where the atomistic/continuum coupling is more exacting. The need for a higher quality coupling may be due to the physics of the system under study, or it may just be that the the FE/MD interface needs to be brought in closer to the center (e.g. closer to the crack tip). In these cases, it is desirable to have a methodology in which the errors introduced by the atomistic/continuum coupling can be estimated and controlled. This is the advantage of an effective model in which the coupling is derived from the underlying atomistics.

As in the far field of the FE/MD/TB CLS model, consider a system of atoms in a solid, crystalline or amorphous, and a coarse-grained (CG) mesh partitioning the solid into cells (cf. Fig. 8). The mesh size may vary, so that in important regions a mesh node is assigned to each equilibrium atomic position, whereas in other regions the cells contain many atoms and the nodes need not coincide with atomic sites. CGMD offers a way to reduce the atomistic coordinates to a much smaller set of degrees of freedom associated with the displacement field at the nodes of the CG mesh, and the equations of motion for this mean displacement field. In particular, the energy for the CG system is defined as a constrained ensemble average of the atomistic energy under fixed thermodynamic conditions. The equations of motion are Hamilton's equations for this conserved energy functional plus dissipative terms.

The classical ensemble is required to obey the constraint that the position and momenta of the atoms are consistent with the mean displacement and momentum fields.

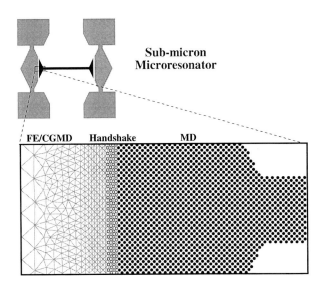

Fig. 8. Schematic diagram showing the mesh refinement used to couple length scales in the micro-resonator. An MD simulation is used in the central region of the device where the strain oscillations are the largest, while a 3D FE or CGMD simulation is used in the periphery where the strain oscillations are small. Both are run simultaneously in lock-step

Let the displacement of atom μ be $\mathbf{u}_\mu = \mathbf{x}_\mu - \mathbf{x}_{\mu 0}$ where $\mathbf{x}_{\mu 0}$ is its equilibrium position. The displacement of mesh node j is a weighted average of the atomic displacements

$$\mathbf{u}_j = \sum_\mu f_{j\mu} \mathbf{u}_\mu, \tag{32}$$

where $f_{j\mu}$ is a weighting function, related to the microscopic analog of FE interpolating functions below. An analogous relation applies to the momenta $\mathbf{p}_\mu$. Since the nodal displacements are fewer or equal to the atomic positions in number, fixing the nodal displacements and momenta does not determine the atomic coordinates entirely. A subspace of phase space remains, corresponding to degrees of freedom that are missing from the mesh. We have defined the CG energy as the average energy of the canonical ensemble on this constrained phase space:

$$\begin{aligned} E(\mathbf{u}_k, \dot{\mathbf{u}}_k) &= \langle H_{\mathrm{MD}} \rangle_{\mathbf{u}_k, \dot{\mathbf{u}}_k} \\ &= \int d\mathbf{x}_\mu d\mathbf{p}_\mu \, H_{\mathrm{MD}} \, e^{-\beta H_{\mathrm{MD}}} \, \Delta / Z, \end{aligned} \tag{33}$$

$$Z(\mathbf{u}_k, \dot{\mathbf{u}}_k) = \int d\mathbf{x}_\mu d\mathbf{p}_\mu \, e^{-\beta H_{\mathrm{MD}}} \, \Delta, \tag{34}$$

$$\Delta = \prod_j \delta\left(\mathbf{u}_j - \sum_\mu \mathbf{u}_\mu f_{j\mu}\right) \delta\left(\dot{\mathbf{u}}_j - \sum_\mu \frac{\mathbf{p}_\mu f_{j\mu}}{m_\mu}\right), \tag{35}$$

where $\beta = 1/(kT)$ is the inverse temperature, Z is the partition function and $\delta(\mathbf{u})$ is a three-dimensional delta function. The delta functions enforce the mean field constraint (32). Note that Latin indices, $j, k, \ldots$, denote mesh nodes and Greek indices, $\mu, \nu, \ldots$, denote atoms. The energy (33) is given below (Eq. (38)).

In CGMD the FE shape functions determine how the underlying forces and masses are weighted in order to compute the mean forces at the nodes. In particular, if $N_j(\mathbf{r})$ is the shape function interpolating the displacement field around node j, then the discrete CGMD analog is $N_{j\mu} = N_j(\mathbf{r}_{\mu 0})$ where $\mathbf{r}_{\mu 0}$ is the equilibrium position of atom μ. Then the weighting function $f_{j\mu}$ is given by

$$f_{j\mu} = \sum_k \left(\sum_v N_{jv} N_{kv} \right)^{-1} N_{k\mu}, \tag{36}$$

where the quantity in the middle is the matrix inverse of NN^T.

Note that when the mesh nodes and the atomic sites are identical, $f_{j\mu} = \delta_{j\mu}$ and the CGMD equations of motion agree with the atomistic equations of motion. As the mesh size increases some short-wavelength degrees of freedom are not supported by the coarse mesh. These degrees of freedom are not neglected entirely, because their thermodynamic average effect has been retained. This approximation is expected to be good provided the system is initially in thermal equilibrium, and changes to the system would only produce adiabatic changes in the missing degrees of freedom. As long as this condition is satisfied, the long wavelength modes may be driven out of equilibrium without problems.

The statistical coarse-graining procedure described above clearly differs substantially from the traditional derivation of finite elements. Part of the reason CGMD is well behaved as the mesh spacing approaches the atomic limit is that no continuum model is used in the derivation. A finite set of coupled ordinary differential equations (ODEs) is transformed into another set of ODEs. There is no need for an intermediate continuum PDE, as in the case of conventional finite elements. The problems that are introduced by such a continuum model result from taking the large N limit (the limit of infinitely many atoms), which involves an asymptotic expansion that is poorly behaved when N is small. CGMD avoids these issues.

The CG Hamiltonian (33) has been computed using standard techniques, both in the linear approximation [3] and retaining the full nonlinear atomistic interactions [4]. Also the dissipative forces have been computed which result from interactions of the coarse-grained displacements with the heat bath of short wavelength degrees of freedom that are missing from the mesh [4]. For the purposes of this Review, we limit our discussion to CGMD for a monatomic harmonic solid. In this case an exact expression for the CGMD energy has been derived. Let the form of the atomistic Hamiltonian be

$$H_{\text{MD}} = \sum_\mu \frac{\mathbf{p}_\mu^2}{2m_\mu} + \sum_{\mu,v} \tfrac{1}{2} \mathbf{u}_\mu \cdot D_{\mu v} \cdot \mathbf{u}_v, \tag{37}$$

where $D_{\mu v}$ is the dynamical matrix. It acts as a tensor on the components of the displacement vector at each site. The CG energy (33) for a monatomic harmonic solid of N atoms coarse grained to N_{node} nodes is given by [3]

$$E(\mathbf{u}_k, \dot{\mathbf{u}}_k) = U_{\text{int}} + \frac{1}{2} \sum_{j,k} \left(M_{jk} \dot{\mathbf{u}}_j \cdot \dot{\mathbf{u}}_k + \mathbf{u}_j \cdot K_{jk} \cdot \mathbf{u}_k \right), \tag{38}$$

where the internal energy of the missing modes is given by

$$U_{\text{int}} = 3(N - N_{\text{node}})kT. \tag{39}$$

This expression indicates the way Eq. (26) was derived. M_{jk} and K_{jk} are defined as follows. The mass matrix is

$$M_{jk} = \left(\sum_{\mu} f_{j\mu} m_\mu^{-1} f_{k\mu} \right)^{-1}$$
$$= m \sum_{\mu} N_{j\mu} N_{k\mu} \quad \text{(monatomic)}, \tag{40}$$

where the second line applies to monatomic solids with atomic mass m. The stiffness matrix is given by

$$K_{jk} = \left(\sum_{\mu\nu} f_{j\mu} D_{\mu\nu}^{-1} f_{k\nu} \right)^{-1}$$
$$= [NN^{\mathrm{T}}(ND^{-1}N^{\mathrm{T}})^{-1}NN^{\mathrm{T}}]_{jk}, \tag{41}$$

where each of the entries on the second line is a matrix. The inverses in (40) and (41) are matrix inverses.

CGMD has also been implemented with a lumped mass approximation. Then the actual mass matrix (40), which is analogous to the distributed mass matrix of finite elements, is replaced with the approximate

$$M_{jk} = \delta_{jk} \sum_{\mu} N_j(\mathbf{x}_{\mu0}) m_\mu \tag{42}$$

which is diagonal (cf. Eq. (22)). This results from using the coarse-graining procedure on the mass rather than on the kinetic energy.

The energy (38) contains terms representing the average kinetic and potential energies, plus the thermal energy term expected from the equipartition theorem for the modes that have been integrated out. This Hamiltonian continues to work for polyatomic solids, in which the optical modes may be coarse grained in various ways to represent different physics.

As with the FE methodology discussed in Section 2.3, the mass and stiffness matrices are computed once at the beginning of the simulation. Then Eqs. (23) and (24) are used to compute the trajectory of the system through time using the velocity-Verlet algorithm (3).

4. Implementation on Parallel Computers

Many of the features of the models described in Sections 2 and 3 have been dictated by the need to perform a meaningful calculation on the current generation of supercomputers. In this section we examine some of these considerations. The crux of this methodology is to allow the study of equilibrium and non-equilibrium dynamics of virtually macroscale systems. In order to take meaningful ensemble averages, or in order to access timescales of use, it is necessary to propagate the system through times on the order of one nanosecond. A typical MD time step is, for chemically bonded systems, approximately 10^{-15} s. Thus, with single user computer allotments typical of current national supercomputer centers, it is necessary to be able to propagate the entire system, on roughly 50 to 100 processors, through one time step in one second of wall clock time. One million time steps then take roughly ten days on a significant, but not unreasonable, fraction of a national machine. Note that this algorithm does not require dedicated use of an entire national machine — hence its utility.

The second point relates to the material chosen to illustrate the methodology: silicon. This is a material for which many proven empirical interatomic potentials [49] already exist. The Stillinger-Weber potential [31] has been used because of its computational simplicity. Silicon is a good choice for other reasons too: its industrial importance for one, its locality of bonding for another. The principles outlined here will work for metallic bonding, but it requires more thought and is the subject of on-going work.

The third point is regarding the choice of tight-binding [34] for illustration of the quantum mechanical coupling. This is dictated mainly by the requirement of computational speed. Tight binding is able to propagate a non-trivial number of atoms under the influence of a quantum mechanical Hamiltonian within one second of wall clock time. Other real-space Hamiltonians could be used in the quantum region, such as the moment methods (bond order methods) advocated by Pettifor and coworkers [50, 51]. TB was chosen for its simplicity and intuitive appeal.

Finally, the implementation of finite elements has been limited to linear elastic theory. Since FE is used only in the far-field region where atoms are perturbed only slightly from equilibrium, there is no need to employ non-linear elasticity theory. The use of a more sophisticated model would allow the FE/MD interface to be moved closer to the region of interest, but as we have explained, it would be appropriate to use a model with improved coupling such as CGMD, before turning to nonlinear FE. The issues of elastic waves and thermodynamics become important before higher order terms in the strain do.

As the discussion in Section 2 implies, and as Figs. 9 and 10 illustrate, each of the three primary algorithms, and the regions of space which they describe, are distributed to different processors. The crack propagation and MEMS simulations were run on the IBM SP2 at the Maui High Performance Computing Center and the Wright-Patterson Major Shared Resource Center. Each FE region was handled by a different processor; the MD region was domain decomposed across many computer nodes and the TB re-

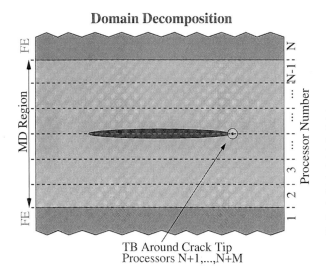

Fig. 9. Illustration of domain decomposition of the pseudo-1D crack system depicting coupling of length scales and the distribution of the computational load across parallel processors. In a typical crack simulation 24 processors were used for molecular dynamics (MD), 8 for tight binding (TB) and 2 for finite elements (FE)

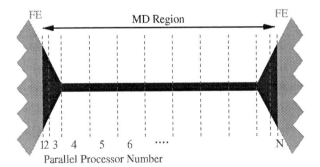

Fig. 10. Illustration of the domain decomposition of the central part of the microresonator system showing coupling of atomistic and continuum length scales, and distribution of the computation across parallel processors. Of the N processors, two are shown designed for the FE regions. In the largest resonator simulation, 40 processors were used for MD and 8 for FE

gion was likewise spread over several processors. The code was written in FORTRAN with MPI for the message passing interface. This kind of code ports to most parallel architecture machines. The advantage of a pseudo-1D topology is that much message passing can be performed using the "shift" operator (MPI_CART_SHIFT). Further, only data within interaction range (defined by the Hamiltonian) need be passed across boundaries (defined by the domain decomposition) between processors. The domain decompositions are shown in Fig. 9 for the fracture system and in Fig. 10 for the MEMS resonator.

5. Applications

5.1 Fracture

The first application we discuss is the rapid brittle fracture of a silicon slab flawed by a microcrack at its center and under uniaxial tension as simulated by Abraham et al. [1]. This example illustrates and validates the hybrid scheme. Figure 1 shows the geometrical decomposition of the silicon slab into the five different dynamic regions of the simulation. The MD region was spatially domain decomposed onto 24 processors. Each FE region was handled by its own processor. The path of the crack was tracked and the center of the TB region was placed at the apex of the crack. This is where bond breaking occurs. It is the region that governs the kinetics of the crack propagation process. For the extended regions of bond rupture (see Fig. 1), a "clover leaf" of eight overlapping TB regions was used, each being cylindrical and assigned to a different processor. The exposed notch faces were xz-planes with (100) faces, with the notch pointed in the $\langle 010 \rangle$ direction. There were 258048 mesh points in each FE region, 1032192 atoms in the MD region, and around 280 unique atoms in the TB region. Each of the eight TB regions is a cylinder with radius of 5.43 Å in the yz-plane with an additional monolayer of silogens to terminate the surface. The lengths of the MD region were 10.9 Å (the slab thickness and periodic), 521 Å (before the pull, in the direction of pull), and 3.649 Å (the primary direction of propagation and periodic). The full pull-length of the combined FE and MD system was 5.602 Å. The entire system including the FE represented 11093376 atoms. The wall clock time for a TB force update was 1.5 s, that for the MD update was 1.8 s and that of the FE was 0.7 s. The size of the FE region could be doubled in order to accomplish complete computational load balancing but without any sacrifice of wall-clock time. The TB region was relocated every 10 time steps.

The slab was initialized at zero temperature, and a constant strain rate was imposed on the outermost FE boundaries defining the opposing horizontal faces of the slab.

Fig. 11. Stress waves from the crack tip propagating through the slab using a finely tuned potential energy color scale after the asymptotic crack speed has been achieved (see Ref. [1]). The figure is shaded according to the stress, with the highest stress at the crack tip, and the lowest stress in the region above and below the center of the crack. Additional low stress regions lie near the plane of the crack but well ahead of it in each direction. The dashed white lines indicate the FE/MD interface. Note the continuity of the stress waves as they cross the interface

Further, a linear velocity gradient was applied within the slab which increased the internal strain with time. The solid failed at the notch tip when it had been stretched by about 1.5%. The imposed strain rate was set to zero at the onset of crack motion. The propagating cracks rapidly achieved a limiting speed of 2770 m/s, equal to 85% of the Rayleigh speed, the sound speed of the solid silicon surface. The distance traveled by the crack as a function of time is very similar regardless of whether MD or TB is used at the crack tip, one indication that the handshaking between the MD region and the TB region was reliable.

A more powerful signature of seamless coupling, one which represents a validation of the method, is depicted in Fig. 11. We note that stress waves pass from the MD region to the FE regions with no visible reflection at the FE/MD interface. Further, there are no obvious discontinuities at the MD/TB interface; this observation remains true even at higher spatial magnifications. Thus the elastic statics and the relevant elastic dynamics confirm that the FE/MD/TB coupling is effectively seamless.

5.2 MEMS

We have simulated micro-resonators over a range of sizes, defect concentrations and temperatures, for comparison. All of the devices have the same geometry and aspect ratio, 25:2:1. The initial configuration of the atoms is taken to be a single crystal of silicon (with some small fraction of the atoms removed at random, if vacancies are to be modeled). To date, the simulations have only utilized the FE/MD coupling of the FE/MD/TB CLS model. The TB coupling would be necessary for a refined model of the behavior of defects, but empirical potentials have proven sufficient to model the principal phenomena of interest: atomistic surface effects.

The largest device that has been simulated consists of a $65 \times 130 \times 1629$ Å^3 bar of silicon with 476012 atoms, with additional material anchoring the bar to the substrate. The bar is taken to have (100) faces. The MD region extends out part of the way into the anchoring tabs at the end of the resonant bar, as shown in Fig. 10. The additional

MD region consists of 564480 atoms at each end, for a total of 1604972 MD atoms. The MD region was distributed across 40 processors. This was coupled to a FE region consisting of a total of 226448 nodes distributed across 8 processors. The mesh grew from the atomic scale at the interface with the MD region, to a cell volume 200 times as large in the far periphery. The mesh was generated based on a 3D generalization of the Sloan mesh generation algorithm [52] appropriate for thin plates.

The motion of the resonator is simulated as follows. The system is brought to thermal equilibrium in 100000 time steps in its undeflected state. Then the normal mode of interest (typically the fundamental flexural mode) is found through a calculation in a 100000 node pure FE system. In particular, the eigenvector is found through inverse iteration. This system uses the CLS FE mesh in the far field, but the mesh is not refined down to the atomic scale. Note that temperature enters into this calculation only through the temperature dependence of the elastic moduli. Then the FE/MD CLS resonator is deflected according to the computed normal mode, using the interpolated displacement field with smoothing, and released. Once released, the thermostat is turned off, and no further energy is put into the system.

Various properties of the resonator have been studied, with a focus on nanoscale phenomena. The Young's modulus has been computed as a function of size for temperature ranging from room temperature down to cryogenic temperatures ($T = 10$ K). It decreases from the bulk value with decreasing system size due to a nanoscale surface effect [7, 53]. This atomistic effect is clearly evident in the simulations for devices less than 0.2 μm in length, even for a single crystal device at cryogenic temperatures. Apart from the surfaces, these devices are essentially a perfect crystal, but it is the surface relaxation which produces deviations from the bulk behavior. Changes in the Young's modulus, E, are accessible in experiment since the resonant frequency of the oscillator is proportional to $E^{1/2}$.

Figure 12 shows how the oscillator rings as a function of time when plucked in flexural mode. Note that relatively large deflections of the resonator are possible, as great as 0.2%, due to the increased compliance of the microscopic devices. The response of the

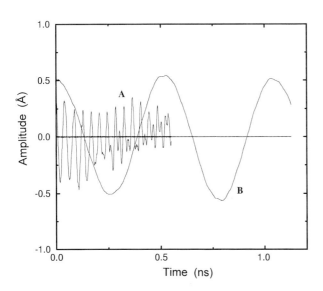

Fig. 12. Plot of the oscillations of two microresonators with different sizes but the same aspect ratio at room temperature. The smaller resonator (A) was 0.017 μm long, and the larger one (B) was 0.14 μm long. The degeneration of the oscillations in A is due to anharmonic mode mixing at the surface of the resonator

oscillator at 300 K shows marked effects of anharmonicity [53]. There is a pronounced frequency doubling effect in smaller oscillators (not shown), and even in the first few periods of the larger oscillator there are clear departures from a sinusoidal oscillation. In studies of smaller devices, the mode mixing is very apparent in the Fourier transform of the oscillations [7, 53]. A significant amount of the first harmonic (as well as a bit of the second) quickly mixes into the spectrum. This is not the case for an identical simulation run at $T = 10$ K, where only the fundamental mode is present (not shown) [53].

We have used simulations such as these to calculate the quality factor, Q, for the various resonators [7, 54]. The simulation time for the large system captures too few oscillations for a direct computation of the Q-value, but a scaling analysis together with a fit to the dominant dissipative processes enables us to estimate this Q-value as well. The MEMS results are presented in greater detail elsewhere [5 to 9, 54].

5.3 CGMD

A variety of calculations have been performed with CGMD in order to validate its effectiveness [3, 4]. It has been shown to provide the lattice statics necessary to represent the long-range strain fields that extend into the FE region. It has also been shown to offer non-pathological lattice dynamics through both improved elastic wave dispersion and reduced scattering off the FE/MD interface. These and other tests of CGMD are described in detail in Ref. [4].

The CGMD phonon spectrum offers a good first test of the model, and we focus on it here. Consider a regular CG mesh, but not necessarily one that is commensurate with the underlying MD lattice; i.e. the nodes do not need to occupy the equilibrium position of atoms. This mesh supports elastic waves with wavelengths greater than the mesh spacing. The spectrum of these waves can be compared to the spectrum given by MD. That is, the energy of the MD elastic waves and the energy of the CGMD elastic waves can be compared at the same wavelength. Similarly, this can be done for finite elements and any other model implemented on the mesh.

It was shown in Ref. [3] that the CGMD phonon spectrum is closer to the true spectrum than that of FE for a one-dimensional chain of atoms with nearest-neighbor interactions. These results carry over to the case of the spectra in the 3D crystal lattice of solid argon [4], and we review these results. Argon forms an f.c.c. solid which is well described by the Lennard-Jones potential

$$V = 4\varepsilon[(\sigma/r)^{12} - (\sigma/r)^6], \tag{43}$$

where $\varepsilon = 1.63 \times 10^{-21}$ J and $\sigma = 3.44$ Å [55]. The corresponding elastic constants are given by $C_{11} = 105.3\varepsilon/\sigma^3$ and $C_{12} = C_{44} = 60.18\varepsilon/\sigma^3$.

We have used the four-fold symmetrized f.c.c. interpolation functions described in Ref. [4] for both CGMD and FE. In their unsymmetrized form, these interpolation functions are a product of one-dimensional linear interpolation functions:

$$N_j(\mathbf{x}) = \prod \tilde{N}(x_a - x_{ja}), \tag{44}$$

$$\tilde{N}(x) = \begin{cases} 1 - |x/L_+| & x > 0 \\ 1 - |x/L_-| & x < 0 \end{cases}, \tag{45}$$

where $L_\pm$ is the internode spacing to either side of node $\mathbf{x}_j$ in the appropriate dimension.

Concurrent Coupling of Length Scales in Solid State Systems

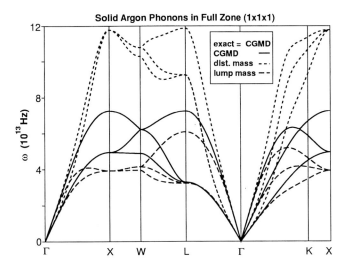

Fig. 13. The phonon spectra for solid argon computed on a mesh in the atomic limit: one atom per node. The CGMD spectrum agrees with the MD spectrum exactly in this limit. The two FE spectra show significant error, but the lumped mass spectrum is better than the distributed mass spectrum, as expected

The spectrum has been computed in each case from secular equation that results from the force equation at a given value of **k**. We consider the spectra at two levels of coarse-graining: the atomic limit ($1 \times 1 \times 1$ or no coarse-graining) and a case approaching the continuum limit ($32 \times 32 \times 32$). The numbers $n_x \times n_y \times n_z$ indicate the number of atoms within a CG cell in each direction. The results are shown in Figs. 13 and 14, respectively.

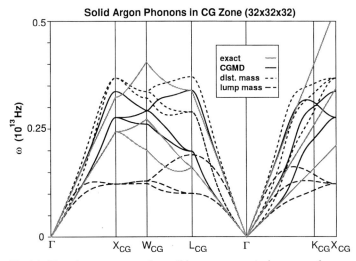

Fig. 14. The phonon spectra for solid argon computed on a mesh approaching the continuum limit: each cell consists of 32768 ($32 \times 32 \times 32$) atoms. The CGMD spectrum agrees with the MD spectrum better than the two FE spectra in this limit as well. Here finite elements with a distributed mass matrix performs somewhat better than the lumped mass approximation, as expected

Consider the spectra in the atomic limit shown in Fig. 13. The CGMD spectrum agrees precisely with the true spectrum. Of the two FE spectra, the lumped mass spectrum is closer to the true spectrum. This is sensible because the mass is localized to the nodes in the atomic limit, since each node represents one atom. Overall, the lumped mass frequencies are lower than the true frequencies, whereas the distributed mass frequencies are higher. This remains true regardless of the level of coarse-graining.

In the continuum limit shown in Fig. 14, the CGMD spectrum no longer agrees exactly with the true spectrum, but it is still a better approximation than either FE spectrum. It is clear that in the continuum limit, the distributed mass produces the better spectrum of the two finite element cases. This again makes sense, because the mass is becoming more evenly spread throughout the CG cell. Still, the CGMD spectrum is significantly better than the distributed mass FE spectrum.

It is interesting to note that all of the spectra agree near the Γ point ($\mathbf{k} = 0$). This is the long wavelength regime where continuum elastic theory is reliable. The fact that all of the spectra agree here indicates that CGMD is finding the correct elastic constants — they determine the slope in the linear dispersion regime.

It is not obvious from the figures, but even in the long wavelength limit, the CGMD spectrum is better than either of the two FE spectra. The relative error for the CGMD spectrum is of order $\mathcal{O}(k^4)$, whereas it is $\mathcal{O}(k^2)$ for the two FE cases. An order $\mathcal{O}(k^2)$ relative error is the naive expectation, since the phonon dispersion relation is linear with the leading corrections of order $\mathcal{O}(k^3)$ due to symmetry. The higher order error for CGMD is due to a subtle cancellation between the contributions from the kinetic and potential energies.

Many other tests of CGMD have been performed, and it has proven to be reliable in lattice statics, lattice dynamics, thermodynamics and non-equilibrium systems. At this point, the implementation of CGMD varies from system to system, because different approximations are made according to the physics that must be represented. Certainly one advantage of an effective model is that these approximations can be controlled. The considerations involved in implementations of CGMD for various systems are discussed elsewhere [4].

6. Conclusions

We have described two algorithms which successfully couple length scales. They are both finite temperature, dynamic, parallel algorithms, and they are seamless in the sense that the effect of the interface cannot be detected in the observables of interest. The general approach is applicable to many material types. The applications of such a methodology will be many and varied. The work described above represents an illustration of what is possible, but much still remains to be done. It is clear that in each of three areas additional work is needed: the identification of new multiscale systems, the extension of multiscale techniques to other applications, and the development of tests to validate their effectiveness. We now discuss some of these possibilities.

Clearly many of the most important multiscale systems are metallic, but substantial further development will be required to treat metals since (a) the bonds are less localized than in silicon, (b) termination of the surface "dangling bonds" of the active TB region with monovalent species is no longer a good approximation, and (c) the ubiquitous dislocations in the real system would quickly cross the atomistic/continuum interface, but they are not supported in conventional finite elements. As we have said, the FE/MD/TB methodology is not tied to TB in particular: other fast quantum mechanical formulations, as long as they meet the speed criterion (one second of wall clock time per time step), may be more appropriate. Indeed, as computational hardware performance improves, the choice of which quantum mechanical scheme to place at the heart of this hybrid scheme will almost certainly change. An example might be the kinetic energy functional [56] local density approximation methods advocated by Carter, Madden and coworkers [57] which require no diagonalization and which parallelize efficiently. Such schemes would require no "clover-leaf" overlapping of quantum mechanical regions as has been, out of necessity, implemented by Abraham et al. A further difficulty for metals involves thermal conductivity and dissipation through mixing of electronic states at the Fermi level; something that an empirical potential cannot capture. Perhaps the empirical potential region will have to be augmented with auxiliary degrees of freedom to enable correct handshaking conditions for this added level of complexity. An effective theory, the analog of CGMD [4] for the MD/TB coupling, would begin to address these issues. Some steps have been taken in this direction. In particular, the development of bond-order potentials by Pettifor and coworkers [50] has shown that empirical potentials can be extracted from quantum mechanical models, but this has yet to be implemented in a seamless, concurrent algorithm.

The handshaking between each neighboring region works well under the conditions of the crack and MEMS examples. We expect long wavelength phonons to propagate with minimal back scattering through the FE/MD interface. However, in systems where significant short wavelength energy is emitted from the central region, we expect that the present handshaking methodology will have to be augmented. The reason is that the shorter wavelength, high frequency vibrations cannot be supported by the larger mesh spacings inherent in the outer part of the FE region and will be scattered back into the MD region. This pathology may be mitigated by adding a random and dissipative heat bath to the FE degrees of freedom. Development of CGMD beyond the linear theory [4] has begun to address these effects through introduction of generalized Nose-Hoover chain thermostats [58] to the FE mesh and specification of the leading non-linear terms for the handshaking region. Indeed, coarse graining provides a natural means to go beyond linear elasticity so that (a) dissipation occurs naturally and (b) constant pressure algorithms in the far-field regions are viable.

The present algorithm dynamically tracks the crack tip with a TB region, assuming that there is one region of breaking bonds. Brittle fracture in silicon is one such example in which it is clear where the TB region should be placed. In the case of dislocation generation at the tip, something indicative of ductility and characteristic of metals, the distribution of computational power must be dynamic, with processors allocated to these areas as needed. Dislocations also pose the challenge of additional structure between the nanoscale and the continuum. Coupling between an efficient representation of this mesoscale and the nanoscale atomistics is an important open problem. This is the subject of on-going work.

In closing, we make the philosophical observation that the algorithms we have described link not only length scales but also disciplines. The TB region focuses on breaking bonds – the traditional realm of chemistry. The MD region describes the statistical mechanics of the system – the forte of physics. And the FE region represents the macroscopic stresses and strains – the historic domain of engineers. This marriage of disciplines and its concomitant dissolution of traditional barriers may well prove to be the true power of this approach.

Acknowledgements J.Q.B. wishes to acknowledge the contributions of Farid Abraham, Tim Kaxiras and Noam Bernstein in developing the crack propagation simulation. R.E.R. would like to thank Daryl Hess and Noam Bernstein for stimulating discussions.

We acknowledge support from ONR and DARPA, and the DoD High Performance Computing Modernization Office in the form of a "Grand Challenge" award of supercomputing resources at the Maui High Performance Computing Center and the Wright-Patterson Major Shared Resource Center.

References

[1] F. F. ABRAHAM, J. Q. BROUGHTON, N. BERNSTEIN, and E. KAXIRAS, Europhys. Lett. **44**, 783 (1998); Comp. Phys. **12**, 538 (1998).
[2] J. Q. BROUGHTON, F. F. ABRAHAM, N. BERNSTEIN, and E. KAXIRAS, to appear in Phys. Rev. B (1999).
[3] R. E. RUDD and J. Q. BROUGHTON, Phys. Rev. B **58**, R5893 (1998).
[4] R. E. RUDD and J. Q. BROUGHTON, submitted to Phys. Rev. B.
[5] D. HESS, N. BERNSTEIN, F. ABRAHAM, and R. E. RUDD, in: Proc. DoD HPCUG Conf., June 1999, Monterey (Ca.).
[6] R. E. RUDD and J. Q. BROUGHTON, in: Proc. Modeling and Simulation of Microsystems, Santa Clara (CA), Computational Publications, Boston 1998 (p. 287).
[7] R. E. RUDD and J. Q. BROUGHTON, in: Proc. Design, Testing and Microfabrication of MEMS, Paris (France), Proc. SPIE **3680**, 104 (1999).
[8] R. E. RUDD and J. Q. BROUGHTON, J. Mod. Sim. Microsys. **1**, 29 (1999).
[9] R. E. RUDD and J. Q. BROUGHTON, in: Proc. DoD HPCUG Conf., June 1998, Houston (Texas).
[10] E. P. WIGNER, Commun. Pure Appl. Math. **13**, 1 (1960).
[11] E. CLEMENTI, Phil. Trans. Roy. Soc. **A326**, 445 (1988).
[12] H. ELBERN, Atmos. Env. **31**, 3561 (1997).
[13] A. F. VOTER, Phys. Rev. Lett. **78**, 3908 (1997).
[14] T. X. NGUYEN, C. K. OH, R. S. SINKOVITS, J. D. ANDERSON, JR., and E. S. ORAN, AIAA J. **35**, 1486 (1997).
[15] J. R. BANAVAR, Continuum Deductions from Molecular Hydrodynamics, invited talk at: Hybrid Computational Methods for Multiscale Modeling of Materials, Workshop, May 12–14, 1999, NIST Gaithersburg (MD), USA.
M. VERGELES, P. KEBLINSKI, J. KOPLIK, and J. R. BANAVAR, Phys. Rev. E **53**, 4852 (1996).
[16] S. KUMAR, D. H. REICH, and M. O. ROBBINS, Phys. Rev. E **52**, R5776 (1995).
[17] D. JACQMIN, Calculation of Two-Phase Navier-Stokes Flows Using Phase-Field Modeling, invited talk at: Hybrid Computational Methods for Multiscale Modeling of Materials, Workshop, May 12–14, 1999, NIST Gaithersburg (MD), USA.
J. R. L. SKARDA, D. JACQMIN, and F. E. MCCAUGHAN, J. Fluid Mech. **366**, 109 (1998).
[18] A. L. GARCIA, J. B. BELL, W. Y. CRUTCHFIELD, and B. J. ALDER, J. Comp. Phys. **154**, 134 (1999).
[19] H. RAFII-TABAR, L. HUA, and M. CROSS, J. Phys.: Condensed Matter **10**, 2375 (1998).
[20] K. G. WILSON and J. B. KOGUT, Phys. Rep. **12c**, 77 (1974).
J. ZINN-JUSTIN, Quantum Field Theory and Critical Phenomena, Clarendon Press, Oxford 1989.
[21] E. B. TADMOR, M. ORTIZ, and R. PHILLIPS, Phil. Mag. **A73**, 1529 (1996).
[22] V. B. SHENOY, R. MILLER, E. B. TADMOR, R. PHILLIPS, and M. ORTIZ, Phys. Rev. Lett. **80**, 742 (1998).
R. MILLER, M. ORTIZ, R. PHILLIPS, V. SHENOY, and E. B. TADMOR, Eng. Frac. Mech. **61**, 427 (1998).
V. B. SHENOY, R. MILLER, E. B. TADMOR, D. RODNEY, R. PHILLIPS, and M. ORTIZ, J. Mech. Phys. Solids **47**, 611 (1999).
D. RODNEY and R. PHILLIPS, Phys. Rev. Lett. **82**, 1704 (1999).
[23] E. KAXIRAS, in preparation.

[24] B. L. HOLIAN and R. RAVELO, Phys. Rev. B **51**, 11275 (1995).
[25] F. F. ABRAHAM, D. BRODBECK, R. RAFEY, and W. E. RUDGE, Phys. Rev. Lett. **73**, 272 (1994).
[26] W. D. MCCOMB, The Physics of Fluid Turbulence, Clarendon Press, Oxford 1990, and references therein.
[27] F. F. ABRAHAM and J. Q. BROUGHTON, Comp. Mater. Sci. **10**, 1 (1998).
[28] F. F. ABRAHAM, D. SCHNEIDER, B. LAND, D. LIFKA, J. SKOVIRA, J. GERNER, and M. ROSENDRANTZ, J. Mech. Phys. Solids **45**, 1461 (1997).
[29] M. E. BACHLECHNER, A. OMELTCHENKO, A. NAKANO, R. K. KALIA, P. VASHISHTA, I. EBBSJO, A. MADHUKAR, and P. MESSINA, Appl. Phys. Lett. **72**, 1969 (1998).
[30] B. L. HOLIAN, P. S. LOMDAHL, and S. J. ZHOU, Physica **A240**, 340 (1997).
[31] F. H. STILLINGER and T. A. WEBER, Phys. Rev. B **31**, 5262 (1985).
[32] A. NAKANO, B. LINGSONG, P. VASHISHTA, and R. K. KALIA, Phys. Rev. B **49**, 9441 (1994).
[33] M. TUCKERMAN, B. J. BERNE, and G. J. MARTYNA, J. Chem. Phys. **97**, 1990 (1992).
[34] W. M. C. FOULKES and R. HAYDOCK, Phys. Rev. B **39**, 12520 (1989).
[35] N. BERNSTEIN and E. KAXIRAS, Phys. Rev. B **56**, 10488 (1997).
[36] F. S. KHAN and J. Q. BROUGHTON, Phys. Rev. B **39**, 3688 (1989).
[37] X-P. LI, R. W. NUNES, and D. VANDERBILT, Phys. Rev. B **47**, 10891 (1993).
[38] M. S. DAW, Phys. Rev. B **47**, 10895 (1993).
[39] P. ORDEJON, D. A. DRABOLD, R. M. MARTIN, and M. P. GRUMBACH, Phys. Rev. B **51**, 1456 (1995).
[40] F. MAURI, G. GALLI, and R. CAR, Phys. Rev. B **47**, 9973 (1993).
[41] C. S. JAYANTHI, S. Y. WU, J. COCKS, Z. L. XIE, M. MENON, and G. YANG, Phys. Rev. B **57**, 3799 (1998).
[42] T. J. R. HUGHES, The Finite Element Method, Prentice-Hall, Englewood Cliffs (NJ) 1987.
[43] I. BABUSKA, J. E. FLAHERTY, W.D. HENSHAW, J.E. HOPCROFT, J.E. OLIGER, and T. TEZDUYAR (Eds.), Modeling, Mesh Generation, and Adaptive Numerical Methods for Partial Differential Equations, in: The IMA Volumes in Mathematics and its Applications, Vol. 75, Springer-Verlag, Berlin/Heidelberg/New York 1995.
[44] K. H. HUEBNER, E. A. THORNTON, and T. G. BYROM, Finite Element Method for Engineers, 3rd ed., Wiley, New York 1995 (p. 228).
[45] H. BALAMANE, T. HALICIOGLU, and W. A. TILLER, Phys. Rev. B **46**, 2250 (1992).
[46] J. R. RAY, Comp. Phys. Rep. **8**, 109 (1988).
[47] O. C. ZIENKIEWICZ and R. L. TAYLOR, The Finite Element Method, Vol. 2, 4th ed., McGraw-Hill, London 1991.
[48] S. KOHLHOFF, P. GUMBSCH, and H. F. FISCHMEISTER, Phil. Mag. **A64**, 851 (1991).
[49] M. Z. BAZANT, E. KAXIRAS, and J. F. JUSTO, Phys. Rev. B **56**, 8542 (1997).
[50] P. ALINAGHIAN, P. GUMBSCH, A. J. SKINNER, and D. G. PETTIFOR, J. Phys.: Condensed Matter **5**, 5795 (1993).
[51] M. AOKI, Phys. Rev. Lett. **71**, 3842 (1993).
[52] S. W. SLOAN, Comp. Struct. **47**, 441 (1993).
[53] J. Q. BROUGHTON, C. A. MELI, P. VASHISHTA, and R. K. KALIA, Phys. Rev. B **56**, 611 (1997).
[54] R. E. RUDD and J. Q. BROUGHTON, in preparation.
[55] J. GRINDLAY and R. HOWARD, in: Lattice Dynamics, Proc. Internat. Conf. Copenhagen (Denmark), Ed. R. F. WALLIS, Pergamon Press, Oxford 1965 (p. 129).
[56] L-W. WANG and M. P. TETER, Phys. Rev. B **45**, 13196 (1992).
[57] S. WATSON, B. J. JESSON, E. A. CARTER, and P. A. MADDEN, Europhys. Lett. **41**, 37 (1998).
[58] G. J. MARTYNA, M. L. KLEIN, and M. TUCKERMAN, J. Chem. Phys. **97**, 2635 (1992).

phys. stat. sol. (b) **217**, 293 (2000)

Subject classification: 63.10.+a; 71.24.+q; 78.30.−j

Electric Fields in Electronic Structure Calculations: Electric Polarizabilities and IR and Raman Spectra from First Principles

K. JACKSON

Department of Physics, Central Michigan University, Mt. Pleasant, MI 48859, USA

(Received August 10, 1999)

By including an external electric field in first-principles electronic structure calculations we are able to explore a variety of properties of atomic clusters. In this paper we briefly discuss the technique and present several recent examples of its use, including the polarizabilities of Si clusters, the IR spectra of hydrogenated Fe particles, and the Raman spectra of chalcogenide glasses. The common feature in all these applications is that they involve predictions of directly observable properties. The value of the electric field calculations is thus that they produce a closer link between theory and experimental measurements.

1. Introduction

Computational studies of materials have become increasingly useful over the past decade due to the development of accurate methods for modeling materials properties and the creation of ever-faster computers. Among the most successful approaches are those based on the density functional theory (DFT) [1]. These combine high accuracy and relative computational efficiency without the use of empirical parameters. A large and growing body of results demonstrates that the theory gives reliable predictions of a number of properties in both the Local Density Approximation (LDA) [2,3] and the Generalized Gradient Approximation (GGA) [4]. Examples include equilibrium atomic structures, vibrational frequencies, and ionization energies and electron affinities [5].

Despite the many successes of DFT methods, it is sometimes difficult to couple theory to experiment. Determining the atomic structure of aperiodic systems presents a good example. In systems like glasses or atomic clusters, for which there are no direct experimental methods that can be used to specify the arrangement of the atoms, vibrational spectroscopy is often used to indirectly probe the structure. In these cases the IR and Raman spectra of the systems can serve as fingerprints for the atomic structure, but the fingerprints must be matched against a well-characterized database for this process of identification to be effective. While DFT calculations can yield accurate vibrational frequencies for a given structure, the fingerprint is incomplete without also having the corresponding spectral intensities of the vibrational modes. To forge an effective partnership between theory and experiment, theory must calculate what experiment actually measures.

In this paper we show how the ability to include external electric fields in DFT calculations can help bridge the gap between theory and experiment. We briefly discuss our method for including an external electric field in DFT calculations and how the technique can be used to calculate IR and Raman spectral strengths [6,7]. We then present three applications: the electric polarizabilities of Si clusters [8], the IR spectrum of $Fe_{13}H_{14}$ [9], and the Raman-active modes of the chalcogenide glass $GeSe_2$ [10].

In the next section we give an overview of the method we use to compute IR and Raman spectral intensities [6] and present the results of benchmark calculations on small silicon clusters [7]. The reader is referred to other papers in this volume for further discussion of the Gaussian-orbital-based DFT methodology. In the following section we turn to the applications. We conclude the paper with a brief section of conclusions.

2. Computational Method

In the DFT approach, the ground-state energy of a many-electron system is written as a functional of the electronic charge density. Representing the density as a sum of one-electron orbital densities and applying the variational method, one arrives at the one-electron Schrödinger-like Kohn-Sham equation [2] that must be solved self-consistently for the orbitals ϕ_i,

$$(-\nabla^2/2 + V_{Nuc} + V_C + V_{xc}) \phi_i = \epsilon_i \phi_i. \qquad (1)$$

Here V_{Nuc} represents the Coulomb attraction of the nuclei, V_C the Coulomb repulsion of the electrons, and V_{xc} the approximate exchange-correlation potential. Most of our calculations are done in the local density approximation (LDA) using the form of V_{xc} developed by Perdew, Burke, and Ernzerhof [3]. The self-consistent orbitals computed from Eq. (1) are used to compute the total energy for the system.

In our calculations Gaussian basis sets are used to represent the electron orbitals. A powerful numerical integration scheme [11] is used to obtain accurate total energies and atomic forces. The forces are the derivatives of the total energy with respect to the nuclear coordinates and can be computed analytically from the self-consistent orbitals [12]. The forces are essential ingredients in gradient-based search algorithms to find the minimum energy configuration of a cluster of atoms. Accurate forces are also required for computing the vibrational modes of a cluster. We compute the modes in the harmonic approximation, using finite differences of the atomic forces to build the dynamical matrix [6]. The vibrational frequencies and corresponding eigenvectors are obtained by diagonalizing the dynamical matrix.

The key feature of the calculations discussed in this paper is the introduction of a spatially uniform, static external electric field $\mathbf{G}$, into the calculation. The field adds a term to the potential seen by the electrons in the Kohn-Sham equations

$$(-\nabla^2/2 + V_{Nuc} + V_C + V_{xc} + V_{ext}) \phi_i = \epsilon_i \phi_i, \qquad (2)$$

where

$$V_{ext} = -e\mathbf{r} \cdot \mathbf{G}. \qquad (3)$$

The orbitals obtained from solving Eq. (2) are used to construct the total energy $E(\mathbf{G})$ as a function of the external field. Including an external field is straightforward in a finite cluster calculation. At worst, V_{ext} breaks the point group symmetry of the cluster, which can add to the computer time required for the calculation. By contrast, V_{ext} breaks the translational symmetry in a calculation based on periodic boundary conditions. Since translational symmetry is the basic assumption on which such methods are built, V_{ext} cannot be simply introduced into periodic calculations.

The cluster electric dipole moment μ and polarizability tensor α_{ij} can be defined in terms of derivatives of the total energy with respect to the external field,

$$\mu_i = -\left.\frac{\partial E}{\partial G_i}\right|_{G=0} \tag{4}$$

and

$$\alpha_{ij} = -\frac{\partial^2 E}{\partial G_i \partial G_j} = \left.\frac{\partial \mu_i}{\partial G_j}\right|_{G=0}. \tag{5}$$

The field derivatives can be evaluated by finite differences (see below). Gas phase measurements of α typically determine an orientationally averaged polarizability, which is proportional to the trace of α_{ij}.

The IR intensity and Raman activity of a vibrational mode depend, respectively, on how the dipole moment and polarizability change with the corresponding atomic oscillations. To lowest order, the spectral strengths are proportional to the derivatives of the dipole moment and polarizability with respect to the vibrational normal modes of the cluster, evaluated at the equilibrium geometry. For example, the IR intensity of the i-th vibrational mode is given by [13]

$$I_i^{IR} = \frac{N\pi}{3c}\left[\frac{d\mu}{dQ_i}\right]^2. \tag{6}$$

Here, N is the number of clusters per unit volume, μ is the cluster dipole moment, and Q_i is the normal coordinate corresponding to the i-th mode. The units of I_i^{IR} are $(D/\text{Å})^2/\text{amu}$.

The differential cross section for Raman scattering in the i-th mode is [14]

$$\left(\frac{d\sigma}{d\Omega}\right)_i = \frac{\omega^4}{c^4}\frac{h}{4\pi\omega_i}\frac{I^{Ram}}{45}, \tag{7}$$

where

$$I^{Ram} = 45(\alpha')^2 + 7(\beta')^2, \tag{8}$$

$$\alpha' = \frac{1}{3}\left(\frac{d\alpha_{xx}}{dQ_i} + \frac{d\alpha_{yy}}{dQ_i} + \frac{d\alpha_{zz}}{dQ_i}\right), \tag{9}$$

$$\beta' = \frac{1}{2}\left[\left(\frac{d\alpha_{xx}}{dQ_i} - \frac{d\alpha_{yy}}{dQ_i}\right)^2 + \left(\frac{d\alpha_{xx}}{dQ_i} - \frac{d\alpha_{zz}}{dQ_i}\right)^2 + \left(\frac{d\alpha_{yy}}{dQ_i} - \frac{d\alpha_{zz}}{dQ_i}\right)^2\right]$$
$$+ 3\left[\left(\frac{d\alpha_{xy}}{dQ_i}\right)^2 + \left(\frac{d\alpha_{xz}}{dQ_i}\right)^2 + \left(\frac{d\alpha_{yz}}{dQ_i}\right)^2\right]. \tag{10}$$

In these equations ω is the frequency of the scattered radiation, ω_i is the frequency of the i-th vibrational mode, and α is the polarizability tensor. I^{Ram} is the Raman scattering activity, typically given in units of $\text{Å}^4/\text{amu}$. The above expression for I^{Ram} includes averaging over all possible cluster orientations and assumes a backscattered experimental arrangement.

It is interesting to translate the definitions for the IR and Raman spectral strengths given above into more physical terms. For a vibrational mode to be IR-active, the dipole moment of a cluster must change as a result of the displacement of the atoms in the mode. Such a mode for Si$_4$ is illustrated in the left panel of Fig. 1, where the atomic displacements can clearly create a charge separation in the vertical direction. By contrast, a Raman-active mode must change the cluster polarizability. Since the polarizability can be related to the volume of the cluster, modes that change the volume tend to be Raman-active. Such a mode is shown on the right in Fig. 1.

To obtain quantitative values for the IR and Raman spectral strengths, we must compute dipole and polarizability derivatives with respect to the normal mode coordinates. These can be viewed as directional derivatives in the space of $3N$ nuclear coordinates and expressed, using the chain rule, in terms of derivatives with respect to atomic coordinates R_k. For an arbitrary function A

$$\frac{dA}{dQ_i} = \sum_{k=1}^{3N} \frac{\partial A}{\partial R_k} X_{ki}, \tag{11}$$

where X_{ki} is the k-th atomic displacement of the i-th normal mode. The necessary derivatives can be expressed in terms of the atomic forces as follows:

$$\frac{\partial \mu_i}{\partial R_k} = -\frac{\partial^2 E}{\partial G_i \, \partial R_k} = \frac{\partial F_k}{\partial G_i}, \tag{12}$$

$$\frac{\partial \alpha_{ij}}{\partial R_k} = -\frac{\partial^3 E}{\partial G_i \, \partial G_j \, \partial R_k} = \frac{\partial^2 F_k}{\partial G_i \, \partial G_j}, \tag{13}$$

F_k is the calculated force on the k-th atomic coordinate. We obtain the derivatives by taking finite differences of the forces from independent self-consistent calculations with an applied electric field. We use a field strength of $\delta G_i = 0.005$ a.u. to evaluate the derivatives. Fields of this size have been shown to yield well-converged results for the derivative [6,15].

The field strength used in the calculations corresponds to about 2.6×10^9 V/m in SI units. This is much larger than strong laboratory fields that are used, for example, to make polarizability measurements [16]. The latter are about 2×10^7 V/m. Despite the

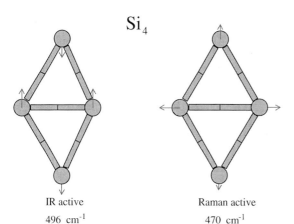

Fig. 1. Atomic displacements for the IR-active and Raman-active modes of Si$_4$. In the IR-active mode, the oscillation of the atoms in the indicated directions creates an oscillating dipole moment for the cluster. In the Raman-active mode, the overall cluster volume, which is loosely related to the cluster polarizability, oscillates with the atoms

fact that the applied field is enormous by laboratory standards, the total energy shifts in the calculations are still very small, typically on the order of 1×10^{-4} Hartree (about 3×10^{-3} eV). Obtaining accurate results for the IR and Raman spectral strengths thus requires good numerical precision in the calculations. We have found that the key ingredients are highly self-consistent wavefunctions based on large basis sets and an accurate scheme for integrating total energies and forces [6,11]. A representative basis set used in calculations for Si clusters had 9 s-type, 8 p-type, and 6 d-type orbitals (9,8,6) on each Si atom, contracted from a set of 15 primitive Gaussians per atom. Extensive tests [17] indicated that the Raman activities are well-converged using this procedure.

The derivatives required for calculating the IR and Raman spectral strengths can all be evaluated using data from a total of 13 electric field calculations, independent of cluster size. For clusters containing 10 or more atoms, this represents a fraction of the computational cost of constructing the dynamical matrix, so that computing IR and Raman intensities adds little additional overhead to the overall calculation.

In an initial application of the DFT-based method, we studied the IR and Raman spectra of small Si clusters [7]. These calculations provided a useful benchmark because both experimental and quantum chemistry (QC) results for the spectra existed for clusters with $N \leq 7$. Thus, we were able to compare the DFT-based technique directly to other results and also to fill in calculated results in cases where they were lacking [17,18].

Table 1 compares the LDA-based Raman spectra for Si_4, Si_6 and Si_7 to the corresponding QC and experimental results [19]. The LDA and QC Raman intensities are both given in relative units, normalized to the same mode. The table shows excellent agreement between both theories and experiment for the positions of the Raman-active

Table 1

Comparison of relative Raman activities calculated within the LDA and by quantum-chemistry techniques (Ref. [19]), along with the observed Raman-active frequencies [19]. The relative units are computed separately for each cluster; absolute activities cannot be inferred when comparing modes from different clusters. Frequencies are in cm^{-1} and activities are in relative units

	LDA		MP2		expt.
	ω	I^{RAM}	ω	I^{RAM}	ω
Si_4	346	1.00	337	1.00	345
	436	0.40	440	0.5	–
	470	3.9	463	5.0	470
Si_6	270	1.0	209	1.0	252
	315	1.3	298	2.6	300
	384	1.0	376	0.6	386
	412	0.8	425	2.0	404
	463	7.9	457	7.7	458
Si_7	301	1.0	300	1.0	289
	347	1.8	339	0.5	340
	347	1.4	346	0.5	340
	362	4.5	352	1.6	358
	448	9.2	441	4.1	435

modes. There are some differences in the calculated intensities, but overall there is good qualitative agreement, and both theories agree on which are the strongest modes. Regarding the accuracy of the DFT intensities in absolute terms, Porezag and Pederson [6] compared calculated intensities against experimentally measured intensities for a small set of molecules. They found typical differences of between 30 and 50%. They also showed that the agreement between DFT and QC intensities depended on the level of theory used in the QC calculations. The differences were smallest for highly correlated calculations [6].

For the calculated IR spectra, shown in Table 2, the LDA- and QC-based spectra are in close agreement in terms of the positions of the IR-active peaks, and these are in within 11 cm^{-1} of the observed frequencies in all cases [7]. The *absolute* IR intensities calculated within the two methods show some differences, however, the QC intensities are roughly a factor of two larger than the LDA intensities for all clusters except Si_5. In that case, the IR intensities calculated by the QC method are much larger than the LDA-based intensities. The QC calculation [20] uses a high level of theory (MP2) to determine relaxed cluster geometries, but then reverts to HF level to compute the IR intensities. Electron correlation effects may therefore be responsible for the differences in the intensities. Li et al. [20], note that the calculated QC bond lengths and vibrational frequencies for Si_5 are particularly sensitive to the treatment of electron correlation, suggesting that the use of HF wavefunctions to compute IR intensities may be a particularly poor approximation in this case. Interestingly, no modes found in the experimental IR spectra of Ref. [20] could be attributed to Si_5. The authors suggested that this could be due to a low concentration of Si_5 clusters in the experiment. (The clusters are not mass-selected before the IR measurements are made. The output of a

Table 2

Comparison of absolute IR intensities calculated within the LDA and by quantum chemistry techniques (Ref. [20]), along with the observed IR-active frequencies [20]. Frequencies ω are in units of cm^{-1} and IR intensities are in units of (D/Å)^2/amu

	LDA		QC		expt.
	ω	IR	ω	IR	ω
Si_3	173	0.09	148	0.21	–
	536	0.80	525	1.41	525
	546	0.20	551	0.47	550.6
Si_4	53	0.12	101	0.05	–
	249	0.03	239	0.19	–
	496	1.34	499	3.85	501
Si_5	174	0.05	160	0.05	–
	412	0.07	382	0.51	–
	444	0.04	436	1.81	–
Si_6	51	0.01	049	0.02	–
	319	0.01	340	0.10	–
	410	0.01	–	–	–
	463	0.84	458	2.07	461
Si_7	240	0.01	249	0.07	–
	430	0.54	421	1.41	422

cluster source is soft-landed onto an inert matrix prior to the IR measurement, producing an array of cluster sizes in the sample.) The LDA-based results suggest a different explanation. All the IR-active modes identified in Ref. [20] have calculated IR intensities ≥ 0.20 (D/Å)^2/amu within the LDA. The largest IR intensity found for Si_5 was 0.07. Thus the LDA results suggest that the Si_5 IR-modes were simply too weak to be seen in the experiment.

The results shown in Tables 1 and 2 show that both the DFT-based and QC methods produce very good agreement with experiment for the small Si clusters. A practical advantage of the DFT approach is its relative computational efficiency. Whereas the complexity of the correlation treatment in the QC methods limits calculations to clusters containing no more than about 10 atoms, the DFT method can treat much larger systems. For example, we have used the method to calculate the spectra of clusters up to Si_{20} using a modest workstation [17,18]. Such calculations may play a key role in identifying the structures of the larger clusters in the future.

3. Applications

3.1 Si cluster polarizabilities

An important goal of research on semiconductor clusters is to understand the evolution of cluster properties as a function of the number of atoms in the cluster. One property of interest is the electric polarizability, which can be viewed as a measure of the metallic nature of the cluster. For bulk silicon, a value for the average polarizability per atom can be obtained from the Clausius-Mossotti relation

$$\alpha = \frac{3}{4\pi}\left(\frac{\epsilon-1}{\epsilon+2}\right) v_{at}, \qquad (14)$$

where α is the polarizability per atom, v_{at} is the volume per Si atom in the Si unit cell, and ϵ is the static dielectric constant of the bulk solid. Taking $\epsilon = 11.8$ and $v_{at} = 19.47$ (the latter from the LDA calculation of Ref. [21]) we obtain a value for the bulk atomic polarizability of 3.64 Å^3/atom. Initial calculations using first-principles methods [15] indicated that the polarizabilities for small Si clusters with $N \leq 10$ were significantly larger than this bulk value, indicating that the clusters are more metallic than the bulk. Recently, new models for cluster structures in the size range from $11 \leq N \leq 20$ have emerged from extensive theoretical searches [18,22]. The new models are consistent with both ion mobility measurements [22] and measured ionization energies [23]. They thus appear to be the cluster structures found in experiment. These models allowed us to track the evolution of α to the intermediate size range [8].

In Fig. 2 we show the variation of α with cluster size for the new clusters. The bulk value of the polarizability has been included on the plot for reference. Note that for clusters larger than $N = 16$ there are two branches to the curve, corresponding to different cluster *shapes*. For these cluster sizes, the searches [22] give two nearly degenerate structures that differ significantly in overall shape. The upper branch corresponds to prolate structures that can be thought of as a stack of cluster subunits along a common axis. The lower branch corresponds to compact structures that have a roughly spherical overall shape. Fig. 3 illustrates the prolate and compact geometries for Si_{20}. The prolate structure is seen to be a stack of two Si_{10} subunits. This stable Si_{10} structure is a basic building block of all the prolate structures [22].

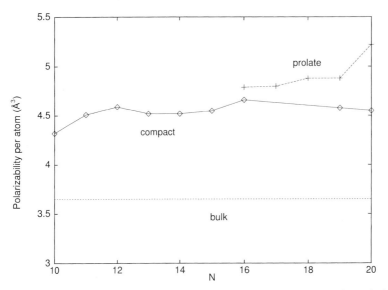

Fig. 2. Calculated average polarizability per atom for Si clusters. The polarizabilities for compact and prolate clusters are compared to the bulk value obtained using the Clausius-Mossotti relation

Fig. 2 suggests the possibility that compact and prolate clusters can be distinguished by their polarizabilities. This is interesting in light of ion mobility experiments [24] showing a shape transition for Si clusters in the intermediate size range. For clusters smaller than $N = 26$, prolate shapes predominate, while for clusters larger than $N = 26$ the compact geometries are favored. The plot shown in Fig. 2 implies that measurements should also show a corresponding decrease in cluster polarizability over this size range.

We found little correlation between α and other properties, such as the electric dipole moment μ or the HOMO-LUMO gap E_g. These properties were found to vary

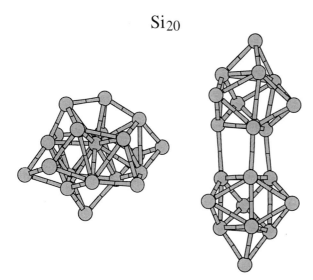

Fig. 3. Comparison of the compact and prolate isomers for Si_{20}. The prolate cluster has a higher average coordination number (larger number of bonds per atom) and slightly longer bonds than the compact cluster

significantly from cluster to cluster across the $10 \leq N \leq 20$ size range, while α can be seen to vary relatively smoothly in Fig. 2. Both μ and E_g appear to be much more sensitive to the local arrangement of the atoms in the cluster [8] than α, which appears to be a measure of a more global property, the cluster shape.

The connection between α and cluster shape can be understood on the basis of the bonding in the two cluster types. The prolate clusters are characterized by higher coordination numbers and longer bond lengths than the compact clusters [8]. Stated differently, the prolate clusters have a larger number of somewhat weaker (and thus longer) bonds than the compact clusters, and the electrons making up those bonds are more easily polarized than the electrons in the stronger bonds of the compact structures.

Experimental measurements of α for small and intermediate-sized Si clusters were published recently [16]. In contrast to our calculations, the measurements suggest that cluster polarizabilities are smaller than the bulk value and approach the bulk value from below with cluster size. There is no evidence of a prolate-compact transition in the experimental data.

The source of the differences between theory and experiment in this case is not clear. It is known that LDA values of the polarizabilities for molecules agree well with experiment. For example, the average polarizabilities calculated by Porezag and Pederson [6] for a small group of hydrocarbons were within 2% of experimental values in all cases. A similar level of agreement should be expected for the Si clusters. Uncertainties in cluster structure cannot be responsible for the magnitude of the differences between theory and experiment, since this is much larger than the compact/prolate difference seen in Fig. 2. The differences may be due to experimental uncertainty, as the error bars quoted in Ref. [16] are very large. While theory and experiment are not in agreement here, it is clear that the ability to directly compare calculated and observed polarizabilities is important and has raised an interesting question that will require further work to settle.

3.2 IR spectrum of $Fe_{13}H_{14}$

Small Fe clusters are known to be reactive toward H_2, with the H_2 undergoing dissociative chemisorption on the cluster surface. The reactivity of the clusters has been found to vary over orders of magnitude as a function of cluster size [25 to 27]. Because of the importance of transition metals in catalysis, it is of interest to understand the nature of the Fe–H bonding in these structures. One approach is to use photodissociation spectroscopy [28]. In experiments in which small Fe_n particles are exposed to molecular hydrogen, H_2 molecules react with the particles to form Fe_nH_m. For a given n and a sufficiently high H_2 concentration, there is a saturation coverage m observed for room temperature clusters [29]. At higher H_2 concentrations (and low temperature), additional H_2 molecules physisorb to the Fe_nH_m particles, forming $Fe_nH_mH_{2p}$ complexes. Photodissociation spectroscopy uses a probe laser to inject energy into the complexes, causing H_2 molecules to be ejected [9]. The probe laser is pulsed and the experiments record the depletion in the number of $Fe_nH_mH_{2p}$ complexes seen with the probe laser on compared to when it is off. The depletion is studied as a function of the wavelength of the probe laser, essentially yielding the IR spectrum of the parent Fe_nH_m cluster. In a recent set of experiments, a tunable CO_2 laser was used to study clusters with $n = 9$

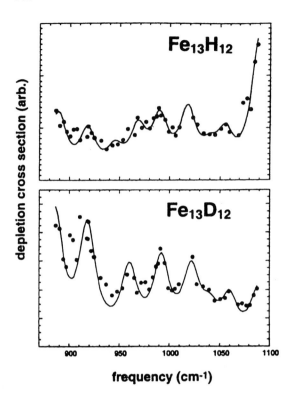

Fig. 4. Photodissociation spectra for $Fe_{13}H_{12}$ and $Fe_{13}D_{12}$ from Ref. [9]

to 20. The laser could be tuned over the range 885 to 1100 cm^{-1}, which nicely brackets the vibrational modes of 1060 and 880 cm^{-1} observed for chemisorbed H atoms on the Fe(110) surface [30].

Fig. 4. shows typical data for the clusters $Fe_{13}H_{12}$ and $Fe_{13}D_{12}$ [9]. Interestingly, both clusters show photodepletion in the laser window. This is surprising, since the larger deuterium mass is expected to shift the frequency of the Fe–D modes down by approximately $1/\sqrt{2}$ relative to the Fe–H modes. For example, the 1060 cm^{-1} mode of Fe(110) would shift to 750 cm^{-1}, well outside the laser window. This raises questions about the origin of the features seen in the spectrum. One possibility is that they are due to modes of the Fe_n cluster core, and not only to the surface Fe–H bonds.

To get a clearer picture of the nature of the vibrational modes in the Fe_nH_m complexes, we used our DFT method to study the $Fe_{13}H_{14}$ cluster shown in Fig. 5 [9]. The detailed structures of the Fe_nH_m clusters are not known, and for a system as large as $Fe_{13}H_{12}$ a systematic search for the equilibrium structure is not practical. We therefore chose to treat the highly symmetric $Fe_{13}H_{14}$ cluster shown in Fig. 5. The Fe core in this model has icosahedral symmetry. Earlier DFT calculations [31] found this to be the lowest in energy among possible high symmetry structures.

Our all-electron calculations used large basis sets for the Fe atoms, featuring a set of 20 even-tempered Gaussians contracted to seven s-type, five p-type, and four d-type orbitals. For the H atoms we used 6 Gaussians contracted to four s-type, three p-type and one d-type orbital. The cluster was found to have a total spin of $S = 7$, corresponding to 14 unpaired spins. Note that the H atoms can be grouped in two classes as shown in the figure: those on twofold sites bridging an Fe-Fe bond ("bridge" atoms) and those on threefold sites ("face" atoms). The Fe–H bond lengths were 1.66 Å for the bridge atoms and 1.76 Å for the face atoms.

In Fig. 6 we show the calculated vibrational density of states (DOS) for $Fe_{13}H_{14}$. The DOS is concentrated in a number of bands which we analyzed by computer animation. The bands between 800 and 1500 cm^{-1} correspond to motion of the H atoms on the cluster surface. All vibrations involving significant motion of the Fe core atoms were

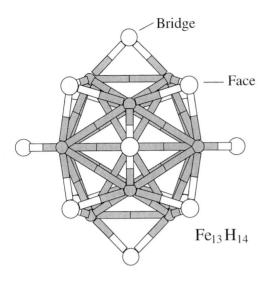

Fig. 5. The $Fe_{13}H_{14}$ cluster. The central core of Fe atoms has icosahedral symmetry. The H atoms are placed on bridge and face sites as shown

confined to the 0 to 400 cm^{-1} range which is not shown. As indicated in the figure, the different bands can be attributed to bond-bending and bond-stretching motion of the face and bridge atoms, respectively. The bond-bending modes feature H atoms moving parallel to the local cluster surface, while in the bond-stretching modes the H atoms move in the direction normal to the surface. The stretching modes are clearly stiffer than the bending modes, and the splittings between the bands corresponding to bridge and face atoms reflect the bonding differences at these sites.

Also shown in Fig. 6 are the IR strengths of the calculated modes. For $Fe_{13}H_{14}$ we find strong IR modes at 850, 1300, and 1460 cm^{-1}, as well as a weak mode at

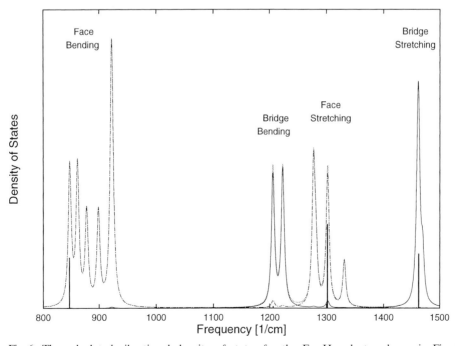

Fig. 6. The calculated vibrational density of states for the $Fe_{13}H_{14}$ cluster shown in Fig. 5. The modes are analyzed in terms of bond-bending and bond-stretching modes of the H atoms on the bridge and face sites as indicated. The IR intensities of active modes are indicated by vertical bars

1205 cm^{-1}. For Fe$_{13}$D$_{14}$, the isotope effect shifts the strong IR-modes to 605, 925, and 1040 cm^{-1}, respectively.

The results of the calculations are in good qualitative agreement with the experiment. The modes of the Fe$_{13}$ core are well-removed from the laser window and we find evidence of strongly IR active modes for both Fe–H and Fe–D in the vicinity of the laser window. In the case of the hydrogenated complexes, the observed modes are bending modes, while for the deuterated complexes they are stretching modes.

The choice of a highly symmetric model for the calculations concentrates the IR intensity into a few modes of the proper symmetry to satisfy IR selection rules. Lowering the cluster symmetry would relax the selection rules and distribute the IR intensity more uniformly across the various bands. Lowering the symmetry would also split degenerate vibrational modes and broaden narrow bands like the one around 1450 cm^{-1}. Both changes would strengthen the agreement between theory and experiment. Another factor that bears on the results is the use of 14 H atoms instead of the experimental saturation coverage of 12 atoms. Reducing the number of H atoms in the model would leave the remaining H atoms somewhat more strongly bound to the core and increase the frequencies of the vibrational modes. For example, in the case of the Fe$_4$H$_6$ and Fe$_4$H$_4$ clusters, where both clusters have a distorted tetrahedral Fe$_4$ core and H atoms occupying bridge sites, we find bond-bending modes in the region 940 to 1000 cm^{-1} for Fe$_4$H$_6$ and bond-stretching modes in the region 1340 to 1493 cm^{-1}. For Fe$_4$H$_4$, the bond-bending modes move up to the range 1047 to 1060 cm^{-1}, while the bond-stretching modes move to 1443 to 1530 cm^{-1}. Both the stretching and bending modes stiffen as the H content decreases. This effect in Fe$_{13}$H$_{14}$ would push the bond-bending modes more squarely into the laser window.

The qualitative agreement between theory and the experimental photodissociation spectroscopy results found for the simple model cluster discussed here suggests the possibility that the photodissociation technique, combined with DFT calculations, could be used to identify cluster structures. As a first step in this direction, it would be interesting to focus on smaller clusters, like Fe$_4$H$_m$, for which theory could be used to systematically search for the minimum energy structure. The calculated spectrum could then be compared quantitatively to experiment.

3.3 The Raman-active modes of GeSe$_2$

Cluster calculations can also be applied to studies of bulk materials if the properties of interest are local, i.e. if they depend only on the local atomic structure and not on long-range effects of the bulk environment. A good example is the Raman spectrum of the chalcogenide glass GeSe$_2$. In addition to some broad bands, the spectrum contains some characteristic peaks at around 200 cm^{-1} [32,33]. The relative sharpness of these peaks implies that they are due to localized vibrational modes in the material that can be accurately treated using cluster calculations.

To study the Raman modes of GeSe$_2$ we constructed cluster models containing structural features of interest. We used H atoms to terminate dangling bonds on the cluster surfaces in order to better model the chemical bonding in the glass. H atoms are useful for this purpose, since the electronegativity of hydrogen is 2.1, close to and intermediate between that of Ge (1.8) and Se (2.4). This means that H atoms will terminate the dangling bonds without a dramatic change in the electron distribution between the Ge and Se atoms. The H atoms are also much lighter than Ge or Se, so vibrational modes involving the H atoms are not expected to mix strongly with Ge-Se modes. For a given model, we relax the positions of all the atoms to the minimum energy configuration and compute the full vibrational spectrum. We then use the technique outlined in Section 2 to compute the IR and Raman intensities for the vibrational modes.

The gross structure of the Raman spectrum of $GeSe_2$ can be understood in terms of the vibrational modes of the tetrahedral $GeSe_4$ building block shown in Fig. 7 [34]. (We have omitted H atoms from all the figures in this section.) To allow reasonable Ge–Se–H bond angles, we lowered the symmetry of the cluster from T_d to D_{2d}. The lower symmetry causes some of the modes that are degenerate in T_d to split slightly, but does not affect the overall form of the spectrum. The T_d-based symmetry labels are used in the figure for convenience.

As shown in Fig. 7, the $GeSe_4$ tetrahedron has vibrational modes at 69 cm^{-1} (E), 201 cm^{-1} (A_1), and 285 cm^{-1} (T_2). The A_1 mode is strongly Raman active, corresponding to the symmetric stretching of all four Ge-Se bonds. This mode is responsible for the Raman peaks near 200 cm^{-1} in the glass spectrum. The E modes are weakly Raman-active, bond-bending modes. Combinations of these modes are responsible for the broad, weak band centered at about 70 cm^{-1} in the experimental spectrum. The T_2 modes feature two bonds stretching while two bonds shrink. This asymmetric mode is Raman-forbidden in T_d symmetry, but gains a very small activity in the lower D_{2d} symmetry. Combinations of these modes are the source of the broad band centered at around 310 cm^{-1} in the glass.

The $GeSe_4$ tetrahedra can be connected in various ways in the glass. Two basic arrangements are corner-sharing (CS), in which neighboring tetrahedra share a single Se atom (one corner) and edge-sharing (ES), in which neighboring tetrahedra share two corners or one edge. We have also considered face-sharing tetrahedra which share three corners, but we find the binding energy of that structure to be much lower than the other arrangements and do not consider that structure further here.

To determine the Raman signatures of CS and ES tetrahedra, we studied the clusters shown in Fig. 8. We can expect the vibrational modes of these structures to consist of linear combinations of the modes of the two component $GeSe_4$ tetrahedra. For the CS model, we found a single strong Raman-active mode at 195 cm^{-1}. This mode, as illustrated by the arrows indicating atomic displacements, is characterized by an even combination of the A_1 breathing modes in the two tetrahedra. A second mode corresponding to an odd combination occurs at 200 cm^{-1}. That mode has essentially no Raman intensity.

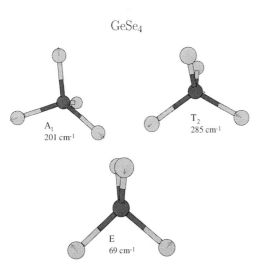

Fig. 7. Characteristic vibrational modes of the $GeSe_4$ tetrahedron

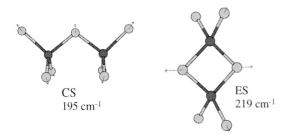

Fig. 8. Corner-sharing (CS) and edge-sharing (ES) combinations of GeSe$_4$ tetrahedra. The displacements of the atoms in the strongly Raman-active mode for the clusters are indicated

This is consistent with the rule of thumb mentioned in Section 2 that Raman-active modes change the volume of the system. In the symmetric breathing mode illustrated in the figure, the overall cluster volume clearly oscillates. In the odd combination, one tetrahedron shrinks while the other grows, leaving the overall volume roughly unchanged.

In the ES cluster, the symmetric mode falls at 219 cm^{-1}, while the odd or anti-symmetric mode lies at 193 cm^{-1}. In this case the frequency of the symmetric mode is shifted up from that of the GeSe$_4$ building block because of increased strain in the fourfold ring. The coupling between the tetrahedral building blocks, as judged by the frequency differences between odd and even modes, is greater in the ES case than in the CS case, because in the latter the two share only a corner, while in the former they share an entire edge. The frequencies of the Raman-active modes in the CS and ES models are in excellent agreement with the positions of the peaks found in the observed spectrum of the glass at 202 and 218 cm^{-1} [32].

The observed spectrum also provides evidence of broken chemical order in GeSe$_2$, i.e. evidence of homopolar Ge-Ge and Se-Se bonds [35]. To determine the Raman signatures of such modes, we studied the clusters shown in Fig. 9. The cluster containing a Ge-Ge bond is labeled FS for face-sharing, adopting terminology used by experimentalists [35]. We show the frequencies and atomic displacements for strongly Raman-active modes associated with these clusters. The position of the FS mode is in close agreement with the experimental mode [32] at 178 cm^{-1}. The Se-Se stretch mode is close to the centroid of a broad band at about 265 cm^{-1} [32].

To assess the relevance of cluster calculations for studying properties of the bulk glass, it is important to understand what effect coupling between the ES and CS units in the bulk might have on the normal mode frequencies. Strong couplings would mix the ES and CS-related modes and shift the positions of the Raman-active modes from those found above. To test this directly, we constructed the large cluster shown in Fig. 9, consisting of a number of ES and CS structures coupled together. Computing the full vibrational spectrum of this cluster would be be computationally very demanding. Instead, we opted to compute only the subset of vibrational modes corresponding to

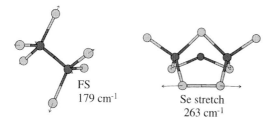

Fig. 9. Raman-active modes associated with homo-polar bonds in GeSe$_2$. The face-sharing (FS) structure features a Ge-Ge bond

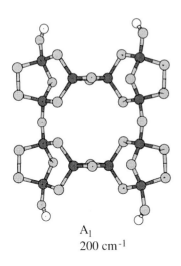

Fig. 10. Atomic displacements for a vibrational mode in the A_1 region of the spectrum for a large model of GeSe$_2$. The model contains four CS units and 2 ES units (see Fig. 8)

A_1
200 cm^{-1}

atomic displacements that do not break the symmetry of the cluster. We construct the dynamical matrix within the space spanned by linear combinations of atomic displacements that preserve the overall cluster symmetry. Diagonalizing gives us the frequencies and eigenvectors of the symmetric oscillation modes. By analyzing the eigenvectors, we can describe the modes for the large cluster in terms of contributions from the various modes of the smaller component clusters.

Focusing our attention on the A_1 region of the spectrum, we find modes at 195 and 200 cm^{-1}, and a third at 211 cm^{-1}. The first two are dominated by motion of the atoms on the CS tetrahedra (the displacements for the 200 cm^{-1} mode are shown in Fig. 10), while the third primarily involves motion in the ES tetrahedra. The A_1 modes of the ES and CS clusters thus do not mix strongly in the vibrations of the cluster shown in Fig. 10, and the frequencies of the modes in the composite cluster are nearly unchanged from the corresponding positions in the smaller ES and CS clusters. We also find a mode at 263 cm^{-1} due almost entirely to stretching of the Se–Se bonds on the cluster edges. Again there is little mixing of these modes with any others, and little shift of the frequencies relative to the small cluster results. The conclusion that can be drawn here is that the coupling between the CS and ES units in this cluster is small even though they are directly bonded. This suggests that the results obtained for the small clusters will be representative of the Raman-active modes of the glass.

We summarize the results of the cluster calculations in Table 3, where we compare the calculated frequencies for the strongly Raman-active modes to the positions of experimental peaks in the GeSe$_2$ spectrum [32]. The agreement between the calculated and measured frequencies is excellent in all cases, giving an atomistic interpretation to the observed Raman peaks.

In Table 3 we also list the calculated Raman intensities for each of the modes and the binding energy per bond of the corresponding clusters. The fact that we can calculate the *absolute* Raman intensities is significant, because it allows us to make a qualitative prediction of the concentrations of the ES, CS, and FS structural units in the glass. The measured Raman intensity of the different modes can be viewed as the product of the Raman activity of the mode in a given cluster times the relative concentration of that structural feature. Using this idea, we can express the ratio of the concentrations of two structural units as follows:

$$\frac{C_i}{C_j} = \frac{I_i^{\text{exp}}}{I_j^{\text{exp}}} \frac{I_j^{\text{th}}}{I_i^{\text{th}}}. \tag{15}$$

Table 3
Frequencies and Raman intensities of the strong Raman-active modes found for the cluster models shown in Figs. 8 and 9. For each unit, the calculated frequency is compared to the position of the centroid of the corresponding peak in the observed spectrum for $GeSe_2$. The binding energy per bond of the various cluster models is also given

	$GeSe_2$			
	ω^{th} (cm^{-1})	ω^{exp} (cm^{-1})	I^{Ram} (Å^4)/amu	E_B (eV/bond)
CS	195	202	48.0	2.91
ES	219	218	40.6	2.90
FS	178	179	49.6	2.82
Se–Se	263		9.8	

If it is further assumed that all the Ge atoms are in either the ES, CS or FS clusters, we can then add the additional requirement that

$$C_{ES} + C_{CS} + C_{FS} = 1. \tag{16}$$

Using the experimental intensities of Ref. [32], i.e. $I_{ES} = 0.43$, $I_{CS} = 1.00$ and $I_{FS} = 0.18$, we obtain the following relative concentrations: $C_{CS} = 60\%$, $C_{ES} = 30\%$ and $C_{FS} = 10\%$. These estimates must be taken as qualitative, given the the simple assumptions that underlie them. For example, no accounting has been made for tetrahedra that have both CS and ES connections. A more quantitative estimate will require further systematic study of the intensities in such mixed systems.

The binding energy E_B per bond values given in the final column of Table 2 can be viewed as a measure of the relative stability of the clusters we have examined and could therefore be expected to be related to the relative concentrations of the various structural features in the amorphous system. For example, we find E_B to be essentially the same for the CS and ES structures (2.91 and 2.90 eV, respectively), suggesting that both should be present in roughly equal amounts in the glass. The concentration estimates given above bear this out.

We are currently using our DFT-based method to investigate $SiSe_2$ and GeS_2. These systems are closely related to $GeSe_2$, but have somewhat different characteristic features in their Raman spectra. It will be interesting to study these systems systematically to gain more insight regarding the details of the spectra and to further test the utility of our method.

4. Conclusions

In this paper we have illustrated a number of applications of using electric fields in electronic structure calculations. The central theme of the work is to calculate quantities that can be directly measured in experiments. Benchmark calculations discussed here and elsewhere [6] demonstrate clearly that DFT-based calculations yield accurate results for cluster polarizabilities, IR intensities and Raman activities. Currently, calculations like the ones described in this paper can be carried out easily on standard workstations for systems containing on the order of 20 to 30 atoms. The prospects for moving to larger cluster sizes is excellent, as processor speeds continue to grow. Central

Michigan University has current plans for a modest parallel network of workstations that will yield about a factor of ten speed up over our current serial machines. Assuming a roughly N^2 scaling for our calculations in this size regime, where N is the number of atoms in a cluster, the implication is that we will be able to extend practical electric field calculations to clusters containing around 60 atoms in the very near future. This will open a variety of exciting new research opportunities.

Acknowledgements This work was supported in part by a grant from the National Science Foundation (NSF-DMR-9972333) and also in part by an award from Research Corporation. Several individuals contributed to the work described here. For the work on Si clusters, the author would like to thank Prof. Dr. Thomas Frauenheim and Alexander Sieck, and Prof. Kai-Ming Ho and Cai-Zhuang Wang; the work on Fe clusters was done in partnership with Dr. Mark Knickelbein and Dr. Geoffrey Koretsky at Argonne. Zoltan Hajnal also contributed to this work. Arlin Briley and Shau Grossman contributed to the work on chalcogenide glasses. The author would also like to thank Prof. Punit Boolchand for many helpful discussions on problems related to these glasses. Finally, the author would like to express his thanks to Dr. Dirk Porezag for his work in developing the technique for IR and Raman intensities, and to Dr. Mark Pederson for his many contributions to all of the work described here.

References

[1] P. HOHENBERG and W. KOHN, Phys. Rev. **136**, B864 (1964).
[2] W. KOHN and L. J. SHAM, Phys. Rev. **140**, A1133 (1965).
[3] J. P. PERDEW, K. BURKE, and M. ERNZERHOF, Phys. Rev. Lett. **77**, 3865 (1996).
[4] J. P. PERDEW, J. A. CHEVARY, S. H. VOSKO, K. A. JACKSON, M. R. PEDERSON, D. J. SINGH, and C. FIOLHAIS, Phys. Rev. B **46**, 6671 (1992).
[5] R. O. JONES and O. GUNNARSON, Rev. Mod. Phys. **61**, 689 (1989).
[6] D. V. POREZAG and M. R. PEDERSON, Phys. Rev. B **54**, 7830 (1996).
[7] K. A. JACKSON, M. R. PEDERSON, D. V. POREZAG, Z. HAJNAL, and TH. FRAUENHEIM, Phys. Rev. B **55**, 2549 (1997).
[8] K. JACKSON, M. PEDERSON, CAI-ZHUANG WANG, and KAI-MING HO, Phys. Rev. A **59**, 3685 (1999).
[9] M. B. KNICKELBEIN, G. M. KORETSKY, K. A. JACKSON, M. R. PEDERSON, and Z. HAJNAL, J. Chem. Phys. **109**, 10692 (1998).
[10] K. A. JACKSON, A. BRILEY, and S. GROSSMAN, unpublished.
[11] M. R. PEDERSON and K. A. JACKSON, Phys. Rev. B **41**, 7453 (1990).
[12] K. A. JACKSON and M. R. PEDERSON, Phys. Rev. B **42**, 3276 (1990).
[13] E. B. WILSON, J. C. DECIUS, and P. C. CROSS, Molecular Vibrations, McGraw-Hill Publ. Co., New York 1955.
[14] M. CARDONA, in: Light Scattering in Solids II, Topics in Applied Physics, Vol. 50, Springer-Verlag, Berlin 1982.
[15] I. VASILIEV, S. OGUT, and J. R. CHELIKOWSKY, Phys. Rev. Lett. **78**, 4805 (1997).
[16] R. SCHÄFER, S. SCHLECHT, J. WOENCKHAUS, and J. A. BECKER, Phys. Rev. Lett. **76**, 471 (1996).
[17] M. R. PEDERSON, K. JACKSON, D. V. POREZAG, Z. HAJNAL, and TH. FRAUENHEIM, Phys. Rev. B **54**, 2863 (1996).
[18] A. SIECK, D. POREZAG, TH. FRAUENHEIM, M. R. PEDERSON, and K. JACKSON, Phys. Rev. A **56**, 4890 (1997).
[19] E. HONEA, A. OGURA, C. A. MURRAY, K. RAGHAVACHARI, W. O. SPRENGER, M. F. JARROLD, and W. L. BROWN, Nature **366**, 42 (1993).
[20] S. LI, R. Z. VAN ZEE, W. WELTNER, JR., and K. RAGHAVACHARI, Chem. Phys. Lett. **243**, 275 (1995).
[21] Y.-M. JUAN, E. KAXIRAS, and R. G. GORDON, Phys. Rev. B **51**, 9521 (1995).
[22] K. M. HO, A. A. SHVARTSBURG, B. PAN, Z.-Y. LU, C.-Z. WANG, J. G. WACKER, J. L. FYE, and M. F. JARROLD, Nature **392**, 582 (1998).

[23] B. Liu, A. A. Shvartsburg, Z.-Y. Lu, B. Pan, C.-Z. Wang, K.-M. Ho, and M. F. Jarrold, J. Chem. Phys. **109**, 9401 (1998).
[24] M. F. Jarrold and V. A. Constant, Phys. Rev. Lett. **67**, 2994 (1992).
[25] M. R. Morse, M. E. Geusic, J. R. Heath, and R. E. Smalley, J. Chem. Phys. **83**, 2293 (1985).
[26] R. L. Whetten, D. M. Cox, D. J. Trevor, and A. Kaldor, Phys. Rev. Lett. **54**, 1494 (1985).
[27] S. C. Richtsmeier, E. K. Parks, K. Liu, L. G. Pobo, and S. J. Riley, J. Chem. Phys. **82**, 3659 (1985).
[28] M. B. Knickelbein, J. Chem. Phys. **104**, 3517 (1996).
[29] E. K. Parks, K. Liu, S. C. Richtsmeier, L. G. Pobo, and S. J. Riley, J. Chem. Phys. **82**, 5470 (1983).
[30] A. M. Baró and W. Erley, Surf. Sci. **112**, 759 (1981).
[31] B. I. Dunlap, Phys. Rev. A **41**, 5691 (1990).
[32] S. Sugai, Phys. Rev. B **35**, 1345 (1987).
[33] Xingwei Feng, W. J. Bresser, and P. Boolchand, Phys. Rev. Lett. **78**, 4422 (1997).
[34] G. Lucovsky, J. P. de Neufville, and F. L. Galeener, Phys. Rev. B **9**, 1591 (1974).
[35] P. Boolchand and J. Grothaus, Phys. Rev. B **33**, 5421 (1986).

phys. stat. sol. (b) **217**, 311 (2000)

Subject classification: 61.46.+w; 64.70.Dv; 71.15.Mb; S2

Ab initio Monte Carlo Investigations of Small Lithium Clusters

S. Srinivas[1]) (a, b) and J. Jellinek (b)

(a) Department of Physics, Central Michigan University, Mt. Pleasant, MI 48859, USA

(b) Chemistry Division, Argonne National Laboratory, Argonne, IL 60439, USA

(Received August 10, 1999)

Structural and thermal properties of small lithium clusters are studied using ab initio-based Monte Carlo simulations. The ab initio scheme uses a Hartree-Fock/density functional treatment of the electronic structure combined with a jump-walking Monte Carlo sampling of nuclear configurations. Structural forms of Li_8 and Li_9^+ clusters are obtained and their thermal properties analyzed in terms of probability distributions of the cluster potential energy, average potential energy and configurational heat capacity all considered as a function of the cluster temperature. Details of the gradual evolution with temperature of the structural forms sampled are examined. Temperatures characterizing the onset of structural changes and isomer coexistence are identified for both clusters.

1. Introduction

A good theoretical account of the temperature dependent behavior of the physical properties of a cluster requires a) an accurate depiction of the underlying interactions of its constituent entities and b) a reliable means of obtaining the representative configurational states of the system at finite temperatures. In practice, first-principles methods or semi-empirical potentials are used for the former and simulation techniques such as Monte Carlo and molecular dynamics methods provide the statistically significant sampling required for the latter. In general, one expects a higher degree of accuracy from first principles methods such as quantum chemistry techniques and density functional theory. Most of the detailed studies on the dynamical and thermal properties of atomic clusters, however, utilize semiempirical potentials which are computationally more efficient. These are typically pair or many-body potentials fitted to reproduce first-principles calculations for small clusters (dimer, trimer) and/or experimental properties of the bulk. Monte Carlo (MC) and molecular dynamics (MD) simulations using such potentials have provided a practical and efficient means of tackling an otherwise intractable problem and extensive information on the manner in which structural and phase transitions occur in clusters has emerged from these studies. First-principles calculations on the other hand, have typically concentrated on the static ($T = 0$ K) ground state properties of clusters, such as isomeric forms, binding energies, ionization potentials and other electronic and spectroscopic properties. While these studies have contributed substantially to our understanding of atomic clusters, there are obvious reasons to go beyond static properties and study structural, dynamical and thermal properties of clusters

---
[1]) e-mail: srinivas@phy.cmich.edu

as well, at a first-principles level. The isomerization processes a cluster undergoes, the barriers between different isomers, the thermal stability of the cluster, the structural and phase transitions that occur, are all important properties of clusters. In cases where two or more isomers of a cluster lie within a very small energy range — a phenomenon that is not rare in atomic, and especially metal, clusters — it is important to examine how the cluster samples its potential energy surface at finite temperatures. Over the past decade considerable effort has gone into the development of first principles-based dynamical [1 to 9] and Monte Carlo [10 to 15] schemes many of which have found applications in the study of atomic clusters.

The large majority of ab initio MD studies have concentrated on issues related to structural forms (isomers) of clusters and changes in these forms (isomerizations). A few have also addressed issues related to the dynamics of these changes. For example, Sung et al. [3], Jones et al. [4] and Gibson and Carter [5] have all studied the pseudo-rotation of Li_5 clusters using Car-Parinello or generalized valence bond and complete active space self-consistent field approaches. Recent ab initio molecular dynamics studies have investigated details of both structural and dynamical characteristics of lithium clusters using Hartree-Fock [6 to 8] calculations for neutral and cationic Li_n clusters ($6 \leq n \leq 11$) and density functional theory [9] for Li_8.

An alternate method that facilitates the investigation of the issues (structural and phase transitions, thermal properties etc.) mentioned above is one that combines an ab initio treatment of electronic structure with a Monte Carlo rather than a dynamical sampling of nuclear configurations. Monte Carlo offers several important advantages over molecular dynamics simulations, especially in cases where structural and thermal properties are of interest. The MC sampling involves exploring the configuration space, without the computationally expensive task of calculating the energy gradients (forces) at each step. An added advantage is the fact that temperature in the MC sampling scheme is the true thermodynamical (as opposed to dynamical) temperature and the canonical Monte Carlo method directly yields the thermal properties of the system. The concept of combining a first-principles calculation of energies with a Monte Carlo scheme is not a novel one. Recently, structural properties of Li_5H have been examined [10] using an MC sampling of the nuclei with a second-order Möller-Plesset (MP2) treatment of electronic structure. Other investigations [11 to 13] have used perturbative or hybrid first-principles methods in classical Monte Carlo simulations to study the interactions of various ions with water clusters. Two recent works [14, 15] use classical and path integral Monte Carlo sampling techniques combined with density functional theory, in one case, to study interactions of HCl with water clusters [14] and in the other, to investigate the quantal effects in small lithium clusters (Li_4 and Li_5^+) [15]. However, none of the studies mentioned above have been concerned with details of the thermodynamic properties of the system under investigation and the length of the MC sampling in these studies have been short ($\approx 10^3$ to 10^4 steps).

Currently, the applicability of ab initio-based MD and MC investigations as tools to probe the finite temperature behavior of clusters is limited to small systems — the calculations involved are too expensive to achieve the level of sampling required for a detailed analysis of larger systems. Small lithium clusters are ideal prototypes for studies such as these with few enough electrons that the calculations are computationally feasible, yet complex enough that interesting information emerges from ab initio investigations [3 to 9, 15 to 24]. An important issue is the task of achieving a balance between the need to treat all the interactions in the system as accurately as possible and the requirement of a sufficiently complete sampling of nuclear configuration space. We present a description of the structural and thermal properties of lithium clusters, at an ab initio level, in a study that we believe satisfies these criteria. The scheme uses a hybrid Hartree-Fock (HF)/density functional theory (DFT) calculation of configurational energies combined with a jump-walking [25] Monte Carlo sampling of configurational space to study the structural and thermal properties of Li_8 and Li_9^+ clusters. The methodology used is described in the next section. The results obtained for the structural and thermal properties are analyzed in Section 3 and summarized in Section 4.

2. Methodology

The two major components of the scheme are the ab initio calculation of the electronic structure followed by a Monte Carlo sampling of the nuclear configuration space. The configurational energy at each step is calculated using a hybrid HF/DFT scheme, where the converged Hartree-Fock electron density $\varrho(\mathbf{r})$ is used in the Lee-Yang-Parr [26] (LYP) correlation functional to obtain the correlation correction to the Hartree-Fock energy. A single numerical integration is performed at each step for the calculation of the correlation energy in this formalism and represents a considerable saving in computational time. The exchange energy used is the exact Hartree-Fock exchange. The validity of using the Hartree-Fock density in a correlation functional has been addressed [27, 28] and applied successfully [29] to study the binding energies of different systems. The basis set used for Li in our calculations is a (411/1) set of contracted Gaussian type orbitals (CGTO) which has been used in previous work [6 to 8, 24]. The parallelization and optimization of the ab initio scheme — the Hartree-Fock calculations of the configurational energies and the Lee-Yang-Parr correlation correction to this energy — is described in detail in our earlier work [24] on Li_8 clusters. This parallelization is an important factor in our ability to perform the Monte Carlo simulations efficiently and obtain the level of sampling that is necessary for examining the thermal averages for the systems under investigation in the present work.

A jump-walking (j-walking) Monte Carlo scheme augments the regular Metropolis moves. The j-walking technique [25] offers a means of increasing the efficiency of sampling of the nuclear configuration space by providing a route through which the system can avoid being trapped in one of the local minima. The problem of getting trapped in a local minimum is a particular concern at low to intermediate temperatures in the canonical Metropolis Monte Carlo [30a] scheme.[2]) It is less probable at high temperatures, where the system samples many different isomeric forms, as well as at very low temperatures, where the global minimum is the most likely one. The j-walking scheme involves occasional sampling from a distribution of configurations generated at a higher temperature by a (previous) Monte Carlo sampling. The j-walker configurations at each temperature are generated in our simulations by storing every 100th configuration during a Monte Carlo sampling at that temperature. The spacing of the temperature intervals for the j-walking Monte Carlo sampling is such that approximately 50% of attempts to jump to the higher temperature j-walker configurations are accepted. The probability of acceptance is appropriately weighted so that the accepted j-walker config-

---

[2]) A detailed description of the Metropolis Monte Carlo scheme is provided in [30b].

uration (originally generated at a higher temperature) follows the correct Boltzmann probability for the sampling temperature. At the highest temperature simulation, the Metropolis Monte Carlo procedure is the one used, while at every subsequent simulation, a j-walking Monte Carlo sampling is used, which in turn generates the j-walker configurations representative of the sampling temperature.

At each step of the Monte Carlo simulation, a randomly chosen atom is moved through random displacements in the x, y and z-directions. The maximum displacement is chosen such that the acceptance rate over the entire run at each temperature is approximately 50%. Independent random number generators with different initial seeds are used for each of the independent quantities to ensure that no correlation exists between these physical quantities. Each simulation performed at a particular temperature generates a representative Boltzmann weighted set of configurations and involves regular Metropolis moves interspersed with jump-walking attempts. Jumps are attemted at random approximately 10% of the time, to randomly chosen stored j-walker configurations. The interaction energy V for each configuration is given as the difference between the total energy of that particular configuration and the sum of the energies of the constituent atoms/ion at infinite separation, all calculated within the HF/LYP scheme. The averages of all the physical quantities are calculated over the entire length of the simulation, which is typically 300 000 to 400 000 steps in our calculations.

The physical quantities of interest in our investigations of the thermal properties of the cluster are the probability distribution function for the potential energy V, the average potential energy $\langle V \rangle$ and the configurational heat capacity (at constant volume) C_V. The latter is calculated using the expression for the heat capacity of a canonical ensemble,

$$C_V = \frac{\langle V^2 \rangle - \langle V \rangle^2}{kT^2}, \tag{1}$$

where V is the potential energy of the cluster, k is the Boltzmann constant, and $\langle \ \rangle$ denotes the averaging over an entire MC sampling at a given temperature T. The configurational heat capacity, which is a measure of the energy fluctuations in the system, is a quantity highly sensitive to small changes in the configurational energies sampled. Changes in C_V herald the onset of structural or phase transitions in the system.

The different isomers of the cluster are obtained through a simulated cooling procedure within the Monte Carlo scheme, using as a starting point, configurations sampled during high temperature simulations. The quenching technique used follows a combination of two scenarios. In the first, the MC sampling is combined with a step-wise reduction of the temperature, decreasing the temperature in steps that vary from about 5% to 15% of the previous temperature at each simulation. The maximum displacement is adjusted to give an acceptance rate of approximately 50%. This is performed repeatedly until a final temperature of 10^{-3} K is reached. In the second scenario, only moves that lower the energy of the cluster are accepted. This process of thermal cooling is performed until the variation in the total configurational energy of the cluster remains less than 10^{-7} Hartree (2.7×10^{-6} eV). All computations were performed using a parallel 12-node cluster of DEC-3000/700 workstations and a parallel 8-node cluster of DEC alpha 500/333 workstations.

3. Results and Discussion

3.1 Li_8

Four isomers are obtained for Li_8 – a lowest energy T_d structure, followed by two nearly degenerate C_{2v} and C_s structures, which are 0.18 and 0.21 eV higher in energy than the T_d structure, and a highest energy ($\Delta E = 0.58$ eV) D_{2d} structure (Fig. 1). We also find that inclusion of correlation in this hybrid scheme marginally lowers the energy spacing between the first three isomers and significantly increases the energy spacing for the highest energy isomer (D_{2d}) in comparison with those obtained at the Hartree-Fock level.

The distribution of the energies sampled as a function of temperature is shown in Fig. 2, where each panel shows the normalized probability distributions for the potential energies sampled by the cluster during each j-walking Monte Carlo simulation. The smoothness of the graphs, which represent histograms obtained for a very fine mesh of energy intervals (0.00136 eV), indicate that our results are well-converged. An examination of Fig. 2 shows that at low temperatures, the probability distributions for the energies are especially narrow and have a high peak centered around lower configurational energies. The narrowness of the distribution is due to the fact that the only configurations sampled up to a temperature of about 120 K are those associated with the lowest energy isomer the T_d structure, which has a binding energy of 6.38 eV. This is verified using simulated rapid cooling from many configurations generated from our sampling. The only structure that emerges in the quenches of numerous configurations associated with the 120 K sampling is the T_d isomer. As the cluster temperature increases, wider ranges of energies are sampled, with more of the configuration space now accessible to the system. Just above 120 K (in fact at 170 K, the next temperature sampled in the MC simulations), the second and third isomers – the C_{3v} and the C_s structures make their first appearance. These two isomers have very similar packing and are very close in energy – the difference in their binding energies is less than 0.03 eV – so they easily interconvert. The fourth and highest energy isomer with the D_{2d} symmetry is only present among the configurations sampled at temperatures above 300 K.

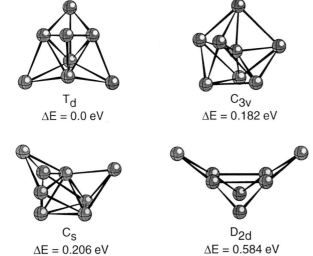

Fig. 1. Isomers of Li_8 obtained within the HF-LYP MC scheme. The numbers are the energy gaps (in eV) between the lowest energy T_d structure and the higher energy isomers

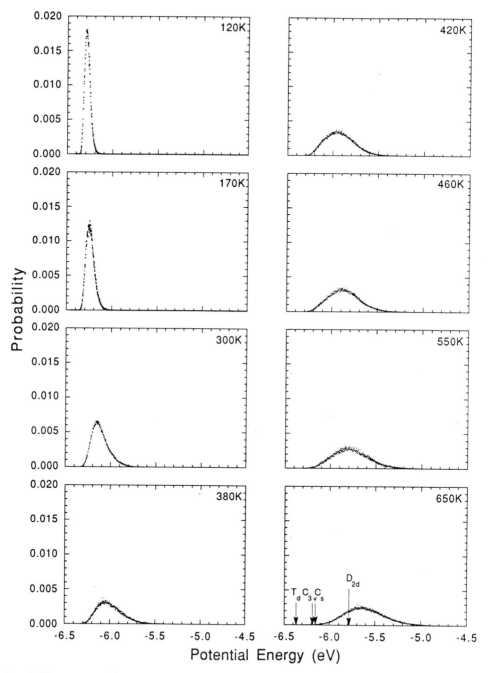

Fig. 2. Histograms of the normalized probability distribution of the potential energy of Li_8 at different temperatures. The probabilities are calculated with an energy box size of 0.00136 eV. The equilibrium energies of the isomers are indicated by arrows

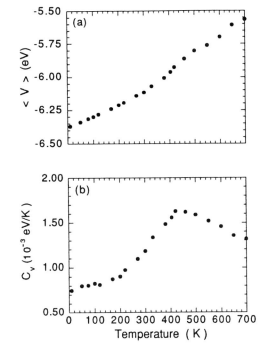

Fig. 3. a) Averaged potential energy of Li_8 as a function of temperature. b) Configurational heat capacity of Li_8 as a function of temperature

Figure 3a shows a graph of the average potential energy of the cluster as a function of the cluster temperature, the caloric curve. The averaged potential energy at each temperature of the caloric curve is obtained by averaging over the entire potential energy distribution (Fig. 2) at that temperature. Figure 3b shows the variation of the configurational heat capacity of the cluster with temperature. Changes in the curvature of the caloric curve and the presence of a peak in the configurational heat capacity signify the occurrence of a phase transition. Clusters are known to exhibit a broad peak in their heat capacity, indicating a gradual, stage-wise melting-like transition [31 to 36], rather than the sharp transitions found in bulk. The presence of an inflection in the caloric curve over a wide temperature region between 180 to 400 K signifies the occurrence of this type of transition. The phenomenon is even more dramatically illustrated by the behavior of the configurational heat capacity, whose graph shows a very broad peak starting at a temperature of around 120 K, with the maximum value of C_v occurring in the region around 420 to 430 K.

3.2 Li_9^+

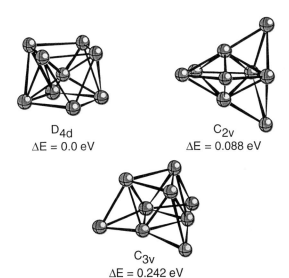

The three isomers obtained for Li_9^+ are shown in Fig. 4. The lowest energy isomer of Li_9^+ is a distorted antiprism-type structure with a central atom and has a D_{4d} symmetry. The next higher energy isomer is a C_{2v} structure ($\Delta E = 0.088$ eV) and

Fig. 4. Isomers of Li_9^+ obtained within the HF-LYP MC scheme. The numbers are the energy gaps (in eV) between the lowest energy D_{4d} structure and the higher energy isomers

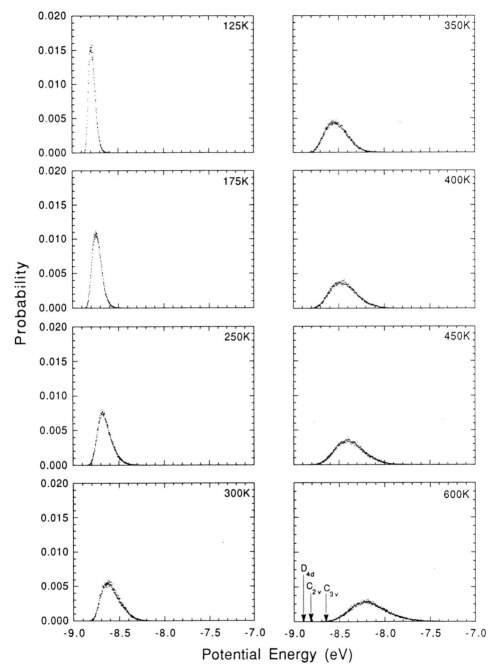

Fig. 5. Histograms of the normalized probability distribution of the potential energy of Li_9^+ at different temperatures. The probabilities are calculated with an energy box size of 0.00136 eV. The equilibrium energies of the isomers are indicated by arrows

the third isomer has a C_{3v} geometry and is 0.242 eV higher in energy than the lowest energy isomer. Our calculations show that in the case of Li_9^+, the energy spacing between the three isomers increases marginally with the inclusion of correlation, as compared with that obtained at a Hartree-Fock level.

The normalized distributions of the potential energies sampled by the cluster in our Monte Carlo simulations for different temperatures are illustrated in Fig. 5. The histograms shown in the above use an energy interval of 0.00136 eV as the sorting box size. As in Li_8, the potential energy distributions are characterized by narrow, high peaks at low temperatures, where the sampling takes place in the region of the lowest energy isomer(s), and broad, low peaks at higher temperatures, as more of the configuration space becomes accessible to the cluster. Our simulations show that below 75 K, the only isomer that is sampled is the lowest energy D_{4d} isomer, which has a binding energy of 8.90 eV in its equilibrium geometry. At around 75 K, the second lowest energy structure (the C_{2v} isomer, which has a binding energy of 8.81 eV) should, in principle, be accessible from energy considerations. However, we see configurations associated with the second isomer being included in the sampling only at temperatures of 125 K and above. By 175 K, all three isomers of Li_9^+ are present in the sampling.

The caloric curve obtained for Li_9^+ is shown in Fig. 6a. The behavior of the configurational heat capacity as a function of the temperature is shown in Fig. 6b. The graph shows an increase in the heat capacity beginning at a temperature of about 125 K. The peak in the configurational heat capacity is a rather "soft" one and occurs at a temperature of about 300 K. The simulations indicate a significant amount of structure present in the configurational heat capacity in the region between 300 and 420 K, which could be a signal of additional changes taking place or could be the consequence of inadequate sampling in this region. We have been unable to resolve this structure in the simulations performed with increased lengths (5 to 6×10^5 steps) of sampling.

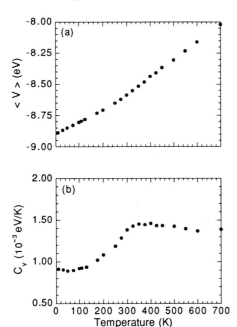

The characteristics of the solid-like to liquid-like transition experienced by a cluster – the temperatures at which individual isomers make their appearance and the temperature range in which the structural and phase transitions occur – are intimately linked to details of the energy landscape, such as the energy ordering of the isomers and the barriers between different isomers. Our results for Li_8 and Li_9^+ clusters illustrate this dependence. The temperatures at which the higher energy isomers become accessible in the sampling are 125 K for the sampling of

Fig. 6. a) Averaged potential energy of Li_9^+ as a function of temperature. b) Configurational heat capacity of Li_9^+ as a function of temperature

the second and third isomers (which are almost degenerate) and 300 K for sampling of all the four isomers of Li_8. In the case of Li_9^+, where the three isomers are more closely spaced, only the lowest energy isomer is sampled up to a temperature of 100 K. The second isomer gets added to the sampling at temperatures above 100 K, and the catchment areas of all three isomers of Li_9^+ are accessible to the cluster by 175 K. The lower characteristic temperatures obtained for Li_9^+ as compared with those for Li_8 are consistent with our observation of smaller energy spacing between the isomers of Li_9^+ present in the sampling as compared with the energy spacing between isomers of Li_8. The melting-like transition, where the cluster goes from a solid-like stage associated with a single isomer to a liquid-like-stage where all structures are sampled, thus occurs over a wide temperature range for both clusters. It should be noted that the melting temperature for bulk lithium is 453 K.

4. Summary

We have presented a brief outline of an efficient ab initio Monte Carlo scheme utilized to study the structural and thermal properties of Li_8 and Li_9^+ clusters. The scheme combines a hybrid Hartree-Fock/density functional treatment of electronic structure with a jump-walking version of the Monte Carlo sampling. Parallelization of the ab initio part of the calculations and the inclusion of j-walking contribute to rendering this scheme very efficient, the former from the viewpoint of the time spent in the calculation of configurational energies and the latter for an adequate sampling of the configuration space. It is worthwhile to comment on some of the differences between this work and previous ones, namely, the HF and DFT-based molecular dynamics studies on lithium clusters [6 to 9]. While the MD investigations have carried out studies that involve very extensive sampling – certainly among the most exhaustive investigations performed to date with ab initio dynamics – there are some unresolved features. For example, it is difficult when comparing these studies, to accurately pinpoint whether the differences in the temperatures characterizing the onset of isomerizations in the cluster within the two approaches, arise from the absence (HF) or presence (DFT) of correlation or from the fact that the "dynamical temperature" of the cluster may not accurately represent the "thermodynamical temperature" of the cluster at those sampling lengths. In the alternate ab initio based Monte Carlo sampling scheme adopted in our work, the cluster temperature is a well-defined thermodynamical parameter regardless of the length of the simulation, and the temperatures characterizing the onset of isomerizations quoted here are meaningful quantities.

At present, when there is little experimental data on the isomer-specific thermal properties of atomic clusters, ab initio theoretical studies such as these provide a means to understand the nature and mechanism of structural and phase transitions in clusters. The study enables us to characterize, at a first-principles level, the temperatures at which a specific cluster undergoes structural transitions, and to pinpoint regions of isomer co-existence. From a practical viewpoint, the characterization of the temperatures at which structural changes occur is expected to aid in the structural identifications of the isomers present in cluster beam experiments carried out at finite temperatures.

Acknowledgement The work presented was performed under the auspices of the Office of Basic Energy Sciences, Division of Chemical Sciences, US-DOE under contract number W-31-109-ENG-38.

References

[1] R. CAR and M. PARRINELLO, Phys. Rev. Lett. **55**, 2471 (1985);
R. CAR and M. PARRINELLO, in Simple Molecular Systems at Very High Density, NATO ASI Series B: Physics, Ed. A. POLIAN, P. LOUBEYRE and N. BOCCARA, Plenum, Vol. 186, New York 1989 (p. 455).

[2] P. BALLONE, W. ANDREONI, R. CAR, and M. PARRINELLO, Phys. Rev. Lett. **60**, 271 (1988).
D. HOHL, R. O. JONES, R. CAR, and M. PARRINELLO, J. Chem. Phys. **89**, 6823 (1988).
R. KAWAI and J. H. WEARE, Phys. Rev. Lett. **65**, 80 (1990).
J.-Y. YI, D. J. OH, and J. BERNHOLC, Phys. Rev. Lett. **67**, 1594 (1991).
U. RÖTHLISBERGER and W. ANDREONI, J. Chem. Phys. **94**, 8129 (1991).
N. BINGGELI, J. L. MARTINS, and J. R. CHELIKOWSKY, Phys. Rev. Lett. **68**, 2956 (1992).
B. HARTKE and E. A. CARTER, Chem. Phys. Lett. **216**, 324 (1993), and references therein.

[3] M.-W. SUNG, R. KAWAI, and J. H. WEARE, Phys. Rev. Lett. **73**, 3552 (1994).
R. KAWAI, J. F. TOMBRELLO, and J. H. WEARE, Phys. Rev. A **49**, 4236 (1994).

[4] R. O. JONES, A. I. LICHTENSTEIN, and J. HUTTER, J. Chem. Phys. **106**, 4566 (1997).

[5] D. A. GIBSON and E. A. CARTER, Chem. Phys. Lett. **271**, 266 (1997).

[6] J. JELLINEK, V. BONAČIĆ-KOUTECKÝ, P. FANTUCCI, and M. WIECHERT, J. Chem. Phys. **101**, 10092 (1994).

[7] P. FANTUCCI, V. BONAČIĆ-KOUTECKÝ, J. JELLINEK, M. WIECHERT, R. J. HARRISON, and M. F. GUEST, Chem. Phys. Lett. **250**, 47 (1996).

[8] V. BONAČIĆ-KOUTECKÝ, J. JELLINEK, M. WIECHERT, and P. FANTUCCI, J. Chem. Phys. **107**, 6321 (1997).

[9] D. REICHARDT, V. BONAČIĆ-KOUTECKÝ, P. FANTUCCI, and J. JELLINEK, Z. Phys. D **40**, 486 (1997); Chem. Phys. Lett. **279**, 129 (1997).

[10] V. KESHARI and Y. ISHIKAWA, Chem. Phys. Lett. **218**, 406 (1994).

[11] T. ASADA and S. IWATA, Chem. Phys. Lett. **260**, 1 (1996).

[12] T. N. TRUONG and E. V. STEFANOVICH, Chem. Phys. Lett. **256**, 348 (1996).

[13] E. V. AKHMATSKAYA, M. D. COOPER, N. A. BURTON, A. J. MASTERS, and I. HILLIER, Chem. Phys. Lett. **267**, 105 (1997).

[14] D. A. ESTRIN, J. KOHANOFF, D. H. LARIA, and R. O. WEHT, Chem. Phys. Lett. **280**, 280 (1997).

[15] R. O. WEHT, J. KOHANOFF, D. A. ESTRIN, and C. CHAKRAVARTY, J. Chem. Phys. **108**, 8848 (1998).

[16] J. GARCIA-PRIETO, W. L. FENG, and O. NOVARO, Surf. Sci. **147**, 555 (1984).

[17] I. BOUSTANI, W. PEWESTORF, P. FANTUCCI, V. BONAČIĆ-KOUTECKÝ, and K. KOUTECKÝ, Phys. Rev. B **35**, 9437 (1987).
V. BONAČIĆ-KOUTECKÝ, P. FANTUCCI, and J. KOUTECKÝ, Chem. Phys. Lett. **146**, 518 (1988).
P. FANTUCCI, S. POLEZZO, V. BONAČIĆ-KOUTECKÝ, and J. KOUTECKÝ, J. Chem. Phys. **92**, 6645 (1990).

[18] O. SUGINO and H. KAMIMURA, Phys. Rev. Lett. **65**, 2696 (1990).

[19] J. BLANC, V. BONAČIĆ-KOUTECKÝ, M. BROYER, J. CHEVALEYRE, PH. DUGOURD, J. KOUTECKÝ, C. SCHEUCH, J. P. WOLF, and L. WÖSTE, J. Chem. Phys. **96**, 1793 (1992).

[20] G. GARDET, F. ROGEMOND, and H. CHERMETTE, J. Chem. Phys. **105**, 9933 (1996).

[21] I. G. KAPLAN, J. HERNÁNDEZ-COBOS, I. ORTEGA-BLAKE, and O. NOVARO, Phys. Rev. A **53**, 2493 (1996).

[22] J. M. PACHECO and J. L. MARTINS, J. Chem. Phys. **106**, 6039 (1997).

[23] R. ROUSSEAU and D. MARX, Phys. Rev. A **56**, 617 (1997).

[24] J. JELLINEK, S. SRINIVAS, and P. FANTUCCI, Chem. Phys. Lett. **288**, 705 (1998).

[25] D. D. FRANTZ, D. L. FREEMAN, and J. D. DOLL, J. Chem. Phys. **93**, 2769 (1990).
D. D. FRANTZ, D. L. FREEMAN, and J. D. DOLL, J. Chem. Phys. **97**, 5713 (1992).

[26] C. LEE, W. YANG, and R. G. PARR, Phys. Rev. B **37**, 785 (1988).
B. MIEHLICH, A. SAVIN, H. STOLL, and H. PREUSS, Chem. Phys. Lett. **157**, 200 (1989).

[27] R. A. HARRIS and L. R. PRATT, J. Chem. Phys. **83**, 4024 (1985).

[28] M. LEVY, Density Matrices and Density Functionals, Eds. R. ERDAHL and V. H. SMITH, JR., Reidel, Boston 1987 (p. 479).

[29] J. CIOSLOWSKI and A. NANAYAKKARA, J. Chem. Phys. **99**, 5163 (1993).
M. CAUSÀ and A. ZUPAN, Chem. Phys. Lett. **220**, 145 (1994).

[30a] N. METROPOLIS, A. W. METROPOLIS, M. N. ROSENBLUTH, A. H. TELLER, and E. TELLER, J. Chem. Phys. **21**, 1087 (1953).

[30b] J. P. Valleau and S. G. Whittington, Statistical Mechanics, Ed. B. Berne, Plenum Press, New York 1977.
[31] R. S. Berry, T. L. Beck, H. L. Davis, and J. Jellinek, in: The Evolution of Size Effects in Chemical Dynamics, Advances in Chemical Physics, Vol. 70, Eds. I. Prigogine and S. A. Rice, Wiley-Interscience, New York 1988 (p. 75).
[32] J. Jellinek, in: Metal–Ligand Interactions, Eds. N. Russo and D. R. Salahub, Kluwer Academic Publishers, Dordrecht 1996 (p. 325), and references therein.
[33] H. L. Davis, J. Jellinek, and R. S. Berry, J. Chem. Phys. **86**, 6456 (1987).
[34] P. Labastie and R. L. Whetten, Phys. Rev. Lett. **65**, 1567 (1990).
[35] C. J. Tsai and K. D. Jordon, J. Chem. Phys. **95**, 3850 (1991).
[36] R. S. Berry, Nature **393**, 212 (1998).

phys. stat. sol. (b) **217**, 323 (2000)

Subject classification: 61.46.+w; 71.15.Mb; S9.11

Structure and Isomerization in Alkali Halide Clusters

C. Ashman (a), S. N. Khanna (a), and M. R. Pederson (b)

(a) Department of Physics, Virginia Commonwealth University,
Richmond, VA 23284-2000, USA

(b) Complex Systems Theory Branch – 6692, Naval Research Laboratory,
Washington D.C. 20375-5000, USA

(Received August 10, 1999)

Theoretical electronic structure calculations on the geometry and stability of $M_4Cl_3^-$ (M = Li, Na, K, Rb, and Cs) have been carried out within a gradient corrected density functional theory. It is shown that all the clusters possess three energetically close structures in the form of a defected cuboid, ladder, and a ring. The three isomers have different electron affinities and can be identified through negative ion photodetachment experiments. It is shown that all the clusters are marked by low energy vibrational modes which can permit interconversion between the structures at high temperatures. The transition (minimum) temperature for observing thermal isomerization increases as one goes from Cs to Li. Further experiments to probe this interesting phenomenon are suggested.

1. Introduction

One of the most fascinating aspects of clusters is the possibility that clusters of the same size and composition can exhibit different physical, electronic [1, 2], magnetic [3] or chemical properties [4]. This arises due to the possibility of geometrical arrangements that are energetically close or degenerate and yet are separated by barriers in the phase space [5]. As the clusters are formed from a vapor phase, the atoms can condense into these different configurations called isomers. Alkali halides are ideal systems for observing isomers. Here, the inter-atomic interactions are marked by ionic character and several energetically competing geometrical arrangements are possible. Indeed, previous theoretical work by Landman and co-workers [6] and others [7] and experiments by Whetten and co-workers [8] have discussed the different modes of electron localization and the possibility of isomers in these clusters almost ten years ago. While these works only considered isomers, a unique situation arises when the energetically close configurations are separated by low barriers or are connected by very long rotational or vibrational pathways. Under these conditions, the clusters can isomerize at ordinary temperatures. Further, if a physical property which varies from one isomer to another can be measured, one can probe the thermal isomerization in real time.

Fatemi et al. have recently reported the first such observations of thermal isomerization in $Cs_4Cl_3^-$ clusters near room temperature [9, 10]. In these experiments, the clusters were generated in beams via a laser vaporization of the solid halide and studied via the negative ion photoelectron spectroscopy. The photoelectron spectra of clusters stabilized at different temperatures show features which change with temperature. The photoelectron spectra at 297 K show a major peak at 0.39 eV followed by minor peaks or humps at 0.72 and 1.01 eV. Upon lowering the temperature to 116 K, the subpeaks at 0.72 and 1.01 eV disappear indicating that the clusters settle into a well defined ground state. Fatemi et al. proposed that the subpeaks at 0.72 and 1.01 eV correspond to two isomers and that the clusters isomerize at ordinary temperatures. As a further evidence to thermal isomerization, they depleted the lowest electron affinity isomer which survives at the low temperature and therefore corresponds to the most stable configuration, from the beam. It repopulated via thermal isomerization from the remaining isomers.

In a recent paper, we carried out theoretical studies of the thermal isomerization in $Cs_4Cl_3^-$ [11]. We had shown that the anion is indeed characterized by three energetically close isomers with defected cuboid, ladder, or an octagonal ring arrangement. Each structure was missing an atom and was consequently distorted. The calculated electron affinities of the three structures and the photoelectron spectrum of the ground state structure were in excellent agreement with experiments. The three structures were shown to be marked by low frequency vibrational modes which enable inter-conversion between different structures at ordinary temperatures. While these studies provided an understanding of the data on $Cs_4Cl_3^-$, such thermal isomerization is not observed for other alkali chlorides. Also, the first major peak in the negative ion photoelectron spectrum of $Na_4Cl_3^-$ is 2.02 eV while it is 0.39 eV for $Cs_4Cl_3^-$. This raises several questions: (1) How do the energetics of the three structures, i.e. cuboid, ring and ladder, which are energetically degenerate for $Cs_4Cl_3^-$ change as we change the alkali atom? (2) Are other alkali chlorides also marked by energetically close geometries? (3) If so, why isn't thermal isomerization observed in other alkali chlorides?

The problem of structural isomerization is also related to another important issue namely the electron pairing and its accomodation. $Cs_4Cl_3^-$ and other alkali halides have one excess alkali atom and an additional electron. These systems are therefore regarded as two excess electron systems. A number of workers have addressed the issue of how the electron pair is accomodated in the small anionic alkali halide clusters [8, 12, 13]. In particular, Bloomfield and co-workers [14] have examined photoelectron spectra of several sodium chloride anions containing two excess electrons and have identified three distinct types of spectra, corresponding to three modes of accomodation of the extra electron. These are: (1) excess electrons are spin paired and localized in a single anion vacancy, (2) electron pair localized on a Na^+ ion to form Na^- anion, and (3) electrons occupy two different anion vacancies as a double F color center. In particular, for $Na_4Cl_3^-$, they propose that the two excess electrons are accomodated as a spin pair in a single anion vacancy forming the cluster equivalent of a F' color center. According to these authors this leads to a high electron affinity as observed experimentally and they propose a cuboid ground state with a missing atom. This argument does have difficulties. Because of electron–electron repulsion, the presence of a localized electron pair would be expected to lead to a low and not a high electron affinity as deduced by these authors. Another puzzling experimental result is that whereas $Na_4Cl_3^-$ has an electron affinity of 2.02 eV, $Cs_4Cl_3^-$ has an electron affinity of only 0.39 eV. Is the change in electron affinity related to a different mode of electron localization and how does it affect the geometrical arrangement of atoms? The structural progressions as one goes from Li to Cs are also interesting from another perspective. The bulk alkali halides form a compact face centered cubic NaCl structure for Li, Na, K and Rb while the Cs halides form a body centered cubic structure. The geometrical progressions in the bulk can be understood in terms of the atomic sizes. How do these size effects play on the geometrical arrangements at small sizes?

In this review we present detailed theoretical investigations of the $M_4Cl_3^-$ (M = Li, Na, K, Rb, and Cs) clusters in order to answer some of the above question. Our studies are based on first principles density functional formalism and have proven predictive capabilities. We show that all the alkali chlorides are marked by three energetically close structures (defected cuboid, ladder, and a ring) whose relative stability changes with the alkali atom. The three structures, while nearly degenerate, are marked by significantly different electron affinities. We find that the excess electron in the anionic cluster and the excess electron due to extra alkali atom do form an electron pair which is delocalized and centered at the vacancy site. We further show that the progressions in the relative stability of the three structures with alkali can be understood in terms of the size of the alkali atom. The low vibrational frequencies and the energy differences between the ladder and the ground state structures increase in going from Cs to Li. As the thermal isomerization proceeds via the ladder structure, we propose that the size, frequencies and the barriers make the thermal isomerization progressively difficult as one goes from Cs to Li.

In Section 2, we briefly outline our theoretical procedure. Section 3 contains the details of the calculations and results. Finally Section 4 is devoted to a discussion of the results and a summary.

2. Theory

The theoretical studies are based on a linear combination of atomic orbitals molecular orbital (LCAO-MO) approach. In this approach the molecular orbitals are expressed in terms of atomic orbitals $\phi(\mathbf{r} - \mathbf{R}_j)$ centered at the atomic sites $\mathbf{R}_j$, i.e.

$$\psi_i = \sum_j C_{ij} \phi_j(\mathbf{r} - \mathbf{R}_j). \tag{1}$$

The atomic orbitals, in turn, are built from a linear combination of the basis functions in the form of Gaussian orbitals. The determination of the optimal basis set is important and we refer the reader to our earlier papers for details [15]. The basis sets were tested for their completeness and basis set superposition errors. For Li, Na, K, Rb and Cs the basis sets had 10, 16, 19, 22, and 25 bare Gaussians, respectively. These were contracted to 5s, 3p, and 1d for Li, 6s, 4p, and 3d for Na, 7s, 5p, and 3d for K, 8s, 6p, and 4d for Rb and 8s, 6p, and 8d for Cs. For Cl, a total of 17 bare Gaussians were contracted to a 6s, 5p and 3d basis set. In all cases, a supplementary d-function was included. The inclusion of d functions in the basis set of Cl is important to accurately describe the Cl atom particularly in view of the charge transfer to the Cl atoms.

The exchange–correlation effects were treated within a density functional formalism by using the form proposed by Ceperley and Adler [16]. The gradient corrections were implemented within a generalized gradient approximation (GGA) using the most recently improved [17] GGA energy functionals. The variational coefficients C_{ij} in Eq. (1) were obtained by solving the one electron Kohn-Sham equations [18]

$$(-\tfrac{1}{2}\nabla^2 + V_{\text{ion}} + V_H + V_{\text{xc}}^\alpha) |\psi_n^\alpha\rangle = \epsilon_n^\alpha |\psi_n^\alpha\rangle \tag{2}$$

self-consistently. In Eq. (2), the first term is the kinetic energy operator, V_{ion} is the ionic potential, V_H the Hartree potential, and V_{xc}^α the exchange–correlation potential depending on the spin α. $|\psi_n^\alpha\rangle$ is the n-th molecular orbital for the spin α. The calculations were carried out at the all-electron level and the details of the calculations are given in earlier papers [15].

As mentioned above, the configuration space of the ionic seven atom clusters considered here are marked by valleys separated by barriers and the determination of the ground state is a difficult task. For each cluster, several geometrical configurations were tried. Of the symmetric structures, we tried a cuboid with a missing atom, a ring, a ladder and a hexagonal capped structure. In each case, the energy was optimised by calculating the conjugate gradients and the possible Jahn-Teller distortions were investigated. In addition, we started with several random configurations and the geometry was optimized to see any possible structures. Various spin multiplicities were tried to optimize the spin. It is clear that it is impossible to guarantee that we have looked at the entire phase space, but we feel confident about our determination of the low energy states.

3. Results

3.1 Atoms and dimers

We start by presenting results on atoms and dimers in order to establish the accuracy of the calculations. In Table 1 we compare the calculated ionization potentials and the electron affinities of the various atoms with the corresponding experimental values. Since the halides are marked by ionic bonds, the most relevant comparisons are the ionization potentials of the alkali atoms and the electron affinity of the Cl. In all cases, the calculated values for these are within few percent of experiment. In addition to atoms, we also calculated the bond length, binding energy and the adiabatic electron affinity of the ionic dimers. Table 2 compares the calculated values with the corresponding experimental numbers. It is quite remarkable that both the calculated bond length and the binding energy are within a few percent of experiment. The bond length monotonically increases as one goes from Li to Cs. This is expected since the atomic size of the alkali increases. The binding energy, on the other hand, shows non-monotonic behavior. LiCl has the highest while NaCl has the lowest binding energy. The binding energy increases as one goes from Na to Cs. This progression can also be easily understood. The binding in dimers is predominantly ionic and the binding is determined by the charge transfer. Unlike the approaches based on Gaussian or other localized basis

Table 1

The ground state multiplicity, ionization potential (IP), and electron affinity (EA) of alkali and Cl atoms. The experimental results are included for comparison

atom	S	IP (eV)		EA (eV)	
		present	exp.	present	exp.
Li	doublet	5.59	5.39	0.35	0.62
Na	doublet	5.36	5.14	0.44	0.55
K	doublet	4.45	4.34	0.44	0.50
Rb	doublet	4.22	4.18	0.42	0.49
Cs	doublet	3.84	3.89	0.48	0.47
Cl	doublet	13.06	12.97	3.68	3.61

Table 2
Equilibrium bond length (BL), binding energy (BE), and electron affinity (EA) for various alkali chloride dimers

dimer	EA (eV)		BL (Å)		BE (eV)	
	present	exp.	present	exp.	present	exp.
LiCl	0.51	0.59	2.03	2.02	4.83	4.86
NaCl	0.77	0.73	2.38	2.36	4.18	4.27
KCl	0.64	0.58	2.71	2.67	4.33	4.49
RbCl	0.54	0.54	2.98	2.79	4.42	4.43
CsCl	0.46	0.46	3.08	2.91	4.69	4.64

where one can estimate the charge transfer via a Mullikan population analysis, it is difficult to partition the charge between atoms in the numerical schemes such as the present. To get an idea about the charge transfer we first calculated the radius of a sphere containing 95% of the charge in a free Cl atom. This radius was 2.6 a.u. We then took a sphere with a radius of 5.2 a.u. and placed it 2.6 a.u. away from the Cl site in the alkali-halide molecule, away from the alkali site. Such a sphere is likely to include most of the charge around the Cl site. By calculating the total charge within such a sphere for the molecule and for case where a free Cl atom is placed at the site occupying the Cl atom in the molecule, we estimated that the Cl gains about 0.7 e^- in various halides. Further, the ionization potential decreases as one goes from Li to Cs. Assuming a point charge model, the binding energy depends on the energy required to remove the charge, the value of the localized charge as well as the interparticle separation which increases with size. The net trend in the binding energy represents the competition between the decrease in ionization potential which facilitates the charge transfer and the increase in interparticle distance with size which reduces the interaction energy. Since the present work concerns the comparison of the negative ion photoelectron spectra with experiment, we also calculated the adiabatic electron affinity (EA) of the molecules. Table 2 compares these numbers with the available experimental data. The agreement is fairly close. Note that the trend in EA is opposite to that in binding energy. This is reasonable considering the fact that a stable neutral cluster is less likely to enhance its stability by acquiring an e^-.

3.2 Geometry and binding in M_4Cl_3 (M = Li, Na, K, Rb, and Cs)

Having established the accuracy of the method, we now present our results on the anionic and neutral $M_4Cl_3^-$ (M = Li, Na, K, Rb, and Cs) clusters. As mentioned earlier, Fatemi et al. [9, 10] had suggested three isomers in $Cs_4Cl_3^-$ to be a hexagonal capped ring, a cuboid (with a missing atom), and an octagon. Our earlier investigations suggested that the ladder structures are also very stable. We therefore focussed on four low energy structures namely a cuboid, a hexagonal ring with a capped atom, a ring, and a ladder structure. For all the structures, the calculations were carried out using the lowest symmetry and possible distortions were included. In each case, the energy was optimized via conjugate gradients. In addition and to further ensure that we are getting all the low energy structures, we started with several random configurations and tried to partly relax them to look for possible low energy candidates. We found that all the

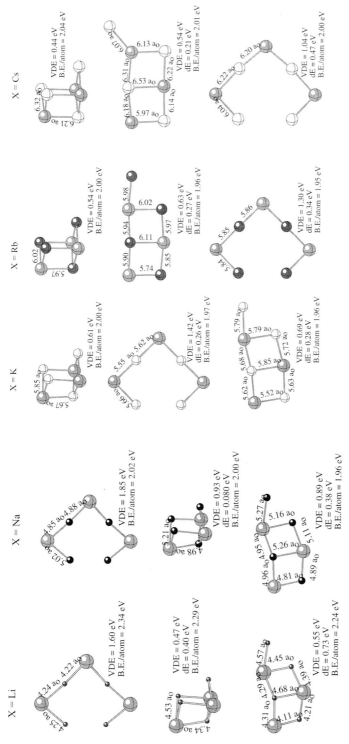

Fig. 1. Geometries, bond lengths, binding energies, electron affinities and relative energies for anionic $X_4Cl_3^-$ clusters (X = Li, Na, K, Rb, and Cs)

Table 3
Binding energy per atom and the relative stability of various structures for $M_4Cl_3^-$ (M = Li, Na, K, Rb and Cs) clusters

geometry	BE/atom (eV)				
	Li	Na	K	Rb	Cs
octagon	2.34	2.02	1.97	1.95	2.00
cuboid	2.29	2.00	2.00	2.00	2.04
ladder	2.24	1.96	1.96	1.96	2.01
	order of stability (energies in eV)				
most stable	oct	oct	cub	cub	cub
	cub (0.40)	cub (0.08)	oct (0.26)	lad (0.27)	lad (0.21)
least stable	lad (0.73)	lad (0.38)	lad (0.28)	oct (0.34)	oct (0.47)

anionic chlorides are marked by three lowest energy structures namely a cuboid with a missing Cl site, an octagonal ring and a ladder with a missing atom. In Fig. 1 we have shown the three structures for all the halides along with the various bond lengths, binding energies per atom and their relative stability. In Table 3 we give the binding energy per atom for all the clusters. It also gives the relative stability of the various structures and their energetics.

For $Li_4Cl_3^-$, the ground state is a ring where the Li atoms are pushed towards the inside. It has a binding energy of 2.34 eV per atom. It is closely followed by a cuboid with a binding energy of 2.29 eV/atom. A distorted ladder with a binding energy of 2.24 eV/atom is the least stable. Note that the binding energies per atom in the three structures differ by less than 0.1 eV/atom which is at the limit of theoretical accuracy. The $Na_4Cl_3^-$ clusters show a similar progression of geometries although the binding energies are smaller and the bond lengths are larger than in case of $Li_4Cl_3^-$. In particular, in case of ring structure, the atoms are only marginally pushed inside. For $K_4Cl_3^-$, $Rb_4Cl_3^-$ and $Cs_4Cl_3^-$ the situation is different. In the case of $K_4Cl_3^-$ the cuboid is the lowest energy structure while the ring is the second lowest structure. The ladder is the least stable although, as before, the changes in binding energy are within 0.05 eV/atom. For $Rb_4Cl_3^-$ and $Cs_4Cl_3^-$ the ground state is a cuboid, but the ladder is more stable than the ring structure. One also notices that as one goes from Li to Cs, the position of the uncoordinated alkali in the ladder changes progressively from one side to the other side and the ladder becomes less distorted. The alkali atoms in the ring structure also relax from inwards for Li to progressively outward for Cs. As expected, the bond lengths continuously increase from 4.2 a.u. in the case of Li to 6.2 a.u. in the case of Cs. This is simply due to increase in the size of the alkali atom. The binding energy per atom continuously decreases as one goes from Li to Cs. Since the bonding is mainly ionic, it is reasonable that an increase in bond length decreases the binding energy. It is interesting to observe that the alkali–Cl bond length in clusters and the binding energy per atom are larger than the corresponding quantities for dimers given in Table 1. The increase of bond length and the binding energy per atom with size is common to metallic and ionic clusters.

In addition to ionic clusters, we also calculated the ground state geometries of neutral clusters. These are shown in Fig. 2. Like the anionic species, the neutral clusters are

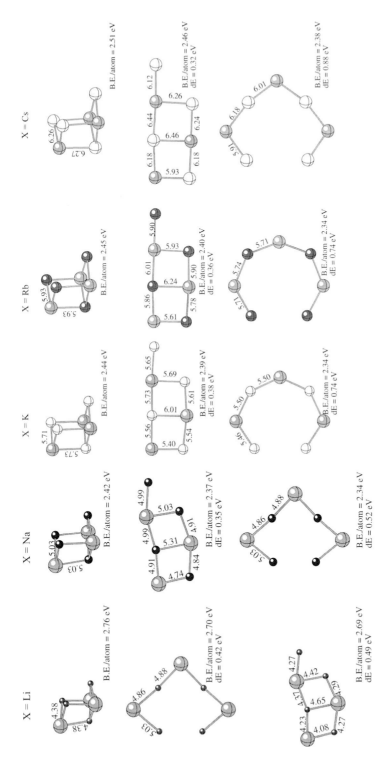

Fig. 2. Geometries, binding energies, and relative stabilities of neutral X_4Cl_3 clusters (X = Li, Na, K, Rb, and Cs)

also marked by three isomers with competing binding energies. The bond lengths in the neutral clusters are generally smaller than in the corresponding anionic clusters. In all cases, the ground state in the neutral clusters is a defected cuboid structure. Except for the case of Li_4Cl_3, all the clusters have a ladder as the next isomer and the octagon as the least stable isomer. It is also interesting to note that the ordering of the three isomers in the anionic and neutral clusters is different except for the case of Cs_4Cl_3 clusters. Since the thermal isomerization has been observed for Cs_4Cl_3 clusters, it raises the issue whether the identical ordering of isomers in the anionic and neutral species facilitates the isomerization? We will come back to this point later. It is also interesting to compare the shapes of the various clusters for the anionic and neutral geometries. For example, the octagon structures for the neutral K, Rb, and Cs chlorides are more buckled outwards than the corresponding anionic species. This can be understood in terms of localization of the extra electronic charge. Our studies show that this charge is localized in the interior of the octagon close to the opening of the ring. This charge attracts the positive alkali ions thereby reducing the buckling. In the cases of Li and Na, the charge is more localized on the outside of the octagon and this effect is not observed.

3.3 Comparison with experimental negative ion photoelectron spectra

We now come to a critical comparison of the present results with the negative ion photoelectron spectra. In these experiments, one generates the anionic clusters in beams and ionizes them by exposing them to photons. The energy distribution of the ensuing electrons provides information about the electronic structure of the neutral clusters. The most significant quantity in these experiments is the location of the main peak which corresponds to the vertical detachment energy (VDE) of the electron from the anion. In cases where the geometry of the neutral is not too different from the anion, the tail of the photoelectron spectra can provide information on the adiabatic electron affinity. Thermal effects can broaden the spectra and this does obscure the determination of the adiabatic electron affinity. In view of this, we will mainly focus on VDE for comparison with experiments.

Table 4 contains a summary of the VDE for all the clusters. As before, Li behaves anomalously. In all cases, VDE increase in going from Li to Na and then monotonically decrease. What is more interesting is that different structures for the same cluster generally have different VDE. It should therefore be possible to identify the structural isomers in well controlled photodetachment experiments. The octagonal ring has the highest VDE while the cuboid and the ladder have comparable VDEs. A study of the electron density shows that the high VDE for the ladder is related to its ability to

Table 4
Vertical detachment energy (VDE, in eV) for various geometries of $M_4C_3^-$ (M = Li, Na, K, Rb and Cs) clusters

geometry	Li	Na	K	Rb	Cs
octagon	1.60	1.85	1.42	1.30	1.04
cuboid	0.47	0.93	0.61	0.54	0.44
ladder	0.55	0.89	0.69	0.63	0.54

accomodate the extra electrons in a diffuse manner. One also notes that the earlier suggestion that the cuboid structures have the highest VDE is not correct. In fact the cuboids have the lowest VDE except for $Na_4Cl_3^-$. Before we discuss charge localization, let us discuss a comparison of the present findings with experiment.

Bloomfield and co-workers [9, 10] have measured the photoelectron spectra of $Na_4Cl_3^-$ and $Cs_4Cl_3^-$ clusters. We begin with the case of $Na_4Cl_3^-$. The spectra is characterized by a single peak around 2.0 eV. From Fig. 1, the most stable structure for $Na_4Cl_3^-$ corresponds to an octagonal ring and from Table 4, it has a VDE of 1.85 eV which is really close to experiment. The experimental spectrum does not have any features around 0.9 eV which shows that the cuboid and the ladder structures are not seen in the photodetachment experiments. This changes as one goes to the case of $Cs_4Cl_3^-$ where the photodetachment spectrum does exhibit features related to isomers. As mentioned before, the photoelectron spectrum of $Cs_4Cl_3^-$ at 273 K shows the main peak at 0.39 eV, minor humps at 0.72 and 1.01 eV and a set of high energy peaks at 1.50, 1.84, and 2.1 eV. For $Cs_4Cl_3^-$, the calculated relative stability of the three isomers are cuboid, ladder and octagon (see Fig. 1). These have VDE of 0.44, 0.54, and 1.04 eV. Taking account of the theoretical accuracy and the experimental uncertainty, it is safe to say that the three low energy peaks seen in experiments correspond to the three structures found here. The experiments also show that as the clusters are cooled to 116 K, the humps at 0.72 and 1.01 eV disappear and the clusters settle into a well defined state with VDE of 0.39 eV. Further the high energy peaks remain intact as the temperature is lowered. As explained in our earlier work, this can be understood via a study of the excited states. Our studies show that the cuboid has a total of nine excited states with energies of approximately 1.45, 1.85 and 2.1 eV which correspond to the major peaks seen in the low temperature experiments. The ring and ladder have first excited states at 2.28 and 1.10 eV, respectively. The 2.28 eV ring excitation coincides with the 2.1 eV excitation peak of the cuboid structure. The 1.10 eV ladder excitation energy is close to the 1.01 eV affinity of the ring structure. The absence of the 1.10 eV excitation peak of the ladder at low temperature further confirms that the ladder is not present at these temperatures. To compare directly to the low-temperature experimental photodetachment spectra we have calculated a joint density of states (JDOS) and show the results in Fig. 3. Also included in the figure is the peak at 0.44 eV which corresponds to the photoionization of the negative ion. For presentational purposes the JDOS is broad-

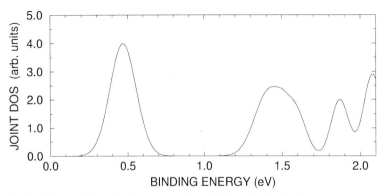

Fig. 3. The joint density of states for the $Cs_4Cl_3^-$ cuboid cluster

ened by 0.25 eV. Comparison of our JDOS to the experimental work of Ref. [9] shows excellent agreement. This further reaffirms that the cuboid is the most stable low-temperature structure and that the three structures isomerize at ordinary temperatures. Bloomfield and co-workers [9, 10] also experimented with other alkali halide clusters. They found no evidence of thermal isomerization in Na and K clusters.

3.4 Thermal isomerization

The above results show that while all the alkali halides are marked by three energetically close isomers, only the Cs halides have been found to exhibit thermal isomerization. This raises the issue why the isomerization is not observed in other clusters. What is even more surprising is that there is no evidence of the isomers in the photodetachment spectra of $Na_4Cl_3^-$. Before we resolve this paradox, one must remember that the isomerization requires that (1) the structures are energetically close, (2) they are separated by low energy barriers and (3) there exist low energy vibrational modes which enable the interconversion of the structures. To this end, we first present the vibrational frequencies of the various clusters.

Using the equilibrium geometries given in Fig. 1, we calculated the vibrational frequencies for all the anion structures. These calculations require a knowledge of the dynamical matrix. The details of the calculations are given in our previous paper [15]. Since we are interested only in low energy modes, we found that each structure has several low energy modes which are close in frequencies. In each case, some of the modes correspond to deformations which will convert one structure into another. To make the presentation easier, we present in Table 5 the lowest vibrational frequency for all the clusters. There are two interesting results. First, all the halides are marked by very low energy vibrational modes. Secondly, the lowest vibrational frequency for cuboid and octagon structures continuously increases in going from Cs to Li. To get an idea about the energy involved, the lowest vibrational frequency in $Cs_4Cl_3^-$ corresponds to an energy of 10^{-3} eV which is really small. While the existence of low frequency modes which can transform structures is the first step, one must also consider the dynamics of the transition. A conversion from the cuboid to the octagon proceeds via the ladder structure [19]. The energy difference between the ground state structure (which is either octagon or cuboid) and the ladder is critical to the isomerization. This energy is 0.21, 0.27, 0.28, 0.38 and 0.73 eV for the case of Cs, Rb, K, Na, and Li, respectively. This shows that the transition temperature to thermal isomerization would continuously increase as we go from Cs to Li. We, therefore, hold the view that it is the barrier to the ladder structure which impedes the isomerization in clusters other than Cs under similar conditions.

Table 5
Lowest vibrational frequency (in cm^{-1}) for various geometries of $M_4C_3^-$ (M = Li, Na, K, Rb and Cs) clusters

geometry	Li	Na	K	Rb	Cs
cuboid	130	80	48	33	32
octagon	61	33	18	20	12
ladder	30	8	11	16	8

4. Conclusion

We have shown that all the seven atom anionic alkali chlorides clusters are marked by low energy cuboid, ladder and ring structures. The relative stabilities of these three structures change with the alkali atom but the binding energies are close and they can be regarded as isomers. The structures, while energetically close, have different VDEs and therefore can be identified through negative ion photoelectron spectroscopy. All the clusters possess low frequency vibrational modes which can allow their interconversion. The lowest vibrational frequency increases from the Li to Cs chlorides indicating that the minimum isomerization temperature increases with decreasing cluster size. For $Cs_4Cl_3^-$ the clusters isomerize at around 270 K but settle into a well defined state at much lower temperatures. For $Na_4Cl_3^-$, thermal isomerization is not observed below 300 K. As discussed above, we suggest that this is because the ladder structure poses a barrier which requires a much higher temperature to cross. It is hoped that the present study will stimulate further experiments to investigate thermal isomerization in alkali halide clusters other than Cs by heating them to higher temperatures.

References

[1] J. Akola, H. Hakkinen, and M. Manninen, Phys. Rev. B **58**, 3601 (1998).
[2] C. Ashman, S. N. Khanna, and M. R. Pederson, Phys. Rev. B (submitted for publication).
[3] L. A. Bloomfield, private communication.
[4] J. L. Elkind, F. D. Weiss, J. M. Alford, R. T. Laaksonen, and R. E. Smalley, J. Chem. Phys. **88**, 5215 (1988).
[5] R. S. Berry, J. Burdett, and A. W. Castleman, Jr. (Eds.), Small Particles and Inorganic Clusters, Z. Phys. **26**, (1993).
[6] D. Scharf, J. Jortner, and U. Landman, J. Chem. Phys. **87**, 2716 (1987).
[7] A. Heidenreich, I. Schek, D. Scharf, and J. Jortner, Z. Phys. D **20**, 227 (1991).
[8] E. C. Honea, M. L. Homer, P. Labastie, and R. L. Whetten, Phys. Rev. Lett. **63**, 394 (1989).
[9] D. J. Fatemi, F. K. Fatemi, and L. A. Bloomfield, Phys. Rev. A **54**, 3674 (1996).
[10] F. K. Fatemi, D. J. Fatemi, and L. A. Bloomfield, Phys. Rev. Lett. **77**, 4895 (1996).
[11] C. Ashman, S. N. Khanna, M. R. Pederson, and D. V. Porezag, Phys. Rev. A **58**, 744 (1997).
[12] G. Rajagopal, R.N. Barnett, A. Nitzan, and U. Landman, Phys. Rev. Lett. **64**, 2933 (1990).
[13] P. Xia, L. A. Bloomfield, and M. Fowler, J. Chem. Phys. **102**, 4965 (1994).
[14] P. Xia and L. A. Bloomfield, Phys. Rev. Lett. **70**, 1779 (1992).
[15] M. R. Pederson and K. A. Jackson, Phys. Rev. B **41**, 7453 (1990).
K. A. Jackson and M. R. Pederson, Phys. Rev. B **42**, 3276 (1990).
[16] D. M. Ceperley and B. J. Adler, Phys. Rev. Lett. **45**, 566 (1980).
[17] J. P. Perdew, K. Burke, and M. Ernzerhof, Phys. Rev. Lett. **77**, 3865 (1996).
[18] W. Kohn and L. J. Sham, Phys. Rev. **140**, A1133 (1965).
[19] A. Heidenreich, J. Jortner, and I. Oref, J. Chem. Phys., **97**, 197 (1992).

phys. stat. sol. (b) **217**, 335 (2000)

Subject classification: 71.15.Mb; 73.61.Jc; 73.61.Tm; 73.61.Wp

Linear Scaling ab initio Calculations in Nanoscale Materials with SIESTA

Pablo Ordejón

Institut de Ciència de Materials de Barcelona (CSIC),
Campus de la U.A.B, E-08193 Bellaterra, Barcelona, Spain

(Received August 10, 1999)

In the context of linear scaling methods for electronic structure and molecular dynamics calculations, SIESTA was developed as a fully first-principles method able to deal with systems with an unprecedented number of atoms, with a modest computational workload. The method has allowed us to study a large variety of problems involving nanoscale materials, such as nanoclusters, nanotubes, biological molecules, adsorbates at surfaces, etc. Here we present a review of such applications.

1. Introduction

The contribution of first-principles calculations to several fields in physics, chemistry, materials sciences, and recently geology and biology is more important than ever. The main reasons are, on one hand, the steady increase in computer power, and on the other, the continuous progress in methodology (both in efficiency and in accuracy of algorithms and approximations). As larger and more complex systems are falling within the range of applicability of these methods, some barriers (formerly believed to be fundamental) are being reached. One of these is the scaling of the computational effort with system size. In the most favorable cases like with Density Functional Theory (DFT) [1,2], quantum mechanical formulations of the electronic structure of atomic systems scale as the cube of the number of atoms (or electrons) of the system [3]. This makes it very difficult to reach system sizes larger than a few hundreds of atoms, and is therefore a huge barrier for the study of problems in nanoscale materials. The way out of this trap was found several years ago, when a number of ideas suggested the possibility of developing approximate although accurate schemes to reduce the computational cost to linear scaling. These so called $O(N)$ methods (for a review, see [4]) have matured since those first proposals, and now constitute a viable route for studying systems with unprecedented size.

The key for achieving linear scaling is the explicit use of locality, meaning by it the insensitivity of the properties of a region of the system to perturbations sufficiently far away from it [5]. A local language will thus be needed for the two different problems one has to deal with in a DFT-like method: building the self-consistent Hamiltonian, and solving it. Most of the initial effort was dedicated to the latter [4,6] using empirical or semi-empirical Hamiltonians. The SIESTA project [7 to 10] started in 1995 to address the former. Atomic-orbital basis sets were chosen as the local language, allowing for arbitrary basis sizes, what resulted in a general-purpose, flexible linear-scaling DFT program [9 to 11]. A parallel effort has been the search for orbital bases that would meet the standards of precision of conventional first-principles calculations, but keeping as small a range as possible for maximum efficiency [10].

In this paper, we give an overview of the applications of SIESTA to problems in different nanoscale systems. Section 2 summarizes the main features and approximations used in SIESTA. We have divided the applications in four groups. In Section 3 we describe work done in carbon nanostructures: fullerenes and nanotubes. Section 4 deals with several metallic nanostructures: transition and noble metals, and nanowires. In Section 5 we deal with some application to biomolecules, and Section 6 is devoted to calculations in surfaces, disordered systems and defects. Finally, the conclusions are given in Section 7.

2. The SIESTA Method

SIESTA is based on DFT, and can use both local-density (LDA) [3] and generalized-gradients (GGA) functionals [12], including spin polarization, collinear and non-collinear [13]. The core electrons are replaced by norm-conserving pseudopotentials [14] factorized in the Kleinman-Bylander form [15], including scalar-relativistic effects, and non-linear partial-core corrections [16]. The one-particle problem is then solved using linear combination of atomic orbitals (LCAO). There are no constraints either on the radial shape of these orbitals (which are treated numerically), or on the size of the basis, allowing for the full quantum-chemistry (QC) know-how [17] (multiple-ζ, polarization, off-site, contracted, and diffuse orbitals). Forces on the atoms and the stress tensor are obtained from the Hellmann-Feynman theorem (including Pulay corrections), and can be used for structure relaxations or molecular dynamics simulations of different types.

The DFT equations are solved using the self-consistent field (SCF) method. For a Hamiltonian, the one-particle Schrödinger equation is solved yielding the energy and density matrix for the ground state. This is performed either by diagonalization (cube-scaling, appropriate for systems under a hundred atoms or for metals) or with a linear-scaling algorithm. These have been extensively reviewed elsewhere [4]. SIESTA implements two $O(N)$ algorithms [6,18] based on localized Wannier-like wavefunctions.

Once the density matrix has been obtained, the SCF procedure continues with the calculation of a new Hamiltonian matrix. The matrix elements of the different terms of the Kohn-Sham Hamiltonian are calculated in one of two different ways [9]. The terms that involve integrals over two atoms only (kinetic energy, overlap, and other terms related with the pseudopotential) are computed a priori as a function of the distance between the centers, and stored in tables to be interpolated later with very little use of time and memory. The other terms are calculated with the help of a uniform grid of points in real space. The smoothness of the integrands determines how fine a grid is needed, and, of course, the finer the grid, the more expensive the calculation. We remark that the use of pseudopotentials, which eliminates the rapidly varying core charge, is essential to provide functions smooth enough to make the grid integration feasible. This fineness is measured by the energy of the shortest wavelength plane-wave that can be described with the grid, in analogy with plane-wave calculations.

The calculation of the Hamiltonian matrix elements sketched above has an $O(N)$ scaling *provided that* the range of overlap between the basis orbitals is finite. To achieve that, we use basis orbitals which strictly vanish beyond a cut-off radius [10] (instead of the usual approach of using decaying orbitals and neglecting matrix elements by whatever criterion). The main advantage is consistency: given a basis, the eigenvalue problem is solved for the *full* Hamiltonian. Thus, the procedure is numerically very stable even for short ranges, in contrast with the usual approach. In this and previous works, the radial parts of the finite-range orbitals were determined in the spirit of the method Sankey and Niklewski [19] who proposed a scheme for minimal (single-ζ) bases that we have generalized to arbitrarily complete sets [10]. The single-ζ orbitals are obtained by solving the DFT atomic problem (including the pseudopotential) with the boundary condition for the orbitals of being zero beyond the cut-off radius, while remaining continuous.[1]) For the efficient generation of larger, more complete basis sets we have used the ideas developed within the QC community over the years, incorporating them into new schemes adapted to numerical, finite-range bases for linear scaling. Numerical multiple-ζ bases are constructed in the split-valence philosophy [9,10]. Our approach also allows polarization orbitals [10] which are obtained by numerically solving the problem of the isolated atom in the presence of a polarizing electric field.

3. Carbon Nanostructures: Fullerenes and Nanotubes

The discovery of the C_{60} fullerene [20] and, later, of the other fullerenes and carbon nanotubes [21], is one of the most stimulating findings in the last decades in chemistry and physics. It triggered a huge amount of work, and is one of the rare cases where theory has been well ahead of experiment in predicting new phenomena and behavior (mostly due to the difficulty of growing and handling samples for experimentation).

3.1 Structure of large carbon fullerenes

One of the first applications of the preliminary version of SIESTA was on the problem of the shape of large hollow carbon fullerenes [7]. Since 1992, structures of nested fullerenes, so called buckyonions, had been observed to be nearly perfectly spherical when irradiated with a high energy electron beam [22]. Soon these structures were proposed to consist of nested icosahedral fullerenes, with sizes ranging from that of C_{60} at the center of the onion, to many nanometers at the outer shells. However, the reason for the origin of the spherical shape remained unclear for several years. Elasticity arguments [23] indicate that *single shell* large icosahedral fullerenes would be faceted in their ground state structures, due to the constraints in the local curvature introduced around the twelve pentagons (except C_{60}, which is perfectly spherical). Simulations using empirical classical potentials reached the same conclusions [24]. However, the results of further tight-binding studies showed contradictory results [25,26] some of which suggested the possibility that the ground state of large icosahedral fullerenes was spherical rather than faceted [25]. In order to contribute to the solution of the controversy, we used SIESTA to calculate the lowest energy structures of icosahedral fuller-

---

[1]) For attaining a desired degree of convergence, the cut-off radii of the different orbitals depend on the species, and also on the specific orbitals themselves. We have proposed a systematic way of finding the different cut-off radii as a function of a single parameter that defines the finite-range approximation: the energy shift [10]. All pseudoatomic orbitals shift upwards in energy when confined to a sphere; defining this energy shift gives all the required radii, guaranteeing a balanced basis.

enes up to C_{540}, and found that they tend to be polyhedral instead of spherical, and that this polyhedral character is more pronounced as the cluster size increases [7]. Our calculations showed that the observed spherical shapes of buckyonions are not a consequence of an intrinsic property of the ground state of the individual fullerenes, but rather must be due to the interaction between shells, or be a non-equilibrium effect due to the electron bombardment. Several theories have been constructed since, following these two lines, the most convincing of which seems to be the formation of defects (a seven- and a fivefold ring adjacent to the original fivefold ring of the icosahedral fullerene, via the release of a C_2 molecule and Stone-Wales rearrangement), which locally remove the curvature of the fivefold ring producing a flat surface around the defect [27]. These structures are only slightly higher in energy than the icosahedral clusters [28], and are therefore likely to be formed during the electron irradiation process.

Other studies of carbon fullerenes include the absorption of C_{60} on Si(111) surfaces, which will be described in Section 6.

3.2 BN fullerenes

The similarities between B–N and C–C bonds makes boron and nitrogen natural candidates for the formation of fullerenes and nanotubes. While pure BN nanotubes have been synthesized [29] since 1995, the synthesis of fullerenes was much more difficult, and it was not until 1998 that pure BN fullerenes [30,31] were obtained. Alexandre et al. [32] used SIESTA to study the structure of these BN fullerenes, and to try to explain the difficulty in their growth compared to BN nanotubes. They developed a structural model for a single-shell BN fullerene which was observed experimentally with dimensions of 9 to 10 Å. The model is shown in Fig. 1. The structure consists of 36 BN pairs distributed in six 4-membered rings and thirty-two 6-membered rings (in contrast to the case of carbon, the presence of fourfold rings in BN structures is more favorable than fivefold rings, due to the absence of "wrong" B–B or N–N bonds). The structure, initially set with octahedral symmetry, was relaxed and a final T_d symmetry was obtained, due to the distortion of the fourfold rings which is caused by the tendency of N to form tetrahedral bonds and of B to form planar bonds. The calculations of Alexandre et al. [32] indicate that the total energy of the $B_{36}N_{36}$ fullerene is higher than that of BN nanotubes with comparable diameter (in particular an infinite (9,0) zigzag BN nanotube with 7 Å of diameter) by 0.48 eV per BN pair. This is considerably larger than the equivalent values for carbon fullerenes and nanotubes (which amounts to 0.31 eV for the C_{60} fullerene). This is an indication that BN fullerenes are not to be obtained as easily as

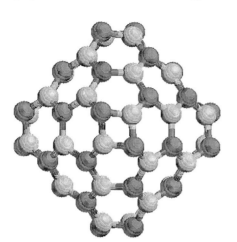

Fig. 1. $B_{36}N_{36}$ fullerene obtained by Alexandre et al. [33] Light (dark) spheres represent B (N) atoms. The back atoms are not shown

their carbon analogues, as is observed experimentally. However, the BN fullerene is quite stable once it is formed, as is inferred from the large calculated HOMO-LUMO gap (1.2 eV larger than that of the BN zigzag nanotube).

3.3 Structure, elastic and vibrational properties of nanotubes

Several studies of nanotubes have been done using SIESTA. In a first work, the structural, elastic and vibrational properties of single wall carbon nanotubes with different radii and chiralities were characterized [33]. The cases studied include (n,n) tubes (with n ranging from 4 to 10), and the (8,4) and (10,0) tubes. The main goal of the study was to monitor the evolution of different physical properties versus the radius and the chirality of the nanotube. Also, our work served to validate the results of more simplistic calculations of these properties (for instance, empirical models and estimates based on extrapolations of graphene results), and to point out their limits of applicability. These methods leave out effects of curvature which may be important when determining the properties of the nanotubes. From the structural point of view, our calculations show that relaxation effects due to the curvature of the tubes are small in general. Decreasing the tube radius increases the difference in the inequivalent bond lengths, and also the radius of the tube compared to the ideal wrapped graphene plane tube. These effects amount to less than 0.5% for tubes larger than (6,6), and therefore are hardly observable experimentally. The elastic properties show a similar trend, with small deviations from the ideal behavior. The ab-initio calculated strain energy follows very closely the α/r^2 behavior predicted by elasticity theory [34] for wrapping a graphene plane (see Fig. 2). Even for tubes as narrow as the (4,4), where deviations could be thought to be large due to the strong curvature and re-hybridization, the elasticity predictions hold very well. The calculated Young modulus does not show any appreciable difference with that of graphene, and does not depend (within the accuracy of the calculation) on the chirality or radius, even down to the narrowest (4,4) tube considered. The Poisson ratio also retains graphitic values, except for a slight reduction for small radii, and some chirality dependence (with armchair tubes displaying somewhat smaller values).

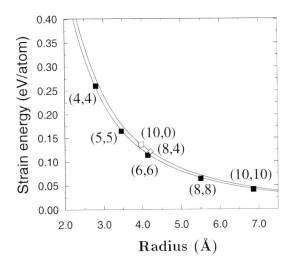

Fig. 2. Strain energy versus tube radius for several carbon nanotubes. The solid lines correspond to least-squares fit to the α/r^2 behavior (one for the (n,n) tubes, and another for the (10,0) and (8,4) tubes)

A detailed analysis of the evolution of the vibrational spectra of nanotubes with chirality and size was also performed [33]. The most important result of these calculations is that the phonon frequencies behave as expected from a zone-folding (ZF) calculation [35] from the graphene dispersion curves, indicating that the effects of curvature are again small. The ZF analysis gives a good qualitative (and sometimes quantitative) picture of the phonon properties, except for the well known deficiencies in low-frequency vibrations such as the breathing modes [35]. These, nevertheless, follow very closely the A/r law predicted from a graphene-derived force-constants calculation [35], indicating again that the effects of curvature on the force constants is small. For the smallest tubes (4,4) and (6,6), the deviations in the vibrational frequencies from the ZF or graphene force-constant results are significant, with a general trend to diminish with decreasing radius by effect of the curvature. This is true both for the low frequency modes (such as breathing modes and twistons) and for the high frequency ones.

3.4 Bent nanotubes

The former nanotube calculations describe the properties of these systems in their free state, or subject to small perturbations (as for the elastic properties) However, their response to large external perturbation is also very important. For instance, carbon nanotubes have been shown to be very flexible and can bend at large angles without fracture [36]. Beyond a critical angle, a kink is formed at the central region of the bend. This behavior has been observed experimentally, and reproduced by means of classical potential simulations [36,37]. Mazzoni and Chacham [38] studied this problem using SIESTA, and found dramatic differences with the results from classical simulations. In particular, they find that the kink is very localized, within a length of about 4 Å (in good agreement with the experimental results). Most importantly, the structure of the kink was predicted to present a considerable restructuring and re-bonding at the bend region. Fourfold rings, very unusual for carbon structures, result from the collapse of carbon hexagons at the highest stressed region of the bend. Part of the atoms at the fourfold rings are also fourfold coordinated, indicating a strong sp^3 character. The electronic properties of the tube (semiconducting in its free state) change in the region of the bend, where the character is metallic.

3.5 Oxidation and opening of nanotubes

In deposited samples, the tips of the nanotubes are always observed to be closed [21]. This represents a serious limitation to the possibility of filling them with a variety of substances, which would open the use of nanotubes to a large number of potential applications. However, several experiments have shown that the nanotube caps could be eliminated by exposing the samples to air and heat, suggesting an oxidation reaction at the ends of the tubes [40, 41]. Similar results were obtained in a CO_2 atmosphere. While the detailed description of the burning and opening process is a very complicated problem, beyond the scope of present day techniques, Mazzoni et al. [45] addressed two relevant questions about this process using SIESTA: (i) why does O_2 preferentially attack the tube caps, and (ii) why and how the tubes remain open after the oxidation takes place (i.e., after the O_2 supply has been cut). The calculations showed that the burning process by O_2 (with the release of CO_2) is very favorable energetically, both in the cap and the walls of the tube. However, the answer for the first question was shown to lay in the release of strain energy accumulated in the tube cap. The opening of a hole by the oxidation process releases part of this energy, even when the smallest holes are starting to open, and is therefore more exothermic than the opening at the wall. The difference in strain energy between the cap and wall of the tubes is quite large, amounting to about 9 eV for the cap of the (6,6) tube. Thus, releasing the strain energy favors the oxidation of the cap versus the wall. With respect to the second question, the structure of the hole is highly stabilized due to the presence of oxygen atoms saturating two carbon dangling bonds. In order to estimate the likelihood of a spontaneous closing of the hole (with the release of CO_2), a series of calculations were done to find an upper bound to the energy barrier for closing. For the (6,6) tube, it was found that the oxidized open tube was slightly more stable (by 0.3 eV) than the closed tube plus three CO_2 molecules if the hole was at the cap, with an upper limit to the energy barrier of 4.2 eV. In contrast, the closing reaction for the wall tube was favorable by 9.2 eV, and no energy barrier was found. Therefore, the picture that arises is the following: when O_2 is present in the atmosphere, the tubes suffer oxidation and holes open both in the walls and the tips, since the burning is a very exothermal process. When the oxygen flow is cut, the wall holes tend to close due to the high energy of the open, oxidized hole with respect to the release of CO_2. However, the oxidized holes at the tip are stable against closing, mainly due to the strain energy accumulated in the closed cap. This also explains the presence of open tips in a CO_2 atmosphere.

3.6 Electron states in finite nanotubes

The electronic properties of nanotubes exhibit a wide range of very interesting phenomena. Some of these are related to the structure and topology of the tubes. For instance, armchair (n,n) tubes are metallic, while zigzag $(n,0)$ tubes can be semiconducting or metallic depending on n [46]. However, there are phenomena which are purely related to the nanoscale size of these materials, such as quantum confinement effects. For instance, a conducting armchair tube cut to a finite length should behave as a finite 1D box in which the conducting electrons are confined. This 1D quantum confinement should then yield electronic states which are standing waves (SW) in contrast to the running waves of an infinite system. The periodicity of the SW is not directly related to the atomic periodicity, but rather to the length of the tube. Since the Fermi level occurs at the wave vector $k_F \approx 2\pi/3a$ (where a is the atomic periodicity along the tube axis), the periodicity of the SWs at the Fermi level corresponds to $\lambda_F \approx 3a$, i.e., three times larger than the atomic periodicity. For these SWs to be observable, the energy resolution must be better than the SW energy separation (which is inversely proportional to the tube length L), and the electron coherence length larger than the tube length. These conditions were matched by the experiments of Venema et al. [47], performed with tubes of $L \approx 30$ nm at a temperature of 4 K deposited on a gold surface. The experiments consisted on Scanning Tunneling Spectroscopy (STS) imaging, a technique that allows to sample the spatial shape of the density of the electronic states with a given energy from the Fermi level. Calculations using SIESTA [48] served to confirm the interpretation of the experiments in terms of standing wave states, and in particular explained several unclear points like: (i) the position and periodicity of the observed peaks, (ii) the small effect of the gold substrate on the observed standing waves. As to the first issue, it was shown that, depending on the length of the tube, and the particular point where the scanning was made, different patterns could be found, explaining the experimental results. A catalog of possible images was developed by J. Soler, based on a simple Hückel model of the SW states. As for the tube–substrate interaction, the calculations indicate that there is a small but significant charge transfer from the substrate to the nanotube, but mainly localized to the atoms closest to the substrate. The binding of the nanotube to the substrate was shown to be strong. However, this does not seem to have a significant effect on the SW states, which resemble very closely those of a free finite nanotube.

4. Metallic Nanostructures: Clusters and Wires

The study of metallic nanostructures has experienced an impressive boost in the last few years, mainly due to the advances in preparation and characterization techniques. Nanomanipulation at the atomic scale is opening the way to a whole new world of nanosize clusters, with properties well distinct from those of the macroscopic solids, and also of small molecules. The nanometer size of these aggregates is precisely the source of this difference in behavior, because that is the critical length at which many properties start to change from molecular-like to solid-like.

The case of metallic nanostructures is specially interesting, but also difficult from the theoretical point of view. In particular, nanoparticles of transition and noble metals pose a tremendous difficulty to theory, due to the large number of atoms involved, their quasi-metallic character, and the presence of d electrons and, at the same time, vacuum regions, which make the computations extremely demanding. SIESTA is specially well suited for the study of these systems. The d electrons are very accurately and efficiently described by means of an atomic-like basis set. At the same time, the description of vacuum around the clusters does not involve a significant computational workload, since the basis set is localized around the cluster region (unlike in plane waves where the vacuum is as expensive as the atomic regions).

4.1 Gold nanoclusters

The structure of nanoclusters is one of their most important characteristics and it is intimately related to the rest of the properties of the cluster. Despite impressive advance in experimental techniques like X-ray diffraction, transmission electron microscopy (TEM) or scanning tunneling microscopy (STM), it is still very difficult to extract precise information on the structure of these systems from experiment. Theory represents an important help in determining the structure of these aggregates. Garzón et al. [49] performed thorough optimizations of intermediate size (1 to 1.5 nm) Au_n ($n = 38, 55, 75$) nanoclusters, using classical (Gupta n-body) interatomic potentials [50]. They used dynamical and genetic-symbiotic (evolutive) optimization methods [51] to make an exhaustive search of possible minima. An interesting trend was observed in the resulting minima found: a set of many (more than a hundred) distinct, but almost degenerate in energy (within 10 meV/atom), stable configurations were found within a very narrow energy range from the ground state. For all cluster sizes, the great majority of these structures are amorphous in nature (based on the features of the pair distribution functions). More intriguing, for clusters with $n = 38$ and 55, the lowest energy configurations are amorphous, instead of the expected crystalline (O_h) or quasicrystalline (I_h) structures. However, the difference in binding energy between the most stable disor-

Table 1
Differences in energy between the lowest amorphous and ordered configurations, for Au_n clusters with $n = 38$, 55 and 75. The SIESTA values correspond to a double-ζ basis with polarization for the s shell, and the GGA approximation

	$E_{am} - E_{ord}$ (eV)		
	Au_{38}	Au_{55}	Au_{75}
Gupta	−0.014	−0.515	−0.419
SIESTA	−0.608	−0.357	−0.214

dered and the lowest energy ordered isomer was very small. For $n = 75$ the situation is reversed, but still with a very small difference between the ordered cluster (D_h symmetry) and the manifold of disordered isomers.

Garzón and coworkers [49,52] tested the validity of the classical potential results using SIESTA to obtain the energy differences between amorphous and ordered structures within DFT. The coordinates of the lowest energy clusters obtained with the Gupta potentials were used as initial configurations in an unconstrained relaxation using the DFT forces (both at the LDA and GGA levels) [49,52]. The results confirm the predictions of Gupta's model, and are shown on Table 1. We see that, within DFT-GGA, for all the cluster sizes (even for $n = 75$) the most stable structure is amorphous. However, as predicted by Gupta's potential, the energy difference is very small in all cases.

In a further effort to understand the origin of this striking situation, Soler et al. [52] developed a model in terms of local stresses which explains the tendency to form amorphous low energy structures. In essence, the tendency of metallic bonds to contract at the surface (due to the decreased coordination) produces a de-stabilization of the compact, ordered structures. The amorphization is further favored by the low energy cost associated to bond length and coordination disorder in metals. These are very general properties of metallic bonding, which are specially important in the case of gold. However, other metallic clusters (such as Pt and Pd) are predicted to behave similarly, based on their metallic properties.

4.2 Gold nanowires

A recent spectacular achievement is the fabrication of monoatomic gold wires, which can be produced in several ways (like the tip of a STM microscope [53], mechanically controllable break junctions [54] or even with simple tabletop set-ups [55]). Even more impressive is the possibility of visualizing these nanometer wires by means of transmission electron microscopy [56]. In these experiments, a bridge of four atoms connecting two gold tips was observed stable for more than 2 min. A striking fact about these observations was the spacing between atoms in the wire, which was of 3.5 to 4.0 Å. Later reports [57] have even increased this distance to about 5 Å, a value much larger than the typical Au–Au distance (2.5 Å in Au_2, and 2.9 Å in bulk gold). In order to explain these unexpectedly large interatomic distances, SIESTA has been used in several works. A first study by Torres et al. [58] has shown that a linear gold chain (both infinite and finite) would break at interatomic distances of less than 3 Å, much smaller than the observed ones. In a further work, Sánchez-

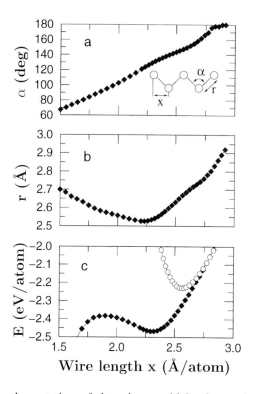

Fig. 3. Results for the a) bond angle and b) bond length of a monoatomic gold chain with zigzag geometry, as a function of its length per atom. c) Binding energy of the zigzag chain (solid symbols) and the linear wire (open symbols)

Portal et al. [59] propose an explanation of the observed interatomic distances in the nanowires. They show that gold monoatomic wires exhibit a zigzag shape instead of a linear one. This shape remains under tension, and only become linear just before breaking. Fig. 3 shows the results of the calculations, where it is clearly seen that, except for wire lengths close to the breaking limit, the ground state of the wire has a zigzag geometry. It was further shown that the barrier for rotation of the zigzag chain was very small (about 60 meV for a seven atom wire suspended between two pyramidal tips). At temperatures higher than 40 K, the rotation of the wire would be faster than the millisecond scale, and therefore the TEM images would represent an average of the rotating wire. This provides a possible explanation for the observed interatomic distances: if the actual wires have an odd number of atoms, with those of the extremes fixed by the contacts, the odd-numbered atoms would stay almost fixed on the same axis, while the even-membered ones could rotate rapidly around the axis. They would then offer only a fuzzy image, that could be missed by the TEM, while the fixed odd-numbered atoms would be clearly visible. Therefore, the observed TEM interatomic distances would correspond to the distance between odd-numbered atoms in the zigzag chain, being thus much larger than the real interatomic distance between first neighbors. Further calculations of the vibrational frequencies were performed, as a possible way to verify experimentally the zigzag geometries.

4.3 Magnetic properties of low-dimensional Fe systems

The interest in low-dimensional magnetic systems stems from the enhancement of the magnetic moment of the material when the atomic coordination is diminished. This opens potential applications for storage devices characterized by high storage density and miniaturization. Iron is the most studied magnetic element, and its a likely candidate for use in such low-dimensional devices. Izquierdo et al. [60] have studied the magnetic moments of several low-dimensional arrangements of Fe atoms, with SIESTA (which predicts a magnetic moment for bulk bcc iron is $2.35\mu_B$, in close agreement with the experimental value of $2.22\mu_B$). In particular, they have considered aggregates of Fe

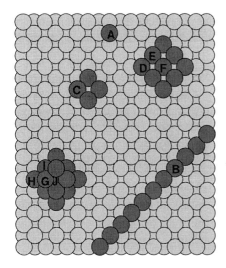

Fig. 4. Labeling of the different inequivalent sites of Fe nanostructures supported on Ag(001)

deposited on the Ag(001) surface. Fig. 4 shows the nanostructures studied: a single adatom, a 1D wire, a 4-adatom and a 9-adatom 2D clusters, a full monolayer and a 13 atom 3D aggregate. The summary of the results for the magnetic moments is shown in Table 2. From these results, we observe a clear tendency towards higher magnetic moments with decreasing iron coordination. We also see that SIESTA predicts magnetic moments very close to those obtained by KKR calculations [61].

4.4 Other transition metal clusters

Several other problems related to clusters of transition metals have been studied using SIESTA. Calleja et al. [62] considered the problem of the ground state structure of the $Ni_{12}Al$ cluster. This was a controversial issue due to the discrepancies found between calculations with different empirical potentials. The results of the DFT calculations showed that the most stable structure is a distorted icosahedron configuration, with the Al atom in the cluster surface. Rey et al. [63] have also studied the structural and magnetic properties of free Cu and Fe clusters.

5. Biomolecules

One of the most exciting possibilities which are being open by the development of efficient linear scaling ab-initio techniques is the study of biological molecules from first

Table 2
Local magnetic moments (in units of μ_B) at the inequivalent sites of several Fe clusters supported on Ag(001). Inequivalent sites are indicated with capital letters in Fig. 4. Available KKR data for the adatom and for the 9 atom cluster are given for comparison

		atom label	SIESTA	KKR [62]
0D	adatom	A	3.36	3.32
1D	wire	B	3.17	
2D	4-atom	C	3.23	
	9-atom	D	3.26	3.22
		E	3.23	3.17
		F	3.17	3.11
	monolayer		3.02	3.00
3D	13-atom	G	3.15	
		H	3.17	
		I	3.04	
		J	2.91	

principles. These systems are extremely complicated for a number of reasons: (i) the large number of atoms involved, which make calculations very demanding; (ii) the complexity of the chemical interactions present, which include widely different kinds of bonding: covalent, ionic, hydrogen bridges, and van der Waals, which must be described with a similar degree of accuracy; (iii) the time scale at which many processes take place (such as conformational changes) are orders of magnitude larger than those available for atomistic simulations. The situation is further complicated by the fact that most of the biological processes occur in solution, and the presence of water molecules and ions is essential to describe these processes. Despite these facts, electronic structure calculations and atomistic simulations can be a unique tool to address the study of many properties of biological molecules. For instance, the conductivity mechanism of DNA molecules is currently being studied experimentally using newly developed techniques, but is very poorly understood (there is even a strong controversy on the interpretation of the experimental results [64]). Ab initio calculations can help resolve these issues. We have started to study several biological systems using SIESTA, beginning to explore the actual possibilities of this approach in biomolecules.

5.1 Hydrogen bonding between nitrogenated bases

As a first step in the study of DNA systems, the more simple problem of the hydrogen bonding between nitrogenated base pairs (adenine, guanine, thymine and cytosine) was considered, as the first building blocks of DNA. Although hydrogen bonds are only one of the many factors which stabilize the DNA helix, it was important to establish the reliability of the approach for the description of this kind of weak bonding. A thorough study [65] of 30 nucleic acid pairs was done, addressing the precision of the approximations (basis sets, grid, etc.) and the accuracy of the GGA functional [12]. It is well known that LDA underestimates the bonding distance in hydrogen bonds, while GGA produces very good results. This was again confirmed by our calculations. In order to assess the accuracy of the SIESTA results, the energies of the 30 base pairs were compared with those obtained by Sponer et al. [66] with second-order Møller-Plesset (MP2) perturbation theory (a correlated quantum chemical method that is considered the best method currently capable to treat for systems of this size). The basis set used was a double-ζ plus polarization in all the atoms (including hydrogen), and with cut-off radii defined by an energy shift of 50 meV (a basis set which was later used in the simulation of the whole DNA helix, as described below). As in the MP2 calculations, the coordinated were those obtained by Sponer and coworkers at the Hartree-Fock level. The results for the binding energy of the base pairs (the difference between the energy of the hydrogen-bonded pair and the free bases) are plotted in Fig. 5. We can see that the correlation between MP2 and SIESTA results is excellent, with an average deviation of less than 0.73 kcal/mol. This is well bellow the accuracy that one can expect for total energies in DFT. This shows that our approach produces accurate values for the binding energies at hydrogen bonds. We also performed structural optimizations with SIESTA, to obtain the first DFT geometries for many of the base pairs under study. Former DFT works had only considered the structures of the common Watson-Crick base pairs, and of the simple cytosine–cytosine (CC) pair (which has a high symmetry and the smallest number of atoms of the set of base pairs). For these systems, our results agree closely (always within 2% for the bonding distances at the hydrogen bonds) with those of previous DFT calculations.

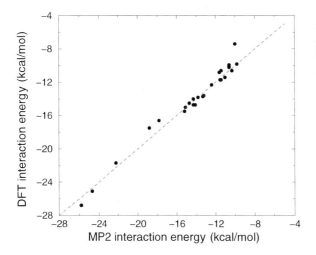

Fig. 5. Binding energies of a set of DNA base pairs, comparing the values obtained with SIESTA [65] with those of MP2 perturbation theory (by Sponer et al. [67]

5.2 DNA strands

Feasibility tests on DNA were performed in the early stages of the SIESTA project, by relaxing a dry B-form poly(dC)-poly(dG) structure with a minimal basis [8,9]. Based on this preliminary tests, and with confidence from the results of the base-pair calculations, a first serious study of a realistic DNA strand was started [67]. In this calculation, Artacho et al. studied a dry form of DNA in vacuum, with no water and counterions. Therefore, they considered the molecule in the A-helix configuration (which is the one that obtains in dry conditions), in its acidic (protonated) form. This choice of extremely idealized conditions were taken because it is a very sensible starting point for the theory, since it provides a clean cut reference for the isolation of environment effects in further studies under more realistic conditions. Moreover, several recent experiments (on DNA conductivity for instance) explore the properties of DNA strands in very similar conditions, which are not relevant for biological processes but provide insight in the physics of transport within the strand.

The DNA strand used in the simulation was an A-helix configuration of a poly(dC)-poly(dG) structure, in which the sequence of one of the strands contains only cytosine and the other guanine. Therefore, all base pairs are equivalent. The initial coordinates for the calculation were obtained from experimental data on X-ray diffraction of long DNA fibers [68]. It must be stressed that the resolution of these X-ray data is fairly poor, and does not allow to resolve the atomic coordinates. Therefore, in obtaining the "experimental" coordinates, there is a process of refinement with empirical force constants involved. Only to that extent can the coordinates be considered as "experimental". We simulate the long helix by using periodic boundary conditions, with a whole turn of the helix in the simulation cell. This contains eleven base pairs, with a total of 715 atoms. We relaxed the structure with a conjugate gradient algorithm using the quantum mechanical DFT forces from SIESTA. In an initial optimization, the basis used was a double-ζ set with no polarization orbitals (except in the P atoms), and rather short confinement radii. The result of this optimization was further relaxed using a better basis set (of quality

Table 3

Bonding distances at H-bonds in a A-DNA helix of poly(dC)-poly(dG). Experimental values from Ref. [69] are compared with the results of a structural optimization using SIESTA

	experiment	SIESTA
N2(H) ... O2	2.86	2.86
N2–H	1.00	1.03
N1(H) ... N3	2.85	2.83
N1–H	1.00	1.04
O6 ... (H)N4	2.86	2.73
H–N4	1.00	1.05

similar to that of the study of the base pairs above: double-ζ plus polarization, and longer orbitals) for the atoms at the hydrogen-bonds. The minimization was continued until the maximum atomic force was smaller than 0.1 eV/Å (with average forces below 0.01 eV/Å). The resulting geometries were then analyzed and compared with the original experimental results. Table 3 shows the atomic distances in the hydrogen bonds, in which overall excellent agreement is observed between the experimental values and those obtained from the geometry optimization. It should be noticed that the positions of the H atoms cannot be extracted from the X-ray diffraction data, and they are therefore located at a nominal distance of 1 Å, as seen in Table 3. The values obtained in the calculation are, in any case, very close to this nominal distance.

Besides the geometry of the H-bonds, other structural parameters are important in the description of the DNA conformation, describing the relative positions of the bases along the helix. Some of them are presented in Table 4, again comparing the experimental and SIESTA values. The meaning of each parameter is shown in Fig. 6. The value of the twist angle is fixed by the supercell geometry, and therefore coincides in the experiment and calculated results. The other structural parameters show some differences, but overall the agreement is quite good.

Besides the structural properties, Artacho et al. [68] have made a detailed study of the electronic properties of the A-helix, analyzing the structure of occupied and empty levels around the gap, the electrostatic potential created by the molecule, the charge density contours, etc. This study is the first ab-initio approach to DNA, and hopefully will be followed by more realistic studies where the effect of water and counterions will be taken into account.

Table 4

Configuration parameters in a A-DNA helix of poly(dC)-poly(dG). Experimental values from Ref. [69] are compared with the results of a structural optimization using SIESTA

	experiment	SIESTA
twist	32.7°	32.7°
roll	8.95°	10.47°
slide	− 1.78 Å	− 1.54 Å
propeller twist	16.07°	11.58°

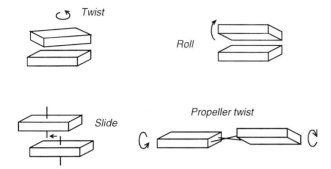

Fig. 6. Scheme of some of the parameters defining the structure of the DNA helix, used in Table 4

5.3 Charge density in crambin

With the recent advances in ultra-high resolution X-ray diffraction, the experimental analysis of the electronic charge density in biomolecules is starting to be possible, even for complex systems. The availability of theoretical results is very useful, since it serves as a guide for the interpretation and the processing of experimental results. Recently, the group of C. Lecomte has measured the charge density maps of crambin, using synchrotron radiation. Crambin is a plant seed hydrophobic protein with 46 residues and 642 atoms. Fernandez-Serra et al. [69] have collaborated closely in the interpretation of the experimental results, performing calculations of the charge density profiles in the different peptidic bonds along the protein, using SIESTA. A thorough study and comparison with the experimental results was done on the average charge density maps in the peptidic bonds, and its dispersion among the different bonds. The results of the calculations are shown in Fig. 7, for the deformation density (difference between the actual density of the molecule and the sum of atomic densities). Positive (negative) regions represent accumulation (depletion) of charge with respect to the sum of atomic densities. Both the average deformation density and the dispersion are shown. Also, an

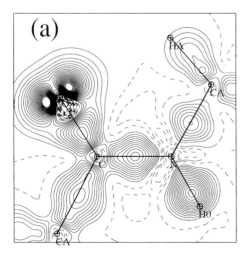
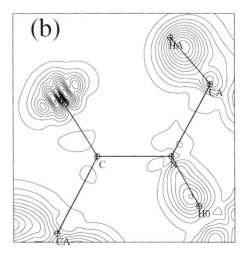

Fig. 7. Deformation charge (difference between the actual charge of the molecule and the sum of atomic charges) at the peptidic bond of crambin. Full (broken) lines indicate positive (negative) contours. a) Average between the different peptidic bonds. b) Dispersion with respect to the average

analysis was made of the differences among peptidic bonds due to the environment (different residues, and different position in the globular protein: surface or inner part).

6. Surfaces and Disordered Systems

The field of surfaces and disordered systems is by no means new. However, the possibility of performing simulations in systems of very large size opens the possibility of exploring problems which were usually beyond the scope of traditional schemes, like absorption and deposition of molecular species in surfaces, or defects and surfaces in disordered systems. These problems require the use of cells with a very large number of atoms, and therefore approach like SIESTA are very much needed for their study. Several works along these lines have been done recently.

6.1 Absorption of C_{60} on Si(111)

The absorption of C_{60} on different surfaces has been studied intensely during the last years, both experimentally and theoretically. The interaction with the surface is very different depending on the type of substrate and its orientation. In particular, for the Si(111) surface there is a very strong interaction due to the presence of highly reactive dangling bonds at the surface. Rebonding and charge transfer are therefore expected, and certainly have been observed experimentally. Pascual and coworkers [70] have studied the absorption sites and configurations of the molecules on the surface with STM, finding two main types of sites: strongly and weakly bonded, respectively. Sánchez-Portal [71] has used SIESTA to study the atomic configurations of these bonding sites. An interesting result of these calculations is the interpretation of the observed STM intermolecular features in terms of molecular orbitals of the free C_{60} molecule [72]. In the

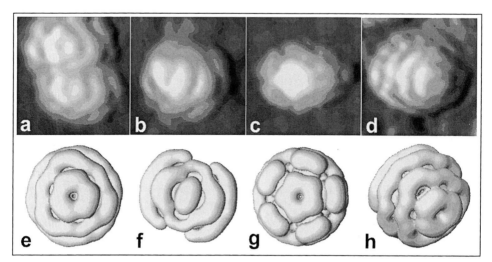

Fig. 8. Comparison of the experimental STM images of individual C_{60} molecules on Si(111) (parts a to d) with the calculated molecular orbitals of the free molecule (parts e to h). In the calculation, a uniaxial deformation of 5% was applied along an axis of three- (e), two- (f), five- (g) and threefold symmetry (h). In parts e) to g), the axis was perpendicular to the plotting plane, whereas in h) it was rotated at an angle of 30°. In all cases, the orbitals plotted are the HOMO

upper panel of Fig. 8 we see some of the experimental STM images. They present very different internal shape, orientation and symmetry. The lower panel of the same figure shows the shape of selected molecular orbitals of the free C_{60} molecule. In all cases, the molecular orbitals were the highest occupied orbital that was obtained by applying a uniaxial deformation of 5% along an axis of three-, two-, five- and threefold symmetry, respectively. This deformation was used to simulate the effect of the substrate interaction, which will split the manifold of degenerate HOMO states of the free C_{60}. We see that the comparison between computed MOs of the distorted molecules and the experimental STM images is excellent, suggesting that the STM is imaging the C_{60} molecular orbitals.

6.2 Absorption and growth of Ba on Si(001)

Thin-film metal silicides have received great attention in recent years, due to their potential applications to microelectronics. Among them, the alkaline-earth-metal silicides are of special interest since the discovery [73] that they are an essential step in the epitaxial growth of crystalline oxides on Si(001). These could eventually eliminate the fundamental limitations of silica-based gate technology in field effects transistors. The growth of the crystalline oxide was shown to depend critically on the prior formation of a sub-monolayer of silicide. Therefore, studying the process of adsorption and growth of this silicide is important to a full understanding of the oxide formation. Wang et al. [74] have used SIESTA to simulate the first stages of the deposition of Ba on Si(001). They studied the interaction between a Ba adatom and the reconstructed Si(001) surface at very low coverage. The preferential absorption site was found to be located in the trough between Si dimer rows, in good agreement with the observations of STM experiments at very low coverage. The bonding was shown to be strongly covalent, in contrast to what would be expected from the large electronegativity difference between Si and Ba. Only a small charge transfer of $0.15e$ was obtained, which explains the core level shifts experimental results of Cheng et al. [75] The problem of the diffusion of Ba on the Si(001) surface was also addressed in view of the energetics results obtained in this work. It was found that diffusion is much more likely (seven orders of magnitude) to occur in the direction of the dimer rows than perpendicular to it. Therefore, the authors predicted a one-dimensional growth behavior, with the formation of Ba chains in the trough between dimer rows, which was later confirmed by STM experiments [76]. Further studies [77] are in progress to explain the experimentally observed phase diagram for higher coverages [78], and the structure and properties of the epitaxially grown oxide.

6.3 Metastable FeSi on Si(111)

Another interesting problem concerning metal silicides on silicon surfaces is that of metastable phases. For FeSi, the stable phase in the bulk (ϵ-FeSi) has a lattice mismatch of -6.4% with the silicon substrate. On the other hand, the CsCl phase, which is not stable in the bulk, has only a 2% lattice mismatch with Si(111), and is therefore metastable when grown epitaxially. Experimentally, FeSi is found to grow in the CsCl structure in different conditions [79,80]. However, the details of the surface structure and termination were unclear. Hinarejos et al. [80] measured the angle-resolved photoemission spectra to shed light in this issue. LEED spectra indicated an unreconstructed 1×1

structure. In a first attempt to interpret the photoemission results, simplified calculations were done for the surface band structures for both the Fe and Si terminated FeSi surfaces, with the apparent conclusion that the surface was Fe terminated. However, Junquera et al. [81] have re-examined the problem using SIESTA, and found that only the Si-terminated surface is compatible with the experimental surface band dispersions. In particular, the photoemission data reveal the presence of a surface band of Λ_1 symmetry at the Γ-point, which can only be reproduced if the surface is Si-terminated. The calculated band position (-3.6 eV from the Fermi level at the Γ-point) agrees very well with the experimental value (-3.5 eV), as well as the shape of the band.

6.4 Liquid silicon surfaces

Liquid silicon is a rather peculiar system. At melting, it transforms from a covalent semiconductor to a liquid metal. This phase, however, is interesting in that its coordination (≈ 6 to 7) is lower than that of typical liquids (≈ 12), due to the persistence of directional bonding in the liquid phase. Very little is known about the structure of its surface, and measurements are extremely difficult due to the high reactivity and melting temperature (1684 K). First principles molecular dynamics calculations were performed with SIESTA by Fabricius et al. [82], to learn about the structure of the liquid silicon surface. The most striking result of these calculations is the observation of a marked atomic layering of the density near the surface. This layering is similar to that observed experimentally for other metals (like Ga and Hg) with low melting temperatures. In the case of silicon, the layering originates from a different effect than in the other metals, with remanent directional bonding playing an essential role. This layering has, however, no major effects on the electronic and dynamical properties of the surface, which are very similar to those of bulk liquid silicon.

6.5 Defects in amorphous semiconductors: a-Si and ta-C

The study of the structure and properties of amorphous solids by means of first principles simulations is difficult due to the large supercell sizes needed to reproduce fairly the lack of translational order in the real materials. However, these simulations can provide essential information on the microscopic characteristics of the materials, which sometimes are not accessible from the experimental point of view. Such is the case with the defects in the amorphous network, for which experiments provide only partial information on their atomistic structure and origin. For instance, electron spin resonance (ESR) experiments give an idea of the concentration of unpaired spin defects in the samples, and their degree of localization and even orbital character, but theory is needed to build a particular model for the structure of the defects. Fedders et al. [83] have studied the problem of defects in a-Si and ta-C performing calculations with SIESTA. For a-Si, they find that the dangling bond configuration is the most likely responsible for the ESR ($g = 2.0055$) signal in that material, and discarding the floating-bonds model. These conclusions are obtained by the analysis of the spin localization in each of the defect configurations. Whereas the dangling bond shows a localization of about 40 to 50% in the central atom of the dangling bond (in good agreement with the experimental estimate of 50 to 80%), the floating bond shows a much more diffuse spin density, with only 55% of the spin in the *five* atoms forming the defect. It is interesting to mention that the spatial extent of spin and charge for the defect state are quite different, the spin being much more localized than the charge of the state. Another conclusion of the work is that, in order to study the properties of defects states, a cell with sufficiently large size and a very small number of defects is required; otherwise, the interaction and hybridization between defect states destroys their localization properties. For ta-C, the study suggests that isolated dangling bonds may not be necessary to produce a ESR signal, but that it can be produced by π-bonded pairs which are known to be present in the materials [84].

7. Conclusions

SIESTA was developed as an approach to compute the electronic properties and to perform atomistic simulations of complex systems from first principles. The developement process required a considerable number of fundamental and technical problems to be addressed and solved. Although this process was completed very recently, SIESTA has already shown a phenomenal potential for the study of very different, large scale problems. Here we have reviewed many of these applications, including carbon nanostructures (fullerenes and nanotubes), metallic nanostructures, biomólecules, surfaces and disordered systems. Many other lines of research are currently open by several groups, in which hopefully SIESTA will be an effective tool for the study of complex systems.

Acknowledgements The author is indebted to the people directly involved in the SIESTA Project: J. M. Soler, E. Artacho, D. Sánchez-Portal, A. García, J. Junquera, M. Machado and J. M. Alonso Pruneda, for the ongoing collaboration during these years. The SIESTA team is grateful for ideas, discussions, and support from José L. Martins, Richard M. Martin, David A. Drabold, Otto F. Sankey, Julian D. Gale, and Volker Heine. PO is the recipient of Sponsored Research Projects from Motorola PSRL and Sumitomo Chemical. This work has been supported by Spain's DGES grant PB95-0202. Support of the Ψ_k network of the ESF is also very much acknowledged.

References

[1] P. Hohenberg and W. Kohn, Phys. Rev. **135**, 864 (1964).
[2] W. Kohn and L. J. Sham, Phys. Rev. **140**, 1133. (1965).
[3] M. Payne, M. Teter, D. Allan, T. Arias, and J. D. Joannopoulos, Rev. Mod. Phys. **64**, 1045 (1992).
[4] P. Ordejón, Comp. Mater. Sci. **12**, 157 (1998).
    S. Goedecker, Rev. Mod. Phys. **71**, 1085 (1999).
[5] W. Kohn, Phys. Rev. Lett. **76**, 3168 (1996).
[6] P. Ordejón, D. A. Drabold, R. M. Martin, and M. P. Grumbach, Phys. Rev. B **51**, 1456 (1995), and references therein.
[7] P. Ordejón, E. Artacho, and J. M. Soler, Phys. Rev. B **53**, R10441 (1996).
[8] P. Ordejón, E. Artacho, and J. M. Soler, Mater. Res. Soc. Symp. Proc. **408**, 85 (1996).
[9] D. Sánchez-Portal, P. Ordejón, E. Artacho, and J. M. Soler, Internat. J. Quantum Chem. **65**, 453 (1997).
[10] E. Artacho, D. Sánchez-Portal, P. Ordejón, A. García, and J. M. Soler, phys. stat. sol. (b) **215**, 809 (1999).
[11] D. Sánchez-Portal, P. Ordejón, E. Artacho, and J. M. Soler, to be published.
[12] J. P. Perdew, K. Burke, and M. Ernzerhof, Phys. Rev. Lett. **77**, 3865 (1996).
[13] T. Oda, A. Pasquarello, and R. Car, Phys. Rev. Lett. **80**, 3622 (1998).
[14] N. Troullier and J. L. Martins, Phys. Rev. B **43**, 1993 (1991).

[15] L. KLEINMAN and D. M. BYLANDER, Phys. Rev. Lett. **48**, 1425 (1982).
[16] S. G. LOUIE, S. FROYEN, and M. L. COHEN, Phys. Rev. B **26**, 1738 (1982).
[17] S. HUZINAGA et. al., Gaussian Basis Sets for Molecular Calculations, Elsevier Sci. Publ. Co., Amsterdam 1984.
R. POIRIER, R. KARI, and R. CSIZMADIA, Handbook of Gaussian Basis Sets, Elsevier Sci. Publ. Co., Amsterdam 1985, and references therein.
[18] J. KIM, F. MAURI, and G. GALLI, Phys. Rev. B **52**,1640 (1995).
[19] O. F. SANKEY and D. J. NIKLEWSKI, Phys. Rev. B **40**, 3979 (1989).
[20] H. W. KROTO, J. R. HEATH, S. C. O'BRIAN, R. F. CURL, and R. E. SMALLEY, Nature **318**, 162 (1985).
W. KRÄTSCHMER, L. D. LAMB, K. FOSTIROPOULOS, and D. R. HUFFMAN, Nature **347**, 354 (1990).
[21] S. IIJIMA, Nature **354**, 56 (1991).
[22] D. UGARTE, Nature **359**, 707 (1992).
[23] J. TERSOFF, Phys. Rev. B **46**, 15546 (1992).
T. A. WITTEN and H. LI, Europhys. Lett. **23**, 51 (1993).
[24] A. MAITI, C. J. BABREC, and J. BERNHOLC, Phys. Rev. Lett. **70**, 3023 (1993).
[25] D. YORK, J. P. LU, and W. YANG, Phys. Rev. B **49** 8526 (1994).
J. P. LU and W. YANG, Phys. Rev. B **49**, 11421 (1994).
[26] S. ITOH, P. ORDEJÓN, D. A. DRABOLD, and R. M. MARTIN, Phys. Rev. B **53**, 2132 (1996).
[27] M. TERRONES and H. TERRONES, Fuller. Sci. Tech. **4**, 517 (1996).
[28] K. R. BATES and G. E. SCUSERIA, Theor. Chem. Acc. **99**, 29 (1998).
[29] N. G. CHOPRA, R. J. LUYKEN, K. CHERREY, V. H. CRESPI, M. L. COHEN, S. G. LOUIE, and A. ZETTLE, Science **269**, 966 (1995).
[30] O. STÉPHAN, Y. BANDO, A. LOISUEAU, F. WILLAIME, N. SHRAMCHENKO, T. TAMIYA, and T. SATO, Appl. Phys. A **67**, 107 (1998).
[31] D. GOLDBERG, Y. BANDO, O. STÉPHAN, and K. KURASHIMA, Appl. Phys. Lett. **73**, 2441 (1998).
[32] S. S. ALEXANDRE, M. S. C. MAZZONI, and H. CHACHAM, Appl. Phys. Lett. **75**, 61 (1999).
[33] P. ORDEJÓN, D. SÁNCHEZ-PORTAL, E. ARTACHO, and J. M. SOLER, Fuller. Sci. Technol., in press.
D. SÁNCHEZ-PORTAL, E. ARTACHO, J. M. SOLER, A. RUBIO, and P. ORDEJÓN, Phys. Rev. B **59**, 12678 (1999).
[34] G. G. TIBBETTS, J. Cryst. Growth **66**, 632 (1983).
[35] R. A. JISHI, L. VENKATARAMAN, M. S. DRESSELHAUS, and G. DRESSELHAUS, Chem. Phys. Lett. **209**, 77 (1993).
[36] S. IIJIMA, C. J. BABREC, A. MAITI, and J. BERNHOLC, J. Chem. Phys. **104**, 2089 (1996).
[37] A. ROCHEFORT, D. R. SALAHUB, and P. AVOURIS, Chem. Phys. Lett. **297**, 45 (1998).
[38] H. S. C. MAZZONI, and H. CHACHAM, submitted to Phys. Rev. Lett.
[39] P. M. AJAYAN and S. IIJIMA, Nature **361**, 333 (1993).
[40] P. M. AJAYAN, T. W. EBBESEN, T. ICHIHASHI, S. IIJIMA, K. TANIGAKI, and H. HIURA, Nature **362**, 522 (1993).
[41] S. C. TSANG, P. J. F. HARRIS, and M. L. H. GREEN, Nature **362**, 520 (1993).
[42] D. UGARTE, A. CHÂTELAIN, and W. A. DE HEER, Science **274**, 1897 (1996).
[43] A. C. DILLON, K. M. JONES, T. A. BEKKEDAHL, C. H. KIANG, D. S. BETHUNE, and M. J. HEBEN, Nature **386**, 377 (1997).
[44] J. LIU, A. G. RINZLER, H. DAI, J. H. HAFNER, R. K. BRADLEY, P. J. BOUL, A. LU, T. IVERSON, K. SHELIMOV, C. B. HUFFMAN, F. RODRIGUEZ-MACIAS, Y. SHON, T. R. LEE, D. T. COLBERT, and R. E. SMALLEY, Science **280**, 1253 (1998).
[45] M. S. C. MAZZONI, H. CHACHAM, P. ORDEJÓN, D. SÁNCHEZ-PORTAL, J. M. SOLER, and E. ARTACHO, Phys. Rev. B **60**, 2208 (1999).
[46] J. W. MINTMIRE and C. T. WHITE, Carbon **33**, 893 (1995).
[47] L. C. VENEMA, J. W. G. WILDÖER, S. J. TANS, J. W. JANSSEN, L. J. HINNE, T. TUINSTRA, L. P. KOUWENHOVEN, and C. DEKKER, Science **283**, 52 (1999).
[48] A. RUBIO, D. SÁNCHEZ-PORTAL, E. ARTACHO, P. ORDEJÓN, and J. M. SOLER, Phys. Rev. Lett. **82**, 3520 (1999).
[49] I. L. GARZÓN, K. MICHAELIAN, M. R. BELTRÁN, A. POSADA-AMARILLAS, P. ORDEJÓN, E. ARTACHO, D. SÁNCHEZ-PORTAL, and J. M. SOLER, Phys. Rev. Lett. **81**, 1600 (1998).
[50] V. ROSSATO, M. GUILLOPE, and B. LEGRAND, Phil. Mag. A **59**, 321 (1989).
[51] K. MICHAELIAN, Amer. J. Phys. **66**, 231 (1998).

[52] J. M. Soler, M. R. Beltrán, K. Michaelian, I. L. Garzón, P. Ordejón, D. Sánchez-Portal, and E. Artacho, Phys. Rev. B, in press.
[53] J. I. Pascual, Phys. Rev. Lett. **71**, 1852 (1993); Science **267**, 1793 (1995).
N. Agraït, J. G. Rodrigo, and S. Vieira, Phys. Rev. B **47**, 12345 (1993).
N. Agraït, G. Rubio, and S. Vieira, Phys. Rev. Lett. **74**, 3995 (1994).
[54] C. J. Muller, J. M. van Ruitenbeek, and L. J. de Jongh, Phys. Rev. Lett. **69**, 140 (1992).
J. M. Krans et al., Nature **375**, 767 (1995).
E. Scheer, Nature **394**, 154 (1998).
[55] J. L. Costa-Krämer, N. García, P. Gacía-Mochales, and P. A. Serena, Surf. Sci. **342**, L1144 (1995).
[56] H. Ohnishi, Y. Kondo, and K. Takayanagi, Nature **395**, 780 (1998).
A. I. Yanson, G. Rubio Bollinger, H. E. van den Brom, N. Agraït, and J. M. van Ruitenbeek, Nature **395**, 783 (1998).
[57] Y. Kondo and K. Takayanagi, Bull. Amer. Phys. Soc. **44**, 312 (1999).
[58] J. A. Torres, E. Tosatti, A. dal Corso, F. Ercolessi, J. J. Kohanoff, F. Di Tolla, and J. M. Soler, Surface Science **426**, L441 (1999).
[59] D. Sánchez-Portal, J. Junquera, P. Ordejón, A. García, E. Artacho, and J. M. Soler, Phys. Rev. Lett. **83**, 3884 (1999).
[60] J. Izquierdo, A. Vega, L. C. Balbás, P. Ordejón, D. Sánchez-Portal, E. Artacho, and J. M. Soler, Recent Advances in Density Functional Methods, Part III, Ed. V. Barone, A. Bencini, and P. Fantucci, World Scientific Publ. Co., Singapore, in press.
[61] V. S. Stepanyuk, W. Hergert, P. Rennert, K. Wildberger, R. Zeller, and P. H. Dederichs, Phys. Rev. B **59**, 1681 (1999).
P. Rennert, V. S. Stepanyuk, W. Hergert, J. Izquierdo, A. Vega, and L. C. Balbás, Advances in Science and Technology, Vol. 19, Surface and Near-Surface Analysis of Materials, Ed. P. Vincencini and S. Valeri, Techna Srl, 1999 (pp. 29 to 36).
[62] M. Calleja, C. Rey, M. M. G. Alemany, L. J. Gallego, P. Ordejón, D. Sánchez-Portal, E. Artacho, and J. M. Soler, Phys. Rev. B **60**, 2020 (1999).
[63] C. Rey, M. Calleja, M. M. G. Alemany, L. J. Gallego, and P. Ordejón, to be published.
[64] D. N. Beratan, S. Priyadarshy, and S. M. Risser, Chemistry and Biology **4**, 3 (1997).
S. Priyadarshy, S. M. Risser, and D. N. Beratan, J. Phys. Chem. **100**, 17678 (1996).
S. O. Kelley and J. K. Barton, Science **283**, 375 (1999).
H.-W. Fink and C. Schönenberger, Nature **398**, 407 (1999).
[65] M. Machado, P. Ordejón, D. Sánchez-Portal, E. Artacho, and J. M. Soler, submitted to J. Chem. Phys., and a-print physics/9908022.
[66] J. Sponer, J. Leszczynski, and P. Hobza, J. Phys. Chem. **100**, 1965 (1996)
[67] E. Artacho, D. Sánchez-Portal, P. Ordejón, and J. M. Soler, to be published.
[68] R. Chandrasekaran and S. Arnott, in: Landolt-Börnstein, New Series, Group VII: Biophysics, Vol. I, Nucleic Acids, Subvolume b: Crystallography and Structural Data II, Ed. W. Saenger, Springer-Verlag, New York 1989.
[69] M. V. Fernandez-Serra, J. Junquera, C. Jelsch, C. Lecomte, and E. Artacho, to be published.
[70] J. I. Pascual, Thesis, Universidad Autónoma de Madrid, March 1998.
J. I. Pascual, J. Gómez-Herrero, and A. Barón, unpublished.
[71] D. Sánchez-Portal, Thesis, Universidad Autónoma de Madrid, June 1998.
[72] J. I. Pascual, J. Gómez-Herrero, A. Baró, D. Sánchez-Portal, E. Artacho, P. Ordejón, and J. M. Soler, to be published.
[73] R. A. McKee, F. J. Walker, and M. F. Chisholm, Phys. Rev. Lett. **81**, 3014 (1998).
[74] J. Wang, J. Hallmark, D. S. Marshall, W. J. Ooms, P. Ordejón, J. Junquera, D. Sánchez-Portal, E. Artacho, and J. M. Soler, Phys. Rev. B **60**, 4968 (1999).
[75] Chiu-Ping Cheng, Ie-Hong Hong, and Tun-Wen Pi, Phys. Rev. B **58**, 4066 (1998).
[76] X. Hu and D. Sarid, private communication.
[77] J. Wang, J. Junquera, and P. Ordejón, to be published.
[78] X. Hu, C. A. Peterson, D. Sarid, Z. Yu, J. Wang, D. S. Marshall, R. Droopad, J. Hallmark, and W. J. Ooms, Surface Sci. **426**, 69 (1999).
[79] H. von Känel, K.A. Mäder, E. Müller, N. Onda, and H. Sirringhaus, Phys. Rev. B **45**, 13807 (1992).

[80] J. J. Hinarejos, G. R. Castro, P. Segovia, J. Alvarez, E. G. Michel, R. Miranda, A. Rodriguez-Marco, D. Sánchez-Portal, E. Artacho, F. Ynduráin, S. H. Yang, P. Ordejón, and J. B. Adams, Phys. Rev. B **55**, 16065 (1997).
[81] J. Junquera, R. Weht, and P. Ordejón, to be published.
[82] G. Fabricius, E. Artacho, D. Sánchez-Portal, P. Ordejón, D. A. Drabold, and J. M. Soler, Phys. Rev. B **60**, 16283 (1999).
[83] P. A. Fedders, D. A. Drabold, P. Ordejón, G. Fabricius, D. Sanchez-Portal, E. Artacho, and J. M. Soler, Phys. Rev. B **60**, 10594 (1999).
[84] D. A. Drabold, P. Stumm, and P. Fedders, Phys. Rev. Lett. **72**, 2666 (1994).

phys. stat. sol. (b) **217**, 357 (2000)

Subject classification: 71.15.Fv; 71.15.Mb; 78.30.Jw; S12

A Self-Consistent Charge Density-Functional Based Tight-Binding Scheme for Large Biomolecules

M. ELSTNER (a, b), TH. FRAUENHEIM (b), E. KAXIRAS (a), G. SEIFERT (b), and S. SUHAI (c)

(a) *Department of Physics, Harvard University, Cambridge MA 02138, USA*

(b) *Theoretische Physik, Universität Paderborn, D-33098 Paderborn, Germany*

(c) *Molekulare Biophysik, Deutsches Krebsforschungszentrum, D-69120 Heidelberg, Germany*

(Received August 10, 1999)

A common feature of traditional tight-binding (TB) methods is the non-self-consistent solution of the eigenvalue problem of a Hamiltonian operator, represented in a minimal basis set. These TB schemes have been applied mostly to solid state systems, containing atoms with similar electronegativities. Recently self-consistent TB schemes have been developed which now allow the treatment of systems where a redistribution of charges, and the related detailed charge balance between the atoms, become important as e.g. in biological systems. We discuss the application of such a method, a self-consistent charge density-functional based TB scheme (SCC-DFTB), to biological model compounds. We present recent extensions of the method: (i) The combination of the tight binding scheme with an empirical force field, that makes large scale simulations with several thousand atoms possible. (ii) An extension which allows a quantitative description of weak-bonding interactions in biological systems. The latter include an improved description of hydrogen bonding achieved by extending the basis set and improved molecular stacking interactions achieved by incorporating the dispersion contributions empirically. In applying the method, we present benchmarks for conformational energies, geometries and frequencies of small peptides and compare with ab initio and semiempirical quantum chemistry data. These developments provide a fast and reliable method, which can handle large scale quantum molecular dynamic simulations in biological systems.

1. Introduction

Biomolecules are challenging systems for computational methods for several reasons:

(i) they usually contain many different types of atoms with very different electronegativities and properties, such as H, C, N, O, P, S as well as several types of metal atoms. The molecules can occur in multiply charged states and the different electronegativities lead to large inter- and intra-molecular charge transfer.

(ii) Biomolecules exhibit a large variety of different bonding types, ranging from covalent and ionic bonding to hydrogen bonding and van der Waals-type interactions. Compared to covalent bond energies, the hydrogen bonds are much weaker, ranging from 2 to 20 kcal/mol, which puts considerable demands on the accuracy of computational methods.

(iii) The potential energy surfaces (PES) of biomolecules covering e.g. H bonded complexes, polypeptides, etc. are highly complex, exhibiting many local minima with small energy differences and often separated by small energy barriers. These minima can be very shallow, where large geometrical changes can occur with only a small change in energy.

(iv) The description of chemical reactions and transfer processes of protons and electrons is of special interest in the theoretical modelling. Therefore, the methods should address issues of bond breaking and formation properly. Further, it is desirable that reaction energies, transition states and reaction pathways are determined with high accuracy, i.e. within a few kcal/mol.

(v) Many biomolecules of interest are very large, while chemical reactions depend sensitively on the protein enviroment and on solvent effects. Therefore, simple model calculations of only one part of the system are often of limitied value. For predictive calculations, realistically large systems containing up to several thousand atoms must be considered. Furthermore, in order to compare to experimental situations, free energies rather than potential energies have to be calculated. This would involve molecular dynamic (MD) simulations over rather long time scales, ranging form several hundreds of picoseconds up to milliseconds. In the latter cases, special sampling methods may allow a considerable reduction in the simulation time. Note, however, that many structure forming processes appear only for long time scales, like the folding of a polypeptide into its three dimensional proteine structure.

Several of these features do not only occur in biomolecules, but are common to many organic and inorganic systems (molecules, clusters, solids, surfaces, adsorbates, ...) addressed by computational methods. Presently, there is no single method available which covers all different chemical bonding types at the desired level of accuracy to satisfy all the demands. For this reason, it is common practice that different methods are used for different purposes.

Methods using classical empirical potential energy functions can perform MD simulations for several thousands of atoms and can reach time-scales in the nano- and microsecond region. However, they usually do not allow for bond breaking and reformation. Further, they are generally parametrized to experimental data for equlibrium conformations of molecules and solids. Little information is included for the regions of the PES far from those equilibrium structures or for that matter for variable chemical environments.

The so called ab initio methods, on the other hand, have been shown to be highly accurate and predictive, but at a very high computational cost. This limits their applications to relatively small systems containing up to 100 atoms and MD time scales of about 1 ps. Even within the ab initio schemes, the accuracy varies with the special method used and the basis set applied. Hartree-Fock (HF) calculations can be applied for certain classes of systems; of more general applicability are post-Hatree-Fock methods like MP2, but they are limited to even smaller systems sizes of about 20 to 40 atoms. Density functional theory (DFT) methods have been increasingly applied to biomolecules yielding an accuracy comparable to MP2 at much lower cost [1], but those are still too computationally demanding for many interesting systems.

Semi-empirical methods provide a compromise: They are less accurate than the ab initio methods, but about two to three orders of magnitude faster, and by the same factor slower than empirical potential methods. Quantum chemical semi-empirical methods have been developed for several decades. The most widely used methods today, the AM1 [2] and PM3 [3] methods, are an approximation to the Hartree-Fock theory: they neglect certain types of integrals and determine the remaining integrals from experimental data and/or by fitting them to reproduce properties of organic molecules, like heats of formation, geometries, dipole moments etc. [4]. These methods have been sucessfully applied to biomolecular systems, and a variety of attempts have been made to further improve their accuracy.

Another type of semi-empirical method, the so-called tight binding (TB) approach, has been developed mostly in the context of solid state theory. The standard TB method works by expanding eigenstates of a Hamiltonian in an orthogonalized basis set of atomic-like wavefunctions, and by representing the exact Hamiltonian operator with a parametrized matrix. A common feature of these methods is the non-self-consistent solution of the correponding eigenvalue problem, which does not take into account the charge redistribution in a molecule. This is of minor importance for systems which contain atoms with similar electronegativies. TB methods are therefore mainly applied to such systems which contain only one or two types of atoms. This approach fails when applied to organic or biological molecules consisting of carbon, hydrogen, oxygen and nitrogen. For a more detailed discussion of these methods, see the article of Frauenheim et al. [5], this volume.

Recently, we have presented the development of a self-consistent-charge tight binding scheme (SCC-DFTB), based on the density functional theory [6, 7]. This method is derived by a second order expansion of the DFT total energy functional with respect to the charge density fluctuations $\Delta\varrho$ around a given reference density ϱ_0. The second order terms in the density fluctuations are approximated by a simple distribution of atom-centered point charges $\Delta q_\alpha = q_\alpha - q_\alpha^0$, estimated by a Mulliken charge analysis, while all other terms maintain the standard functional form of the tight-binding approach. Hence, this method can be applied in a straight-forward manner to any tight binding scheme in which the total energy is written in terms of the band structure energy $E_{bs} = \sum_i^{occ} \sum_{\mu\nu} c_\mu^i c_\nu^i H_{\mu\nu}[\varrho_0]$ and a short-range repulsive pair potential, E_{rep}, both determined at the reference density ϱ_0. The approximate DFT energy functional, explicitly containing the second order terms becomes

$$E_{tot} = \sum_i^{occ} \sum_{\mu\nu} c_\mu^i c_\nu^i H_{\mu\nu}[\varrho_0] + E_{rep}[\varrho_0] + \frac{1}{2} \sum_{\alpha\beta} \Delta q_\alpha \Delta q_\beta \gamma_{\alpha\beta}. \qquad (1)$$

The third term on the right hand side represents the longe-range Coulomb interactions between point charges at different sites and includes the self-interaction contributions of the single atoms [6]. Making this approximate Kohn-Sham functional subject to a variational principle within a LCAO representation for the single-particle electronic states, we derive a Kohn-Sham equation by minimizing the second order energy functional which is self-consistent in the Mulliken charge distribution through the modification of the Hamitonian matrix [6, 5]. We implemented this self-consistent charge extension into our density-functional based tight-binding method [8], where the $H_{\mu\nu}[\varrho_0]$ are calculated within DFT-GGA in a two-center approximation using a minimal basis of atomic-like wavefunctions ϕ_μ. In order to apply the method to biological molecules, we have derived tight-binding integrals as a function of distance between atom types S, O, N, C, and H. The parametrization for P, Zn and other metals like Mg or Na, relevant to biological processes, is in progress.

Before an approximate method, like AM1, PM3 or SCC-DFTB, can be applied to biological systems, its reliability has to be tested. A first test is to benchmark the method for properties of small organic molecules, to examine reaction energies, structures, vibrational frequencies etc. Such tests for the SCC-DFTB method have been presented elsewhere [6, 7]. They show that this method has an accuracy which is comparable to that of fully self-consistent DFT methods. However, the performance for the properties of small organic molecules is not a sufficient test of the method for larger, biologically relevant structures. Important interactions and chemical situations, present in larger molecules, as e.g. H bonding and molecular stacking interactions, are not covered by the tests on small molecules. Further, the structure and energetics of larger molecules can be determined by the energy landscape given, e.g., by rotations around covalent bonds. These can be very different from the ground state of small organic molecules. Additionally, a different chemical environment can alter the energy landscape. This might be the environment in a protein, or the effect of aqueus solution. Typical examples are polypeptides, where different conformations like α- and 3_{10}-helical structures or β-sheets, result from rotations around covalent bonds with low energy barriers. Therefore, to examine the reliability of an approximate method, one must test it for molecules and molecular models that represent typical situations in biological systems. If a method performs well for representatives of a certain class of molecules, it can be used for a prediction of yet unknown properties. The approximate methods should also be examined for typical reactions, transition states, etc.

Another issue is the computational cost. Although the approximate methods mentioned above are two to three orders of magnitude faster than ab initio methods, they are still limited to system sizes of several hundred atoms. There are basically two approaches suitable for reaching larger system sizes in the framework of quantum mechanical methods: The first is to circumvent the computationally costly solution of the generalized eigenvalue problem, which exhibits a cubic scaling with increasing system size; the second consists of combining quantum mechanical (QM) methods with empirical molecular mechanical (MM) force fields (QM/MM).

The outline of the paper is as follows: In Section 2, we describe the combination of the SCC-DFTB with a MM method. In Section 3, we discuss the challenges of describing H bonding within an approximate method and summarize the results of the SCC-DFTB method. In Section 4, we focus on the investigation of typical peptide structures and examine the performance of the SCC-DFTB in comparison with ab initio methods for energetic, structural and vibrational properties. Molecular stacking interactions, as they appear for example in the DNA double helix, seem to be even more challenging to theoretical modelling. They are poorly described even within more elaborate methods, such as DFT. Therfore, since the SCC-DFTB method is an approximation to DFT, we included these interactions empirically into the SCC-DFTB scheme; this is described in Section 5. Finally, Section 6 contains conclusions and outlook.

2. Hybrid SCC-DFTB/Molecular-Mechanical Coupling

The computational cost for large molecules (containing more than 100 atoms) within the SCC-TB method is determined by the diagonalization of the Hamitonian matrix, which exhibits N^3 scaling with increasing system size N. The method allows routine application to systems containing several hundred atoms on a workstation. However, for extended molecular dynamics runs or for the study of very large systems (containing several 1000 atoms), the SCC-DFTB method is computationally too costly. One way to deal with this is through the so called O(N) methods, which circumvent the matrix diagonalization for the solution of the generalized eigenvalue problem. Such methods are described and applied to large scale simulations in the articles by Galli and Ordejon, in Refs. [9, 10], this Special Issue. Here we focus on an alternative approach, which has become popular in quantum chemistry in the last few years: the combination of a quantum mechanical method with an empirical force field. The idea behind the combined quantum mechanical/empircal force field methods (QM/MM) is to describe a part of the molecule quantum mechanically and the rest of the system within the computationally much faster empirical force field approach. In this approach, the total energy is usually written as

$$E_{tot} = E_{QM} + E_{MM} + E_{QM-MM}, \quad (2)$$

where E_{QM} is the energy of the QM part of the subsystem represented by the SCC-DFB energy Eq. (1), E_{MM} is the energy of the MM subsystem given by the energy function of the empirical force field, and E_{QM-MM} describes the coupling of the two subsystems.

If the boundary of the QM and MM regions intersects a covalent bond, the combination of those methods is not straight-forward. Several suggestions have been made to tackle this problem. A popular approach is the so called link-atom approach, where the quantum system is saturated with a fictitious atom for the QM calculation only, while the bond across the QM/MM boundary is modeled by the bonding interaction of the empirical force field. QM/MM approaches have been reviewed recently [11], and we will not discuss details of their implementation. We will only present the main ideas for the SCC-DFTB/MM-coupling [12, 13].

E_{QM-MM} consists of Coulomb and van der Waals (vdW) interactions between the two subsystems. The vdW interaction is modeled by the interaction terms present in the empirical force field method, while the Coulomb term is approximated by the interactions of the point charges between the subsystems, where the QM charges are given by the Mulliken charges Δq_α of the SCC-DFTB method and the MM charges Q_β are given by the force field parameters:

$$E_{QM-MM} = -\sum_{\alpha \in QM, \beta \in MM} \frac{\Delta q_\alpha Q_\beta}{R_{\alpha\beta}} + E_{vdW}. \quad (3)$$

The total energy is

$$E_{tot} = E_{QM} + E_{MM} - \sum_{\alpha \in QM, \beta \in MM} \frac{\Delta q_\alpha Q_\beta}{R_{\alpha\beta}} + E_{vdW}. \quad (4)$$

Applying the variational principle to this energy expression, we arrive at the generalized eigenvalue problem

$$\sum_\nu c^i_{n\nu}(H_{\nu\mu} - \epsilon_i S_{\nu\mu}) = 0 \quad (5)$$

with the matrix elements

$$H_{\nu\mu} = H^0_{\nu\mu} + \tfrac{1}{2} S_{\nu\mu} \sum_\delta (\gamma_{\nu\delta} + \gamma_{\mu\delta}) \Delta q_\delta - \tfrac{1}{2} S_{\nu\mu} \sum_{\beta \in MM} \left(\frac{1}{R_{\nu\beta}} + \frac{1}{R_{\mu\beta}} \right) Q_\beta. \quad (6)$$

$S_{\nu\mu}$ is the overlap matrix, $R_{\mu\beta}$ the distance between the corresponding QM and MM atoms. The generalized eigenvalue problem therefore has to be solved in the presence of the "external" charges Q_β. We have tested this method extensively for H bonded compounds, where one molecule is treated quantum mechanically and the other is treated with the force field method. The results are very promising: geometries and energies compare well with higher level calculations and the relative ordering of the energies of several conformers is well reproduced [13].

The effects of external electrical fields can be included similarly in an approximate way, i.e. making use of the monopole approximation for the SCC-DFTB charges and coupling the electric field to the SCC-DFTB charges [12]. We have used this extension to study proton transport in linear water filaments, driven by the external electric field [14].

3. The Description of H Bonding

Hydrogen bonds are a common bonding pattern in biolocial structures which play a crucial role for determining the geometries, the energetics and other properties of biomolecules. Two or more molecules can be bound together by an inter-molecular H bond, or the conformation of one molecule can be stabilized by intra-molecular H bonds. Intra-molecular H bonds e.g. stabilize the three-dimensional structure of polypeptides and proteins. In these systems, different conformations may exhibit different H bonding patterns, as will be discussd in the next section.

H bonds can be described as van der Waals (vdW)-type interactions, as a combination of electrostatic attraction, charge transfer effects and repulsion and dispersion interactions. The dispersion interaction contributes significantly to the binding energy and strongly reduces the H bond lengths.

Although ab initio methods are computationally too demanding for many biomolecules of interest, they can prove helpful in gaining insight into the physics of the H bond by studying small model systems. They can also be used to build up a data set for either parametrizing the approximate methods or for testing their performance.

Most ab initio studies of H bonded complexes have been performed at the HF and post-HF (MP2 and higher) level of theory using localized basis functions. In calculating H bond energies, three major sources of error have to be considered [15]. The first is called the basis set superposition error (BSSE), while the second is referred to as the basis set incompleteness error (BSIE) and the third is related to an insufficient treatment or even neglect (HF) of the dispersion interaction. The first error stems from the manner in which binding energies are evaluated by subtracting the energies of the monomers from the energy of the compound. By doing this, the energy of the monomers is calculated in the monomer basis sets, while in the calculation of the compound energy, each monomer of the compound is described in the basis set of the whole compound. This results in lowering the monomer enery in the compound relative to the energy of the isolated one due to the variational principle. Therefore, BSSE errors lead to an overestimation of the binding energy; typical errors are in the range of 0.5 to 3 kcal/mol, depending on the system and the basis set. The BSSE error decreases with increasing size of the basis set, because when a monomer is described in a satisfactory basis the additional basis functions from neighboring molecules have neglegible effect on its total energy. There are several ways to account for the BSSE error; the most widley used method is given by the counterpoise procedure (CP) of Boys and Bernardi [16] where the monomer energies are calculated in the basis set of the compound, that is the bonded and isolated monomers are treated within the same basis set. The BSIE also leads to an overestimation of the binding energy; the error again decreases for larger basis sets. The effect of correlation is also to increase the binding energy as a function of the size of the basis set.

These three factors seem to show typical trends for a wide class of H bonded systems and have been examined in detail for the water dimer [15] (see Fig. 1). As a result of these three factors, calculations at lower theoretical levels can lead accidentally to good binding energies. For example, the binding energy of the water dimer is 3.6 kcal/mol and 4.9 kcal/mol at the HF and MP2 level within the basis set limit. The experimental value is 5.4 kcal/mol [17]. Because the neglect of correlation lowers the binding energy and the BSIE overestimates it, a HF calculation using singly polarized basis sets like 6-31G(d,p) finds a surprising accurate value of 5.5 kcal/mol, while it would be overestimated by 30% at the MP2 level using the same basis set [15].

While the H bonding energies are clearly overestimated within the the DFT-LDA approach, gradient corrected DFT methods seem to perform very well for H bonded systems too: the binding energy of the water dimer is slightly underestimated by about 0.5 to 1 kcal/mol, depending on the basis set and the exchange–correlation functional used [18].

Concerning the approximate (semi-empirical) methods it is difficult to account properly for the sources of errors as discussed above for the ab initio methods. The lack of extended basis sets or appropriate consideration of correlation effects (dispersion interaction) is partially compensated by a proper choice of empirical parameters. Semi-empirical methods show the overall tendency to understimate H bond strenghs (for a review, see [19]).

The same tendency is found for the SCC-DFTB method [20, 13]. The geometries, especially the H bond distances, are in very good agreement with MP2 and DFT calculations, with the ordering in energy of different conformations well reproduced. But the SCC-DFTB binding energies of weakly bonded complexes are consistently underestimated by 1 to 2 kcal/mol when compared to ab initio methods.

Several attempts to refine the semi-empirical methods for H bonded systems have been made. One strategy is to modify the core–core repulsion term. This term not only contains the ion–ion repulsion, but also an additional effective energy contribution. In the AM1 method, this part has been modified compared to its predecessor MNDO [21], which improves H bonding significantly. However, an accurate description of H bonds seems to require a specific parametrization for these systems, as has been done in the MNDO/M [22] approach. In that approach, new functions for the core–core repulsion energy have been introduced and explicitly parametrized to reproduce binding energies and geometries of simple H bonded complexes. A different strategy has been followed in the framework of the SINDO1 model [23], where polarization func-

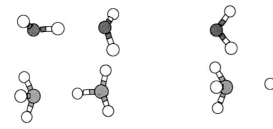

Fig. 1. Linear and bifurcated water dimer configurations (upper left and upper right), linear ammonia dimer (lower left) and ammonia–water complex (lower right)

tions have been introduced for the hydrogen atoms to increase the inter-molecular bonding. But this approach did not succeed without introducing an additional empirical function, which damps the effect of the p-functions for distances smaller and larger than the typical hydrogen bond distance.

To improve the description of H bonding in the SCC-DFTB model, we extended the clearly very limited basis for hydrogen (only 1s) by including also 2p basis functions. This should improve the description of the charge transfer effects in the H bonding region (≈ 2 Å X–H distance) and increase the H bonding energy. However, since the 2p atomic wavefunction is quite diffuse, the long range decay of the corresponding matrix elements had to be modified that they approach smoothly to zero at about 5 to 6 Å.

A large variety of hydrogen bonding complexes have been studied with this extension of the method. We compared the results of different levels of theory, including empirical force-field and semiempirical methods, DFT, HF and post-HF methods and the minimal basis DFTB approach. Here we focus on discussing only a few weakly bonded complexes with interaction energies ranging from 2 to 6 kcal/mol, which are the least satisfactorily described systems within SCC-DFTB (see Fig. 1).

The global minimum of the water dimer is a linear structure, where one H bond is formed between the water monomers (structure 1 in Table 1). The bifurcated structure, where both hydrogens of one water molecule form H bonds with the oxygen of the other water molecule (structure 2), is about 2 kcal/mol lower in energy at the MP2/6-311+G** level of theory [24]. The NH_3 dimer considered here forms one N–H H bond, and in the conformer NH_3–H_2O-1, the water molecule acts as a donor forming one H bond, while in conformer NH_3–H_2O-2 both hydrogens form H bonds with the nitrogen, similar to the case in the conformer H_2O–H_2O-2. In the H_3COH–H_2O-1 complex, the H_3COH acts as a donor, while in the H_3COH–H_2O-2 the reverse is true. With the exception of the NH_3–H_2O-1 conformer, where the binding energy is underestimated by more than 1 kcal/mol, the results in Table 1 show that the SCC-DFTB method agrees quite well with the higher level calculations.

SCC-DFTB geometries compare very well with higher level results, and the energetic ordering of different conformers of the complexes (not discussed here in detail) is also reproduced (which is also true for the SCC-DFTB method applying the minimal basis set only [13]). H bond lengths are not given here in detail, but they deviate with respect to the DFT and MP2 results by about 0.05 to 0.1 Å. For example, the O–O

Table 1

H bonding energies (kcal/mol) of H bond complexes; min denotes SCC-DFTB calculations with the minimal LCAO basis set. AI denotes ab initio calculations as dicussed in the text

complex	SCC-DFTB	min	AI	complex	SCC-DFTB	min	AI
H_2O–H_2O-1	5.0	3.3	5.4[1]	H_3CNH_2–H_2O	4.9	3.3	6.5[4]
H_2O–H_2O-2	3.6	2.1	3.5[1]	NH_3–H_2O-1	4.9	3.4	5.8[3]
NH_3–NH_3	3.4	1.8	3.1[2]	NH_3–H_2O-2	3.3	1.4	3.2[3]
H_3COH–H_2O-1	5.3	3.5	5.6[4]	$HCOOH$–H_2O	9.0	7.1	10.8[4]
H_3COH–H_2O-2	5.1	3.1	5.6[4]				

[1]) MP2/6-311+G** [24], [2]) HF/6-31+G(2d,2p) [29], [3]) MP2/6-311+G** corrected for BSSE [13], [4]) HF/6-31G* [28]

distance in H_2O-H_2O-1 is 2.85 Å, whereas it is 2.9 Å at the MP2 and MP4 level of theory [25].

The performance of semi-empirical methods like AM1, PM3 and MNDO/M for H-bonded complexes has been evaluated recently [19]. Most semi-empirical methods underestimate H-bond strengths when there is no special emphasis on H bonds in the parameter determination procedure, as e.g. in the MNDO/M method. The performance of the MNDO/M method seems to be very good for H-bonding interactions: it has been shown to reproduce the stabilization energies of DNA H-bonded base pairs very well [27], while for other H-bonding complexes it shows a tendency to overestimate interaction enthalpies [19]. AM1 often does not predict the right ground state structure, e.g. bifurcated H bonds are favored against linear ones [19]. Another example is given by the $H_3O^+-H_2O$ complex, which is symmetric, with the proton centered between the oxygens. Both, AM1 and PM3 fail to reproduce this global minimum: PM3 yields an asymmetric structure and AM1 predicts bifurcated binding while SCC-DFTB reproduces the correct symmetric structure.

The H_2O-OH^- compound has an asymmetric structure with two different O–H bond lengths at the HF, MP2 and MP4 levels of theory, while DFT predicts this to be a symmetric structure [26]. The SCC-DFTB follows here the DFT results, also predicting a symmetric conformation. The energy difference between the symmetric and asymmetric conformation is only a few tenths of kcal/mol at the post-HF level of theory.

Finally, we compare the interaction energies for H bonded DNA base pairs with MP2 results [27], as we have already done for the SCC-DFTB method without H-p type orbitals [20]. Hobza et al. [27] investigated the H-bonding energies of 26 base pair conformations of the bases adenine (a), cytosine (c), guanine (g) and thymine (t). In those compounds the interaction energies are underestimated in the SCC-DFTB (with the H-s basis only) compared to the MP2 values, yielding a mean average error of 2.8 kcal/mol compared to MP2. Empirical force fields and the semi-empirical methods like AM1 [2], PM3 [3] and MNDO/M [22] have been tested for this molecule set as well [27]. The empirical force fields show mean average errors of 0.9 to 2.4 kcal/mol, whereas the semi-emprical methods have mean average errors of 7.3 kcal/mol (AM1), 6.3 kcal/mol (PM3) and 2.5 kcal/mol (MNDO/M). The results for the SCC-DFTB including H-p type orbitals as given in Table 2 (for the abreviations see [27]) show similar trends as with s-basis only [20]: For instance, the interaction energy of the gg1 base pair is higher than that of the gcwc base pair. The tt and gt base pair interaction energies were nearly equal to the MP2 values in the H-s only basis, whereas other base pair interactions were underestimated on average by 3 kcal/mol. These base pair interaction energies are higher than the corresponding MP2 values. With the new H-p basis, we find a mean average error of 1.5 kcal/mol with respect to the MP2 values.

Clearly, none of the approximate methods, including SCC-DFTB, are able to produce highly accurate results for all the H bonded systems considered. But many systems also put high computational demands on the ab initio methods, where correlation, large basis sets and correction for BSSE have to be taken into account to provide the needed high accuracy. This makes the computations extremely demanding, so that any application to larger molecular complexes will be prohibitive. The SCC-DFTB scheme is able to reproduce geometries and energies very well, showing deviations from higher level calculations which are of the same order as those at the ab initio level, when different methods and basis sets are compared. The compounds tested so far (not all of which

Table 2
Interaction energies (kcal/mol) of H bonded base pairs, as described in the text. For the notation identifying the base pairs see Ref. [27]

base pair	SCC-TB	MP2	base pair	SCC-TB	MP2
gcwc	24.0	25.4	atwc	11.6	12.4
gg1	24.4	24.0	atrwc	11.5	12.4
cc	16.2	18.8	aa1	9.7	11.5
gg3	15.0	17.1	ga4	9.7	11.1
ga1	14.7	15.7	tc2	11.2	11.8
gt1	16.2	14.7	tc1	11.0	11.6
gt2	15.8	14.3	aa2	9.0	11.0
ac1	12.2	14.3	tt2	11.9	10.6
gc1	12.2	13.9	tt1	11.9	10.6
ac2	11.8	14.1	tt3	11.9	10.5
ath	11.3	13.3	ga2	9.0	10.4
ga3	12.8	15.2	gg4	8.6	10.3
atrh	11.3	13.2	aa3	8.3	10.0

are discussed here), were chosen to be representative of patterns that occur in biomolecules. Since these compounds are described satisfactorily, the SCC-DFTB method is expected to be successful in describing H bonding in biological systems.

4. Polypeptides: Conformational Energies and Geometries

The three-dimensional structure of peptides and proteins is given by the spacial arrangement of simple, so-called secondary structural elements, like α-helices, β-sheets (shown in Fig. 2) or turn structures. The pure sequence of amino acids, like glycine,

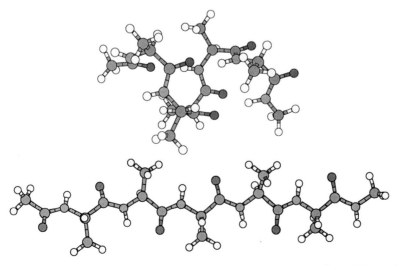

Fig. 2. α-helical and β-sheet (extended) conformations of a polypeptide model containing five alanine residues

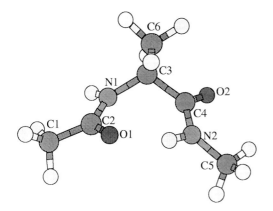

Fig. 3. The C_7^{ax} conformer of NA-LA-NMA (see text)

alanine etc., which build up secondary and tertiary structure, is called the primary structure. Since protein folding from a theoretical viewpoint can be understood as a hierarchical process, where secondary structure elements are formed first and then assembled to build up the three dimensional system [30], it is of great interest to study the fundamental structure and energetics of such elements with theoretical methods.

Molecules like those shown in Fig. 2, which contains five alanine residues, truncated by the "capping" groups $COCH_3$ at the left end and $NHCH_3$ (at the right end), are still very large for extensive ab initio studies. This is why smaller model peptides, like the N-acetyl-L-alanine-N'-methylamide (NA-LA-NMA) (Fig. 3) molecule have been studied at the HF, MP2 and DFT levels of theory in order to understand the geometric structure and the detailed energetics of polypeptides.

The NA-LA-NMA peptide model is derived from the alanine molecule by subsituting the H at N1 in the alanine molecule by the $C2O1C1H_3$ (acetyl) group and the OH group at C4 in the alaline molecule by the $N2HC5H_3$ (methylamide) group. These substitutions establish the Ψ and Φ dihedral angles, which characterize the peptide backbone conformations in polypeptides and proteins. The Φ angle is defined by rotations about the N1–C3 bond, Φ(C2–N1–C3–C4), whereas the Ψ angle is defined by rotations of the C3–C4 bond, Ψ(N1–C3–C4–N2), see Fig. 3.

In glycine based peptides the $C6H_3$ group is substiuted by a hydrogen atom. Other types of aminoacid residues result from substitution of this group by other organic restgroups. These side-chains can themselves have rotational degrees of freedom, which are important for the tertiary structure, since they may allow for additional binding between amino-acid residues, as for example in sulfid bridges. Here, we will concentrate only on alanine based polypeptides.

NA-LA-NMA has six stable conformers on the DFT-B3LYP/6-31G* and MP2/6-31G* potential energy surfaces (PES) [31], which have been taken as starting structures for further geometry optimizations with the SCC-DFTB, AM1 and PM3 methods. The relative energies of these conformers for different methods are shown in Table 3. The three lowest energy conformers form internal H bonds, whereas the higher energy conformers do not. The C_7^{ax} structure is shown in Fig. 3. The C_7^{ax} and C_7^{eq} conformations differ in the orientation of the methyl group attached to the central C3 atom. In the C_7^{ax} conformer, this methyl group is perpendicular to the seven membered ring which forms the H bond, whereas in the C_7^{eq} this methyl group is in the plane of the seven mem-

Table 3

Relative energies (kcal/mol) of the different conformers of NA-LA-NMA for different methods as described in the text. DFT-GGA refers to the method B3LYP. The geometries at the DFT, MP2 and HF level have been determined with the 6-31G* basis set

conf.	DFT-GGA	MP2	HF	SCC-DFTB	PM3	AM1
C_7^{eq}	0.00	0.00	0.00	0.00	0.00	0.00
C_5^{ext}	1.43	1.76	0.41	0.99	−1.55*)	1.72*)
C_7^{ax}	2.58	2.61	2.82	1.03	0.87	0.72
β_2	3.18	3.37	2.58	2.20+)	–	–
α_L	5.82	4.60	4.72	3.70	3.53	–
α_P	6.85	6.34	5.74	4.78	1.10	3.29

+) The β_2 conformer is not stable within the SCC-DFTB, but the maximum force at the B3LYP geometry is very small. The energy is given for a geometry, in which the forces are smaller than 0.00065 a.u.
*) Distorted structure.

bered ring. In the C_5^{ext} conformer, the hydrogen attached to N1 forms a H bond with O2, leading to a five membered ring.

The α-helical structures are predicted to be high energy conformers on the PES. Since these two structures form no internal hydrogen bonds, the smaller energy difference with respect to the ground state could be due to the fact that the energy of the H bonds may be considered to be underestimated in the SCC-DFTB method. Therefore, the stabilization of the C_7 and C_5 structures is underestimated compared to the α-helical structures. The β_2 conformer is found to be unstable in the SCC-DFTB model, but the maximum force at the dihedral angles (as shown in Table 4) is very small. This conformer may be stabilized due to internal H bonds in larger polypeptides.

Both, the AM1 and PM3 methods find a very distorted C_5^{ext} conformation, where the internal H bond is broken, as can be seen from the dihedral angles in Table 4. The β_2 conformer is also not stable in both methods and the α_L is unstable in the AM1 model.

The effects of solvation and peptide length are expected to play a crucial role for stabilizing the different conformers. Therefore, relative stabilities of the various conformers estimated within quantum mechanical calculations may change when the effects of solvents are included and longer peptides are considered. By applying a quantum che-

Table 4

Dihedral angles (in degrees) of the NA-LA-NMA conformers for different methods as described in the text. DFT-66A refers to the B3LYP method

conformer angles	C_7^{eq}		C_7^{ax}		C_5^{ext}		β_2		α_L		α_P	
	Φ	Ψ	Φ	Ψ	Φ	Ψ	Φ	Ψ	Φ	Ψ	Φ	Ψ
DFT-66A	−81.9	72.3	73.8	−60.0	−157.3	165.3	−135.9	23.4	68.5	24.5	−169.4	−37.8
SCC	−81.3	72.0	74.6	−66.1	−153.2	176.6	−136.7	24.9+)	65.6	13.0	−172.5	−51.1
AM1	−84.4	68.5	76.6	−64.0	−117.7	141.5	–	–	–	–	−115.5	−55.2
PM3	−71.4	77.7	68.8	−67.9	−93.9	147.9	–	–	62.3	39.6	−137.6	−60.5

+) The β_2 conformer is not stable within the SCC-DFTB method. The Φ, Ψ values refer to a conformation, where the maximum force is lower than 0.00065 a.u.

mical reaction field model at the RHF/6-31G* level of theory [32], the α_R conformation is shown to be considerably stabilized relative to the C_7^{eq} conformer. However, the conformation is still not a local minimum on the PES. Recently, it was shown that the α_R conformer will become stabilized only by explicitly including water molecules on a quantum theory level [33], supporting free energy calculations with emprirical force fields which also stabilize this conformer in solution [34]. We have shown that the SCC-DFTB model is also able to describe the changes of the PES due to the solvent effects [13].

While NALANMA does not show secondary structure motifs, like α-helical or turn-like conformers, such structural elements may be stabilized for larger polypeptides due to internal H bond formation. In addition, the relative stability of extended versus helical structures is expected to change due to cooperative effects which are much more pronounced in helical conformations. This has been examined with ab initio calculations, confirming large cooperative effects for the α_R helix compared to corresponding linear structures [35]. This in turn may result in an increased stability of helical conformers compared to extended ones with increasing peptide size.

We discuss next the tripeptide model NA-LA$_2$-NMA, containing two amino acids "capped" with the acetyle and methylamide groups, where the formation of turn structures is possible. The so-called β turn structures reverse the direction of a polypeptide via four amino acids and are therefore a common structural motif in proteins. In the tripeptide model (see Fig. 4), the oxygen atom of the acetyle goup can form a H bond with the nitrogen atom of the methylamide group, a so called $i \rightarrow i+3$ H bond, since this H bond connects the i-th residue (represented by the acetyle group) along a polypeptide chain with the residue $i+3$ (represented by the methylamide group). Protein α-helices form $i \rightarrow i+4$ H-bond patterns, therefore they cannot appear in this tripeptide model. Longer polypeptides, starting from the NA-LA$_3$-NMA model, would allow the formation of such H bonds.

Turn structures have been studied at the HF/3-21G [36], HF and MP2 (6-31G*) level of theory [37, 38]. Here we focus on the C_7^{eq}, C_5^{ext} linear repeat conformers and two β turn structures, the type I (βI) and type II (βII). These are classified by idealized backbone dihedral angles [39]. The dihedral angles are defined, as above in the NA-LA-NMA molecule, as rotations around the C_α–N and C_α–C bonds, where C_α is defined as the C atom to which the side chains (in the case of alanine the CH_3 group) are attached. A βI type turn structure is characterized by the ideal values $\Phi_1 = -60°$ and $\Psi_1 = -30°$, $\Phi_2 = -90°$ and $\Psi_2 = 0°$ respectively. A βII type turn is classified by the ideal dihedral

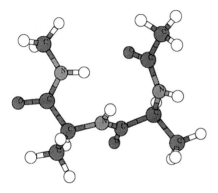

Fig. 4. The βI conformer of NA-LA$_2$-NMA (see text)

angles $\Phi_1 = -60°$, $\Psi_1 = 120°$, $\Phi_2 = 80°$ and $\Psi_2 = 0°$. A βIII type turn structure is characterized by the ideal values $\Phi_1 = -60°$ $\Psi_1 = -30°$ $\Phi_2 = -60°$ and $\Psi_2 = -30°$, but this turn is not a stable conformer in this tripeptide model. All these turn structures form $i \rightarrow i + 3$ H bonds, but only the type βIII turns are usually referred to as 3_{10} helices. Table 5 shows the dihedral angles evaluated at the SCC-DFTB and B3LYP/6-31G* (using the GAUSSIAN98 program package [43]) levels of theory for the four conformations. The dihedral angles of SCC-DFTB compared to B3LYP show deviations of up to 20° as in the case of the βII conformer. However, the PES for these molecules are very shallow and, consequently, deviations in this range can be expected even when comparing ab initio calculations at various levels of theory. The relative energy differences are consistently underestimated in the SCC-DFTB model, but the relative stabilities are reproduced well.

In the SCC-DFTB method, dipole moments are calculated by using the SCC-DFTB Mulliken charges. Despite this approximation, the SCC-DFTB dipole moments compare quite well with those at the B3LYP level of theory, as shown in Table 5.

We also have investigated larger polyalanine molecules, containing up to 11 alanine residues at the B3LYP/6-31G*, SCC-DFTB, AM1 and PM3 levels of theory [41, 44]. For the SCC-DFTB model we found similar trends as described above: the relative energetic ordering of different conformers is reproduced well, although the relative energy differences are underestimated. Geometries are in good agreement with B3LYP results, showing similar deviations as dicussed above. The SCC-DFTB model, therefore, seems to give a reliable description of structures and a semi-quantitative estimate of the energetics. Secondary structural motives, like β-sheets, helices and turn structures, are predicted to be stable conformers in agreement with B3LYP results. For approximate methods, this good performance cannot be expected a priori. It has been shown that some empirical force field methods, AM1 and even HF/3-21G calculations, do not predict correctly turn structures to be stable conformers [37]. Further, semi-empirical methods like AM1 and PM3 do not reproduce satisfactorily the relative ordering in energy and structural properties of these secondary structural motifs [41, 44]. β-sheet structures show large distortions in both models. While PM3 is not able to describe helical structures, it unwinds helices and breaks the internal H bonds, AM1 seems to favor helices which are in between the 3_{10} and α_R conformations. In the latter case bifurcated H bonds (of both $i \rightarrow i + 3$ and $i \rightarrow i + 4$ type) are formed.

Table 5

Dihedral angles (in degrees), relative energies ΔE (in kcal/mol) and dipole moments (in Debeye) of NA-LA$_2$NMA at different levels of theory. DFT denotes B3LYP/6-31G* calculations

conformer method	C_7^{eq}		C_5		βI		βII	
	DFT	SCC-DFTB	DFT	SCC-DFTB	DFT	SCC-DFTB	DFT	SCC-DFTB
Φ_1	−82.7	−81.3	−158.7	−157.4	−74.7	−70.6	−60.8	−59.2
Ψ_1	69.5	68.8	165.5	175.7	−12.3	−5.4	128.8	110.2
Φ_2	−84.6	−83.3	−159.4	−160.5	−105.5	−111.0	69.8	64.1
Ψ_2	70.0	68.7	165.0	178.5	13.1	18.1	15.1	20.9
dipole	5.8	5.3	6.3	5.8	8.1	7.5	6.6	6.3
ΔE	0.0	0.0	2.13	1.51	2.59	1.76	4.32	2.71

Recently, we also studied the vibrational frequencies, infrared absorption (IR) and vibrational circular dichroism (VCD) intensities of NA-LA-NMA with the SCC-DFTB method in comparision to HF, B3LYP and MP2 calculations [45]. In this work, in order to estimate the intensities at the SCC-DFTB level we used an SCC-DFTB – DFT hybrid approach. Ground state geometries and second energy derivatives were calculated with the SCC-DFTB method while the dipole derivatives and VCD tensors were calculated with the B3LYP method by using the SCC-DFTB geometries. Frequencies for the C_7^{eq} and C_5 conformers at the B3LYP, MP2, HF (6-31G*) and SCC-DFTB level of theory are given in Ref. [45] in detail. The SCC-DFTB values compare very well with the higher level calculations; this also holds for the C_7^{ax}, α_L and α_P structures, but these results will not be given here explicitly. As one example, for the C_7^{eq} and C_5 conformers the standard deviation from experimental frequencies is 3.0%, 4.4% and 6.7% for the MP2, B3LYP and SCC-DFTB methods respectively. The IR and VCD intensities estimated with the hybrid approach compare satisfactorily with those of the higher level calculations [45] for the two conformers investigated, the C_7^{eq} and C_5^{ext}.

This good agreement encouraged us to calculate the IR intensities fully at the SCC-DFTB level of theory. As exploited above, the dipole moments from the SCC-DFTB model compare reasonably well with the full DFT results. We used the derivatives of the SCC-DFTB dipolmoments (calculated from Mullikan charges) with respect to the atomic coordinates as an approximate way to calculate the IR intensities.

Important modes for the characterization of peptide conformations are the N–H stretch (amide A) mode, the C=O stretch (amide I) and the N–H bend (in combination with C–N stretch) (amide II) modes, located around 3400 to 3500 cm^{-1}, 1700 cm^{-1} and 1500 to 1550 cm^{-1}, respectively (for the assignment of experimental frequencies see Ref. [31] and references therein).

The N–H and C=O bonds occur twice in NA-LA-NMA (Fig. 3), but the corresponding vibrational modes are not degenerate, since only one of the two N–H and C=O bonds is involved in a H bond. This alters the vibrational frequency and leads to a splitting of the amide A, I and II modes. The two modes corresponding to each bond type will be labeled by A_a and A_b, I_a and I_b etc. In Fig. 5 we show the IR spectra of the C_7^{eq} conformer of NA-LA-NMA estimated on the SCC-DFTB and the B3LYP/6-31G* level of theory.

The SCC-DFTB mainly overestimates the intensities of the more intense modes, i.e. the C=O stretch, relative to the less intense modes. This is mainly due to the approximation of the dipole derivatives, since in the hybrid SCC-DFTB/B3LYP approach the relative intensities compare much better with the full ab-initio data [45]. As can be seen from Fig. 5, the SCC-DFTB reproduces the frequency splitting of the N–H stretch mode at 3500 cm^{-1} and of the C=O stretch mode at 1750 cm^{-1} reasonably well. Only for the N–H bend mode the splitting is smaller than at the B3LYP level of theory, so that only one line occurs at 1600 cm^{-1} in the SCC-DFTB spectrum.

Next, we analyze the frequency splitting for the five stable conformers at the SCC-DFTB potential energy surface in more detail. Table 6 shows the values for the splitting of the amide A, I and II modes (experimental values and assingnement and B3LYP/6-31G* data are from Ref. [31]), which are calculated as the difference of the two frequencies of the modes A, I and II, e.g. $\Delta \nu_A = \nu_{A_a} - \nu_{A_b}$ (ν_{A_a} is larger than ν_{A_b}). As can be seen from the values for the C_7^{eq} and C_5 conformers, both theoretical methods show deviations from the experimental values of up to 30 cm^{-1}. Both methods

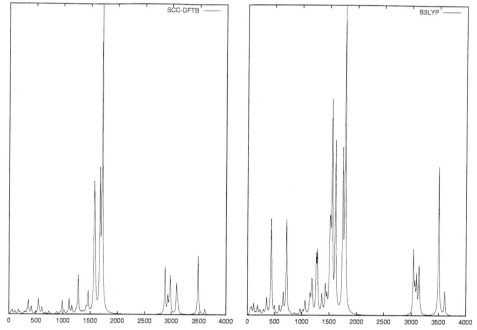

Fig. 5. IR intesities for the C_7^{eq} conformer of NA-LA-NMA at the SCC-DFTB (left) and B3LYP level of theory, see text. Intensities are in arbitray units, frequencies in cm^{-1}

reproduce the trends in the splitting for the amide A and I modes when going from the C_7^{eq} to the C_5 conformer. For the amide II mode, the SCC-DFTB does not reproduce the experimental trend, wheras B3LYP does. To discuss relative intensities of the a and b modes in more detail, we take the ratio of the intensities I_a^X and I_b^X:

$$Q = \frac{I_a^X}{I_b^X},$$

Table 6

Splitting of the amide A, I and II modes (in cm^{-1}) for the NA-LA$_1$NMA conformers at the B3LYP and SCC-DFTB level of theory respectively in comparison to experiment (see text)

	B3LYP	SCC-DFTB	EXP	B3LYP	SCC-DFTB	EXP	B3LYP	SCC-DFTB
	C_7^{eq}			C_5			α_P	
A	105	141	115	43	89	71	12	23
I	40	50	25	16	11	17	3	5
II	60	12	42	41	40	34	34	62
	C_7^{ax}			α_L				
A	45	33		34	20			
I	37	46		7	12			
II	59	13		27	46			

Table 7
Relative intensities of amide A (Q_A), I (Q_I) and II (Q_{II}) modes for the NA-LA$_1$NMA conformers at the B3LYP, SCC-DFTB and SCC-B3LYPB/B3LYP hybride (HYB) levels of theory, respectively (see text)

	B3LYP	SCC-DFTB	HYB	B3LYP	SCC-DFTB	HYB	B3LYP	SCC-DFTB
	C_7^{eq}			C_5			α_P	
A	1:6.4	1:9.7	1:6.5	1:4.0	1:6.3	1:3.6	1:1.2	1:1.1
I	2.1:1	2.1:1	2.2:1	1:4.7	1:1.3	1:1.1	1:1.1	1:1.0
II	1:1.3	1:1.8	1:1.6	1:2.7	1:3.0	1:2.9	1.2:1	1:1.7
	C_7^{ax}			α_L				
A	1:9.8	1:17.1		1.7:1	1.7:1			
I	2.7:1	3.1:1		1.1:1	1:1.2			
II	1.2:1	1:1.2		1.3:1	1:1.6			

where X labels the modes (X = A, I and II). The Q values are given in Table 7 (values for B3LYP and the hybrid SCC-DFTB/B3LYP are taken from Ref. [31] and Ref. [45], respectively). Compared to B3LYP the relation of the intensities is reversed for some modes. Further, the amide I mode of the C_5 conformer shows a much smaller ratio at the SCC-DFTB level than at the B3LYP level of theory, which is in agreement with MP2 results [45]. However, at the HF/6-31G* level of theory deviations in these ratios from the B3LYP or MP2 results are similar, whereas, when smaller basis sets are used in HF, like 4-31G, even larger deviations are obtained [45, 31]. The higher ratios of the intense modes at the SCC-DFTB level indicate that more intense modes are indeed overestimated compared to the modes with lower intensity. This shows that there is a systematic overestimation of the intense modes (or underestimation of the modes with low intensity), which possibly might be corrected by scaling the intensities.

5. DNA: H Bonding and Stacking

Base pair stacking is an even more challenging test than H bonding (see Fig. 6), since correlation is responsible for the stacking stabilization energies to a large extent and very diffuse functions are of primary importance for a proper description. Further, HF has been shown to reproduce stacking energies poorly and MP2 seems to overestimate correlation contributions by 15 to 30% [46]. Recently, there have been suggestions for constructions of density functionals to include the van der Waals interaction (see Ref. [47] and references therein).

Since within the SCC-DFTB model the dispersion interaction is clearly left out, we have chosen to include it empirically. To our knowledge, a first attempt in this dircetion was based on the Slater-Kirkwood approximation [48] as implemented by Lewis and Sankey [49]. We choose to add the London dispersion formula

$$E_{\text{dis}} = \sum_{i,j} \frac{(I_i \alpha_i)(I_j \alpha_j)}{(I_i + I_j) R_{ij}^6} \quad (7)$$

to the total energy Eq. (1). I_i and α_i are experimental values for the ionization potential and the polarizability of atom i, R_{ij} is the distance between atoms i and j. However,

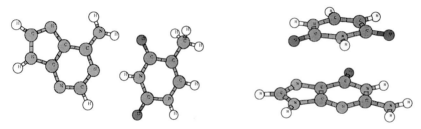

Fig. 6. The H-bonded ac base pair and the gu stacked base pair

in the intermediate distance region in the vicinity of the potential minimum (of the base–base interaction) the London formula is no longer valid due to the overlap of the charge densities, i.e. the interaction becomes too attractive. Therefore, we have chosen a scaling function in order to make E_{dis} vanish for small distances compared to the base pair equilibrium distance. Details will be given in a future publication [51]. The expression of E_{dis} has the advantage that it can be applied to all atoms in the system, and does not need to be restricted to inter-base pair interactions. Since analytic expressions for atomic forces are easily evaluated, geometry optimizations and molecular dynamic simulations can be performed in the standard way.

Stacking energies for DNA-base pairs have been evaluated at the MP2 level of theory with an a posteriori correction for BSSE using geometries resulting from empirical force field optimizations [40]. Empirical force fields are capable of satisfactorily describing the stacking interactions, although there are discrepancies between the force fields themselves and with respect to the MP2 results of up to 100% in the interaction energies [27]. The AM1 and PM3, as well as the MNDO/M method, have been shown to be unable to reproduce the attractive stacking interactions; for these methods, the interactions are erroneously found to be repulsive (2 to 10 kcal/mol), leading to a destabilization of the stacked base pairs [27].

At the SCC-DFTB level of theory, the interaction energies are attractive, although significantly underestimated [20]. The empirical correction leads to interaction energies which compare well with the MP2 results, see Table 8. We also investigated the radial and torsional dependence of the stacking interactions. The empirically extended SCC-DFTB model is able to reproduce these dependencies very accurately compared to the MP2 values [42].

Table 8

Stacking energies (kcal/mol) of base pairs for the SCC-DFTB, MP2 and SCC-DFTB with empirical inclusion of dispersion (see text)

	SCC-DFTB	MP2	SCC-DFTB + dispersion		SCC-DFTB	MP2	SCC-DFTB + dispersion
ga	2.6	11.2	9.0	gg	5.6	11.3	11.1
gu	6.5	10.6	11.4	aa	1.5	8.8	6.9
ac	3.0	9.5	8.1	cc	2.6	8.3	8.9
gc	6.0	9.3	10.7	uu	3.3	6.5	5.1
au	3.8	9.1	8.9	cu	5.3	8.5	9.3

6. Conclusions

We have presented several extensions of an approximate DFT scheme (SCC-DFTB) in order to improve the accuracy for biologically relevant molecular interactions with the aim to develop a highly efficient method for accurate large-scale simulations of biomolecules. The SCC-DFTB has been implemented into a QM/MM scheme, which has been shown to yield satifactory results for intermolecular H-bonded compounds. We have also extended the SCC-DFTB minimal basis set to include p-type basis functions on the hydrogen atoms. Tests for a large set of compounds relevant to biological molecules show a clear improvement in the description of H bonding energies, which now are in quantitative agreement with higher level ab initio results. The SCC-DFTB method has been benchmarked for different conformers of small model peptides. The geometries are shown to be reliably determined and the relative energies show the right ordering compared to B3LYP/6-31G* reference calculations, although the energy differences are slightly underestimated. We further discussed the performance of the SCC-DFTB method for vibrational frequencies and IR absorption for the conformers of the model peptide NA-LA-NMA. The determination of IR intensities is based on the Mulliken approximation for atomic point charges, which shows an overestimation of the more intense peaks. Trends in the absorption spectra are reproducible to some extent and a correction of the intensities via a scaling might be possible. Finally, we described briefly the inclusion of the dispersion interaction in an empirical manner. This extension leads to interaction energies comparable to those obtained at the MP2 level of theory. In summary, we have shown that the SCC-DFTB method is able to describe biologically relevant structures with high accuracy, while it is several orders of magnitude faster than ab initio methods.

References

[1] A. St-Amant, Density Functional Methods in Biomolecular Modeling, in: Reviews in Computational Chemistry, Vol. 7, Eds. K. B. Lipkowitz and D. B. Boyd, New York 1996 (p. 217).
[2] J. S. Dewar, E. Zoebisch, E. F. Healy, and J. J. P. Stewart, J. Amer. Chem. Soc. **107**, 3902 (1985).
[3] J. J. P. Stewart, J. Comput. Chem. **10**, 209, 221 (1989).
[4] M. C. Zerner, Semiempirical Molecular Orbital Methods, in: Reviews in Computational Chemistry, Eds. K. B. Lipkowitz and D. B. Boyd, New York 1990 (p. 45).
[5] Th. Frauenheim et al., phys. stat. sol. (b) **217**, 41 (2000).
[6] M. Elstner, D. Porezag, G. Jungnickel, J. Elsner, M. Haugk, T. Frauenheim, S. Suhai, and G. Seifert, Phys. Rev. B **58**, 7260 (1998).
[7] M. Elstner, D. Porezag, G. Jungnickel, T. Frauenheim, S. Suhai, and G. Seifert, in: Tight-Binding Approach to Computational Materials Science, Eds. P. Turchi, A.Gonis, and L. Colombo, Mater. Res. Soc. Symp. Proc. **491**, 131 (1998).
[8] D. Porezag, T. Frauenheim, T. Köhler, G. Seifert, and R. Kaschner, Phys. Rev. B **51**, 12947 (1995).
[9] G. Galli, phys. stat. sol. (b) **217**, 231 (2000).
[10] P. Ordejón, phys. stat. sol. (b) **217**, 335 (2000).
[11] J. Gao, Methods and Applications of Combined Quantum Mechanical and Molecular Mechanical Potentials, in: Reviews in Computational Chemistry, Vol. 7, Eds. K. B. Lipkowitz and D. B. Boyd, New York 1996 (p. 119).
[12] M. Elstner, Ph.D. Thesis, University Paderborn, Paderborn (Germany) 1998.
[13] W. Han, M. Elstner, K. J. Jalkanen, T. Frauenheim, and S. Suhai, submitted to Internat. J. Quant. Chem.
[14] M. Elstner, S. M. Lee, Y. H. Lee, E. Kaxiras, and T. Frauenheim, in preparation.

[15] J. G. C. M. van Duijneveldt-van de Rijdt and F. B. van Duijneveldt, Ab initio Methods Applied to Hydrogen-Bonded Systems, in: Theoretical Treatment of Hydrogen Bonding, Ed. D. Hadzi, Wiley, New York 1997.
[16] S. F. Boys and F. Bernardi, Mol. Phys. **19**, 553 (1970).
[17] L. A. Curtiss, D. J. Frurip, and M. J. Lander, Chem. Phys. **71**, 2703 (1979).
[18] H. Guo, S. Sirois, E. I. Proynov, and D. R. Salahub, Density Functional Theory and Its Application to Hydrogen-Bonded Systems, in: Theoretical Treatment of Hydrogen Bonding, Ed. D. Hadzi, Wiley, New York 1997.
[19] D. Hadzi and J. Koller, Hydrogen Bonding by Semi-Empirical Molecular Orbital Methods, in: Theoretical Treatment of Hydrogen Bonding, Ed. D. Hadzi, Wiley, New York 1997.
[20] M. Elstner, D. Porezag, T. Frauenheim, S. Suhai, and G. Seifert, in: Multiscale Modelling of Materials, Eds. T. Diaz de la Rubia, T. Kaxiras, V. Bulatov, N. M. Ghoniem, and R. Phillips, Mater. Res. Soc. Symp. Proc. **538**, 243 (1999).
[21] M. J. S. Dewar and W. J. Thiel, J. Amer. Chem. Soc. **99**, 4899, 4907 (1977).
[22] A. A. Voityuk and A. A. Blizniuk, Theor. Chim. Acta **72**, 223 (1987).
[23] K. Jug and G. Geudtner, J. Comput. Chem. **14**, 639 (1993).
[24] B. K. Smith, D. J. Swanton, J. A. Pople, H. F. Schaefer, and L. Randon, J. Chem. Phys. **92**, 1240 (1990).
[25] S. Suhai, J. Phys. Chem. **99**, 1172 (1995).
[26] R. V. Stanton and K. M. Merz, J. Chem. Phys. **101**, 6658 (1994).
[27] P. Hobza et al., J. Comput. Chem. **18**, 1136 (1997).
[28] Y.-J. Zheng and K. M. Merz, J. Comput. Chem. **13**, 1151 (1992).
[29] J. E. Del Bene, J. Comput. Chem. **10**, 603 (1998).
[30] M. Karplus and D. L. Weaver, Protein Sci. **3**, 650 (1994).
[31] K. J. Jalkanen and S. Suhai, Chem. Phys. **208**, 81 (1996).
[32] K. Rommel-Möhle and H.-J. Hofmann, J. Mol. Struc. (Theochem) **285**, 211 (1993).
[33] W. Han, K. J. Jalkanen, M. Elstner, and S. Suhai, J. Phys. Chem. B **102**, 2587 (1998).
[34] C. L. Brooks III and D. A. Case, Chem. Rev. **93**, 2487 (1993).
[35] P. T. van Duijnen and B. T. Thole, Biopolymers **21**, 1749 (1982).
[36] A. Perczel, M. A. McAllister, P. Csaszar, and I. C. Csizmadia, J. Amer. Chem. Soc. **115**, 4849 (1993).
[37] H.-J. Böhm, J. Amer. Chem. Soc. **115**, 6152 (1993).
[38] K. Möhle, M. Gussmann, A. Rost, R. Cimiraglia, and H.-J. Hofmann, J. Phys. Chem. **101**, 8571 (1997).
[39] J. Richardson, Adv. Protein Chem. **43**, 167 (1991).
[40] J. Sponer, J. Leszczynski, and P. Hobza, J. Phys. Chem. **100**, 5590 (1996).
[41] K. Jalkanen, M. Elstner, S. Suhai, and T. Frauenheim, to be published.
[42] P. Hobza, M. Elstner, T. Frauenheim, E. Kaxiras, and S. Suhai, to be published.
[43] M. J. Frisch, G. W. Trucks, H. B. Schlegel, G. E. Scuseria, M. A. Robb, J. R. Cheeseman, V. G. Zakrzewski, J. A. Montgomery, Jr., R. E. Stratmann, J. C. Burant, S. Dapprich, J. M. Millam, A. D. Daniels, K. N. Kudin, M. C. Strain, O. Farkas, J. Tomasi, V. Barone, M. Cossi, R. Cammi, B. Mennucci, C. Pomelli, C. Adamo, S. Clifford, J. Ochterski, G. A. Petersson, P. Y. Ayala, Q. Cui, K. Morokuma, D. K. Malick, A. D. Rabuck, K. Raghavachari, J. B. Foresman, J. Cioslowski, J. V. Ortiz, B. B. Stefanov, G. Liu, A. Liashenko, P. Piskorz, I. Komaromi, R. Gomperts, R. L. Martin, D. J. Fox, T. Keith, M. A. Al-Laham, C. Y. Peng, A. Nanayakkara, C. Gonzalez, M. Challacombe, P. M. W. Gill, B. Johnson, W. Chen, M. W. Wong, J. L. Andres, C. Gonzalez, M. Head-Gordon, E. S. Replogle, and J. A. Pople, Gaussian 98, Revision A.5, Gaussian, Inc., Pittsburgh (PA) 1998.
[44] K. Jalkanen, M. Elstner, T. Frauenheim, and S. Suhai, submitted to Chem. Phys.
[45] H. G. Bohr, K. Frimand, K. J. Jalkanen, M. Elstner, and S. Suhai, Chem. Phys. **246**, 13 (1999).
[46] J. Sponer and P. Hobza, Chem. Phys. Lett. **267**, 263 (1997).
[47] J. F. Dobson and J. Wang, Phys. Rev. Lett. **82**, 2123 (1999).
M. Lein, J. F. Dobson, and E. K. U. Gross, J. Comput. Chem. **20**, 12 (1999).
[48] T. A. Halgren, J. Amer. Chem. Soc. **114**, 7827 (1992).
[49] J. P. Lewis and O. F. Sankey, Biophys. J. **69**, 1068 (1995).

phys. stat. sol. (b) **217**, 377 (2000)

Subject classification: 68.35.Bs; 68.35.Md; 71.15.Hx; S10.1

First-Principles Calculations of α-Alumina (0001) Surfaces Energies with and without Hydrogen

P. D. Tepesch[1]) (a) and A. A. Quong (b)

(a) Sandia National Laboratories, Livermore, CA, USA

(b) Lawrence Livermore National Laboratory, Livermore, CA, USA

(Received August 10, 1999)

We used first-principles, mixed-basis (pseudo-atomic and plane-wave basis functions) pseudopotential calculations to study the energetics and structure of the α-alumina (0001) surface for three different terminations as well as two different hydrogen covered surfaces. The calculations show that over the range of chemical fields where α-alumina is stable, the stoichiometric (1-Al terminated) surface is the most stable with a surface energy of 2.13 J/m^2. The lowest values of the 2-Al-terminated and oxygen terminated surface energies are 2.7 and 3.5 J/m^2, respectively. We find that hydrogen bonds weakly to the stoichiometric surface with a binding energy of 0.5 eV/bond and prefers to sit directly on top of the terminating Al atoms. Finally, we find that an oxygen terminated surface with one hydrogen atom per surface oxygen has a lower surface energy over the relevant range of chemical fields.

1. Introduction

Alumina continues to be a material of significant interest in science and technology: a search in the SciSearch database for alumina or Al_2O_3 gave a list of more than 2000 papers several reasons. First, the preparation of ceramic materials is usually through a solid-state sintering process, in which the starting material is in powder form. Understanding the properties of the surfaces will lead to better understanding of this sintering process. Second, surfaces of single-crystal alumina and, in particular, the (0001) α-alumina surface, are used as substrates for thin-film deposition. Finally, alumina surfaces are of interest as catalyst materials.

The atomic structure of the (0001) α-alumina surface has been the subject of many experimental studies. It is known that, in vacuum at moderate temperatures (<1100 °C), the surface is unreconstructed (1×1) and that the surface goes through three reconstructions at higher temperature which are proposed to be accompanied by changes in surface composition [1, 2]. For the unreconstructed surface, there are several proposals for composition and structure. In Fig. 1, the different surface terminations are illustrated, with a cut along the line labeled S yields two identical stoichiometric surfaces (1-Al-terminated). The cut along the line NS yields two different non-stoichiometric surfaces, one being aluminum rich, the so-called 2-Al-terminated surface, and the oxygen rich O-terminated surface. On the basis of experiments on ultra-thin layers of alumina on metal surfaces, it has been proposed that the surface is oxygen terminated [3]. On the basis of growth experiments, it has been proposed that there are two different

---

[1]) Current address: Corning, Inc., Cellular Ceramics Research, Science & Technology, SP-DV-1-9, Corning, NY 14831, USA.

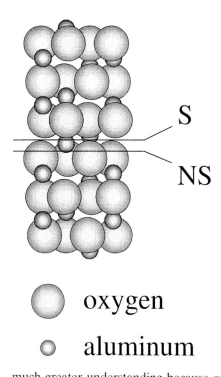

Fig. 1. The crystal structure of α-alumina viewed along the [2̄110] direction. The oxygen atoms are in a distorted h.c.p. stacking and the aluminum atoms occupy 2/3 of the distorted octahedral interstitial sites. There are two distinct places to cut this crystal perpendicular to the [0001] direction, which are labeled in the figure by S, for stoichiometric, and NS, for non-stoichiometric types of surfaces with different composition [4]. Finally, ion-scattering experiments suggest that there is only a single, stoichiometric surface structure [5]. The ion-scattering experiments also show significant concentrations of hydrogen that are proposed to come from the bulk. In some sense, it is not surprising that different experiments lead to different conclusions about the surface structure and termination since the surface environment can vary significantly depending on experimental conditions, surface preparation, growth conditions, etc.

Unfortunately, the theoretical work on the (0001) α-alumina surface does not lead to a much greater understanding because of complexities neglected in the calculations.

The first calculations carried out on this surface were shell-model calculations on the stoichiometric surface (Fig. 2), which predict a large inward relaxation of the topmost Al layer and a surface energy of 2.03 J/m^2 [6]. This has been largely reconfirmed with later first-principles DFT calculations, which predict a surface energy of 1.76 J/m^2 [7, 8] with similar results for the relaxations. Although results have been published with rigid-ion models for non-stoichiometric surfaces, there is some question about how to interpret those results without reference to chemical field terms (see Section 2). The only first-principles calculations for the clean surface have been performed for the stoichiometric surface. One calculation has been done for a hydrogen-covered, oxygen-terminated surface, but the results are difficult to interpret, given that the unphysical result of a negative surface energy is reported [9].

In this paper, we will first describe the methods we apply, including the thermodynamic framework in which we will interpret total-energy calculations and the first-principles technique we use to compute the necessary energies. We will then present the results for the three types of surfaces derived from cutting the bulk crystal and show that the stoichiometric surface is the thermodynamically stable surface at 0 K. Finally, we show the results of our calculations for the hydrogen-covered oxygen-terminated and stoichiometric surfaces, which indicate that the former is thermodynamically stable (at 0 K).

2. Modeling Methods

In this section we first present the thermodynamic framework in which we compute surface or, more generally, interface energies. We then describe the details of the mixed-basis pseudopotential calculations.

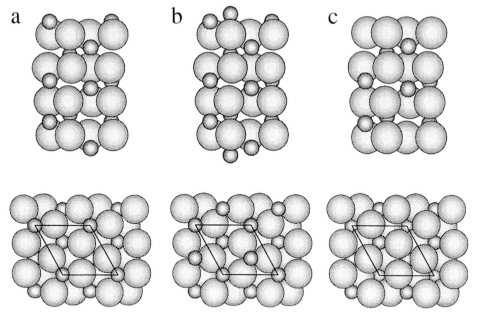

Fig. 2. The three distinct surfaces that can be obtained by cutting α-alumina perpendicular to the [0001] direction shown along [2$\bar{1}$10] (top) and [0001] (bottom). a) The stoichiometric, or 1-Al-terminated surface which contains 1 Al atom on the surface per unit cell. This surface is obtained by cutting with the plane labeled S in Fig. 1. b) The oxygen-rich, or oxygen-terminated surface. c) The aluminum-rich, or 2-Al-terminated surface. The oxygen- and aluminum-rich surfaces are obtained by cutting with the plane labeled NS in Fig. 1

2.1 General thermodynamic considerations

We are interested in understanding the equilibrium surfaces of a bulk (semi-infinite) crystal at constant temperature (T) and pressure (P). In a system with n_c components (atom types) and one phase, Gibbs' phase rule gives the number of degrees of freedom to be $n_c - 1$. One is free to set the chemical potentials, μ_i, of $n_c - 1$ components in the system. The remaining chemical potential is determined by the condition of equilibrium. A useful thermodynamic potential under these conditions is

$$\omega(T, P, \mu'_1, \mu'_2, \ldots, \mu'_{n_c-1}) = u - Ts + Pv - \sum_{i=1}^{n_c-1} \mu'_i c_i + \frac{1}{n_c} \sum_{i=1}^{n_c-1} \mu'_i = \mu, \quad (1)$$

where u is the internal energy, s is the entropy, v is the volume, $\mu'_i = \mu_i - \mu_{n_c}$, c_i is the concentration of species i, and the average value of the chemical potential is

$$\mu = \frac{1}{n_c} \sum_{i=1}^{n_c} \mu_i. \quad (2)$$

We will refer to μ'_i as the chemical field of species i.

To obtain surface energies from first-principles calculations, we use a supercell approach at $T = P = 0$. For a supercell containing two non-interacting surfaces, N_i^s atoms of type i, N^s total atoms, surface area A, and total energy E^s,

$$A(\sigma_1 + \sigma_2) = E^s - N^s e^b + \sum_{i=1}^{n_c-1} \mu'_i (N^s c_i^b - N_i^s). \quad (3)$$

where σ is the surface energy, c_i^b is the concentration of component i in the perfect crystal, and e^b is the energy per atom of the perfect crystal. In cases where there is a symmetry element that relates the two surfaces, $\sigma_1 = \sigma_2$, the surface energy can be uniquely determined by Eq. (3). This is the case for all surfaces in crystals with inversion symmetry, including α-alumina.

Because we are interested in equilibrium surface energetics of a certain crystalline phase b, we need to know the constraints on the chemical fields which define the thermodynamic stability range of this phase. These constraints are determined by the condition that, at equilibrium, $\omega^b < \omega^j$, for all other phases j. At $T = P = 0$, these constraints give

$$\sum_{i=1}^{n_c-1} \mu_i'(c_i^j - c_i^b) < e^j - e^b, \qquad (4)$$

where c_i^j and c_i^b are the concentrations of species i in phases j and b, respectively. The "tightest" constraints will be those determined by the phases that can be in equilibrium with b. In a two-component system, there are at most two relevant constraints. For systems with more than two components, there is no fundamental limit on the number of relevant constraints.

2.2 Total energy method

To calculate the energetics of this system, we use a first-principles electronic structure method. The electron–ion interaction is taken as a norm-conserving pseudopotential and the electronic wavefunctions are expanded in a linear combination of plane waves and pseudo-atomic orbitals. Exchange-correlation is treated in the local density approximation and we use the Perdew-Zunger [10] parametrization of the energy. Sampling of the Brillouin zone is performed by summing over k points generated by the method of Monkhorst and Pack [11], and in the case of metallic systems, the states are smeared by a Gaussian. The pseudopotentials are generated using the method of Troullier and Martins [12]. The oxygen pseudopotential would require a plane wave cut-off of 60 Ry (to obtain 1 mRy/atom convergence of energy differences) in a pure plane wave calculation. In the present method, however, the presence of the pseudoatomic orbitals allows us to reach this level of convergence with a 10 Ry cut-off. The pseudoatomic orbitals have been generated by the use the LDA eigenstates of the neutral pseuodoatom.

3. Results and Discussion

3.1 Alumina (0001) surface termination

As discussed in the Introduction, there have been reports in literature suggesting that all three surface terminations of α-alumina, shown in Fig. 2, can be found. Only the stoichiometric surface has been treated within a first-principles calculation. Empirical calculations have been reported [13], but without reference to the necessary chemical field terms (Eq. (3)). Here we show the results of first-principles calculations for the relative stability of these three surfaces at $T = 0$.

Equation (3) for α-alumina can be written as

$$2A(\sigma) = E^s - N^s e^b + \mu'_{Al}(N^s c_{Al}^b - N^s_{Al}). \qquad (5)$$

Calculations of α-Alumina (0001) Surfaces Energies

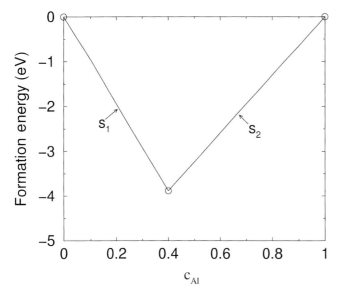

Fig. 3. Illustration of the determination of the constraints on chemical field for the stability of α-alumina

To compare the surface energies for surface configurations with different compositions we need the bulk energy e^b, the supercell energies E^s, and the values of the chemical field μ'_{Al}, for which the crystal is thermodynamically stable. In the two-component Al–O system, there are only two relevant constraints on μ'_{Al}. The two constraints are determined by applying Eq. (4) to the thermodynamically stable phases closest in composition at higher and lower values of c_{Al}. To illustrate this, Fig. 3 shows the formation

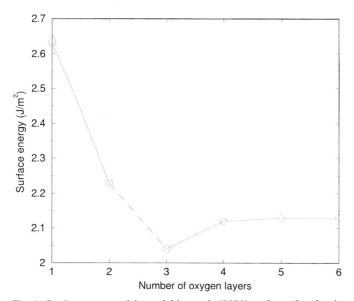

Fig. 4. Surface energy of the stoichiometric (0001) surface of α-alumina as a function of slab thickness

energies of three structures (fcc aluminum, α-alumina, and isolated O_2 molecules) in the Al–O system as a function of aluminum concentration. The reference energies are taken to be fcc aluminum and isolated O_2 molecules. The slopes S_1 and S_2 are the constraints on μ'_{Al}. Since α-alumina is the only stable compound observed in the Al–O system, the constraints are determined by equilibrium conditions with the pure elements. We estimate the $T = 0$ energy for oxygen to be half the energy of an isolated O_2 molecule. If one were considering surface structures of pure Al, one of the limits would be $\mu'_{Al} < \infty$.

For the calculation of the surface energy of the stoichiometric surface, we used supercells containing one to six close-packed oxygen layers (see Fig. 4). There is only a 0.01 J/m² change in surface energy going from four-layer to six-layer slabs. For comparing the surface energies of the three different terminations, we used slabs with six oxygen layers, corresponding to unit cells with 28, 30, and 32 atoms per unit cell for the oxygen-rich, stoichiometric, and aluminum-rich surfaces, respectively. The unit cell is hexagonal with an in-plane lattice parameter of 4.731 Å and a vacuum region of 5.5 Å. All geometries were constrained with inversion and threefold rotational symmetry. The coordinates and interplanar spacings at the minimum energy configurations for half of each slab are given in Tables 1 to 4 along with the unrelaxed (bulk) coordinates and interplanar spacings.

Table 1

Unrelaxed (bulk) coordinates and interplanar spacings of the (0001) surface of α-alumina. Atoms related by the threefold and inversion symmetries are not shown

type	x (fractional)	y (fractional)	Z (Å from slab center)	spacing (Å)
Al	0.333333333	0.666666666	6.677211417	
Al	0.666666666	0.333333333	6.196589416	0.480622
O	0.639333333	0.666666666	5.364083683	0.832500
Al	0	0	4.531578141	0.832500
Al	0.333333333	0.666666666	4.050956275	0.480622
O	0.306000000	0.306000000	3.218450350	0.832500
Al	0.666666666	0.333333333	2.385944616	0.832500
Al	0	0	1.905322750	0.480622
O	0.972666666	0.333333333	1.072817016	0.832500
Al	0.333333333	0.666666666	0.240311475	0.832500

Table 2

Relaxed coordinates and interplanar spacings of the 1-Al-terminated (0001) surface of α-alumina. Atoms related by the threefold and inversion symmetries are not shown

type	x (fractional)	y (fractional)	Z (Å from slab center)	spacing (Å)
Al	0.666666666	0.333333333	5.516612984	
O	0.655556699	0.682278703	5.371175279	0.1454377
Al	0	0	4.503531058	0.8676442
Al	0.333333333	0.666666666	4.229791717	0.2737393
O	0.310398203	0.310318699	3.237579796	0.9922119
Al	0.666666666	0.333333333	2.362669895	0.8749099
Al	0	0	1.919678200	0.4429917
O	0.971538999	0.331964424	1.072658318	0.8470199
Al	0.333333333	0.666666666	0.254032345	0.8186260

Table 3
Relaxed coordinates and interplanar spacings of the 2-Al-terminated (0001) surface of α-alumina. Atoms related by the threefold and inversion symmetries are not shown

type	x (fractional)	y (fractional)	Z (Å from slab center)	spacing (Å)
Al	0.333333333	0.666666666	6.751376302	
Al	0.666666666	0.333333333	6.212652647	0.53872366
O	0.641716196	0.669477599	5.363135291	0.84951736
Al	0	0	4.516749494	0.84638580
Al	0.333333333	0.666666666	4.058870142	0.45787935
O	0.305806220	0.305998969	3.212420776	0.84644937
Al	0.666666666	0.333333333	2.383109919	0.82931085
Al	0	0	1.900281495	0.48282842
O	0.972149121	0.333424942	1.069617177	0.83066431
Al	0.333333333	0.666666666	0.240579978	0.82903719

Table 4
Relaxed coordinates and interplanar spacings of the O-terminated (0001) surface of α-alumina. Atoms related by the threefold and inversion symmetries are not shown

type	x (fractional)	y (fractional)	Z (Å from slab center)	spacing (Å)
O	0.630012008	0.675629035	5.372858513	
Al	0	0	4.600442952	0.772415561
Al	0.333333333	0.666666666	4.126819480	0.473623472
O	0.308133950	0.304941390	3.233394766	0.893424714
Al	0.666666666	0.333333333	2.406242044	0.827152722
Al	0	0	1.936374157	0.469867887
O	0.973301152	0.333801243	1.081004623	0.855369534
Al	0.333333333	0.666666666	0.235551158	0.845473466

Figure 5 shows the surface energies of the three terminations of bulk alumina as a function of μ'_{Al}. It is clear that over the range of μ'_{Al} for which α-alumina is stable, the stoichiometric surface is the lowest energy surface relative to the oxygen-terminated, or 2-Al-terminated surfaces. This result is in agreement with recent ion-scattering results by Ahn and Rabalais et al. [5], although in their experiments there was evidence for a significant hydrogen concentration on the surface. Our results also indicate that the stoichiometric surface is the most favorable at least at low temperatures, and that the oxygen-terminated surface is unlikely to be thermodynamically stable under any conditions.

3.2 H on Alumina (0001)

There are several reasons for considering the effect of hydrogen on the surface structure of alumina. Alumina is typically derived from hydrated minerals such as gibbsite, and diaspore. Several useful metastable high surface area phases of alumina are likely to have significant surface hydrogen concentrations. There is evidence in the ion-scattering experiments by Ahn and Rabalais et al. [5] of significant hydrogen concentrations on the surface, which may effect the surface structure (such as interplanar spacings). Finally, in typical surface processing there is either water in direct contact with the surface, or there is significant water concentration in the air, making it interesting and

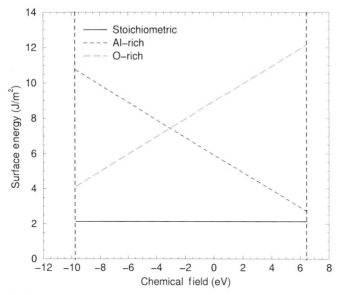

Fig. 5. Surface energies of the three terminations of α-alumina as a function of chemical field, μ'_{Al}. The limits on μ'_{Al} corresponding to equilibrium with f.c.c. aluminum and isolated O_2 molecules are shown as vertical lines. The stoichiometric surface is stable over all μ'_{Al} for which α-alumina is stable

important to understand which changes in surface structure might occur in presence of water. There has been one calculation of the structure and energetics of a hydrogen-saturated, oxygen-terminated surface, using a combination of empirical potentials and quantum-chemistry techniques [9]. However, because of the negative surface energy reported, the results are difficult to interpret.

For calculations of hydrogen-covered surfaces, we used cells with four close-packed oxygen layers. For the stoichiometric surface there was one hydrogen atom per surface aluminum atom and minimizations were started with the hydrogen in the starting positions shown in Fig. 6. In all cases, the minimum energy was achieved with the hydrogen

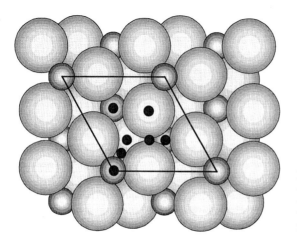

Fig. 6. Starting positions for hydrogen on the stoichiometric surface. The out-of-plane coordinate was chosen to make the distance from the nearest oxygen ion 2 Å

directly on top of the surface aluminum atom. For the oxygen-rich surface there was one hydrogen atom per surface oxygen atom (three hydrogen atoms per surface unit cell). The hydrogen atoms were first started directly on top of the surface oxygen atoms. The final geometry for this situation is similar to previous results [9]. If the hydrogen atoms were then perturbed as to destroy the threefold symmetry, the final

Table 5

Relaxed coordinates and interplanar spacings of the hydrogen-covered, 1-Al-terminated (0001) surface of α-alumina. Atoms related by the threefold and inversion symmetries are not shown

type	x (fractional)	y (fractional)	Z (Å from slab center)	spacing (Å)
H	0.358634601	0.656141410	5.459355026	
Al	0.333333333	0.666666666	3.828399506	1.630955520
O	0.324755696	0.324755696	3.183518892	0.644880614
Al	0.666666666	0.333333333	2.360536973	0.822981919
Al	0	0	2.092027672	0.268509301
O	0.974228135	0.331065308	1.073662684	1.018364988
Al	0.333333333	0.666666666	0.243136242	0.830526442

Table 6

Relaxed coordinates and interplanar spacings of the hydrogen-covered, O-terminated (0001) surface of α-alumina with threefold symmetry constraints. Atoms related by the threefold and inversion symmetries are not shown

type	x (fractional)	y (fractional)	Z (Å from slab center)	spacing (Å)
H	0.176074990	0.792144266	3.724828913	
O	0.317751930	0.291390151	3.211611262	0.513217651
Al	0.666666666	0.333333333	2.340218114	0.871393148
Al	0	0	1.967579538	0.372638576
O	0.972292192	0.331502047	1.078696969	0.888882569
Al	0.333333333	0.666666666	0.261104914	0.817592055

Table 7

Relaxed coordinates and interplanar spacings of the hydrogen-covered, O-terminated (0001) surface of α-alumina without threefold symmetry constraints. Atoms related by inversion symmetry are not shown. Spacings for atoms marked with an asterisk (∗) are computed by averaging the positions of atoms nearly in the same plane above

type	x (fractional)	y (fractional)	Z (Å from slab center)	spacing (Å)
H	0.93995475	0.74031752	4.037292121	
H	0.24518617	0.34617165	3.983419730	
O∗	0.69744417	0.00452243	3.211713626	0.798642300
O∗	0.00131668	0.67391513	3.207433178	0.802922748
O∗	0.32027027	0.30098277	3.149093741	0.861262185
H∗	0.47388908	0.80905682	3.204442384	−0.015028860
Al∗	0.67243443	0.32923613	2.337357347	0.852056168
Al	0.00019806	0.99703876	1.979544479	0.357812868
O	0.35990317	0.02560272	1.084284695	0.895259784
O	0.97443535	0.33214473	1.077437558	0.902106921
O	0.67035981	0.64028155	1.071703903	0.907840576
Al∗	0.33397973	0.66549468	0.252220263	0.825588456

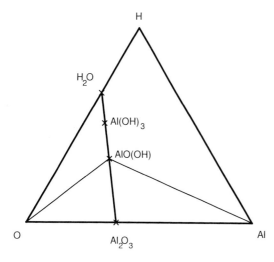

Fig. 7. Map of ground-state structures in the Al–O–H system

positions were slightly different (one of the hydrogen atoms moves slightly below the topmost plane of oxygen atoms near the position normally occupied by the terminating aluminum atom on the stoichiometric surface) with a surface energy less than 0.01 J/m² lower than with the threefold symmetry imposed. The relaxed geometries for the hydrogen-covered surfaces are give in Tables 5 to 7.

To compare the surface energies in the ternary Al–O–H system, we need to consider another chemical field variable, $\mu'_H = \mu_H - \mu_O$. Experimentally, as illustrated in Fig. 7, it is found that the only compounds outside of the Al–O binary lie along the line between Al_2O_3 and H_2O. The ternary compound closest in composition to Al_2O_3 along the Al_2O_3–H_2O line is the mineral diaspore, with the chemical formula AlO(OH). The energy of this compound has been determined and with Eq. (4) gives the only ternary constraint for the stability of α-alumina. The final results for the stability of the alumina surfaces considered in this paper are given in Fig. 8. The hydrogen-covered stoichiometric surface is not stable in comparison to either the clean surface, or the hydrogen-covered, O-terminated surface.

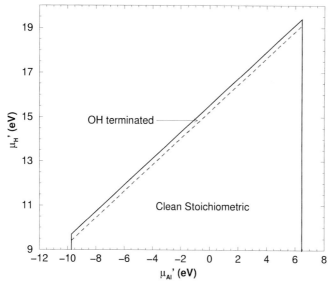

Fig. 8. Stability map of (0001) α-alumina surface structures as a function of μ'_{Al} and μ'_H

4. Conclusions

We computed the surface energies for the three surface structures that can be derived by cutting an α-alumina crystal perpendicular to (0001). We find that, at $T = P = 0$, over the range of chemical field for which the crystal is thermodynamically stable, the stoichiometric surface is the energetically favored. At values of the chemical field near equilibrium with fcc Al, the 2-Al-terminated surface is within 0.6 J/m^2 of the stoichiometric surface. The oxygen-terminated surface is not in close competition energetically at any reasonable value of chemical field. The presence of hydrogen atoms on the stoichiometric surface increases the topmost Al–O interplanar spacing, consistent with the ion-scattering experiments of Ahn and Rabalais et al. [5] However, we find that the hydrogen-covered, *oxygen-terminated* surface is energetically much more favorable than the hydrogen-covered, stoichiometric surface.

Acknowledgements This work was supported by the Office of Basic Energy Sciences of the US DOE, Division of Materials Science under contract No. DE-AC04-94AL85000 (PDT) and was performed under the auspices of the US Department of Energy by Lawrence Livermore National Laboratory under contract W-7405-Eng-48 (AAQ).

References

[1] M. GAUTIER et al., J. Amer. Ceram. Soc. **77**, 323 (1994).
[2] T. M. FRENCH and G. A. SOMORJAI, J. Phys. Chem. **74**, 2489 (1970).
[3] R. M. JAEGER, Surf. Sci. **259**, 235 (1991).
[4] M. W. BENCH, P. G. KOTULA, and C. B. CARTER, Surf. Sci. **391**, 183 (1997).
[5] J. AHN and J. W. RABALAIS, Surf. Sci. **388**, 121 (1997).
[6] W. C. MACKRODT, J. Chem. Soc. Faraday Trans. II **85**, 54 (1989).
[7] I. MANASSIDIS, A. D. VITA, and M. J. GILLAN, Surf. Sci. Lett. **285**, L517 (1993).
[8] C. VERDOZZI, D. R. JENNISON, P. A. SCHULTZ, and M. P. SEARS, submitted.
[9] M. A. NYGREN, D. H. GAY, and R. A. CATLOW, Surf. Sci. **380**, 113 (1997).
[10] J. PERDEW and A. ZUNGER, Phys. Rev. B **23**, 5048 (1981).
[11] H. J. MONKHORST and J. D. PACK, Phys. Rev. B **13**, 5188 (1976).
[12] N. TROULLIER and J. L. MARTINS, Phys. Rev. B **43**, 1993 (1991).
[13] S. BLONSKI and S. H. GAROFALINI, Surf. Sci. **295**, 263 (1993).

phys. stat. sol. (b) **217**, 389 (2000)

Subject classification: 71.15.Mb; 71.15.Pd; 73.20.Hb; S1.3; S5.11

Ab initio Molecular Dynamics Simulations of Reactions at Surfaces

A. Gross

Physik-Department T30, Technische Universität München, D-85747 Garching, Germany

(Received August 10, 1999)

In general the statistical nature of reactions on surfaces requires the calculation of a very large number of trajectories in order to determine reaction rates. We show that even in massively parallel schemes a sufficient number of trajectories determined from first principles can only be obtained in a approach in which first the potential energy surface (PES) on which the nuclei move is determined and then the dynamical calculations on an appropriate representation of the PES are performed. The PES can nowadays be evaluated in great detail by first-principles methods based on density-functional theory. These electronic structure calculations also allow the investigation of the factors that determine the reactivity of a particular system. We discuss different methods to represent an ab initio PES and present a massively parallel ab initio quantum dynamics approach for the dissociation of hydrogen on metal surfaces.

1. Introduction

Modern ab initio algorithms based on density-functional theory (DFT) allow the determination of the high-dimensional potential energy surface (PES) and the potential gradients for reactions on surfaces at many different configurations [1 to 6]. This is a prerequisite for the ab initio description of reactions due to the complexity of the high-dimensional PES. However, in order to assess the reactivity of a particular system it is necessary to perform calculations of the reaction dynamics [1, 7]. "Traditional" ab initio molecular dynamics methods (AIMD) perform a complete total-energy calculation for each step of the numerical integration of the equations of motion. We will demonstrate that even in massively parallel approaches the number of trajectories that can be calculated by this approach is still well below 100 [8, 9]. It will be shown that there are special cases in which the crucial trajectories originate from a small portion of the relevant phase space so that already from a small number of trajectories useful information can be extracted [9]. But usually this traditional approach does not allow the determination of a sufficient number of trajectories for obtaining reliable reaction rates.

We have therefore proposed a three-step approach for performing ab initio molecular dynamics calculations [10]: First a sufficient number of ab initio total-energy calculations is performed. Then an interpolation scheme is used to fit the ab initio energies and to interpolate between the actual calculated points. And finally the dynamics calculations are performed on this continuous representation of the ab initio PES. In this way easily 100.000 ab initio trajectories can be determined [10] on a workstation. In addition, on such a continuous representation also quantum dynamical calculations can be performed [11 to 13]. Quantum dynamical simulations can actually be less CPU time consuming than classical trajectory calculations for the determination of reaction rates because the averaging over initial conditions is done automatically in quantum mechanics by choosing the appropriate initial quantum states [10].

In this paper, we will illustrate the three-step approach for the example of dissociation of hydrogen on clean and adsorbate-covered metal surfaces. We will also show that the electronic structure calculations can be used in order to understand the factors determining the reactivity of a particular surface. Furthermore, we will introduce a massively parallel implementation of the very stable coupled-channel scheme [14] we have used to solve the time-independent Schrödinger equation of the interaction of hydrogen with metal surfaces. Because of the use of curvelinear reaction path coordinates this scheme requires the diagonalisation of non-symmetric matrices. Due to the lack of massively parallel public domain diagonalisation schemes for general matrices we implemented a second order pertubation diagonalisation scheme. We will present results concerning the performance and scaling properties of this ab initio quantum dynamics scheme.

2. AIMD with the Determination of the Forces "on the Fly"

The first ab initio molecular dynamics study of reactions at surfaces with the determination of the forces "on the fly" was an investigation of the adsorption of Cl_2 on Si(111)-2×1 [8]. Only five trajectories were determined in this study so that the information about the reaction dynamics gained from this study was rather limited.

Here we focus on a more recent example, the desorption of hydrogen from Si(100). The interaction of hydrogen with silicon surfaces is of strong technological relevance. On the one hand, hydrogen is used to passivate silicon surfaces, on the other hand, hydrogen desorption from silicon is an important step in the chemical vapor deposition (CVD) growth of silicon substrates.It is a well-studied system [1, 15], but still it is discussed very controversely, as far as experiment [16 to 18] as well as theory is concerned [9, 19]. One of the debated issues is the role of the surface rearrangement of the silicon substrate degrees of freedom upon the adsorption and desorption of hydrogen.

In Fig. 1 this surface rearrangement is illustrated for the Si(100) surface. While at the hydrogen covered monohydride surface the outermost silicon atoms form symmetric dimers (Fig. 1a), at the clean surface these dimers are buckled (Fig. 1c). Consequently,

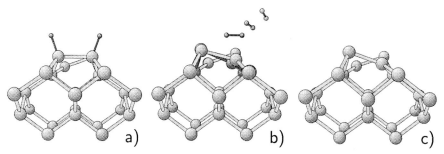

Fig. 1. a) Hydrogen covered Si(100) surface (monohydride). b) Snapshots of a trajectory of D_2 desorbing from Si(100) starting at the transition state with the Si atoms initially at rest [9]. The dark Si atoms correspond to the Si positions after the desorption event. c) Clean anti-buckled Si(100) surface [9]

the silicon surface atoms participate in the hydrogen adsorption and desorption process. In order to investigate the energy redistribution among the different hydrogen and silicon substrate degrees of freedom upon the desorption of hydrogen from Si(100) we have performed AIMD calculations [9] using the Generalized Gradient Approximation (GGA) for the treatment of the exchange and correlation effects. The forces necessary to integrate the equations of motion were determined by DFT calculations for every step of the numerical integration routine. The electronic wave functions were expanded in a plane-wave basis set with a cutoff of 40 Ry, and we used two k-points in the irreducible part of the Brillouin zone. The calculations had been performed using the massively parallel version of the fhi96md code [20]. We chose a time step of 1.2 fs in the numerical integration of the motion which took about 20 min on 64 nodes of a Cray T3D. In total 40 trajectories of D_2 desorbing form Si(100) have been determined in that fashion. Fig. 1b shows some snapshots of such a trajectory. It illustrates how the silicon atom beneath the desorbing D_2 molecule relaxes after the desorption thereby gaining a kinetic energy of about 0.1 eV.

However, the number of 40 trajectories is usually much too small to determine any reaction probabilities. Only in certain cases as the hydrogen desorption from Si(100) where the crucial trajectories originate from a small portion of the relevant phase space one can still get reasonable information out of a small number of trajectories. The results of the AIMD calculations were in good quantitative agreement with the experiment [9], except for the experimentally observed low kinetic energy of desorbing D_2 molecules [16] which is still highly debated [1].

Usually the calculation of reaction probabilities requires the determination of the order of 10^3 to 10^6 trajectories or a quantum dynamical scheme. Ab initio molecular dynamics simulations with the determination of the forces "on the fly" are still far away from fulfilling this requirement, as was just shown. The calculation of ab initio reaction probabilities can only be achieved by a three-step approach which will be presented in the next section.

3. Three-Step Approach to AIMD

3.1 Determination of the ab initio potential energy surface

The first step in the general scheme for determining the ab initio dynamics of reactions at surfaces is represented in Fig. 2, namely the determination of the ab initio PES by density-functional theory calculations [4]. It has turned out that it is crucial to treat the exchange-correlation effects in the DFT calculations within the generalized gradient approximation (GGA) in order to obtain realistic barrier heights for the hydrogen dissociation on surfaces [2]. For the hydrogen dissociation on close-packed metal surfaces usually the surface rearrangement upon hydrogen adsorption is negligible. Still total energies for several hundred different configurations have to be determined in order to gain sufficient information about the PES as a function of the molecular coordinates. For the PES of the interaction of hydrogen with Pd(100) total energies of approximately 250 different configurations were calculated. In a later study energies for more than 750 different configurations were computed [21] which resulted in a better agreement of the dynamical calculations based on these ab initio input points with the experiment [7].

As Fig. 2 shows, the PES of the interaction of hydrogen with Pd(100) has non-activated paths towards dissociative adsorption and no molecular adsorption well. However, the majority of pathways towards dissociative adsorption has in fact energy barriers with a rather broad distribution of heights and positions, i.e. the PES is strongly anisotropic and corrugated. That is the reason why so many DFT calculations are needed.

The DFT-GGA calculations can also be used in order to understand the electronic factors that determine the reactivity of a surface [22 to 24]. We will illustrate this for the case of the H_2 dissociation at the (2×2) sulfur-covered Pd(100) surface. The pre-

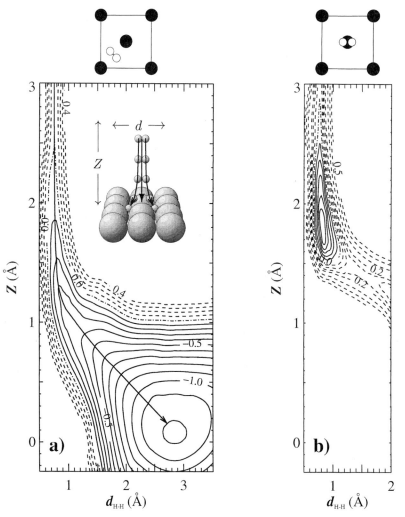

Fig. 2. Contour plots of the PES along two two-dimensional cuts through the six-dimensional coordinate space of H_2/Pd(100), so-called elbow plots, determined by GGA calculations [4]. The coordinates in the figure are the H_2 center-of-mass distance from the surface Z and the H–H interatomic distance d. The lateral H_2 center-of-mass coordinates in the surface unit cell and the orientation of the molecular axis are depicted above the elbow plots. Energies are in eV per H_2 molecule. The contour spacing in a) is 0.1 eV, while it is 0.05 eV in b)

sence of an adsorbate on a surface can profoundly change the surface reactivity. An understanding of the underlying mechanisms and their consequences on the reaction rates is of decisive importance for, e.g., designing better catalysts. Sulfur is known to reduce the reactivity of the Pt-based car exhaust catalyst, hence it is important to analyse the reasons for this so-called poisoning. Hydrogen adsorption on Pd(100) can be used as a model system because on Pd(100) sulfur preadsorption also leads to the poisoning, i.e., the hydrogen dissociation is no longer non-activated as on the clean Pd(100) surface.

This is demonstrated in Fig. 3 where we have collected four elbow plots of the hydrogen dissociation on the (2×2) sulfur covered Pd(100) surface determined by GGA-DFT calculations [22, 24]. These calculations show that hydrogen dissociation on sulfur-covered Pd(100) is still exothermic, however, the dissociation is hindered by the formation of energy barriers in the entrance channel of the PES. The minimum barrier, which is shown in Fig. 3a, has a height of 0.1 eV and corresponds to a configuration in which the H_2 center of mass is located above the fourfold hollow site. This is the site which is farthest away from the sulfur atoms in the surface unit cell. The closer the hydrogen molecule is to the sulfur atoms on the surface, the larger the barrier towards dissociative adsorption becomes. Directly over the sulfur atoms the barrier has a height of 2.5 eV. To our knowledge, this is the most corrugated surface for dissociative adsorption studied so far by ab initio calculations.

In order to understand the origins for the formation of this huge variety in the barrier heights, we have analysed the density of states (DOS) for the H_2 molecule in these different geometries. The information provided by the density of states alone is often not sufficient to assess the reactivity of a particular system. It is also essential to know the character of the occupied and unoccupied states. For the dissociation the occupation of the bonding σ_g and the anti-bonding σ_u^* H_2 molecular levels *and* of the bonding and anti-bonding states with respect to the surface–molecule interaction are of particular importance. We will see that the barrier distribution of the H_2 dissociation over (2×2)S/Pd(100) can be understood by a combination of direct and indirect electronic effects.

The DOS for the situation without any interaction between molecule and surface, i.e, when the H_2 molecule is still far away from the surface, is shown in Fig. 4a. However, the electronic states of the adsorbed sulfur, in particular the p orbitals, are strongly hybridized with the Pd d states. The d band at the surface Pd atoms is broadened and shifted down somewhat with respect to the clean surface due to the interaction with the S atoms [24]. The intense peak in the hydrogen DOS at -4.8 eV which corresponds to the σ_g state is degenerate with the sulfur related bonding state at -4.8 eV. This degeneracy, however, is accidental, as will become evident immediately.

When the molecule comes closer to the surface, the σ_g state starts interacting with the Pd d band. At the minimum barrier position of Fig. 3a the σ_g state has shifted down to -7.1 eV, as Fig. 4b shows. On the other hand, the sulfur state at -4.8 eV remains almost unchanged. This indicates that there is no direct interaction between hydrogen and sulfur. Furthermore, we find a broad distribution of hydrogen states with a small, but still significant weight below the Fermi level. These are states of mainly H_2-surface antibonding character [22, 24] which become populated due to the sulfur induced downshift of the Pd d band. These H_2-surface antibonding states lead to a repulsive interaction and thus to the building up of the barriers in the entrance channel of the PES [24]. It is therefore an indirect interaction between sulfur and hydrogen that is responsible for

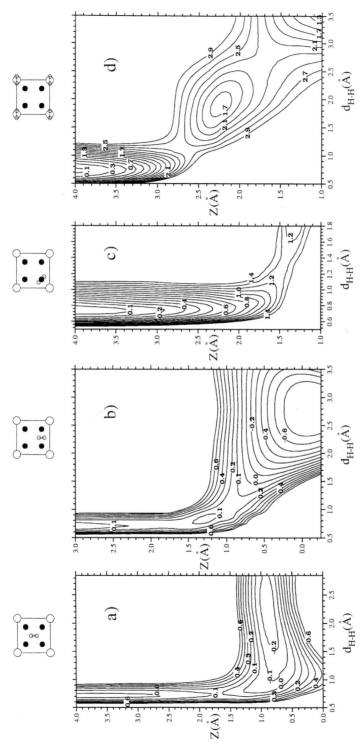

Fig. 3. Cuts through the six-dimensional potential energy surface (PES) of H_2 dissociation over $(2 \times 2)S/Pd(100)$ at four different sites with the molecular axis parallel to the surface: a) at the fourfold hollow site; b) at the bridge site between two Pd atoms; c) on top of a Pd atom; d) on top of a S atom. The energy contours, given in eV per molecule, are displayed as a function of the H–H distance, d_{H-H}, and the height Z of the center-of-mass of H_2 above the topmost Pd layer. The geometry of each dissociation pathway is indicated in the panel above the contour plots. The large open circles are the sulfur atoms, the large filled circles are the palladium atoms (from [22])

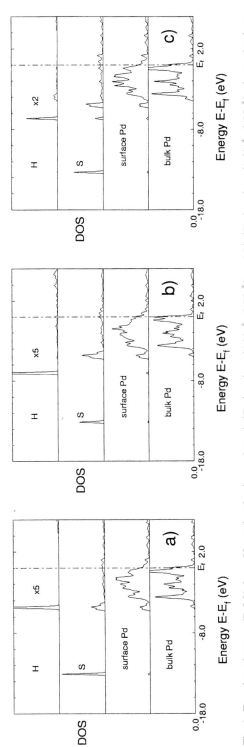

Fig. 4. Density of states (DOS) for a H_2 molecule situated at a) $(Z, d_{H-H}) = (4.03$ Å, 0.75 Å$)$ and b) $(Z, d_{H-H}) = (1.61$ Å, 0.75 Å$)$ above the four-fold hollow site which corresponds to the configuration depicted in Fig. 3a, and for a H_2 molecule situated at c) $(Z, d_{H-H}) = (3.38$ Å, 0.75 Å$)$ above the sulfur atom which corresponds to the configuration depicted in Fig. 3d. Z and d_{H-H} denote the H_2 center-of-mass distance from the surface and the H–H interatomic distance, respectively. Given is the local DOS at the H atoms, the S adatoms, the surface Pd atoms, and the bulk Pd atoms. The energies are given in eV (from [22])

the barriers at this site. A similar picture explains why for example noble metals are so unreactive for hydrogen dissociation: The low-lying d bands of the noble metals cause a downshift and a substantial occupation of the antibonding H_2-surface states resulting in high barriers for hydrogen dissociation [23].

The situation is entirely different if the molecule approaches the surface above the sulfur atom. This is demonstrated in Fig. 4c. The center of mass of the H_2 molecule is still 3.38 Å above the topmost Pd layer, but already at this distance the hydrogen and the sulfur states strongly couple. The intense peak of the DOS at -4.8 eV has split into a sharp bonding state at -6.6 eV and a narrow anti-bonding state at -4.0 eV. Thus it is a direct interaction of the hydrogen with the sulfur related states that causes the high barriers towards hydrogen dissociation close to the sulfur atoms.

In conclusion, the poisoning of hydrogen dissociation on Pd(100) by adsorbed sulfur is due to a combination of an indirect effect, namely the sulfur-related downshift of the Pd d bands resulting in a larger occupation of H_2-surface antibonding states, with a direct repulsive interaction between H_2 and S close to the sulfur atoms.

3.2 Representing the ab initio PES

The second step in the ab initio dynamics scheme is the interpolation of the ab initio data. For the hydrogen dissociation on close-packed metal surfaces the surface rearrangement due to the impinging hydrogen molecules can be neglected due to the large mass mismatch between hydrogen and the metal substrate. This allows to describe the dissociation dynamics within the six-dimensional PES only considering the molecular degress of freedom. In six dimensions it is still possible to fit the ab initio data to an analytical expression, which has be done successfully for the H_2 dissociation on the clean [10, 11] and sulfur-covered Pd(100) surface [22, 25] and also for the H_2/Cu(100) PES [26]. The case of the H_2/Cu(100) PES, however, shows how difficult still the fitting of a six-dimensional PES is. Due to errors in the fitting an artificial well was introduced in the fitted PES which influenced the results [27].

However, once a reliable analytical fit is found, it is computationally inexpensive to calculate the potential gradients at arbitrary configuration which is needed, e.g, to perform molecular dynamics simulations of reactions. However, it becomes very cumbersome to find an appropriate analytical form if more degrees of freedom like, e.g., surface degrees of freedom, have to be considered. Neural networks offer a very flexible interpolation scheme which has been used for fitting an ab initio PES of chemical reactions [28 to 30]. On the one hand, they require no assumptions about the functional form of the underlying problem, but on the other hand, their parameters have no physical meaning. For that reason a relatively large number of ab initio input points is needed for an accurate description of a whole PES. As an alternative approach, recently a genetic programming scheme has been proposed which searches for both the best functional form and the best set of parameters [31]. This method has so far only been used for three-dimensional potentials so that a proof of its applicability for higher-dimensional problems is still missing.

All the interpolations schemes mentioned so far allow a fast determination of the fitted PES at arbitrary configurations. However, these methods usually require a rather large number of training points in order to reproduce the input PES within a sufficient accuracy. As a rough estimate of the neccessary number of input points, at least three points are needed for each degree of freedom. Consequently, in six dimensions about 10^3 and in twelve 10^6 ab initio total energy calculations as an input are required. The ab initio determination of such a large number of total energies for molecule–surface systems is still computationally very expensive. Therefore, an intermediate step is needed.

Tight-binding methods with parameters derived from first-principles calculations offer such an intermediate approach. Recently, it has been shown that a non-orthogonal tight-binding total-energy (TBTE) method that so far had successfully been used for material properties [32, 33] can also be applied for fitting an ab initio PES of the dissociation of molecules at surfaces [34]. The parameters of this tight-binding scheme had been fitted to reproduce the ab initio PES for the $H_2/Pd(100)$ system. Figure 5 shows the $H_2/Pd(100)$ PES determined with the tight-binding Hamiltonian. This PES should be compared with the ab initio results of Fig. 2. The comparison reveals that indeed the tight-binding method is able to accurately reproduce an ab initio PES. Moreover, due

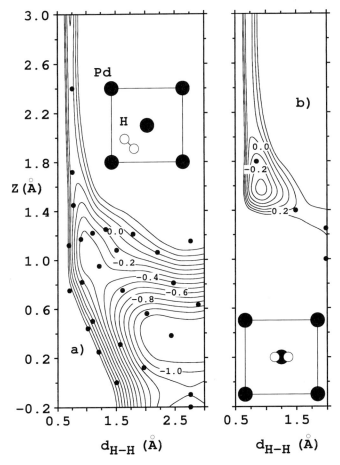

Fig. 5. Contour plots of the TB-PES along two two-dimensional cuts through the six-dimensional coordinate space of $H_2/Pd(100)$, determined by a tight-binding Hamiltonian adjusted to ab initio calculations [34]. The notation corresponds to Fig. 2. The dots denote the points that have been used to obtain the fit. Energies are in eV per H_2 molecule. The contour spacing is 0.1 eV

to the fact that the quantum mechanical nature of bonding is taken into account properly by tight-binding methods [35], and that the parameters of the TBTE method, the Slater-Koster integrals [36], have a physical meaning, only a moderate number of input total energies is needed for a reliable global description of the PES.

The computational effort to determine total energies with a TBTE method is larger compared to an analytical representation since it requires the diagonalization of matrices. However, it is still much faster by about 2 to 3 orders of magnitude than ab initio total energy schemes. Either one can use the TB method directly to perform tight-binding molecular dynamics simulations, or one can use the information obtained by the TB total energies to adjust the parameters of a neural network which allows a fast evaluation of total energies, but needs at large number of input points for adjusting the parameters. In the first application of the TBTE method for the dissociation of hydrogen on Pd(100) [34] the tight-binding parameters describing the Pd–Pd interaction had been taken from an independent calculation [33, 37]. These parameters already reproduce Pd bulk and surface properties such as elastic constants and phonon energies. Hence this set of parameters can be used in simulations where the substrate atoms are no longer kept rigid but are treated as dynamical variables. This will allow to assess the influence of surface motion upon the hydrogen adsorption.

3.3 Performing ab initio dynamics simulations

The third step in the ab initio dynamics scheme involves the determination of the reaction dynamics. Once a reliable interpolation scheme is found, dynamics simulations can be carried out. Quantum mechanically this is done by plugging in the PES in a suitable form into the Hamiltonian and then solving either the time dependent [14] or the time-independent Schrödinger equation [38]. For solving the classical equation of motion, on the other hand, the gradient of the potential energy surface is needed, and then the equation of motion can be integrated numerically.

The first quantum dynamical treatment of hydrogen dissociation on surfaces in which all hydrogen degrees of freedom were treated dynamically was actually done by solving the time-independent Schrödinger equation in an efficient coupled-channel scheme. For details of the coupled-channel scheme we refer to Ref. [14].

Just recently this coupled-channel scheme has been implemented on a massively parallel computer, the Cray T3E. The code has been rewritten by using public domain libraries. However, the description of the dissociation in a time-independent method requires the use of curvelinear coordinates. Due to the use of curvelinear coordinates a non-symmetric matrix V has to be diagonalized which in the notation of Ref. [14] was denoted by q^2. Apparently, there are no massively parallel public domain diagonalisation schemes for general matrices available yet. For the particular problem of the molecular dissociation on surfaces we have therefore used a second order pertubation diagonalisation scheme which will be briefly sketched in the following. This schemes uses the fact that the anti-symmetric part of the matrix is small and can be treated as a pertubation.

First we decompose the matrix V_{nm} into a symmetric and an anti-symmetric part,

$$V_{nm} = \frac{1}{2}(V_{nm} + V_{mn}) + \frac{1}{2}(V_{nm} - V_{mn})$$
$$\equiv V_{nm}^{\text{sym}} + V_{nm}^{\text{asym}}. \tag{1}$$

The anti-symmetric part of the matrix which is related to the curvature of the curve-linear axis is in general sparse and small for the case of hydrogen dissociation on metal surfaces,

$$V_{nm}^{\mathrm{asym}} \leq 0.1 V_{nm}^{\mathrm{sym}}. \tag{2}$$

This allows to treat V^{asym} as a pertubation. First the symmetric part V^{sym} is diagonalised and the transformation matrix U determined. This matrix is used to transform V^{asym},

$$\tilde{V}^{\mathrm{asym}} = U^{\mathrm{T}} V^{\mathrm{asym}} U. \tag{3}$$

The new eigenvalues are determined in a second order pertubation diagonalisation scheme via

$$E_n = \varepsilon_n + \sum_m{}' \frac{|\langle n|\tilde{V}^{\mathrm{asym}}|m\rangle|^2}{\varepsilon_m - \varepsilon_n}. \tag{4}$$

Here the ε_m are the eigenvalues of the symmetric matrix. The prime denotes that the terms with $m = n$ are excluded from the sum.

Equivalently, the new eigenvectors are computed,

$$|N\rangle = |n\rangle + \sum_m{}' |m\rangle \frac{\langle m|\tilde{V}^{\mathrm{asym}}|n\rangle}{\varepsilon_n - \varepsilon_m}$$
$$+ \sum_m{}' \sum_k{}' |m\rangle \frac{\langle m|\tilde{V}^{\mathrm{asym}}|k\rangle \langle k|\tilde{V}^{\mathrm{asym}}|n\rangle}{(\varepsilon_n - \varepsilon_k)(\varepsilon_n - \varepsilon_m)}. \tag{5}$$

This scheme assumes that the eigenvalues are non-degenerate. This requirement is fulfilled because $\tilde{V}^{\mathrm{asym}}$ only couples different harmonic oscillator eigenstates which are always non-degenerate.

We have carefully checked the accuracy of this diagonalisation scheme. The error in the reaction probabilities due to the second order pertubation diagonalisation scheme is less than 0.5%. In Table 8 we have collected results concerning the performance of this second order diagonalisation scheme on a Cray T3E. This table shows that this scheme produces a reasonable speedup for up to 32 or 64 processors.

Figure 6 presents six-dimensional quantum dynamical calculations of the sticking probability using the coupled-channel method [14] as a function of the kinetic energy of a H_2 beam under normal incidence on a Pd(100) surface together with the integrated

Table 1
Performance of the massively parallel version of the coupled-channel quantum dynamics method [14] using a second order diagonalisation scheme on a Cray T3E

No. of PE	time (s)	speedup	MFlops/PE	total MFlops
1	4393	1.00	127	127
4	1035	4.24	132	530
16	282	15.60	124	1989
32	168	26.15	108	3464
64	109	40.30	80	5118
128	86	51.08	51	6487

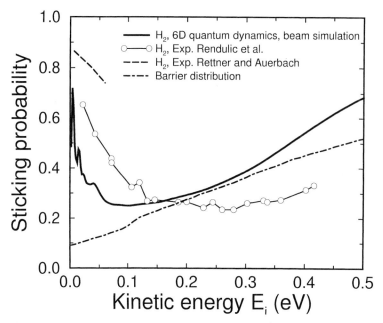

Fig. 6. Sticking probability versus kinetic energy for a hydrogen beam under normal incidence on a Pd(100) surface. Theory: six-dimensional results for H_2 molecules with an initial rotational and energy distribution adequate for molecular beam experiments (solid line) [11]; integrated barrier distribution: dash-dotted line. H_2 molecular beam adsorption experiment under normal incidence (Rendulic et al. [40]): circles; H_2 effusive beam scattering experiment with an incident angle of $\theta_i = 15°$ (Rettner and Auerbach [43]): long-dashed line

barrier distribution [10, 11]. The system H_2/Pd(100) is an experimentally well-studied system [39 to 43]. In Fig. 6 the results of H_2 molecular beam experiment by Rendulic et al. [40] and Rettner and Auerbach [43] are also plotted.

The integrated barrier distribution corresponds to the sticking probability in the classical sudden approximation or the so-called hole model [44]. Fig. 6 demonstrates that the static information gained from the barrier distribution is not sufficient in order to assess the reactivity of the H_2/Pd(100) system: At low kinetic energies the sticking probability is more than five times larger than what one would have estimated from the barrier distribution.

The high sticking probability at low kinetic energies, which agrees with the experiment, is caused by the steering effect: A slow molecule moving on a PES with non-activated as well as activated paths towards dissociation can be steered efficiently towards non-activated paths to adsorption by the forces acting upon the molecule even if the molecule approaches with an unfavorable initial configuration. This mechanism becomes less efficient at higher kinetic energies because then the molecule is too fast to be diverted significantly. Furthermore, Fig. 6 shows that at high kinetic energies the incoming molecules are still slightly steered since the sticking probability is larger than the integrated barrier distribution. However, in the intermediate range of $E_i \approx 0.25$ eV the sticking probability and the barrier distribution are rather close. At first sight this seems to be paradoxical. But a detailed analysis of swarms of classical trajectories on the same PES reveals that this behavior is caused by "negative" steering. Far away from the surface the H$_2$ molecules are first steered towards the on top site. There they will eventually encounter a barrier (see Fig. 2b). At very low energies the molecules are then further steered to the bridge site, but at higher energies they are too fast, they hit the barrier at the top site and scatter back into the gas phase.

These results demonstrate the power or high-dimensional ab initio molecular dynamics calculations. Before this six-dimenensional study it was generally believed that a sticking probability decreasing with increasing kinetic energy is *always* caused by a precursor mechanism [40]. In this mechanism the molecules are assumed to be trapped molecularly in a precursor well before dissociation, and this trapping probability decreases with increasing kinetic energy. Now it is generally accepted that for hydrogen dissociation at reactive transition metal surfaces it is not necessary to invoke the precursor mechanism.

There is a dynamical property that allows to distinguish between steering and the precursor mechanism: the influence of the sticking probability on the rotational motion of the molecule. While steering is suppressed by additional rotational motion of the molecule [45], trapping in the molecular adsorption well in the presursor mechanism does not show any significant dependence on the initial rotational motion of the molecules [46].

Six-dimensional dynamical calculations have also been performed on the analytical representation of the ab initio PES of H$_2$ at S(2×2)/Pd(100) in order to assess the dynamical consequences of the sulfur adsorption on the hydrogen dissociation [25]. The results of these quantum and classical calculations for the H$_2$ dissociative adsorption probability as a function of the incident energy are compared with experiment [39, 40] in Fig. 7.

First of all it is evident that the calculated sticking probabilities are significantly larger than the experimental results. Only the onset of dissociative adsorption at $E_i \approx 0.12$ eV is reproduced by the calculations. This onset is indeed also in agreement with the experimentally measured mean kinetic energy of hydrogen molecules desorbing from sulfur covered Pd(100) [39], which is denoted by the arrow in Fig. 7. We believe that those large differences between theory and experiment might be caused by the existence of subsurface sulfur which was, however, not discussed in the experimental studies. While the DFT calculations yield that the poisoning is caused by the building up of barriers hindering the dissociation, the vanishing hydrogen saturation coverage for roughly a quarter monolayer of adsorbed sulfur [42] suggests that any attractive adsorption sites for hydrogen have disappeared due to the presence of sulfur. These seemingly contradicting results and also the discrepancy between calculated and measured molecular beam sticking probabilities could be reconciled if subsurface sulfur plays an important role for the hydrogen adsorption energies. Subsurface sulfur is not considered in the calculations but might well be present in the experimental samples. The possible influence of subsurface species on reactions at surfaces certainly represents a very interesting and important research subject for future investigations.

Except for this open question, there are further interesting results obtained by the dynamical calculations. The calculated sticking probabilities are not only much larger than the experimental ones, they are also much larger than what one would expect from integrated barrier distribution, which corresponds to the sticking probability in the hole model [44]. This demonstrates that steering is not only operative for potential energy surfaces with non-activated reaction paths like for H$_2$/Pd(100), but also for activated systems as H$_2$/S(2×2)/Pd(100). The huge corrugation of this system leads to an

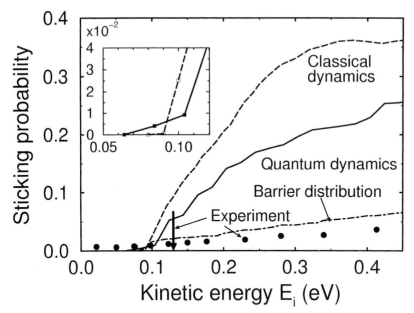

Fig. 7. Sticking probability versus kinetic energy for a H_2 beam under normal incidence on a $S(2 \times 2)/Pd(100)$ surface. Full dots: experiment (from Ref. [40]); the arrow denotes the barrier towards adsorption deduced from desorption experiments [39]. Dashed-dotted line: integrated barrier distribution; solid line: quantum mechanical results for molecules initially in the rotational and vibrational ground-state; dashed line: classical results for initially non-rotating and non-vibrating molecules. The inset shows the quantum and classical results at low energies

enhancement of the sticking probability with respect to the hole model by a factor of three to four.

Figure 7 shows in addition that the classical molecular dynamics calculations overestimate the sticking probability of H_2 at $S(2 \times 2)/Pd(100)$ compared to the quantum results. At small energies below the minimum barrier height the quantum calculations still show some dissociation due to tunneling, as the inset of Fig. 7 reveals, whereas the classical results are of course zero. But for higher energies the classical sticking probability is up to almost 50% larger than the quantum sticking probabilities. This suppression is also caused by the large corrugation and the anisotropy of the PES. The wave function describing the molecule has to pass narrow valleys in the PES in the angular and lateral degrees of freedom in order to dissociate. This leads to a localization of the wave function and thereby to the building up of zero-point energies which act as additional effective barriers. While the vibrational H–H mode becomes softer upon dissociation so that the zero-point energy in this particular mode decreases, for the system $H_2/S(2 \times 2)/Pd(100)$ this decrease is over-compensated by the increase in the zero-point energies of the four other modes perpendicular to the reaction path, i.e., the sum of *all* zero-point energies increases upon adsorption [25]. Therefore the quantum particles experience an effectively higher barrier region causing the suppressed sticking probability compared to the classical particles. Interestingly enough, if the sum of all zero-point energies remains approximately constant along the reaction path as in the system $H_2/Pd(100)$, then these quantum effects almost cancel out [10].

Furthermore, at energies slightly above the minimum energy barrier in Fig. 7, the quantum results are rather close to the classical results. This is caused by the fact that in this energy regime steering is much more efficient in the quantum than in the classical dynamics leading to a compensation of the hindering zero-point effects [47].

A final remark concerning the dynamical calculations: It is a common belief that classical trajectory calculations are less CPU-time consuming than quantum dynamical calculations. For the determination of a single trajectory this is certainly true, however, if it comes to the determination of reaction probabilities, then quantum methods can become more efficient. This is due to the fact that the evaluation of reaction probabilities requires averaging over initial conditions which is done automatically in quantum dynamics by choosing the appropriate initial quantum state. However, owing to the scaling properties with the number of atoms and the relative large memory requirement of quantum methods, one still has to rely on classical methods if dynamical simulations involving, say, more than ten degrees of freedom are to be performed.

4. Conclusions and Outlook

The last years have seen a tremendous step forward in the understanding of the interaction of molecules with surfaces. Based on advances in density functional theory algorithms, the potential energy surface of a molecule interacting with a surface can be mapped out in great detail. This development has motivated an increased effort in the dynamical simulation of processes on surfaces. The paradigm for simple reactions on surfaces – the dissociation of hydrogen on metal surfaces – seems to be understood to a large extent now, as this brief review has shown, although there are still open question. Consequently there is a lot of room for further investigations. In particular the importance of electronic transitions like electron–hole pair excitations upon adsorption is unknown yet. Currently also systems including oxygen – either in the substrate (oxide surfaces) or in the molecule (oxidation reactions) – are being addressed where electronic transitions are probably even more important.

The next great challenge is the description of more complex reactions and processes on surfaces. The study of these reactions is important not only for achieving one of the classical goals of surface science, which is a better understanding of heterogeneous catalysis. It is also directly relevant for a wide range of applications as the corrosion or passivation of surfaces, lubrication, growth properties for building better devices, the hydrogen storage in metals; and one even starts to address the microscopic description of biological systems. To investigate these reactions theoretically it is crucial not only to evaluate the potential energy surface on which the reaction takes place, but also to perform dynamical simulations on these potential energy surfaces to actually obtain reaction rates and probabilities. Since in many problems surface processes occur on a long time scale and/or on a large length scale, also appropriate statistical and analytical methods to deal with these different scales have to be developed and applied. The theoretical treatment of reactions on surfaces is certainly a growing research field.

Acknowledgements It is a pleasure to acknowledge the co-workers who have made this work possible. I like to thank Michel Bockstedte, Wilhelm Brenig, Mike Mehl, Dimitri Papaconstantopoulos, Matthias Scheffler, Ching-Ming Wei and Steffen Wilke. Special thanks go to Jakob Pichlmeier for creating the massively parallel version of the coupled-channel scheme.

References

[1] A. GROSS, Surf. Sci. Rep. **32**, 291 (1998).
[2] B. HAMMER, M. SCHEFFLER, K.W. JACOBSEN, and J.K. NØRSKOV, Phys. Rev. Lett. **73**, 1400 (1994).
[3] J.A. WHITE, D.M. BIRD, M.C. PAYNE, and I. STICH, Phys. Rev. Lett. **73**, 1404 (1994).
[4] S. WILKE and M. SCHEFFLER, Phys. Rev. B **53**, 4296 (1996).
[5] A. EICHLER, G. KRESSE, and J. HAFNER, Phys. Rev. Lett. **77**, 1119 (1996).
[6] C. STAMPFL and M. SCHEFFLER, Phys. Rev. Lett. **78**, 1500 (1997).
[7] A. EICHLER, J. HAFNER, A. GROSS, and M SCHEFFLER, Phys. Rev. B **59**, 13297 (1999).
[8] A. DE VITA, I. ŠTICH, M.J. GILLAN, M.C. PAYNE, and L.J. CLARKE, Phys. Rev. Lett. **71**, 1276 (1993).
[9] A. GROSS, M. BOCKSTEDTE, and M. SCHEFFLER, Phys. Rev. Lett. **79**, 701 (1997).
[10] A. GROSS and M. SCHEFFLER, Phys. Rev. B **57**, 2493 (1998).
[11] A. GROSS, S. WILKE, and M. SCHEFFLER, Phys. Rev. Lett. **75**, 2718 (1995).
[12] M. KAY, G.R. DARLING, S. HOLLOWAY, J.A. WHITE, and D.M. BIRD, Chem. Phys. Lett. **245**, 311 (1995).
[13] G.J. KROES, E.J. BAERENDS, and R.C. MOWREY, Phys. Rev. Lett. **78**, 3583 (1997).
[14] W. BRENIG, T. BRUNNER, A. GROß, and R. RUSS, Z. Phys. B **93**, 91 (1993).
[15] K.W. KOLASINSKI, Internat. J. Mod. Phys. B **9**, 2753 (1995).
[16] K.W. KOLASINSKI, W. NESSLER, A. DE MEIJERE, and E. HASSELBRINK, Phys. Rev. Lett. **72**, 1356 (1994).
[17] K.W. KOLASINSKI, W. NESSLER, K.-H. BORNSCHEUER, and E. HASSELBRINK, J. Chem. Phys. **101**, 7082 (1994).
[18] P. BRATU, W. BRENIG, A. GROSS, M. HARTMANN, U. HÖFER, P. KRATZER, and R. RUSS, Phys. Rev. B **54**, 5978 (1996).
[19] A.J.R. DA SILVA, M.R. RADEKE, and E.A. CARTER, Surf. Sci. **381**, L628 (1997).
[20] M. BOCKSTEDTE, A. KLEY, J. NEUGEBAUER, and M. SCHEFFLER, Comput. Phys. Commun. **107**, 187 (1997).
[21] A. EICHLER, G. KRESSE, and J. HAFNER, Surf. Sci. **397**, 116 (1998).
[22] C.M. WEI, A. GROSS, and M. SCHEFFLER, Phys. Rev. B **57**, 15572 (1998).
[23] B. HAMMER and M. SCHEFFLER, Phys. Rev. Lett. **74**, 3487 (1995).
[24] S. WILKE and M. SCHEFFLER, Phys. Rev. Lett. **76**, 3380 (1996).
[25] A. GROSS, C.M. WEI, and M. SCHEFFLER, Surf. Sci. **416**, L1095 (1998).
[26] G. WIESENEKKER, G.J. KROES, and E.J. BAERENDS, J. Chem. Phys. **104**, 7344 (1996).
[27] G.J. KROES, E.J. BAERENDS, and R.C. MOWREY, J. Chem. Phys. **110**, 2738 (1999).
[28] T.B. BLANK, S.D. BROWN, A.W. CALHOUN, and D.J. DOREN, J. Chem. Phys. **103**, 4129 (1995).
[29] K.T. NO, B.H. CHANG, S.Y. KIM, M.S. JHON, and H.A. SCHERAGA, Chem. Phys. Lett. **271**, 152 (1997).
[30] S. LORENZ, A. GROSS, and M. SCHEFFLER, APS Bull. **43**, 235 (1998).
[31] D.E. MAKAROV and H. METIU, J. Chem. Phys. **108**, 590 (1998).
[32] R.E. COHEN, M.J. MEHL, and D.A. PAPACONSTANTOPOULOS, Phys. Rev. B **50**, 14694 (1994).
[33] M.J. MEHL and D.A. PAPACONSTANTOPOULOS, Phys. Rev. B **54**, 4519 (1996).
[34] A. GROSS, M. SCHEFFLER, M.J. MEHL, and D.A. PAPACONSTANTOPOULOS, Phys. Rev. Lett. **82**, 1209 (1999).
[35] G.M. GORINGE, D.R. BOWLER, and E. HERNÁNDEZ, Rep. Progr. Phys. **60**, 1447 (1997).
[36] J.C. SLATER and G.F. KOSTER, Phys. Rev. **94**, 1948 (1954).
[37] WWW address: http://cst-www.nrl.navy.mil/bind
[38] G.R. DARLING and S. HOLLOWAY, Rep. Progr. Phys. **58**, 1595 (1995).
[39] G. COMSA, R. DAVID, and B.-J. SCHUMACHER, Surf. Sci. **95**, L210 (1980).
[40] K. D. RENDULIC, G. ANGER, and A. WINKLER, Surf. Sci. **208**, 404 (1989).
[41] L. SCHRÖTER, H. ZACHARIAS, and R. DAVID, Phys. Rev. Lett. **62**, 571 (1989); Surf. Sci. **258**, 259 (1991).
[42] M.L. BURKE and R.J. MADIX, Surf. Sci. **237**, 1 (1990).
[43] C.T. RETTNER and D.J. AUERBACH, Chem. Phys. Lett. **253**, 236 (1996).
[44] M. KARIKORPI, S. HOLLOWAY, N. HENRIKSEN, and J.K. NØRSKOV, Surf. Sci. **179**, L41 (1987).
[45] A. GROSS, S. WILKE, and M. SCHEFFLER, Surf. Sci. **357/358**, 614 (1996).
[46] M. BEUTL, K.D. RENDULIC, and G.R CASTRO, Surf. Sci. **385**, 97 (1997).
[47] A. GROSS, J. Chem. Phys. **110**, 8696 (1999).

phys. stat. sol. (b) **217**, 405 (2000)

Subject classification: 68.35.Md; 71.15.Mb; S1; S1.1; S1.3

Metal Surfaces:
Surface, Step and Kink Formation Energies

J. Kollár (a), L. Vitos (b), B. Johansson (b), and H. L. Skriver (c)

(a) Research Institute for Solid State Physics, H-1525 Budapest, P.O. Box 49, Hungary

(b) Condensed Matter Theory Group, Department of Physics, Uppsala University, Box 530, S-75121 Uppsala, Sweden

(c) Center for Atomic-scale Materials Physics and Department of Physics, Technical University of Denmark, DK-2800 Lyngby, Denmark

(Received August 10, 1999)

We review the surface, step, and kink energies in monoatomic metallic systems. A systematic comparison is given between the theoretical results based on density functional theory and available experimental data. Our calculated values are used to predict the equilibrium shapes of small metal particles, monoatomic surface islands, and the instability of different surface geometries.

1. Introduction

Solid state physics as well as the other natural sciences are based on the doctrine that a well designed experiment will eventually be used to find the true value of any physical quantity in which we are interested. This doctrine is often supplemented by the statement that a theory which does not predict but only explain is not a theory at all. The obvious question that comes to mind is: What if the quantities we are interested in are difficult or perhaps even impossible to measure? One possible answer is that modern computational physics based on density functional theory has reached a level of accuracy where the atomic numbers of the chemical species involved are sufficient to yield values of physical quantities with an accuracy equal to or better than experiments. The surface, step, and kink energies of metals to be reviewed in the following are all examples of quantities which are difficult to measure but may be calculated with great accuracy.

2. The Surface Energy

The surface energy γ defined as the surface excess free energy per unit area of a particular crystal facet is one of the basic quantities in surface physics. It determines the equilibrium shape of mesoscopic crystals, it plays an important role in faceting, roughening, and crystal growth phenomena, and may be used to estimate surface segregation in binary alloys. Most of the experimental surface energy data [1, 2] stems from surface tension measurements in the liquid phase extrapolated to zero temperature. Although these data at present form the most comprehensive experimental source of surface energies they include uncertainties of unknown magnitude and correspond to an isotropic crystal. Hence, these data do not yield information as to the surface energy of a particular surface facet, and if one needs this information one must, except for the classical direct measurements on Pb and In surface facets [3], turn to density functional calculations.

2.1 Calculation of the surface energy from first principles: The surface-energy database

At $T = 0$ K the surface excess free energy per two-dimensional (2D) unit cell of area A_{2D} may be calculated as the difference

$$E_{\text{surf}} \equiv \gamma A_{2D} = E_{2D}(N) - NE_{3D} \tag{1}$$

between the total energy of N atoms in the surface region (2D) and N times the total energy per atom in the bulk (3D). In the actual calculations the surface region includes a number of empty spheres simulating the semi-infinite vacuum. The total energy of the bulk, E_{3D}, and of the surface regions, E_{2D}, is obtained by means of the full charge density (FCD) method based on density functional theory [4], and the Kohn-Sham one-electron equations are solved by means of the linear muffin-tin orbitals (LMTO) method using the atomic sphere approximation (ASA) for the potential and electron density [5]. The complete non-spherically symmetric charge density $n(\mathbf{r})$ constructed from the output of a self-consistent LMTO-ASA calculation is used to evaluate the FCD energy functional within the local density approximation (LDA) or generalized gradient approximation (GGA) for the exchange-correlation term. Thereby, the FCD method retains most of the simplicity and the computational efficiency of the LMTO-ASA method, but attains an accuracy comparable to that of the full potential methods [6 to 8]. Details of the FCD method can be found in Refs. [7, 9 to 11].

In the surface calculations the LMTO equations are solved by means of the surface Green's function technique developed by Skriver and Rosengaard [12] which correctly accounts for the semi-infinite nature of a surface by means of the principal layer technique [13]. The Dyson equation describing the relaxation of the electronic structure of the surface region is set up and solved within the ASA, taking into account the electrostatic mono and dipole contributions to the spherically symmetric one-electron potential [12]. However, the Green's function technique neglects the relaxation of the atomic positions. According to the first-principles work by Feibelman and co-workers [14, 15] and by Mansfield and Needs [16] the effect of the relaxation of the surface atomic positions on the calculated surface energy of a particular crystal facet may vary from 2 to 5%, depending on the roughness. Further, the semi-empirical results by Rodriguez et al. [17] show that surface relaxation typically affects the anisotropy by less than 2%.

Recently Vitos et al. [18] used the technique described above to calculate the surface energies for the low-index surfaces of 60 metals with the aim of generating a consistent database that may be used to establish models of a range of surface science phenomena. A comparison with available full potential calculations [19] for the f.c.c. surfaces of the 4d metals, see Fig. 1, shows a mean deviation in the calculated anisotropies $\gamma_{100}/\gamma_{111}$ and $\gamma_{110}/\gamma_{111}$ of $\approx 5\%$. We therefore expect the database to reflect the true density functional result with sufficient accuracy, also so for those metals, the majority in fact, for which there are no full potential calculations.

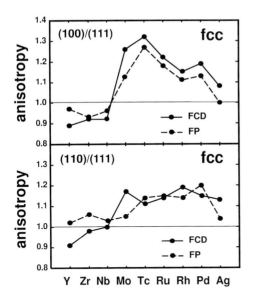

Fig. 1. Surface energy anisotropy defined as $\gamma_{(100)}/\gamma_{(111)}$ and $\gamma_{(110)}/\gamma_{(111)}$ for the (hypothetical) f.c.c. surfaces of the 4d elements from the LMTO-FCD-LDA calculation (FCD) and from the full-potential LMTO calculation (FP) by Methfessel et al. [19]. Note that the so defined anisotropy includes the surface area per atom which may be found in Table I in the database [18]

2.2 Surface energy of transition and noble metals

In Fig. 2 we show a series of surface energies taken from the database [18] together with the measured surface tensions reduced to 0 K by de Boer et al. [2]. The calculated values correspond to the most close-packed surfaces of the 4d metals in their ground state crystal structure. We immediately observe a perfect agreement between the two sets of data for most of the 4d metals and take this agreement as a measure of the validity of the measurements and the subsequent reduction to 0 K and of the accuracy of the calculations. However, there are substantial deviations between the two sets of data for Mo, Tc, and Ru which reflects the fact that the experimental data are valid for isotropic materials while the theoretical data correspond to the surface energy of specific surface facets of metals with specific crystal structures. Hence, if one wants to understand or, more ambitiously, predict the shapes of small metal clusters for which one needs the surface energy anisotropy, i.e., the kind of data shown in Fig. 1, one must turn to first-principles calculations for help.

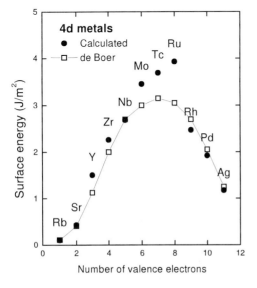

Fig. 2. The calculated surface energies for the close-packed surfaces of the 4d metals including Rb and Sr [18] compared with the experimentally derived values by de Boer et al. [2]

2.3 The Friedel model

Before we discuss more complicated surface properties we would like to point out that the variation of the surface energy across a transition metal series, such as the 4d series shown in Fig. 2, has a very simple physical origin. In the Friedel model [20] of cohesion in transition metals one considers the one-electron contribution from the d-electrons and writes the total energy in the form

$$E_{\text{tot}} \approx \int_B^{E_F} (E - \epsilon_d)\, D_d(E)\, dE, \qquad (2)$$

where $D_d(E)$ is the d state density, B the bottom of the valence band, ϵ_d the center of gravity of the state density, here assumed to be equal to the atomic d-level, and E_F the Fermi level. Further, each metal is characterized by its d-occupation number n_d or by the filling fraction

$$f = \frac{n_d}{10} = \frac{1}{10} \int_B^{E_F} D_d(E)\, dE. \qquad (3)$$

To compute the integral (2) the state density is approximated by a rectangle containing ten electrons, i.e.,

$$D_d(E) = \begin{cases} \dfrac{10}{W} & \text{for } B \leq E \leq A \\ 0 & \text{otherwise}, \end{cases} \qquad (4)$$

where $W = A - B$ is the bandwidth, and one obtains for the total energy, which in this case is the cohesive energy, the well-known Friedel parabola

$$E_{\text{coh}} = 5Wf(1-f). \qquad (5)$$

The experimental cohesive energies of the transition metals follow the prediction of the Friedel model quite well, especially when they are corrected for atomic effects [21].

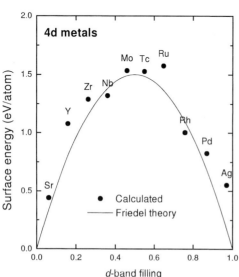

Within tight-binding theory the bandwidth W which enters the expression for the cohesive energy varies as $\sqrt{z}$, where z is the coordination number. As a result the bandwidth for a transition-metal surface is typically reduced by $\approx 20\%$ and the surface energy is given by

$$E_{\text{surf}} = Wf(1-f). \qquad (6)$$

From the comparison in Fig. 3 it is clear that the general behaviour of the experi-

Fig. 3. The calculated surface energies of the 4d metals compared to the Friedel parabola, $E_{\text{surf}} = Wf(1-f)$, for $W = 6$ eV

mental as well as the calculated surface energies for the 4d metals is captured by the Friedel model. However, it is also clear that the model can neither describe the dependence on crystal structure which is included in the first-principles calculations shown in the figure, nor describe surface energy anisotropies.

2.4 Anisotropy of the surface energy

The surface energy anisotropy or the orientation dependence of the surface energies can be characterized by the ratio between the surface energies of facets with increasing roughnesses. The anisotropies of the f.c.c. and b.c.c. metals, for example, may be described by $\gamma_{100}/\gamma_{111}$ and $\gamma_{100}/\gamma_{110}$, respectively. In the case of the h.c.p. metals, there are two facets $(10\bar{1}0)_A$ and $(10\bar{1}0)_B$ with the same roughness. Therefore, for these metals we define the anisotropy as the average between $\gamma_{10\bar{1}0_A}/\gamma_{0001}$ and $\gamma_{10\bar{1}0_B}/\gamma_{0001}$. These anisotropies are defined in terms of the surface energy per unit area rather than per atom, and, therefore, they include the effect of the different area per atom on different facets [18].

In Fig. 4 we summarize the surface energy anisotropies as defined above and calculated from the database [18]. Depending on the nature of the metallic bonds three different behaviours can be distinguished. When most of the valence electrons are localized around the atoms the surface energy contribution from the interstitial region is negligible. In this case the surface energy per atom is, to a good approximation, constant, and the surface energy per area decreases with increasing roughness. This group

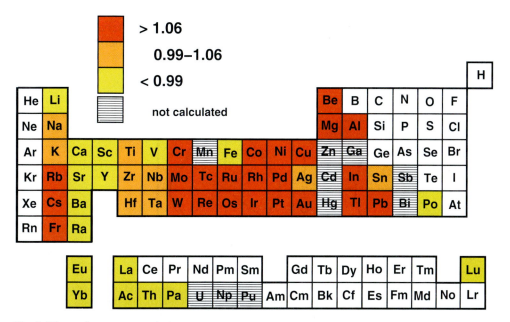

Fig. 4. The surface energy anisotropy

includes the light alkali, the alkaline earths and the first transition metals in each series.

In the second case a sizable valence density is distributed uniformly in the interstitial region, giving rise to a surface energy per atom proportional to the surface area per atom, and, therefore, to a very week anisotropy. This behaviour is typical for the Ti and V group elements.

In the third case the metallic bonds have an important covalent character, i.e., the charge density on a spherical surface of radius close to the half interatomic distance is strongly anisotropic with maxima in the nearest neighbour directions. This covalent character of the metallic bond is reflected in a surface energy per atom determined mainly by the number of broken bonds. The middle and the late transition and p metals may be characterized by this kind of strong anisotropy. The b.c.c. Fe presents an anomalous anisotropy as a consequence of its ferromagnetic ground state. The magnetic contribution to the surface energy [22] of the (100) Fe surface lowers the surface energy of this facet relative to that of the most closely packed (110) surface.

2.5 *The equilibrium shape of the crystals*

The equilibrium shape of a small cluster of nanometer scale at constant volume is that which minimizes the total surface free energy. At $T = 0$ K the shape is therefore completely determined by the anisotropy of the surface energy [23]. Hence, in the case of weak anisotropy, e.g., early transition metals, the equilibrium crystal shape is a sphere, while in the case of strong anisotropy, e.g., late transition and p metals, it may be a complex polyhedron.

The equilibrium shape may be determined by the so-called Wulff construction [24], which assumes a complete knowledge of the orientation dependence of the surface energy $\gamma(\mathbf{n})$. In the present application of the Wulff construction we use for the low-index surfaces the surface energies from Tables IV to VII of the database [18], and for the high-index surfaces we use a cluster expansion of the surface energy based on the low-index first-principles results.

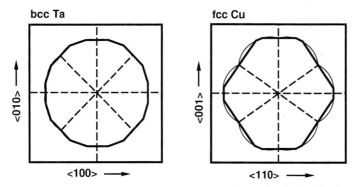

Fig. 5. The equilibrium shape of the b.c.c. Ta cluster in the (001) plane, and of the f.c.c. Cu cluster in the (110) plane. The dashed lines point to the directions for which first-principles calculations were performed. The thin lines denote the results by the cluster expansion and the heavy lines the theoretical equilibrium crystal shapes

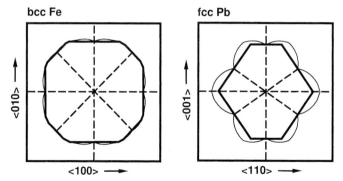

Fig. 6. The equilibrium shape of the b.c.c. Fe and f.c.c. Pb clusters in the (001) and (110) planes, respectively. The dashed lines point to the directions for which first-principles calculations were performed. The thin lines denote the results by the cluster expansion and the heavy lines the theoretical equilibrium crystal shapes

In a cluster expansion one expresses the total energy of a given crystal as a linear combination of one-site, two-site, etc., interatomic potentials [25]. In the application to transition metal surfaces it turns out that one needs only include pairwise interactions to describe the higher-index surfaces with a resonable accuracy [23]. We therefore use the expansion

$$E_{\text{surf}}(hkl) = \sum_{s}^{N_s} n_s(hkl)\, V_s^{(2)}, \qquad (7)$$

where $n_s(hkl)$ is the number of broken pair-bonds in the s-th coordination-shell for a surface of index (hkl), and N_s the number of shells considered in the expansion. The pair interactions, $V_s^{(2)}$, listed in Table II of Ref. [26], were found for each metal by fitting (7) to the relevant low-index surface energies listed in the database [18].

The polar plots of the shapes of b.c.c. Ta in the (001) plane and f.c.c. Cu cluster in the (110) plane are shown in Fig. 5 while Fig. 6 shows the polar plots of b.c.c. Fe in the (001) plane and f.c.c. Pd in the (110) plane. Further polar plots may be found in Ref. [18]. In the figures the dashed lines point to the directions for which first-principles calculations were performed. The thin lines correspond to the results obtained by the cluster expansion and the heavy lines indicate the theoretical equilibrium crystal-shapes.

The almost spherical shape of the b.c.c. Ta cluster seen in the figure reflects the weak surface energy anisotropy in this metal. The equilibrium shape of Cu exhibits mainly the (111) and (100) facets, but due to the relatively low value of γ_{110} a small fraction of the (110) facet is also formed. The particular squarish shape of b.c.c. Fe clusters is a magnetic effect which causes an anomalous surface energy anisotropy whereby the equilibrium area of an individual (100) facet is larger than that of a (110) facet. We note, that in a paramagnetic calculation the anisotropy for Fe is reversed, and in that case the equilibrium shape corresponds to that shown in Fig. 6 but rotated by 45°. In recent TEM experiments [27] one finds that an Fe cluster has a nearly cubic shape oriented in complete agreement with the predictions in Fig. 6.

As a last example we show the equilibrium shape of f.c.c. Pb which, as a result of the strong anisotropy, forms in a slightly distorted hexagonal shape. The only facets present are the (111) and (001) facets. A very similar equilibrium shape was observed experimentally by Heyraud and Metois [3] at T above 473 K.

3. Formation Energies for Mono-Atomic Surface Steps

The energetics of mono-atomic steps on otherwise perfect metal surfaces are basic in the description of surface morphology. Unfortunately, a direct experimental determination of the energy needed to create these steps does not appear to be feasible and most of the low temperature experimental step-energy data [28] stem from measurements of the orientation dependence of the surface free energy at high temperatures extrapolated to 0 K. At present, there are only very few measurements of this type and a theoretical determination of step energies is therefore of vital importance.

Step energies have been calculated in the past either by approximate methods, such as the embedded atom method (EAM) [29] and the effective medium theory (EMT) [30] or from first-principles using density functional theory [31, 32] for a few selected metals. Recently a database of step energies for all f.c.c. and b.c.c. transition and noble metals has been established by Vitos et al. [26].

3.1 Calculation of step energies: Vicinal surfaces and steps

A mono-atomic step the face of which has Miller indices $(h'k'l')$ on a perfect (hkl) crystal surface may be characterized by the excess free energy β per unit length along the edge of the step. This step energy can be calculated as the difference in the surface energy γ of a vicinal surface specified by $p(hkl) \times (h'k'l')$ (for notation see [33]) and the surface energy γ_0 of the flat (hkl) surface, in the limit where the number of 2D unit cells on the terrace $p \to \infty$, i.e.,

$$\beta = \gamma w - \gamma_0 w_0 . \qquad (8)$$

Here, w is the distance between adjacent step edges and w_0 the width of the (hkl) terrace exposed between a step bottom and the next step edge. Assuming that the step–step interaction is short-ranged, in practical calculations one may use (8) to good accuracy with finite, not too large, Miller indices for the vicinal surfaces [31, 32], or equivalently, relatively small values of p.

To analyse the calculated step energies and the contributions from different coordination shells we insert the cluster expansion (7) for the vicinal

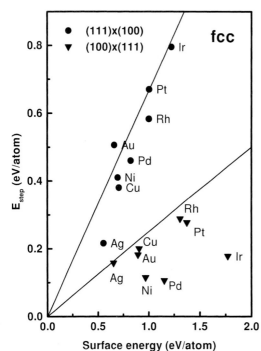

Fig. 7. The calculated step energies plotted as functions of the surface energies of the terraces [18]. The lines of slope 2/3 and 1/4 are the results of the nearest neighbor model for $(111) \times (100)$ and $(100) \times (111)$, respectively

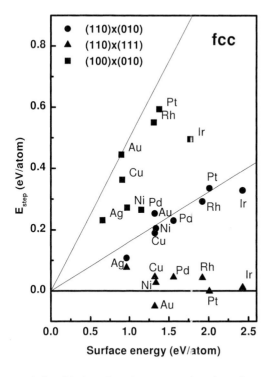

Fig. 8. The calculated step energies plotted as functions of the surface energies of the terraces [18]. The lines of slope 1/2, 1/6, and 0 are the results of the nearest neighbor model for $(100) \times (010)$, $(110) \times (010)$, and $(110) \times (111)$, respectively

and the flat surface into the expression for the step energy to obtain

$$E_{\text{step}} \equiv \beta d = \sum_{s}^{N_s} n_{\text{step},s}^{\text{eff}} V_s^{(2)}, \quad (9)$$

in terms of an effective number of disrupted bonds per shell. Here d is the repeat distance along the step edge.

3.2 Correlation with surface energies

If we restrict the range of pair interactions to nearest neighbors, the step energy in Eq. (9) may be written as

$$E_{\text{step}} = n_{\text{step},1}^{\text{eff}} V_1^{(2)} \quad (10)$$

and should therefore be proportional to the surface energy of the terrace on which it is formed with a constant of proportionality equal to $n_{\text{step},1}^{\text{eff}} / n_1(hkl)$. To examine how well this correlation suggested by Bonzel [28] is fulfilled for the cubic transition metals, we plot in Figs. 7 to 10 the calculated step energies as functions of the surface energy of the appropriate terrace. From the figures it is seen that the steps with the longest interatomic distance along the edge on the most close-packed surfaces have the highest step energies, e.g., f.c.c. $(111) \times (100)$ and b.c.c. $(110) \times (001)$. This is clearly in accordance with the fact that for these steps the terrace has a minimal number of broken bonds

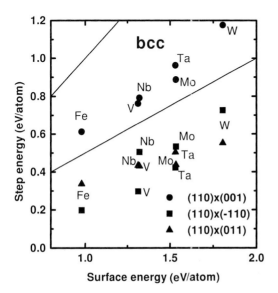

Fig. 9. The calculated step energies plotted as functions of the surface energies of the terraces [18]. The lines of slope 1 and 1/2 are the results of the nearest neighbor model for $(110) \times (001)$ and $(110) \times (\bar{1}11)$, $(110) \times (011)$, respectively

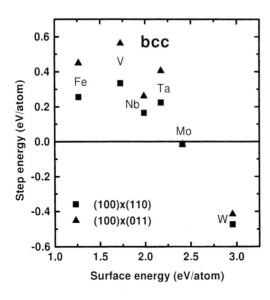

Fig. 10. The calculated step energies plotted as functions of the surface energies of the terraces [18]. The line of slope 0 is the result of the nearest neighbor model

while the step has a maximum [26]. It is further seen that the proportionality between step and surface energy is well obeyed by the f.c.c. transition metals, see Figs. 7 and 8, reflecting the fact that here the nearest neighbor pair interaction is the dominant term in the expansion. Owing to the importance of the second nearest neighbor interactions the b.c.c. transition metals do not exhibit the slope predicted by the nearest neighbor model although there still is a clear correlation between step and surface energy, Figs. 9 and 10.

3.3 Step energy anisotropy: Equilibrium island shapes

The anisotropy of the step energies is connected with the formation of kinks and therefore has consequences for the shapes of islands formed on the surface. For the f.c.c. (100) surface the anisotropy of the step energy is found to be $\beta_{(100)\times(010)}/\beta_{(100)\times(111)} = 1.1$ to 1.9 and by means of a 2D Wulff construction (see, e.g. [24]) we predict islands in the shape of squares[1]) with sides parallel to the $\langle 110 \rangle$ directions. This may be compared with the experimental findings: For the Ag(100) surface a preferential formation of $(100) \times (111)$ steps is observed [34], and in the case of the Ir(100) surface the experimental equilibrium island-shape is a square with $\langle 110 \rangle$ sides [35]. On the f.c.c. (110) surface for most of the transition metals we have $\beta_{(110)\times(010)} \approx \beta_{(110)\times(\bar{1}11)}$ and $\beta_{(110)\times(111)}/\beta_{(110)\times(010)} \ll 1$. Therefore, on this facet islands of mono-atomic heigth, if they are formed, should have long edges parallel to the $\langle 110 \rangle$ direction.

For the b.c.c. (110) surface the anisotropy of the step energies obtained for the two main directions is found to be $\beta_{(110)\times(001)}/\beta_{(110)\times(\bar{1}10)} = 1.1$ to 1.7, except for Fe where this ratio increases to 2.2. Therefore, on the Fe (110) surface the formation of the b.c.c. $(110) \times (001)$ type steps is less probable. From the step energy ratios for the (110) facets using the 2D Wulff construction we predict slightly distorted octagonal equilibrium island shapes in the case of V, Nb, and Ta, and hexagonal island shapes for Mo and W. For the b.c.c. (100) surfaces, except Mo and W, the step energy anisotropy is $\beta_{(100)\times(011)}/\beta_{(100)\times(110)} = 1.1$ to 1.3, which leads to octagonally shaped islands.

---

[1]) When $\beta_{(100)(100)}/\beta_{(110)(100)} > \sqrt{2}$ the equilibrium shape is a perfect square, otherwise it is octogonal.

4. The Kink Formation Energies on the Mono-Atomic Surface Steps

The kink formation energy is defined as the excess free energy of a kink in an otherwise perfect one atom high step and may be determined from the anisotropy of the step energy. Starting from the cluster expansion for the vicinal surfaces (7) the expression for the kink energy may be written in a form similar to (9), i.e.,

$$E_{\text{kink}} = \sum_{s}^{N_s} n_{\text{kink},s}^{\text{eff}} V_s^{(2)}. \tag{11}$$

The effective numbers of broken bonds per shell $n_{\text{kink},s}^{\text{eff}}$ in the formation of a kink along a particular step edge are listed in the last column of Table I in Ref. [26]. It turns out that for any given step there is only one shell, i.e., one $n_{\text{kink},s}^{\text{eff}} \neq 0$, that contributes to the kink formation energy which therefore is reduced to one of the pair potentials listed in Table II of Ref. [26]. The formation energies of kinks in mono-atomic steps on transition and noble metal surfaces may be derived from Tables I and II in Ref. [26].

For the f.c.c. transition metals the calculated kink energies have their largest values for steps such as $(110) \times (111)$ where the interatomic distance along the edge is shortest. For more open steps, e.g., $(100) \times (010)$ and $(110) \times (010)$, the kink energies are substantially smaller, and in the case of Ni, Pd, Ir, Pt, and Au they even become negative indicating an instability of these steps. It is worth noting that the kink energy for the $(110) \times (111)$ step on Pt and Au is positive, indicating that this step which should form on account of its negative step energy will be stable in accordance with the missing row reconstruction observed experimentally in these metals [36].

Based on the pair potentials in Ref. [26] the kink formation energies for all the b.c.c. transition metals should be positive except for Mo and W, where only the $(110) \times (011)$ steps have positive kink energy. This means, that not only are the (100) facets of these two metals unstable against the formation of $(100) \times (110)$ and $(100) \times (011)$ steps but also the steps themselves are unstable against the formation of kinks. This is clearly connected with the 2×2 reconstruction observed experimentally for the (100) surface of W [37].

5. Stability of Transition and Noble Metal Surfaces

5.1 Steps, kinks, and reconstruction

Using the databases of surface, step, and kink formation energies [18, 26] one may analyse the stability of transition and noble metal surfaces against the formation of mono-atomic high steps or surface reconstruction.

For the steps with the shortest interatomic distance along the edge on open surfaces, i.e., f.c.c. $(110) \times (111)$, b.c.c. $(100) \times (110)$, and b.c.c. $(100) \times (011)$, there are no extra broken nearest neighbor bonds, and the step energy is therefore expected to be governed by the next nearest pair interactions. For the f.c.c. transition metals these pair interactions $V_2^{(2)}$ are small, and for that reason the f.c.c. $(110) \times (111)$ step usually has a low step energy. If we include three nearest neighbor coordination shells in Eq. (9) the energy of the f.c.c. $(110) \times (111)$ step may be written in terms of the surface energies per atom of the two facets involved, i.e., $E_{(110) \times (111)} = E_{(111)} - E_{(110)}/2$. Consequently, the condition for the stability of the f.c.c. (110) surface against the formation of f.c.c. $(110) \times (111)$ type steps is $E_{(110)} < 2E_{(111)}$ or equivalently $\gamma_{(110)}/\gamma_{(111)} < \sqrt{3/2}$. From the tables of Refs. [18, 26] it is seen that this stability condition is violated for Pt and Au, a fact which is clearly related to the negative values of the calculated step energies and to the missing row reconstruction observed experimentally in these metals [36].

The two b.c.c. transition metals Mo and W also exhibit negative step energies in those cases where the next nearest neighbor interactions dominate, i.e. for b.c.c. $(100) \times (110)$ and b.c.c. $(100) \times (011)$ steps, as seen in Fig. 10. Using Eq. (9) with four nearest neighbor coordination shells, the condition for the stability of the b.c.c. (100) surface against the formation of b.c.c. $(100) \times (110)$ type steps may be written in terms of the surface energies as $E_{(100)}/E_{(310)} < 2/3$ or $\gamma_{(100)}/\gamma_{(310)} < \sqrt{10/9}$. The surface energies of the (310) facet in Mo and W are 3.601 and 4.338 eV/atom, respectively [18]. On the basis of these results one would predict a preferential formation of these steps on the Mo and W (100) surfaces. However, in these cases also the $(100) \times (110)$ and $(100) \times (011)$ steps are unstable. Therefore, the reconstruction should be more complex than a missing row reconstruction, in agreement with the 2×2 reconstruction observed experimentally for the (100) surface of W [37].

5.2 Rippled surfaces

Recently an interesting surface reconstruction into a sequence of (331) and $(33\bar{1})$ facets is reported to occur on the (110) surface of Ir [38], and the existence of this mesoscopic "rippling" shows that in this case the formation of (331) vicinal facets is preferred to the formation of (111) vicinal facets which, in the one-atom wide terrace limit, is the (1×2) missing row reconstruction observed in Pt and Au (110) surfaces [36]. The negative formation energies for the f.c.c. $(110) \times (111)$ type steps on the (110) surfaces of Pt and Au calculated in Ref. [26] clearly indicate the instability of these surfaces against the formation of close-packed (111) microfaceted steps and, presumably, the appearance of the (1×2) missing row reconstruction. For all other f.c.c. transition and noble metals (including Ir), however, the calculations [26] show that the (110) surface is energetically favourable relative to the formation of f.c.c. $(110) \times (111)$ type steps.

In Ref. [39] we examined the stability of the f.c.c. (110) transition and noble metal surfaces againts a "rippled" reconstruction formed by vicinal surfaces with Miller indices $(2\lambda + 1, 2\lambda + 1, 1)$ and $(2\lambda + 1, 2\lambda + 1, \bar{1})$ where $\lambda = 0, 1, \ldots$ Below, we present the main results of this analysis.

The stability of the (110) surface against the reconstruction into "rippled" vicinal facets is described by the ratio

$$f_{hkl} \equiv 1 - \frac{E_{110}}{E_{hkl}} = 1 - \frac{\gamma_{110}}{\gamma_{hkl}} \cos \theta_{hkl}, \qquad (12)$$

where E_{hkl} is the total surface energy of a "rippled" surface consisting of a sequence of vicinal facets with indices (hkl) and $(hk\bar{l})$, E_{110} the total surface energy of the corresponding flat (110) surface, and θ_{hkl} the angle between the normals of the flat (110) and of the vicinal surfaces. If $f_{hkl} > 0$ the (110) surface is stable, otherwise the "rippled" surface reconstruction occurs [39].

The stability ratio (12) may be calculated by noting that the surface-energy database [18] includes in addition to the surface energies of the f.c.c. (110) facet the surface energies of vicinal surface with $\lambda = 0$. For the vicinal surface with Miller indices corresponding to $(\lambda \geq 1)$ one uses the cluster expansion (7).

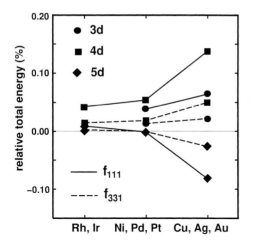

Fig. 11. The relative stability of the f.c.c. (110) transition metal surfaces against the reconstruction into a sequence of vicinal surfaces with Miller indices $(2\lambda+1, 2\lambda+1, 1)$ and $(2\lambda+1, 2\lambda+1, \bar{1})$. f_{111} and f_{331} denote the $\lambda = 0$ and $\lambda = 1$ cases, respectively

In Fig. 11 we show the f_{111} and the f_{331} ratios for the 3d, 4d and 5d f.c.c. metals. One immediately realizes that none of the 3d and 4d metals undergoes the "rippled" surface reconstruction. Moreover, for all the 3d and 4d f.c.c. metals we have $f_{331} < f_{111}$. Hence, for these metals the formation of (331) vicinals would be more stable than the formation of (111) vicinals. For Pt and Au both f_{111} and f_{331} are negative and $f_{331} > f_{111}$, which is in agreement with the missing row reconstruction observed experimentally in these metals [36].

Ir is the only one among the 5d f.c.c. metals for which $f_{111} > f_{331}$ and in this case a surface with (331) and (33$\bar{1}$) vicinals is expected to be more stable than that with (111) and (11$\bar{1}$) vicinals. According to the cluster expansion approximation of the surface energy for the vicinal (331) facet remains positive for $f_{331} < f_{111}$, suggesting the stability of the (110) flat surface in disagreement with experiment. However, a first-principles calculation [39] obtained by the exact muffin-tin orbitals method [40] for the surface energies $\gamma_{111}, \gamma_{100}, \gamma_{110}$, and γ_{331} in Ir leads to a small negative value for both f_{111} and f_{331} with $f_{331} < f_{111}$, in perfect agreement with the experimental observation [38]. Consequently, for Ir, in contrast to the other f.c.c. transition and noble metals with $f_{331} < f_{111}$, the pair potentials from the second and third coordination shell approximately cancel, i.e., $V_2^{(2)} + 2V_3^{(2)} \approx 0$, and the unusual (331)-type reconstruction is stabilized by the higher order pair- and multisite potentials.

Acknowledgements The Swedish Natural Science Research Council is acknowledged for financial support. Center for Atomic-scale Materials Physics is sponsored by the Danish National Research Foundation. Part of this work was supported by the research project OTKA 23390 of the Hungarian Scientific Research Fund.

References

[1] W. R. TYSON and W. A. MILLER, Surf. Sci. **62**, 267 (1977).
[2] F. R. DE BOER, R. BOOM, W. C. M. MATTENS, A. R. MIEDEMA, and A. K. NIESSEN, Cohesion in Metals, North-Holland, Amsterdam 1988.
[3] J. C. HEYRAUD and J. J. METOIS, Surf. Sci. **128**, 334 (1983); **177**, 213 (1986).
[4] R. M. DREIZLER and E. K. U. GROSS, in: Density Functional Theory, Springer-Verlag, Berlin 1990.

[5] O. K. Andersen, O. Jepsen, and M. Sob, in: Electronic Band Structure and Its Applications, Ed. M. Yussouff, Springer Lecture Notes, Springer-Verlag, Berlin 1987.
[6] J. Kollár, L. Vitos, and H. L. Skriver, Phys. Rev. B **55**, 15353 (1997).
[7] L. Vitos, J. Kollár, and H. L. Skriver, Phys. Rev. B **55**, 13521 (1997).
[8] L. Vitos, J. Kollár, and H. L. Skriver, in: NATO ASI Series B: Physics, Stability of Materials, Eds. A. Gonis, P. E. A. Turchi, and J. Kudrnovsky, Plenum Press, New York 1996 (p. 393).
[9] L. Vitos, J. Kollár, and H. L. Skriver, Phys. Rev. B **49**, 16694 (1994).
[10] L. Vitos, J. Kollár, and H. L. Skriver, Phys. Rev. B **55**, 4947 (1997).
[11] J. Kollár, L. Vitos, and H. L. Skriver, in: Ab-initio electronic properties of solids, Lecture Notes in Physics, Springer-Verlag, Berlin 2000.
[12] H. L. Skriver and N. M. Rosengaard, Phys. Rev. B **43**, 9538 (1991).
[13] B. Wenzien, J. Kudrnovsky, V. Drchal, and M. Sob, J. Phys.: Condensed Matter **1**, 9893 (1989).
[14] P. J. Feibelman, Phys. Rev. B **46**, 2532 (1992).
[15] P. J. Feibelman and D. R. Hamann, Surf. Sci. **234**, 377 (1990).
[16] M. Mansfield and R. J. Needs, Phys. Rev. B **43**, 8829 (1991).
[17] A. M. Rodriguez, G. Bozzolo, and J. Ferrante, Surf. Sci. **289**, 100 (1993).
[18] L. Vitos, A.V. Ruban, H. L. Skriver, and J. Kollár, Surf. Sci. **411**, 186 (1998).
[19] M. Methfessel, D. Henning, and M. Scheffler, Phys. Rev. B **46**, 4816 (1992).
[20] J. Friedel, in: The Physics of Metals, Ed. J. M. Ziman, Cambridge University Press, Cambridge 1969 (p. 494).
[21] M. S. S. Brooks and B. Johansson, J. Phys. F **13**, L197 (1983).
[22] M. Aldén, H. L. Skriver, S. Mirbt, and B. Johansson, Surf. Sci. **315**, 157 (1994).
[23] Siqing Wei and M. Y. Chou, Phys. Rev. B **50**, 4859 (1994).
[24] M. C. Desjonquères and D. Spanjaard, Concepts in Surface Physics, Springer-Verlag, Berlin/Heidelberg 1996.
[25] J. A. Moriarty and R. Phillips, Phys. Rev. Lett. **66**, 3036 (1991).
[26] L. Vitos, H. L. Skriver, and J. Kollár, Surf. Sci. **425**, 212 (1999).
[27] L. Theil Hansen, private communication.
[28] H. P. Bonzel, Surf. Sci. **328**, L571 (1995).
[29] S. V. Khare and T. L. Einstein, Surf. Sci. **314**, L857 (1994).
[30] P. Stoltze, J. Phys.: Condensed Matter **6**, 9495 (1994).
[31] P. J. Feibelman, Phys. Rev. B **52**, 16845 (1995).
[32] G. Boisvert, L. J. Lewis, and M. Sheffler, Phys. Rev. B **57**, 1881 (1998).
[33] B. Lang, R. W. Joyner, and G. A. Somorjai, Surf. Sci. **30**, 454 (1972).
[34] Ch. Teichert, Ch. Ammer, and M. Klaua, phys. stat. sol. (a) **146**, 223 (1994).
[35] Chonglin Chen and Tien T. Tsong, Surf. Sci. **336**, L735 (1995).
[36] H. Hörnis, J. R. West, E. H. Conrad, and R. Ellialtioglu, Phys. Rev. B **47**, 13055 (1993) and references therein.
[37] R. A. Barker, P. J. Estrup, F. Jona, and P. M. Marcus, Solid State Commun. **25**, 375 (1978).
[38] J. Kuntze, S. Speller, and W. Heiland, Surf. Sci. **402**, 764 (1998).
[39] L. Vitos, B. Johansson, H. L. Skriver, and J. Kollár, submitted to Comput. Mater. Sci. (1999).
[40] O. K. Andersen, O. Jepsen, and G. Krier, in: Lectures on Methods of Electronic Structure Calculations, Eds. V. Kumar, O. K. Andersen, and A. Mookerjee, World Scientific Publ. Co., Singapore 1994 (pp. 63 to 124).
L. Vitos, H. L. Skriver, B. Johansson, and J. Kollár, submitted to Comput. Mater. Sci. (1999).

phys. stat. sol. (b) **217**, 419 (2000)

Subject classification: 71.38.+i; 74.25.Kc; 74.70.Ad; S2; S5

Linear-Response Studies of the Electron–Phonon Interaction in Metals

A. Y. Liu

Department of Physics, Georgetown University, Washington DC 20057-0995, USA

(Received August 10, 1999)

Applications of the first-principles linear-response method for calculating electron–phonon coupling parameters are presented. Calculations show that the electron–phonon coupling in the low-temperature 9R phase of Li is significantly weaker than in the room-temperature b.c.c. phase, which helps explain the observed absence of a superconducting transition down to the mK regime. Results for the electron–phonon interaction in compressed S indicate that the surprisingly high superconducting T_c of 17 K observed at 160 GPa arises from a combination of a moderately large electron–phonon mass enhancement parameter and a very high phonon energy scale for the compressed lattice.

1. Introduction

Effects of the electron–phonon interaction in metals are evident in many experimentally accessible quantities [1]. For example, the electron–phonon interaction causes the enhancement of the effective electron mass as measured in the electronic heat capacity, it produces a finite phonon lifetime arising from electron–phonon scattering, it contributes to electrical and thermal resistivities, and, of course, it plays a prominent role in superconductivity. Within the Migdal-Eliashberg theory of strong-coupling superconductivity, the superconducting transition temperature T_c and other properties depend on the Coulomb parameter μ^* and the electron–phonon spectral function $\alpha^2 F(\omega)$ [2 to 4]. The spectral function measures the effectiveness of phonons of energy $\hbar\omega$ to scatter electrons from one part of the Fermi surface to another. It is similar to the phonon density of states $F(\omega)$, but is weighted by an average of the square of the electron–phonon matrix element.

Experimentally, low-temperature specific-heat measurements and quasiparticle tunneling are two common methods for obtaining information about the electron–phonon interaction in metals and superconductors [1]. The thermal effective mass in the electronic contribution to specific heat is given by $m_{th} = (1 + \lambda) m_b$, where m_b is the band mass and λ is the dimensionless electron–phonon mass enhancement parameter. The mass enhancement parameter is proportional to the inverse-frequency moment of $\alpha^2 F(\omega)$. Since determining λ from specific-heat data requires knowledge of the quasiparticle band mass, there can be large uncertainties in the values of λ extracted this way. Another probe of the electron–phonon interaction is provided by tunneling measurements across metal–insulator–superconductor junctions. Structure in the tunneling conductance reflects structure in the superconducting gap function that arises from the interaction of electrons and phonons. In the McMillan-Rowell tunneling inversion procedure [5], the Eliashberg equations are solved iteratively to find an $\alpha^2 F(\omega)$ that accurately reproduces the measured tunneling conductance. Tunneling experiments have provided valuable information about the electron–phonon interaction in many metals. However, the method is not amenable to all systems, particularly those for which fabrication of high-quality junctions is problematic and those in which the coupling is so weak that the phonon structure in the tunneling data is too subtle to extract [6].

First-principles calculations can play an important role in providing information about the electron–phonon interaction, especially in materials where limited experimental data are available. Calculation of electron–phonon coupling parameters requires knowledge of the low-energy electronic excitation spectrum, the complete phonon spectrum, and the self-consistent response of the electronic system to lattice vibrations. In the past, ab initio calculations of electron–phonon coupling parameters have proceeded along two distinct lines. In the rigid ion (RI) [7] and rigid muffin tin (RMT) [8] schemes, the change in the potential due to ionic displacements is approximated by a rigid shift of the potential within atomic spheres and assumed to be negligible elsewhere. While these non-self-consistent approximations may be adequate for many transition metals [9], their validity has been questioned in some cases, especially for anisotropic or low-density-of-states materials in which the electronic screening may not be effective in confining changes in the potential to the atomic spheres [10]. An alternative to the RI and RMT methods is the frozen-phonon total-energy method [11]. In this approach, the electron–phonon matrix elements are evaluated using the self-consistently screened potentials corresponding to frozen-in phonon displacements. The primary drawback of the frozen-phonon approach is that only phonon wavevectors that are commensurate with the lattice and that correspond to reasonably sized supercells can be considered. This makes it difficult to determine accurately quantities that involve integrations over the phonon wavevector **q** throughout the Brillouin zone. These include, for example, the mass enhancement parameter λ, the phonon density of states $F(\omega)$, and the spectral function $\alpha^2 F(\omega)$.

Recently, linear-response theory within the framework of density-functional calculations has been shown to be a powerful alternative to the frozen-phonon method for calculating lattice dynamical properties [12] and electron–phonon coupling parameters [13] in solids. Atomic displacements are treated as perturbations, and the electronic response to the perturbation is calculated self-consistently. Perturbations of arbitrary wavevector **q** can be treated without constructing supercells. The linear-response method has been implemented with a variety of different basis sets for representing the electronic wavefunctions, and it has been successfully applied to the study of lattice dynamics in a wide range of materials [12, 14, 15]. The method has also been used to investigate the electron–phonon interaction in a variety of superconductors [13, 16 to 18]. Superconducting metals for which good tunneling data are available have been used as test cases for demonstrating the accuracy of the method. In the case of In, for example, it has been shown that the calculated $\alpha^2 F(\omega)$ reproduces the measured tunneling conductance to better than one part in a thousand [18].

In this paper, applications of the plane-wave-based density-functional linear-response method to the calculation of electron–phonon coupling parameters [16] in Li and in high-pressure phases of S are presented. The structure dependence of the electron–phonon coupling strength in Li is examined to try to resolve a long-standing puzzle regarding the lack of an observed superconducting phase of this material [19]. Results are also presented for S, which is an insulating molecular solid at ambient pressures, but which metallizes and becomes superconducting with a surprisingly high transition temperature under Megabar pressures [20]. For both these systems it is difficult to obtain detailed information about the electron–phonon interaction from experiments. In the case of Li, specific-heat measurements provide some estimates for λ [21], but with no superconducting phase, tunneling experiments are not applicable. For S, the fact that the superconducting phases exist only within diamond anvil cells severely limits the types of experimental probes possible.

2. Formalism

For notational simplicity, we consider the case of a single atom of mass M per unit cell. The electron–phonon matrix element for scattering of an electron from a Bloch state $n\mathbf{k}$ to another Bloch state $n'\mathbf{k}+\mathbf{q}$ by a phonon of frequency $\omega_{\mathbf{q}\nu}$ is

$$g(n\mathbf{k}, n'\mathbf{k}+\mathbf{q}, \nu) = \left(\frac{\hbar}{2M\omega_{\mathbf{q}\nu}}\right)^{1/2} \langle n'\mathbf{k}+\mathbf{q}|\hat{\epsilon}_{\mathbf{q}\nu}\cdot\nabla_{\mathbf{R}}V_{\text{sc}}|n\mathbf{k}\rangle, \qquad (1)$$

where $\hat{\epsilon}_{\mathbf{q}\nu}$ is the phonon polarization vector, and $\nabla_{\mathbf{R}}V_{\text{sc}}$ is the gradient of the self-consistent potential with respect to atomic displacements [1]. The scattering of phonon $\mathbf{q}\nu$ by electrons gives rise to a finite phonon linewidth that can be determined from the Golden Rule,

$$\gamma_{\mathbf{q}\nu} = \frac{2\pi}{\hbar}\sum_{\mathbf{k},n,n'} |g(n\mathbf{k}, n'\mathbf{q}+\mathbf{k}, \nu)|^2 (f_{n'\mathbf{k}+\mathbf{q}} - f_{n\mathbf{k}})\,\delta(E_{n'\mathbf{k}+\mathbf{q}} + \hbar\omega_{\mathbf{q}\nu} - E_{n\mathbf{k}}) \qquad (2)$$

$$= 2\pi\omega_{\mathbf{q}\nu}[N(E_{\text{F}})]^2 \langle\langle|g|^2\rangle\rangle, \qquad (3)$$

where $N(E_{\text{F}})$ is the density of states at the Fermi level, $f_{n\mathbf{k}}$ is the Fermi distribution function, and $\langle\langle|g|^2\rangle\rangle$ is a Fermi-surface average defined by

$$\langle\langle|g|^2\rangle\rangle = \frac{\sum_{\mathbf{k},n,n'}|g(n\mathbf{k}, n'\mathbf{k}+\mathbf{q}, \nu)|^2\,\delta(E_{n\mathbf{k}}-E_{\text{F}})\,\delta(E_{n'\mathbf{k}+\mathbf{q}}-E_{\text{F}})}{\left[\sum_{\mathbf{k},n}\delta(E_{n\mathbf{k}}-E_{\text{F}})\right]^2}. \qquad (4)$$

Equation (3) follows from Eq. (2) with the usual assumption that the temperature and phonon energies are much smaller than the electronic energy scale.

The Eliashberg spectral function is given by a sum over contributions to the coupling from each phonon mode,

$$\alpha^2 F(\omega) = \frac{1}{2\pi N(E_{\text{F}})}\sum_{\mathbf{q}\nu}\delta(\omega-\omega_{\mathbf{q}\nu})\frac{\gamma_{\mathbf{q}\nu}}{\hbar\omega_{\mathbf{q}\nu}}. \qquad (5)$$

The dimensionless electron–phonon mass enhancement parameter also involves a sum over modes and is equal to twice the first inverse-frequency moment of the spectral function,

$$\lambda = 2\int d\omega\, \alpha^2 F(\omega)/\omega \qquad (6)$$

$$= \frac{1}{\pi\hbar N(E_{\text{F}})}\sum_{\mathbf{q}\nu}\frac{\gamma_{\mathbf{q}\nu}}{\omega_{\mathbf{q}\nu}^2}. \qquad (7)$$

In this work, the electronic wavefunctions $|n\mathbf{k}\rangle$ and eigenvalues $E_{n\mathbf{k}}$ are calculated using the ab initio pseudopotential local-density-approximation (LDA) formalism. The electron–ion interaction is represented by soft separable pseudopotentials [22], and the single-particle wavefunctions are expanded in a plane-wave basis set. The Wigner [23] and Perdew-Zunger [24] forms of the exchange and correlation functional are employed for Li and S, respectively, and the partial core correction is used to handle the nonlinearity of the exchange and correlation interaction between the core and valence charge densities [25].

The phonon frequencies and polarization vectors are calculated using linear-response theory. The second-order change in the total energy, and hence the dynamical matrix, depends only on the first-order change in the electronic charge density. The linear response of the electronic density to atomic displacements is determined self-consistently by solving a Bethe-Salpeter equation as discussed in Ref. [14]. The electron–phonon matrix elements g are computed from the first-order change in the self-consistent potential.

The doubly-constrained Fermi surface sums in Eq. (4) are performed using dense meshes of order a thousand or more $\mathbf{k}$ points in the irreducible Brillouin zone (IBZ),[1]) and the δ-functions in energy are replaced by Gaussians, typically of width 0.01 to 0.03 Ry. Because of the large number of $\mathbf{k}$ points sampled, the results are not very sensitive to the Gaussian width. Dynamical matrices are usually computed on coarser meshes of up to a hundred or so $\mathbf{q}$ points in the IBZ.[2]) For some applications, it is useful to perform a Fourier interpolation of the dynamical matrices onto a larger set of $\mathbf{q}$ points: the dynamical matrices are Fourier transformed to yield the real-space force constants, which can then be used to construct the dynamical matrix at arbitrary $\mathbf{q}$ points. The accuracy of this interpolation procedure depends on the decay of the force constants in real space. With the same scheme, the dissipative part of the dynamical matrix,

$$\Gamma_{\alpha\beta}(\mathbf{q}) = \frac{\hbar}{2M\omega_{\mathbf{q}\nu}} \sum_{\mathbf{k},n,n'} \langle n\mathbf{k}|\nabla_\alpha V^{\mathrm{sc}}|n'\mathbf{k}+\mathbf{q}\rangle \langle n'\mathbf{k}+\mathbf{q}|\nabla_\beta V^{\mathrm{sc}}|n\mathbf{k}\rangle \delta(E_{n\mathbf{k}} - E_{\mathrm{F}}) \delta(E_{n'\mathbf{k}+\mathbf{q}} - E_{\mathrm{F}}),$$
(8)

can also be interpolated onto arbitrary $\mathbf{q}$ points. Here α and β are labels combining the atomic site and the direction of displacement. The phonon linewidth is simply related to this matrix by $\gamma_{\mathbf{q}\nu} = \hat{\epsilon}_{\mathbf{q}\nu} \cdot \boldsymbol{\Gamma}(\mathbf{q}) \cdot \hat{\epsilon}_{\mathbf{q}\nu}$ [3].

To study superconducting properties, the Coulomb parameter μ^* is also needed. A realistic calculation of μ^* is extremely difficult since it requires knowing the full frequency- and momentum-dependent dielectric response function of the system. Fortunately, μ^* tends not to vary much from material to material, generally lying in the range of 0.1 to 0.15. An upper bound on μ^* can be found by letting the unrenormalized Coulomb parameter μ become infinite. Because of the retarded nature of the electron–phonon interaction compared to the Coulomb interaction, μ^* is reduced in value from μ to $\mu/[1 + \mu \ln(E_{\mathrm{F}}/\hbar\omega_{\max})]$, where $\omega_{\max}$ is the maximum phonon frequency. Thus μ^* is bounded from above by $[\ln(E_{\mathrm{F}}/\hbar\omega_{\max})]^{-1}$, which is typically around 0.2.

---

[1]) For Li, the Fermi-surface integrals were performed using 728 and 650 irreducible $\mathbf{k}$ points for the b.c.c. and 9R structures, respectively. For S, 9800 and 2600 $\mathbf{k}$ points were sampled in the β-Po and simple cubic IBZs, respectively.

[2]) Dynamical matrices were computed for 285 and 98 irreducible phonon wavevectors in b.c.c. and 9R Li, respectively, while for S, 32 and 35 phonon wavevectors were sampled in the β-Po and simple cubic phases, respectively.

3. Applications

3.1 Structure dependence of electron–phonon coupling in Lithium

The lack of a superconducting transition in Li has been a long-standing puzzle. Frozen-phonon and RMT calculations [26], as well as specific-heat experiments [21], have suggested that the electron–phonon mass enhancement parameter in b.c.c. Li is $\lambda \approx 0.4$, which is similar to that in Al. This would suggest a transition temperature on the order of 1 K if a value of $\mu^* \approx 0.12$ is assumed. Experimentally, however, no transition is observed, at least down to 6 mK [27].

An important point to note is that Li undergoes a martensitic phase transition around 80 K, transforming from the room-temperature b.c.c. structure to the 9R phase, a close-packed structure with a nine-layer stacking sequence [28]. Earlier calculations had focused on the b.c.c. structure because of uncertainty about the crystal structure of the low-temperature phase. This naturally raises the question of how the electron–phonon interaction in Lithium varies with crystal structure.

The spectral functions calculated for b.c.c. and 9R Li are shown in Fig. 1. The b.c.c. phase has enhanced spectral weight at low frequencies compared to the close-packed 9R structure. Since λ is proportional to the inverse-frequency moment of $\alpha^2 F$, this difference at low frequencies leads to significant differences in λ. As can be seen in Table 1, we find $\lambda = 0.45$ for b.c.c. Li, which is in good agreement with earlier first-principles calculations [26]. The electron–phonon coupling in 9R Li is found to be significantly weaker, with $\lambda = 0.34$. These results can be compared to low-temperature specific-heat measurements which yield an effective mass of $m_{th} = 1.22m$, where m is the bare electron mass [21]. Although λ is calculated to be smaller in the 9R phase compared to the b.c.c. phase, the LDA band mass in 9R Li is calculated to be about 7% larger than in b.c.c. Li. These differences tend to offset each other in the expression for m_{th}, and the result is that the thermal masses computed for 9R and b.c.c. Li ($2.25m$ and $2.24m$, respectively) are both consistent with the thermodynamic data.

The importance of low-frequency modes in enhancing the electron–phonon coupling in the b.c.c. phase can be seen in a comparison of ω_{ln} for the two phases. A logarithmic moment of $\alpha^2 F$, this parameter can be expressed as $\omega_{ln} = \exp\left[\sum_{\mathbf{q}\nu} \lambda_{\mathbf{q}\nu} \ln(\omega_{\mathbf{q}\nu}) \Big/ \sum_{\mathbf{q}\nu} \lambda_{\mathbf{q}\nu}\right]$. As shown in Table 1, although the average phonon frequencies for the two structures

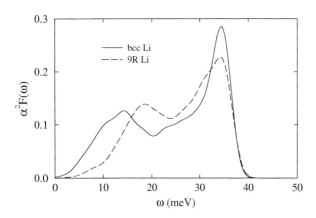

Fig. 1. Electron–phonon spectral functions calculated for b.c.c. and 9R Li

Table 1
Electronic and electron–phonon related properties computed for b.c.c. and 9R Li: the electronic density of states at the Fermi level $N(E_F)$, the average phonon frequency ω_{ave}, a logarithmic average of the phonon frequencies ω_{ln}, the mass enhancement parameter λ, and the minimum value of the Coulomb pseudopotential, μ^*_{min}, that is consistent with the observed lack of a superconducting transition above 6 mK

	$N(E_F)$ (states/Ry/spin/atom)	ω_{ave} (meV)	ω_{ln} (meV)	λ	μ^*_{min}
b.c.c.	3.33	25.0	14.9	0.45	0.27
9R	3.57	25.2	20.6	0.34	0.19

are nearly the same, ω_{ln} is almost 30% smaller in b.c.c. Li than in the close-packed phase. This indicates that some low-frequency phonon modes in b.c.c. Li couple strongly to the electrons, thereby dramatically lowering the λ-weighted logarithmic average of the phonon frequencies.

The modes that contribute most strongly to λ in b.c.c. Li can be identified in Fig. 2, which shows the phonon dispersion curves and the wavevector- and branch-dependent phonon linewidths and electron–phonon coupling parameters along high-symmetry directions. Figure 2c shows that the low-lying $T_1[\xi\xi 0]$ phonon modes, which correspond to shearing of (110) planes, give anomalously large contributions to λ. Since the linewidths

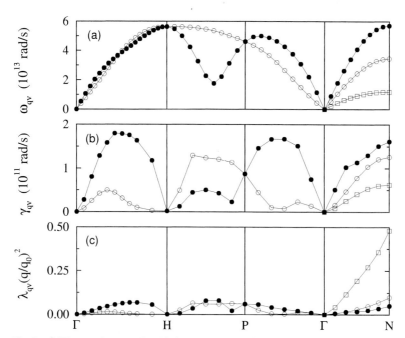

Fig. 2. a) Phonon frequencies, b) linewidths, and c) mass enhancement parameters calculated along high-symmetry directions in b.c.c. Li. Solid circles correspond to longitudinal modes, while open symbols indicate transverse modes. Along the $[\xi\xi 0]$ direction (Γ to N), the open squares mark the T_1 branch. The coupling parameters $\lambda_{\mathbf{q}\nu}$ are weighted by a phase-space factor of $(q/q_0)^2$, where $q_0 = 2\pi/a_0$

for these modes are comparable in magnitude to those for transverse modes along other directions, and since $\lambda_{\mathbf{q}\nu}$ is proportional to $\gamma_{\mathbf{q}\nu}/\omega_{\mathbf{q}\nu}^2$, where $\gamma_{\mathbf{q}\nu}$ does not depend explicitly on $\omega_{\mathbf{q}\nu}$, the large values of $\lambda_{\mathbf{q}\nu}$ associated with the $T_1[\xi\xi 0]$ modes arise primarily from the small phonon frequencies rather than from Fermi-surface nesting or other properties related to the electronic structure. In the 9R structure, there are no comparable low-lying phonon branches that contribute anomalously to λ.

Using the calculated $\alpha^2 F$ for b.c.c. and 9R Li, we have solved the Eliashberg equations to find T_c as a function of μ^*. For b.c.c. Li, a value of $\mu^* = 0.27$ is needed in order to suppress T_c below the experimental limit of 6 mK. For 9R Li, the transition temperature is found to be consistent with the experimental limit if $\mu^* = 0.19$. For both phases, the upper limit on μ^* of $[\ln(E_F/\omega_{\max})]^{-1}$ is about 0.23. This confirms that the large value needed to suppress T_c below the experimental limit in b.c.c. Li is most likely unphysical. For the 9R structure the electron–phonon interaction is weak enough to be consistent with the lack of a transition above 6 mK, provided that the Coulomb pseudopotential is somewhat larger than is usually assumed. In a recent paper, static momentum-dependent dielectric matrices calculated within the LDA were used to estimate the Coulomb repulsion parameters in b.c.c. and 9R Li [29]. The estimates of $\mu^* \approx 0.16$ for both b.c.c. and 9R Li indicate that the standard values assumed for μ^* may indeed be underestimates for this material.

3.2 Superconductivity in compressed Sulfur

Sulfur, an insulating molecular solid under ambient conditions, undergoes a series of structural phase transitions when compressed, becoming metallic at pressures above 90 GPa [30]. Recent experiments have found metallic phases of compressed S to be superconducting, with transition temperatures that increase from 10 K at 93 GPa to 14 K at 157 GPa [20]. Near 160 GPa, the crystallographic transformation to the rhombohedral β-Po structure is accompanied by a jump in T_c to 17 K, which is the highest T_c observed among elemental solids. Earlier first-principles calculations had predicted that β-Po S would transform to the b.c.c. structure at 550 GPa, and that this b.c.c. phase would be superconducting with T_c around 15 K [31].

Here we present results for the electron–phonon coupling in S in an intermediate pressure range. Parameters characterizing the phonon spectrum and electron–phonon coupling in β-Po S at three pressures are listed in Table 2. At 160 GPa, where T_c has been measured to be 17 K, we find the electron–phonon mass enhancement parameter to be only moderately large at $\lambda = 0.76$. This is significantly smaller than experimental and theoretical estimates for λ in Nb, for example, which has a lower T_c of 9.2 K [32]. It is also smaller than estimates of λ for In, which superconducts at 3.4 K [18]. Our result for β-Po S, however, is consistent with the measured transition temperature because the scale of phonon energies in compressed S is very high, e.g., approximately a factor four (eight) that of Nb (In). We have solved the Eliashberg equations using the calculated spectral functions and find that in order to reproduce the measured value of T_c at 160 GPa, we need to assume $\mu^* = 0.11$, which is well within the range of typical values for this parameter.

The mass enhancement parameter λ for β-Po S does not vary much in going from 160 GPa to 200 GPa. The effect on λ of an overall increase in frequencies is offset by increases in the linewidths, due in part to an increase in $N(E_F)$ (see Eqs. (3) and (7)).

Table 2

Calculated values of electron–phonon coupling parameters for the β-Po and simple cubic phases of S at various pressures. The transition temperatures are determined by solving the Eliashberg equations with μ^* fixed at 0.11, which gives the measured $T_c = 17$ K in the β-Po structure at 160 GPa

P (GPa)	$N(E_F)$ (states/Ry/spin)	ω_{ave} (meV)	ω_{ln} (meV)	λ	T_c (K)
β-Po					
160	2.01	46.4	37.7	0.76	17.0
200	2.12	50.6	40.2	0.78	19.2
280	2.07	57.2	42.3	0.66	12.8
simple cubic					
280	2.20	71.8	48.1	0.53	7.4

At higher pressures, the average coupling strength falls due to a continued rise in the frequencies and a small decrease in $N(E_F)$. The weak pressure dependence of $N(E_F)$ suggests that it is reasonable to approximate μ^* as constant. With this assumption, we estimate that over the range of stability of the β-Po phase, T_c first follows the increase in ω_{ln}, which sets the scale for T_c, then decreases as the coupling weakens at higher pressures.

Contrary to earlier theoretical work [31], we predict that S does not transform directly from the β-Po to the b.c.c. structure near 550 GPa, but rather transforms from the β-Po phase to the simple cubic structure around 280 GPa, and then to the b.c.c. phase near 550 GPa [33]. Our calculations show that λ decreases to 0.53 upon transformation to the simple cubic structure. The weaker coupling in the simple cubic phase results primarily from a dramatic change in the phonon spectrum. The phonon density of states for β-Po and simple cubic S at 280 GPa are plotted in Fig. 3. The shift to higher frequencies in the simple cubic phase is due primarily to the openness of the lattice. Since very little charge occupies the large interstitial regions in the simple cubic lattice, the charge density along the bonds connecting neighboring atoms is large, producing a very stiff lattice. The reduction in λ in the simple cubic phase is attributable to the large overall increase in phonon frequencies. As can be seen from Table 2, the

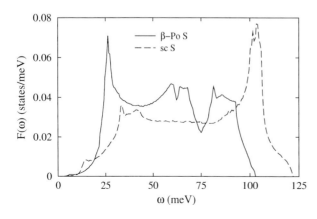

Fig. 3. Phonon density of states calculated for β-Po and simple cubic S at 280 GPa. The dynamical matrices were computed for 32 (35) **q** points in the β-Po (simple cubic) IBZ, and then interpolated onto denser grids. The phonon density of states was calculated using the interpolated results

increase in $\omega_{\ln}$ in going from the β-Po to the simple cubic structure is much smaller than the jump in the average phonon frequency. This is because $\omega_{\ln}$ is most strongly influenced by low-frequency modes of which there are fewer in the simple cubic phase. The net result is that the superconducting T_c is expected to fall upon transformation to the predicted simple cubic phase.

4. Conclusions

In summary, two applications of the accurate and efficient linear-response method for calculating electron–phonon coupling parameters from first principles have been presented. For Li, we find that the electron–phonon mass enhancement parameter is about 30% larger in the room temperature b.c.c. phase than in the low-temperature 9R phase. The low-lying shear modes along the [$\xi\xi 0$] direction have been identified as the primary source of the enhanced coupling in the b.c.c. phase. The coupling in the 9R structure is weak enough to be consistent with current experimental limits on T_c. In the case of compressed S, calculated phonon spectra and electron–phonon coupling parameters are consistent with the large superconducting transition temperature of 17 K that has been measured for β-Po S at 160 GPa. Since the coupling itself is only moderately strong, the large phonon energy scale plays a crucial role in raising T_c. Over the range of stability of the β-Po phase, we predict that the superconducting transition temperature initially rises slightly with pressure, then decreases down to about 13 K near the structural transition to the simple cubic phase. The transition temperature is expected to drop further upon transformation to the simple cubic structure.

Acknowledgements I am grateful to A. A. Quong, S. P. Rudin, R. Bauer, J. K. Freericks and E. Nicol for their contributions to this work. This work was supported by the Clare Boothe Luce Fund, the National Science Foundation under grant number DMR-9627778, and the NPACI.

References

[1] G. GRIMVALL, The Electron-Phonon Interaction in Metals, North-Holland, Amsterdam 1981.
[2] G. M. ELIASHBERG, Zh. Eksper. Teor. Fiz. **38**, 966 (1960) [Soviet Phys. – JETP **11**, 696 (1960)].
[3] P. B. ALLEN, Phys. Rev. B **5**, 2577 (1972); in: Dynamical Properties of Solids, Eds. G. K. HORTON and A. A. MARADUDIN, North-Holland, New York 1980.
  P. B. ALLEN and B. MIKOVIC, in: Solid State Physics, Eds. H. EHRENREICH, F. SEITZ, and D. TURNBULL, Academic Press, New York, 1982.
[4] J. P. CARBOTTE, Rev. Mod. Phys. **62**, 1027 (1990).
[5] W. L. MCMILLAN and J. M. ROWELL, Phys. Rev. Lett. **14**, 108 (1965); in: Superconductivity, Vol. 1, Ed. R. PARKS, Dekker, New York 1969.
[6] E. L. WOLF, Principles of Electron Tunneling Spectroscopy, Oxford University Press, New York 1985.
[7] L. J. SHAM and J. M. ZIMAN, Solid State Physics, Vol. 15, Eds. H. EHRENREICH and D. TURNBULL, Academic Press, New York 1963.
[8] G. D. GASPARI and B. L. GYORFFY, Phys. Rev. Lett. **28**, 801 (1972).
[9] B. M. KLEIN and W. E. PICKETT, in: Superconductivity in d- and f-Band Metals, Eds. W. BUCKEL and W. WEBER, Kernforschungszentrum, Karlsruhe 1982.
[10] H. WINTER, J. Phys. F **11**, 2283 (1981).
  D. GLOTZEL, D. RAINER, and H. R. SCHOBER, Z. Phys. B **35**, 317 (1979).
  H. KRAKAUER, W. E. PICKETT, and R. E. COHEN, Phys. Rev. B **47**, 1002 (1993).
[11] P. K. LAM, M. M. DACOROGNA, and M. L. COHEN, Phys. Rev. B **34**, 5065 (1986).

[12] S. Baroni, P. Giannozzi, and A. Testa, Phys. Rev. Lett. **58**, 1861 (1987).
[13] S. Y. Savrasov, D. Y. Savrasov, and O. K. Andersen, Phys. Rev. Lett. **72**, 372 (1994).
[14] A. A. Quong and B. M. Klein, Phys. Rev. B **46**, 10734 (1992).
    A. A. Quong, A. Y. Liu, and B. M. Klein, in: Materials Theory and Modeling, Eds. J. Broughton, P. Bristow, and J. Newsam, MRS Symp. Proc. No. 291, Mater. Res. Soc., Pittsburgh 1993.
[15] P. Giannozzi, S. de Gironcoli, P. Pavone, and S. Baroni, Phys. Rev. B **43** 7231 (1991).
    A. A. Quong, Phys. Rev. B **49**, 3226 (1994).
    S. Y. Savrasov, Phys. Rev. Lett. **69**, 2819 (1992).
    S. de Gironcoli, Phys. Rev. B **51**, 6773 (1995).
    R. Yu and H. Krakauer, Phys. Rev. Lett. **74**, 4067 (1995).
[16] A. Y. Liu and A. A. Quong, Phys. Rev. B **53**, R7575 (1996).
[17] R. Bauer, A. Schmid, P. Pavone, and D. Strauch, Phys. Rev. B **57**, 11276 (1998).
[18] S. P. Rudin, R. Bauer, A. Y. Liu, and J. K. Freericks, Phys. Rev. B **58**, 14511 (1998).
[19] A. Y. Liu, A. A. Quong, J. K. Freericks, E. J. Nicol, and E. C. Jones, Phys. Rev. B **59**, 4028 (1999).
[20] V. V. Struzhkin, R. J. Hemley, H.-K. Mao, and Y. A. Timofeev, Nature **390**, 382 (1997).
[21] N. E. Phillips, CRC Crit. Rev. Solid State Sci. **2** 467 (1971).
[22] N. Troullier and J. L. Martins, Phys. Rev. B **43**, 1993 (1991).
[23] E. Wigner, Phys. Rev. **46**, 1002 (1934).
[24] J. Perdew and A. Zunger, Phys. Rev. B **23**, 5048 (1981).
[25] S. G. Louie, S. Froyen, and M. L. Cohen, Phys. Rev. B **26**, 1738 (1982).
[26] D. A. Papaconstantopoulos, L. L. Boyer, B. M. Klein, A. R. Williams, V. L. Moruzzi, and J. F. Janak, Phys. Rev. B **15**, 4221 (1977).
    T. Jarlborg, Physica Scripta **37**, 795 (1988).
    A. Y. Liu and M. L. Cohen, Phys. Rev. B **44**, 9678 (1991).
[27] T. L. Thorp, B. B. Triplett, W. D. Brewer, M. L. Cohen, N. E. Phillips, D. A. Shirley, J. E. Templeton, R. W. Stark, and P. H. Schmidt, J. Low Temp. Phys. **3**, 589 (1970).
    K. M. Lang, A. Mizel, J. Mortara, E. Hudson, J. Hone, M. L. Cohen, A. Zettl, and J. C. Davis, J. Low Temp. Phys. **114**, 445 (1999).
[28] A. W. Overhauser, Phys. Rev. Lett. **53**, 64 (1984).
[29] Y. G. Jin and K. J. Chang, Phys. Rev. B **57**, 14684 (1998).
[30] H. Luo, S. Desgreniers, Y. K. Vohra, and A. L. Ruoff, Phys. Rev. Lett. **67**, 2998 (1991).
    H. Luo, R. G. Greene, and A. L. Ruoff, Phys. Rev. Lett. **71**, 2943 (1993).
[31] O. Zakharov and M. L. Cohen, Phys. Rev. B **52**, 12572 (1995).
[32] P. B. Allen, in: Handbook of Superconductivity, Ed. C. P. Poole, Academic Press, San Diego 1999.
[33] S. P. Rudin and A. Y. Liu, Phys. Rev. Lett. **83**, 3049 (1999).

phys. stat. sol. (b) **217**, 429 (2000)

Subject classification: 61.50.Lt; 68.45.Da; 68.45.Kg; 71.15.Mb; S5; S10.1

Modelling Carbon for Industry: Radiolytic Oxidation

P. Leary (a), C. P. Ewels (a), M. I. Heggie (a), R. Jones (b), and P. R. Briddon (c)

(a) *Department of Chemistry, Physics and Environmental Science,*
University of Sussex at Brighton, Falmer, Brighton, Sussex, BN1 9QJ, UK

(b) *Department of Physics, University of Exeter, Exeter, EX4 4QL, UK*

(c) *Department of Physics, University of Newcastle, Newcastle, NE1 7RU, UK*

(Received August 10, 1999)

An *ab initio* density functional technique (AIMPRO) has been employed to investigate the structure, vibrational properties, and dissociation mechanisms of CO_3^0, the important radical anion CO_3^- and the interaction of this species with the graphite basal plane. The results are discussed in the context of the radiolytic oxidation of graphite: a process of relevance to the British nuclear industry, which relies for the most part on graphite-cored, CO_2-cooled reactors. The radiation field splits coolant molecules and produces, amongst other things, a very reactive radical anion CO_3^-, which has been suggested as the main agent for the accelerated oxidation of graphite. This paper shows that CO_3^- binds strongly to graphite after combining with an electronic hole and forming a long and strong ionic bond. It still remains mobile on the basal plane and can diffuse to a graphite edge and oxidize it.

1. Introduction

The radiolytic oxidation of graphite leads to the abstraction of basal carbon atoms by an active gaseous species [1, 2]. A random distribution of vacancies is formed by the oxidation of a basal C atom by a particularly aggressive gaseous species. Although some gaseous species are able to remove basal carbon atoms forming vacancies, the dominant mechanism for mass loss was found to be via the reactive edge sites (edge recession) and vacancies (vacancy ring expansion). Both the concentration of these vacancies, and the vacancy ring-radii are proportional to the gas pressure [1]. Due to the chemical similarity of vacancy rings and step-edges, the ring expansion and step-erosion rates are found to be almost identical. Significantly, on areas segregated by a number of parallel steps, the basal erosion rate was found to be proportional to the exposed area of that basal plane; this observation suggested the following: an active species, responsible for edge oxidation was first trapped on the basal plane before migrating over the surface to the reactive sites (edges, vacancies) [1, 3]. The hypothesis that the step erosion occurs as a result of a surface reaction involving a species diffusing on the basal plane, and not from a direct attack from the gas phase, was confirmed by careful measurements of the edge positions following the reaction [1]. Therefore, at least for this edge mechanism the oxidizing species should be highly mobile on the graphite surface. Both CO and CO_2 are desorbed in the gasification reaction, and these desorptions are found to occur at well defined energies [4]. For the reaction in the presence of radiolysed CO_2, the CO and CO_2 desorption energies were found to occur at energies of 132 and in the range 65 to 132 kJ/mol, respectively.

In summary, during the course of radiolytic oxidation, a plane one-layer thick is oxidized by a mechanism which involves the surface diffusion of the oxidizing species. This species is first trapped on the basal surface, before migrating to edges (or vacancy rings) where the abstraction of basal edge carbon atoms (gasification) takes place.

This paper is concerned with the adsorption of carbon-trioxide on the (0001) graphite surface. The motivation for this work relates to the radiolytic oxidation of graphite, in which the CO_3^- radical anion is believed to play a crucial role [5]. This adsorbate-substrate system is not only of profound technological importance; as will be illustrated in the following sections, but it also exhibits a novel and unique bonding mechanism.

Before considering the bound system, we must summarize the previous work on both isolated CO_3^0 and CO_3^-. It is also informative to briefly discuss CO_3^q as an adsorbate and graphite as a substrate, to outline the type of bonding which can arise. For these purposes, we choose three model systems which have been considered in the past, and are well understood: the interaction of CO_3 on alkali-metal surfaces is briefly described, followed by the adsorption of oxygen and then potassium on graphite. To varying extents, these systems exhibit similar chemical and physical bonding effects to the system we go on to study – CO_3 on graphite.

1.1 Isolated CO_3

Carbon trioxide (CO_3) and its radical anion CO_3^- play important roles in atmospheric chemistry. In particular, the latter was found to be present in the D-region [6, 7], and has a dominant role in the negative ion chemistry of the ionosphere [8]. The CO_3^- anion is formed from the interactions of O^- with CO_2,

$$CO_2 + O^- \rightarrow CO_3^-. \tag{1}$$

More recent work has implied that CO_3^- is crucial in the radiolytic oxidation of the graphite moderator of CO_2 cooled nuclear reactors [5]. In this process the O^- arises from γ-irradiation photodissociation of CO_2.

A large amount of work, both experimental and theoretical, has been performed on molecular $CO_3^{-/0}$ in the past, and a more complete investigation by our methods is the subject of another paper [9]. Brief comments pertaining to the previous results are made here.

The neutral molecule was first detected directly by vibrational mode data through Fourier Transform Infra-Red Spectroscopy (FTIR) [10]; the molecule being trapped in a solid CO_2 matrix. Five vibrational frequencies were observed, and a detailed study of the isotope shifts of these suggested a C_{2v} structure [11]. This neutral molecule has a relatively low stability, dissociating into CO_2 and O with an activation energy of (0.4 ± 0.2) eV [12]. The vibrational frequencies of CO_3 are detailed in Table 1.

A number of theoretical investigations at various levels of theory have been performed on the neutral CO_3 molecule [13 to 18], and all favor either a D_{3h} or C_{2v} structure. An in-depth investigation into the relative energies of the D_{3h} and C_{2v} structures has shown that the energies of the closed shell states $^1A_1$ (C_{2v}) and $^1A_1'$ (D_{3h}) are very sensitive to correlation effects [16]; the ground state could switch between the two with various levels of theory. At the highest level of theory, the C_{2v} structure was the ground state; nevertheless, the energy differences between the C_{2v} and D_{3h} structures were very small (0.2 eV). The vibrational frequencies of CO_3 have also been calculated (given in

Table 1

The observed and calculated vibrational frequencies (in cm^{-1}) for the isolated CO_3 and CO_3^- molecules ($^{16}O^{16}O^{12}C^{16}O$)

symmetry	ref. [11]	ref. [19]	ref. [16]	ref. [23]
A_1	1981.1	2138	2069	1307
A_1	1073.4	1144	1086	–
B_1	971.9	754	1099	1494
A_1	593.2	720	662	–
B_1	568.2	559	559	–
B_2	–	630	665	–

Table 1) for the C_{2v} structure [19, 16], and the results compare favorably to the experimental ones [11].

The first observation of CO_3^- was in the solid state, after γ-irradiation of $KHCO_3$ crystals [20], and their results suggested that the symmetry was lower than D_{3h}. Other work on CO_3^- trapped in a solid matrix supported the distorted (C_{2v}) structure [21 to 23]. In particular, the IR spectroscopic data of CO_3^- in solid Ar [23] unambiguously implied a C_{2v} structure. These modes of vibrations are given in Table 1, and are very different to those of the neutral molecule.

CO_3^- in the gas phase was first detected in drift tube studies [24], and three body reaction experiments have established that reaction 1 occurs extremely rapidly even at 200 K [25, 26]. The reaction between O_2^- and CO has also been observed, but occurs at a much slower rate [27]. Many investigations into the photodissociation of CO_3^- via the following reaction:

$$CO_3^- + h\nu \rightarrow CO_2 + O^-, \qquad (2)$$

have also been reported [28 to 31], and an accurate measurement of (2.258 ± 0.008) eV has been made for this process [29]. The magnitude of this dissociation energy compared to that of the neutral molecule implies that the capture of an electron significantly stabilizes the molecule. This result implied an electron affinity of (3.26 ± 0.17) eV for CO_3, significantly different to the value of 2.69 eV determined in [32].

Early calculations on CO_3 yielded a C_{2v} structure [33, 34], in agreement with the experimental findings. Later work at higher levels of theory focused on the structure and dissociation mechanism of CO_3^- [31]. A D_{3h} structure was obtained, along with a dissociation energy of 2.6 to 2.88 eV in good agreement with the experimental value of 2.27 eV. The clustering of CO_2 around CO_3^- has recently been considered [35]; there the ground state for the isolated anion possessed C_{2v} symmetry.

In summary, the ground state structure of the gas phase $CO_3^{0/-}$ molecules is still a matter of contention. The structure obtained from experiment is of C_{2v} symmetry in both charge states, with the unique OCO angle being <120° for the neutral defect, and >120° for the negative one. The theoretical calculations are not in agreement with one another, with both C_{2v} and D_{3h} structures being found as the ground state in both charge states (the energy differences often being close, and the sign depending on the level of theory). Experimentally, the structure of gas phase $CO_3^{0/-}$ is not known: the C_{2v} structure for both charge states came from solid state measurements, and it has been suggested that crystalline imperfections, and induced polarization of the CO_3^- by the nearby cations may result in a distortion of the molecule. This polarization induced splitting has been shown to be the case for NO_3^- which is known to be a D_{3h} molecule, but appears C_{2v} in a solid argon matrix [36]. There, the splitting in the two stretching fundamentals is 171 cm^{-1}, very close to the splitting observed in CO_3^- of 187 cm^{-1}.

1.2 CO_3 as an adsorbate

The model system we choose to illustrate the properties of CO_3^q as an adsorbate is the trapping/formation of this complex over alkali-metal surfaces. This process does not follow from a direct attack by a charged CO_3 species, as is hypothesized for radiolytic oxidation, but instead from the activation of CO_2. The process is described thoroughly in the work of Hoffmann et al. [37] and here we briefly summarize their results.

Neutral CO_2 is adsorbed on alkali-metal surface at temperatures lower than 100 K, and is relatively weakly bound. An increase in the binding is achieved by the formation of an ionically bound CO_2^-; a result of the electron donating properties of the substrate. The interaction of a further CO_2 molecule with a bound CO_2^- ultimately leads to the formation of an ionically bound complex, $C_2O_4^{2-}$ with the capture of a second electron. This species is stable to around 300 K, and penetrates into the alkali-metal layer. In this sense, the complexes formed here are neither truly surface or bulk bound, since they do not wholly penetrate into the surface of the alkali-metal. Following the dimerization of the CO_2^- ions, forming the oxalate ion $C_2O_4^{2-}$, conversion to a surface carbonate can proceed via the ejection of CO from the surface: $C_2O_4^{2-} \rightarrow CO_3^{2-} + CO$. This process has only a very small reaction barrier; and the resultant CO_3^{2-} is strongly bound and stable to around 650 K. The bound complexes formed were identified using FTIR spectroscopy, and the vibrational spectra obtained for the carbonate agreed very well with those of gas phase CO_3^{2-} and solid K_2CO_3. A combination of the experimental results and theoretical considerations led to the conclusion that the bonding of CO_3^{2-} was non-directional, with the molecules described as ionic-alkali coordination complexes whose interaction was predominantly electrostatic.

The bonding of CO_3^q to this substrate can thus be understood in terms of the electron affinity of the molecule, and the ionization potential of the metal. The net result of the transfer is an extremely stable adsorbate which has a very strong electrostatic attraction to the metal surface.

1.3 Graphite as a substrate

Many previous investigations have considered adsorbates on graphite, since the (0001) surface is ideal in many respects: the material has a chemically inert and large adsorption surface, which is extremely homogeneous and has a low concentration of defects. The task of discussing chemisorption on the semi-metallic graphite is basically reduced to two possible bonding mechanisms: (i) a covalently bonded system, where one or more of the surface C atoms becomes sp^3 like, forming strong covalent bonds; this has the effect of reducing the resonance energy of the system, and (ii) an ionically bonded system, whereby electronic charge is transferred to the adsorbate from the graphite π levels, or from the adsorbate into the graphite π^* levels, with a corresponding electrostatic attraction between the ion and its image charge in the basal plane.

For the first bonding mechanism, we describe the previous theoretical considerations of the adsorption of atomic oxygen on graphite. This represents an ideal system since

this complex has implications for our results which will be discussed in the later sections. An understanding of the ionic bonding to graphite also facilitates an appreciation of the later results, and for this purpose, the binding of potassium ad-atoms to the basal plane is briefly discussed.

1.3.1 Oxygen on the basal plane

The interaction of graphite with both O and O_2 ultimately forming graphite oxide is a very diverse subject. Here we restrict ourselves to a brief discussion of the previous investigations into the structure and binding of atomic oxygen on the basal plane.

Previous theoretical calculations into this system [38, 39] (based on MINDO/3 and DFT-LDA, respectively) both find an epoxy structure where the oxygen atom forms two covalent bonds with adjacent C atoms. The C–C bond is not broken, the COC atoms forming a triangle, with the O atom being around 2.00 Å from the basal plane, corresponding to C–O bonds of length 1.42 Å [39]. In both calculations, the binding energy of O was very high, 2.41 and 3.18 eV, respectively. Spin dependencies were neglected in the latter, and referring the binding energy to the triplet ground state of atomic oxygen would lead to a substantial reduction in the reported binding.

When the oxygen lays above one of the C atoms, an energy increase of 0.07 [38] and 0.36 eV [39] was obtained. This relatively small value suggests that the barrier for O migration over the surface is low, and therefore the atoms could be mobile on the basal plane around 430 °C. The migration to defect sites or edges can result in the emission of CO/CO_2 and erosion of the material.

1.3.2 Potassium on the basal plane

There is currently a large amount of interest in the adsorption of alkali-metals on graphite, and the intercalation compounds which can be formed; despite the chemical similarities of these in their bulk phases, alkali-metals display a range of differing properties on graphite. Of these, the sub-monolayer of potassium has been extensively studied both experimentally and theoretically.

At relatively large coverages, an ordered phase termed (2×2) is formed, having one K atom per eight C atoms; with the K atoms residing over the "hollow" sites at the center of graphitic hexagons [40]. The spacing between atoms in this phase is $2a_0$, where a_0 is the basal lattice parameter. At very low coverages, a dispersed phase exists, with a K–K distance of ≈ 60 Å [41]. These lower coverages are more typical of an *isolated* ad-atom, and again, the K atom resides over the hollow site. The experimental results suggested that a considerable amount of charge, $0.7e$, was transferred from the K atom to the graphite resulting in the formation of an ionically bound system [41]. It was also noted that the charge transfer to the substrate decreased as the coverage increased to that of the (2×2) phase, which is more reminiscent of a quasi-metal.

Recent theoretical calculations [42, 39] support the above picture. In these calculations a single graphite monolayer was considered, although some calculations [39] were extended to multiple layered systems, where the results were very similar. These calculations were based on LDA-DF supercell techniques, and coverages of $\frac{1}{4}$ and $\frac{1}{16}$ K atoms per C atom were considered. The lowest energy configuration was with the K atom above the hollow site, as observed experimentally. For the (4×4) phase, the K atom was a distance of 2.72 Å (2.80 Å) from the basal plane, with a corresponding binding energy of ≈ 0.9 (≈ 0.5) eV from [42] ([39]). The binding energy and height above the basal plane were fairly insensitive to the adsorption site, the largest fluctuations occurring at the bridge site (between two C atoms) where the energy was 0.05 eV above the ground state, and the difference in ad-atom height was ≈ 0.1 Å [39]. Clearly, this large distance (in good agreement with the observed value of around 2.70 Å for the (2×2) phase) and the binding energy is consistent with the formation of an ionic bond. An estimate of $q = 0.28e$ ($0.40 \, e$) was obtained for the charge transfer in the (4×4) phase broadly consistent with the experimental results [42] ([39]).

The (0001) surface of graphite is very versatile, taking part in a range of bonding mechanisms with different adsorbates; a result of the semi-metallic nature of the material.

2. Theoretical Method

The calculations presented here, are based on ab initio local density functional cluster theory utilizing a Gaussian basis expansion AIMPRO [43, 44], and for a qualitative theory of resonance: simple Hückel theory [45].

The atomic structure of both CO_3^q and graphite, as well as the complexes formed between the two, is investigated using a first principles cluster method. For most calculations, a single layer polyaromatic sheet was used to model graphite, the justification for the use of the graphene sheet is given in the following section.

The first principles calculations are performed under the local density approximation (LDA), and norm-conserving pseudopotentials of [46] are used, removing the need to treat the core electrons explicitly. Our total energies therefore only relate to the "valence" electrons due to the omission of the chemically inert core electrons.

A basis consisting of 32 Gaussian functions for each CO_3 carbon and oxygen atom was used to model the molecular wavefunctions (8 each of s-, p_x-, p_y- and p_z-), and 8 s-type Gaussian functions were used on all atoms for the valence charge density. In addition to this, 2s-, p_x-, p_y- and p_z-like Gaussian functions were placed at the center of every bond in the wavefunction basis with 2 s-like Gaussian functions being used for the charge density basis.

The self-consistent energy E and the force on each atom are calculated and the atoms moved by a conjugate gradient algorithm until equilibrium was attained. In some cases the second derivatives of E between selected atoms were evaluated in the way described previously [43]. From these, the dynamical matrix can be set up, and the quasi-harmonic local vibrational modes found.

The theoretical method employed here has been successfully applied to a wide range of problems in chemistry and solid state physics: examples include the modeling of fullerenes [47], diamond surfaces [48], and oxygen related point defects in crystalline silicon [49].

In the following section, a comparison of the electronic structure of the graphene molecules from both the ab initio calculations and the Hückel theory is made.

3. Graphite

In this section, we demonstrate that the cluster model is a reasonable approximation for the graphite basal plane, and discuss in general terms the effect of cluster topology on electronic structure.

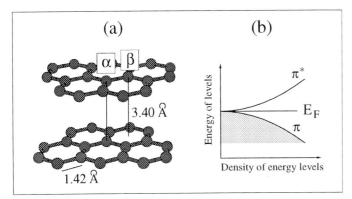

Fig. 1. A simple illustration of the physical properties of graphite: a) the structure and structural parameters are shown in a two layered portion of graphite. The α and β sites for the inequivalent carbon atoms in the AB stacking are marked, and b) a schematic diagram of the density of states close to the Fermi level, E_F, in bulk graphite which illustrates the semi-metallic character

The carbon atoms in graphite adopt a planar layered structure, as illustrated in Fig. 1a. Strong covalent bonds are formed between the atoms within each layer; each carbon atom possessing three nearest neighbors in an hexagonal arrangement. The C–C bond lengths are 1.42 Å, and the sp^2 bonding results in a delocalized π-cloud from atom centered orbitals orthogonal to the plane, p$_z$. Bulk graphite is built up by adding many layers parallel to one another. The distance between layers of 3.35 Å is much longer than the intra-layer bonding distance, and is a result of the much weaker van der Waals like bonding between the planes, or to a weak band structure term [50].

The chemically important aspects of the electronic structure of graphite, and the polycyclic aromatic hydrocarbons we employ, can be simply understood in terms of the sp^2 hybridization and linear combinations of the unhybridized p-orbital (we label p$_z$) within simple Hückel theory (Section 3.1). Such a theory neglects the σ and σ* bands which lie well below and above E_F, respectively. Delocalized π-orbitals dictate the chemistry of the molecule or graphite: for an infinite (2D) single layer of graphite, the electronic structure corresponds to a zero-band gap semiconductor, with a zero density of states at E_F; for crystalline graphite, the small interaction between the layers results in a limited overlap of the π and π* bands, and therefore in a weakly conducting material (semi-metal). The graphene density of states is illustrated in Fig. 1b.

Our cluster model for graphite is a single layered polycyclic aromatic hydrocarbon C_XH_Y. The H-atoms passivate the dangling bonds of σ character, leaving only the extended π system contributing to the density of states around E_F. Typically our calculations involve clusters where $X \approx 60$ and $Y \approx 20$. These finite clusters inevitably break down the π-delocalization over the sheet, and introduce a tendency toward alternating double–single bonds most notably at the edges. For the $C_{62}H_{20}$ and $C_{61}H_{21}$ cluster (considered using AIMPRO) the extent to which this occurs can be seen in the bond lengths across the sheet; after relaxation the core C–C bond lengths are 1.41 Å (within ±0.005 Å) whereas at the sheet edge (i.e. C bonded to a H) the C–C bonds vary between 1.35 and 1.41 Å. Typically, C–C bonds on the *boat* edge are the shortest, consistent with a high bond-order for these edge carbon dimers. Nonetheless, the average C–C length within the cluster remains at 1.41 Å (in good agreement with the experi-

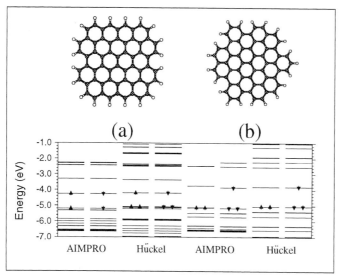

Fig. 2. The two clusters used in most of the calculations presented here: a) $C_{62}H_{20}$ and b) $C_{61}H_{21}$. The Kohn-Sham electronic levels (both spin up and down) and those calculated from the Hückel theory are also given to illustrate their nature, and to demonstrate the similarity of the results of both theories when applied to such hydrocarbons. For a direct comparison, we rigidly shifted all the Hückel levels down so that the Fermi levels from both calculations matched (implying $\alpha \approx -3.8$ eV)

mental value of 1.42 Å, an error of less than 1%). This variation can also be observed in the bond orders from the Hückel theory.

Another consequence of the use of finite clusters is that our "graphite" clusters are semiconducting or metallic depending on the radical nature of the sheet. In particular, throughout this work we predominantly focus on two clusters: $C_{62}H_{20}$ and $C_{61}H_{21}$, shown in Fig. 2a, b. A critical difference exists between the two clusters — for the neutral charge state, the HOMO is fully occupied in the former and half-filled in the latter. There are two reasons for duplicating the calculations in two clusters: 1. graphite is a semi-metal — our two clusters are semiconducting and metallic, respectively, and their behavior might be expected to bracket the behavior of graphite. Obviously, as the number of atoms in the cluster approaches infinity, the density of states approaches that shown in Fig. 1b. 2. in modeling erosion of graphite by removal of one carbon atom at a time, reactions will occur alternately on semiconducting and radical substrates. Either substrate could give rise to a rate limiting step.

Several of the physical properties of graphite are very well reproduced using our method. As well as the bond lengths which have already been discussed, the calculated electron affinity and ionization potential of $C_{62}H_{20}$ are 2.1 and 5.5 eV, in fair agreement with the known work-function (5 eV) of graphite. Similarly, using the radical sheet $C_{61}H_{21}$ values of 2.5 and 5.1 eV, respectively, are determined. Comparing the total energies of two polyaromatics which differ only in the number of carbon atoms, a cohesive energy of ≈ 7.4 eV has been determined for graphite [48], and is in good agreement with the experimental value of 7.34 eV [51].

In making the one-layer approximation, the inequivalence on the α and β sites is also neglected i.e. alternate carbons in one plane reside above carbon atoms and hexagonal sites in the adjacent plane (in AB stacking). However, the resulting slight inequivalence of the basal carbon atoms is likely to be small.

3.1 The electronic properties of polycyclic aromatic hydrocarbons

The electronic nature of the polycyclic aromatic sheets is an interesting topic in itself, depending both on cluster size and shape of the sheet. The π-electron properties of large condensed polyaromatic sheets have been previously examined using Hückel theory [52]. There, the authors investigated the electronic structure of hexagonal sheets with various edge structures; the [1$\bar{1}$00] *zig-zag* and [2$\bar{1}\bar{1}$0] *boat* edge being the two extremes. Here we summarise our work embracing general cluster shapes and various compositions of the two basis edge types.

We find: (i) The HOMO-LUMO gap diminishes more rapidly with cluster size for clusters with predominantly the *zig-zag* edge, (ii) sheets possessing exclusively the *zig-zag* edge generally give rise to a finite density of states at E_F, and the localization of the HOMO/LUMO charge density is on that edge, (iii) sheets possessing the *boat* edge only have the widest HOMO-LUMO gap for any given size, and no edge states at or around E_F exist; such clusters have the greatest resonance stabilization. Sheets of mixed edge composition give rise to an electronic structure intermediate between the two extremes although the HOMO and LUMO are still exclusively localized on the *zig-zag* portions of the edge.

The following therefore represents the HOMO-LUMO gap size dependence for polycyclic aromatic hydrocarbon clusters: for a given cluster size, the HOMO-LUMO gap is smallest (often zero) when the ratio n[zig-zag]/n[boat] is largest; conversely if n[zig-zag]/n[boat] ≈ 0 then the gap is largest for the particular cluster size (the ratio can never equal zero or one due to the corners).

The results for hexagonal clusters differ from the above trends in that there is no peak in the density of states at E_F for *small* molecules possessing only the *zig-zag* edge. Indeed, even when $N \approx 1000$ (N denoting the number of C atoms) an appreciable gap of 0.5 eV exists. This result was also noted in [52] and a graph of the HOMO-LUMO gap with cluster size was presented. In contrast to this result we find that for other topologies (triangular clusters for example) a level, or band of levels, *always* exists at E_F even as $N \to 0$. However, in small clusters a large energy difference still exists between the level/band at E_F and the next nearest level; and this closes very slowly with increasing cluster size. Nevertheless, even the smallest triangular clusters possessing the *zig-zag* edge remain metallic in character.

Elaborating further on the these novel triangular clusters, the smallest ones in the series: $C_{13}H_9$, $C_{22}H_{12}$, $C_{33}H_{15}$, $C_{46}H_{18}$, ... — all have states at E_F, and the degeneracy of this state increases with increasing cluster size. The next nearest state to that at E_F is at least 2 eV away for these clusters and this difference decreases slowly with cluster size. For small triangular molecules with a *boat* edge termination the gap size is very large and again decreases slowly with cluster size.

The localized state for the *zig-zag* edge, and the peak in the density of states at E_F is undoubtedly a feature of that edge type, and not a result of cluster topology. Both these features in the electronic activity have been reported previously in the literature for large hexagonal clusters [52] and for graphene ribbons [53].

The electronic levels and localizations of the states near E_F are also compared for both Hückel theory and AIMPRO. This is illustrated for the two clusters we proceed to use in further sections — $C_{62}H_{20}$ and $C_{61}H_{21}$ — illustrated in Fig. 2, and the results are almost identical (after adjustment of the Hückel Coulomb integral). This was also true of every other cluster in which we made the comparison, whether wide gap semiconducting or metallic. These results therefore show that finite single-layered sheets with a range of electronic properties can be constructed by varying the edge composition and that Hückel theory is a robust and quick test of sheet properties.

Due to a small interaction between different layers in graphite, there is a slight overlap between the π and π^* bands. Such an interaction cannot be investigated using simple Hückel theory, and we therefore revert to first principles calculations. These show that the electronic levels of bi-layered sheets can be understood in terms of those of each layer in isolation (and therefore Hückel theory), provided an allowance for electron–electron interactions between the sheets is made. If one superimposes the eigenvalues of the two separated sheets, and allows a moderate splitting (up to 0.4 eV) of the levels (to account for e^-–e^- interactions), then the resulting levels closely match those of the bi-layered system. This approach is valid for any number of layers.

Consequently, when two-layer clusters are used, the following represents their behavior: when two semiconducting clusters are brought up to one another, the resultant double-layer model is still semiconducting, and the band-gap is slightly smaller than the isolated sheet; on the other hand, if two radical (metallic) clusters are joined, then the double-layer cluster formed is also semiconducting, but possesses a much smaller band gap. As far as the basic electronic properties are concerned, for clusters of the size we can consider there is no real advantage in using a multiple layered cluster; indeed the results of [39] demonstrate that the essential properties of both ionic and covalent mechanisms are well described by one sheet.

4. Isolated CO_3 and CO_3^-

4.1 Geometry of CO_3 and CO_3^-

A full geometry optimization of CO_3 and CO_3^- from several starting structures has been performed, both with and without symmetry and bond length constraints. Considering the neutral molecule first, two stable atomic structures were found, possessing D_{3h} and C_{2v} symmetries. The first of these represented the ground state, having three equivalent C–O bonds of length 1.256 Å. The metastable structure having C_{2v} symmetry was 0.076 eV less stable, and had one short and two long bonds of length 1.175 and 1.328 Å, respectively and a unique OCO bond angle of 77.8°. The global minimum of CO_3^- was the D_{3h} structure, and no local minima were found. The C–O bond lengths were 1.279 Å, slightly longer than those of the neutral D_{3h} molecule. Calculations with constrained bond angles confirmed that the above structures were the only minima on the potential energy surface, the variations in energy are illustrated in Fig. 3. From these results, it is apparent that the neutral molecule is relatively floppy, and the molecule stiffens significantly upon the addition of an electron.

The difference in total energy between CO_3^0 and CO_3^- is 4.83 eV, which represents the electron affinity of CO_3^0. This result is in relatively poor agreement with the experimental value of 2.7 to 3.3 eV [29, 32]; but preliminary work with higher levels of theory suggests the experimental values are low. The dissociation of the molecules in the gas phase can be trivially calculated by comparing the energies of the molecule and the

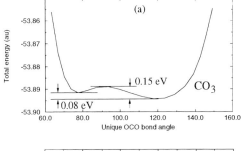
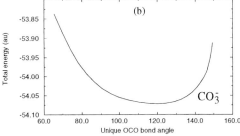
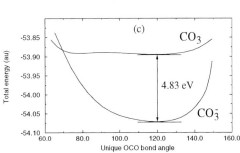
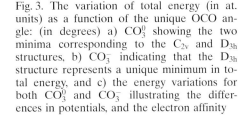

Fig. 3. The variation of total energy (in at. units) as a function of the unique OCO angle: (in degrees) a) CO_3^0 showing the two minima corresponding to the C_{2v} and D_{3h} structures, b) CO_3^- indicating that the D_{3h} structure represents a unique minimum in total energy, and c) the energy variations for both CO_3^0 and CO_3^- illustrating the differences in potentials, and the electron affinity

reaction fragments (atomic O and O_2 being in their triplet ground states). For CO_3^0, the lowest energy dissociation path is $CO_3 \rightarrow CO_2 + O$, which is endothermic by 0.508 eV, commensurate with the experimental estimate (0.2 to 0.4 eV, [12]). Similarly, the lowest energy dissociation pathway for CO_3^- was found to be $CO_3^- \rightarrow CO_2 + O^-$, which required 3.14 eV. This is the same as the observed photodissociation mechanism of CO_3^- where the bond energy was 2.27 eV [29]. Our result is in fair agreement with the experimental one, and such apparent overbinding is not untypical of LDA. In general our results agree well with experiment as far as the energetics are concerned: CO_3 has a relatively small binding energy, and a large reduction in energy is achieved by the capture of an electron: The formation of the anion significantly stiffens and strengthens the molecule.

4.2 Vibrational frequencies of CO_3^0 and CO_3^-

The vibrational frequencies of CO_3^0 [11] and CO_3^- [23] as measured in solid matrix of CO_2 and Ar:K, respectively, are given in Table 1 along with those calculated previously for CO_3^0.

Here, we calculate the vibrational frequencies of CO_3^0 (C_{2v} and D_{3h}) and CO_3^- (D_{3h}). Our calculated isotope shifts are to be detailed in a separate paper [9]. Considering the results of the C_{2v} CO_3^0 molecule first (Table 2) it can be seen that these results agree very well (most within 20 cm^{-1}) with the experimentally measured ones: 1981.1 A_1, 1073.4 A_1, 971.9 B_1, 593.2 A_1, 568.2 A_1 (experiment, cm^{-1}); and 1940.3 A_1, 1091.5 A_1, 951.6 B_1, 886.0 A_1, 586.2 B_2, 540.0 A_1 (calculated, cm^{-1}). The frequencies associated with the neutral D_{3h} molecule (our ground state) are also listed, and have not been observed. It remains possible that our calculated ground state is incorrect, since the

Table 2

The calculated vibrational frequencies (AIMPRO) (in cm^{-1}) for the isolated CO_3 and CO_3^- molecules ($^{16}O^{16}O^{12}C^{16}O$)

symmetry	CO_3 (C_{2v})	CO^3 (D_{3h})	CO_3^- (D_{3h})
A_1	1940.3	1507.1	1358.6
B_1	951.6	1507.1	1358.6
A_1	1091.5	1119.7	1047.1
A_1	886.0	1032.5	932.4
B_1	586.2	335.1	489.9
B_2	540.0	335.1	489.9

energy differences are small, and it has already been noted that the energies are extremely sensitive to different levels of theory. Additionally, the solid CO_2 matrix may affect the potential energy surface in such a way as to favor the C_{2v} structure.

An apparent discrepancy arises for the vibrational frequencies of the CO_3^- radical anion. Experimentally, the molecule appears C_{2v}, with the two stretching vibrations appearing at 1494.0 cm^{-1} B_1 and 1307.2 cm^{-1} A_1 (Table 1). Our calculated structure for CO_3^- has D_{3h} symmetry, and therefore leads to a degenerate mode at 1358.6 cm^{-1} (Table 2). The potential energy surface was very steep around D_{3h} CO_3^- and no C_{2v} structure was found from our calculations; however, it has already been noted that interactions between the molecules and the solid matrix may result in a lowering of the molecular symmetry and hence the observed splitting.

5. Atomic Oxygen on the Basal Plane

In light of the tendency of atomic O to form covalent bonds with C, and the results of the previous investigations into atomic O on the graphite basal plane; we investigate the two most likely structures formed when $O_{(g)}$ adsorbs onto the basal plane, namely: a) epoxy oxygen, and b) above-atom oxygen. The atomic structure of both these complexes is illustrated in Fig. 4a, b respectively.

We perform calculations for both adsorbed O and O$^-$ on the basal plane; modeled by the $C_{62}H_{20}$ and $C_{61}H_{21}$ aromatic sheets. For the neutral adatom, the binding energy is referred to that of the isolated sheet and triplet atom, respectively.

For the neutral adatom, our calculated binding energies (Table 3) show that the epoxy site represents the ground state, with a binding energy of around 1.4 eV. This value is different from that quoted in many previous calculations, however, we note that if energies are incorrectly referred to $O(S=0)$ then our binding energy rises to

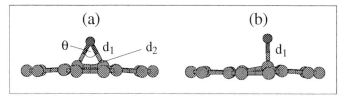

Fig. 4. The two atomic configurations for the chemisorbed O atom on graphite: a) epoxy structure, b) the above atom structure

Table 3
The binding energies (in eV) of atomic oxygen and O^- on the basal plane. For the neutral atom, the energies are referred to the triplet ground state; the binding energy with respect to the singlet state of atomic O are given in parenthesis

	O (epoxy)	O (atom)	migration barrier
$C_{62}H_{20}$	1.36 (3.04)	1.13 (2.81)	0.22
$C_{61}H_{21}$	1.44 (3.12)	1.28 (2.96)	0.16
	O^- (epoxy)	O^- (atom)	migration barrier
$C_{62}H_{20}$	1.21	1.54	0.33
$C_{61}H_{21}$	1.65	1.62	0.03

around 3.1 eV which is typical of previously quoted values. Such an error will be implicit in a spin-averaged LDA calculation. The above-atom atomic configuration represents a local minimum, being ≈0.2 eV higher in energy that the epoxy ground state. A similar value was noted in the previous modeling. While this cannot automatically be taken as the saddle point for O motion, it is indicative that basal oxygen should be highly mobile at elevated temperatures ($T \approx 300\,°C$), a result also noted in [39]. Basal oxygen (when present) is therefore likely to be of importance in various graphite oxidation mechanisms, although as we will demonstrate in the next section, this species can be solvated by CO_2.

For the epoxy structure, equivalent CO bonds of 1.54 Å are made, with a COC bond angle of 56.6 °C. Further details regarding the structure are given in Table 4. For both structures, the C atom(s) bonded to the O are raised by distances of 0.2 to 0.4 Å above the basal plane.

For the negative ion, O^-, the binding energy is still large, around 1.6 eV (Table 3). The difference in energy between the epoxy and above-atom structures remains small, and the results are qualitatively the same as those of the neutral atom: a strong binding is present, and the barrier to migration is very low.

Table 4
Structural parameters (in Å and degrees) for the epoxy and above-atom adsorbed oxygen on the basal plane. Refer to Fig. 4 a, b for the details of the labeling

epoxy	d_1	d_2	θ
$C_{62}H_{20}O$	1.539	1.458	56.5
$C_{62}H_{20}O^-$	1.585	1.446	54.3
$C_{61}H_{21}O$	1.537	1.463	56.9
$C_{61}H_{21}O^-$	1.539	1.467	56.9
atom	d_1		
$C_{62}H_{20}O$	1.555		
$C_{62}H_{20}O^-$	1.504		
$C_{61}H_{21}O$	1.503		
$C_{61}H_{21}O^-$	1.504		

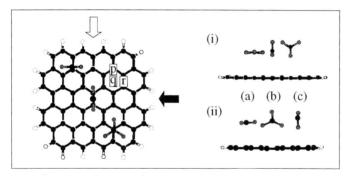

Fig. 5. Some possible atomic structures for chemisorbed CO_3 on the basal plane. Three structures are shown schematically over a $C_{62}H_{20}$ cluster. These are as follows: a) the parallel structure, b) the two-down structure, and c) the one-down structure. The sub-figures (i) and (ii) are views from the directions of the white [1$\bar{1}$00] and black [2$\bar{1}\bar{1}$0] arrows, respectively. In addition to the structures shown, the orientation of the molecule with respect to both the surface normal (rotations) and the lateral site (p, above-atom, q, above hexagon, and r, above bond-center) can also be varied

6. The Interaction between Graphite and CO_3

6.1 The structure of CO_3 on graphite

Before describing in detail the results for chemisorbed CO_3^- on graphite, a number of structures and bonding mechanisms which we investigate are outlined. Given the ionic nature of the radical anion, a bonding mechanism similar to that of K on graphite might be expected, and we investigate a number of structures where the molecule resides several Å above the basal plane (varying both lateral position of the molecule and its orientation with respect to the surface). Some possible ionic structures are illustrated in Fig. 5. The formation of covalent bonds between the oxygen atom(s) from the molecule to the basal plane is also considered, along with the breakup of the bound molecule to covalently bonded O and $CO_{2(g)}$.

The binding energy of the CO_3 molecule was investigated over a number of polycyclic aromatic sheets, and again we focus on the results from $C_{62}H_{20}$ and $C_{61}H_{21}$, denoted G′ and G″ respectively. The binding energies of the complex were investigated in two charge states $[(G', G'')-CO_3]^{(0/-)}$.

Upon relaxing the molecule over the basal plane, starting from a number of configurations, both bonded into the basal plane and residing over it, a remarkable bonding mechanism is found: from all starting structures, an ionically bonded system is formed where a CO_3^- molecule resides above an "image hole" in the basal plane. In all cases the molecule resides nearly 3 Å above the surface (defined as the smallest z-coordinate of any CO_3 atom above a sheet in the xy-plane).

Considering the molecule with one CO bond aligned along z and the O neighbouring a basal plane C atom (the "one-down" configuration) Fig. 5, we find that for G′CO_3 and G′CO_3 the bonding is very strong at 2.15 and 2.62 eV, respectively. These energies relate to the following reaction: $GCO_3 \rightarrow G + CO_3$, (G = G′, G″). For the desorption of CO_3^- (leaving G⁺), higher energies of 2.81 and 2.88 eV, respectively, were found for G′ and G″. The difference in the CO_3 binding energies for the two hydrocarbons comes from the non-radical (G′) and radical (G″) nature of the sheets; for the latter the

HOMO is singly degenerate and half-occupied, the wavefunction associated with this level is a non-bonding orbital characteristic of odd-alternant systems. As a consequence of this, the G″ HOMO is higher than that of G′, and its corresponding ionization potential is lower. A lower ionization potential results in a stronger G–CO_3 ionic bond, this trend was confirmed using a variety of aromatic hydrocarbons. An analysis of the Mulliken populations reveals that the charge on the molecule is $1e^-$, with the graphene sheet being positively charged. Therefore, CO_3 is a powerful oxidant (Lewis acid) on graphite and the bond formed between the two is purely electrostatic. However, it must be noted that the radiolytic oxidation proceeds via CO_3^- attack on graphite.

For the negative complexes $(G'CO_3)^-$ and $(G''CO_3)^-$ the binding energy of the CO_3^- is much reduced at 0.76 and 0.79 eV, respectively. In these systems, the net charge on the molecule is still $1e$, but the graphene sheet has no charge. $(G''CO_3)^-$ is in an $S=1$ state. The smaller binding in this charge state comes from a redistribution of the electronic charge in the basal plane. In our finite sheets (which are used to model bulk graphite) it is important to consider both charge states, in the real semi-metallic system charge is free to flow on and off the basal plane, and the formation of an "image hole" below the molecule is always possible. The strength of the binding energy of the molecule to bulk graphite will be dependent on E_F. The addition or removal of charge from our clusters merely represents a change in E_F, although it must be pointed out that this varies discontinuously with occupation number due to the discrete nature of the electronic levels in these small polycyclic clusters. The results for the two charge states demonstrate that the binding energy of CO_3^- is greater when the Fermi-level is deeper. When the Fermi-level is deep (applicable to the neutral complex) dissociation of the complex giving both CO_3 and CO_3^- is possible although the dissociation energy is high; for shallower values of E_F (corresponding to an increase in the bulk electron density) emission of CO_3^- becomes more favorable and its corresponding binding energy is significantly lower.

Regarding the structure of the "one-down" atomic configuration, the CO_3 molecule bond lengths were around 1.27 Å in all cases; this length is intermediate between those of isolated CO_3 and CO_3^- (D_{3h}), respectively. The closest the molecule approaches the basal plane (C(G)–OCO_2) in this orientation is approximately 2.8 Å, the distance between the molecular and basal C atoms being 4.1 Å. Although strongly bound, the molecule resides well above the basal plane (as was the case in the K–graphite interaction).

Having established that the bonding is ionic in nature, and discussed in detail the "one-down" atomic structure our attention now turns to other orientations of the molecule over the basal plane. Clearly there is a number of structures that the molecule might adopt over the basal plane, both in the lateral site of the molecule (over a C atom, C–C bond, or hexagon center) and in the orientation of the molecule itself either with the plane of the molecule parallel to surface, or with the CO bond along z (pointing toward ("one-down") or away from ("two-down") the surface). Several structures incorporating all the above degrees of freedom were investigated, the resultant energies were remarkably insensitive to the structure: the energies of all structures were within 0.1 eV of each other. This statement applies to both the neutral and negatively charged $(G–CO_3)^q$ complexes. This consistency in the total energy has profound consequences for the migration of the molecule over the basal plane, the barrier for migration must be extremely small (< 0.1 eV) and is independent of the orientation of the molecule

Table 5

The calculated vibrational frequencies (AIMPRO) (in cm^{-1}) for the isolated CO_3^- molecule and when the molecule is bound to graphite ($^{16}O^{16}O^{12}C^{16}O$).

symmetry	CO_3^- (D_{3h})	GCO_3 one-down^0	$(GCO_3)^-$ one-down$^-$
A_1	1358.6	1438.5	1428.6
B_1	1358.6	1375.3	1359.9
A_1	1047.1	999.4	1055.7
A_1	932.4	926.0	898.5
B_1	489.9	497.4	484.1
A_1	489.9	442.4	464.6

with respect to the surface. When the molecule is oriented parallel to the basal plane, it resides a distance of 3.4 Å above the surface. In all orientations, the distance between the molecule and the basal plane is similar. Again, this distance is defined as the molecule's smallest atomic z-coordinate above a sheet in the xy-plane. Therefore the molecule is essentially confined to a 2D plane, since the potential energy surface in the (xy) directions is very smooth. The magnitude of the binding energy coupled with its insensitivity to structure again reflects the ionic nature of the bonding.

6.2 Vibrational frequencies of bound CO_3

The local vibrational frequencies of the bound CO_3 molecule, in both the one-down and parallel atomic configurations have been calculated and are given in Table 5. The precise structures are: one-down, with the unique O over a basal C atom, and two equivalent O atoms symmetrically placed on the ($2\bar{1}10$) axis; and parallel where the molecular C atom is over a basal C atom, and the O atoms are placed over the centers of hexagons. One important point to note is the large splitting of the highest modes in the first of these structures. This results from the inequivalence of the O atoms in this orientation, a combination of an induced polarization of the molecule and a slight distortion to C_{2v} symmetry. In the parallel orientation, the O atoms are equivalent, and the two highest A_1 and B_1 modes should be degenerate. We find a slight splitting presumably from the D_{2h} symmetry of the cluster substrate. In both cases, the differences between the CO_3 vibrations in GCO_3 and $(GCO_3)^-$ is small, and both lie close to those of the isolated CO_3^- molecule, again consistent with the results from the Mulliken analysis which implied a net charge of $1e$ on the bound molecule. These calculations represent two of the extremes for the vibrational frequencies. Since the bound molecule is free to adopt any orientation over the basal plane, and this causes changes in the frequency, then this system will adsorb/emit in a very broad range of modes around 1400 cm^{-1}. It is unlikely that any unambiguous experimental identification of these modes could be made since they lie in the region of C–C and C–O vibrations from the basal plane itself, and the absolute concentration of basal CO_3 molecules is presently unknown.

6.3 The breakup of CO_3 over graphite

Given that CO_3^- is believed to play a crucial role in the radiolytic oxidation of graphite, the breakup of CO_3 over graphite is an important reaction to consider. Since $CO_3^{(0/-)}$ is known to dissociate in the following way: $CO_3^{(0/-)} \rightarrow CO_2 + O^{(0/-)}$, we investigate the reaction enthalpies for the conversion of bound CO_3 to bound O and CO_2, for both the neutral and negative complexes with G' and G''. All reactions are endothermic, the emission of CO_2 costing 1.3, 1.7, 2.3, and 2.2 eV, from G'CO_3, G''CO_3, G'CO_3^-, and G''CO_3^- respectively. These energies suggest that the ionically bound CO_3 molecule will not breakup into covalently bound O and $CO_{2(g)}$; in-fact CO_2 will solvate surface oxygen forming bound CO_3^-. We must point out that these energies refer to the formation of basal oxygen, as discussed in the previous section, which is relatively weakly bound because of loss of π bonding upon formation of covalent bonds with basal C-atoms. Although not investigated here, the reaction of bound CO_3 with an sp^2 hybrid (at a vacancy or an unsaturated edge) will be strongly exothermic and this will control the edge oxidation shown to be typical of radiolytic oxidation by Feates [1].

7. Conclusions

The work presented here outlines the strong ionic bond formed between the radical anion, CO_3^-, and the graphite basal plane.

Graphite has been approximated by finite polyaromatic clusters in these calculations, and Section 3 is dedicated to a justification of this approximation. It is shown that the physical properties of graphite are well reproduced by our method, although the finite structure of the graphene sheets results in a discrete spectrum for the electronic levels. Interestingly, the nature of these electronic levels (for a given cluster size) depends on the topology of the hydrocarbon, and it is found that the magnitude of the HOMO-LUMO gap is a function of the edge composition of the sheet.

In Section 4 we calculate the structure, vibrational properties, and dissociation mechanisms of isolated CO_3^0 and CO_3^-, and our results compare favorably with the previous work on these complexes. Our calculations suggest that both have a D_{3h} ground state, which is unique for the negatively charged complex. For the neutral molecule, a metastable C_{2v} structure exists, being 0.07 eV above the D_{3h} one. Our structures are in apparent conflict with those determined from experiment; however, these were derived from measurements of the vibrational frequencies in the solid state. The calculated frequencies for the C_{2v} CO_3 molecule, agree extremely well with those observed; our D_{3h} structures give rise to degenerate high modes for CO_3 and CO_3^- and those of the latter lie between the ones observed experimentally. It remains possible that the crystalline environment perturbed the structures in those measurements; to date no information on the gas phase species is available. A more in-depth discussion of this problem is presented elsewhere [9]. Finally, the dissociation reactions have been calculated for these molecules, and the results are in good agreement with experiment.

Having established the reliability of the method, and the use of graphene as a model for the basal plane, Sections 5 and 6 are concerned with the main thesis of this paper, the interaction of CO_3^- with graphite. The bonding in this system is remarkable: in spite of the fact that the molecule is strongly bound (>2 eV but dependent on the Fermi-level) the variation of energy with respect to both the molecular orientation and lateral position is less than 0.1 eV. The smallest atomic z-coordinate of the molecule above a sheet in the xy-plane is always approximately 3 Å and the potential energy surface in the lateral directions in extremely flat. Therefore, the molecule will be highly mobile, even at relatively low temperatures, while still being restricted to a plane. The bond formed between the molecule and the basal plane is strongly ionic, with CO_3^- residing above an image hole in the basal plane. It is the interaction between these two charges which dictates the strength of the bonding, and therefore, the binding energy decreases with increasing Fermi level (higher bulk electron density). Technologically, this adsorbate–substrate system is important, since CO_3^- is believed to be responsible for the radiolytic oxidation of graphite. Additionally, it is interesting in its own right through the long yet strong ionic bond.

Acknowledgements We acknowledge support from the UK Engineering and Physical Sciences Research Council (K28350, L04955, K4230) and the Higher Education Funding Council for England (JR-SUHE). M.I.H., C.P.E. and P.L. acknowledge support from British Energy Generation Ltd. and the Industry Management Committee of the UK Health and Safety Executive Fruitful discussions with Prof. A. M. Stoneham, the late Dr. M. Fujita and Dr. M. Bradford are gratefully acknowledged.

References

[1] F. S. FEATES, Trans. Faraday Soc. **64**, 3093 (1968).
[2] F. S. FEATES, Trans. Faraday Soc. **65**, 211 (1969).
[3] I. Y. ADAMSON, I. M. DAWSON, F. S. FEATES, and R. S. SACH, Carbon **3**, 393 (1966).
[4] F. S. FEATES and C. W. KEEP, Trans. Faraday Soc. **66**, 3156 (1970).
[5] J. V. BEST, W. J. STEPHEN, and A. J. WICKHAM, Progr. Nucl. Energy **16**, 127 (1985).
[6] R. S. NARCISI, A. D. BAILEY, L. DELLA LUCCA, C. SHERMAN, and D. M. THOMAS, J. Atmos. Terr. Phys. **33**, 1147 (1971).
[7] F. ARNOLD, J. KISSEL, D. KRANKOWSKY, H. WIEDER, and J. ZAEHRINGER, J. Atmos. Terr. Phys. **33**, 1169 (1971).
[8] L. THOMAS, P. M. GONDHALEKAR, and M. R. BOWMAN, J. Atmos. Terr. Phys. **35**, 397 (1973).
[9] M. I. HEGGIE, C. P. EWELS, P. LEARY, R. JONES, and P. R. BRIDDON, in preparation.
[10] N. G. MOLL, D. R. CLUTTER, and W. E. THOMPSON, J. Chem. Phys. **45**, 4469 (1966).
[11] M. E. JACOX and D. E. MILLIGAN, J. Chem. Phys. **54**, 919 (1971).
[12] S. W. BENSON, *Thermochemical Kinetics*, 2nd ed., Wiley, New York 1976.
[13] J. F. OLSEN and L. J. BURNELLE, J. Amer. Chem. Soc. **91**, 7286 (1969).
[14] J. A. POPLE, U. SEEGER, R. SEEGER, and P. VON R. SCHLEYER, J. Comput. Chem. **1**, 199 (1980).
[15] S. CANUTO and G. H. F. DIERCKSEN, Chem. Phys. **120**, 375 (1988).
[16] M. A. CASTRO, S. CANUTO, and A. M. SIMAS, Chem. Phys. Lett. **177**, 98 (1991).
[17] R. D. J. FROESE and J. D. GODDARD, J. Phys. Chem. **97**, 7484 (1993).
[18] W. J. VAN DE GUCHTE, J. P. ZWART, and J. J. C. MULDER, J. Mol. Struct. **152**, 213 (1987).
[19] J. S. FRANCISCO and I. H. WILLIAMS, Chem. Phys. **95**, 373 (1985).
[20] G. W. CHANTRY, A. HORSEFIELD, J. R. MORTON, and D. H. WHIFFEN, Mol. Phys. **5**, 589 (1962).
[21] R. A. SERWAY and S. A. MARSHALL, J. Chem. Phys. **46**, 1949 (1967).
[22] R. A. SERWAY and S. A. MARSHALL, J. Chem. Phys. **47**, 868 (1967).
[23] M. E. JACOX and D. E. MILLIGAN, J. Mol. Spectroscopy **52**, 363 (1974).
[24] J. I. MORUZZI and A. V. PHELPS, J. Chem. Phys. **45**, 4617 (1966).
[25] D. K. BOHME, D. B. DUNKIN, F. C. FEHSENFELD, and E. E. FERGUSEN, J. Chem. Phys. **51**, 863 (1969).
[26] D. A. PARKES, J. Chem. Soc. Faraday Trans. I **68**, 627 (1972).
[27] D. A. PARKES, J. Chem. Soc. Faraday Trans. I **69**, 198 (1973).
[28] J. T. MOSELY, R. A. BENNETT, and J. R. PETERSON, Chem. Phys. Lett. **26**, 288 (1974).
[29] J. F. HILLER and M. L. VESTAL, J. Chem. Phys. **72**, 4713 (1980).
[30] D. E. HUNTON, M. HOFFMAN, T. G. LINDEMANN, and A. W. CASTLEMAN, JR., J. Chem. Phys. **82**, 134 (1985).
[31] J. T. SNODGRASS, C. M. ROEHL, P. A. M. VAN KOPPEN, W. E. PALKE, and M. T. BOWERS, J. Chem. Phys. **92**, 5935 (1990).
[32] S. P. HONG, S. B. WOO, and E. M. HELMY, Phys. Rev. A **15**, 1563 (1977).

[33] J. F. Olsen and L. Burnelle, J. Amer. Chem. Soc. **92**, 3659 (1970).
[34] S. P. So, J. Chem. Soc. Faraday Trans. II **72**, 646 (1976).
[35] K. Hiraoka and S. Yamabe, J. Chem. Phys. **97**, 643 (1992).
[36] D. Smith, D. W. James, and J. P. Delvin, J. Chem. Phys. **54**, 4437 (1971).
[37] F. M. Hofmann, M. D. Weisel, and J. Paul, Surf. Sci. **316**, 277 (1994).
O. Axelsson, J. Paul, M. D. Weisel, and F. M. Hoffmann, J. Vac. Sci. Technol. A **12**, 150 (1994).
Y. Shao, J. Paul, O. Axelsson, and F. M. Hoffmann, J. Phys. Chem. **97**, 7652 (1993).
O. Axelsson, Y. Shao, J. Paul, and F. M. Hoffmann, J. Phys. Chem. **99**, 7028 (1995).
[38] T. Fromherz, C. Mendoza, and F. Ruette, Mon. Not. Astron. Soc. **263**, 851 (1993).
[39] D. Lamoen and B. N. J. Persson, J. Chem. Phys. **108**, 3332 (1998).
[40] N. J. Wu and A. Ignatiev, J. Vac. Sci. Technol. **20**, 896 (1982).
[41] K. M. Hock, J. C. Barnard, R. E. Palmer, and H. Ishida, Phys. Rev. Lett. **71**, 641 (1993).
[42] F. Ancilotto and F. Toigo, Phys. Rev. B **47**, 13713 (1993).
[43] R. Jones, Phil. Trans. Roy. Soc. Lond. A **350**, 189 (1995).
[44] R. Jones and P. R. Briddon, in: *Identification of Defects in Semiconductors*, Ed. M. Stavola, Vol. 51 A, Chap. 6, Ser. Semiconductors and Semimetals, Academic Press, Boston 1998 (p. 287).
[45] J. N. Murrell, S. F. A. Kettle, and J. M. Tedder, *Valence Theory*, Chap. 15, Wiley, New York 1965.
[46] G. B. Bachelet, D. R. Hamann, and M. Schlüter, Phys. Rev. B **26**, 4199 (1982).
[47] B. R. Eggen, M. I. Heggie, G. Jungnickel, C. D. Latham, B. Jones, and P. R. Briddon, Science **272**, 87 (1996).
[48] C. D. Latham, M. I. Heggie, and R. Jones, Diamond and Relat. Mater. **2**, 1493 (1993).
[49] C. P. Ewels, R. Jones, S. Öberg, J. Miro, and P. Deak, Phys. Rev. Lett. **77**, 865 (1996).
[50] J.-C. Charlier, X. Gonze, and J.-P. Michenaud, Europhys. Lett. **28**, 403 (1994).
J.-C. Charlier, X. Gonze, and J.-P. Michenaud, Carbon **32**, 289 (1994).
J.-C. Charlier, X. Gonze, and J.-P. Michenaud, Phys. Rev. B **43**, 4579 (1991).
[51] P. Badziag, W. S. Verwoerd, W. P. Ellis, and N. R. Greiner, Nature **343**, 244 (1990).
[52] S. E. Stein and R. L. Brown, J. Amer. Chem. Soc. **109**, 3721 (1987).
[53] K. Nakada, M. Fujita, G. Dresselhaus, and M. S. Dresselhaus, Phys. Rev. B **54**, 17954 (1996).

phys. stat. sol. (b) **217**, 449 (2000)

Subject classification: 71.15.Fv; 71.20.Nr; 73.20.Dx; 78.66.Fd; S7.12; S7.15

Calculation of Electronic States in Semiconductor Heterostructures with an Empirical spds* Tight-Binding Model

R. SCHOLZ (a), J.-M. JANCU (b), F. BELTRAM (b), and F. BASSANI (b)

(a) *Institut für Physik, Technische Universität Chemnitz, D-09107 Chemnitz, Germany*

(b) *Scuola Normale Superiore and Istituto Nazionale per la Fisica della Materia, Piazza dei Cavalieri 7, I-56126 Pisa, Italy*

(Received August 10, 1999)

We illustrate how the tight-binding formalism can be used to accurately compute the electronic states in semiconductor quantum wells and superlattices. To this end we consider a recently developed empirical tight-binding model which carefully reproduces ab initio pseudopotential calculations and experimental results of bulk semiconductors. The present approach is particularly suited both for short-period superlattices and large and complex unit cells where the transferability of the hopping parameters is required. First applications for ultrathin GaAs/AlAs superlattices and InGaAs/AlAs heterostructures are presented and discussed.

1. Introduction

Availability of very powerful computers and improvements in theoretical techniques have vastly improved our ability to understand and predict the properties of semiconductor materials from first principles and empirical calculations. While first principles techniques like density functional theory (DFT) in the local density approximation (LDA) are well suited to predict structural properties and electronic bands over a large energetic region [1 to 3], the calculation of excited-state properties is still unsatisfactory. The known LDA band gap problem for instance results from the difference between the local density single particle potential and the self-energy operator that describes the quasi-particle of the system. It can be overcome by the use of the GW approximation for the self-energy [4].

Even though it is already known for a long time how to circumvent the self-interaction problem inherent in standard local density theory [5], the application of self-interaction-relaxation corrections (SIRC) to electronic bands in solids is a rather recent development [6]. Using SIRC methods, it was shown that the band gap can be improved substantially, at a computational cost far below the GW quasi-particle correction. However, the precision of both the GW and SIRC approaches still remains insufficient for a prediction of the conduction band energies within typical experimental uncertainties achieved in optical measurements.

The necessity to accurately interpret the excited-state properties of semiconductors has led to the development of various empirical approaches, ranging from empirical pseudopotentials [7, 8] over $\mathbf{k} \cdot \mathbf{p}$ models [9, 10] to tight-binding (TB) methods [11, 12]. Most of them use the precise experimental knowledge of band positions for well-controlled input parameters determining the quantities entering the model.

From interpretations of the electronic wavefunctions calculated with pseudopotential methods in terms of atomic symmetries [13, 14], it became obvious that d-symmetric contributions play a crucial role both for the valence band maximum at Γ and for the conduction states at X and L. This has led to basis extensions of earlier sp^3 [11] and sp^3s^* [15] TB models including parts of the d-states [16]. Finally, it turned out that the inclusion of all five atomic d-states becomes necessary for a reliable description of the valence bands and the first two conduction bands around the Brillouin zone [17], defining the empirical $sp^3d^5s^*$ TB model applied in the present work.

The outline of the paper is as follows. In Section 2 we discuss the general features of the present TB bulk parametrization and apply them in Section 3 to short-period superlattices. Intersubband transitions of InGaAs/AlAs quantum wells are investigated in Section 4, and the paper is concluded in the last section.

2. Bulk Parametrization

2.1 Basis size

In free atoms of group III and V, an sp^3 basis is physically complete for the valence shell. However, when a zincblende III–V semiconductor is formed, these valence states interact via matrix elements not much smaller than the atomic level spacing [11], so that interactions of the same magnitude beetwen the valence orbitals and higher-lying atomic states cannot be excluded. Therefore, the completeness of the atomic valence shell for the occupied states in the tetrahedral group T_d becomes questionable, and it has to be rediscussed in terms of the representations Γ_i of T_d. In order to improve the numerical completeness, we require a further basis orbital for each occupied valence state Γ_i. While for Γ_1, the inclusion of a second s-symmetric basis state called s^* is the obvious issue, the Γ_4-symmetric valence maximum would allow for an $\{p, p^*\}$ or an $\{p, d\}$ basis. Analyzing the degenerate sets of reciprocal lattice vectors close to the direct gap, the decomposition into irreducible representations $\langle 1, 1, 1\rangle \to 2\Gamma_1 + 2\Gamma_4$ and $\langle 2, 0, 0\rangle \to \Gamma_1 + \Gamma_3 + \Gamma_4$ allows to decide that the inclusion of a full set of five atomic d-states is the better choice, as these cover $d \to \Gamma_3 + \Gamma_4$. This completeness argument for the occupied states coincides with the necessity to use d-symmetric basis orbitals known from the projection of conduction band wavefunctions on the atomic symmetries [13, 14].

2.2 Determination of tight-binding parameters

While the above considerations regarding numerical completeness provide arguments for a reasonable basis choice, they give no indication concerning the size of the on-site energies and the interaction matrix elements of the one-electron Hamiltonian. Therefore, a starting parametrization for the $sp^3d^5s^*$ nearest-neighbour model is required. We illustrated previously how universal TB parameters can be obtained from an analysis of the free electron energy bands for the face-centred cubic lattice [17]. Based on this parametrization of the empty lattice, we deduced the TB parameters for semiconductors as follows: First, the splitting $E_p - E_s$ is fixed to the atomic energy difference while E_d and E_{s^*} are taken from the values derived for the free electrons, second, the interaction matrix elements are modified in order to achieve the correct band ordering, especially at Γ, and third, the deviations from known band positions [18] are minimized

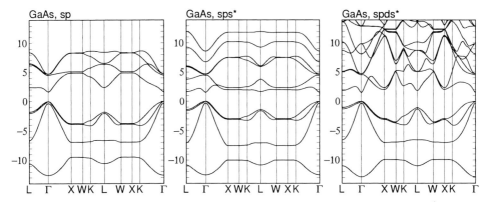

Fig. 1. Comparison of GaAs TB band structures obtained for different basis sizes: sp^3 (left) [11], sp^{3}s* (middle) [15], and sp^{3}d^{5}s* (right) [17]. The energy reference is the valence band maximum, and the energy unit is eV

within a numerical procedure optimizing the off-diagonal matrix elements. When comparing the resulting parametrization of a real semiconductor with the free electron reference, e.g. for germanium, the average change in the TB parameters is only 1/3 of their absolute mean, indicating that the order of magnitude for the interactions was correctly reproduced with the analysis of the free electrons. In a final step, spin–orbit splitting of p-states is included with procedures known from smaller bases [19], leading to the present 40-band TB model.

2.3 Comparison of different basis sizes

In Fig. 1, we illustrate the quality of the sp^{3}d^{5}s* Hamiltonian by comparing the calculated band structure of GaAs with previous TB models. In the smallest sp^3 basis, the valence band dispersion is quite reasonable, in contrast to the conduction bands where only the direct surroundings of Γ_{6c} are reliable [19]. The nearest-neighbor sp^{3}s* model gives the correct positions of the L_{6c}, X_{6c}, and X_{7c} conduction band minima, but the transverse masses at L and X-points are not reproduced [15]. Even though some improvement of the band dispersion can be achieved with interaction parameters between more distant atoms, the erroneous projection on the atomic symmetries cannot be corrected [20]. Moreover, the difference between electron and light-hole effective masses is not correctly reproduced in sp^{3}s* calculations [21], and our nearest-neighbour sp^{3}d^{5}s* Hamiltonian corrects this deficiency. Concerning the transverse mass at X, it was shown analytically that pd interactions are required for the dispersion of the conduction bands along $X \to W$ [17]. The resulting band structure is reliable all around the Brillouin zone up to about 5 eV above the valence maximum, and at Γ even the next two band positions are well described, the higher one being the lowest state with 100% d(Γ_3)-symmetry.

2.4 Band structure of arsenides

For the binary arsenides discussed in the present work, the resulting TB band positions and reduced masses are given in Table 1, and the agreement with experimental data [18] is quite good. The precise value for the reduced electron mass, especially at Γ, will

Table 1

Comparison of energetic positions and masses obtained in the present work (TB) with experimental values (exp.) [18] and pseudopotential calculations in the GW approximation (PP) [22], for AlAs, GaAs, and InAs. All energies are in eV, and the reference energy is taken at the maximum of the valence band. Bands are assigned with representations in the double-group notation, and the origin is chosen on the anion site

	AlAs			GaAs			InAs		
	TB	exp.	PP	TB	exp.	PP	TB	exp.	PP
Γ_{6v}	−12.020		−12.41	−12.910	−13.1	−13.03	−12.188	−12.3	−12.10
Δ_0	−0.300	−0.30	−0.27	−0.340	−0.341	−0.34	−0.380	−0.38	−0.38
Γ_{6c}	−3.130	3.13	2.88	1.519	1.519	1.22	0.418	0.418	0.31
Γ_{7c}	−4.569	4.54	5.14	4.500	4.53	4.48	4.252	4.39	4.51
Γ_{8c}	4.725	4.69	5.14	4.716	4.716	4.48	4.580	4.39	4.51
X_{6v}	−2.760	−2.41	−2.44	−3.109	−2.88	−2.91	−2.654	−2.4	−2.49
X_{7v}	−2.565	−2.41	−2.44	−2.929	−2.80	−2.91	−2.546	−2.4	−2.49
X_{6c}	2.223	2.229	2.14	1.989	1.98	1.90	2.176		2.01
X_{7c}	2.584	2.579	3.03	2.328	2.35	2.47	2.441		2.50
L_{6v}	−1.191		−0.99	−1.330	−1.42	−1.28	−1.124	−0.9	−1.13
$L_{4,5v}$	−0.983		−0.99	−1.084	−1.20	−1.28	−0.830	−0.9	−1.13
L_{6c}	2.581	2.54	2.91	1.837	1.85	1.64	1.691		1.43
L_{6c}	5.069		5.59	5.047	5.47	5.40	4.723		5.32
$m_c(\Gamma)$	0.156	0.15		0.067	0.067		0.024	0.023	
$m_{lh}(\Gamma)$	0.19	0.21		0.090	0.094		0.029	0.026	
$m_t(X_{6c})$	0.237	0.19		0.237	0.27		0.278		
$m_t(L_{6c})$	0.155			0.117	0.075		0.110		

become important in the calculation of the heterostructure band positions. For the GaAs and AlAs band structures shown in Fig. 2, the band positions of Γ_{6c} agree with the experimental reference within 1 meV, and the X_{6c} positions within 10 meV. The X_{6c} wavefunction is the lowest band at a high symmetry point containing a large admixture of bonding d-states, compare Table 2. For a reliable calculation of the folded bands in

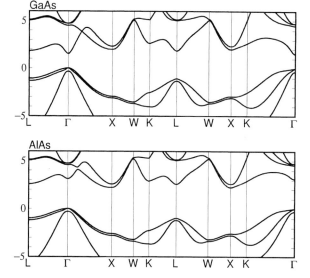

Fig. 2. GaAs and AlAs bulk band structures obtained with the empirical $sp^3d^5s^*$ TB calculation (energy in eV)

Table 2
Decomposition of GaAs wavefunctions at different points of the Brillouin zone, calculated with the sp^{3}d^{5}s* TB model. The contributions of different atomic symmetries agree favorably with the corresponding projections derived from empirical pseudopotential calculations [14]. Entries vanishing due to symmetry restrictions are denoted with —

	Γ_{6v}	Γ_{6c}	X_{6c}	L_{6c}	Γ_{8v}	Γ_{8c}	X_{6v}	$L_{4,5v}$
E (eV)	−12.910	1.519	1.989	1.837	0.0	4.716	−2.929	−1.084
s_a	0.564	0.411	0.029	0.145	—	—	—	—
s_c	0.303	0.456	—	0.297	—	—	—	—
s_a^*	0.065	0.128	0.025	0.043	—	—	—	—
s_c^*	0.068	0.005	—	0.003	—	—	—	—
p_a	—	—	—	0.147	0.553	0.290	0.580	0.589
p_c	—	—	0.458	0.239	0.234	0.510	0.416	0.350
d_a	—	—	0.325	0.090	0.084	0.169	0.001	0.022
d_c	—	—	0.163	0.037	0.129	0.032	0.002	0.038
$d_a(\Gamma_3)$	—	—	0.325	—	—	—	—	0.007
$d_c(\Gamma_3)$	—	—	—	—	—	—	—	0.013
$d_a(\Gamma_4)$	—	—	—	0.090	0.084	0.169	0.001	0.015
$d_c(\Gamma_4)$	—	—	0.163	0.037	0.129	0.032	0.002	0.025

(100)-grown heterostructures, a good description of the Γ and X conduction states and their surroundings is essential, both in terms of energetic positions and symmetry character of the wavefunctions.

2.5 Deformation potentials

The experimental deformation potentials can be reproduced in our TB model by including a distance dependence for the interaction parameters [17]. It was shown in detail that the erroneous sign for the uniaxial and hydrostatic deformation potentials of the Γ_{4c}-states and the X conduction-band valleys in smaller TB models is corrected due to the large bonding d-component of the corresponding wavefunctions. Even though the strain in GaAs–AlAs heterostructures is quite small, some details of the heterostructure band positions could not be reproduced without reasonable values for the bulk deformation potentials, see below.

3. Zincblende Superlattices: GaAs/AlAs

The first application of our TB model to heterostructures concerns the case of (GaAs)$_n$(AlAs)$_n$ superlattices (SL) grown lattice-matched on GaAs substrate [23]. The deformation of the AlAs regions is treated as usual with classical elasticity, resulting in homogeneous tetragonal strain, and the modifications of the AlAs TB parameters are accounted for by including the distance dependences derived from the bulk deformation potentials, compare Tables IX and X of [17].

The direct applicability of our bulk parameters to heterostructures is obviously restricted to nearest-neighbour TB models: In an earlier sp^3 parametrization applied to superlattices, the next-nearest neighbour parameters needed especially for the correct transverse mass at X had to be changed in the presence of an interface [20].

Except for the small changes of our TB parameters due to the tetragonal strain, the only further input required is the valence band offset, chosen in accordance with experimental data as $\Delta E_v = 0.55$ eV [24], i.e. we do not attempt to calculate this value, as this would require more sophisticated ab initio methods [25]. For all atoms not on the interface, the on-site energies are shifted in accordance with this valence band offset, and for the As layer forming the interface, half this shift is applied. Except for small strain corrections, all interactions are parametrized as in the corresponding bulk materials.

Figure 3 shows the gap energies of $(GaAs)_n(AlAs)_n$ SL resulting in our TB model together with experimental data. The gaps shown correspond to free electron–hole recombination, neglecting any excitonic effects. As these exciton corrections are rather small compared to the magnitude of the heterostructure quantization energies displayed in Fig. 3, the main results are not affected. We found the $n = 1$ SL is characterized by a L-derived conduction band minimum (CBM) with nearly half of the wavefunction localized on the Ga sublattice ($s(Ga) + p(Ga) + d(Ga) \approx 40\%$), in agreement with first-principle calculations [29]. Both in the present TB approach and in ab initio and empirical pseudopotential (PP) investigations [29 to 31], this energy level is found to oscillate strongly with the SL period n. For $n = 2$, the TB CBM derives from the bulk AlAs $X_{x,y}$-states because tetragonal strain shifts the X_z-level above $X_{x,y}$ by about 20 meV for $n = 1$ and 2, in agreement with experiment [26]. For larger n, the quantum confinement of X_z according to the large longitudinal mass $m_l(X)$ becomes smaller than for the $X_{x,y}$-states with quantization energy based on the much lower transverse mass $m_t(X)$, resulting in a crossover $X_{x,y} \rightarrow X_z$ for the CBM between $n = 2$ and 3 in our calculations, in good agreement with experiment, where this crossover occurs between $n = 3$ and 4 [26]. For $3 \leq n \leq 13$, our absolute CBM derives from the bulk AlAs X_z-states folded to the center of the tetragonal Brillouin zone. Since the valence band maximum (VBM) remains a Γ-like state localized in the GaAs well for all SL periods n, the inter-

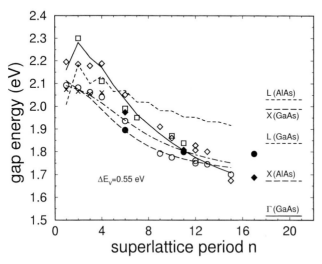

Fig. 3. Comparison of experimental and tight-binding gap energies in $(GaAs)_n(AlAs)_n$ superlattices. States deriving from Γ: TB (solid), experimental (open diamonds: low T [26], filled diamonds: RT [27], open squares: compilation of different temperatures [28]); X_z: TB (long-dashed), experimental (open circles: low T [26], filled circles: RT [27]); $X_{x,y}$: TB (dash-dotted), experimental ($\times$, low T [26]); L_{6c}: TB (short-dashed). The asymptotic band positions for large n are indicated on the right-hand side

band transition is then pseudodirect in momentum space and indirect in real space (type-II). The quantum size effects associated with the period n and the reduced masses $m_l(X)$ and $m(\Gamma)$ induce a type-II → type-I transition between $n = 13$ and 14 monolayers, and for $n \geq 14$ the CBM is a Γ-like state localized mainly in the GaAs well. This calculated value is in good agreement with the experimental results for the crossover thickness of $n_c \approx 11$ at room temperature [27] and of $n_c \approx 14$ at low temperature [26]. The overall agreement between theory and experiment is excellent and clearly demonstrates the transferability of the TB parameters to quantum structures with energy offsets at the interface region. Interestingly, the best empirical pseudopotential approaches use As pseudopotentials depending on the number of neighbouring Ga and Al sites [30, 31], in qualitative correspondence with our TB parametrization of the interface. However, the advantage of the TB model is that we can better distinguish the bonds towards the Ga and Al sites: They are parametrized as in the corresponding (strained) bulks.

The present TB model is in better agreement with experiment than recent empirical pseudopotential calculations [31], and most of our improvements can be related to a more precise parametrization of the bulk band structures, especially the precise values for the AlAs Γ_{6c} energy and for the GaAs reduced electronic mass at Γ, compare Table 1. Furthermore, our value of 0.55 eV for the valence band offset yields better results than lower values, as has been discussed elsewhere in more detail [23]. The calculated inversion of the X_z and $X_{x,y}$-states for small n requires the inclusion of tetragonal strain in the model calculation, so that it cannot be reproduced in approaches where this strain is neglected [30, 31].

Our nearest-neigbour $sp^3d^5s^*$ calculation and an older sp^3 TB model with next-nearest neighbour interactions [20] give similar results for larger SL with $n \geq 4$, in very good agreement with experimental data. In both cases, the quality of the SL band positions is directly related to a good fit of the bulk band structure, including the curvatures at Γ and X. However, for ultrathin SL with $n \leq 3$ where the interface region becomes a large fraction of the total volume, the present results are more reliable. This indicates that for ultrathin SL, the correct localization of the eigenstates on certain atomic planes is required for a reliable description of intervalley mixing and the resulting SL energy levels. Obviously, these mixing effects cannot be obtained in $\mathbf{k} \cdot \mathbf{p}$ models expanded around the Γ-point only [31], and generalizations of the $\mathbf{k} \cdot \mathbf{p}$ approach including basis states around Γ and X are required for investigations of Brillouin zone folding [32].

4. Intersubband Transitions in InGaAs/AlAs Quantum Wells

4.1 InGaAs/AlAs grown on GaAs substrate

For lattice constant and elastic properties, the InGaAs alloys are treated with a linear interpolation in the spirit of the virtual crystal approximation, e.g. for the lattice constant a_0,

$$a_0(\text{In}_x\text{Ga}_{1-x}\text{As}) = xa_0(\text{InAs}) + (1-x)a_0(\text{GaAs}). \tag{1}$$

Applying the same interpolation laws to the tight-binding parameters, the dependence of important quantities like the direct gap shows a close to linear dependence on the In concentration x, contrary to the experimentally determined pronounced parabolic gap bowing [18]. Therefore, we interpolate all tight-binding parameters linearly, but allow for a parabolic contribution of two parameters determining the band positions around the direct gap,

$$s_a^* s_c \sigma(\text{In}_x\text{Ga}_{1-x}\text{As}) = x \, s_a^* s_c \sigma(\text{InAs})$$
$$+ (1-x) \, s_a^* s_c \sigma(\text{GaAs})$$
$$+ x(1-x) \, b(s_a^* s_c \sigma),$$
$$p_c d_a \sigma(\text{In}_x\text{Ga}_{1-x}\text{As}) = x \, p_c d_a \sigma(\text{InAs})$$
$$+ (1-x) \, p_c d_a \sigma(\text{GaAs})$$
$$+ x(1-x) \, b(p_c d_a \sigma). \qquad (2)$$

The first of these bowing parameters, $b(s_a^* s_c \sigma) = -0.401$ eV, has to be rather large and influences the Γ_{6c} position in $\text{In}_{0.5}\text{Ga}_{0.5}\text{As}$ by nearly 0.2 eV, compare also Table XIII of [17] for the relative importance of different TB parameters. The second interaction where we allow for a bowing, $b(p_c d_a \sigma) = -0.061$ eV, influences on the position of the valence band maximum only in the meV range; it is just needed to conserve the valence band maximum as our energy reference. The above assignment of the required gap bowing to only two TB parameters is not unique, but as we shall be concerned in the following with heterostructures where the properties of InGaAs are needed mainly around the direct gap, this simple interpolation scheme was found to be most convenient.

For heterostructures containing $\text{In}_x\text{Ga}_{1-x}\text{As}$ grown on GaAs substrates, the much larger lattice constant of InAs compared to the GaAs substrate produces a considerable strain, even for low indium molar fractions, and the distance dependence of the TB parameters is interpolated again linearly between the exponents derived for InAs and GaAs, compare Table IX of [17]. The use of AlAs as barrier material allows for quite high barriers and therefore large energy differences between consecutive Γ conduction subbands. A drawback of AlAs is the low-lying X-like state in the barrier which can become the absolute conduction minimum for thin enough well widths, producing a type-II behaviour as dicussed in the previous section for GaAs/AlAs superlattices.

As was shown recently in more detail, the TB calculations are in quantitative agreement with observed transition energies: For a 5ML type-II sample with 30% indium content, both experiment and TB theory gave a pseudo-direct hh1–c1 gap energy of 1730 meV, and for a 8 ML type-I sample, the direct hh1–c1 transition was observed at 1635 meV and calculated to be 1640 meV, the difference corresponding roughly to excitonic corrections neglected in the calculation [33]. For the latter 8 ML sample, a room temperature intersubband transition c1–c2 occurred at 675 meV, a value reproduced again with the TB calculation (680 meV) [33].

The intersubband transition energy in the 8 ML sample with 30% indium is already close to the region required for tele-communication ($\lambda = 1.55$ μm or 0.8 eV), and therefore, a systematic study of this phenomenon as a function of well width and indium concentration could give an indication for useful device parameters. In Fig. 4, we show the calculated intersubband transition energies for indium concentrations and well widths giving type-I superlattices. Albeit all the points shown give an absolute conduction band minimum at Γ, this does not yet guarantee strong intersubband absorption between the lowest Γ-like states: The X-point with its large electronic density of states

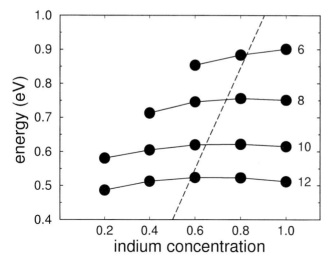

Fig. 4. Calculated intersubband transition energies for $In_xGa_{1-x}As$ quantum wells between AlAs barriers, grown lattice matched on GaAs substrate (dots). The dashed line represents the critical thickness for strained $In_xGa_{1-x}As$ quantum wells calculated from a mechanical equilibrium model [34]. Only well thicknesses (in ML) resulting in type-I behaviour are included

leads to a leakage of carriers $\Gamma \to X$, even if X is slightly above the lowest Γ conduction state.

The window between the type-II – type-I crossing and the mechanical stability limit turns out to be quite small. Therefore, a device design with lower strain in the well region would be favourable, and it should be combined with a systematic reduction of carrier leakage to the X-point, as will be discussed below.

4.2 InGaAs/AlAs grown on InP substrate

Changing the substrate to InP with its larger lattice constant, the growth of $In_xGa_{1-x}As$ quantum wells with lower strain becomes possible. The sample suggested in the follow-

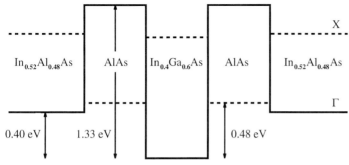

Fig. 5. Band alignment of a $In_{0.52}Al_{0.48}As/AlAs/In_{0.4}Ga_{0.6}As/AlAs/In_{0.52}Al_{0.48}As$ heterostructure grown pseudomorphically on InP [35]. The calculated Γ electron confinement barrier is 1.33 eV in the $In_{0.4}Ga_{0.6}As$ layers whereas the X electrons are confined in the AlAs layers. The potential offset between the Γ band edge of the central strained $In_{0.4}Ga_{0.6}As$ well and the X_z band edge of the strained AlAs barrier is 0.48 eV

ing consists of an $In_{0.4}Ga_{0.6}As$ quantum well with AlAs barriers, confined between buffer layers of $In_{0.52}Ga_{0.48}As$, compare Fig. 5. While the buffer layers are close to lattice-matched with the substrate, both the AlAs barriers and the $In_{0.4}Ga_{0.6}As$ are well characterized by tensile strain, lowering the Γ and X_z band edges in both materials. For the calculation of the band profile, we use now the strain-modified conduction to valence band offset of 67:33, resulting in a discontinuity between the $In_{0.52}Ga_{0.48}As$ and $In_{0.4}Ga_{0.6}As$ Γ edges of 0.40 eV, in agreement with experiment [36]. For the 40% indium concentration used in the central quantum well, the Γ confinement energy of 1.33 eV is somewhat larger than for the $In_{0.3}Ga_{0.7}As$ samples discussed before, and from Fig. 4 it can be deduced that an intersubband transition energy of about 0.8 eV (1.55 μm) can be achieved with a well width of 7 ML. There are two further important device criteria to be fulfilled: First, the AlAs barriers should be thin enough to allow for an efficient population of the lowest Γ-like state in the central $In_{0.4}Ga_{0.6}As$ well by tunneling, and second, the X-like states in the AlAs barriers should be higher than the lowest Γ-like state in order to avoid carrier leakage $\Gamma \to X$. Both requirements are met with barrier thicknesses below about 10 ML, compare Fig. 6a, and the heterostructure quantization energy of the X-states in the AlAs region shifts them upwards for decreasing barrier thickness. However, a very thin barrier does not provide an efficient confinement of the second Γ-like state to the central quantum well, resulting in reduced intersubband transition strengths for too thin barriers, compare Fig. 6b. A compromise between high-lying X-states in Fig. 6a and a high intersubband transition strength must be found, and a barrier thickness of 7 ML appears to be sufficient for a saturation of the transition strength in Fig. 6b [35]. It should be noted that this system provides strong optical activity in the transparency window of optical fibres, $\lambda = (1.55 \pm 0.02)$ μm, and

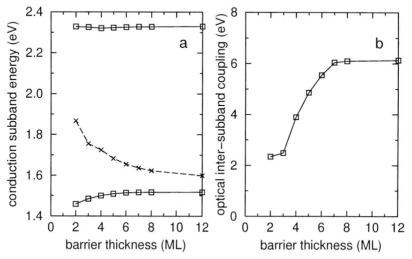

Fig. 6. a) Subband energies for the structure of Fig. 5 at $k = 0$, for 7 ML $In_{0.4}Ga_{0.6}As$ wells, as a function of AlAs barrier thickness. The energy is measured from the $In_{0.52}Al_{0.48}As$ valence-band maximum. The lowest two Γ-like quantum-well states are shown as solid lines with squares, the X_z-like barrier states as a dashed line with crosses. b) Squared dipole matrix element for the intersubband transition between the two lowest Γ conduction states, as a function of AlAs barrier thickness

due to the low strain, it is fully compatible with good crystalline quality. As the hh1–c1 gap has a much larger value, no interband absorption is expected around $\lambda = 1.55$ μm, and the lower valence band mobility as a limiting factor for the device performance can be avoided completely.

5. Conclusion and Outlook

In the present paper, we have shown that the empirical $sp^3d^5s^*$ tight-binding model is a highly predictive tool for the design of heterostructures with specific properties. This was achieved with a reliable tight-binding description of the wavefunctions in terms of atomic symmetries, a quantitative modelling of the energetic positions of the bulk electronic bands and the inclusion of strain in the Hamiltonian describing the heterostructure. Actually, the proposed method can also be used to assess the quality of different interface boundary conditions in $\mathbf{k} \cdot \mathbf{p}$ models commonly used for large quantum wells [37]. A further important application of tight-binding methods concerns the simulation of devices under external bias voltages, including self-consistent electric fields and charge densities due to electrons or holes localized in some region of the heterostructure. An application of the present $sp^3d^5s^*$ empirical TB model to self-consistent free-carrier screening of piezoelectric fields in wurtzite heterostructures is discussed elsewhere in this volume [38, 39].

References

[1] P. Hohenberg and W. Kohn, Phys. Rev. **136**, B864 (1964).
[2] W. Kohn and L.J. Sham, Phys. Rev. **140**, A1133 (1965).
[3] L.J. Sham and W. Kohn, Phys. Rev. B **145**, 561 (1965).
[4] M.S. Hybertsen and S.G. Louie, Phys. Rev. B **34**, 5390 (1986); B **37**, 2733 (1988).
[5] J.P. Perdew and A. Zunger, Phys. Rev. B **23**, 5048 (1981).
[6] D. Vogel, P. Krüger, and J. Pollmann, Phys. Rev. B **52**, R14316 (1995); B **54**, 5495 (1996); B **55**, 12836 (1997).
[7] M.L. Cohen and T.K. Bergstresser, Phys. Rev. **141**, 789 (1966).
[8] M.L. Cohen and J. Chelikowsky, Electronic Structure and Optical Properties of Semiconductors, 2nd ed., Springer-Verlag, Berlin 1989.
[9] E.O. Kane, J. Phys. Chem. Solids **1**, 249 (1957).
[10] M. Cardona and F.H. Pollak, Phys. Rev. **142**, 530 (1966).
[11] D.J. Chadi and M.L. Cohen, phys. stat. sol. (b) **68**, 405 (1975).
[12] W.A. Harrison, Electronic Structure and Properties of Solids, Freeman, San Francisco 1980.
[13] S.L Richardson, M.L. Cohen, S.G. Louie, and J.R. Chelikowsky, Phys. Rev. B **33**, 1177 (1986).
[14] P. Boguslawski and I. Gorczyca, Semicond. Sci. Technol. **9**, 2169 (1994).
[15] P. Vogl, H.P. Hjalmarson, and J.D. Dow, J. Phys. Chem. Solids **44**, 365 (1983).
[16] Y.C. Chang and D.E. Aspnes, Phys. Rev. B **41**, 12002 (1990).
[17] J.-M. Jancu, R. Scholz, F. Beltram, and F. Bassani, Phys. Rev. B **57**, 6493 (1998).
[18] Semiconductors: Group IV Elements and III–V Compounds, Ed. O. Madelung, Springer-Verlag, Berlin 1991; Landolt-Börnstein: Numerical Data and Functional Relationships in Science and Technology, Ed. O. Madelung, Group III, Vol. 17a, Springer-Verlag, Berlin 1982 and Vol. 22a, Springer-Verlag, Berlin 1987.
[19] D.J. Chadi, Phys. Rev. B **16**, 790 (1977).
[20] Y.-T. Lu and L.J. Sham, Phys. Rev. B **40**, 5567 (1989).
[21] T.B. Boykin, Phys. Rev. B **56**, 9613 (1997).
[22] X. Zhu and S.G. Louis, Phys. Rev. B **43**, 7840 (1991).
[23] R. Scholz, J.-M. Jancu, and F. Bassani, MRS Symp. Proc. **491**, 383 (1998).
[24] J. Batey and S.L. Wright, Surf. Sci. **174**, 320 (1986).
[25] S.-H. Wei and A. Zunger, Phys. Rev. Lett. **59**, 144 (1987).

[26] M. Holtz, R. Cingolani, K. Reimann, R. Muralidharan, K. Syassen, and K. Ploog, Phys. Rev. B **41**, 3641 (1990).
W. Ge, W.D. Schmidt, M.D. Sturge, L.N. Pfeiffer, and K.W. West, J. Lum. **59**, 163 (1994).
M. Nakayama, K. Imazawa, K. Suyama, I. Tanaka, and H. Nishimura, Phys. Rev. B **49**, 13564 (1994).
[27] G. Li, D. Jiang, H. Han, Z. Wang, and K. Ploog, Phys. Rev. B **40**, 10430 (1989).
[28] R. Cingolani, L. Baldassare, M. Ferrara, M. Lugarà, and K. Ploog, Phys. Rev. B **40**, 6101 (1989).
[29] S.-H. Wei and A. Zunger, J. Appl. Phys. **63**, 5794 (1988).
[30] K.A. Mäder and A. Zunger, Phys. Rev. B **50**, 17393 (1994).
[31] D.M. Wood and A. Zunger, Phys. Rev. B **53**, 7949 (1996).
[32] L.-W. Wang, A. Franceschetti, and A. Zunger, Phys. Rev. Lett. **78**, 2819 (1997).
[33] J.-M. Jancu, V. Pellegrini, R. Colombelli, F. Beltram, B. Mueller, L. Sorba, and A. Franciosi, Appl. Phys. Lett. **73**, 2621 (1998).
[34] J.W. Matthews and A.E. Blakeslee, J. Cryst. Growth **27**, 118 (1974).
[35] J.-M. Jancu, R. Scholz, A. di Carlo, and F. Beltram, Superlattices and Microstructures **25**, 351 (1999).
[36] J.-H. Huang, T.Y. Chang, and B. Lalevic, Appl. Phys. Lett. **60**, 10 (1992).
[37] S. De Franceschi, J.-M. Jancu, and F. Beltram, Phys. Rev. B **59**, 9691 (1999).
[38] A. Di Carlo, phys. stat. sol. (b) **217**, 705 (2000).
[39] F. Della Sala, A. Di Carlo, P. Lugli, F. Bernardini, V. Fiorentini, R. Scholz, and J.-M. Jancu, Appl. Phys. Lett. **74**, 2002 (1999).

phys. stat. sol. (b) **217**, 461 (2000)

Subject classification: 61.43.Dq; 64.70.Dv; 71.15.Pd; 71.23.Cq; S5.11

Constant-Pressure Molecular Dynamics of Amorphous Si

H. M. Urbassek and P. Klein[1])

Fachbereich Physik, Universität Kaiserslautern, Erwin-Schrödinger-Straße, D-67663 Kaiserslautern, Germany

(Received August 10, 1999)

Metastable materials like amorphous silicon (a-Si) are created under non-equilibrium conditions, such as quenching from the melt. Therefore, in contrast to the equilibrium phases, careful control of the external thermodynamic variables (temperature and pressure) is important, as they influence the quality of the material produced. We present a scheme for incorporating temperature and pressure control in tight-binding molecular dynamics. It is applied to a-Si formation from the melt under zero external pressure. The influence of the control mechanisms is studied. The characteristics of the prepared a-Si specimen are discussed.

1. Introduction

Amorphous silicon (a-Si) can be considered as a prototype of a metastable material. It has been intensely investigated [1, 2] in particular due to its importance in applications such as photovoltaics. But also from a fundamental point of view, as an elementary amorphous material, it is well suited to demonstrating the complexities of this non-equilibrium phase of materials: in contrast to equilibrium phases, its properties are not uniquely defined, but can fluctuate locally, and in particular between different samples. More importantly, its properties depend on the production method. We concentrate here on the most frequently applied method, i.e., rapid quenching from the molten phase. However, we note that a-Si production by ion irradiation [3] of crystalline silicon (c-Si) offers an interesting alternative.

These complexities have their counterpart on the computational side. One problem – which is shared by all computational materials studies – is certainly the proper identification of a suitable interatomic potential. Due to the large interest in a-Si, a large number of conputational studies have been performed on it, ranging from classical empirical potentials [4, 5] to so-called ab initio density functional studies [6]. Intermediate in terms of computer-time requirements is the so-called tight-binding (TB) molecular dynamics method, which is the subject of these proceedings. As a common rule, however, these studies employ molecular dynamics methods, and simulate the quench of a liquid silion (l-Si) sample to the amorphous phase. A problem which all theses methods have to face is that the Hamiltonians applied obviously must be able to describe both the metallic liquid and the semiconducting amorphous phase of Si. Both phases are characterized by rather different local structures; thus, the average coordination in l-Si is around 6, while in a-Si it is around 4. Hence the general problem of potential transferability is a particularly stringent requirement for the Si system.

---

[1]) Present address: ITWM, Erwin-Schrödinger-Straße, D-67663 Kaiserslautern, Germany.

But also the careful control of the external thermodynamic variables under which the quench from the molten phase is performed is important in calculation, as it is in experiment. While this point has been discussed in terms of the quench speed [5, 7] – that is the control of the temporal decrease of the specimen temperature – as a rule the external pressure has not been controlled in simulations, while of course, in experiment it is well defined at 1 bar, and that is virtually zero in comparison to the relevant elastic moduli of Si. Rather, in the simulations usually the density of the specimen is prescribed; as an example, in the ab initio study [6] the density is scaled linearly during the quench from the experimental density of l-Si to the experimental density of a-Si. However, for the potential employed, the experimental density needs not to be the equilibrium density in the computation – neither in the liquid nor in the amorphous phase. Thus, we found recently [8] that in the well known and often employed Goodwin-Skinner-Pettifor (GSP) [9] parametrization of a Si TB scheme, the zero-pressure equilibrium density in the simulated liquid deviates by 10% from the experimental density. Hence, the quench procedure described above exerts considerable external stress during the quench and has a definite influence on the quality of the material produced. In particular, GSP a-Si produced from the experimental density is almost metallic, displaying no band gap.

Hence we report here on a scheme which combines temperature and pressure control in TB molecular dynamics, and illustrate its use by preparing an a-Si specimen by quench from the melt.

2. Method

We perform our simulations using a tight-binding molecular dynamics scheme in the formulation of Frauenheim et al. [10]. However, we go beyond the minimal sp-basis set used in that work, by including d-orbitals [11]. The radial d-wave function has been optimized to obtain good agreement to the ab initio band structure of crystalline Si. The matrix elements of the Hamiltonian and of the wave function overlap matrix are given elsewhere [12]. $N = 128$ Si atoms move in a rectangular supercell with periodic boundary conditions.

In contrast to computations in insulating systems, we use a finite electron temperature T_e in order to describe the liquid metallic phase; otherwise strongly discontinuous forces would result as soon as an energy eigenvalue crosses the Fermi surface. We set T_e equal to the external temperature of the system, although the exact value of $T_e > 0$ does not much influence the result.

2.1 Pressure

Let us consider as a first step a classical system of N particles, interacting via a pair potential $V_{cl}(r_{ij})$. Here, $r_{ij} = |\mathbf{r}_i - \mathbf{r}_j|$ denotes the distance between particles i and j. The pressure tensor consists of a kinetic and a potential contribution. Its components $P_{cl}^{\alpha\beta}$ – upper Greek indices denote Cartesian components of vectors and tensors ($\alpha, \beta = 1, 2, 3$) – can be written as

$$P_{cl}^{\alpha\beta} \Omega = \sum_i \frac{p_i^\alpha p_i^\beta}{m} + \sum_{i<j} (r_i^\alpha - r_j^\alpha) F_{i,j}^\beta. \tag{1}$$

Here, Ω denotes the supercell volume, m is the atomic mass, and $\mathbf{p}_i$ the momentum of atom i. The classical force which atom i experiences from atom j amounts to

$$\mathbf{F}_{i,j} = -\frac{\partial}{\partial \mathbf{r}_i} V_{\mathrm{cl}}(r_{ij}). \tag{2}$$

In tight-binding molecular dynamics, in addition to classical (short-range repulsive) forces $\mathbf{F}_{\mathrm{cl}}$, which can be treated as above, valence electron contributions have to be added. Thus, the total energy can be written as

$$V_{\mathrm{tot}} = \sum_{i<j} V_{\mathrm{cl}}(r_{ij}) + \sum_{\kappa} \gamma_\kappa \epsilon_\kappa, \tag{3}$$

where γ_κ is the occupation number of electron state κ, and ϵ_κ its one-electron eigenenergy. Here and in the following, lower Greek indices denote valence electrons ($1 \le \kappa \le 9N$). ϵ_κ is obtained in the tight-binding formalism by the solution of

$$(H - \epsilon_\kappa S) c_\kappa = 0. \tag{4}$$

Here H is the tight-binding Hamiltonian, S the overlap matrix, and c_κ is the electronic eigenvector of state κ.

At $T_e = 0$, the Hellmann-Feynman theorem can be used to derive the quantum-mechanical force $\mathbf{K}_i$ on atom i from the band-structure energy $\sum_\kappa \gamma_\kappa \epsilon_\kappa$,

$$\mathbf{K}_i = -\frac{\partial}{\partial \mathbf{r}_i} \sum_\kappa \gamma_\kappa \epsilon_\kappa = -\sum_\kappa \gamma_\kappa c_\kappa^\dagger \left(\frac{\partial H}{\partial \mathbf{r}_i} - \epsilon_\kappa \frac{\partial S}{\partial \mathbf{r}_i} \right) c_\kappa. \tag{5}$$

The total pressure hence is the sum of the classical part, Eq. (1), and a quantum-mechanical part,

$$P^{\alpha\beta}\Omega = P_{\mathrm{cl}}^{\alpha\beta}\Omega - \sum_{i,j}(r_i^\alpha - r_j^\alpha)\sum_\kappa \gamma_\kappa c_\kappa^\dagger \left(\frac{\partial H_{ij}}{\partial r_i^\beta} - \epsilon_\kappa \frac{\partial S_{ij}}{\partial r_i^\beta} \right) c_\kappa. \tag{6}$$

Here, $H_{ij} = P_i H P_j$, where P_i denotes the projection on the valence basis states at atom i, and S_{ij} analogously; $c_\kappa^\dagger$ is the transpose of c_κ.

Two complications have to be taken into account:

(i) Since we use a supercell with periodic boundary conditions, we have to decide which particles we allow to interact. We do it in the usual way [13]. Let us concentrate on a simulation box shaped as a rectangular parallelepiped and oriented along a cartesian system. Its sides extend from $-l^\alpha$ to $+l^\alpha$ along each axis α and its volume is $\Omega = 8l^1 l^2 l^3$. As we always use the nearest image of a particle to calculate its interaction partners, the particle distance $\mathbf{r}_{ij} = \mathbf{r}_i - \mathbf{r}_j$ has to be modified in the above equations to

$$(r_i^\alpha - r_j^\alpha)_p = \begin{cases} r_i^\alpha - r_j^\alpha, & \text{if } |r_i^\alpha - r_j^\alpha| < l^\alpha, \\ r_i^\alpha - r_j^\alpha + 2l^\alpha, & \text{if } r_i^\alpha - r_j^\alpha < -l^\alpha, \\ r_i^\alpha - r_j^\alpha - 2l^\alpha, & \text{if } r_i^\alpha - r_j^\alpha > l^\alpha. \end{cases} \tag{7}$$

(ii) For the reasons described above, we use a finite electron temperature T_e and hence the Fermi distribution $F(\epsilon_i; T_e)$ enters into the description, such that the occupation number is $\gamma_i = 2F(\epsilon_i; T_e)$. The pertinent Fermi energy is calculated self-consistently. Hence the Hellmann-Feynman forces, Eq. (5), now read

$$K_i^\alpha = -\sum_\nu \left(2F(\epsilon_\nu; T_e) + 2\sum_\mu \epsilon_\mu \frac{\partial F(\epsilon_\mu; T_e)}{\partial \epsilon_\nu} \right) \frac{\partial \epsilon_\nu}{\partial r_i^\alpha}, \tag{8}$$

where

$$\frac{\partial \epsilon_\nu}{\partial r_i^\alpha} = \sum_{kk'} c_\nu^\dagger \left(\frac{\partial H_{kk'}}{\partial r_i^\alpha} - \epsilon_\nu \frac{\partial S_{kk'}}{\partial r_i^\alpha} \right) c_\nu. \tag{9}$$

As a consequence, the total system pressure finally reads

$$P^{\alpha\beta}\Omega = P_{cl}^{\alpha\beta}\Omega$$
$$- \sum_{i,j} (r_i^\alpha - r_j^\alpha)_p \sum_\nu \left(2F(\epsilon_\nu; T_e) + 2\sum_\mu \epsilon_\mu \frac{\partial F(\epsilon_\mu; T_e)}{\partial \epsilon_\nu} \right) \sum_{kk'} c_\nu^\dagger \left(\frac{\partial H_{kk'}}{\partial r_i^\alpha} - \epsilon_\nu \frac{\partial S_{kk'}}{\partial r_i^\alpha} \right) c_\nu. \tag{10}$$

As a non-trivial point, we mention that the calculation of the derivative $\partial F(\epsilon_\mu; T_e)/\partial \epsilon_\nu$ appears to work best in a central-difference scheme: We define (slightly) shifted 'eigenvalues' $\epsilon_\nu^\pm = \epsilon_\nu \pm \delta$ (with $\delta = kT_e/20$), and then calculate self-consistently the accordingly shifted Fermi energies. This allows us to evaluate the Fermi function F slightly below and above ϵ_ν, and hence its derivative.

2.2 Control

The equations derived in Refs. [8, 14] read

$$\dot{r}_i^\alpha = \frac{1}{m} p_i^\alpha + r_i^\alpha \frac{\pi^\alpha}{l^\alpha}, \tag{11}$$

$$\dot{p}_i^\alpha = F_i^\alpha + K_i^\alpha - p_i^\alpha \frac{\pi^\alpha}{l^\alpha} - p_i^\alpha \frac{\xi^\alpha}{N}, \tag{12}$$

$$\dot{l}^\alpha = \pi^\alpha, \tag{13}$$

$$\dot{\pi}^\alpha = \frac{1}{Ml^\alpha} (P^{\alpha\alpha} - P_{ext})\Omega, \tag{14}$$

$$\dot{\xi}^\alpha = \frac{1}{Q} k(T^{\alpha\alpha} - T_{ext}), \tag{15}$$

$$\dot{\eta} = \eta \sum_\alpha \xi^\alpha. \tag{16}$$

These (non-Newtonian) dynamical equations have been derived using the 'extended system' concept [13], and combine elements of the Nosé-Hoover thermostat [15, 16] and of Andersen's [17] pressure control.

If the normal stress, $P^{\alpha\alpha}$, on any side of the simulation box does not coincide with the exterior pre-determined pressure P_{ext}, the corresponding length l^α is changed, Eqs. (13), (14). In these equations, π is the extension speed of the simulation box, and M is the inertial mass that regulates the response speed of the system on pressure deviations. In this work, we choose M equal to the total mass of the system. We note that the more general (classical) Rahman-Parrinello approach [18] allows to change the shape of the simulation box in addition to its side lengths. This does, however, provide no advantage in non-crystalline systems, since the lack of anisotropy in the shear moduli — and in particular their vanishing in the case of liquids — leads to strong fluctuations in the box shape, which are irrelevant to the physics of the problem and even disturbing.

Analogously, we introduce in Eqs. (15) and (16) a temperature control which serves to counteract any deviation of the 'temperature components',

$$T^{aa} = \frac{1}{Nk} \sum_i \frac{(p_i^a)^2}{m}, \tag{17}$$

from the prescribed external temperature T_{ext}. Here k denotes the Boltzmann constant. The quantity ξ has the dimension of a frequency, and acts in a friction-like force – which may, however, have either sign – damping or exciting particle motion in order to achieve the required kinetic energies, cf. Eq. (12). Q is a 'temperature inertia' controlling the reaction speed of the system to temperature deviations. The careful choice of its value is crucial for obtaining an optimal performance of the algorithm, cf. the discussion in Section 3.1. The quantity η is not needed for the formulation of the dynamical equations, but will be used below in the construction of a conserved enthalpy, Eq. (18). It is conjugated to temperature, and shares some properties with an entropy of the system.

Equation (11) takes account of the fact that any length scaling of the simulation volume influences the particle speeds. Finally, Eq. (12) sums all 'forces' that change the particle momenta: these comprise besides the classical Newtonian forces $\mathbf{F}_i$, Eq. (2), and the Hellmann-Feynman forces $\mathbf{K}_i$, Eq. (5), also the friction-like forces determined from the temperature control, Eq. (15). Finally, also changes in the box size exert forces on the particles, which may be understood in analogy to (adiabatic) temperature changes upon contraction or expansion of the system.

It can be shown that the so-called generalized enthalpy

$$\mathcal{H} = \sum_i \frac{\mathbf{p}_i^2}{2m} + V_{\text{tot}} + Q\xi^2/2 + M\pi^2/2 + kT_{\text{ext}}\ln\eta + P_{\text{ext}}\Omega \tag{18}$$

is constant along any system trajectory. We shall use this function to check the numerical accuracy of our simulation, cf. Section 3.2. We note that as integration routine, we use algorithm II of Ref. [14].

We finally mention that for applications of pressure and temperature control such as they are of interest here – viz. the isobaric quench of a material – a partitioning of the supercell volume into an 'outer' and an 'inner' part may prove useful: In the inner part, atoms follow a 'free', i.e., iso-energetic dynamics, while in the outer part, pressure control and temperature control with a (time-dependent) external temperature $T_{\text{ext}}(t)$ is prescribed. This partitioning is achieved by changing the control parameters smoothly whithin the supercell volume; for details see Ref. [14]. An application of this partitioning scheme has been recently given in Ref. [19].

3. Results

The liquid Si structure compares well with available experimental and ab initio data [20]. In particular, its mass density amounts 2.58 g/cm³, that is 10% above the c-Si density, as in experiment. The liquid is characterized in detail in Ref. [11]. We summarize the small deviations to the experimental observations, which are found on top of the general good agreement: A slightly too small average coordination is found, accompanied by a deficit of bond angles at 60°. The phonon spectrum shows a slightly too large weight of high-frequency vibrations, corresponding to the optical phonons of the crystalline phase; simultaneously the velocity autocorrelation function shows – albeit moderately – the so-called caging effect, which is absent in the ab initio data.

3.1 Preparation: dependence on quench rate

We display in Fig. 1 the characteristics of the phase transition of the liquid Si to the amorphous phase. We note that the data have been smoothed over a time period of 0.6 ps to suppress noise. Three different quenches have been performed, characterized by different values of the temperature inertia Q, which has been chosen as $Q_i = 1\ (8, 16) \times 10^6$ at. units for quench number $i = 1\ (2, 3)$. Fig. 1a shows that a large value of Q corresponds to a smaller quench rate, as is evident. Furthermore, it is seen that for a too fast quench rate, no real phase transition occurs; for the smaller quench rates, on the other hand, the temperature has a plateau around 1500 K, indicating the latent heat which

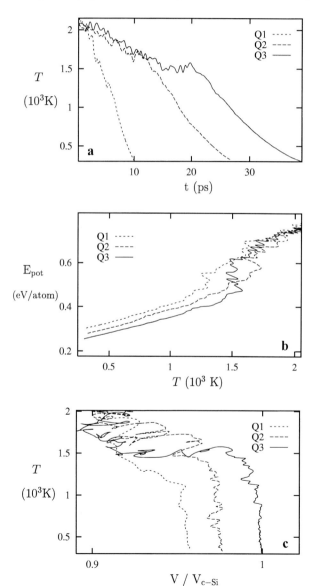

Fig. 1. a) Variation of temperature T vs. time t during quench of liquid Si to the amorphous phase at zero pressure. b) Potential energy E_{pot} relative to the potential energy of c-Si at 300 K, and c) simulation volume V, normalized to the volume V_{c-Si} of crystalline silicon, vs. temperature during quench. The different data sets apply to different quench rates, impemented by the temperature inertias Q_i with values given in the text

has to be delivered during the phase transition. Below 1500 K, temperature decreases continuously with time. The latent heat release can also be seen in Fig. 1b as the loss in potential energy per atom occurring at the phase-transition temperature. Here, however, it becomes furthermore evident that with a smaller quench rate the 'best' amorphous structures are created, in the sense that they have the least potential energy. However, at 300 K our best structure is still 0.25 eV above the c-Si energy, while in experiment, the energetic difference between a-Si and c-Si, while depending on the preparation conditions, is below 0.18 eV/atom [21, 22]. We note that the ab initio calculations of [6] give an even larger energetic difference of 0.28 eV/atom. This exemplifies that small quench rates are needed to simulate experimental a-Si structures. Finally, Fig. 1c demonstrates that the small quench rate is also important for obtaining the correct density, which is in experiment within 1% equal to the c-Si density [3]. We achieved this value with our small quench rate. Obviously a too fast quench conserves liquid-like coordination structures in the a-Si specimen, which consequently has a too high density. This is corroborated by an analysis of atomic coordination, which shows 10% or more over-coordinated atoms in the amorphous structure for the faster quenches. The rather distinct dependence of the a-Si density on the quench rate, even for rather low quenches, is astonishing.

3.2 Numerical accuracy: the conserved enthalpy

Figure 2 shows the time evolution of the generalized enthalpy during the quench. For our pressure- and temperature-controlled scheme, it is a conserved quantity. Due to the finite time step employed in the integration algorithm, it fluctuates slightly. Note that these fluctuations, which are of the order of 2 meV/atom at most, are negligible in comparison to temperature, which corresponds to a kinetic energy of 26 meV/atom at the end of the quench. Fig. 2 hence demonstrates that the time step chosen (80 at. units) is adequate, and also that our integration algorithm works correctly; in particular, the conserved enthalpy shows no drift.

3.3 Characteristics of a-Si

In the following we describe some of the characteristics of the a-Si specimen prepared by quenching from the melt using the smallest quench rate. We note that this structure shows several deviations from the a-Si structure on which we reported in Ref. [11] and which has been obtained using the same Hamiltonian. The reason here-

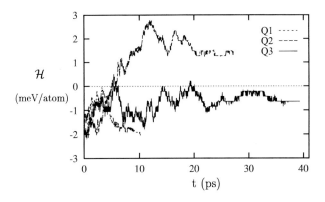

Fig. 2. Generalized enthalpy $\mathcal{H}$ (with arbitrary zero) as a function of time t during the amorphisation process. The different data sets apply to different quench rates, implemented by the temperature inertias Q_i with values given in the text

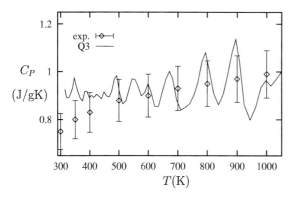

Fig. 3. Specific heat at constant pressure, C_P, vs. temperature T of a-Si, as prepared by the slowest quench rate, and compared to experimental data [25]

to is mainly that the present structure is based on a 128-atom supercell, while the previous study employed only 64 atoms; therefore we consider the present results as more trustworthy.

Immediately from Fig. 1a, we can obtain the specific heat for constant pressure ($p = 0$) during the quench. The data obtained below the phase transition temperature are reproduced in Fig. 3. They agree satisfactorily with experimental data. The decrease of the experimental data below about 500 K is due to quantum effects which start affecting the high-frequency vibrations below this temperature, and hence cannot be modelled in our simulation.

After the quench was finished and the specimen had attained 300 K, we equilibrated it for 5 ps, during which time interval also the data displayed below have been taken. The pair correlation function, Fig. 4, is in good agreement with the ab initio data, both in the location and width of the first peak, and in the first minimum. 96.9% of the atoms are on average fourfold coordinated; we find only 1.6% fivefold and 1.5% threefold coodinated atoms. This compares well with the available ab initio data [6], where 96.4% fourfold coordination was found. Note that the superior quality of the more slowly quenched a-Si specimen with respect to the quicker quench shows up mainly in a broader nearest-neighbor peak and, concomitantly, a reduced gap to the second-nearest neighbor shell. Also the bond-angle distribution, Fig. 5, shows good agreement to the ab initio data. The slight secondary maximum at 60° bond angle parallels the one found in the ab initio description, and must be interpreted as a remnant from the liquid phase. This 60°-contribution is considerably more pronounced for the fast quench. While our

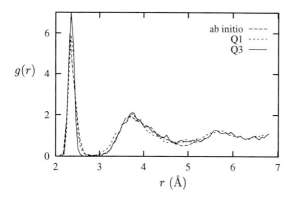

Fig. 4. Pair distribution function $g(r)$ of a-Si, as prepared by the slowest quench rate, compared to ab initio data [6]

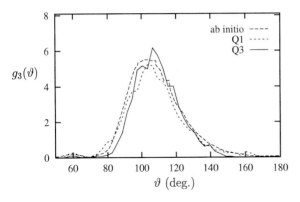

Fig. 5. Bond-angle distribution $g_3(\vartheta)$ of a-Si, as prepared by the slowest quench rate, compared to ab initio data [6]

mean bond angle appears to be slightly larger than in the ab initio simulation, the width of our distribution is very close to the ab initio width; the experimental [23] width (10°) is considerably smaller than the width of either simulation. Note that the quench rate in the ab initio simulation amounted to only 17 ps in comparison to the 40 ps of our slow quench (cf. Fig. 1a); this fact may explain the slightly stronger tails of the ab initio bond angle distribution.

The vibrational power spectrum is displayed in Fig. 6 and compared to experimental and ab initio data. While the experiment shows almost equal weight for optical and acoustical phonon contributions, the tight-binding results display a slight overestimation of the high-frequency optical contribution, which appears furthermore shifted to higher frequencies. So the tight-binding Si structure appears somewhat too hard in comparison to ab initio, and also to real a-Si.

The electronic density of states shows a clear gap at the Fermi energy. Instead of displaying these data, we plot immediately in Fig. 7 the optical excitation gap, which is presented here in the form of a Tauc plot, showing the imaginary part of the dielectric constant ϵ_2 as a function of the optical excitation frequency ω. ϵ_2 is calculated according to Ref. [24]. The calculated optical gap is evaluated as usual [24] from a straight-line fit to the high-frequency data. This procedure is based on the assumption of parabolic band shapes for delocalized states and hence ignores the −optically inactive − localized states at the band edge.

We also include in this figure the results of our faster quenches, which allows us to read off the considerably superior electronic quality of our slow-quench a-Si specimen.

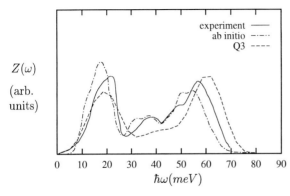

Fig. 6. Vibrational power spectrum $Z(\omega)$ of a-Si, compared to experimental [26] and ab initio data [6]. The $Z(\omega)$ data sets have been normalized to equal area

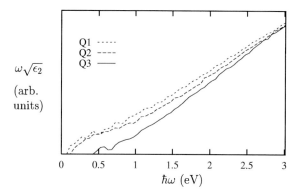

Fig. 7. Tauc plot: imaginary part of dielectric constant ϵ_2 vs. optical excitation frequency ω. The different data sets apply to different quench rates, implemented by the temperature inertias Q_i with values given in the text

Its optical gap amounts to around 0.8 eV, which is considerably larger than the gaps obtained by the faster quenches. Note, however, that the experimental value [1] amounts to 1.3 eV. We attribute this discrepancy to the fact that simulational a-Si is not sufficiently relaxed with respect to experimental a-Si; this must of course show up in a larger number of localized states, which tend to shrink the band gap.

4. Conclusions

Amorphous materials are an example of metastable phases; they are produced – in the calculation as well as in the experiment – by non-equilibrium processes. A careful control of the external thermodynamic variables (temperature and pressure) during the production process is important. For example, in the GSP-TB scheme [9], a-Si produced from l-Si at the experimental density shows quite different properties than when produced under zero pressure [8]: it is under considerable stress and exhibits almost metallic electronic properties with a vanishing band gap.

We presented a scheme for temperature and pressure control for application in tight-binding molecular dynamics calculations. It is straightforward to implement. As a benefit, we could derive a generalized enthalpy, which is conserved under the non-Newtonian time evolution of the system in our scheme. This enthalpy can be used to check the accuracy of the integration routines used.

By way of example, we prepared an 128 atom a-Si specimen by quenching from the melt. Here, we used an spd-TB Hamiltonian which predicts l-Si at zero pressure at the correct experimental density. The characteristics of the quenched a-Si specimen are in favorable agreement with experimental data. By varying the appropriate control parameter (the so-called temperature inertia), and hence the quench rate in the algorithm, the considerable influence of the quench rate on the a-Si characteristics could be demonstrated.

Acknowledgement We are grateful to Th. Frauenheim for discussions and for making available the Si Hamiltonian to us and acknowledge financial support by the Deutsche Forschungsgemeinschaft.

References

[1] H. FRITZSCHE (Ed.), Properties of Amorphous Silicon, 2nd ed., INSPEC, London 1989.
[2] H. FRITZSCHE, Amorphous Silicon and Related Materials, World Scientific Publ. Co., Singapore 1989.

[3] D. L. Williamson, S. Roorda, M. Chicoine, R. Tabti, P. A. Stolk, S. Acco, and F. W. Saris, Appl. Phys. Lett. **67**, 226 (1995).
[4] M. D. Kluge, J. R. Ray, and A. Rahman, Phys. Rev. B **36**, 4234 (1987).
[5] W. D. Luedtke and U. Landman, Phys. Rev. B **40**, 1164 (1989).
[6] I. Štich, R. Car, and M. Parrinello, Phys. Rev. B **44**, 11092 (1991).
[7] G. Servalli and L. Colombo, Europhys. Lett. **22**, 107 (1993).
[8] P. Klein and H. M. Urbassek, phys. stat. sol. (b) **207**, 33 (1998).
[9] L. Goodwin, A. J. Skinner, and D. G. Pettifor, Europhys. Lett. **9**, 701 (1989).
[10] T. Frauenheim, F. Weich, T. Köhler, S. Uhlmann, D. Porezag, and G. Seifert, Phys. Rev. B **52**, 11492 (1995).
[11] P. Klein, H. M. Urbassek, and T. Frauenheim, Comput. Mater. Sci. **13**, 252 (1999).
[12] P. Klein, Ph.D. Thesis, University Kaiserslautern, 1998.
[13] D. C. Rapaport, The Art of Molecular Dynamics Simulation, Cambridge University Press, Cambridge 1995.
[14] P. Klein, Modelling Simul. Mater. Sci. Engng. **6**, 405 (1998).
[15] S. Nose, J. Chem. Phys. **81**, 511 (1984).
[16] W. G. Hoover, Phys. Rev. A **31**, 1695 (1985).
[17] H. C. Andersen, J. Chem. Phys. **72**, 2384 (1980).
[18] M. Parrinello and A. Rahman, Phys. Rev. Lett. **45**, 1196 (1980).
[19] P. Klein, H. M. Urbassek, and T. Frauenheim, Phys. Rev. B **60**, 5478 (1999).
[20] I. Štich, R. Car, and M. Parrinello, Phys. Rev. B **44**, 4262 (1991).
[21] S. Roorda, S. Doorn, W. C. Sinke, P. M. L. O. Scholte, and E. van Loenen, Phys. Rev. Lett. **62**, 1880 (1989).
[22] S. Roorda and W. C. Sinke, Mater. Res. Soc. Symp. Proc. **205**, 9 (1992).
[23] J. Fortner and J. S. Lannin, Phys. Rev. B **39**, 5527 (1989).
[24] G. A. N. Connell, in: Amorphous Semiconductors, Vol. 36, Topics Appl. Phys., Ed. M. H. Brodsky, Springer-Verlag, Berlin 1979 (p. 73).
[25] J. C. Brice, in: Properties of Amorphous Silicon, 2nd ed., Ed. H. Fritzsche, INSPEC, London 1989 (p. 480).
[26] W. A. Kamitakahara, C. M. Soukoulis, H. R. Shanks, U. Buchenau, and G. S. Grest, Phys. Rev. B **36**, 6539 (1987).

phys. stat. sol. (b) **217**, 473 (2000)

Subject classification: 61.72.Ji; 66.30.Jt; 68.35.Bs; 71.15.Mb; S7.12; S7.14

Structures, Energetics and Electronic Properties of Complex III–V Semiconductor Systems

M. HAUGK (a), J. ELSNER (a), TH. FRAUENHEIM (a), T.E.M. STAAB (b),
C.D. LATHAM (c), R. JONES (c), H.S. LEIPNER (b), T. HEINE (a), G. SEIFERT (a),
and M. STERNBERG (a)

(a) *Universität GH Paderborn, Fachbereich Physik, D-33098 Paderborn, Germany*

(b) *Martin-Luther-Universität, Fachbereich Physik, Friedemann-Bach Platz 6, D-06108 Halle, Germany*

(c) *Department of Physics, University of Exeter, Exeter, EX4 4QL, UK*

(Received August 10, 1999)

A parallel implementation of the selfconsistent-charge density-functional based tight-binding (SCC-DFTB) method is used to examine large scale structures in III–V semiconductors. We firstly describe the parallel implementation of the method and its efficiency. We then turn to applications of the parallel code to complex GaAs systems. The geometries and energetics of different models for the $\sqrt{19} \times \sqrt{19}$ reconstruction at the $(\bar{1}\bar{1}\bar{1})$ surface are investigated. A structure containing hexagonal rings of As at the surface consistent with STM experiments is found to be stable under Ga-rich growth conditions. We then examine voids in the bulk material which are mainly caused by the movement of dislocations. Void clusters of 12 missing atoms are found to be energetically favorable. This is in very good agreement with recent positron annihilation measurements. Additionally, we investigate the diffusion of C in p-type material and suggest a diffusion path with an activation energy of less than 1 eV which is consistent with experimental studies. Finally, focusing on GaN we provide atomistic insight into line defects in wurtzite GaN threading along the growing c-axis. We highlight the stability and electronic properties of screw and edge dislocations, discuss reasons for the formation of nanopipes and relate the yellow luminescence observed in highly defected materials to deep acceptors, V_{Ga} and $V_{Ga}-(O_N)_n$, trapped at threading edge dislocations.

1. Introduction

In the last few years a lot of progress has been made in developing *ab initio* methods to compute the total energy, electronic structure and elastic properties of molecules, clusters and solid state systems. In particular selfconsistent field density-functional theory (SCF-DFT) based methods have been made more efficient by involving pseudopotentials and implementing the code to parallel machines [1, 2]. This has lead to a remarkable reduction in the computing time, so that on parallel machines it is now possible to include more than ≈ 200 atoms in the calculations. Large surfaces [3] or clusters [4] can now be investigated by SCF-DFT. There is, however, a need for methods capable of examining much larger structures within a reasonable accuracy at a high predictive level. To this end a selfconsistent-charge density-functional based tight-binding (SCC-DFTB) formalism has recently been developed (see [5] and references therein). This method can be applied in the standard non-selfconsistent-charge approach to systems with small interatomic charge transfer as well as in the selfconsistent-charge mode to investigate systems where the delicate balance of interatomic charge transfer between atoms having different electronegativity plays an important role. It has been used to examine systems with sizes ranging from small clusters [6 to 8] over fullerenes [9, 10], surfaces and interfaces [11 to 13] to extended defects [14] of a variety of semiconductors and provides an accuracy comparable to SCF-DFT calculations. However, due to memory restrictions and computation time, at the moment on standard workstations the SCC-DFTB method is limited to structures up to ≈ 500 atoms. This size is still too small to model some interesting technologically relevant problems. Moreover, a speed-up of the SCC-DFTB method is desirable for investigations of medium size structures with metallic character or systems where a large configuration space requires many structures to be examined. To meet these demands we have implemented SCC-DFTB on parallel computers making use of the ScaLAPACK, PBLAS and BLACS libraries. The code is efficient and portable to most of the parallel platforms.

This paper is organized as follows: In Section 2 we briefly describe the SCC-DFTB scheme and the way it can be implemented in an efficient way on parallel machines. In the following sections we apply the parallel SCC-DFTB code to problems which could previously not be examined due to their large system size, and/or metallic behaviour and/or large configuration space. The examples chosen are the $\sqrt{19} \times \sqrt{19}$ reconstruction at the GaAs($\bar{1}\bar{1}\bar{1}$) surface, voids in bulk GaAs and the diffusion of carbon in GaAs are discussed.

2. The SCC-DFTB Method and Its Parallel Implementation

2.1 The SCC-DFTB method

The SCC-DFTB method is described in detail in Elstner et al. [5, 15]. It is based on a second-order expansion of the Kohn–Sham total energy functional with respect to charge density fluctuations at a given reference density n_0, $n(\mathbf{r}) = n_0(\mathbf{r}) + \delta n(\mathbf{r})$. Retaining only the leading monopolar term of atom-centred charge fluctuations calculated by a Mulliken-charge analysis one finally obtains an approximate tight-binding-(TB)-like expression for the Kohn–Sham energy functional,

$$E_2^{TB} = \sum_i^{occ} \langle \Psi_i | \hat{H}_0 | \Psi_i \rangle + E_{rep}(n_0) + \frac{1}{2} \sum_{\alpha,\beta}^{N} \gamma_{\alpha\beta} \Delta q_\alpha \Delta q_\beta. \qquad (1)$$

The first term represents the leading matrix element of the many-atom Hamiltonian $\hat{H}_0$ and the second term stands for a short-range repulsive pair interaction, both taken at the reference density n_0. The latter beside the ion–ion repulsion includes the double-counting Hartree and exchange correlation contributions. While just these two terms are usually considered within standard non-selfconsistent TB theory now in our new approach a third term in (1) is added, explicitly introducing long-range Coulomb interactions correcting the total energy due to fluctuations in the charges $\Delta q_\alpha = q_\alpha - q_\alpha^0$ centred at atom-sites α (q_α is the actual electronic charge of atom α, q_α^0 is the number of electrons of an isolated atom). Assuming exponentially decaying spherical charge densities the interaction integral $\gamma_{\alpha\beta}$ can be evaluated analytically [5], yielding the pure Coulomb interaction between two charge distributions located at $\mathbf{R}_\alpha, \mathbf{R}_\beta$ in the limit of large distances $|\mathbf{R}_\alpha - \mathbf{R}_\beta|$. For $\alpha = \beta$ the functional $\gamma_{\alpha\beta}$ gives the self-interaction contribution of atom α.

Within our tight binding scheme we expand the one-electron wave functions Ψ_i in the expression for the total energy (1) into contracted, atom centred, valence orbitals

φ_ν obtained in an atomic SCF-LDA calculation [6],

$$\Psi_i(\mathbf{r}) = \sum_\nu^M c_{\nu i}\varphi_\nu(\mathbf{r} - \mathbf{R}_\nu). \tag{2}$$

Since we apply the Γ-point approximation ($e^{i\mathbf{k}\cdot\mathbf{r}} = 1$), the basis set expansion (2) can also be used for periodic systems. It is well known that a Γ-point approximation may lead to incorrect results if the supercell has small admeasurements in one or more dimensions. However, the parallel code has been designed for structures consisting of hundreds to thousands of atoms, where the supercells are large enough so that it is nearly always sufficient to include only the Γ-point in the calculations. We have checked the convergence of energy with respect to the size of the supercell for all results we present in this paper. Only supercells which are converged in energy are used for further calculations.

For future applications where it might be necessary to include more **k**-points, the calculation of energies for each of these points will be distributed to a set of processors.

2.1.1 Eigenvalue problem

By applying the variational technique for minimizing the total energy expression (1) within our basis set (2) the secular equation for the determination of the wave function expansion coefficients $c_{\nu i}$ and the eigenvalues ϵ_i is derived,

$$\sum_\nu^M c_{\nu i}(H_{\mu\nu} - \varepsilon_i S_{\mu\nu}) = 0; \qquad \forall \mu, i, \tag{3}$$

with Hamilton matrix elements

$$H_{\mu\nu} = \langle\varphi_\mu|\hat{H}_0|\varphi_\nu\rangle + \tfrac{1}{2}S_{\mu\nu}\sum_K^{N_{\text{nuc}}}(\gamma_{IK} + \gamma_{JK})\Delta q_K = H^0_{\mu\nu} + H^1_{\mu\nu}, \tag{4}$$

and overlap matrix elements

$$S_{\mu\nu} = \langle\varphi_\mu|\varphi_\nu\rangle. \tag{5}$$

M is the number of basis functions. Equation (3) has to be solved within an iterative cycle until selfconsistency for the charge fluctuations $\Delta q_I = q_I - q_I^0$ (q_I is the electronic charge of atom I, q_I^0 is the number of electrons of an isolated atom) is reached.

2.1.2 Mulliken charges

To evaluate the atomic charges q_I we employ the Mulliken charge analysis

$$q_I = \tfrac{1}{2}\sum_{\mu\in I}^M\sum_\delta\sum_i^{\text{occ}} n_i(c^*_{\mu i}S_{\mu\delta}c_{\delta i} + c^*_{\delta i}S_{\delta\mu}c_{\mu i}). \tag{6}$$

We note that for non-periodic systems the eigenstates $c_{\nu i}$ are real. Large periodic systems which shall be considered here are very well described within a Γ-point approximation resulting again in real eigenstates. Therefore using the symmetry of the overlap matrix **S** (6) simplifies to

$$q_I = \sum_{\mu\in I}^M\sum_\delta\sum_i^{\text{occ}} n_i c_{\mu i}S_{\mu\delta}c_{\delta i}. \tag{7}$$

2.1.3 Atomic energies

For many problems, like the calculation of surface energies or defect formation energies, it is very useful to know the energy contribution of the single atoms. Therefore this atomic energy is derived from (1) as the energy contribution of a single atom I,

$$E_I = \sum_{\mu \in I}^{M} \sum_{\nu} \sum_{i} n_i c_{\mu i}^* c_{\nu i} H_{\mu\nu}^0 + \frac{1}{2} \sum_{J}^{N_{\text{nuc}}} (q_I - q_I^0)(q_J - q_J^0) \gamma_{IJ} + \sum_{J}^{J \neq I} \Phi(I, J). \tag{8}$$

2.1.4 Forces

Finally an analytic expression for the interatomic forces follows by taking the derivative of the final SCC-DFTB energy (1) with respect to the nuclear coordinates:

$$\mathbf{F}_I = -\sum_{i}^{\text{occ}} n_i \sum_{\mu\nu}^{M} c_{\mu i}^* c_{\nu i} \left(\left[\frac{\partial H_{\mu\nu}^0}{\partial \mathbf{R}_I} + \frac{1}{2} \frac{\partial S_{\mu\nu}}{\partial \mathbf{R}_I} \sum_{L}^{N_{\text{nuc}}} (\gamma_{KL} + \gamma_{JL}) \Delta q_L \right] - \varepsilon_i \frac{\partial S_{\mu\nu}}{\partial \mathbf{R}_I} \right)$$

$$- \Delta q_I \sum_{L}^{N_{\text{nuc}}} \frac{\partial \gamma_{IL}}{\partial \mathbf{R}_I} \Delta q_L - \frac{\partial E_{\text{rep}}}{\partial \mathbf{R}_I}; \quad \mu \in K, \quad \nu \in J. \tag{9}$$

2.2 Parallelization of the SCC-DFTB method

The problems described by Equations (3), (6), (8) and (10) scale with M^3, where M is the number of basis functions, and are therefore computationally expensive. Recent works [16 to 20] show that in contrast to single processor machines, where the expressions in these equations can be implemented in a straight forward manner, for parallel machines specifically developed linear algebra routines (PBLAS) should preferably be used in order to reduce the time consuming transport of data between processors. Therefore we transform the expressions into linear algebra operations, which can then be solved using the available linear algebra packages for parallel computers.

2.2.1 Eigenvalue problem

The generalized eigenvalue problem in (3) reads in matrix notation:

$$\mathbf{H}\mathbf{C}_i = \varepsilon_i \mathbf{S}\mathbf{C}_i, \tag{10}$$

where $\mathbf{H}$ is the Hamilton matrix, $\mathbf{S}$ the overlap matrix and $\mathbf{C}_i$ are the eigenvectors.

2.2.2 Mulliken charges

If $\mathbf{O}$ denotes the diagonal matrix containing the occupation numbers n_i, then we can transform (7) and write the Mulliken charges as a diagonal matrix $\mathbf{Q}$ with the charges per orbital q_μ in the diagonal,

$$\mathbf{Q} = \mathbf{S} \cdot \mathbf{C} \cdot (\mathbf{C} \cdot \mathbf{O})^T = \mathbf{S} \cdot \mathbf{C} \cdot \mathbf{O} \cdot \mathbf{C}^T. \tag{11}$$

2.2.3 Atomic energies

The first term of Equation (8) can be rewritten in analogy to the Mulliken charges (again $\mathbf{O}$ is a diagonal matrix containing the occupation numbers),

$$\mathbf{E} = \mathbf{H} \cdot \mathbf{C} \cdot (\mathbf{C} \cdot \mathbf{O})^T = \mathbf{H} \cdot \mathbf{C} \cdot \mathbf{O} \cdot \mathbf{C}^T, \tag{12}$$

where **E** is a diagonal matrix with the energy contributions of single orbitals. The energy of a single atom can be obtained by summing over the energy contributions of all orbitals belonging to this atom plus the atomspecific second and third terms in Equation (8).

2.2.4 Forces

To calculate the forces we define the matrices $\mathbf{DH}^{\mu s}$ and $\mathbf{DS}^{\mu s}$,

$$DH^{\mu s}_{\delta \nu} = \left[\frac{\partial H^0_{\delta \nu}}{\partial \mathbf{R}_{\mu s}} + \frac{1}{2} \frac{\partial S_{\delta \nu}}{\partial \mathbf{R}_{\mu s}} \sum_L^{N_{\text{nuc}}} (\gamma_{KL} + \gamma_{JL}) \Delta q_L \right]; \qquad \delta \in K, \ \nu \in J,$$

and

$$DS^{\mu s}_{\delta \nu} = \frac{\partial S_{\delta \nu}}{\partial \mathbf{R}_{\mu s}}, \tag{13}$$

where $s = x, y, z$ and $\partial/\partial \mathbf{R}_{\mu s}$ denotes the contribution to the derivative of orbital μ in the direction s. To get the gradient per atom α one simply sums over the orbitals belonging to the atom. Then the first term in (10) transforms into

$$\sum_i n_i \mathbf{C}_i^\mathsf{T} (\mathbf{DH}^{\mu s} - \varepsilon_i \mathbf{DS}^{\mu s}) \mathbf{C}_i. \tag{14}$$

The second and the third term in Equation (9) are not computationally expensive and are only evaluated on the processor where the corresponding atom is located. We now reduce the matrix–vector multiplications with the nearly empty matrices $\mathbf{DS}^{\mu s}$ and $\mathbf{DH}^{\mu s}$ in Equation (14) to more efficient expressions. Therefore we substitute $\mathbf{X} = \mathbf{DH}^{\mu s}$ to simplify the notation. In the case of $\mathbf{X} = \mathbf{DS}^{\mu s}$ the following steps are identical. Only in the μ-th row and μ-th column $\mathbf{X}$ has non-vanishing elements, since all other elements are independent of the position of orbital μ, i.e. $X_{\delta \nu} = 0$ for $\delta \neq \mu$ and $\nu \neq \mu$. We then have

$$\sum_i n_i \mathbf{C}_i^\mathsf{T} \begin{pmatrix} & & X_{1\mu} & & \\ \mathbf{0} & \dots & & \mathbf{0} & \\ X_{\mu 1} & \dots & X_{\mu \mu} & \dots & X_{\mu M} \\ \mathbf{0} & & & \dots & \mathbf{0} \\ & & X_{M\mu} & & \end{pmatrix} \mathbf{C}_i = \sum_i n_i \mathbf{C}_i^\mathsf{T} \begin{pmatrix} X_{1\mu} c_{\mu i} \\ \dots \\ \sum_\delta X_{\mu \delta} c_{\delta i} \\ \dots \\ X_{M\mu} c_{\mu i} \end{pmatrix}_{X_{\mu \mu} = 0}$$

$$= \sum_i n_i \left(\sum_\delta c_{i\delta}^\mathsf{T} X_{\delta \mu} c_{\mu i} + c_{i\mu}^\mathsf{T} \sum_\delta X_{\mu \delta} c_{\delta i} \right) = 2 \sum_i n_i c_{\mu i} \sum_\delta (X_{\delta \mu} c_{\delta i} + X_{\mu \delta} c_{\delta i})$$

$$= 2 \sum_i n_i c_{\mu i} X_{\mu *} c_{* i} = 2 \sum_i (\mathbf{X} \cdot \mathbf{C})_{\mu i} (\mathbf{C} \cdot \mathbf{O})_{\mu i}$$

$$= 2 (\mathbf{X} \cdot \mathbf{C})_{\mu *} \cdot (\mathbf{C} \cdot \mathbf{O})_{* \mu}^\mathsf{T} = 2 (\mathbf{X} \cdot \mathbf{C} \cdot \mathbf{O} \cdot \mathbf{C}^\mathsf{T})_{\mu \mu}.$$

It is therefore sufficient to evaluate the diagonal elements of the matrix product

$$\hat{\mathbf{X}}^s \cdot C \cdot O \cdot C^\mathsf{T} \tag{15}$$

for each direction $s = x, y, z$. Here $\hat{\mathbf{X}}^s$ is a full, i.e. dense, matrix. For $1 \leq \mu \leq M$ the μ-th row of $\hat{\mathbf{X}}^s$ contains the derivatives of matrix elements of $\mathbf{H}$ which involve the μ-th orbital. These derivatives are formed by moving the μ-th orbital in the s direction,

$$(\hat{X}^s)_{\mu \nu} = \frac{\partial \langle \mu | \mathbf{H} | \nu \rangle}{\partial R_{\mu s}}. \tag{16}$$

The same transformation can be done for $\mathbf{X} = \mathbf{DS}^{us}$ with

$$(\hat{X}^s)_{\mu\nu} = \frac{\partial \langle \mu | \mathbf{S} | \nu \rangle}{\partial R_{\mu s}} . \tag{17}$$

Considering the factor ε_i in Eq. (17) we get analogously to Eq. (18),

$$\hat{\mathbf{X}}^s \cdot \mathbf{C} \cdot \mathbf{O} \cdot \mathbf{V} \cdot \mathbf{C}^T , \tag{18}$$

where $\mathbf{V}$ denotes the diagonal matrix containing the eigenvalues.

2.3 Benchmarking

In a previous paper [5] we have already shown that applied to GaAs surface reconstructions the SCC-DFTB method gives surface energies and geometries in very good agreement with plane-wave SCF-LDA calculations [3]. In view of possible applications of the SCC-DFTB method to large periodic systems containing surface steps we test the performance of the parallel code for a GaAs(100) surface cell with an α2(2 × 4) reconstruction which we increase laterally to obtain larger systems. Ga and As atoms are modeled by an sp-basis (4 basis functions per atom), the pseudo hydrogen atoms saturating the bottom of the surface slabs are expressed by one s wave function. Usually the geometry is optimized using the conjugate gradient relaxation technique. In this case and for a semiconducting system on average five SCC-iterations per SCC-cycle are needed until the charges are converged.

Fig. 1 illustrates the time scaling depending on the system size for one SCC-iteration. The discrete data points were fitted to cubic polynomials (the problem scales with M^3). As can be seen run on a parallel machine the SCC-DFTB method allows to treat large systems within a reasonable time. We now compare the behavior of the fitted polynomials for 4, 16 and 64 PEs in the limit of large numbers of basis functions. The relationships of the leading coefficients of the polynomials for 4 to 16 (16 to 64) PEs are 16.4/5.1 = 3.2 (5.1/1.4 = 3.6). They are only slightly below the optimum value of 4 indicating that for large systems the degree of parallelization is quite high, i.e. the time lost in interprocessor communication has become small. However, as can be seen in

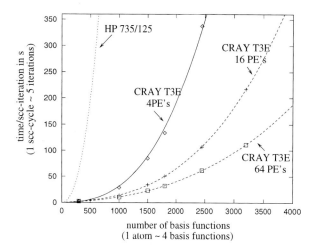

Fig. 1. Time scaling of the SCC-DFTB method for a supercell with a GaAs(100) surface

Fig. 1 the total amount of time used for small and medium size problems (matrix sizes up to 3000×3000) diminishes only by a factor of ≈ 2 or less. Indeed we have found that an efficient parallelization is achieved for problem sizes with matrices greater than $\approx 500 \times 500$/node. This corresponds to total matrix sizes of 1000×1000, 2000×2000 and 4000×4000 for 4, 16 and 64 PEs respectively.

The required memory per PE scales as $C \sim M^2/\sqrt{n_{PE}}$ where n_{PE} is the number of PEs. Assuming an sp basis and 64 nodes on the T3E at the moment a system of ≈ 2000 atoms may be examined before exceeding the available memory of 128 Mbyte/node.

In the following we will apply the parallel implementation of the SCC-DFTB method to investigate the geometries, energetics and electrical properties of complex GaAs structures.

3. The $\sqrt{19} \times \sqrt{19}$ Reconstruction at the GaAs($\bar{1}\bar{1}\bar{1}$) Surface

A lot of progress has been made during the last years in determining the geometries of semiconductor surface reconstructions. In particular, total energy calculations have been used to suggest atomic models for the most important reconstructions of the GaAs(110) [21], (111) [22] and (100) [23] surfaces which are consistent with experimental data obtained by low energy electron diffraction (LEED) [24] and scanning tunneling microscopy (STM) [25].

A common feature of all these energetically favorable models is that they obey the electron counting rule (ECR). This rule demands the energetically high Ga derived dangling bonds to be emptied in favour of filling the As derived dangling bonds having energies close to the valence band maximum (VBM). The charge transfer is often accompanied by a rehybridisation yielding sp^2 (p^3) bonded Ga(As) atoms in order to reduce the surface energy and to make the surface semiconducting. This rule is a useful tool for the construction of energetically promising surface models. To explain the 2×2 periodicity observed at the GaAs($\bar{1}\bar{1}\bar{1}$) surface in an As-rich environment, the semiconducting As trimer model was suggested by the ECR as a possible candidate, which was then confirmed by total energy calculations [26]. Unfortunately, the electron counting rule can not be used to predict any semiconducting model for the $\sqrt{19} \times \sqrt{19}$ periodicity reported in a Ga-rich environment [26 to 30]. Due to symmetry constraints this surface has to be metallic in its neutral charge state. Moreover, it is not clear whether energetically favorable metallic surfaces could be found or explained by a model as simple as the ECR. Several structures have already been suggested to explain the $\sqrt{19} \times \sqrt{19}$ periodicity [26, 31]. However, a total energy calculation by standard SCF-LDA schemes is still computationally too expensive.

Here we present calculations for the formation energies of various new models for the GaAs ($\bar{1}\bar{1}\bar{1}$) $\sqrt{19} \times \sqrt{19}$ reconstruction together with previously examined models for 2×2 reconstructions [26, 31].

3.1 Details of calculation and analysis

The ($\bar{1}\bar{1}\bar{1}$) surfaces were modeled by slabs consisting of ten monolayers with periodic boundary conditions in two dimensions. The first six monolayers were allowed to relax, while the remaining atoms were fixed to preserve the bulk lattice spacing. In order to prevent artificial charge transfer between the bottom of the slab and the surface, we saturated the Ga dangling bonds on the bottom with pseudo–hydrogen having a charge of 1.25 e. This charge corresponds to the charge per bond contributed from a tetrahedrally bound As atom [32].

The relative stabilities of two structures having different numbers of Ga and As atoms depend on the reservoir with which the atoms are exchanged in the structural transition. Questions of thermodynamic stability are therefore posed within the context of atomic chemical potentials [33]. It can be shown that the surface energy can be expressed as a function of the atomic chemical potential of one species, which we take to be μ_{Ga},

$$\gamma_0 = [E_{tot} - \mu_{GaAs}^{cryst.} N_{As} - \mu_{Ga}(N_{Ga} - N_{As})]/F, \quad (19)$$

where F denotes the area of the surface cell and $N_{As(Ga)}$ the total number of As(Ga) atoms. The allowed range of this chemical potential is then

$$\mu_{Ga(bulk)} - \Delta H_f \leq \mu_{Ga} \leq \mu_{Ga(bulk)},$$

ranging from the As-rich ($\mu_{Ga} = \mu_{Ga(bulk)} - \Delta H_f$) to the Ga-rich ($\mu_{Ga} = \mu_{Ga(bulk)}$) environment. ΔH_f is the heat of formation for GaAs, which has been determined from enthalpy measurements to be 0.74 eV [34]. To obtain absolute surface energies we use an energy density formalism [35] in a modified version for TB [36].

3.2 Results

We first investigate the surfaces with 2×2 periodicity, the As vacancy, the Ga adatom and the Ga and As trimer models. For illustrations of these structures see Ref. [3]. The absolute surface energies for the most stable 2×2 reconstructions depending on the Ga chemical potential are shown in Fig. 2. The values are very similar to those calculated by Moll et al. [3] and can be compared to the relative formation energies of Biegelsen et al. [26]. In agreement with those works we find the As trimer as the by far the most stable structure under As-rich growth conditions, whereas the Ga trimer has very high surface energy in any environment.

We now turn to models with $\sqrt{19} \times \sqrt{19}$ periodicity. This type of reconstructions can be observed by heating [27, 28] or annealing [26] the sample at about 500 °C. A signifi-

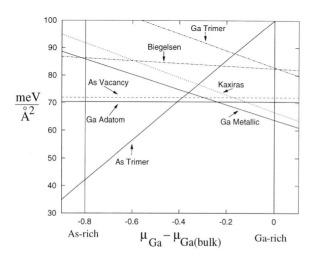

Fig. 2. Surface energies plotted versus the Ga chemical potential of different models for the 2×2 and $\sqrt{19} \times \sqrt{19}$ periodicities at the GaAs($\bar{1}\bar{1}\bar{1}$) surface. The part on the left (right) of the diagram corresponds to As(Ga) rich growth conditions

cant desorption of surface As is reported at the transition from 2×2 to $\sqrt{19} \times \sqrt{19}$ periodicity. Moreover, Woolf et al. [29] found that the $\sqrt{19} \times \sqrt{19}$ periodicity does not exist under a strong As$_4$ flux. These results clearly show that the $\sqrt{19} \times \sqrt{19}$ surface should exist in a Ga-rich environment.

Kaxiras et al. [31] suggested a model consisting of Ga trimers and threefold coordinated As atoms at the surface. If one counts all As and Ga dangling bonds in this structure (see Ref. [31] for a figure) all As lone pairs can be filled leaving 3/4 of an electron for being placed into Ga derived dangling bonds. These, however, have high energies causing the structure to relax from this configuration.

Analyzing the bonding configuration of this specific surface it is possible to describe its stability in a manner that is similar to the ECR. In contrast to semiconducting surfaces, there are fourfold coordinated Ga atoms which are not in a typical sp^n-hybridisation state. They are therefore considered to exhibit metallic Ga bonding character like in the bulk Ga crystal. Hence, the electrons could be distributed in a way that all As dangling bonds are filled, all Ga dangling bonds are emptied and excess charge is placed in the Ga bonds with metallic character.

However, the surface energy is high, see Fig. 2, predicting this model to be energetically unfavorable. This can be understood by noting that surface Ga trimers are far from being sp^2 coordinated and have thus a high surface energy, see Fig. 2.

High resolution STM images performed by Biegelsen et al. [26] furthermore indicate that the top layer As atoms should be arranged within a hexagonal ring. To match the STM images Biegelsen et al. proposed a model, where six top layer As atoms are bound to six threefold and six twofold coordinated Ga atoms. A figure of the structure before geometrical optimization was performed can be found in Ref. [26]. Again, all As lone pairs are filled. Due to the six twofold coordinated Ga atoms 3.75 electrons would have to occupy Ga derived dangling bonds. It is therefore not surprising that the atoms relax towards a configuration with only two twofold coordinated Ga atoms and yielding four additional Ga–Ga bonds. The Ga–Ga bonding is achieved by forcing the Ga atoms into nearly linear Ga–Ga–As chains, which are energetically costly. The resulting high surface energy, shown in Fig. 2, suggests this model not to correspond to the observed STM picture.

We then investigated the stability of a variety of different structures with preferably threefold coordinated Ga and As surface atoms. However, we found high surface energies for all of these structures. We then constructed models where Ga–Ga bonds could form, favouring fourfold coordinated Ga atoms.

The energetically most auspicious configuration obtained is shown in Fig. 3. In agreement with the previously discussed model, suggested on the basis of STM pictures by Biegelsen et al. [26], six top layer As atoms form a hexagonal ring. They all adopt a p^3 hybridisation with doubly occupied s-like orbitals. Fifteen threefold coordinated and three fourfold coordinated Ga atoms build the second layer. The threefold coordinated Ga atoms move to an sp^2-like hybridisation. But the most important feature is the existence of the three fourfold coordinated Ga atoms, labeled 1, 2, 3 in Fig. 3. These Ga atoms exhibit weak metallic Ga–Ga bonds with a bond length of ≈ 2.9 Å which is slightly larger than in bulk Ga (≈ 2.7 Å). As discussed for the Ga metallic bonds in the $\sqrt{19} \times \sqrt{19}$ Ga trimer model proposed by Kaxiras [31], this structure allows the non-ECR-compensated excess charge to be placed into Ga–Ga bonds before Ga derived dangling bonds would have to be occupied. All As lone

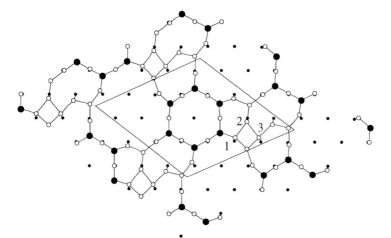

Fig. 3. Top view of the Ga metallic model. Large (small) filled circles represent top (third) layer As, empty circles second layer Ga atoms

pairs are filled. Therefore, equivalent to the $\sqrt{19}\times\sqrt{19}$ Ga trimer structure, the rule similar to the ECR may be used to explain the stability of this surface configuration. Hence, this rule can be called a modified electron counting rule. Additional to the fulfilling of the modified ECR we found another feature that favours this surface to be more stable than the other ones. All surface Ga atoms are either sp^2 hybridised (3-fold coordinated) or exhibit metallic bonding character (4-fold coordinated) and consequently do not have to adopt energetically unfavourable configurations, like the Ga trimer or the Ga–Ga–As linear chain structures. These two facts explain why the formation energy of this structure is low, see Fig. 2. It is worth noting that reconstruc-

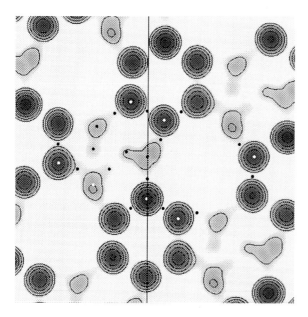

Fig. 4. Horizontal plot of charge density in 1 Å distance from the surface atoms for an energy of (1.8 ± 0.1) eV. Black and white dots mark the atom positions in the first two layers

tions of polar GaN surfaces, where metallic Ga is believed to be at the surface could be driven by a similar mechanism.

As can be seen from Fig. 2 that the proposed model is energetically favorable under Ga-rich growth conditions and could thus be a candidate for explaining the observed periodicity.

Analysing the structure in Fig. 3, one might expect the Ga metallic bonds in the most stable structure to become a remarkable feature appearing in an STM experiment. We therefore calculated the charge density within an energy range of (1.8 ± 0.1) eV below the Fermi level at 1 Å above the surface atoms (Fig. 4).

The density is plotted in a range of four orders of magnitude, from dark regions with high charge density to white regions with nearly no charge density. It can clearly be seen that the As hexagonal ring is the dominant feature which was identified in the STM measurements of Biegelsen et al. [26]. The influence of the Ga metallic bonds in the charge density distribution is very small compared to that of the dominant markers arising from the As lone pairs. The symmetry of the As hexagon is not disturbed significantly.

4. Vacancy Aggregates in GaAs

Vacancy agglomerates are well known to be produced in semiconductors by neutron irradiation [37], plastic deformation [38, 39], or saw-cutting [40]. In addition to the occurrence of voids in damaged materials, vacancy agglomerates are present even in as-grown silicon [41]. Vacancy-like defects are not only electrically active defects in semiconductors, moreover, they are regarded to play an important role in the brittle or ductile fracture of the material [42]. When excess vacancies are mobile but cannot find sinks (like surfaces or dislocations) to anneal, they are likely to agglomerate – either in a plane, hence forming a dislocation loop of the vacancy-type, or into stable three-dimensional vacancy clusters.

The primary mechanism of vacancy formation during plastic deformation is supposed to be related to dislocation motion (see Fig. 5).

Mott [43] was the first to suggest a model of emission or absorption of intrinsic point defects by screw dislocations containing jogs. Jogs can be generated by the intersection of dislocations belonging to different slip systems. Jogs on screw dislocations are not glissile. They are forced to follow the glide motion of the screw due to the line tension of the dislocation. During this nonconservative motion of the jog, vacancies or interstitials are emitted (Fig. 5). Primarily, one would expect the formation of chains of point defects as a result of this jog dragging. But there is no experimental evidence for such chains. Hence, it has been supposed [44] that the clustering of vacancies to stable agglomerates is a primary process of jog dragging. Vacancy chains should thus collapse immediately during deformation to stable clusters. This could happen by atomic re-arrangement of vacancies during or immediately after the climb step of the jog.

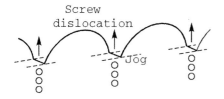

Fig. 5. Formation of point defects (vacancies and interstitials according to the sign of the jog) by the dragging of jogs at screw dislocations

Positron annihilation spectroscopy (PAS) is a suitable tool to detect small vacancy clusters. Since positron lifetime spectroscopy (POLIS) measures the electron density at the trapping site of the positron, this method is sensitive to the size of open-volume defects, i.e. the positron lifetime increases with increasing open volume [45]. Hence, it is possible to distinguish between monovacancies, vacancy chains, and three-dimensional agglomerates. Vacancy agglomerates are usually detected by POLIS in deformed metals or semiconductors [38, 39, 46]. Additionally, one finds a positron lifetime component which is typical for trapping into monovacancies (GaAs: 260 ps) or divacancy-type defects [38, 39, 44]. It has been attributed to vacancy-type defects bound to the dislocation core. The long positron lifetime component of about 500 ps measured in GaAs is an indication of large vacancy agglomerates created during deformation. The positron bulk lifetime is 230 ps.

Only in covalently bound materials there exist some indications that stable vacancy clusters of certain sizes should be present. The atomic configurations which have a low number of unsaturated or dangling bonds are considered to be favourable. This led Chadi and Chang [47] to the conclusion that there should be so called *magic numbers* of vacancies in diamond-like homonuclear structures like silicon. They proposed stable vacancy clusters for $N = 6, 10$, and 14 vacancies. This may be expanded in a straightforward manner to $N = 4i + 2$, $i = 1, 2, 3, \ldots$ since one always has to remove another four atoms in addition to hexagonal ring to end up with a surface consisting of hexagonal rings (minimum number of dangling bonds). It may now be interesting to investigate whether there exist stable clusters of certain sizes (magic numbers) in GaAs. In contrast to homonuclear systems, the existence of dangling bonds of the different atomic species make reconstruction and relaxation of the surrounding atoms much more important. Without theoretical calculations, such a conclusion cannot be drawn from experiments such as PAS alone. Therefore, we examine the formation energy for three-dimensional agglomerates as well as for vacancy chains. Furthermore, we calculated the corresponding positron lifetimes to be able to compare our data to experimental results. To avoid defect–defect interactions we use much larger supercells than ever before: 512 instead of widely used 64 atoms. Since this cannot be achieved by fully SCF methods, one has to apply a more approximative scheme.

4.1 Details of calculation

The method gives very good results for surface reconstructions of GaAs [48] compared to SCF-LDA calculations. Thus SCC-DFTB may also provide a suitable tool for describing the properties of the inner surface of extended vacancy aggregates. Additionally we carried out test calculations for monovacancies and antisite defects and obtained defect formations energies which are in very good agreement with SCF-LDA results by Northrup et al. [49], see Table 1. This indicates that SCC-DFTB allows calculations of defects with an accuracy comparable to SCF-LDA calculations.

Table 1
Defect formation energies E'_D (in eV, $\Delta\mu = 0$, $\mu_e = 0$) for selected intrinsic point defects

Method	V_{Ga}^0	V_{Ga}^{3-}	As_{Ga}^{2+}	Ga_{As}^0	V_{As}^{1+}
LDA [49]	4.55	5.26	0.93	2.74	2.97
SCC-DFTB	4.54	5.58	1.03	2.77	3.09

In our calculations we used supercells containing 512 atoms. Stable configurations of vacancies or vacancy aggregates in GaAs can be charged, depending on the position of the Fermi level. However, we restricted our calculations to vacancy clusters consisting of an equal number of missing arsenic and gallium atoms. The justification comes from the generation mechanism of vacancies by jog dragging: An equal number of V_{Ga} and V_{As} must be produced to recover the dislocation configuration. Due to the electron counting rule, this should lead to configurations which are charge-neutral over a wide range of the position of the Fermi level. Additionally, the defect formation energies are then independent of the relative chemical potential and have an absolute value. This allows a direct comparison of the defect formation energies of the vacancy aggregates.

The formation energies Ω_n for neutral defects with an equal number of Ga and As vacancies have been obtained using [50],

$$\Omega_n = E_{tot} - \mu_{GaAs}^{bulk}(n_{Ga} + n_{As}), \qquad (20)$$

where n_{Ga} (n_{As}) denotes the number of Ga(As) atoms in the supercell and μ_{GaAs}^{bulk} is the chemical potential of the Ga and As pair in bulk GaAs, which we obtained in a SCC-DFTB total energy calculation for the GaAs single crystal.

The positron lifetimes for the perfect lattice and for different vacancy cluster configurations are calculated using the superimposed-atom model by Puska and Nieminen [51] in the semiconductor approach [52]. The lattice relaxations under the influence of the trapped positron can be neglected for large open volume defects (cf. e. g. [53, 54]). We took the unrelaxed atomic positions as well as those determined by the SCC-DFTB method. The theoretical data will be scaled to the experimental bulk lifetime (a detailed description can be found in [55]).

For the zincblende structure, we tested the number of atoms and the grid points used for discretizations of the problem. With the system size of 512 atoms and the number of grid points used the results are converged within the employed method up to 12 vacancies in the cluster. For cluster sizes larger than 14, the positron results have to be taken with care.

To be consistent with the deformation experiments in GaAs, the vacancy chains were oriented in a $\langle 112 \rangle$ direction, since this is the direction the jogs are dragged by the gliding screw dislocation in a $\{111\}$ slip plane. In this densely packed direction, zig-zag chains of alternating Ga and As vacancies in nearest neighbour positions can be formed.

4.2 Results

Our calculations show that the formation energy for chains of vacancies is much higher than that for vacancy clusters (1.4 eV, 3.6 eV and 5.9 eV for four, six and eight missing atoms, i.e. 2, 3, 4 pairs of V_{Ga} and V_{As}). Hence, if the vacancies are able to move along this vacancy chain, they will re-arrange from the initial chain configuration and aggregate as 3D clusters, which have a much lower formation energy. The re-arrangement of vacancies should require less energy than a vacancy diffusion in the perfect crystal since less bonds have to be broken for vacancy migration along a vacancy chain. This means that clustering can happen even below the temperature of vacancy diffusion as discussed in the introduction. Therefore, the generation of vacancy clusters as a result of jog dragging and collapsing of vacancy chains is observed during plastic deformation.

We then examined different three-dimensional vacancy aggregates in order to determine the stability of clusters with different numbers of missing atoms. Since a huge number of possible candidates for clusters consisting of up to 14 vacancies exists, we used a straightforward way to construct these vacancy aggregates and checked only a small number of different configurations for some vacancy complexes. In our approach of constructing the clusters we followed a recently published work by Hastings et al. [56], who found hexagonal rings to be the most stable vacancy configuration in silicon. Accordingly, we tried to form hexagonal ring-like structures. One configuration is just the hexagonal ring consisting of $3V_{Ga}$ and $3V_{As}$. The sequence in which the atoms were removed in larger vacancy clusters can be seen in Fig. 7. Other configurations which do not follow this straightforward way of construction were checked and found to be significantly higher in energy. They are consequently not considered here.

To compare the energies of the different clusters we calculated the energy gained by adding an isolated divacancy to the most stable aggregate consisting of $n-2$ vacancies. In our calculations this is

$$\Delta E_n = \Omega_n - (\Omega_{n-2} + \Omega_2),$$

which sets the energy zero at ΔE_2. Note that ΔE_n is the negative dissociation energy of V_n into $V_{n-2} + V_2$.

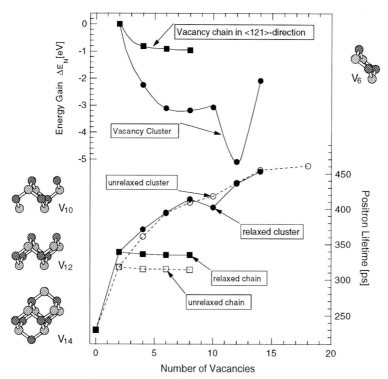

Fig. 6. Energy gained by adding a divacancy to an aggregate of $(N-2)$ vacancies (upper part) and the corresponding positron lifetime (lower part). Structures of some V_n are shown beside the graphic (As atoms dark grey, Ga Atoms light gray)

This energy is plotted in Fig. 6 for the most stable aggregates of vacancies. The picture is very similar to that for Si [56], except that the smallest stable structure, with respect to dissociation, is found to be V_{12} in GaAs, whereas in Si it is V_6.

We assumed in our calculations that the vacancy aggregates grow by addition of divacancies. Of course these aggregates can dissociate into other fragments. However, SCC-DFTB predicts the lowest dissociation energies for $V_n \rightarrow V_{n-2} + V_2$.

A general feature of all relaxed structures is that threefold coordinated Ga atoms undergo a sp^2 hybridisation by emptying all dangling bonds, whereas all threefold coordinated As atoms show p^3 bonding configuration with filled dangling bonds. This is typical for GaAs and can be illustrated by the electron counting rule. Following this rule a Ga atom contributes 3/4 electrons per bond, whereas As contributes 5/4 electrons. Since the energy states created by Ga dangling bonds are higher than the energy states by As dangling bonds, the electrons are transferred from gallium to arsenic atoms. As a result the atoms will change the hybridisation state to further minimize the total energy.

It can be seen from Fig. 6 that V_{12} is by far the most stable configuration, which can be understood by looking at the relaxed structure of V_{12} in Fig. 7. The dominant feature of this structure is the formation of an As–As and a Ga–Ga dimer. This reduces the number of dangling bonds and, therefore, lowers the total energy. Additionally, it can be seen that the threefold coordinated As atom (a in Fig. 7) is in a p^3 configuration state, whereas the Ga atom (d in Fig. 7) prefers a planar structure with sp^2-character. Furthermore, Fig. 7 shows that the formation of dimers is already possible for V_{10}. However, in contrast to V_{12} some Ga atoms are not able to form an energetically favorable threefold coordinated sp^2 structure and, therefore, V_{10} is less stable than V_{12}. Furthermore, V_{12} has a bandgap which is larger than that for the other structures except the divacancy. This additional fact supports the stability of this structure. Similar to Si, where one hexagonal ring is the first stable structure, the vacancies in GaAs are arranged here in two hexagonal rings placed one upon the other (see Fig. 7).

The dimer formation of inner surface atoms is responsible for the lower energy for certain configurations of vacancy agglomerations in compound semiconductors – called

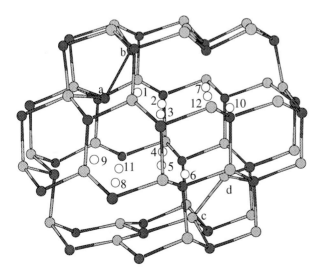

Fig. 7. Structure consisting of 12 vacancies. Dark grey balls denote As and light grey balls Ga atoms. The white spheres are the vacancies. The numbers give the order in which the atoms have been removed, starting from V_2 to V_{12}. Atoms a and d are removed to get V_{14}

magic number clusters. Since in GaAs – in contrast to Si – the energy of a structure can already be lowered by transferring electrons from Ga dangling bonds to As dangling bonds, a remarkable reduction of energy can only be achieved when dangling bonds are saturated by dimer formation. This explains why in Si the hexagonal ring with a minimal number of dangling bonds is the first stable cluster, whereas in GaAs it is the double hexagonal (V_{12}) structure, which shows an energetically favorable dimer configuration.

After examining the structure and energetics of the different types of vacancy aggregates we calculated the positron lifetimes for the unrelaxed and the relaxed structures. For small clusters (V_2 and V_4) and for the vacancy chains the positron lifetime is higher for relaxed clusters compared to unrelaxed ones. This is consistent with an outward relaxation of the atoms surrounding the vacancy cluster (decrease of the electron density). For clusters consisting of more than four vacancies no significant change in positron lifetime before and after relaxation can be observed. An exception is V_{10}, where the unfavorable configuration of some Ga atoms, mentioned above, leads to an increase in the electron density within the open volume. Therefore, we find a reduced positron lifetime after relaxation.

The calculated positron lifetime related to trapping into a zig-zag chain of alternating Ga and As vacancies is roughly that of a divacancy, since the electron densities within the open volume are similar.

The increase in the positron lifetime due to the larger open volume of the defects seems to saturate for about 12 to 14 vacancies agglomerated. So, it is difficult to tell from the experimentally observed positron lifetimes if there are even bigger clusters. While the calculations reach their validity limit, the experimental error is certainly roughly 50 ps.

5. The Diffusion of Carbon in GaAs

At low concentrations ([C] < 10^{19} cm^{-3}) the diffusion rate for isolated substitutional C_{As} acceptors in GaAs is some two orders of magnitude less than other p-type dopants such as Be or Zn [57]. Under As-rich conditions, measurements of the activation energy for migration of C_{As} cover the range of 2.8 to 3.1 eV [58 to 60]. (The rate is lower under Ga-rich conditions.) Hence it possible to create carbon-doped regions with very abrupt p–n boundaries which do not degrade over time. Unfortunately, the advantage which carbon has is lost when the concentration rises above [C] $\approx 5 \times 10^{19}$ cm^{-3} [61]. In secondary ion mass spectrometry (SIMS) and microstructural studies the smearing of a carbon-doped region with initially sharp boundaries was observed [62]. In these experiments a considerable quantity of carbon interstitials were found $\approx 25\%$ of [C]), which appeared to be very mobile even at growth temperatures, and therefore reduced the sharpness of the p–n junction [63]. Moreover, it was observed, that the annealing of highly C-doped GaAs at temperatures between 650 and 850 K leads to a drastic reduction in the hole concentration [p] [64 to 66]. This cannot be explained by interstitial diffusion, but instead the formation of compensating defects is required. In the experiments previously cited, the initial hole concentration before any heat treatment was nearly equal to the existing C concentration, i.e. nearly all C atoms were activated as acceptors, thus the concentration of interstitials was negligible. The loss of holes is in fact seen across the whole spectrum of $Al_xGa_{1-x}As$ alloys for $x = 0$ to 1 [67].

The dominant defect responsible for hole loss has been identified in material grown by solid source molecular beam epitaxy (MBE) as some form of carbon pairs [68]. Raman scattering observations supported by *ab initio* theoretical modelling have shown that new dicarbon defects consisting of a pair of carbon atoms lying at an arsenic lattice site, $(C-C)_{As}$, or interstitial dicarbon defects, $(C-C)_i$, are formed when annealing heavily doped material [69, 70]. These defects are deep donors, or double donors respectively, hence three or four holes are lost for each dicarbon complex formed (provided that all C atoms were active as acceptors intitially). Their formation implies the activation energy for carbon diffusion has a relatively low value in material where $[C] \gtrsim 5 \times 10^{19}$ cm^{-3}. In high-temperature annealing experiments on heavily carbon-doped GaAs, Fushimi and Wada have measured a value of ≈ 1 eV for the activation energy of the hole loss process [71]. At present, no details have yet emerged of the reaction mechanism for carbon migration, however, it is widely thought that it involves a 'kick-out' process, where highly mobile arsenic interstitials (As_i) displace C_{As} atoms. This was first suggested by H. M. You et al. [59].

In strongly p-doped material it would be expected that any As_i present exists in the triply ionised state, As_i^{3+}. Northrup and Zhang have calculated the formation energy in GaAs of As_i^{3+} to be $3.15 \text{ eV} + 3\mu_e + \Delta\mu/2$, and $\Delta H = -1.05$ eV [72]. Given that $-\Delta H \leq \Delta \mu \leq \Delta H$, and in strongly p-doped material $\mu_e \approx 0$ eV, the minimum formation energy of As_i^{3+} is ≈ 2.6 eV. The measured activation energy for the diffusion of carbon (2.8 to 3.1 eV) lies ~ 0.2 to 0.5 eV above this which is consistent with the formation of As_i^{3+} being the slowest step.

Thus the activation energy for the hole loss process measured by Fushimi and Wada is lower than the formation energy for As_i^{3+} (and hence activation energy for creation). This suggests that there is a source of As_i present in Fushimi and Wada's specimens, and the C_{As} atoms, in effect represents a sink for them.

It has also been suggested that hydrogen may play a role, and the possibility of a vacancy-assisted process is not excluded [73]. The work referred to here (more by Fushimi and Wada) draws these conclusions from experiments performed on p$^+$–n junction structures similar to those used in heterojunction bipolar transistors (HBTs). In the case of hydrogen, however, it is not essential for this to be present for degradation to occur; dicarbon complexes form in material which has been prepared so as to contain as little hydrogen as possible. Moreover, if hydrogen were present in significant quantities, then it would interact with the dicarbon complexes to form new defects containing hydrogen and carbon such as the $(C_{As})_2H$ aligned defect complex identified by Ying Cheng et al. in as-grown GaAs epitaxial layers [74]. It might also be expected that C_{As}–H would also be present (this has an infrared active C–H stretch mode at 2635 cm^{-1}) [75, 76]. No such defects have been observed in the hydrogen-free material. In the case of the possibility of a vacancy-assisted process, it is difficult to see how this could result in dicarbon defects. Moreover, production of C_{Ga} defects appears to be inevitable with this mechanism; these have never been observed. For a recent review of diffusion in GaAs and related compounds see Ref. [77].

In the following sections we will next examine the 'kick-out' process by using local density functional based theoretical methods to explore the potential energy surface for a single interstitial carbon atom in GaAs, and hence calculate reaction and activation energies for diffusion which may be compared with experiment.

5.1 Calculational details

The potential energy surface for interstitial atoms in GaAs is expected to be very complex with many local minima. There is also no information, experimental or theoretical, available from other sources about the migration path for a carbon atom through a GaAs crystal. These combined facts make it extremely difficult to choose relevant structures and constraints for modelling the diffusion mechanism directly with a fully ab initio method. The approach adopted here, therefore, was to first conduct a survey of many possible ideas using the approximate but fast SCC-DFTB method for a 216 atom supercell. This enabled us to identify the main features along the migration path, which could then be examined in detail with a much higher degree of confidence using a more accurate but time-consuming method, AIMPRO [2].

In order to search for reaction pathways using the SCC-DFTB method, constraints were placed on the relaxation of the atomic coordinates in the conjugate-gradient energy minimisation algorithm it uses. To simulate the migration of an atom between a chosen pair of fully relaxed local minima in energy, the total energy was calculated at points along the vector defining the direct trajectory connecting the two metastable states, while simultaneously the movement of selected atoms was restricted to the plane to which the vector is perpendicular at each point. The selection of atoms generally included (but not only) those which appeared to be bonded to different neighbours in each local minimum. Some experimentation is required to obtain a satisfactory result. If too many atoms are included in the selection there is a risk that the system is overconstrained which will result in an overestimate of the activation energy. On the other hand, if the system is underconstrained because too few atoms are selected, the energy plotted as a function of distance along the trajectory vector will not be smooth, and atoms make unreasonably large jumps in position at some point along the migration path. A strength of the SCC-DFTB method in this respect is that it is quick enough to try many different selections until the best one is found.

Once a low energy trajectory has been found (including its ends), it is then apparent how the atomic movements are coordinated, which bonds are exchanged, and so on. The AIMPRO method can then be employed to examine the process in detail and give more reliable estimates of energies. The system of constraints described above can be applied to the AIMPRO method, however, for reactions where an exchange of bonds can be identified the difference of the squares of the bond lengths of bond pairs being exchanged is a better choice of reaction coordinate. For each pair of bonds being exchanged, the total energy as a function of the quantity $r_a^2 - r_b^2$ is calculated, where r_a and r_b are the bond lengths. Provided the correct pair or pairs of bonds have been chosen, this yields a smooth curve for one pair, or a saddle-like surface for two pairs. Although the method is completely general, it is not practical to model systems where more than two pairs of bonds are exchanged – N pairs require $O(x^N)$ complete energy minimisations. In the AIMPRO method the structures are modeled with 135 atom clusters saturated by pseudo hydrogen.

5.2 Results

Heavily carbon-doped GaAs containing nearly all the carbon in the form of C_{As} acceptors is very strongly p-type, therefore the Fermi-level lies close to the top of the valence band, and any donors present will be ionised. A single carbon interstitial atom C_i, in-

Structures, Energetics and Electronic Properties of Complex III–V Systems

Table 2
Calculated energies (in eV) of the $(C-As)_{As}$ and Ga–C–As defects in GaAs relative to $(C-Ga)_{Ga}$

	AIMPRO	SCC-DFTB
$(C-As)_{As}$	+0.38	+0.34
Ga–C–As	+0.30	−0.25

serted into the ideal GaAs lattice has two electrons occupying a level deep in the gap, hence in p-type material its normal state will have a +2 charge.

With the SCC-DFTB method it was very quickly established that the minimum energy location for C_i^{2+} is at the centre of a Ga–As bond, i.e., a linear Ga–C–As configuration with C_{3v} symmetry. The next lowest energy minimum, about 0.25 eV above the bond-centred structure, was found to be a $(C-Ga)_{Ga}$ split interstitial with an approximately [100]-aligned C–Ga bond. In third place, the corresponding $(C-As)_{As}$ split interstitial was also metastable, and had a relative energy of about 0.6 eV.

When these three structures were tested with the AIMPRO method, the energy ordering was found to be different. The $(C-Ga)_{Ga}$ split interstitial was lowest at 0.3 eV below the linear Ga–C–As configuration, which was in turn 0.1 eV below the $(C-As)_{As}$ split interstitial. Assuming these figures are more reliable, it appears therefore that while the relative energies of the split interstitials agree to within 0.04 eV, the bond-centred structure is overbound with the SCC-DFTB method. A possible explanation for this difference is that the carbon atom experiences a very different bonding configuration in the split-interstitial and bond-centred sites, and the SCC-DFTB method being only a minimal basis approach and not fully self-consistent is not very reliable in these circumstances. The slightly harder TB repulsive potential yielded longer bond lengths, thus bringing their values into very close agreement with the AIMPRO calculations. It is quite possible that the agreement may be improved by further optimisation of the basis. Table 2 summarises the main results.

Having found these three structures, and noting that there are only relatively modest differences in energy between each of the split interstitials and the bond-centred site, it was now apparent how carbon might diffuse through the crystal. Starting from one of the two split interstitial structures, the path would take it to the other via the bond-

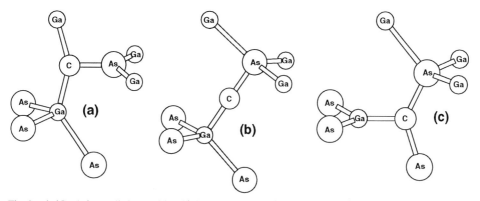

Fig. 8. a) $(C-As)_{As}$ split interstitial, b) bond-centred C interstitial, and c) $(C-Ga)_{Ga}$ split interstitial

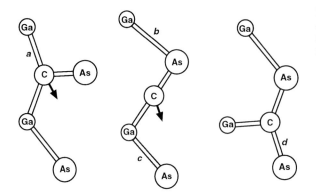

Fig. 9. Diagram showing which pairs of bonds are exchanged for the diffusion of C_i. First, 'a' breaks and 'b' forms, then 'c' breaks and 'd' forms

centred structure (see Fig. 8). By performing many simulations at the SCC-DFTB level of theory in conjunction with the first method for constraining atoms we were able to examine this process and identify the main features of the atomic movements involved. The calculations, however, are not reliable enough to give meaningful energies in this situation where atoms are undergoing large changes in their bonding configuration far from those used in the initial fitting process.

The SCC-DFTB simulations showed that one pair of bonds is exchanged when the carbon moves from $(C-As)_{As}$ to the bond-centred site, and the second pair when it subsequently moves to $(C-Ga)_{Ga}$. The bonds which break and form are illustrated by Fig. 9. Starting from $(C-As)_{As}$, the C-Ga bond labelled 'a' breaks, then the Ga-As bond 'b' forms. Next, the Ga-As bond 'c' breaks, and finally the C-As bond 'd' forms. Hence there are two reactions for which we can define coordinates as described previously in terms of the difference of the squares of the lengths of the exchanging pairs of bonds. For the first reaction, $(C-As)_{As} \rightarrow Ga-C-As$, this is, $r_a^2 - r_b^2$; and in the case of the second reaction, $Ga-C-As \rightarrow (C-Ga)_{Ga}$, it is $r_c^2 - r_d^2$.

Figs. 10 and 11 show how the energy varies as a function of the reaction coordinates. Starting from a $(C-As)_{As}$ state, a migrating carbon atom faces a small barrier of only 0.09 eV to reach a Ga-As bond centre. From here it then encounters a somewhat higher, but still relatively small barrier of 0.39 eV when it moves to a $(C-Ga)_{Ga}$ state. In the opposite direction, the activation energy is 0.71 eV, thus this is the largest barrier an interstitial carbon atom must overcome to move via this path.

Two alternative diffusion paths were also considered in addition to the above. At first glance it looks viable for a carbon atom to move through an interstitial cage from one

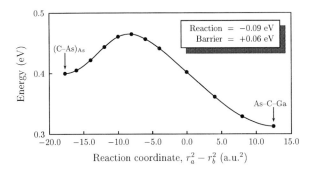

Fig. 10. Energy relative to the $(C-Ga)_{Ga}$ complex plotted as a function of the reaction coordinate $r_a^2 - r_b^2$ described in the main text and Fig. 9. The reaction energy and barrier height are relative to the initial $(C-As)_{As}$ state

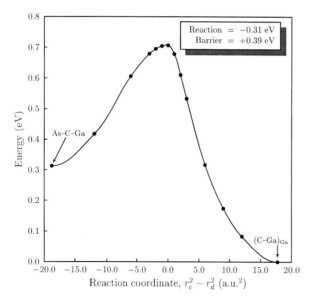

Fig. 11. Energy relative to the (C–Ga)$_{Ga}$ complex plotted as a function of the reaction coordinate $r_c^2 - r_d^2$ described in the main text and Fig. 9. The reaction energy and barrier height are relative to the intermediate Ga–C–As state

(C–As)$_{As}$ or (C–Ga)$_{Ga}$ site to the nearest neighbouring one of the same type. Efforts to simulate this always failed. In the case of (C–As)$_{As}$ to (C–As)$_{As}$, the carbon atom jumped to form a (C–Ga)$_{Ga}$ when it was pushed into the interstitial space away from the arsenic atom. Similar jumping behaviour also occurred in the case of diffusion from (C–Ga)$_{Ga}$ to (C–Ga)$_{Ga}$. The only path which could be found with smoothly varying energy along the chosen coordinate without discontinuous steps in position was the 'bond-centred' one.

6. Threading Dislocations and Nanopipes in GaN

GaN has recently been the subject of considerable interest due to its optoelectronic properties. In particular the wide band gap (3.4 eV for wurtzite GaN) makes blue light applications feasible. Frequently, sapphire substrates are used to grow device quality wurtzite-(α) GaN with the metal-organic chemical vapour phase deposition (MOCVD) technique. In this case, growth proceeds along the c-axis. Figure 12 shows a cross-sectional TEM weak beam image of a typical sample. The large lattice misfit between GaN and the sapphire substrate of 13% results in dislocation tangles near the interface. In addition to these geometric misfit dislocations which have dislocation lines in the basal plane, also isolated threading dislocations with dislocation lines parallel to **c** and Burgers vectors **c**, **a** and **c** + **a** persist beyond the interface [79 to 81]. Since many of them penetrate the entire epilayer from the substrate to the surface they are called threading dislocations.

An unexpected finding [82, 83] is that inspite of their high density (typically $\sim 10^9$ cm^{-2}) and the fact that they cross the active region of the devices (starting typically ≈ 0.5 μm above interface) these threading dislocations in GaN do not lead to a pronounced reduction in the device lifetime of the light-emitting diodes [84] or blue lasers [85]. This can be contrasted with GaAs where radiation enhanced dislocation motion [86] readily occurs and leads to an increase in non-radiative processes. It is therefore of consider-

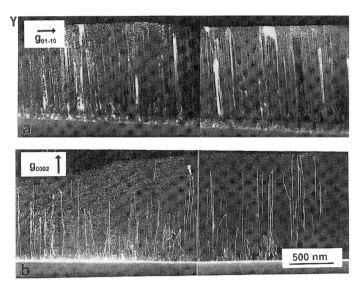

Fig. 12. Dislocation arrangement in a GaN sample grown on sapphire by MOCVD: a) Cross-sectional TEM (g/3g) weak beam image, $\mathbf{g} = (01\bar{1}0)$. Screw dislocations with $\mathbf{b} = \pm[0001]$ are out of contrast. Dislocations with a $\mathbf{b}$-component in the interface, i.e. $\mathbf{b} = \pm\frac{1}{3}[\bar{1}2\bar{1}0]$ are visible. b) Cross-sectional TEM (g/3g) weak beam image, $\mathbf{g} = (0002)$. Screw dislocations with $\mathbf{b} = \pm[0001]$ are visible. Dislocations with a $\mathbf{b}$-component in the interface are out of contrast. Christiansen et al. [78]

able interest to understand the structural and electrical properties of threading dislocations in GaN and to compare them with those of dislocations in more traditional semiconductors.

Threading dislocations are often associated with the appearance of long nanopipes which are parallel to **c** and have hexagonal cross sections with constant diameters ranging from 20 to 250 Å [87 to 90]. Nanopipes degrade the material quality. In particular, they can get filled by metal during the formation of contacts with GaN and have already caused short circuits in laser devices. Frank [91] predicted that a dislocation whose Burgers vector exceeds a critical value should have a hollow tube at the core. The equilibrium radius is achieved by balancing the elastic strain energy released by the formation of a hollow core against the energy of the resulting free surfaces. Liliental-Weber et al. [92] suggested another possible mechanism for the formation of nanopipes. They found the density of nanopipes to be increased with the impurity concentration and proposed that impurities poison the walls of the nanopipes which prevents the nanopipes from growing out.

Finally, the origin of defect-induced electronic states, which lie deep in the GaN band gap and can thus significantly alter the optical performance, is possibly related to threading dislocations. Especially in laser devices deep gap states are of concern since parasitic components in the emission spectrum are highly undesirable. The most commonly observed emission in unintentionally doped n-type GaN, the yellow luminescence (YL), is centred at 2.2 to 2.3 eV with a line width of ≈ 1 eV. Several models for the origin of the YL in GaN have been proposed. Most of them assume the transition to be between a shallow donor and a deep acceptor [93] or a deep donor and a shallow acceptor [94]. Recent work has however found evidence for the deep acceptor model [95]. Cathodoluminescence (CL) studies of the yellow luminescence have shown that the YL is spatially non-uniform. A possible reason for this non-uniform distribution of the YL could be related to threading dislocations, which are non-uniformly distributed throughout the epilayer and might be electrically active. Indeed, also atomic force microscopy (AFM) in combination with CL has led to the conclusion that threading dislocations act as non-radiative recombination centres and degrade the luminescence efficiency in the blue light spectrum of the epilayers [96]. However, the type of dislocations involved in the YL is not clear: Christiansen et al. [78] suggest that the YL arises from threading dislocations with a screw component whereas Ponce et al. [97] localise the YL at low angle grain boundaries which predominantly contain threading edge dislocations. Moreover, it is not clear whether dislocations in the pure, i.e. impurity free form or defects trapped in the stress field of dislocations are responsible for the YL.

In this section, the atomic geometries, electrical properties and line energies of threading screw and edge dislocations with full and open cores are investigated [14]. The results are interpreted by comparing elements of the dislocation cores with non-polar ($10\bar{1}0$) surfaces. Possible mechanisms for the formation of nanopipes are then examined [98]. Finally, we also explore the segregation of gallium vacancies and oxygen as well as related defect complexes to threading edge dislocations and discuss their implication for the yellow luminescence [99].

6.1 Screw dislocations

Threading screw dislocations in wurtzite material have a Burgers vector parallel to the dislocation line [0001]. The smallest screw dislocations have thus elementary Burgers vector ± **c**. Screw dislocations occur at a density $\sim 10^6$ cm^{-2} in α-GaN grown by MOCVD on (0001) sapphire. Since they nucleate in the early stages of growth at the sapphire interface and thread to the surface of the crystallites (see Fig. 12b), screw dislocations are believed to arise from the collisions of islands during growth [80]. At a screw dislocation the surface is rough and has a high energy which favours the nucleation of islands. They are thus vital for the growth process. In GaN, screw dislocations are unusual in often being associated with nanopipes [100]. However, full core screw dislocations [101] and screw dislocations with a very narrow opening of ≈ 8 Å [102] are also reported.

6.1.1 Full core screw dislocations

Screw dislocations with a full core have been observed by Xin et al. [101] using high resolution Z-contrast imaging (see Fig. 13).

Within the SCC-DFTB method the dislocations are modelled in 210 atom clusters periodic along the dislocation line with periodicity **c** and in 576 atom ($12 \times 12 \times 1$) supercells. Because of the large lateral extension of the supercell (12×12), only *k*-points for the sampling along the *c*-direction are necessary. Two *k*-points parallel to this direction were found sufficient to carry out the sum over the Brillouin zone: using four *k*-points gave only a difference of ≤ 0.02 eV/Å in the dislocation line energy. In the AIMPRO case, relaxations were carried out in 392 atom stoichiometric clusters.

Both methods found heavily distorted bond lengths for the full core screw dislocation (see Fig. 14 and Table 3) yielding deep gap states ranging from 0.9 to 1.6 eV above the

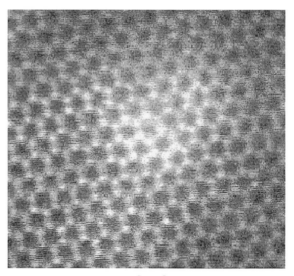

Fig. 13. Left: Low magnification angular dark field image along [0001]. Threading dislocations shown as bright dots due to their strain field. Right: High resolution Z-contrast image of an end-on pure screw dislocation showing a full core. Xin et al. [101]

valence band maximum, VBM, and shallow gap states at ≈0.2 eV below the conduction band minimum, CBM. An analysis of these gap states revealed that the states above the VBM are localised on N core atoms, whereas the states below CBM are localised on core atoms but have mixed Ga and N character. Therefore the full core screw dislocation is electrically active and could act as a non-radiative centre [14]. Similarly one could expect that dislocations of mixed type would also have deep states in the gap as a result of the distortion arising from their screw component. Indeed, CL experiments have related the yellow luminescence centred at 2.2 eV to screw dislocations [78]. In addition, atomic force microscopy in combination with CL imaging has shown that

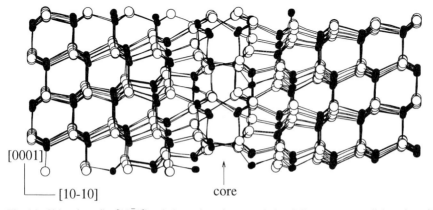

Fig. 14. Side view (in [11$\bar{2}$0]) of the relaxed core of the full-core screw dislocation (**b** = [0001]). The atoms at the dislocation core adopt heavily distorted configurations (see Table 3) yielding deep gap states

Table 3
Bond lengths (min–max (and average)) and bond angles (min–max) for the most distorted atoms at the core centre of the full-core screw dislocation (**b** = [0001])

atom	bond lengths (Å)	bond angles (°)
1 ($Ga_{4\times coord.}$)	1.85–2.28 (2.14)	68–137
2 ($N_{4\times coord.}$)	1.89–2.28 (2.13)	71–136

threading dislocations with a screw component act as nonradiative combination sites [96]. A calculation in a supercell containing a screw dipole consisting of two dislocations with **b** = [0001] and −[0001], which are symmetrically equivalent, confirmed these results and gave a high line energy of 4.88 eV/Å. This is mainly the core energy of each screw dislocation together with the elastic energy stored in a cylinder of diameter roughly equal to the distance between the cores, 19.1 Å.

6.1.2 Screw dislocations with a narrow opening

We now investigate whether the line energy of the full core screw dislocation is reduced if material is taken from the core. Accordingly, calculations were then carried out using the same supercell as for the full core screw dislocations, but with the hexagonal core of each screw dislocation removed leading to a pair of open-core dislocations with diameters $d \approx 7.2$ Å. The relaxed structure (Fig. 15) preserved the hexagonal core character, demonstrating that the internal surfaces of the dislocation cores shown in Fig. 16 are similar to {100} type facets except for the topological singularity required by a Burgers circuit.

It is instructive to compare the distortions of the atoms situated at the wall of the open-core (Table 4) with the corresponding atoms at the $(10\bar{1}0)$ surface (Table 5). In both cases, the three fold coordinated Ga (N) atoms adopt an sp^2–(p^3)-like hybridisa-

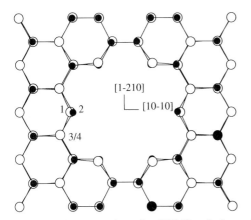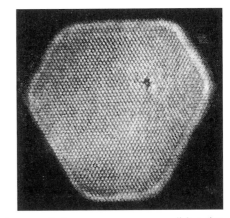

Fig. 15. Left: Top view (in [0001]) of the relaxed core of the open-core screw dislocation (**b** = [0001]). The three-fold coordinated atoms 1 (Ga) and 2 (N) adopt a hybridisation similar to the $(10\bar{1}0)$ surface atoms. Right: TEM image of a nanopipe containing a dislocation with a screw component. During growth the nanopipe closes leaving the dislocation with an opening of three rows (≈ 8 Å wide; see black arrangement within the nanopipe). Liliental-Weber [102]

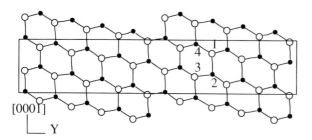

Fig. 16. Projection of the wall of the open-core ($d = 7.2$ Å) screw dislocation ($\mathbf{b} = [0001]$). The three-fold coordinated atoms 1 (Ga) and 2 (N) adopt a hybridisation similar to the ($10\bar{1}0$) surface atoms

tion which lowers the surface energy and cleans the band gap [103]. Indeed, we find that unlike the full-core screw dislocation, the gap is free from deep states [14].

There are, however, in contrast to the ($10\bar{1}0$) surface, energetically shallow gap states. Calculations were carried out for a distorted ($10\bar{1}0$) surface, i.e. a ($10\bar{1}0$) surface in a unit cell where the unit cell vectors were modified to give a distorted surface corresponding to that of the wall of the open-core screw dislocation with diameter $d = 7.2$ Å. We find that the distorted ($10\bar{1}0$) surface has a spectrum with shallow states very similar to those of the open-core screw dislocation with $d = 7.2$ Å. We also calculated the spectrum for a nanopipe with $d = 7.2$ Å but without a dislocation core. This, like the undistorted ($10\bar{1}0$) surface, possesses a gap free from deep states, although there are N (Ga) derived surface states lying slightly below (above) the VBM (CBM). These results indicate that the shallow states in the open-core screw dislocation with diameter $d = 7.2$ Å can be attributed to the distortion arising from the dislocation Burgers vector. Calculations for a series of different distortions of the ($10\bar{1}0$) surface corresponding to open-core screw dislocations with different diameters also suggest that open-core screw dislocations with diameters greater than ~ 20 Å should have no gap states at all. As can be seen in Table 4 the distortion in the open-core screw dislocation is signifi-

Table 4

Bond lengths (min–max (and average)) and bond angles (min–max (and average)) for the most distorted atoms at the wall of the open core screw dislocation ($\mathbf{b} = [0001]$). Atom numbers refer to Figs. 15 and 16

atom	bond lengths (Å)	bond angles (°)
1 ($Ga_{3\times coord.}$)	1.86–1.89 (1.88)	107–123 (117)
2 ($N_{3\times coord.}$)	1.88–2.05 (1.96)	102–111 (108)
3 ($Ga_{4\times coord.}$)	1.89–2.07 (1.96)	100–122
4 ($N_{4\times coord.}$)	1.93–2.03 (1.97)	98–120

Table 5

Bond lengths (min–max (and average)) and bond angles (min–max (and average)) for the top two layers of atoms at the corresponding undistorted GaN($10\bar{1}0$) surface

atom	bond lengths (Å)	bond angles (°)
1 ($Ga_{3\times coord.}$)	1.83–1.88 (1.86)	116–117 (117)
2 ($N_{3\times coord.}$)	1.83–1.92 (1.89)	107–111 (108)
3 ($Ga_{4\times coord.}$)	1.91–2.02 (1.94)	107–112
4 ($N_{4\times coord.}$)	1.88–2.03 (1.93)	99–115

cantly less than that in the full-core screw dislocation (see Table 3). It is therefore not surprising that the calculated line energy of 4.55 eV/Å is lower than the line energy of the full-core screw dislocation. The energy required to form the surface at the wall is compensated by the energy gained by reducing the strain. However, a further opening gave a higher line energy and we conclude that the equilibrium diameter is ≈7.2 Å. This opening has also been reported by Liliental-Weber [102] who found some of the screw dislocations to have holes which are three atomic rows wide (see Fig. 15).

A theoretical approach to predict the opening of a screw dislocation was deduced by Frank [91]. By balancing the elastic dislocation strain energy released by the formation of a hollow core against the energy of the resulting free surfaces, he showed that, for isotropic linear elasticity and a cylindrical core, the equilibrium core radius r_{eq} is

$$r_{eq} = \frac{\mu \mathbf{b}^2}{8\pi^2 \gamma} \,, \qquad (21)$$

where γ is the surface energy, μ is the shear modulus and $\mathbf{b}$ is the Burgers vector. For a rough estimate of r_{eq}, we use the theoretical value for the surface energy of $\{10\bar{1}0\}$ facets which we found to be $\gamma = 121$ meV/Å^2 = 1.9 Jm^{-2}. Taking $\mu = 8 \times 10^{10}$ Nm^2 as an upper limit and $\mathbf{b} = 0.5$ nm for the Burgers vector of an elementary screw dislocation yields $r_{eq} \approx 0.2$ nm. It is unlikely, that isotropic elasticity theory can describe the severely distorted full core dislocation which limits the usefulness of Frank's expression (21) concerning the precise quantitative value of the equilibrium diameter. Our calculated value of ≈7.2 Å and Frank's value are reasonably close since the relatively small line energy difference found between full-core and open-core ($d \approx 7.2$ Å) screw dislocations suggests a shallow minimum which probably allows all intermediate structures to exist. In our calculations only structures constructed by removing entire hexagons, but not those obtained by removing single rows were considered. Calculating the latter ones, may lead to slightly lower energies.

In summary, it can be concluded from our calculations and from Frank's theorem that in GaN screw dislocations with an elementary Burgers vector $\mathbf{c}$ can exist with a full core and with a narrow opening up to ≈7.2 Å. The full core screw dislocation is electrically active whereas the screw dislocation with a hexagonal opening has only shallow gap states. These states are induced by the distortion arising from the Burgers vector.

6.2 The formation of nanopipes

Nanopipes in α-GaN thread along the c-axis and have hexagonal cross sections, i.e. they are inclosed by $\{10\bar{1}0\}$ type walls (see Fig. 17). Nanopipes are commonly observed in MOCVD grown epilayers on sapphire [87 to 89]. However, they have also been reported in samples grown by MBE on SiC [90]. Nanopipes occur at a density up to $\sim 10^8$ cm^2 and have constant diameters ranging from 20 to 250 Å. The first suggestion was that they were the manifestation of screw dislocations with empty cores as discussed by Frank a long time ago [91]. However, as shown above ab initio calculations as well as Frank's theorem do not support the idea that in GaN the core of a screw dislocation with Burgers vector equal to $\mathbf{c}$ is open with such a large diameter. Pirouz [104] has therefore argued that superscrew dislocations with Burgers vectors $n\mathbf{c}$ where $n > 1$ are formed during growth by the collision of islands. Clearly, if n were big enough, then the cores would be open. However, there is at present no

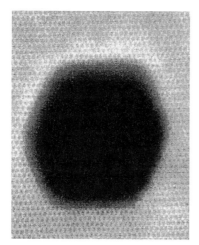
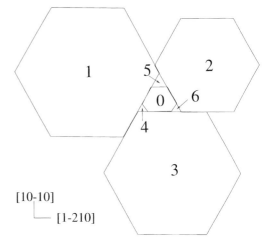

Fig. 17. Left: High resolution Z-contrast image along [0001] of a nanopipe (Y. Xin, unpublished). Right: Suggested mechanism for the formation of a nanopipe (area No. 0). Three hexagons (No. 1,2,3) are growing together. As the surface to bulk ratio at ledges (No. 4,5,6) is very large, they grow out quickly leaving a nanopipe (area No. 0) with $\{10\bar{1}0\}$ type facets

microscopic evidence for such dislocations [105]. We also note that in a typical sample some nanopipes (<10%) were observed which could not be associated with a screw dislocation [105]. Another possibility is that the $\{10\bar{1}0\}$ type surface walls of the nanopipes are coated by hydrogen which is always present during MOCVD growth. This might result in a very low surface energy explaining the large diameter of the nanopipes via formula (21). The adsorption of H on $(10+10)$ surfaces has been investigated by Northrup and Neugebauer [106]. They found however, that at a usual MOCVD growth temperature of $\approx 1000\,°C$ the energy of the $(10\bar{1}0)$ surface is not lowered by the adsorption of H and concluded that hydrogen is not responsible for the formation of nanopipes. This has been confirmed by the recent work of Liliental-Weber et al. [90] who detected nanopipes also in MBE grown material where the concentration of hydrogen is negligible. On the other hand, Liliental-Weber et al. [90] found the diameters and densities of nanotubes to be increased in the presence of impurities, e.g. O, Mg, In and Si, and argued that these impurities decorate the $\{10\bar{1}0\}$ walls of the nanotubes inhibiting overgrowth. O being the main source of unintentional doping in GaN, we will now discuss how O can cause the formation of nanopipes.

From Fig. 18 it can be seen that the surface walls of nanopipes are $(10\bar{1}0)$ surfaces which are predominantly of type I and flat, i.e. they usually have one dangling bond per atom and only little irregularities caused by surface steps. GaN samples usually contain a considerable concentration of gallium vacancies and oxygen which as our calculations show, have both a tendency to diffuse to $(10\bar{1}0)$ surfaces. It is therefore very likely that many gallium vacancies and oxygen atoms have segregated to the nanopipe walls where they can form $V_{Ga}-(O_N)_3$ defect complexes. In Ref. [108] it was shown that $V_{Ga}-(O_N)_3$ are very stable defect complexes on $(10\bar{1}0)$ surfaces of type I. Moreover, overgrowth was determined to be difficult as oxygen atoms would drift away diffusing to the new surface. Since $V_{Ga}-(O_N)_3$ defect complexes do not lead to any

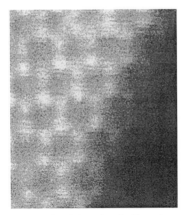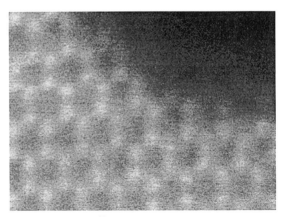

Fig. 18. Straight edge (left) and corner (right) at the $\{10\bar{1}0\}$ type wall. Xin et al. [107]

noticeable change of the atomic positions at the surface they are consistent with the HRTEM image in Fig. 18. Unfortunately at present there seems to be no direct way for detecting V_{Ga}–$(O_N)_n$ at nanopipe walls by experiments.

To explain the formation of nanopipes, we suppose [98, 109] that oxygen atoms constantly diffuse to the $\{10\bar{1}0\}$ type surfaces. Within the framework of Stranski-Krastanow growth, the internal $\{10\bar{1}0\}$ type surfaces between GaN islands are shrinking along with the spaces between colliding GaN islands (see Fig. 17). Therefore, the O coverage and density of V_{Ga}–$(O_N)_3$ defects is expected to increase. The maximum concentration of this defect would be reached if 50% (100%) of the first (second) layer N atoms were replaced by O, see Ref. [108]. It is, however, likely that far lower concentrations are necessary to stabilise the surface and make further shrinkage of the inter-island spaces impossible thus leaving a nanopipe. Provided oxygen could diffuse to the surface fast enough, the diameter and density of the holes would be related to the density of oxygen atoms in the bulk. This has indeed been observed by Liliental-Weber et al. [92] who found that as the concentration of oxygen in the material changed by about an order of magnitude the number of nanopipes increased by a factor of about three and the diameter of the nanopipes changed from (3 to 10) nm to (6 to 12) nm. A more detailed prediction of the radii and density of nanopipes depending on the oxygen partial pressure would require thermodynamic equilibrium for the formation of the nanopipes. However, as can be seen in the large distribution of nanopipe radii, this is obviously not reached.

It is also necessary to explain why the tubes have $\{10\bar{1}0\}$ type surface walls. The other low index surface perpendicular to the growth direction which could become poisoned by O impurities and thus be responsible for the formation of nanopipes is the $(11\bar{2}0)$ surface. $\{11\bar{2}0\}$ type surfaces are not observed presumably because of their higher absolute surface energy and the much higher formation energies of V_{Ga}–(O_N) complexes compared to the $\{10\bar{1}0\}$ surface [108]. Moreover we suggested that because of the different surface topologies V_{Ga}–$(O_N)_3$ is likely to form on $(10\bar{1}0)$ surfaces but not on $(11\bar{2}0)$ surfaces during growth.

Finally, we point out that our arguments are still valid if each nanopipe is associated with a screw dislocation since the walls of the tube with a dislocation are locally equivalent to a (10$\bar{1}$0) surface which is distorted to form a helix (see Section 6.1.2). We therefore conclude that rather than being responsible for the formation of nanopipes screw dislocations are attracted to nanopipes in order to reduce the elastic energy.

6.3 Threading edge dislocations

Pure edge dislocations lie on {10$\bar{1}$0} planes and have a Burgers vector $\mathbf{b} = \mathbf{a} = [1\bar{2}10]/3$. They are a dominant species of dislocation, occurring at extremely high densities of $\sim 10^8$ to 10^{11} cm^{-2} in α-GaN grown by MOCVD on (0001) sapphire (Fig. 12a) and in analogy to screw dislocations are thought to arise from the collisions of islands during growth [80].

Within the SCC-DFTB method threading edge dislocations are modelled in 210 atom clusters periodic along the dislocation line with periodicity $\mathbf{c}$ and in 576 atom ($12 \times 12 \times 1$) supercells containing a dislocation dipole. In analogy to the models for the screw dislocations, two k-points parallel to $\mathbf{c}$ were used to carry out the sum over the Brillouin zone. In the AIMPRO case, relaxations were carried out in 286 atom stoichiometric clusters.

The relaxed core of the threading edge dislocation is shown in Fig. 19. The corresponding bond-lengths and bond angles of the most distorted atoms are given in Table 6. With respect to the perfect lattice the distance between columns (1/2) and (3/4) [and the equivalent on the right] are 9% contracted while the distance between columns (9/10) and (7/8) [and the equivalent on the right] are 13% stretched. This atomic geometry for the threading edge dislocation has recently been confirmed by Xin et al. [101] using atomic resolution Z-contrast imaging (see Fig. 20). Consistent with our calculation they determined a contraction (stretching) of $15 \pm 10\%$ of the distances between the columns at the dislocation core. Our calculations show that in a manner identical to the (10$\bar{1}$0) surface, the three-fold coordinated Ga (N) atoms (no. 1 and 2 in Fig. 19) relax towards sp^2 (p^3) leading to empty Ga dangling bonds pushed towards the CBM, and filled lone pairs on N atoms lying near the VBM. Thus we find threading edge dislocations to be electrically inactive [14].

From a supercell calculation, we obtain a line energy of 2.19 eV/Å for the threading edge dislocation. We note that this line energy is considerably lower than the one

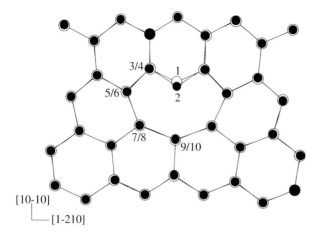

Fig. 19. Top view (in [0001]) of the relaxed core of the threading edge dislocation ($\mathbf{b} = \frac{1}{3}[1\bar{2}10]$). The three-fold coordinated atoms 1 (Ga) and 2 (N) adopt a hybridisation similar to the (10$\bar{1}$0) surface atoms. The distance between columns (1/2) and (3/4) are by 9% contracted while the distance between columns (7/8) and (9/10) is by 13% stretched

Table 6

Bond lengths (min–max (and average)) and bond angles (min–max (and average)) for the most distorted atoms at the core of the threading edge dislocation ($\mathbf{b} = \frac{1}{3}[1210]$). Atom numbers refer to Fig. 19

atom	bond lengths (Å)	bond angles (°)
1 (Ga$_{3\times\text{coord.}}$)	1.85–1.86 (1.85)	112–118 (116)
2 (N$_{3\times\text{coord.}}$)	1.88–1.89 (1.86)	106–107 (106)
3/4 (Ga/N$_{4\times\text{coord.}}$)	1.86–1.95 (1.91)	97–119
5/6 (Ga/N$_{4\times\text{coord.}}$)	1.92–2.04 (1.97)	100–129
7/8 (Ga/N$_{4\times\text{coord.}}$)	1.94–2.21 (2.06)	94–125
9/10 (Ga/N$_{4\times\text{coord.}}$)	1.95–2.21 (2.11)	100–122

found for the screw dislocation with a narrow opening. This can be interpreted by noting that the edge dislocation has a smaller number of three-fold coordinated atoms than the open-core screw dislocation as well as a smaller elastic strain energy arising from the smaller Burgers vector. This last energy is proportional to kb^2. Here, b is the magnitude of the Burgers vector and the constant k is equal to unity for the screw dislocation, and $1/(1-\nu)$ for the edge dislocations, where ν is Poisson's ratio (0.37 for GaN [110]). Thus the ratio of the elastic energies is $E_{\text{screw}}/E_{\text{edge}}$ which is approximately 1.66. Our calculations give the ratio of the line energies, which includes the core energies, to be 2.08. This could explain why threading edge dislocations occur at a higher density than threading screw dislocations.

In analogy to the open-core screw dislocations we have investigated whether the energy of the threading edge dislocation could be lowered by removing the most dis-

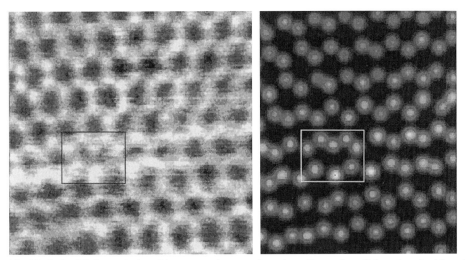

Fig. 20. Left: High-resolution Z-contrast image of a threading edge dislocation looking down [0001]. The bright dots are atomic columns of alternating Ga and N atoms. The dislocation core is shown in the boxed region. Right: Maximum entropy image showing most probable column positions. The distance between the column of three-fold coordinated atoms and the columns on the left (and right) is found to be by $15 \pm 10\%$ contracted. The distance between the column below the three-fold coordinated atoms and the neighbouring columns is found to be by $15 \pm 10\%$ stretched. These results are consistent with our calculations. Xin et al. [101]

torted core atoms (see Fig. 19). However, removal of either the columns of atoms (9,10), or the columns (1,2), (3,4), (5,6), (7,8) and their equivalents on the right, leads to considerably higher line energies. This implies that, in contrast with screw dislocations, which as discussed above can exist with a variety of cores, the threading edge dislocations should exist with a full core.

6.4 Deep acceptors trapped at threading edge dislocations: V_{Ga} and $V_{Ga}-(O_N)_n$

In the previous section we showed that in the defect free form the threading edge dislocation has a band gap free from deep lying states, hence implying that the pure dislocation cannot be responsible for the yellow luminescence detected in n-type GaN. However, as can be seen from Fig. 19 and Table 6 the core atoms adopt a very particular geometry with atoms 1 and 2 being three-fold coordinated and atoms 9 and 10 having very stretched bonds with bond-lengths ranging from 2.0 to 2.2 Å. This geometry differs considerably from a position in bulk-like material and thus gives rise to a stress field which could act as a trap for intrinsic defects and impurities. Gallium vacancies (V_{Ga}) have been detected by positron annihilation studies in bulk GaN and their concentration was found to be related to the intensity of the YL [111]. The relevant transition level in *n*-type GaN is at the centre of the YL spectrum ($E^{2-/3-} \approx 1.1$ eV referenced to the top of the valence band [112]). As a triple acceptor the gallium vacancy is threefold negatively charged in n-type GaN and can attract up to three positively charged donors. Recent experimental [113 to 115] and theoretical [116] works suggest that oxygen at a nitrogen site (O_N) is the main cause of unintentional n-type conductivity in GaN. V_{Ga} forms defect complexes with O_N which sits as a next neighbour of V_{Ga} to reduce the Coulomb energy [112, 117]. V_{Ga} related defect complexes in GaN were found to have electrical properties dominated by the Ga vacancy [112], i.e. they are acceptors and exhibit gap states above the top of the valence band arising from the N dangling bonds surrounding V_{Ga}. Furthermore, Youngman and Harris [118] studied the violet luminescence (VL) in AlN, which is believed to have essentially the same origin as the YL in GaN [117]. They found the VL in AlN to be correlated with the oxygen incorporation and extended defects which are also known to contain substantial amounts of oxygen [119]. Hence, in analogy to the VL in AlN it has been suggested that the YL in *n*-type GaN is caused by O related defect complexes.

6.4.1 Benchmark calculations for V_{Ga}, O_N and $V_{Ga}-(O_N)$ in bulk material

In bulk material the $(V_{Ga}-O_N)^{2-}$ defect complex (see Fig. 21) as well as its constituents, V_{Ga}^{3-} and O_N^+, have previously been investigated by Neugebauer et al. [112, 120] and Mattila et al. [117] using plane wave SCF-LDA methods. As a benchmark they are now investigated by the SCC-DFTB method where the defects are modelled in 128 atom wurtzite supercells using two *k*-points to sample the Brillouin zone. As in references [112, 117, 120], formation energies are evaluated assuming Ga-rich growth conditions, which are common in many growth techniques, O in equilibrium with Ga_2O_3, corresponding to an upper limit for the O concentration [120], and n-type material, i.e. the Fermi level is pinned close to the conduction band minimum. The atomic geometry of the triply charged Ga vacancy is characterised by a strong outward relaxation of the

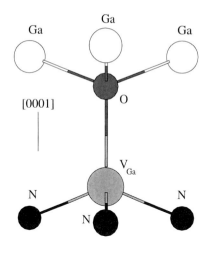

Fig. 21. Schematic view of the V_{Ga}–O defect complex. Substituting further three-fold coordinated N by O leads to $V_{Ga}-(O_N)_2$ and $V_{Ga}-(O_N)_3$

surrounding N atoms. The three equivalent N atoms relax by 10.2% (11.8% in Ref. [112]) outwards and the remaining N atom moves 9.5% in [0001] (10.6% in Ref. [112]). The formation energy is low (1.6 eV in this work, ≈1.3 eV in Ref. [112] and ≈1.5 eV in Ref. [117]). Oxygen on a nitrogen site has slightly larger Ga-O bonds than the Ga-N bond length in bulk GaN (1.95 Å). We obtain again a low formation energy of 1.7 eV (≈1.7 eV in Ref. [112] and ≈ 1.6 eV in Ref. [117]). Bringing V_{Ga}^{3-} and O_N^+ together, one gets $(V_{Ga}-O_N)^{2-}$. We find the distance between the vacancy core and the O (N) increased by 13.5% (8.9%) which is close to the values of 14.9% (9.8%) given by Neugebauer et al. [112]. Furthermore, we determined the energy ΔE for the reaction

$$(V_{Ga} - O_N)^{2-} \rightarrow V_{Ga}^{3-} + O_N^+$$

to be 2.2 eV in good agreement with the plane wave methods (1.8 eV in Ref. [112] and ≈2.1 eV in Ref. [117]). We thus get an absolute formation energy of ≈1.1 eV which is again very close to the plane wave values (≈1.1 eV in Ref. [112] and ≈ 0.9 eV in Ref. [117]) implying a high equilibrium concentration of $\sim 10^{18}/cm^3$ [112, 117] at a usual MOCVD growth temperature of $\sim$1300 K.

The good agreement of the SCC-DFTB method for V_{Ga}^{3-}, O_N^+ and $(V_{Ga}-O_N)^{2-}$ with SCF-LDA plane wave calculations suggests that SCC-DFTB allows a valid description of oxygen in GaN.

6.4.2 Properties of V_{Ga}, O_N and $V_{Ga}-(O_N)_n$ ($n = 1, 2, 3$) in the stress field of the threading edge dislocation

In the following, the geometries, electrical properties and formation energies of V_{Ga}, O_N and $V_{Ga}-(O_N)_n$ ($n = 1, 2, 3$) are investigated in the stress field of threading edge dislocations [99]. In the SCC-DFTB case the dislocations are modelled in a 312 atom supercell containing a dislocation dipole (see Fig. 22). In order to reduce the interaction between the point defects we doubled the 312 atom supercell along the dislocation line, i.e. in [0001], to obtain a 624 atom supercell. In the AIMPRO case, we used 286 atom stoichiometric clusters with one dislocation.

First we place the point defects into a bulk-like position, i.e. a position with a very small stress field, far away from the dislocation core in the supercell (position L in Fig. 22). At position L in this cell we find the atomic geometries and formation energies of the point defects to be in good agreement with the values obtained in the 128 atom perfect lattice supercell (see first two lines in Table 7). We now put the defects at different positions (columns (1/2), (5/6), (9/10) in Fig. 19) in the dislocation stress field and

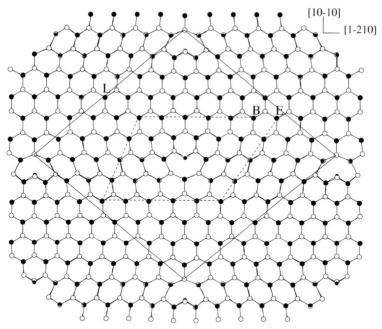

Fig. 22. View along the dislocation line ([0001]) of a wurtzite supercell containing a dipole of threading edge dislocations with Burgers vector $\mathbf{b} = \mathbf{BE} = \pm\frac{1}{3}[1\bar{2}10]$. The cell contains 312 atoms and has a periodicity of [0001] along the dislocation line. Cutting out one dislocation and saturating the dangling bonds with pseudo-hydrogens gives a cluster periodic along the dislocation line (see text). Position L is used as a bulk reference position to model point defects (see Section 6.4)

evaluate the formation energies and electrical properties. As will be seen, some of the formation energies are negative suggesting that under equilibrium conditions the corresponding position would certainly be adopted by the defect. However, since gallium vacancies and oxygen are not necessarily in equilibrium with the dislocation stress field, the precise concentration of defect complexes in the dislocation stress field depends on the history of the sample.

V_{Ga}^{3-} is trapped in the dislocation stress field, in particular, at the dislocation core (pos. 1 in Fig. 19) and at pos. 9 (see Fig. 19). Ga atoms in these positions would have high energies, caused by the under-coordination in pos. 1 or by the strongly strained

Table 7

Formation energies (in eV) of V_{Ga}^{3-}, O_N^+, $(V_{Ga}-O_N)^{2-}$, $(V_{Ga}-(O_N)_2)^{1-}$ and $V_{Ga}-(O_N)_3$ in a 128 atom bulk cell and at the threading edge dislocation (see Figs. 19 and 22). Ga-rich growth conditions, O in equilibrium with Ga_2O_3 and n-type material are assumed

position	V_{Ga}^{3-}	O_N^+	$(V_{Ga}-O_N)^{2-}$	$(V_{Ga}-O_{2N})^-$	$V_{Ga}-O_{3N}$
bulk cell	1.8	1.7	1.1	0.7	0.8
pos. L (bulk-like)	1.7	1.5	1.0	0.9	0.7
pos. (1,2) (core)	−0.2	0.2	−2.3	−2.5	−3.0
pos. (5,6)	0.3	1.0	−1.0	−1.0	−0.8
pos. (9,10)	−0.3	1.3	−0.6	−0.3	−0.3

bonds in pos. 9 (2.11 Å average bond length). This makes the formation of vacancies at these positions energetically favourable (see Table 7). It should be noted that at pos. 1 a Ga vacancy creates a two-fold coordinated N atom at pos. 2 which would result in a high energy. However, since Ga atoms at pos. 7 and its equivalent at the right are quite close to the N atom at pos. 2, this N atom forms a bond (2.00 Å) with one of these Ga atoms and thus achieves three-fold coordination. The new configuration has a distorted core and looks like a first step of a kink formation. This suggests that V_{Ga} play an important role in the dislocation motion.

Oxygen atoms sit preferentially two- or three-fold coordinated. This explains why O_N^+ is by 1.3 eV more stable at the dislocation core (pos. 2) where it replaces a three-fold coordinated N atom than in a bulk-like region (pos. 0) where it is four-fold coordinated (see Table 7).

The high stabilities of V_{Ga}^{3-} and O_N^+ at the dislocation core imply also a very low formation energy for $(V_{Ga}-O_N)^{2-}$ (−3.3 eV below the energy for pos. L, see Fig. 22) and hence a high concentration. Here O sits two-fold coordinated in a bridge position with very strong Ga–O bonds (1.72 Å). Due to these strong bonds and the high complex binding energy of 2.3 eV at the dislocation core we expect $(V_{Ga}-O_N)^{2-}$ to be immobile. Finally, we investigated $(V_{Ga}-(O_N)_2)^-$ and $V_{Ga}-(O_N)_3$, which in analogy to $(V_{Ga}-O_N)^{2-}$ are found to be particularly stable at the core of the threading edge dislocations where they are likely to be immobile. See Table 7 for the detailed formation energies. All these results suggest that $(V_{Ga}-(O_N)_n)^{(3-n)-}$ ($n = 1, 2, 3$) defect complexes increase the oxygen concentration near to threading edge dislocations and in particular at the dislocation core. Threading edge dislocations may therefore be used as a trap for undesired impurities. This has been suggested by Nakamura et al. [121] who proposed that during the initial stages of GaN growth threading edge dislocations should be permitted to clean the sample from impurities which emerge from the substrate. In a following step, a very thin SiO mask is then used to reduce the number of threading edge dislocations in the following region of the epilayer which will be used as the active region of the devices.

Concerning the electrical properties the SCC-DFTB calculations reveal that at bulk positions V_{Ga}^{3-}, $(V_{Ga}-O_N)^{2-}$, $(V_{Ga}-(O_N)_2)^-$ and $V_{Ga}-(O_N)_3$ defects are deep acceptors with gap states around 1.0 to 1.2 eV above VBM. In order to obtain information about the contribution of these defects to the YL we then calculated the difference of the formation energies depending on the charge states relevant to the transition in n-type material. The results referenced to VBM are given in Table 8. Subtracting them from the band gap (≈ 3.4 eV) gives an estimate for the transition energies in n-type material. Since the energies for the different charge states are derived from total energies asso-

Table 8
Transition energies (in eV) of V_{Ga}, $V_{Ga}-O_N$, $V_{Ga}-(O_N)_2$ and $V_{Ga}-(O_N)_3$ at the threading edge dislocation (see Figs. 19 and 22) referenced to VBM

position	$(V_{Ga})^{2-/3-}$	$(V_{Ga}-O_N)^{1-/2-}$	$(V_{Ga}-O_{2N})^{0/1-}$	$(V_{Ga}-O_{3N})^{1+/0}$
pos. L (bulk-like)	1.4	1.0	0.7	0.9
pos. (1/2) (core)	0.8	1.0	0.7	0.4
pos. (5/6)	0.8	1.4	1.0	0.9
pos. (9/10)	0.4	0.3	0.6	0.8

ciated with fully relaxed atomic configuration, the calculated energy differences correspond to zero-phonon transitions. As can be seen, at a variety of positions the defects could contribute to the yellow luminescence. It is interesting to note that in a bulk-like position $V_{Ga}-(O_N)_3$ has a deep gap state (≈ 1.0 eV above VBM) which comes from a three-fold coordinated nitrogen atom in a bulk position. At the dislocation core (column (1/2)), however, for $V_{Ga}-(O_N)_3$ all three-fold coordinated nitrogens surrounding the Ga vacancy are replaced by oxygen. At this position $V_{Ga}-(O_N)_3$ adopts the same configuration as at the $(10\bar{1}0)$ surface [108] and does not induce deep states in the band gap.

7. Conclusion

We described a selfconsistent charge density-functional based tight-binding scheme (SCC-DFTB) which is shown to determine the geometries, energetics and electrical properties of semiconducting systems with an accuracy comparable to fully selfconsistent methods. The method is implemented in an efficient way on parallel computers and applied to GaAs and GaN structures which due to their large size and/or metallic behaviour and/or large configuration space could not be investigated before.

References

[1] M. BOCKSTEDTE, A. KLEY, J. NEUGEBAUER, and M. SCHEFFLER, Comput. Phys. Comm., **107**, 187 (1997).
[2] R. JONES and P. R. BRIDDON, The ab initio Cluster Method and the Dynamics of Defects in Semiconductors, in: Semicond. and Semimet., Vol. 51A, Chap. 6, Ed. M. Stavola, Academic Press, Boston 1998.
[3] N. MOLL, A. KLEY, E. PEHLKE, and M. SCHEFFLER, Phys. Rev. B, **54**, 8844 (1996).
[4] J. WAGNER, R.C. NEWMAN, B.R. DAVIDSON, S.P. WESTWATER, T.J. BULLOUGH, T.B. JOYCE, C.D. LATHAM, R. JONES, and S. ÖBERG, Phys. Rev. Lett. **78**, 74 (1997).
[5] M. ELSTNER, D. POREZAG, G. JUNGNICKEL, J. ELSNER, M. HAUGK, G. JUNGNICKEL, TH. FRAUENHEIM, S. SHUHAI, and G. SEIFERT, Phys. Rev. B **58**, 7260 (1998).
[6] D. POREZAG, TH. FRAUENHEIM, TH. KÖHLER, G. SEIFERT, and R. KASCHNER, Phys. Rev. B **51**, 12947 (1995).
[7] J. ELSNER, M. HAUGK, G. JUNGNICKEL, and TH. FRAUENHEIM. J. Mater. Chem. **6**, 1649 (1996).
[8] A. SIECK, D. POREZAG, TH. FRAUENHEIM, M.R. PEDERSON, and K. JACKSON, Phys. Rev. A **56**, 4890 (1997).
[9] D. POREZAG, M.R. PEDERSON, TH. FRAUENHEIM, and T. KÖHLER. Phys. Rev. B **52**, 14963 (1997).
[10] D. POREZAG, G. JUNGNICKEL, TH. FRAUENHEIM, G. SEIFERT, A. AYUELA, and M.R. PEDERSON, Appl. Phys. A **64**, 321 (1997).
[11] M. KAUKONEN, P.K. SITCH, G. JUNGNICKEL, R.M. NIEMINEN, S. POYKKO, D. POREZAG, and TH. FRAUENHEIM, Phys. Rev. B **57**, 9965 (1998).
[12] M. HAUGK, J. ELSNER, M. STERNBERG, and TH. FRAUENHEIM. J. Phys.: Condensed Matter **10**, 4523 (1998).
[13] M. STERNBERG, W.R.L. LAMBRECHT, and TH. FRAUENHEIM, Phys. Rev. B **56**, 1568 (1997).
[14] J. ELSNER, R. JONES, P.K. SITCH, D. POREZAG, M. ELSTNER, TH. FRAUENHEIM, M.I. HEGGIE, S. ÖBERG, and P.R. BRIDDON, Phys. Rev. Lett. **79**, 3672 (1997).
[15] M. ELSTNER, TH. FRAUENHEIM, E. KAXIRAS, G. SEIFERT, and S. SUHAI, phys. stat. sol. (b) **217**, 357 (2000).
[16] L.S. BLACKFORD, J. CHOI, A. CLEARY, J. DEMMEL, I.D. HILLON, J. DONGARRA, S. HAMMARLING, G. HENRY, A. PETITET, K. STANLEY, D.W. WALKER, and R.C. WHALEY, Proc. Supercomputing '96, 1996.
[17] SCALAPACK USER'S GUIDE, SIAM, 3600 University City Science Center, Philadelphia 1997.
[18] M. ABOELAZE, N. CHRISOCHOIDES, and E. HOUSTIS, Tech. Rep. CSD-TR-91-007, Purdue University, West Lafayette 1991.

[19] J. Choi, J. Dongarra, S. Ostrouchov, A. Petitet, D. Walker, and R.C. Whaley, Computer Science Tech. Rep. CS-95-292, University of Tennessee, Knoxville 1995.
[20] J. Choi, J. Dongarra, and D. Walker, Concurrency: Practice and Experience **8**, 517 (1996).
[21] G.-X. Qian, R.M. Martin, and D.J. Chadi, Phys. Rev. B **37**, 1303 (1988).
[22] E. Kaxiras, Y. Bar-Yam, J.D. Joannopoulos, and K.C. Pandey, Phys. Rev. B **35**, 9625 (1987).
[23] J.E. Northrup and S. Froyen, Phys. Rev. B **50**, 2015 (1994).
[24] S.Y. Tong, G. Xu, and W.N. Mei, Phys. Rev. Lett. **52**, 1693 (1984).
[25] T. Hashizume, Q.K. Xue, J. Zhou, A. Ichimiya, and T. Sakurai, Phys. Rev.Lett. **74**, 3177 (1995).
[26] D.K. Biegelsen, R.D. Bringans, J.E. Northrup, and L.E. Schwartz, Phys. Rev. Lett. **65**, 452 (1990).
[27] A.Y. Cho and I. Hayashi, Solid State Electronics **14**, 125 (1971).
[28] R. Arthur, Surf. Sci. **43**, 449 (1974).
[29] D. A. Woolf, D.I. Westwood, and R.H. Williams, Appl. Phys. Lett. **62**, 1371 (1993).
[30] H.-W. Ren and T. Nishinaga, Phys. Rev. B **54**, R11054 (1996).
[31] E. Kaxiras, Y. Bar-Yam, J.D. Joannopoulos, and K.C. Pandey, Phys. Rev. B **35**, 9636 (1987).
[32] K. Shiraishi, J. Phys. Soc. Jap. **59**, 3455 (1990).
[33] G.-X. Qian, R.M. Martin, and D.J. Chadi, Phys. Rev. B **38**, 7649 (1988).
[34] CRC Handbook of Chemistry and Physics, 67th Edn., Ed R.C. West, Chemical Rubber Company Press, Boca Raton 1986.
[35] N. Chetty and R.M. Martin, Phys. Rev. B **45**, 6074 (1992).
[36] M. Haugk, J. Elsner, and Th. Frauenheim, J. Phys.: Condensed Matter **9**, 7305 (1997).
[37] Y.-H. Lee and J.W. Corbett, Phys. Rev. B **8**, 2810 (1973).
[38] R. Krause-Rehberg, M. Brohl, H.S. Leipner, Th. Drost, A. Polity, U. Beyer, and H. Alexander, Phys. Rev. B **47**, 13266 (1993).
[39] R. Krause-Rehberg, H.S. Leipner, A. Kupsch, A. Polity, and T. Drost, Phys. Rev. B **49**, 2385 (1994).
[40] F. Börner, S. Eichler, A. Polity, and R. Krause-Rehberg, Appl. Surf. Sci. (1999).
[41] T.Y. Tan, P. Plekhanov, and U.M. Gösele, Appl. Phys. Lett. **70**, 1715 (1997).
[42] A.M. Cuitiño and M. Ortiz, Acta Mater. **44**, 427 (1996).
[43] N.S. Mott, Trans. MS AIME **218**, 962 (1960).
[44] H.S. Leipner, C.G. Hübner, T.E.M. Staab, M. Haugk, and R. Krause-Rehberg, phys. stat. sol. (a) **171**, 377 (1999).
[45] M.J. Puska and R.M. Nieminen, Rev. Mod. Phys. **66**, 841 (1994).
[46] J.G. Byrne, Metals Trans. **10A**, 791 (1979).
[47] D.J. Chadi and K.J. Chang, Phys. Rev. B **38**, 1523 (1988).
[48] M. Haugk, J. Elsner, and Th. Frauenheim, J. Phys.: Condensed Matter **9**, 7305 (1997).
[49] J.E. Northrup and S.B. Zhang, Phys. Rev. B **47**, 6791 (1993).
[50] S.B. Zhang and J.E. Northrup, Phys. Rev. Lett. **67**, 2339 (1991).
[51] M.J. Puska and R.M. Nieminen, J. Phys. F (Metal Phys.) **13**, 333 (1983).
[52] M.J. Puska, S. M äkinen, M. Manninen, and R.M. Nieminen, Phys. Rev. B **39**, 7666 (1989).
[53] M. Saito and A. Oshiyama, Phys. Rev. B **53**, 7810 (1996).
[54] M. Hakala, M.J. Puska, and R.M. Niemienen, Phys.Rev. B **57**, 7621 (1998).
[55] F. Plazaola, A.P. Seitsonen, and M.J. Puska, J. Phys.: Condensed Matter **6**, 8809 (1994).
[56] J.L. Hastings, S.K. Estreicher, and P.A. Fedders, Phys. Rev. B **56**, 10215 (1997).
[57] B.T. Cunningham, L.J. Guido, J.E. Baker, J.S. Major, N. Holonyak, and G.E. Stillman, Appl. Phys. Lett. **55**, 687 (1989).
[58] T.H. Chiu, J.E. Cunningham, J.A. Ditzenberger, W.Y. Jan, and S.N.G. Chu, J. Crystal Growth **111**, 274 (1991).
[59] H.M. You, T.Y. Tan, U.M. Gösele, S.T. Lee, G.E. Höfler, K.C. Hsieh, and N. Holonyak, J. Appl. Phys. **74**, 2450 (1993).
[60] J.A. Zhou, C.Y. Song, J.F. Zheng, M. Stavola, C.R. Abernathy, and S.J. Pearton, Mater. Sci. Forum **258**, 1293 (1997).
[61] S.P. Westwater and T.J. Bullough, J. Crystal Growth **170**, 752 (1997).
[62] G.E. Höfler, J.N. Baillargeon, K.C. Hsieh, and K.Y. Cheng, Appl. Phys. Lett. **60**, 1990 (1992).
[63] G.E. Höfler and K.C. Hsieh, Appl. Phys. Lett. **61**, 327 (1992).
[64] G.E. Höfler, H.J. Höfler, N. Holonyak, and K.C. Hsieh, J. Appl. Phys. **72**, 5318 (1992).
[65] T.J. Delyon, J.M. Woodall, M.S. Goorsky, and P.D. Kirchner, Appl. Phys. Lett. **56**, 1040 (1990).

[66] K. WATANABE and H. YAMAZAKI, Appl. Phys. Lett. **59**, 434 (1991).
[67] J.D. MACKENZIE, C.R. ABERNATHY, S.J. PEARTON, and S.N.G. CHU, Appl. Phys. Lett. **66**, 1397 (1995).
[68] D.L. SATO, F.J. SZALKOWSKI, and H.P. LEE, Appl. Phys. Lett. **66**, 1791 (1995).
[69] J. WAGNER, R.C. NEWMAN, B.R. DAVIDSON, S.P. WESTWATER, T.J. BULLOUGH, T.B. JOYCE, C.D. LATHAM, R. JONES, and S. ÖBERG, Phys. Rev. Lett. **78**, 74 (1997).
[70] B.R. DAVIDSON, R.C. NEWMAN, J. WAGNER, C.D. LATHAM, R. JONES, C.C. BUTTON, and P.R. BRIDDON, forthcoming.
[71] H. FUSHIMI and K. WADA, J. Appl. Phys. **82**, 1208 (1997).
[72] J.E. NORTHRUP and S.B. ZHANG, Phys. Rev. B **47**, 6791 (1993).
[73] H. FUSHIMI and K. WADA, IEEE Trans. Electron Devices **44**, 1996 (1997).
[74] Y. CHENG, M. STAVOLA, C.R. ABERNATHY, S.J. PEARTON, and W.S. HOBSON, Phys. Rev. B **49**, 2469 (1994).
[75] B. CLERJAUD, F. GENDRON, M. KRAUSE, and W. ULRICI, Phys. Rev. Lett. **65**, 1800 (1990).
[76] R. JONES and S. ÖBERG, Phys. Rev. B **44**, 3673 (1991).
[77] U. GÖSELE, T.Y. TAN, M. SCHULTZ, U. EGGER, P. WERNER, R. SCHOLZ, and O. BREITENSTEIN, Defect and Diffusion Forum **143**, 1079 (1997).
[78] S. CHRISTIANSEN, M. ALBRECHT, W. DORSCH, H.P. STRUNK, C. ZANOTTI-FREGONARA, G. SALVIATI, A. PELZMANN, M. MAYER, M. KAMP, and K.J. EBELING, MRS Internet J. of Nitride Res., 1, 1996.
[79] X.H. WU, L.M. BROWN, D. KAPOLNEK, S. KELLER, B. KELLER, S.P. DENBAARS, and S.J. SPECK, J. Appl. Phys. **80**, 3228 (1996).
[80] X.J. NING, F.R. CHIEN, and P. PIROUZ, J. Mater. Res. **11**, 580 (1996).
[81] F.A. PONCE, D. CHERNS, W.T. YOUNG, and J.W. STEEDS, Appl. Phys. Lett. **69**, 770 (1996).
[82] S.D. LESTER, F.A. PONCE, M.G. CRANFORD, and D.A. STEIGERWALD, Appl. Phys. Lett. **66**, 1249 (1996).
[83] L. SUGIURA, J. Appl. Phys. **81**, 1633 (1997).
[84] S. NAKAMURA, T. MUKAI, and M. SENOH, J. Appl. Phys. **76**, 8189 (1994).
[85] S. NAKAMURA, M. SENOH, S. NAGAHAMA, N. IWASA, T. YAMADA, T. MATSUSHITA, H. KIYOHU, and Y. SGUIMOTO, Jpn. J. Appl. Phys. **35**, L74 (1996).
[86] K. MAEDA and S. TAKEUCHI, J. Physique **44**, C4–375 (1983).
[87] W. QIAN, G.S. ROHRER, M. SKOWRONSKI, K. DOVERSPIKE, L.B. ROWLAND, and D.K. GASKILL, Appl. Phys. Lett. **67**, 2284 (1995).
[88] W. QIAN, M. SKOWRONSKI, K. DOVERSPIKE, L.B. ROWLAND, and D.K. GASKILL, J. Cryst. Growth **151**, 396 (1995).
[89] P. VENNÉGUÈS, B. BEAUMONT, M. VAILLE, and P. GIBART, Appl. Phys. Lett. **70**, 2434 (1997).
[90] Z. LILIENTAL-WEBER, Y. CHEN, S. RUVIMOR, and J. WASHBURN, Phys. Rev. Lett. **79**, 2835 (1997).
[91] F.C. FRANK, Acta. Cryst. **4**, 497 (1951).
[92] Z. LILIENTAL-WEBER, Y. CHEN, S. REVIMOV, and J. WASHBURN, Mater. Sci. Forum **258**, 1659 (1997).
[93] T. OGINO and M. AOKI, Jpn. J. Appl. Phys. **19**, 2395 (1980).
[94] E.R. GLASER, Phys. Rev. B **51**, 13326 (1995).
[95] M. GODLEWSKI et al., Mater. Sci. Forum **258/263**, 1149 (1997).
[96] S.J. ROSNER, E.C. CARR, M.J. LUDOWISE, G. GIRLAMI, and H.I. ERIKSON, Appl. Phys. Lett. **70**, 420 (1997).
[97] F.A. PONCE, D.B. BOUR, W. GÖTZ, and P.J. WRIGHT, Appl. Phys. Lett. **68**, 57 (1996).
[98] J. ELSNER, R. JONES, M.I. HEGGIE, M. HAUGK, TH. FRAUENHEIM, S. ÖBERG, and P.R. BRIDDON, Phil. Mag. Lett. (1998).
[99] J. ELSNER, R. JONES, M.I. HEGGIE, P.K. SITCH, M. HAUGK, TH. FRAUENHEIM, S. ÖBERG, and P.R. BRIDDON, Phys. Rev. B **58**, 12571 (1998).
[100] D. CHERNS, W.T. YOUNG, J.W. STEEDS, F.A. PONCE, and S.S. NAKAMURA, J. Cryst. Growth **178**, 201 (1997).
[101] Y. XIN, S.J. PENNYCOOK, N.D. BROWNING, P.D. NELIST, S. SIVANANTHAN, F. OMNÈS, B. BEAUMONT, J.-P. FAURIE, and P. GIBART, Appl. Phys. Lett. **72**, 2680 (1998).
[102] Z. LILIENTAL-WEBER, private communications at EDS 1998.
[103] J.E. NORTHRUP and J. NEUGEBAUER, Phys. Rev. B **53**, 10477 (1996).
[104] P. PRIOUZ, Phil. Mag. A **78**, 727 (1998).
[105] D. CHERNS, private communication, Bristol 1997.

[106] J.E. NORTHRUP, R. DI FELICE, and J. NEUGEBAUER, Phys. Rev. B **56**, R4325 (1997).
[107] Y. XIN, S.J. PENNYCOOK, N.D. BROWNING, P.D. NELLIST, S. SIVANANTHAN, J.P. FAURIE, and P. GIBART, in: Nitride Semiconductors, Eds. F.A. PONCE, S.P. DENBAARS, B.K. MEYER, S. NAKAMURA, and S. STRITE, Mater. Res. Soc. Proc. **482**, 781 (1998).
[108] J. ELSNER, R. JONES, M. HAUGK, R. GUTIERREZ, and TH. FRAUENHEIM, Appl. Phys. Lett. **73**, 3630 (1998).
[109] J. ELSNER, R. JONES, M. HAUGK, R. GUTIERREZ, M. HEGGIE, TJ. FRAUENHEIM, S. ÖBERG, and P.R. BRIDDON, Appl. Phys. Lett. **73**, 3530 (1998).
[110] V.A. SAVASTENKO and A. U. SHELOG, phys. stat. sol. (a) **48**, K135 (1978).
[111] K. SAARINEN, T. LAINE, S. KUISMA, J. NISSILÄ, P. HAUTOJÄRVI, L. DOBRZYNSKI, J.M. BARANOWSKI, K. PAKULA, R. STEPNIEWSKI, M. WOJDAK, A. WYSMOLEK, T. SUSKI, M. LESZCYNSKI, I. GRZEGORYY, and S. POROWSKI, Phys. Rev. Lett. **79**, 3030 (1997).
[112] J. NEUGEBAUER and C. VAN DE WALLE, Appl. Phys. Lett. **69**, 503 (1996).
[113] C. WETZEL, T. SUSKI, J.W. AGER, E.R. WEBER, E.E. HALLER, S. FISCHER, B.K. MEYER, R.J. MOLNAR, and P. PERLIN, Phys. Rev. Lett. **78**, 3923 (1997).
[114] C. WETZEL et al., in: ICPS 23, World Scientific, Singapore 1996 (p. 2929).
[115] P. PERLIN et al., Mater. Res. Soc. Proc. **449**, 519 (1997).
[116] C.G. VAN DE WALLE and J. NEUGEBAUER, Mater. Sci. Forum **258/263**, 19 (1997).
[117] T. MATTILA and R.M. NIEMINEN, Phys. Rev. B **55**, 9571 (1997).
[118] R.A. YOUNGMAN and J.H. HARRIS, J. Amer. Ceram. Soc. **73**, 3238 (1990).
[119] A.D. WESTWOOD et al., J. Mater. Res. **10**, 1270 (1995).
[120] J. NEUGEBAUER and C.G. VAN DE WALLE, Festkörperprobleme **35**, 25 (1996).
[121] S. NAKAMURA, in: Nitride Semiconductors, Eds. F.A. PONCE, S.P. DENBAARS, B.K. MEYER, S. NAKAMURA, and S. STRITE, Mater. Res. Soc. Proc. **482**, 1145 (1998).

phys. stat. sol. (b) **217**, 513 (2000)

Subject classification: 61.72.Bb; 61.72.Ji; 63.20.Pw; 66.30.–h; 71.15.–m; S5.11

Structure and Dynamics of Point Defects in Crystalline Silicon

S. K. Estreicher

Texas Tech University, Lubbock TX 79409-1051, USA
Tel.: (806) 742-3723; Fax: -1182; e-mail: stefan.estreicher@ttu.edu

(Received August 10, 1999)

Ab-initio theoretical methods are increasingly being used to study the properties of defects in covalent materials such as silicon. Static properties include potential energy surfaces, binding energies, and electronic structures. The dynamics involve diffusion, vibrational properties, and defect reactions. This paper describes a powerful combination of methods used to study the interactions involving vacancies, self-interstitials, and impurities in Si. The methods are Hartree-Fock in molecular clusters and density-functional based molecular-dynamics simulations in periodic supercells. Three examples of applications are discussed: 1. the dissociation of interstitial H_2 molecules by vacancies and self-interstitials, and the formation of H_2^*, 2. the aggregation of vacancies leading to the formation of the ring-hexavacancy (V_6), and 3. the trapping of interstitial copper at V_6 and the origin of the electrical activity of copper precipitates.

1. Introduction

Most electrical and optical properties of semiconductors such as crystalline Si are determined by the type and concentration of defects they contain. Many defects are present in the as-grown crystal. They include *intrinsic defects* (vacancies, self-interstitials, and their precipitates) and *extrinsic defects* or *impurities* which come from the source material, the ambient, the crucible, the heating element, etc. More defects are introduced during the processing of devices: Etching injects vacancies as well as hydrogen (if H-containing chemicals are used), the growth of surface layers (oxides, nitrides, metallizations, etc.) inject self-interstitials or vacancies into the bulk, ion implantation creates damage, the slurry used in chemomechanical polishing often contains Cu, and the list goes on and on.

All these defects modify the electronic structure of the material. For example, dopants add charge carriers (electrons e^- to the conduction band or holes h^+ to the valence band), while most transition metals are associated with $e^- - h^+$ recombination centers, and therefore reduce carrier lifetimes. In general, defects or impurities disrupt the periodicity of the crystal, introduce local strain, and result in new energy levels which are often somewhere in the gap. Depending on the Fermi level and the position of defect levels, some defects may exist in several charge states. Further, defects can be mobile, leading to a wide range of reactions between them. Such reactions create new, more complicated, defects with different electrical and optical properties than the original ones. Defect diffusion and reaction is enhanced by thermal treatments, and the properties of a sample often depend on its history.

As the size of the active area of devices shrinks, traces of unwanted defects can affect their properties, and it is becoming critical to understand how they diffuse, interact with the crystal, with each other, and with other defects. Such an understanding can only be achieved by a combination of experimental and theoretical studies. Experimentalists need to probe microscopic properties. They use techniques such as Fourier-transform infra-red absorption (FTIR) or Raman spectroscopy to look for local vibrational modes, electron paramagnetic resonance (EPR) to find unpaired electron spins, or deep-level transient spectroscopy (DLTS) to probe the gap for defect levels. More macroscopic methods such as secondary-ion mass spectrometry (SIMS) or electron-beam induced current (EBIC) also provide crucial information.

Theorists use 'ab-initio' or 'first-principles' methods. In Hartree-Fock theory, 'ab-initio' means that all the required four-center integrals have been calculated. There are about N^4 of them, where N is the number of orbitals. This says nothing about the size of the basis set, the use of pseudopotentials, and other variables. As for 'first-principles', the expression has never been precisely defined. The fancy terminology usually means that no parameter adjusted to experimental data has been used. In contrast, 'semi-empirical' methods fit parameters such as potentials, ionization energies, or bond lengths to measured values. These methods have transferability problems. While they are excellent at predicting quantities they were parameterized to calculate, their predictive power is unknown for other situations.

However, it is important to remember that all the methods contain some kind of parameters. These include the basis set size and type (plane waves versus atomic-like functions), the amount of electron correlation included (local density approximation, generalized gradient approximation, or Moller-Plesset corrections to HF theory), the size of the cluster or periodic supercell which represents the host crystal, the cutoff radii for ab-initio pseudopotentials, the number of k-points in supercell calculations, etc. However, none of those parameters are fitted to experimental data.

In the past fifteen years or so, theory has achieved considerable progress. Better modeling techniques and algorithms have been developed, much experience has been gained, and the computers themselves have tremendously increased in power, input/output speed, and memory size. The predictive power of theory is such that experimentalists often collaborate with theorists to help them refine models and/or rule out energetically unfavorable defect structures. Some predictions are nearly quantitative, for example equilibrium configurations, binding energies, diffusion paths and barriers, or vibrational frequencies for various charge states of isolated defects and small complexes. On the other hand, the defect levels and wavefunctions are qualitative, and few or no properties associated with excited (conduction-band) states are calculated.

This paper deals with localized defects in Si, where 'localized' means a size of one nanometer or less. Extended defects such as dislocations or grain boundaries are not discussed. Issues of interest include the formation of small aggregates of vacancies or self-interstitials, the trapping of impurities at internal voids, impurity–impurity interactions leading to the formation of pairs or larger precipitates, and problems of that nature. There have been many results published in this area in the past few years, and recent reviews are available on hydrogen-related defects [1], oxygen and its precipitates [2, 3], transition metal impurities [4, 5], and deep-level defects in general [6].

My purpose here is to focus on a few examples which illustrate how theory contributes to our understanding of defects in semiconductors. They are from the research done in my group. Section 2 describes the methods, Section 3 deals with intrinsic defects and impurity–defect reactions, Section 4 with the aggregation of vacancies and the trapping of impurities at vacancy aggregates. Section 5 concludes with a summary and a discussion.

2. Methodology

The two commonly used ab-initio techniques are based on density-functional (DF) and Hartree-Fock (HF) theories, both of which are discussed elsewhere in this text.

DF theory replaces the many-electron problem with a single-particle one via an effective potential which is not fully known but is quite well approximated either within the local density approximation (LDA) or the generalized gradient approximation (GGA). The method gives excellent total energies and total (spin and charge) densities, but provides no one-electron wavefunctions. The basis sets used are either plane waves [7] or localized pseudo-atomic orbitals [8]. Planes waves are computationally efficient, but enormous numbers of them are required to describe highly localized densities, such as for atomic H. Localized basis sets of Gaussians or Slater orbitals are much more intuitive because of the local nature of the interactions under study. However, they are computationally more cumbersome.

HF theory attacks the many-electron problem up front, and the calculations are much more time-consuming than the DF ones. However, the HF output includes plenty of chemical information, such as overlap populations, degrees of bonding, bond indices, single-electron wavefunctions, etc. Basis sets of localized orbitals are always used. The basis set is called 'minimal' if a single orbital per occupied orbital is used (e.g. 1s, 2sp, 3sp for each Si atom) and 'double-zeta' or 'split-valence' if two orbitals per valence orbital are used. 'Polarization functions' are orbitals with symmetry corresponding to a higher quantum number (2p for H or 3d for Si).

Molecular-dynamics (MD) simulations are used to study the diffusion of defects, local vibrational modes, as well as defect reactions. In these calculations, the temperature is introduced classically via the kinetic energy of the nuclei, but the electrons remain at 0 K in most applications. The electronic problem is solved quantum mechanically (most often within the DF theory), and the ionic motion is treated within classical mechanics. The gradient of the total energy at the time t gives the force on each nucleus via the Hellman-Feynman theorem, and Newton's laws of motion are then solved to obtain the velocity and position of each ion at the time $t + \Delta t$, where Δt is of the order of the femtosecond. The entire problem is solved again, and simulations lasting many thousands of time steps are performed.

There are many types of MD simulations, with varying degrees of speed and accuracy. They are discussed in a recent review [9]. The methods allow both constant-temperature simulations (to calculate vibrational frequencies, defect diffusion, or reactions) and simulated quenching (to probe potential energy surfaces). In the latter case, some fraction of the nuclear kinetic energy is removed every few time steps to mimic the loss of temperature. Slow or fast quenching forces the convergence toward global or local minima of the potential energy surface. Of course, the reverse process (simulated annealing) is also possible.

The host crystal is approximated in one of two ways. Molecular clusters ('clusters') are a fraction of the crystal with surface dangling bonds tied up with H atoms ('saturators'). This approach works quite well because the wavefunctions associated with the Si-H surface bonds are very localized. However, the wavefunctions of a shallow defect tend to be artificially confined by the presence of the surface. Typical cluster sizes range from 10 to 100 host atoms ($Si_{10}H_{16}$ to $Si_{100}H_{72}$). The alternative consists in using periodic supercells ('cells'). Here, a number of host atoms (typically 64) are placed in a box to which periodic boundary conditions are applied. There is no surface to worry about, but the defect is periodic, and interactions between defects in adjacent cells turn defect levels into defect bands. The cell is always kept electrically neutral, even when studying charged defects, by the use of a neutralizing background. This artificially localizes the wavefunctions, and dipole–dipole interactions between cells are likely to occur.

Thus, a number of approximations are present, even in 'state-of-the-art' calculations. However, the results may still be remarkably accurate because many properties depend on total energy *differences* and the systematic errors cancel out. Thus, equilibrium geometries, relative energies of metastable states, binding energies, and sometimes vibrational frequencies, can be calculated accurately. However, the details of a defect wavefunction are more tricky to obtain, as are those quantities which depend on excited states. No information on the conduction band are calculated, and quantitative predictions of the positions of gap levels is risky, although a hybrid approach (scaled ab-initio ionization energies) has recently been proposed [10]. In any case, energy eigenvalues well inside the gap indicate the presence of localized states, but the precise position and number of such states depends on the basis set size and type, the size of the cell or cluster, the amount of electron correlation included in the calculation, and maybe other factors as well. The best vibrational frequencies that can be calculated are within 3 to 5% of the experimental ones. For Si–H stretching modes, this means that predictions are within $100 \, \text{cm}^{-1}$ from the experimental ones, making it hard to differentiate between the numerous lines commonly observed in the 1900 to $2200 \, \text{cm}^{-1}$ range.

In most cases, the results discussed below have been obtained using a combination of methods, in order to make sure that the results are independent of how the host crystal is represented and the Schrödinger equation solved. Typically, we optimize geometries independently using HF calculations in clusters (44 to 100 Si atoms) and MD simulations in supercells (64 and 216 Si atoms).

At the HF level, the geometries are first optimized with gradient techniques and no symmetry assumption using the 'approximate ab-initio' method of Partial Retention of Diatomic Differential Overlap (PRDDO) [11, 12]. In this approach, clever basis set transformations reduce the number of four-center integrals from $\sim N^4$ to $\sim N^3$ with little loss of accuracy. The method reproduces the results of ab-initio HF minimal basis set calculations at a fraction of the computational expense without introducing semiempirical parameters. The geometries are very accurate and are used as inputs for single-point ab-initio HF calculations with large basis sets [13]. Performing geometry optimizations at the ab-initio HF level in large clusters and low symmetry would be computationally prohibitive. Pseudopotentials [14, 15] are used to remove the core electrons from the calculations.

The MD simulations involve fast quenches (about 500 time steps, typically from liquid nitrogen temperature down to a fraction of 1 K) starting with a variety of guess configurations. The calculations are done with the 'ab-initio tight-binding' method developed by Sankey and co-workers [16 to 18]. This DF-based method uses a basis set of pseudo-atomic orbitals which are cut off at some distance from each nucleus. In the case of Si, this distance is $5.0 \, a_B$, and two Si atoms further than $10.0 \, a_B$ from each other do not overlap. The cutoff radius of H is $3.6 \, a_B$. Ab-initio pseudopotentials [19] are used, and the exchange-correlation potential is that of Ceperley-Adler [20] as parameterized by Perdew and Zunger [21]. The time step ranges from 0.2 to 2.0 fs, depending on whether H is in the calculation or not. The method is fast and contains no semiempirical parameters.

3. Interactions Involving V and I

3.1 The isolated vacancy or self-interstitial

The vacancy (V) and the self-interstitial (I) in Si have been studied for a long time [6] and issues dealing with the equilibrium sites and the diffusion properties have been contentious for many years (see e.g., the heated arguments at ICDS-13, Ref. [22]).

These defects are important because many processing steps generate them. They diffuse very fast when minority charge carriers are present (that is, *during* the non-equilibrium phase of the process), and can either kick-out substitutional impurities (in the case of I) or trap interstitial ones (in the case of V). Thus, they affect the electrical properties and the diffusivity of many impurities. They also form precipitates which may be electrically active themselves or serve as gettering sites for transition metals and other impurities.

Little is known about the early stages of the precipitation of intrinsic defects. However, vacancies do aggregate, and self-interstitials form {311} platelets. During heat treatments, vacancy aggregates trap all sorts of impurities, and interstitials may be released from platelets, possibly resulting in the transient-enhanced diffusion of shallow acceptors [23 to 25]. Thus, even though the equilibrium concentrations of V and I are low [26], these defects play key roles while the material is not in equilibrium.

The vacancy diffuses very fast, with activation energies for diffusion [27] of 0.45 eV (V^0), 0.33 eV (V^{++}), and 0.17 eV (V^{--}). In all but the ++ charge state, the vacancy is an orbital triplet in T_d symmetry and distorts to tetragonal symmetry. The four Si nearest-neighbors (NNs) to the vacancy pair up two by two, moving simultaneously away from the vacancy and toward each other [28, 29]. Our calculated formation energy [30] of the neutral vacancy is 4.0 eV, a value consistent with that obtained by other theorists (Kelly et al. [31] calculated 3.5 eV, Blöchl et al. [32] 4.1 eV, Zhu et al. [33] 3.7 eV, Nelson [34] 4.0 eV), and experimentalists (Watkins et al. [35] measured (3.6 ± 0.5) eV and Dannefaer et al. [36] (3.6 ± 0.2) eV).

The self-interstitial has never been observed in Si. Under conditions when I are generated in substantial concentrations, minority carriers are present as well, and I diffuse athermally. This could be due to recombination-enhanced migration [37 to 39]. In any case, as long as minority carriers are present, I flow freely and can reach a surface, trap at impurities, recombine with V, or precipitate. Since these processes occur during a non-equilibrium phase, it has been impossible to observe them experimentally. As a result, the details of the interactions are not known, and the role of theory becomes even more important.

Our calculations [40] of the lowest-energy configuration of I^0 predict a split-$\langle 110 \rangle$ configuration similar to that found by other authors [41 to 44]. Other charge states of I exist but were not considered so far.

3.2 Interactions of V and I with H_2, formation of H_2^*

Hydrogen is the most versatile impurity in semiconductors [1, 45]. It modifies the electrical and optical activity of the material by passivating many electrically active impurities, changes the electrical activity of others, and forms a range of defects visible by FTIR, Raman spectroscopy, and/or photoluminescence (PL). Hydrogen catalyzes the diffusion of interstitial oxygen in Si, passivates grain boundaries and (at least some) dislocations. Hydrogen forms monatomic centers at the tetrahedral interstitial (T) site and the relaxed bond-centered (BC) site. It also forms at least two kinds of dimers, interstitial H_2 molecules at the T site and the H_2^* complex. The latter consists of two Si–H bonds on the same trigonal axis, replacing a single Si–Si bond: The first H is near a BC site, bound to an almost sp^3-hybridized Si atom, and the second H is in an anti-bonding (AB) configuration, bound to an almost sp^2-hybridized Si atom.

Interstitial H_2 molecules, predicted by theory some 15 years ago [46, 47], have recently been observed. A Raman study by Murakami et al. [48] yielded a broad band at 4158 cm^{-1} in plasma-hydrogenated samples. The authors initially attributed this frequency to interstitial H_2 molecules. However, the signal proved [49] to be associated with molecules trapped in voids associated with platelets. Leitch et al. [50] found the Raman signature of isolated interstitial H_2 in similarly hydrogenated samples. At room temperature, the line is at 3601 cm^{-1}, some 500 cm^{-1} below that of free molecules. At 10 K, the line shifts to 3618 cm^{-1}. No ortho/para splitting was detected, in contrast to similar data in GaAs [51].

Molecular hydrogen was also seen by IR spectroscopy at 10 K in 17 mm thick samples soaked at high temperatures in a H_2 gas. In phosphorus-doped CZ-Si [52], two bands at 3789 and 3731 cm^{-1} were assigned to H_2 molecules trapped near interstitial O, and a third band at 3618 cm^{-1} to interstitial H_2. This IR line is present [53] up to at least 350 °C, and is also seen in boron-doped FZ-Si [54]. In these samples, over 90% of the hydrogen initially incorporated is in interstitial H_2 form. This accounts for the large fraction of 'hidden hydrogen' – the difference between the easily detected hydrogen in $\{B,H\}$ pairs and the total amount seen by SIMS [54].

Recent ab-initio calculations agree that the stretching frequency of interstitial H_2 in Si is substantially reduced from the free-molecule value [55 to 57]. But questions remain regarding the orientation of the molecule in the crystal and its barrier for rotation [56, 58, 59].

The existence of H_2^* was predicted by theory [60, 61] before the complex was detected [62] by IR spectroscopy in proton-implanted samples. The stretch modes are at 2062 cm^{-1} (for the H near the BC site) and 1838 cm^{-1} (for the H near the AB site). Both IR lines anneal out simultaneously around 200 °C.

Theorists disagree on which of H_2 or H_2^* is the more stable dimer, but agree that they are within a few tenths of an eV of each other. Both complexes should therefore coexist in H-rich samples. They do not. Further, since H_2 is still observed at 350 °C while H_2^* anneals out at 200 °C, one could expect a thermally-induced transition from H_2^* to H_2. Such a transition was never seen. Instead, H_2^* is only seen in irradiated material [54], and the irreversible conversion from H_2 to H_2^* induced by electron irradiation has now been reported [54]. A transition from 'hidden hydrogen' to H_2^* following 2 MeV electron irradiation has first been found in [63]. Experimental evidence shows that a transition from H_2 to states involving Si–H bonds results from radiation damage [64]. Such states include $\{V,H,H\}$ (a vacancy decorated with two H atoms [65]) and $\{I,H,H\}$ (two H atoms trapped at a self-interstitial [66], Fig. 1), and others.

MD simulations of H_2–V (and H_2–I) interactions show [30] that the strain associated with V (and I) readily dissociates the molecules resulting in the formation of the

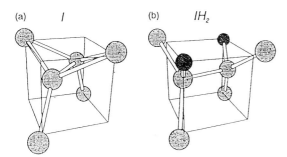

Fig. 1. The split-$\langle 110 \rangle$ configuration of a) the neutral self-interstitial and b) the $\{I, H, H\}$ complex, from Ref. [65]

$\{V, H, H\}$ (and $\{I, H, H\}$) complexes. Both complexes have been observed [65, 66] in proton-implanted samples. The reaction $V + H_2 \rightarrow \{V, H, H\}$ releases 4.0 eV, while $I + H_2 \rightarrow \{I, H, H\}$ releases 1.7 eV. At the same level of theory, the formation energy of the V is 4.0 eV and that of I is 4.2 eV. Thus, 8.2 eV are available from V–I recombination. Since the energy gain from the reaction of either V or I with H_2 is substantially less than 8.2 eV, one would expect[1]) that $\{I, H, H\}$ could further react with V, and $\{V, H, H\}$ with I. Both reactions could lead to the formation of H_2^*. Total energy differences indeed predict that the two recombination reactions release substantial amounts of energy: $\{I, H, H\} + V \rightarrow H_2^* + 7.1$ eV and $\{V, H, H\} + I \rightarrow H_2^* + 4.8$ eV. By themselves, these numbers do not prove that the reactions occur.

However, MD simulations [67] show that the radiation-damage-induced conversion from H_2 to H_2^* follows from two successive interactions equivalent to the V–I recombination at the H_2 molecule. During the irradiation, both V and I diffuse rapidly through the material even at low temperatures. We did observe $\{I, H, H\} + V \rightarrow H_2^*$ in a 4000 time steps MD simulation at $T = 1000$ K. This completes the cycle $H_2 + I \rightarrow \{I, H, H\}$ followed by $\{I, H, H\} + V \rightarrow H_2^*$, as observed in e^--irradiated samples. In an MD simulation at room temperature, the same reaction had barely begun after 3000 time steps. We believe that $\{V, H, H\} + I$ also occurs, but not within the few picoseconds (real time) that are available to us (at 1200 K, the reaction had not begun after 8000 time steps).

Fig. 2 shows six steps from the simulation. The initial configuration (a) is a local minimum of the potential energy. It consists of a vacancy (schematically represented by a dashed circle) at a third nearest-neighbor site from $\{I, H, H\}$. The latter complex is the one observed by IR in proton-implanted Si [66]. At the end of the simulation, the V–I recombination has taken place, and only H_2^* remains. In (b), atom 1 is on its way toward the vacancy, pushed by atom 2 which will take its place. In (c), atom 1 is in place and V–I recombination is essentially achieved. In (d), we are left with a pair of Si–H bonds parallel to each other, but not on the same $\langle 111 \rangle$ axis. Between (e) and (f), the hydrogen atom attached to atom 2 jumps to the correct trigonal axis and H_2^* is formed. Note that the last step shown (f) still corresponds to a temperature of 1000 K (it is not a quenched configuration) and the two Si–H bonds oscillate around the trigonal axis. A quench of configuration (f) gives H_2^*.

These results show that interstitial H_2 in Si reacts with V and I to form IR-active centers, as observed experimentally [54, 63, 64]. V and/or I are present in the sample in above-equilibrium concentrations during a variety of processes such as electron irradia-

---

[1]) The idea was suggested to us by R.C. Newman.

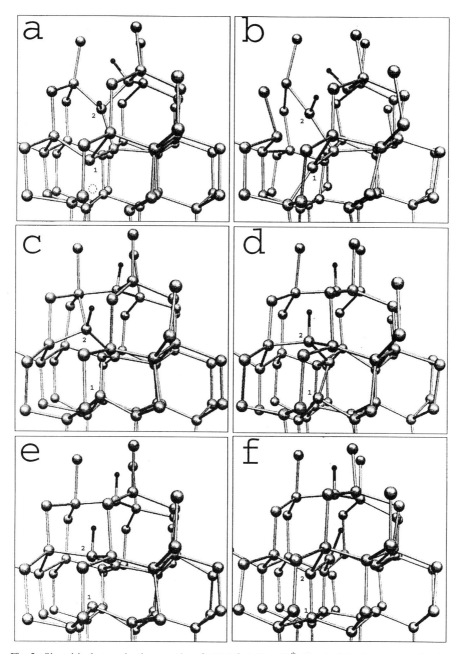

Fig. 2. Six critical steps in the reaction $\{I, H, H\} + V \rightarrow H_2^*$. Most of the host atoms in the cell are not shown. The initial configuration (a) is a local minimum of the potential energy, and the vacancy is shown as a dashed circle. The last configuration (f) is H_2^* (at 1000 K, the two H atoms are oscillating around the same trigonal axis). During the reaction, the atom labeled 1 fills the vacancy while the self-interstitial (atom 2) takes its place, completing the V–I recombination. Note that between the steps (e) and (f), one of the H atoms jumped from one trigonal axis to another

tion, ion implantation, etching, or the deposition of various surface layers. During such processes, V and I diffuse very rapidly and interact with defects. The interactions with H_2 are particularly energetic. In a sample containing 'hidden hydrogen', now identified as interstitial H_2 molecules, the interactions result in the formation of Si–H bonds in centers such as $\{V, H, H\}$ (when V dominate, such as during etching or nitridation), $\{I, H, H\}$ (when I dominate, such as the deposition of Au, O, or n^+ layers), or H_2^* (when both V and I are abundant, such as during electron irradiation).

At temperatures higher than 200 °C, the H_2^* complex thermally dissociates and the net result of the interaction of H_2 with radiation damage is the transformation of the molecule into interstitial H, which suddenly appears in the bulk of the material. There have been reports of anomalously high diffusivity of H, assumed to be vacancy-enhanced [68, 69]. The present results show that if hidden H_2 is *already present* in the bulk, it will become electrically and optically active following the injection of V and/or I. This will make it appear that H diffused very fast from the surface, while in fact only V and I diffused.

4. Vacancy–Vacancy Interactions

4.1 Vacancy aggregates

Vacancies in Si attract each other and form aggregates which affect the electrical and optical properties of the material in a variety of ways. The aggregates themselves can have deep levels in the gap and be electron–hole recombination centers. They can also be the precursors of extended defects and/or gettering centers for a range of impurities. Further, it is believed that small vacancy aggregates are found at the core of dislocations and are responsible for much of their electrical activity [70, 71]. The most important aggregates are those which are the most stable, as these are the most likely ones to survive high-temperature treatments.

The configurations and binding energies of such aggregates have been calculated [72, 73] as follows. At first, all the symmetrically inequivalent sets of $n = 1, \ldots, 6$ Si atoms have been removed from a 64 Si atom supercell. Only one configuration of the heptavacancy was calculated. In each case, the cell was briefly thermalized then rapidly quenched, forcing the convergence toward the nearest local minimum of the potential energy. The energy of all these vacancy aggregates were compared, and the geometry of the lowest-energy aggregates of n vacancies, $\{V_n\}$, was re-optimized in clusters at the PRDDO level, providing inputs for single-point calculations at the ab-initio HF level. This approach allows us to double-check the binding energies of the various aggregates, obtain the energies of the metastable configurations, and gain insights into the chemical details of the rebonding taking place.

For small n, the lowest-energy configuration of $\{V_n\}$ is realized when Si atoms are successively removed from a hexagonal ring in the crystal. Fig. 3 shows the binding energies of the various vacancy aggregates, defined as the energy gained by adding a vacancy to an aggregate of $n - 1$ vacancies:

$$E(V_{n-1}) + E(V_1) \longrightarrow E(V_n) + \Delta E_n .$$

The dots show the energies of the various metastable configurations.

The central result [72] is the prediction that one of the aggregates, the ring-hexavacancy (V_6), is substantially more stable than all the other aggregates. Although it has

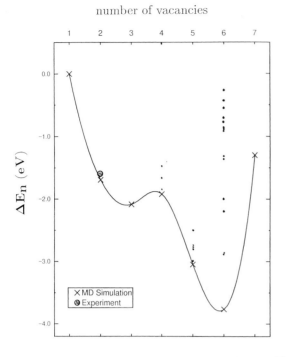

Fig. 3. Binding energy (see text) of a vacancy to the vacancy cluster $\{V_n\}$. The crosses show the lowest-energy aggregates and the dots show the energies of the metastable configurations. Note that even the lowest-lying metastable configuration of V_6 is well above the ground state — the ring-hexavacancy. The open circle is the experimental binding energy of the divacancy

several metastable states, constant-temperature MD simulations starting with a metastable aggregate result in a rearrangement of the vacancies within a short time, strongly suggesting that the defect will quickly collapse into the ring structure [73]. Thus, there should be only one hexavacancy.

Fig. 4 shows the structure of V_6. It is an ellipsoidal void with trigonal symmetry. Its thickness is ≈ 4 Å and diameter ≈ 8 Å. It has two sets of six equivalent Si NNs plus two 'special' Si atoms on the trigonal axis which also participate in the reconstruction.

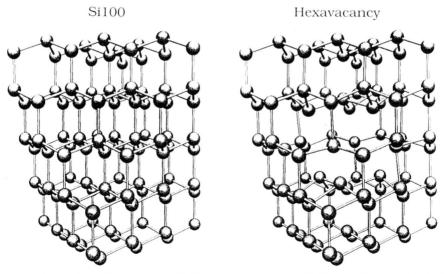

Fig. 4. Comparison of the perfect 100 Si atom cluster to the 94 Si atom cluster, with the V_6 at its center. The H saturators are not shown for clarity

The energy spectra associated with the most stable $\{V_n\}$ aggregates are shown in Fig. 5. The $\{V_6\}$ defect reconstructs so efficiently that it has no localized levels in the gap. Theory therefore predicts not only that is the most stable of all the small vacancy aggregates but also that it is electrically inactive except for a shallow (0/1) level near the conduction band minimum. Such a defect should form following radiation treatments (including ion implantation), survive the subsequent anneal, and remain in the crystal, mostly invisible. However, as will be discussed below, it is likely to be an efficient gettering center for a range of impurities. Such precipitates could well be electrically and/or optically active. Further, V_6 is a plausible nucleus or precursor of extended defects, in particular those which, like V_6, are trigonal and planar.

To date, there is no experimental proof that this defect exists, but there are numerous indications that some invisible defect survives heat treatments. A short list can be found in Ref. [73]. Recent theoretical studies [74] have associated the B_{41} and B_{71}^1 PL bands with two configurations of $\{V_6, H, H\}$, one of which has both H atoms on the trigonal axis, leading to a defect with D_{3d} symmetry. Were this assignment correct, the J

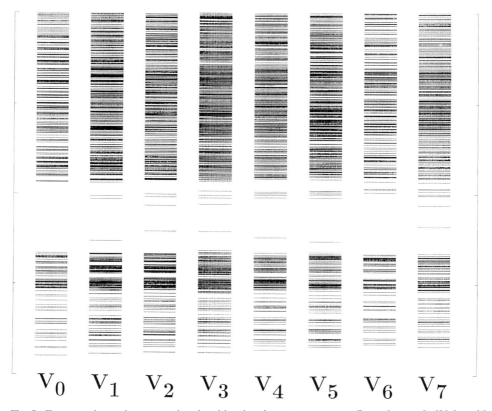

Fig. 5. Energy eigenvalues associated with the lowest-energy configurations of $\{V_n\}$, with $n = 1, \ldots, 7$, calculated at the ab-initio HF level. While these are *not* DLTS levels, the energy eigenvalues deep inside the 'gap' are associated with localized states. Note that all the vacancy aggregates have localized levels except V_6, which only shows some band-tailing from the 'conduction band'. V_0 denotes zero vacancy, that is the perfect cluster

PL band would be V_6. Finally, a recent EPR (spin 1) spectrum, also with D_{3d} symmetry, has been tentatively assigned [75] to $\{V_6, H, H\}$.

4.2 Precipitation of Cu at internal voids

As mentioned above, an internal void such as V_6 is likely to be a gettering center for a range of impurities. Calculations [76] show that H, O, C, and probably other impurities comfortably fit inside this void at a substantial gain in energy relative to the isolated interstitial impurity. The example that follows is that of copper [77].

Copper is a common impurity in silicon [4, 5, 78, 79]. As the interstitial Cu_i^+ ion, it is the fastest-diffusing impurity known to date [80, 81] and reacts with a range of defects and impurities. In particular, copper has a strong tendency to precipitate at defects such as dislocations [4, 71], grain boundaries [82], nanocavities [83 to 85], stacking faults [86], or radiation-damaged regions [87, 88]. Such precipitates are known to reduce the lifetime of charge carriers [89, 90]. A range of gap levels observed by DLTS have been assigned to copper precipitates. These levels tend to form defect bands [4, 79].

Much of what is known about copper–defect interactions in Si deals with large precipitates – visible for example by cross-sectional transmission electron microscopy (XTEM) or electron-beam induced current (EBIC) – or long-range interactions such as the dissociation of copper–acceptor pairs observed in transient-ion drift (TID) experiments. These types of experiments reveal little about the *local* (atomic-level) interactions. The microscopic nature of the interactions involving Cu in Si is poorly understood.

Not surprisingly, most of the open questions are related to nearby interactions. What is the chemistry of copper in lattice defects of nanometer size or larger? Where does Cu fit in a void? What are the binding energies and electronic structures? What is the origin of the electrical activity of copper precipitates?

Such questions can be addressed at the *ab-initio* HF level using the ring-hexavacancy as a model internal void [77]. The activation energy for diffusion [81] of Cu_i^+ and the structure of $\{Cu, B\}$ pairs [91] have been successfully calculated at the same level of theory. Geometry optimizations have been performed for up to five Cu impurities trapped in V_6, with no symmetry assumptions. The optimizations included the 14 Si atoms which participate in the reconstruction of V_6 and all the Cu atoms involved. There are two issues to consider when calculating the precipitation of Cu_i^+.

First, the defect cannot build up too much charge. If one assumes that several Cu_i^+ precipitate and no electron is ever trapped, the defect rapidly builds up excessive positive charge and becomes a long-range repulsive center for additional Cu_i^+. This was proposed [92] as the reason why copper precipitates much easier in n- than in p-type Si. However, since copper precipitates are also observed in p-type material, electrons must trap at the defect to maintain a small or zero net charge. Such a charge neutralization process is assumed here.

Second, only total energy differences of *closed-shell* (spin 0) configurations are directly comparable at the level of theory employed here. In principle, restricted open-shell calculations (ROHF) for the spin 1/2 configurations could be carried out and produce energies that are comparable to the closed-shell (RHF) ones. However, optimizing geometries at the ROHF level for the complexes studied here is far too computer-intensive, especially since PRDDO is ill-equipped to perform such calculations. Therefore, total energy differences involving only closed-shell configurations are included below. The precipitation of Cu_i^+ at V_6 was handled as follows.

1. Cu^+ is placed inside V_6, the geometry optimized, and the gain in energy ΔE_1 relative to Cu_i^+ is defined as

$$Cu_i^+ + V_6 \rightarrow \{Cu_1, V_6\}^+ + \Delta E_1 \, .$$

The calculated gain is $\Delta E_1 = 3.36$ eV. It is then assumed that $\{Cu_1, V_6\}^+$ captures an electron, then traps Cu_i^+. This defect traps a second electron and the geometry of $\{Cu_2, V_6\}^0$ is optimized. This defect, a closed shell, is the trap for the next Cu_i^+.

2. Cu^+ is placed inside $\{Cu_2, V_6\}^0$, the geometry optimized, and the gain in energy ΔE_3 relative to Cu_i^+ is defined as

$$Cu_i^+ + \{Cu_2, V_6\}^0 \rightarrow \{Cu_3, V_6\}^+ + \Delta E_3 \, .$$

The calculated gain is $\Delta E_3 = 4.94$ eV. It is then assumed that $\{Cu_3, V_6\}^+$ captures an electron, then traps Cu_i^+. This defect traps a second electron and the geometry of $\{Cu_4, V_6\}^0$ is optimized. This defect, a closed shell, is the trap for the next Cu_i^+.

3. Cu^+ is placed inside $\{Cu_4, V_6\}^0$, the geometry optimized, and the gain in energy ΔE_5 relative to Cu_i^+ is defined as

$$Cu_i^+ + \{Cu_4, V_6\}^0 \rightarrow \{Cu_5, V_6\}^+ + \Delta E_5 \, .$$

The calculated gain is $\Delta E_5 = 2.01$ eV.

The process could continue for a larger void, but the hexavacancy is filling up. This complex could continue to grow, for example involving sites for Cu outside V_6. This possibility was not considered.

The geometrical configurations of $\{Cu_n, V_6\}^0$ and $\{Cu_n, V_6\}^+$ are very similar to each other. The electron removed from the neutral complex comes from a molecular orbital rather delocalized around V_6 and does not affect the geometry much at all. The optimized configurations of the $\{Cu_n, V_6\}$ complexes with $n = 0, \ldots, 5$ are shown in Fig. 6.

A careful analysis of the orbital populations of the Cu impurities trapped in V_6 reveals that the copper atoms do their best to bind to the inner surface of the void using the nine valence orbitals at their disposal. No two Cu atoms are identically hybridized in the void. Yet several remarkable trends are worth noticing.

1. In a void, the Cu impurities remain *as far apart as possible* from each other (the description of the structure is given below).

2. *Cu–Si covalent overlap* is preferred to Cu–Cu overlap. This is consistent with the bond strengths in diatomics [93] (2.3 eV for Cu–Si versus 1.8 eV for Cu–Cu). As long as there is plenty of room in the void, the Cu atoms bind to different Si atoms. As the void fills up, some Cu–Cu overlap becomes unavoidable, and a few Si atoms bind to two Cu atoms.

3. Each Cu forms weak covalent bonds to *precisely four* Si atoms. This observation is not simply based on a compilation of Cu–Si internuclear distances, but on the requirement that, for a given $\{Cu, Si\}$ pair, the degree of bonding be non-negligible and the overlap population positive (that is, a bonding overlap). The "degree of bonding" [94] is a measure of the amount of covalent overlap between two atoms. It is not propor-

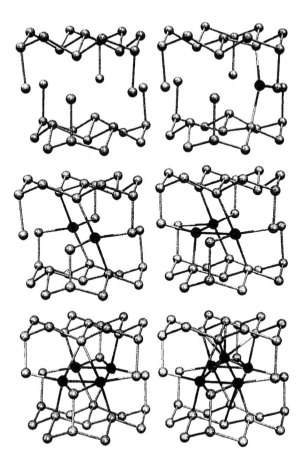

Fig. 6. Energy-optimized configurations of the $\{Cu_n, V_6\}$ complexes for $n = 0,\ldots,5$. The copper atoms are in black (see text)

tional to the bond strength but is defined in such a way that it is 2.0 for a four electron covalent bond (e.g., O_2), 1.0 for a two electron covalent bond (e.g., H_2), and 0.0 if the bond is purely ionic (e.g., NaCl). In the case of the $\{Cu_n, V_6\}$ complexes, the degrees of bonding involving copper are all in the 0.4 to 0.8 range, that is Cu always forms several weak bonds, and even the strongest ones are quite a bit less than a true two electron bond (degree of bonding 1.0).

4. The sum of the degrees of bonding associated with each Cu in any of the $\{Cu_n, V_6\}$ complexes varies only in the *narrow range 2.3 to 2.6*. Thus, the total number of electrons participating in the covalent bonding of Cu to an internal void is *always close to five*. These electrons are distributed among four Cu–Si bonds (plus any Cu–Cu bonds).

5. The maximum strength of each of the Cu–Si bonds is 0.9 eV (that is, less than the width of the gap). This estimate is obtained by dividing the binding energy of Cu to $\{Cu_n, V_6\}$ (relative to $\{Cu_{n-1}, V_6\} + Cu_i^+$) by the number of Cu–Si bonds involved. Since each copper forms four inequivalent Cu–Si bonds with degrees of bonding ranging from 0.4 to 0.8, some of the bonds are necessarily weaker than 0.9 eV. This has the following two consequences.

5a. At least some of the energy eigenvalues associated with the bonding/antibonding orbitals of the weak Cu–Si (and Cu–Cu) bonds are in the gap. In the HF calculations, many energy eigenvalues are clearly visible in the gap of the energy spectrum for all the $\{Cu_n, V_6\}$ complexes. Plots of the wavefunctions associated with the highest-occupied and lowest-unoccupied energy eigenvalues show that they are associated purely with Cu–Si bonding and anti-bonding orbitals, respectively, with no apparent component in the reconstructed Si–Si bonds. This is not a proof, but at least a strong hint that the electrical activity of copper precipitates in internal voids is caused by the energy levels associated with the many weak Cu–Si bonds. This is consistent with the interpretation [95, 96] that the electrical activity of copper precipitates originates at the interface between Si and the precipitates.

5b. The binding energy of a single hydrogen to a vacancy [29] or vacancy aggregate [97] is of the order of 3 eV (this results from the formation of one strong Si–H bond). Thus, if interstitial hydrogen is present, it will easily displace Cu from internal voids, since replacing even the strongest Cu–Si bond by a Si–H bond results in a gain of the order of 2 eV. Such an effect has indeed been reported [84].

A description of the complexes in Fig. 6 is as follows.

In $\{Cu_1, V_6\}$, copper binds to four Si atoms on the inner surface of the void. The four Cu–Si distances range from 2.38 to 2.68 Å, and the degrees of bonding from 0.6 to 0.4. The binding energy relative to Cu_i^+ is 3.36 eV.

In $\{Cu_2, V_6\}$, the two copper atoms are as far from each other as possible (4.55 Å), at opposite inner surfaces of V_6. The Cu–Si bond lengths vary from 2.38 to 2.62 Å, and the degrees of bonding from 0.7 to 0.5.

In the case of $\{Cu_3, V_6\}$, the three copper atoms form an almost perfect isosceles triangle in the plane of V_6. The Cu–Cu distances are 2.75, 2.77, and 4.08 Å, respectively. One Cu is bound to the two other Cu atoms, but the overlap is very small (degree of bonding ≤ 0.2 and overlap population ≈ 0.1). This Cu is mostly bound to only two Si atoms, at 2.28 and 2.33 Å, with degrees of bonding of 0.7 and 0.8. The other two coppers form the usual four Cu–Si bonds, with bond lengths varying from

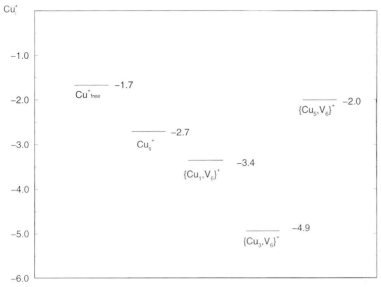

Fig. 7. Energy (in eV) of Cu_{free}^+, Cu_s^+, $\{Cu_1, V_6\}^+$, $\{Cu_3, V_6\}^+$, and $\{Cu_5, V_6\}^+$, all relative to Cu_i^+

2.38 to 2.71 Å and degrees of bonding from 0.6 to 0.4. The binding energy of Cu_i^+ to this complex is the highest in the series, 4.9 eV. The reason why this value is so large is not clear at this point. However, it indicates a tendency for Cu to precipitate since, given a choice between trapping at V_6 or at $\{Cu_2, V_6\}$, copper would prefer the latter.

The copper atoms in $\{Cu_4, V_6\}$ form a perfect rectangle in the plane of V_6. The rectangle has width 2.36 Å and length 3.67 Å. There are two weak Cu–Cu bonds (degree of bonding 0.4), that is each of the four Cu atoms is bound to one Cu and four Si atoms. These bonds are similar to those described above.

Finally, $\{Cu_5, V_6\}$ is a slightly distorted version of $\{Cu_4, V_6\}$, with the fifth Cu forming the tip of a pyramid with a rectangular base, the tip being on the trigonal axis of V_6. The fifth Cu is eight-fold coordinated (!) but the degrees of bonding vary only from 0.2 to 0.4. The other coppers are roughly as described above for $\{Cu_4, V_6\}$. The binding energy relative to $\{Cu_4, V_6\}$ and Cu_i^+ is 2.01 eV.

Note that this value is consistent with the measured [84] dissociation energy of copper from internal voids, (2.2 ± 0.2) eV. However, our results show that some of the Cu atoms trapped in a void are more strongly bound than others. The calculated binding energies associated with Cu_i^+ are summarized in Fig. 7.

5. Conclusions

In the past decade, ab-initio calculations have been performed by several authors and many defect problems in various semiconductors have been addressed. The host crystal is typically simulated by clusters or supercells ranging in size from 50 to 216 host atoms. The key results are mostly independent of cluster or cell size, and this dependence can (and should) be checked for larger defect complexes. The geometries and relative energies of the stable and metastable configurations of isolated impurities, defects, and small complexes can be calculated at a variety of theoretical levels. In general, the calculated features which depend on total energy differences such as configurations, binding energies, and activation energies for diffusion, are independent of the (ab-initio) methodology used. HF calculations, which are slower than DF ones, provide a wealth of very useful chemical information.

The examples discussed above illustrate how ab-initio theory is able to complement experimental studies of defects in semiconductors such as Si. Predictions related to the structure (in particular the symmetry) of point defects and small complexes, their vibrational properties, the spin density distributions, the binding energies, the activation energies for diffusion, and other experimentally accessible properties can be calculated and provide direct connections to experiment. Further, ab-initio theory can describe situations which cannot be measured, such as defect reactions or the details of the chemistry of a defect complex.

MD simulations provide unique insights into the dynamics of defects, and the outputs of such calculations are routinely made into 'movies' which show, in real time, the details of processes that cannot be seen experimentally. The calculations of defect reactions such as the radiation-induced transformation of H_2 into H_2^* illustrate how theory contributes to our understanding of processes that occur during the non-equilibrium phases of device processing.

Theory is now capable to handle much larger defects than just a decade ago. The case of the ring-hexavacancy is particularly interesting for a number of reasons. First, its stability suggests that it should be present in the material and survive high-temperature treatments, even though it could be difficult to detect. Its presence or absence in a given sample can clearly affect how that sample will react to subsequent treatment, providing one example of how the history of a sample affects its behavior. Second, it is a high-symmetry (D_{3d}) internal void that could well be the nucleus of a range of defects, in particular those that are trigonal and planar such as H plasma-induced platelets. Third, since it appears to be a powerful gettering center, it provides theorists with a most useful tool to study the early stages of impurity precipitation. In the example discussed above, copper was the test impurity: at least five Cu interstitials can trap inside V_6, a number large enough to uncover general features and trends.

The interplay between theory and experiment will likely become more important in the years to come. As theorists gain experience, fine-tune their methods, and as computers increase in power and availability, the role of ab-initio theory will increase. MD simulations are rapidly becoming the tool of choice to address issues too complicated for static methods. The simulations can handle any number of degrees of freedom, and allow the theorist to avoid the assumptions that are often needed to do static calculations. For example, many diffusion paths calculated in the past have assumed that the motion occurs in a particular lattice plane (quite often, the {110} plane). Other assumptions dealt with how much the crystal has time to relax while a light interstitial such as H diffuses through it. Clearly, assuming a frozen crystal underestimates the reaction of the Si atoms to the presence of H, while allowing the host atoms to adjust instantaneously to any motion of H overestimates this reaction. MD simulations need no such inputs and handle the dynamics remarkably well. However, they are still limited (in most cases) to the electronic ground state and to classical nuclear motion.

As for the weaknesses of theory, there are still many. Some are likely to be overcome by the increase of computer power. For example, larger cells or clusters will become routine, allowing the study of larger defects. Bigger basis sets, more k-points, and a better treatment of electron correlation will improve the accuracy of the predictions. The calculation of vibrational spectra from the Fourier transform of the velocity–velocity autocorrelation function will improve the predictions of local vibrational modes, including the isotope- and temperature-dependence of the observable modes. Other weaknesses are likely to persist. For example, the quantum motion of impurities (such as hydrogen or muonium) is only beginning to be discussed. Another example deals with conduction-band-related issues. In particular, the ab-initio prediction of the electrical properties of defects and impurities is still qualitative at best. Let the funding continue …

Acknowledgements This work is supported by the grant D-1126 from the R.A. Welch Foundation and the contract XAD-7-17652-01 from the National Renewable Energy Laboratory.

References

[1] S.K. ESTREICHER, Mat. Sci. Engr. R **14**, 319 (1995).
[2] F. SHIMURA (Ed.), Oxygen in Silicon, Semicond. and Semimet., Vol. 42, Academic Press, Boston 1994.
[3] R. JONES (Ed.), Early Stages of Oxygen Precipitation in Silicon, Kluwer Academic Publ., Dordrecht 1996.
[4] A.A. ISTRATOV and E.R. WEBER, Appl. Phys. A **66**, 123 (1998).

[5] A. MESLI and T. HEISER, Defect and Diffusion Forum **131/132**, 89 (1996).
[6] S.T. PANTELIDES (Ed.), Deep Center in Semiconductors, Gordon & Breach, New York 1986.
[7] C.G. VAN DE WALLE, Phys. Rev. B **49**, 4579 (1994).
[8] R. JONES and P.R. BRIDDON, in: Identification of Defects in Semiconductors, Semicond. and Semimet., Vol. 51A, Ed. M. STAVOLA, Academic Press, Boston 1998 (p. 287).
[9] S.K. ESTREICHER and P.A. FEDDERS, in: Computational Studies of New Materials, Eds. D.A. JELSKI and T.F. GEORGE, World Scientific, Singapore 1999 (p. 27).
[10] A. RESENDE, R. JONES, S. ÖBERG, and P.R. BRIDDON, Phys. Rev. Lett. **82**, 2111 (1999).
[11] A. DERECSKEI-KOVACS and D.S. MARYNICK, Int. J. Quant. Chem. **58**, 193 (1996).
[12] A. DERECSKEI-KOVACS, D.E. WOON, and D.S. MARYNICK, Int. J. Quant. Chem. **61**, 67 (1997).
[13] M.W. SCHMIDT, K.K. BALDRIDGE, J.A. BOATZ, S.T. ELBERT, M.S. GORDON, J.H. JENSEN, S. KOSEKI, K.A. NGUYEN, S. SU, T.L. WINDUS, M. DUPUIS, and J.A. MONTGOMERY, JR., J. Comp. Chem. **14**, 1349 (1993).
[14] P.J. HAY and W.R. WADT, J. Chem. Phys. **82**, 270 (1985).
[15] W.J. STEVENS, H. BASH, and M. KRAUSS, J. Chem. Phys. **81**, 6026 (1984).
[16] O.F. SANKEY and D.J. NIKLEVSKI, Phys. Rev. B **40**, 3979 (1989).
[17] O.F. SANKEY, D.J. NIKLEVSKI, D.A. DRABOLD, and J.D. DOW, Phys. Rev. B **41**, 12750 (1990).
[18] A.A. DEMKOV, J. ORTEGA, O.F. SANKEY, and M.P. GRUMBACH, Phys. Rev. B **52**, 1618 (1995).
[19] D.R. HAMMAN, M.SCHLÜTER, and C. CHIANG, Phys. Rev. Lett. **43**, 1494 (1979).
G.B. BACHELET, D.R. HAMANN, and M. SCHLÜTER, Phys. Rev. B **26**, 4199 (1982).
[20] D.M. CEPERLEY and B.J. ADLER, Phys. Rev. Lett. **45**, 566 (1980).
[21] S. PERDEW and A. ZUNGER, Phys. Rev. B **32**, 5048 (1981).
[22] See the following papers in Proc. 13th Int. Conf. Defects Semicond., Eds. L.C. KIMERLING and J.M. PARSEY, JR., The Metallurgical Society of AIME, 1985: S.T. PANTELIDES (p. 151), A. SEEGER and W. FRANK (p. 159), and J.C. BOURGOIN (p. 167).
[23] H.G.A. HUIZING, C.C.G. VISSER, N.E.B. COWERN, P.A. STOLK, and R.C.M. DE KRUIF, Appl. Phys. Lett. **69**, 1211 (1996).
[24] J. LIU, V. KRISHNAMOORTHY, H.-J. GOSSMAN, L. RUBIN, M.E. LAW, and K.S. JONES, J. Appl. Phys. **81**, 1656 (1997).
[25] P.B. GRIFFIN, M. CAO, P. VANDE VOORDE, Y.-L. CHANG, and W.M. GREENE, Appl. Phys. Lett. **73**, 2986 (1998).
[26] M. JACOB, P. PICHLER, H. RYSSEL, and R. FALSTER, J. Appl. Phys. **82**, 182 (1997).
[27] G.D. WATKINS, in: Deep Center in Semiconductors, Ed. S.T. PANTELIDES, Gordon & Breach, New York 1986.
[28] M.A. ROBERSON and S.K. ESTREICHER, Phys. Rev. B **49**, 17040 (1994).
[29] Y.K. PARK, S.K. ESTREICHER, C.W. MYLES, and P.A. FEDDERS, Phys. Rev. B **52**, 1718 (1995).
[30] S.K. ESTREICHER, J.L. HASTINGS, and P.A. FEDDERS, Phys. Rev. B **57**, R12663 (1998).
[31] P.J. KELLY and R. CAR, Phys. Rev. B **45**, 6543 (1992).
[32] P.E. BLÖCHL, E. SMARGIASSI, R. CAR, D.B. LAKS, W. ANDREONI, and S.T. PANTELIDES, Phys. Rev. Lett. **70**, 2435 (1993).
[33] J. ZHU, L.H. YANG, C. MAILHIOT, T. DIAZ DE LA RUBIA, and G.H. GILMER, Nuclear Instrum. and Methods Phys. Res. **102**, 29 (1995).
[34] J. NELSON, private communication.
[35] G.D. WATKINS and J.W. CORBETT, Phys. Rev. **134**, A1359 (1964).
[36] S. DANNEFAER, P. MASCHER, and D. KERR, Phys. Rev. Lett. **56**, 2195 (1986).
[37] J.C. BOURGOIN and J.W. CORBETT, Phys. Lett. **38A**, 135 (1972); Radiat. Eff. **36**, 167 (1978).
[38] L.C. KIMERLING, Solid State Electron. **21**, 1391 (1978).
[39] G.D. WATKINS, A.P. CHATTERJEE, R.D. HARRIS, and J.R. TROXELL, Semicond. and Insulators **5**, 321 (1983).
[40] M. GHARAIBEH, S.K. ESTREICHER, and P.A. FEDDERS, Proc. 20th Int. Conf. Defects Semicond., Physica B, in print.
[41] R. CAR, P. BLÖCHL, and E. SMARGIASSI, Mater. Sci. Forum **83/87**, 433 (1992).
[42] D.J. CHADI, Phys. Rev. B **46**, 9400 (1992).
[43] A. MAINWOOD, Mater. Sci. Forum **196/201**, 1589 (1995).
[44] M. TANG, L. COLOMBO, J. ZHU, and T. DIAZ DE LA RUBIA, Phys. Rev. B **55**, 14279 (1997).
[45] S.J. PEARTON, J.W. CORBETT, and M.J. STAVOLA, Hydrogen in Crystalline Semiconductors, Springer-Verlag, Berlin 1992.

[46] A. Mainwood and A.M. Stoneham, Physica B **116**, 101 (1983); J. Phys. C (Solid State Phys.) **17**, 2513 (1984).
[47] J.W. Corbett, S.N. Sahu, T.S. Shi, and L.C. Snyder, Phys. Lett. **93A**, 303 (1983).
[48] K. Murakami, N. Fukata, S. Sasaki, K. Ishioka, M. Kitajima, S. Fujimura, K. Kikuchi, and H. Haneda, Phys. Rev. Lett. **77**, 3161 (1996).
[49] A.W.R. Leitch, V. Alex, and J. Weber, Mater. Sci. Forum **258/263**, 241 (1997).
[50] A.W.R. Leitch, V. Alex, and J. Weber, Phys. Rev. Lett. **81**, 421 (1998).
[51] J. Vetterhöffer, J. Wagner, and J. Weber, Phys. Rev. Lett. **77**, 5409 (1996).
[52] R.E. Pritchard, M.J. Ashwin, J.H. Tucker, and R.C. Newman, Phys. Rev. B **57**, R15048 (1998).
[53] R.E. Pritchard, private communication.
[54] R.E. Pritchard, J.H. Tucker, R.C. Newman, and E.C. Lightowlers, Semicond. Sci. Technol. **14**, 77 (1999).
[55] Y. Okamoto, M. Saito, and A. Oshiyama, Phys. Rev. B **56**, R10016 (1997).
[56] B.H. Hourahine, R. Jones, S. Öberg, R.C. Newman, P.R. Briddon, and E. Roduner, Phys. Rev. B **57**, R12666 (1998).
[57] C.G. Van de Walle, Phys. Rev. Lett. **80**, 2177 (1998).
[58] B.H. Hourahine, R. Jones, S. Öberg, and P.R. Briddon, Mater. Sci. Engr. B, in print.
[59] R. Jones and B. Hourahine, Semicond. Sci. Technol., in print.
[60] P. Deak, L.C. Snyder, and J.W. Corbett, Phys. Rev. B **37**, 6887 (1988).
[61] K.J. Chang and D.J. Chadi, Phys. Rev. Lett. **62**, 937 (1989).
[62] J.D. Holbech, B. Bech Nielsen, R. Jones, P. Stich, and S. Öberg, Phys. Rev. Lett. **71**, 875 (1993).
[63] M.J. Binns, S.A. McQuaid, R.C. Newman, and E.C. Lightowlers, Semicond. Sci. Technol. **8**, 1908 (1993).
[64] T.S. Shi, G.R. Bai, M.W. Qi, and J.K. Zhou, Mater. Sci. Forum **10/12**, 597 (1986).
[65] M. Budde, B. Bech Nielsen, P. Leary, J. Goss, R. Jones, P.R. Briddon, S. Öberg, and S.J. Breuer, Phys. Rev. B **57**, 4397 (1998).
[66] B. Bech Nielsen, L. Hoffmann, M. Budde, R. Jones, J. Goss, and S. Öberg, Mater. Sci. Forum **196/201**, 933 (1995).
[67] S.K. Estreicher, J.L. Hastings, and P.A. Fedders, Phys. Rev. Lett. **82**, 815 (1999).
[68] J.W. Corbett, J.L. Lindström, and S.J. Pearton, Mater. Res. Soc. Symp. Proc. **104**, 229 (1988).
[69] B.L. Sopori, K. Jones, and X.J. Deng, Appl. Phys. Lett. **61**, 2560 (1992).
[70] C. Kisielowski-Kemmerich, phys. stat. sol. (b) **161**, 11 (1990).
[71] J. Weber, Solid State Phenom. **37/38**, 13 (1994).
[72] S.K. Estreicher, J.L. Hastings, and P.A. Fedders, Appl. Phys. Lett. **70**, 432 (1997).
[73] J.L. Hastings, S.K. Estreicher, and P.A. Fedders, Phys. Rev. B **56**, 10215 (1997).
[74] B. Hourahine, R. Jones, A.N. Safonov, S. Öberg, P.R. Briddon, and S.K. Estreicher, to be published.
[75] Y.V. Gorelkinskii, private communication.
[76] S.K. Estreicher, unpublished.
[77] S.K. Estreicher, Phys. Rev. B **60**, 5375 (1999).
[78] E.R. Weber, Appl. Phys. A **30**, 1 (1983).
[79] T. Heiser, A.A. Istratov, C. Flink, and E.R. Weber, Mater. Sci. Engr. B, in print.
[80] A.A. Istratov, C. Flink, H. Hieslmair, E.R. Weber, and T. Heiser, Appl. Phys. Lett. **81**, 1243 (1998).
[81] D.E. Woon, D.S. Marynick, and S.K. Estreicher, Phys. Rev. B **45**, 13383 (1992).
[82] J.-L. Maurice and C. Colliex, Appl. Phys. Lett. **53**, 241 (1989).
[83] J. Wong-Leung, C.E. Ascheron, M. Petravic, R.G. Elliman, and J.S. Williams, Appl. Phys. Lett. **66**, 1231 (1995).
[84] S.M. Myers and D.M. Follstaedt, J. Appl. Phys. **79**, 1337 (1996).
[85] M. Zhang, C. Lin, X. Duo, Z. Lin, and Z. Zhou, J. Appl. Phys. **85**, 94 (1999).
[86] M. Kaniewska, J. Kaniewski, and A.R. Peaker, Mater. Sci. Forum **83/87**, 1457 (1992).
[87] S. McHugo, E.R. Weber, S.M. Myers, and G.A. Petersen, Appl. Phys. Lett. **69**, 3060 (1996).
[88] S. Koveshnikov and O. Kononchuk, Appl. Phys. Lett. **73**, 2340 (1998).
[89] A. Rohatgi, J.R. Davis, R.H. Hopkins, P. Rai-Choudhury, P.G. McMullin, and J.R. McCormick, Solid State Electron. **23**, 415 (1980).

[90] A.A. Istratov, C. Flink, H. Hieslmair, T. Heiser, and E.R. Weber, Appl. Phys. Lett. **71**, 2121 (1997).
[91] S.K. Estreicher, Phys. Rev. B **41**, 5447 (1990).
[92] A.A. Istratov, O.F. Vyvenko, C. Flink, T. Heiser, H. Hieslmair, and E.R. Weber, Mater. Res. Soc. Symp. Proc. **510**, 313 (1998).
[93] D.R. Lide (Ed.), Handbook of Chemistry and Physics, CRC Press, Boca Raton 1994.
[94] D.R. Armstrong, P.G. Perkins, and J.J.P. Stewart, J. Chem. Soc. Dalton Trans. 838 (1973).
[95] A.A. Istratov, H. Hedemann, M. Seibt, O.F. Vyvenko, W. Schröter, T. Heiser, C. Flink, H. Hieslmair, and E.R. Weber, J. Electrochem. Soc. **145**, 3889 (1998).
[96] M. Seibt, M. Griess, A.A. Istratov, H. Hedemann, A. Sattler, and W. Schröter, phys. stat. sol. (a) **166**, 171 (1998).
M. Seibt, H. Hedemann, A.A. Istratov, F. Riedel, A. Sattler, and W. Schröter, phys. stat. sol. (a) **171**, 301 (1999).
[97] S.K. Estreicher, J.L. Hastings, and P.A. Fedders, Mater. Sci. Engng. B **58**, 31 (1999).

phys. stat. sol. (b) **217**, 533 (2000)

Subject classification: 62.20.Qp; 71.15.Mb; S7.14; S10.1

Superhard Materials

J. E. Lowther

Department of Physics, University of the Witwatersrand, Johannesburg, South Africa

(Received August 10, 1999)

Computer modelling is playing an increasing role in predicting new and novel forms of matter. Ab-initio techniques that employ state-of-the-art density functional and pseudopotential methods in particular are proving quite reliable and these with various semi-empirical approaches can give great insight into materials preparation and possible synthesis procedures. Some properties of recently postulated or observed structures of nitrides and oxides are described. The examples considered show how modelling can suggest possible paths interrelating different structures and how it can shed light on potential problems involved with synthesizing such structures.

1. Introduction

Over the last decade computer modelling has proved capable of giving quite accurate predictions of a variety of material properties. Defects and impurities in crystals as well as the intrinsic features of crystals are only some properties [1] that have been investigated. The computational methods used when studying intrinsic properties of materials essentially can be split into two categories, ab-initio or semi-empirical. The ab-initio route [2] usually relies strongly on techniques of density functional theory [3], at varying levels especially the local density approximation (LDA) or generalized gradient approximation (GGA), whereas the semi-empirical approach [4] involves interatomic potentials or force fields fitted to known structures. In both cases transferability of the approach among different structures is important, especially when it comes to a predictive modelling of potentially new superhard crystalline phases.

The transferability of the ab-initio approach is mainly associated with the choice of pseudopotentials that are employed. There are various formal ways of testing and generating such pseudopotentials [5,6] but, and especially when using a plane wave approach, some balance has to be met computationally by using a soft pseudopotential that essentially restricts the number of plane waves needed. With the semi-empirical route transferability between systems will always be a worrying feature yet in systems like diamond or silicon carbide an analytic three-body Tersoff potential [7] has met with considerable success. Likewise for nominally ionic materials such as silica or other related oxides, analytic rigid pair potentials with (among others) a Buckingham component have often been employed [8,9].

When attempting to relate structural properties of materials to their hardness, it is confusing to quantify directly the precise meaning of hardness. The latter is often measured through some kind of indentation approach with, for example, a Knoop Hardness quantifying the strength of different materials. This Knoop hardness is approximately directly related to the bulk or shear modulus of the material so that either of these latter properties may be used to signify those materials that could have a potential superhard character. The bulk modulus is mainly used for the reason that it is cheaper to calculate from the practical usage of computer time. But the hardness concept is always one of contention and often labored, all that really can be said is that a larger value of the bulk modulus may indicate that the related material has superhard properties. Given this point, materials with a bulk modulus exceeding a value of ≈ 250 GPa can be considered as falling within a region where they could have a potential superhard character.

2. Carbides and Nitrides

Diamond has a hardness that is associated with a three-dimensional network of strong covalent bonds. It is not unreasonable to expect that similar structures can be formed from different chemical components. Boron nitride, in the cubic form, is now well established as one possible structure with silicon carbide also being used as a hard material. Thus, materials with B, C, N or Si hold potential for superhard properties.

Nitrides of silicon or carbon have recently received much attention because of their potential superhard application with suggestions that C_3N_4 may have a hardness exceeding that of diamond [10]. A practical realization of a possible superhard nitride material has been Si_3N_4 and it is likely that modification of this material with carbon will produce even harder materials in the future.

2.1 Silicon nitride

Silicon nitride is a commercially used hard material presently employed in several applications because of its weight saving compared with other materials such as steel. Early calculations using ab-initio approaches on either α-Si_3N_4 or β-Si_3N_4 [10,11] forms first suggested its potential as a hard material and more recently there have been several other calculations. The structure of β-Si_3N_4 is $P3_1c$ with 14 atoms in the unit cell whereas α-Si_3N_4 has a structure $P6_3m$ with 24 atoms in the unit cell. The β-structure is shown in Fig. 1.

Besides the ab-initio approaches, semi-empirical techniques using a variety of potentials have been used. The simplest of these employed only a two-centre Buckingham and Leonard-Jones interaction each describing Si–Si, Si–N and N–N interactions [12]. However, different potentials were needed to adequately describe each of the α or β-phases possibly suggesting that such a potential could have difficulties in dealing with unspecified structures. Likewise a three-body Tersoff potential has also been employed [13] but to date only one structure (namely the β-phase) has been considered. Both the

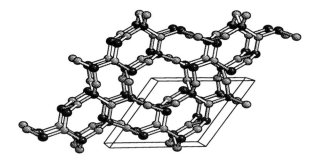

Fig. 1. β-Si_3N_4

Table 1
Structural details of the β phase of Si_3N_4 obtained using different computational approaches

property	pair potential [12]	Tersoff [13]	ab-initio [11]	experiment [15 to 17]
a (Å)	7.761	7.513	7.65	7.595, 7.606, 7.766
c (Å)	5.628	–	5.68	5.628
B (GPa)	270	240	252 to 282	258 to 273

pair and three-body forms of the potential allow for possible phase segregation although another rigid potential that does not explicitly include any separate chemical significance of Si or N has been used in large scale molecular dynamics simulations of the mechanical properties of nonophase Si_3N_4 [14]. In Table 1 we compare results of some approaches, as can be seen there is a consistent agreement with experiment. These results show that although ab-initio approaches give a far more realistic description of the properties of β-Si_3N_4, a far less demanding approach can give quite satisfactory results. For softer materials this appears to be generally the case, polytypes of SiO_2 can also be modelled using such an approach, as we shall discuss later.

2.2 Carbon nitride

Several forms of C_3N_4 have been suggested for crystalline structures with superhard forms being β-C_3N_4 [10,18,19] (hexagonal structure and space group $P6_3/m$), cubic or defect zincblende C_3N_4 [18] ($P\bar{4}3m$) and another cubic form [20] ($I\bar{4}3d$) suggested to be

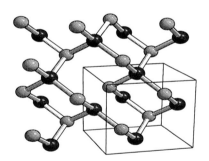

the hardest as measured by the calculated bulk modulus. In each of these structures, carbon atoms are in fourfold coordination by nitrogens, the nitrogens in threefold coordination by carbons. All structures are predicted to be stable with unit cell dimensions given in Table 2. The harder cubic structures are shown in Fig. 2 and the β structure is quite similar to that shown in Fig. 1.

As can be seen each C atom is surrounded in fourfold coordination by N atoms, but the C–N bond length is small: 1.45 Å in β-C_3N_4 and 1.48 Å in cubic (zincblende) C_3N_4.

Graphitic phases consisting of hexagonal planes of bonded C–N atoms with a missing C on one of the conventional planar hexago-

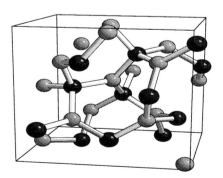

Fig. 2. Cubic phases of C_3N_4. The top structure is commonly referred to as the defect zincblende stucture

Table 2
Structural details of some predicted superhard crystalline phases of C_3N_4 using LDA

phase	unit cell (Å)	bulk modulus (GPa)
β-C_3N_4 ($P6_3/m$)	$a = 6.41$ to 6.44, $c = 2.404$ to 2.467	425 to 451 [18, 21, 20]
cubic-C_3N_4($P\bar{4}3m$) (zincblende)	$a = 3.42$ to 3.44	425 to 448 [18, 22, 20]
cubic C_3N_4($I\bar{4}3d$)	$a = 5.40$	480 to 496 [22, 20]
α-C_3N_4($P3_1c$)	$a = 6.466$ $c = 4.710$	425 [20]

nal lattice sites have also been considered Various sites are possible for the ordering of the graphitic planes depending on the relative location of the missing C atom from one plane to another [23] and another P6m2 structure recently considered by Teter and Helmley [20] also contains 14 atoms in the unit cell but is different to the other graphitic structures as here the two possible C–N bond lengths are in the hexagonal plane. The low energy P6m2 graphite phase may change to one of the other higher energy graphite phases following compression well before the superhard β-phase is reached [23].

The possible properties of the various crystalline, graphitic and amorphous C_3N_4 systems [24] are given in Table 3 and the relative energies of many phases of C_3N_4 are shown in Fig. 3.

2.3 Silicon carbon nitride

Following the earlier theoretical results on β-Si_3N_4 or β-C_3N_4, replacement of the Si atoms with C is being speculated. One of the hardest forms of SiC_2N_4 is, in line with both Si_3N_4 or C_3N_4 expected to be β-SiC_2N_4. This is obtained by replacing two C atoms in the 14 atom unit cell of β-C_3N_4 with Si atoms maintaining lowest C–N bond lengths [11]. In Fig. 4 we show the structure of β-SiC_2N_4 as viewed along the c-axis of the hexagonal unit cell with specific details of the structure being presented in Table 4. This structure is not unlike that for the β-Si_3N_4 and β-C_3N_4 structures apart from a lowering in symmetry to P2/m.

Quite recently a low density phase of SiC_2N_4 has also been found [27] with a possible cubic structure(Pn3m) shown in Fig. 5.

Table 3
Predicted structural properties of C_3N_4

phase	shortest C–N distances(Å)	interplanar distance (Å)	B (GPa)
β	1.45		432
cubic (defect zincblende)	1.48		428
P6m2 graphite	1.32 to 1.45	3.29	253
p-graphite (odd-phase stacking)	1.37	3.37	205
p-graphite (even-phase stacking)	1.37	3.72	198
amorphous			128

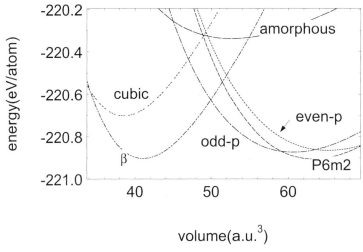

Fig. 3. Relative energies of some possible phases of C_3N_4

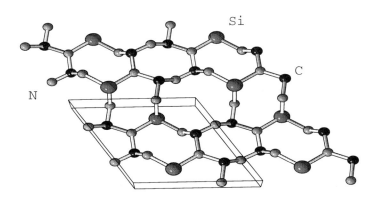

Fig. 4. β-SiC_2N_4

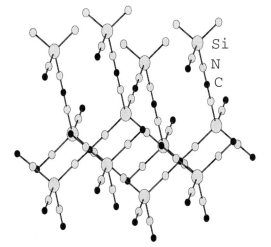

Fig. 5. Possible precursor phase of SiC_2N_4

Table 4

Calculated structure and bulk modulus of some phases of SiC_2N_4 [24]. Structures b and c are approximate structures that have been also considered (Refs. [26] and [25])

phase	unit cell (Å)	B (GPa)
β-SiC_2N_4(P2/m)	$a = 6.78$, $b = 6.92$; $c = 2.58$	330
a (Pn3m)	$a = 6.64$	106
b	$a = 6.65$	95
c	$a = 6.32$	32

In such a structure, Si surrounded tetrahedrally by four N atoms is the essential building block and a linear Si–N–C–N–Si bridging unit. The measured unit cell constant is 6.18 Å, and Riedel et al. [27] noted that this size of cell would have a short interatomic spacing between the C and N atoms of 1.19 Å. They therefore suggested that the measured cell constant may correspond to an effective value relating to a random distribution of N atoms about the Si–C–Si unit. Kroll et al. [26] and Lowther [25] have suggested other structures for the phase essentially involving bending of the N–C–N bond that leads to an overall softening of the structure. As yet the several cubic structures have not been related to the potentially superhard β-$SiC_2N_4C_2N_4$ phase experimentally but the calculated equation of state shown in Fig. 6 suggests that the cubic phase could be a promising precursor phase for β-SiC_2N_4.

3. Oxides

Oxygen has proven to be a clear competitor to C-based superhard materials essentially because of chemical differences. Silica based structures in particular have been extensively examined [28] and an important phase of the material, stishovite, has emerged exhibiting superhard properties. The essential basis for the enormous number, yet quite different, polytypes of silica relates to the nearest neighbor coordination about the Si. In quartz and related structures such as cristobalite, coesite etc. [16], the Si is in four-

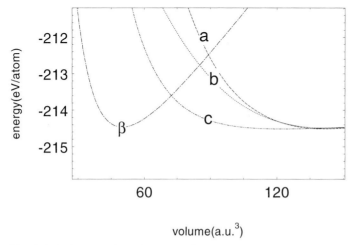

Fig. 6. Relative energies of some possible phases of SiC_2N_4

Table 5

Calculated properties of the relatively soft quartz (fourfold coordination) and potentially superhard stishovite (sixfold coordination) phases of silica. The ab-initio results reflect both LDA and GGA calculations

	α-quartz (P3$_2$21)			stishovite (P4$_2$mnm)		
	pair-potential [29]	ab-initio [29]	expt. [34,35]	pair-potential [29]	ab-initio [29,33]	expt. [36]
a (Å)	5.02	4.84 to 5.02	4.9160	4.26	4.16 to 4.23	4.18
c (Å)	5.53	5.38 to 5.51	5.4054	2.75	2.65 to 2.72	2.667
B (GPa)	39	31 to 48	34 to 37	299	286 to 324	313

fold coordination with O atoms and it is the relative inclination of the SiO$_4$ tetrahedra that ultimately gives rise to very many SiO$_2$ polytypes. In attaining the superhard stishovite phase the coordination increases to six and a rutile type structure is formed. Higher coordinated phase of silica have not been reported to date, however when a metal substitutes for the Si some high coordinated structures emerge which hold quite valuable potential as superhard materials.

3.1 Low oxygen coordination: silica

Several two-body [8,29] and three-body rigid potentials [30] have been used to represent force fields of SiO$_2$ all claiming to be capable of predicting elastic and structural properties to a high degree of accuracy and presume that the rigid potentials are transferable which, of course, ab-initio calculations challenge especially for high pressure forms of the material. Using both a pair-potential in a Buckingham form as well as the local density pseudopotential approximation Keskar and Chelikowsky [29] and other authors [31] to [33] first showed the various phases of SiO$_2$ lie extremely close in energy with quartz being lower in energy than stishovite. Some structural results for the softer and harder phases of SiO$_2$ are given in Table 5 and in Fig. 7 we show the structure of α-quartz and stishovite.

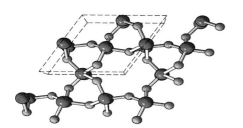

3.2 High oxygen coordination: zirconia and hafnia

Experimental measurements on both ZrO$_2$ and HfO$_2$ have revealed that the ambi-

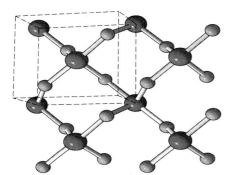

Fig. 7. The α-quartz (top) and stishovite (bottom) structures

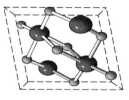

P2$_1$/c

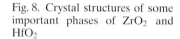

Fig. 8. Crystal structures of some important phases of ZrO$_2$ and HfO$_2$

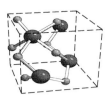

Pbc21

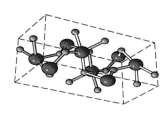

Pbca

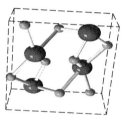

Pnma

ent phase of both materials is a monoclinic baddeleyite phase. Under pressure both dioxides transform to a series of orthorhombic phases, yet the manner and route are slightly different. In the case of ZrO$_2$, the first orthorhombic phase starts at an applied pressure of about 3 GPa depending upon the grain size of the material and is observed to exist up to about 22 GPa when another orthorhombic structure sets in. The first orthorhombic phase is now identified as having its own orthorhombic-I structure-type with Pbca symmetry [37] and the second a cotunnite structure with Pnma symmetry. On the other hand, for HfO$_2$ the transformation to the first orthorhombic phase from baddeleyite now sets in at about 4 to 10 GPa and under pressure the second orthorhombic structure is observed at about 28 GPa and which exists up to about 40 GPa [38]. The structure of quenched samples of the first orthorhombic phase prepared under high-pressure, high-temperature conditions is of the orthorhombic-I Pbca type [39]; however, other structures, Pbcm for example, have been suggested [40] based on in situ, high-pressure X-ray diffraction studies. The Pbc21 structural model can readily be related to that of Pbcm. Somewhat simplistically, the 12-atom Pbc21 unit cell can nearly twin under a reflection with itself to double up giving the 24-atom Pbca phase. The structures of some of these phases are shown in Fig. 8 and the properties using ab-initio LDA calculations [41,42] in Table 6.

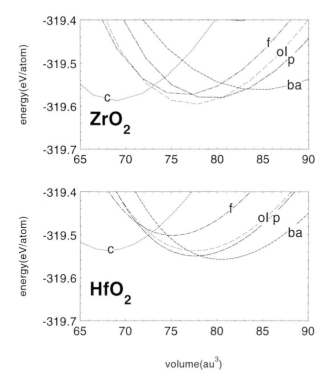

Fig. 9. Calculated energies of ZrO_2 and HfO_2 phases. ba: baddeyelite (P2$_1$/c), p: Pbc21, oI: orthorhombic-I (Pbca), f: fluorite (Fm3m), c: cottunite (Pnma)

In Fig. 9 we show the calculated energies of the various phases for both oxides. As is typical with most hard systems they all lie quite close to each other, but for ZrO_2 the Pbca phase is more stable whilst for HfO_2 the Pbc21 phase is only slightly lower in energy than Pbca. In fact, the three phases of HfO_2, P2$_1$/c, Pbc21 and Pbca are seen to lie within a very close energy range (≈ 10 meV) of each other. This closeness in energy suggests that the transformation between these phases will be quite sensitive and easily affected by, among other factors, temperature and remnant lattice stresses as observed. Therefore, it is not surprising that there are difficulties in identifying the first orthorhombic phase boundaries of HfO_2 as has been observed experimentally [43,44]. A transformation to the second orthorhombic structure is quite clearly seen for ZrO_2 as involving the structures Pbca into Pnma, although again the subtle Pbc21/Pbca interrelation again plays a sensitive role in HfO_2. In both ZrO_2 and HfO_2 however the

Table 6
Some important phases of zirconia and hafnia together with predicted bulk moduli (from Refs. [41] and [42]) using LDA

phase	coordination	ZrO_2 B (GPa)	HfO_2 B (GPa)
baddeleyite (P2$_1$/c)	7	157	251
Pbc21	7	264	272
orthorhombic-I (Pbca)	8	272	256
fluorite (Fm3m)	8	267	280
cottunite (Pnma)	9	305	306

second orthorhombic phase is clearly Pnma, which agrees with what is found experimentally [44,45,46]. The calculated bulk moduli for these phases, 305 and 306 GPa, respectively, are in good agreement with the experimental values of 332 GPa for ZrO_2 and 340 GPa for HfO_2 [46]. Both theory and experiment thus concur that these phases are good candidates for use as superhard materials.

4. Conclusion

In this paper, we have presented an overview of the applications of computer modelling to some materials that could have potentially superhard applications. The theoretical challenge underlying all of these structures is that the phases all lie quite close in energy yet at different volumes (per molecular formula unit) making an exact evaluation of the structural energetics and hence the equation of state quite a formidable task. In both ab-initio and semi-empirical approaches good techniques of molecular dynamics are paramount in understanding these structures as, under pressure, atoms can relax within the unit cell and thus impact on measured elastic properties. Electron correlation is also quite clearly important in understanding superhard materials yet this now seems to be quite well represented through density functional approaches.

The role played by the semi-empirical approaches has possible shortcomings when dealing with superhard materials. As we have seen in the case of both Si_3N_4 and SiO_2 at first sight there is a good prediction of the structural properties using rigid potentials. Yet there is danger in extending such potentials to compressed compact structures of materials because it is precisely in this region that electron correlation will be important. Yet the efficiency of rigid potential codes and their low cost factor will continue to be an attraction especially in giving a quantitative insight into material behavior.

Density functional techniques have undoubtedly been very important in recent developments of electronic structure calculations yet the various functionals available still await further investigation. For some materials especially carbides and nitrides the LDA is preferential than GGA, yet with the oxides it is likely that both could afford similar importance.

Unfortunately, there is some cynicism regarding commercial synthesis of some predicted new materials yet the wisdom of the computational arguments upon which the various structures of the material have been predicted seems firm and have shown to be capable of making valuable insight in many other areas of materials science. Experiences gained regarding the commercial synthesis of diamond and cubic BN could probably lead ways that new materials can be synthesized, yet other more novel routes also cannot be excluded, and here computer modelling continues to play an important role.

References

[1] M. L. COHEN and J.R. CHELIKOWSKY, Electronic Structure and Optical Properties of Semiconductors, Springer-Verlag, Heidelberg 1989.
[2] M. C. PAYNE, M. P. TETER, D. C. ALLAN, T. A. ARIAS, and J. D. JOANOPOULOS, Rev. Mod. Phys. **64**, 1045 (1992).
[3] D. P. JOUBERT (Ed.), Density Functional Theory, Springer-Verlag, Heidelberg 1998.
[4] J. H. HARDING, Rep. Progr. Phys. **53**, 1403 (1990).
[5] X. GONZE, R. STUMPF, and M. SCHEFFLER, Phys. Rev. B **44**, 8503 (1991).
[6] N. TROULLIER and J. L. MARTINS, Phys. Rev. B **43**, 1993 (1991).
[7] J. TERSOFF, Phys. Rev. Lett. **61**, 632 (1988).

[8] B. VAN BEEST, A. P. J. JANSEN, and R. A. VAN SANTEN, Phys. Rev. Lett. **64**, 1955 (1990).
[9] R. NADA, C. R. A.. CATLOW R. DOVESI, and C. PISSANI, Phys. Chem. Minerals **17**, 353 (1990).
[10] A. Y. LIU and M. L. COHEN, Science **245**, 841 (1989).
[11] CHENG-ZHANG WANG, EN-GE WANG, and QINGYUN DAI, J. Appl. Phys. **83**, 1975 (1998).
[12] WAI-YIM CHING, YONG-NIAN XU, J. D. GALE, and M. RUHLE, J. Amer. Ceram. Soc. **81**, 3189 (1998).
[13] F. DE BRITO MOTA, J. F. JUSTO, and A. FAZZIO, Phys. Rev. B **58**, 8323 (1998).
[14] R. K. KALIA, A. NAKANO, A. OMELTCHENKO, K. TSURTA, and P. VASHISHTA, Phys. Rev. Lett. **78**, 2144 (1997).
[15] I. KOHASTU and J. W. MCCAULY, Mater. Res. Bull. **9**, 917 (1974).
[16] R. W. G. WYCKOFF, Crystal Structures, Kreigler, Malabar 1986.
[17] O. BORGEN and H. M. SEIP, Acta. Chem. Scand. **15**, 1789 (1961).
[18] A. M LIU and R. M. WENTZCOVITCH, Phys. Rev. B **50**, 10362 (1994).
[19] J. V. BADDING, Adv. Mater. **11**, 877 (1997).
[20] D. M. TETER and R. J. HEMLEY, Science **271**, 53 (1996).
[21] A. Y. LIU and M. L. COHEN, Phys. Rev. B **41**, 10727 (1990).
[22] J. MARTIN-GILL, F. MARTIN-GILL, M. SARLKAYA, M. QIAN, M. JOSE-YACAMAN, and A. RUBIO, J. Appl. Phys. **81**, 2555 (1997).
[23] J. E. LOWTHER, Phys. Rev. B **59**, 11683 (1999).
[24] J. E. LOWTHER, Phys. Rev. B **57**, 5724 (1998).
[25] J. E. LOWTHER, Phys. Rev. B **60**, 11943 (1999).
[26] P. KROLL, R. RIEDEL, and R. HOFFMAN, Phys. Rev. B **60**, 3126 (1999).
[27] R. RIEDEL, A. GREINER, G. MICHE, W. DRESSLER, H. FUESS, J. BILL, and F. ALDINGER, Agnes. Chem. Internat. Ed. Engl. **36**, 603 (1997).
[28] P. J. HEANEY, G. V. GIBB, and C. T. PREWITT (Eds.), in: Reviews of Mineralogy, Vol. 29, Silica-Physical Behavior, Geochemistry and Materials Applications, Mineral Soc. America, Washington DC, 1994.
[29] N. R. KESKAR and J. R. CHELIKOWSKY, Phys. Rev. B **46**, 1 (1992).
[30] W. JIN, K. KALIA, P. VASHISHTA, and J. P. RINO, Phys. Rev. B **50**, 118 (1994).
[31] D. M. TETER, G. V. GIBBS, M. B. BOISEN, JR., and M. P. TETER, Phys. Rev. B **52**, 8064 (1995).
[32] J. K. DEWHURST, J E. LOWTHER, and L. MAZWARA, Phys. Rev. B **55**, 111003 (1997).
[33] TH. DEMUTH, Y. JEANOINE, J. HAFNER, and J. G. ANGGAN, J. Phys. C **11**, 3833 (1999).
[34] M. SUGIYAMA, S. ENDO, and K. KOTO, Mineral J. Japan **13**, 1987 (455).
[35] N. L. ROSS, T. F. SHU, R. HAZEN, and T. GASPARIK, Amer. Mineral **75**, 739 (1986).
[36] L. LEVIEN and C. T. PREWITT, Amer. Mineral **66**, 324 (1981).
[37] O. OHTAKA, T. YAMANAKA, S. KUME, N. HARA, H. ASANO, and F. IZUMI, Proc. Jpn. Acad. B **66**, 193 (1990).
[38] A. JAYARAMAN, S. Y. WANG, S. K. SHARMA-SK, and L. C. MING, Phys. Rev. B **48**, 9205 (1993).
[39] O. OHTAKA, T. YAMANAKA, S. KUME, N. HARA, H. ASANO, and F. IZUMI, J. Amer. Ceram. Soc. **78**, 233 (1995).
[40] D. M. ADAMS, S. LEONARD, D. R. RUSSELL, and R. J. CERNIK, J. Phys. Chem. Solids **52**, 1181 (1991).
[41] J. E. LOWTHER, J. K. DEWHURST, J. M. LEGER, and J. HAINES, Phys. Rev. B **60**, 14485 (1999).
[42] J. K. DEWHURST and J. E. LOWTHER, Phys. Rev. B **54**, R1234 (1997).
[43] J. M. LEGER, A. ATOUF, P. E. TOMASZEWSKI, and A.S. PEREIRA, Phys. Rev. B **47**, 14075 (1993).
[44] J. TANG, M. KAI, Y. KOBAYASHI, S. ENDO, O. SHIMOMURA, T. KIKEGAWA, and T. ASHIDA, in: Properties of Earth and Planetary Materials at High Pressure and Temperature, Eds. M. H. MANGHNANI and T. YAGI, American Geophysical Union, Washington 1998.
[45] J. HAINES and J. M. LEGER, J. Superhard Mater. **2**, 3 (1998).
[46] J. HAINES, J. M. LEGER, S. HULL, J. P. PETITET, A. S. PEREIRA, C. A. PEROTTONI, and J. A. H. DA JORNADA, J. Amer. Ceram. Soc. **80**, 1910 (1997).

phys. stat. sol. (b) **217**, 545 (2000)

Subject classification: 61.72.Lk; 62.20.–x; 71.15.–m; S5.11

Modeling Brittle and Ductile Behavior of Solids from First-Principles Calculations

U. V. WAGHMARE (a), E. KAXIRAS (a), and M. S. DUESBERY[1]) (b)

(a) *Department of Physics, Harvard University, Cambridge, MA 02138, USA*

(b) *Fairfax Materials Research, Fairfax, VA 22153, USA*

(Received August 10, 1999)

A broad classification of solids in terms of their mechanical behavior would characterize them as brittle or ductile. While there is no doubt that ultimately this behavior is due to processes at the atomistic level, the link between these processes and their macroscopic manifestation is difficult to establish. Phenomenological theories that try to address this link must rely on microscopic parameters, the values of which are beyond their scope. Here we review recent efforts to employ first-principles electronic structure calculations in order to determine important physical quantities which, in conjunction with phenomenological theories, can provide insight into brittle versus ductile behavior. We apply this approach to cases of intrinsic interest, such as silicon, the prototypical brittle solid, as well as in cases of practical interest, such as the improvement of ductility in molybdenum disilicide, a material of potential usefulness in improved turbine blades. Current indications are that this combination of techniques can serve as powerful qualitative predictive tool for the dependence of mechanical behavior on the microscopic structure and chemical composition of a solid.

1. Introduction

Mechanical properties are at the core of many uses of materials in traditional as well as high technology applications. Solids are typically classified as brittle or ductile, according to how they respond to external loads. For example, brittle failure is a singular limiting factor in the application of advanced materials in components ranging from electronic devices to turbine engine blades. In brittle solids, the imposition of external stress results in the extension of pre-existing cracks: the stress concentration at the crack tip (which actually diverges for a sharp tip, when calculated within continuum elasticity theory [1]), leads to bond-breaking and cleavage. In contrast to this, in a ductile substance the large stress at the crack tip is absorbed by generation and motion of dislocations which blunts the crack; the net effect is plastic deformation of the material but no breaking. In advanced electrical and mechanical devices, as the size of critical components shrinks, and the demands for high performance and reliability increase, a detailed understanding of the microscopic processes responsible for ductile or brittle behavior becomes increasingly important. Understanding such processes at the atom-

---

[1]) We are saddened to report that Dr. M. S. Duesbery passed away during the preparation of this manuscript. Dr. Duesbery's contributions to the understanding of mechanical properties of solids have left an indelible mark on the field. The two other authors feel indebted to Dr. Duesbery for generously sharing with them his deep knowledge of this field and for his invaluable contributions to this collaboration.

istic level and connecting them to the macroscopic behavior of real materials will provide unprecedented level of control in the design of new applications.

What would it take to actually simulate at the microscopic level the processes responsible for brittle or ductile behavior? We would need to include in the simulation a large enough number of atoms to capture the different types of defects (cracks, dislocations, surfaces, grain boundaries, point defects) and their response to external loading, at finite temperature. For a solid of linear size one micron this requires the simulation of the dynamic behavior of $\approx 10^{12}$ atoms in a variety of local environments and for times that are very long compared to the time scale of atomic vibrations (which sets the scale for each step in the simulation). This is impossible with current computational capabilities. But even if it were possible, the sheer volume of data generated by following the evolution of 3×10^{12} degrees of freedom for millions of simulation steps, would make it difficult to analyze and comprehend the results. At present, supercomputers can be employed to simulate the behavior of up to 10^9 atoms, using the simplest type of interatomic interaction (referred to as classical potentials). In this approximation, the complex interaction between atoms through the valence electrons responsible for the cohesion of solids, is reduced to a few-body (typically two- and three-body) interatomic potential which depends on atomic distances and the angles between such distances. It is not clear to what extent this simplified description can capture realistically the atomistic processes responsible for bond breaking and cleavage (in the case of a brittle solid) or dislocation nucleation and motion (in the case of a ductile solid). Even if this can be successfully done for a solid consisting of a single element (where a sophisticated interatomic potential might have a chance at performing reasonably), when many types of atoms are present and chemical interactions become dominant at the microscopic level, this approach would quickly run up against insurmountable difficulties.

Alternatively, one can use sophisticated quantum mechanical methods for describing the properties of solids at the atomic level. When these methods are entirely self-reliant, i.e. they do not depend on any parameters that cannot be determined within the theory (other than fundamental physical constants), they are referred to as first-principles approaches. For instance, quantum mechanical methods based on Density Functional Theory [2] have proven to be powerful and unbiased tools for microscopic studies of a wide range of materials properties. Direct application of these methods yields an accurate description of the electronic and atomic structure of bulk crystals, their simple surfaces and interfaces, of small clusters of atoms, etc. Properties that represent small perturbations from the equilibrium crystal structure of the solid, such as the phonon spectrum, elastic moduli, optical, dielectric and magnetic constants, can also be obtained with reasonable accuracy. This is possible because all these properties can be modeled by a small number of atoms, since these calculations are limited to systems of order a few hundred atoms. At present it is entirely out of the question (and, if feasible, might represent a waste of valuable computational resources) to try to simulate systems with millions or billions of atoms using first-principles quantum mechanical calculations.

It is clear from the above discussion that successful modeling of physical phenomena which involve atomistic processes and their implications for macroscopic behavior, must rely on a combination of methodologies. A very active field of "multiscale modeling" of materials behavior has emerged recently from this realization. Several ways to implement multiscale modeling, with different strengths and weaknesses, have been put forward. It is beyond the scope of the present article to review these approaches. Here we aim to present one of the simplest ways of linking information between the microscopic and macroscopic realms, and explore its potential to address real problems in materials behavior. Although admittedly simple and limited, this approach offers some intriguing insight to complex phenomena such as brittle versus ductile failure, and can have predictive capabilities within a well defined scope of applications. The essence of this approach is to use results from phenomenological macroscopic theories (based on continuum elasticity) coupled with first-principles quantum mechanical calculations to determine the values of any physical parameters that enter in the phenomenological description.

The article is organized as follows: Section 2 covers the theoretical background behind the model we use to assess ductile and brittle response from a macroscopic phenomenological perspective. The first-principles quantum mechanical methods which we use to obtain the model parameters are reviewed in Section 3. In Section 4, we discuss applications of this approach to mechanical behavior of real materials: silicon, the prototypical brittle solid, and molybdenum disilicide and its substitutional alloys, which represent a class of materials with potentially important applications. We conclude in Section 5 with a discussion of the approximations inherent in our approach and the consequent limitations.

2. Brittle versus Ductile Behavior: Macroscopic Theory

In this section, as a preamble to subsequent discussion, we first introduce the concept of the generalized stacking fault energy surface (γ-surface) and indicate how it relates to dislocation nucleation and mobility. We also review theories of the Ductile–Brittle Transition (DBT), particularly emphasizing their connection with dislocation nucleation and mobility, and therefore the γ-surface.

2.1 The generalized stacking fault energy

The concept of the γ-surface, originally introduced by Vitek [3], is central to many macroscopic theories of plastic deformation. Consider a crystal cut into two halves parallel to the (hkl) crystallographic plane and suppose that one half is displaced relative to the other by a vector **v**. The plane (hkl) is called the fault plane and $\mathbf{v} = 0$ corresponds to the ideal crystal, the minimum energy configuration. The change in energy per unit area of the crystal as a function of **v**, which is varied on the (hkl) plane to scan a unit cell, is called the γ-surface. If the atoms in the crystal are allowed to relax for an arbitrary **v**, the fault (or γ-surface) is called relaxed. The γ-surface contains a wealth of information about the crystal: at the atomistic level, it can be used to predict dislocation core properties [4]; at the macroscopic level it can yield the stress intensity at which dislocations are nucleated at the crack tip [5]. The γ-surface is a fundamental material property which can be obtained from atomistic calculations.

2.2 Dislocation motion

When dealing with dislocations from the macroscopic perspective, some phenomenological description consistent with the continuum picture of a solid must be invoked. A useful description of this type was put forward by Peierls [6] and Nabarro [7]. The

Peierls-Nabarro model represents the crystal dislocation as a continuous, planar distribution of infinitesimal disregistry of atomic planes on either side of a fault plane. The density of this disregistry, $\varrho(x)$ is given in terms of the displacement field $\mathbf{u}(x)$ as $\varrho(x) = |\mathrm{d}\mathbf{u}/\mathrm{d}x|$. The balance between the forces among the infinitesimal segments of the dislocations and the elastic restoring force of the crystal leads to the Peierls-Nabarro integro-differential equation

$$K \int_{-\infty}^{\infty} \mathrm{d}x' \frac{\mathrm{d}\mathbf{u}(x')}{\mathrm{d}x'} \frac{1}{(x-x')} = -\nabla_{\mathbf{u}}(\gamma(\mathbf{u}(x))) \,, \tag{1}$$

subject to the normalization condition

$$\int_{-\infty}^{\infty} \frac{\mathrm{d}\mathbf{u}(x)}{\mathrm{d}x} \mathrm{d}x = \mathbf{b} \,. \tag{2}$$

Here, K is an elastic constant (its exact value depends on the nature of the dislocation) and $\mathbf{b}$ is the Burgers vector of the dislocation (it is assumed for simplicity that we are dealing with a straight infinite dislocation and the disregistry is only along one direction x, perpendicular to the dislocation line). Eq. (1) relates the stress on the lattice at a position x due to the infinitesimal dislocation distribution at that point $\mathbf{u}(x)$ to the force exerted by the lattice due to its distortion, given by the gradient of the γ-surface. Eq. (2) normalizes the total misfit to the Burgers vector (for more details see, for example, Hirth and Lothe [8]).

In order to determine the effects of dislocation motion, we need to consider what will happen within this picture when this distribution of lattice disregistry is moved by a distance equivalent to the lattice periodicity and compute the change in energy as a function of x. This energy cost can then be related to the external stress required to move the dislocation. This energy is referred to as the Peierls energy $W_P(x)$ and its maximum gradient is the Peierls stress for translation of rigid dislocations at zero temperature. While extremely useful as a conceptual tool, this model has many limitations. The method has been shown to predict Peierls stresses much larger than those determined from a fully atomistic calculation [9, 10]. One of the important limitations, for example, is the neglect of any change in $\varrho(x)$ with position during motion through lattice; if this constraint is removed, the agreement with atomistic methods is improved substantially [11]. For a rough estimation of the Peierls stress, the step in which the distribution is translated through the lattice can be bypassed: it has been shown that, in the limit of narrow dislocations, the Peierls stress is given directly by the maximum gradient of the γ-surface along an extremal path (the lowest energy path between endpoints corresponding to ideal crystal) over which the displacement vector $\mathbf{v}$ varies continuously from 0 to $\mathbf{b}$ [12].

2.3 Ductility versus brittleness

As mentioned in the introduction, the competing processes which lead to brittle or ductile behavior are the extension of the crack by creation of fresh surfaces (brittle response) or the generation of dislocations that exert a back stress which reduces, and thus *shields* the stresses by blunting the crack tip (ductile response). Most solids can undergo a transition between brittle and ductile response, which can be very sharp as a function of temperature (in bulk silicon the transition takes place for a change in temperature of a couple of K!). In brittle failure, the energy required for an incremental advance of the crack front is

$$G = 2\gamma_s, \qquad (3)$$

where G is the energy release rate (the mechanical energy released per unit area swept by the crack front) and γ_s is the surface energy (the energy needed to create a unit area of fresh surface). This is known as the Griffith criterion [13]. For ductile response a more complex picture needs to be invoked, taking into account dislocation nucleation and motion, in the neighborhood of the crack tip.

The first attempt to rationalize the distinction between brittle and ductile behavior was made by Kelly and coworkers [14], who postulated that a material would be ductile if the crack tip stress exceeded the theoretical shear stress before the theoretical tensile stress was reached. This work did not address the DBT. A direct linkage to the DBT was made later by Rice and Thomson [15], who proposed that the onset of ductile behavior occurred when spontaneous emission of dislocations at the crack tip became feasible. In the RT model, dislocations which move more than a critical distance a_c from the crack tip are repelled by the crack and are hence considered to be nucleated. If a_c is smaller than the dislocation core radius a_0, the material is considered to be ductile at all temperatures; on the other hand, if $a_c > a_0$, the material is brittle at low temperatures. The linkage of the RT model with dislocation energy naturally introduces a temperature-dependene DBT. Unfortunately, it predicts rather large activation energies for dislocation nucleation in brittle materials, precluding a DBT significantly below the melting temperature T_m. Experimentally, the DBT is found to occur at much lower temperatures, for example, at $(2/3)T_m$ for Si.

There are three more recent theories to describe the DBT. In a modern variant of the RT model, Rice [16] introduced the unstable stacking fault energy γ_{us} as the maximum energy barrier encountered along the extremal path (introduced in the previous subsection) connecting 0 and **b** on a plane (hkl). Rice showed that γ_{us} is a measure of the nucleation energy for a dislocation of Burgers vector **b** on the (hkl) plane. Within the Rice model, the criterion for dislocation nucleation at the crack tip, and therefore ductility, is reached when

$$G = \alpha\gamma_{us}, \qquad (4)$$

where α is a constant which depends on the geometry of the crack, but is of the order unity. We point out a connection between γ_{us} and the Peierls stress (the former being the maximum energy and the latter being the maximum slope along the extremal path). If the energy has a sinusoidal form, the maximum slope will scale with the maximum energy, implying that the Peierls stress should scale directly with γ_{us}.

In an alternative model of the DBT, Hirsch and Roberts (HR) [17] postulate that the governing mechanism is the motion, rather than the nucleation of dislocations. The crack tip stresses are presumed to activate internal sources, with dislocations of one sign moving into the bulk and of the opposite sign being absorbed into the crack. Of course, absorption of dislocations of one sign is equivalent to emission of dislocations with opposite sign. The HR theory therefore connects the DBT with dislocation mobility. This, as shown above scales with the unstable stacking fault energy γ_{us}.

Yet a different model, due to Khantha, Pope, and Vitek (KPV) [18], argues that the shielding dislocations are nucleated in the vicinity of the crack tip as a self-screening cloud of dipoles by a Kosterlitz-Thouless type mechanism [19]. In this case, the key quantity is the total dislocation energy, as in the RT model. However, the elastic energy contribution is much smaller due to the smaller screening length of the KPV model, while the contribution of the dislocation core energy is the same as in the RT model. It is precisely the dislocation core region in which the misfit assumes the magnitude at which the γ-surface reaches the unstable stacking fault value; i.e. the core energy depends on the unstable stacking fault energy. Thus, although the specific details of the KPV model are quite different from either the Rice or the HR model, γ_{us} can be considered to be the key parameter for dislocation related processes of ductility in all three models.

To summarize, the conditions for both brittle fracture and the nucleation and motion of dislocations can be expressed in terms of features of the γ-surface. If we adopt the view that dislocation nucleation is a necessary precursor to dislocation motion, the conditions become particularly simple, depending only on the two energies γ_s and γ_{us}. It is tempting to try to use these simple arguments to characterize the tendency of a solid to behave as a brittle or a ductile substance. Brittle behavior is the consequence of the condition of eq. (3) being satisfied before the condition of eq. (4); if the converse is true, the material will be ductile. More specifically, we can define a "disembrittlement parameter"

$$D = \frac{\gamma_s}{\gamma_{us}}, \qquad (5)$$

the value of which will determine to which class a solid belongs. The difficulty with using this criterion is that values for γ_s and γ_{us} are not provided by any analysis based on continuum arguments of the type invoked in this section. They have to be established from accurate microscopic calculations. Even then, it is not clear for which slip system (i.e. the combination of fault plane and Burgers vector of dislocation) or for which cleavage plane, these quantities must be evaluated. Moreover, the value of D that corresponds to the DBT is not known with any precision. This critical value has been estimated [16] to be between 1 and 10. One might hope, however, that *changes* in the value of D due to changes in the microscopic structure or chemical composition of a solid will correlate with changes in the solid's tendency to behave as a brittle or a ductile substance. It is this idea that we pursue below.

3. First-Principles Methods

The minimal empirical input for the study of equilibrium properties of a solid is its composition and its atomic structure. At $T = 0$, the optimal structure can be determined by calculating the total energy and finding the minimum energy structure. First-principles methods can provide an accurate determination of the total energy as a function of structure and composition. They can therefore be employed to determine either the optimal structure given the composition and certain external conditions (like pressure), or to determine the energy difference between specific structures of interest. Below we will use such methods to obtain accurate values for γ_s and γ_{us}.

First-principles methods are typically based on the adiabatic approximation (or Born-Oppenheimer approximation), by treating electrons in their quantum mechanical ground state for a given configuration of nuclei $\{\mathbf{R}_i\}$. As a result, the energy of a given state is given by the sum of the electronic ground state energy $E_{0-\text{el}}(\{\mathbf{R}_i\})$ and the ion–ion interaction (Coulomb) energy E_{ion} obtained by an Ewald summation:

$$E_{\text{tot}} = E_{0-\text{el}}(\{\mathbf{R}_i\}) + E_{\text{ion}}. \tag{6}$$

Detailed descriptions of many of the commonly used first-principles methods are presented in other parts of the present collection of review articles (see [20]). Here, we briefly highlight some of the essential features of the methods based on Density Functional Theory, and discuss their use in studying mechanical behavior through total energy calculations. More than 99% of the computational effort is spent in the calculation of the electronic ground state.

The problem of finding the quantum mechanical ground state of electrons in solids is a many-body problem. While highly accurate methods exist for solving this many-body problem, such as quantum Monte Carlo, these can handle only very small systems of order tens of atoms at most. For larger systems of interest, approximate but reasonably accurate schemes that make calculations feasible with existing computational facilities are mostly based on density functional theory (DFT) [2]. The central theorem of DFT, proven by Hohenberg and Kohn, states that the ground state energy of an electronic system is a unique functional of its charge density $\varrho(\mathbf{r})$ and is variational (minimum) with respect to the charge density: for the ground state of a system of electrons in an external potential $V_{\text{ext}}(\mathbf{r})$,

$$E[\varrho(\mathbf{r})] = \int d\mathbf{r} V_{\text{ext}}(\mathbf{r})\varrho(\mathbf{r}) + \frac{1}{2}\int\int d\mathbf{r}d\mathbf{r}' \frac{\varrho(\mathbf{r})\varrho(\mathbf{r}')}{|\mathbf{r}-\mathbf{r}'|} + F[\varrho(\mathbf{r})] \tag{7}$$

is a minimum with respect to variations in $\varrho(\mathbf{r})$, under the constraint that

$$\int d\mathbf{r}\varrho(\mathbf{r}) = N, \tag{8}$$

where N is the total number of electrons. The functional $F[\varrho(\mathbf{r})]$ contains the kinetic and exchange-correlation energies for interacting electrons. This functional is not known exactly and its use in practice generally involves an approximation called the local density approximation (LDA): the ground state energy is a local functional of the total charge density. This functional is derived using results of more elaborate approaches (such as the Random Phase Approximation or Quantum Monte-Carlo simulations) for the homogeneous electron gas [21]. Kohn and Sham [2] expressed the charge density in terms of single particle wavefunctions and mapped the ground state problem in DFT onto a single particle problem in an effective potential that must be determined self-consistently.

The basis set used to represent the Kohn-Sham wavefunctions and the electronic charge density in this work consists of plane waves with kinetic energy smaller than a fixed energy cutoff determined by the problem of interest. We make use of the translational symmetry of bulk crystals through Bloch's theorem, so that the wave vector $\mathbf{k}$ in the Brillouin zone (BZ) labels an electronic eigenfunction, and the cell periodic part of the wavefunction can be expressed in terms of a discrete set of plane waves generated as multiples of the reciprocal space vectors. The plane wave basis set allows for efficient calculation of the various operator matrix elements, including the energy and the forces on the ions (obtained through the Hellmann-Feynman theorem) and an unbiased representation of charge densities.

Another simplifying approximation is the use of pseudopotentials [22] to represent the ion–electron interaction. Thus the core electrons in an atom are frozen and are replaced by an effective potential which is what the valence electrons experience. With this, the full nuclear Coulomb potential Z/r, with Z the atomic number and r the radial distance from the nucleus, is replaced by a pseudopotential which behaves as Z_v/r outside a cutoff radius r_c but remains finite as $r \to 0$, with Z_v the valence charge. The pseudopotential is constrained to ensure that: 1. the lowest s, p, and d orbital energy eigenvalues of the pseudoatom are identical to the valence electron eigenvalues of the full atom, 2. the normalized radial eigenfunctions of these levels are identical beyond r_c to those of the valence electrons in the full atom, 3. the logarithmic derivatives of the pseudo-wavefunctions, evaluated at r_c and the valence electron eigenvalue, are the same as those of the valence wavefunctions of the real atom. These constraints are designed to make the lowest energy states of the pseudoatom identical to those of the valence states of the real atom outside the cutoff radius, which is the region where valence wavefunctions overlap to produce bonding between the atoms in a solid.

The first advantage of using a pseudopotential is the elimination of core electrons, which hardly affect the bonding in solids. A second advantage is that the eigenfunctions of the valence electrons are replaced by nodeless pseudo-wavefunctions, whose representation in a plane wave basis is easier. In certain cases, like first row elements and transition metals, for which the valence p and d wavefunctions are already nodeless and localized, additional consideration is required. The constraints on a pseudopotential mentioned above are not sufficient to completely determine its form for $r < r_c$. This freedom can be used to optimize a pseudopotential, so that the corresponding pseudo-wavefunctions can be represented with a basis set truncated with a relatively low energy cutoff [23]. This class of pseudopotentials, called "optimized pseudopotentials", is used for first-row elements and transition metals in the present work. As a test of these pseudopotentials, our calculations of ideal crystal properties for these elements give lattice constants within 2% of their experimental values.

For the plane wave basis, an energy cutoff up to 60 Ry for solids containing transition metals and oxygen and 12 Ry for Si has been used in the present calculations; these computational parameters ensure convergence of energy differences between different structures to within 1 mRy/atom. A Fermi-Dirac broadening scheme with $k_B T = 0.04$ eV is used to represent the Fermi surface discontinuity. We sample the Brillouin zone using the Monkhorst-Pack scheme [24] with k-point grids of $6 \times 6 \times 6$ for bulk Si (an f.c.c. lattice with a 2-atom basis) and $7 \times 7 \times 7$ for bulk $MoSi_2$ (a b.c.t. lattice with a 3-atom basis), the other brittle solid studied. For the calculation of the γ_s and γ_{us} we use slab configurations (see Fig. 1), with multiples of the unit cell along the slab direction and proportionately smaller grids of k-points in the corresponding direction in reciprocal space. For efficient calculations on metallic systems involving large unit cells and surfaces, we use a preconditioned conjugate gradient algorithm [25] to iteratively diagonalize the Kohn-Sham Hamiltonian and the Kerker charge density mixing scheme [26] in the self-consistent procedure to avoid sloshing of charge between the vacuum and the slab region. This scheme mixes small wave vector components of the charge density gradually and thereby damps oscillations during the self-consistency cycle.

Surface energies γ_s are obtained from the difference between the total energy E_{tot} of a crystal cleaved across a given plane and of the bulk (see Fig. 1). We consider a periodic supercell containing several units of the ideal crystal cell (6 for Si, 2 to 4 for

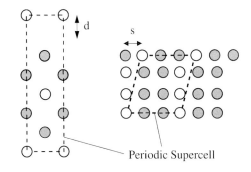

Fig. 1. Geometry of supercells in a typical calculation of surface energy γ_s and unstable stacking fault energy γ_{us}

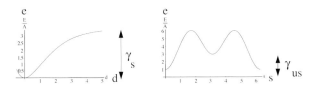

MoSi$_2$) and calculate the total energy as a function of d, the distance between two atomic layers separated by the desired cleavage plane in the supercell. These energies are fit to the universal binding energy function [27]

$$e(d) = \frac{E_{\text{tot}}(d)}{A} = e_\infty - 2\gamma_s(1+f)e^{-f}, \qquad (9)$$

where $f = (d - d_0)/\lambda$, e_∞ is the energy per unit area (A) of the cleaved crystal, d_0 is the inter-planar separation in the ideal bulk crystal and λ a fitting parameter. The unstable stacking fault energy is obtained from the total energy of a supercell containing the fault plane with the two halves of the cell displaced in the direction of the fault vector with respect to each other (see Fig. 1).

4. Applications: Ductile versus Brittle Behavior

4.1 Silicon

In addition to its extensive use in technological applications (especially in the area of electronic devices), silicon also serves as the theorist's prototypical, covalently bonded, brittle solid. Si undergoes a brittle to ductile transition with increasing temperature, at 873 K. A remarkable feature of this transition is its sharpness (it has a width of approximately 2 K). It would be desirable to understand this transition at a microscopic level. This, however, may be beyond current capabilities of simulating all the necessary features, which include the nucleation and motion of dislocations near crack tips in various geometries, such as the formation of dislocation loops near the kinks of cracks, etc. Instead, we will discuss here recent efforts to establish some of the important characteristics of Si as obtained from first-principles atomistic scale calculations, like the values of γ_{us} on different fault planes. As discussed above, these quantities are relevant to macroscopic phenomenological theories of brittle versus ductile behavior.

In Si, the relevant fault plane is the {111} slip and cleavage plane. The crystal structure of Si, the diamond cubic lattice, allows for two distinct placements of the slip

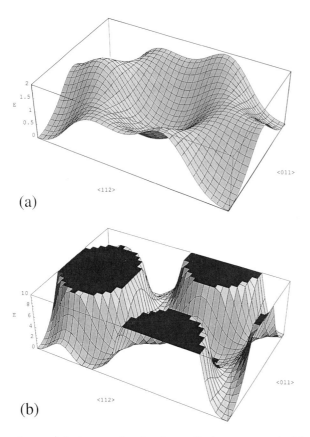

Fig. 2. Generalized stacking fault energy surface for a) the shuffle and b) the glide plane of Si

(a)

(b)

plane: a) between atomic planes that are separated by a distance equal to the nearest neighbor distance (that is, bisecting the Si–Si bond along the [111] crystallographic direction), b) between atomic planes that are separated by one third of the nearest neighbor distance. The two slip planes between a) widely spaced and b) narrowly spaced atomic layers are referred to as "shuffle" and "glide" planes in the context of dislocation motion [8].

Kaxiras and Duesbery [28] obtained the generalized stacking fault energy surface for both shuffle and glide planes using first-principles calculations in the DFT/LDA framework. The generalized stacking fault energy surfaces for these two planes are reproduced in Fig. 2a and b. We summarize their results for the unstable stacking fault energies in Table 1. While the values of γ_{us} are comparable in magnitude for the two slip

Table 1
Energy of unstable stacking faults, γ_{us} (in J/m^2), at various levels of relaxation, for the shuffle and glide planes in Si

	no relaxation	atomic relaxation at ideal volume	atomic relaxation and volume relaxation
a) shuffle	1.84	1.81	1.67
b) glide	2.51	2.02	1.91

planes, the energy scales are vastly different. It is clear from Table 1 that atomic relaxation affects the glide plane energetics more than the shuffle plane. This is not surprising because sliding on the glide and shuffle planes involves breaking of three and one bonds, respectively, per pair of atoms on either side of the slide plane. This indicates that at zero temperature the sessile mode will be favored. The same work attempted to generalize the concepts outlined in Section 2 to finite temperature based on the Vineyard [29] transition state theory for diffusion. With access to the entire generalized stacking fault (GSF) energy surface for slip, the problem can be mapped to that of a classical particle moving on a two-dimensional potential energy surface. This mapping involves folding all the degrees of freedom of the many-atom system to two collective degrees of freedom describing motion on the slip plane. The formulas for calculating entropy associated with the diffusion of a particle in transition state theory can then be used to obtain the entropy associated with the slip process, and from that the associated free energy at finite temperature. It was found that although the glide plane has a higher γ_{us} value, it also has higher entropy. Since the entropy term enters into the free energy with a negative sign, it is possible that under the proper thermodynamic conditions the free energy of the glide plane can become *lower* than that of the shuffle plane. Thus, as the temperature is increased beyond a critical point, a transition from sessile shuffle mode to glissile glide mode can occur. Kaxiras and Duesbery suggested that this change of preferred sliding mode could be related to the ductile to brittle transition in Si. The entire GSF energy surface can also provide useful information about the shape and Peierls stress associated with various dislocations [12, 30]. Juan and Kaxiras [30] calculated various GSF energy surfaces for silicon from first principles and obtained its dislocation properties.

While Rice's criterion for ductile versus brittle behavior addresses the issue of dislocation nucleation and motion, it does not include any information about the surfaces created due to dislocation emission at the crack tip. Juan, Sun, and Kaxiras (JSK) [31] introduced a correction to account for the surface traction effects, which they called the ledge surface contribution. Since the surface exposed near the tip of the crack upon nucleation of a dislocation is not equivalent to a free surface, including these effects in a criterion for brittle versus ductile behavior is quite demanding and can be done accurately only with first-principles calculations. For the case of Si, JSK were able to identify a geometry that allowed the calculation of the ledge surface contribution: they found that the energy associated with ledge surface creation is approximately 60% of the energy of the free surface. Assuming an evanescent force law, they incorporated the ledge effects into Rice's original theory within continuum mechanics, which showed that in silicon the increase in the necessary loading for dislocation emission due to surface traction can be up to 20%.

4.2 Molybdenum disilicide and its substitutional alloys

Experimental background: Molybdenum disilicide, $MoSi_2$, is a metalloceramic material with the strength of a ceramic and the toughness of a metal. It has a high melting point that makes it a promising candidate for high temperature turbine applications. The negative side to its use is that it becomes brittle below 1000 °C and loses its toughness. There are ongoing experimental efforts to disembrittle $MoSi_2$ at low temperatures. Several attempts to improve the low-temperature fracture toughness of $MoSi_2$ by forming composite materials have been made. The addition of ceramic particles or whiskers such as SiC [32] or ZrO_2 [33] inhibits fracture and increases the fracture toughness to a level achieved by incorporation of a ductile second phase. For example, Mo in $MoSi_2$ [34] increases the fracture toughness (up to ≈ 10 MPa $m^{1/2}$). This is believed to be due to the crack-slowing effects of large matrix-particle interfacial stresses. The addition of carbon [35] scavenges silica from the grain boundaries, changing the fracture mode from intergranular to transgranular and increasing fracture toughness to 12 MPa $m^{1/2}$. The addition of carbon-lubricated silicon carbide continuous fibers increases the fracture toughness [36] to 35 MPa $m^{1/2}$ for cracks normal to the fiber length. Unfortunately, the toughness for cracks running parallel to the fibers is unaffected. This process is costly and not easily extended to three dimensions, making it an unsuitable candidate for practical use. The effect of structural modification on the DBT in $MoSi_2$ has also been examined, but the conclusions are conflicting.

There has also been some experimental work, similarly inconclusive, on the effect of chemical alloying. There are many choices to be tried for alloying, both in terms of choice of chemical element and of the concentration. Theoretical analysis can complement experimental efforts and indicate which type of alloying may *intrinsically* improve the material. Even using theory to eliminate some of the possible options can be very helpful in making experimental efforts cost-effective. Here, we present our analysis of ductile versus brittle behavior of $MoSi_2$ based on first-principles calculations.

Crystal structure: $MoSi_2$ crystallizes in a body centered tetragonal structure as shown in Fig. 3a, formed by the alternate stacking of single Mo and double Si (001) layers. Both Mo and Si atoms are highly coordinated, with 10 nearest neighbors; as expected with this level of coordination, $MoSi_2$ has metallic character. It has been found experimentally that many slip systems in $MoSi_2$ remain active at low temperatures [37]. Brittle failure in single crystals occurs only when the stress axis is close to [001] and is

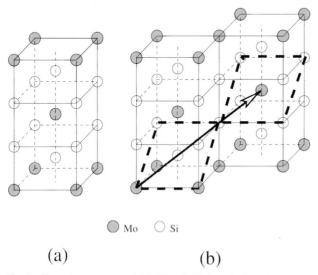

Fig. 3. Crystal structure of $MoSi_2$. a) Unit cell for the body centered $C11_b$ crystal; solid circles represent Mo atoms and open circles represent Si atoms. b) (013) plane and the Burgers vector for the slip system (013)[331]

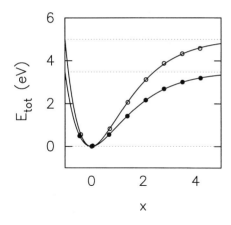

Fig. 4. First-principles energies (symbols) fitted to the universal equation of state (solid lines), for cleavage separating Mo–Si (open circles) and Si–Si (filled circles) planes of MoSi$_2$. x is the distance between the two adjacent atomic layers in the units of the length scale parameter λ (see text for details)

caused by increasing difficulty in the operation of the slip system $\{013\}\langle 331\rangle$, shown in Fig. 3b. This slip system involves motion of dislocations with $\frac{1}{2}[331]$ Burgers vector on (013) planes. To assess ductility trends, we calculate γ_s for (001) cleavage planes and γ_{us} for the $\{013\}\langle 331\rangle$ slip system from first principles.

Cleavage: In MoSi$_2$, there are two types of (001) cleavage planes: one separating adjacent layers of Si atoms and another separating adjacent layers of Mo and Si atoms. We use a supercell consisting of six atomic layers in the (001) direction (the conven-

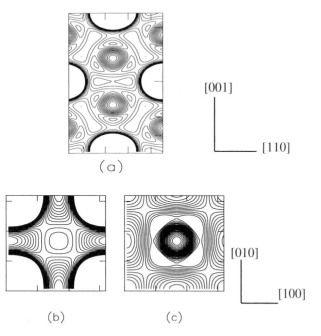

Fig. 5. Electronic charge densities of MoSi$_2$ in a) the (110) plane which contains both Si and Mo atoms, b) the (001) plane of Mo atoms, and c) the (001) plane of Si atoms. In a), Mo atoms are at the center of closely spaced large circular contours and Si atoms at the center of sparsely spaced elliptic contours. In b), Mo atoms are at the corner of the square unit cell and in c), Si atom is at the center of the square unit cell

tional tetragonal unit cell with 2 MoSi$_2$ formula units) in the calculation of the γ_s. In Fig. 4, we show total energies for the two (001) surfaces fitted to eq. (9). The asymptotic value of the energy curve in the limit of infinite interplanar separation, taking as the zero of the energy scale the minimum energy (bulk energy), gives the surface energy γ_s. Our results show that binding between (001) Si planes ($\gamma_s = 2.74$ J/m^2) is weaker than that between (001) Mo and Si planes ($\gamma_s = 3.94$ J/m^2). Since the surface relevant to brittle failure is that with the smaller γ_s, we shall focus on the energetics of cleavage between (001) silicon planes in the alloy calculations below.

To obtain a microscopic picture of bonding and better understanding of the above results, we examine the electronic charge densities on various planes of MoSi$_2$. In Fig. 5a, we show contour plots of the electronic charge density on a (110) plane, which contains all the bonds that are broken for cleavage on the (001) plane. Due to the contribution from core 4s and 4p electrons, there is large electronic density at Mo sites, indicated by circular contours which are very closely spaced. Clearly, there is directional bonding between nearest neighbor Mo and Si atoms which is evident from the triangular (anisotropic) regions, indicating local concentration of electrons that form the covalent part of the bond. In contrast, there is no significant covalent bonding between Si–Si and Mo–Mo atoms. This naturally results in lower cleavage energies for (001) planes that separate Si planes. We also display charge densities on (001) planes in Figs. 5b and c. For both Mo and Si (001) layers, the charge densities are nearly isotropic, which, in addition to the small variation in charge density on Si layers, indicates the metallic character of bonding.

Slip systems: The atomic arrangement on the (013) plane is shown in Fig. 6a with the two smallest Burgers vectors, $\mathbf{b}_0 = \frac{1}{2}[1\bar{3}1]$ and $\mathbf{b} = \frac{1}{2}[331]$. This plane is treated as a basal plane of a supercell used in the calculation of the stacking fault energy surface. We considered supercells with two and three (013) atomic planes to check convergence of the results with respect to supercell size, for a few points on the γ-surface. All the results for the γ-surface presented here have been obtained with a 3-layer supercell,

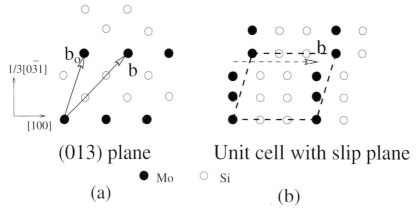

Fig. 6. a) Atomic arrangement in the (013) plane and the two smallest Burgers vectors. b) Schematic geometry of the 3-layer supercell (side view) used in the calculation of slip energies. The dashed arrow indicates displacement of the upper half relative to the lower half of the crystal along a Burgers vector. The thick dashed line indicates the boundary of the supercell corresponding to a finite relative slip of the two halves

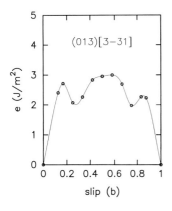
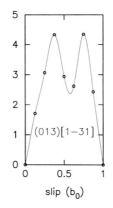

Fig. 7. Energy per unit area of slips on the (013) plane of MoSi$_2$ in (131) and (331) directions, corresponding to the two vectors shown in Fig. 6a. The lines through the points are cubic spline fits

shown in Fig. 6b. In Fig. 7 we show cross-sectional branches of the generalized stacking fault energy surface along the two Burgers vectors. We find $\gamma_{us} = 4.33$ J/m^2 for the slip system $\{013\}\langle 131\rangle$ and $\gamma_{us} = 2.99$ J/m^2 for the slip system $\{013\}\langle 331\rangle$, showing that the nucleation of dislocations on the latter system is energetically favored, despite the longer Burgers vector. The intermediate energy minima along these curves indicate possible stable stacking faults. The connection between these results and possible dislocation dissociation and anti-phase boundary faults in MoSi$_2$ is the subject of Ref. [38].

Substitutional alloys: So far we determined the properties of the cleavage and slip systems relevant to the ductile–brittle transition in MoSi$_2$ using a combination of first-principles calculations and experimental information for the relevant planes. In order to study the effects of substitutional alloying on ductility, we calculate the surface and unstable stacking fault energies for these planes using ordered supercells with a few of the Si or Mo atoms substituted by other elements. The goal here is to examine the effects of a specific alloying element on the bonding in MoSi$_2$ through changes it induces to the disembrittlement parameter D, while ignoring the effects of disorder in such alloys by necessity since quantum mechanical calculations can be performed only for ordered structures with relatively small unit cells.

First we study the effects of Al substitution for Si in detail. We replace a single (001) plane of silicon atoms with aluminium and calculate the surface energy of various (001)

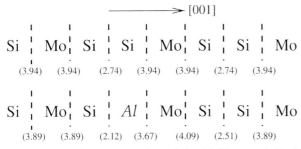

Fig. 8. Schematic representation of the MoSi$_2$ crystal. Each element symbol represents an entire (001) plane of atoms of this type, while the dashed lines represent cleavage planes. The numbers in parentheses are the calculated surface energies (γ_s) for cleavage on those planes. The representation and numbers at the bottom are for a crystal with an entire Si plane near the middle of the slab substituted by Al

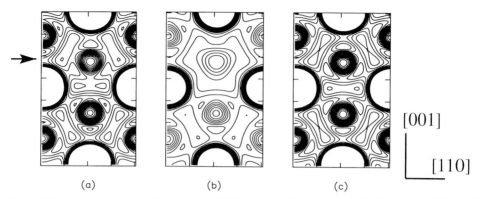

Fig. 9. Electronic charge densities on the (110) plane of Al-substituted $MoSi_2$ with a) 50% and b) 100% planar substitution (Al atoms are substituted in the third atomic layer from the top indicated by an arrow). For comparison, we display the charge density of monolithic $MoSi_2$ in c)

planes, shown schematically in Fig. 8. We find that γ_s for most of the (001) cleavage planes is reduced by Al substitution. The effects on the γ_s for cleavage at Si–Si planes are much stronger than those for cleavage at Mo–Si planes. These effects are strongly localized, decaying rapidly with the distance away from the plane of Al substitution. To obtain a microscopic picture, we examine the electronic charge densities for Al substitution (at 50% and 100% of a plane) which are shown in Fig. 9. This amount of Al substitution corresponds to 16% and 33% substitution by volume or 4.4% and 8.9% substitution by weight, respectively. Comparison to the ideal $MoSi_2$ crystal, included in Fig. 9c, reveals that the changes in charge densities are localized near the plane of Al substitution. These effects become stronger with concentration of planar substitution. The covalent bonding of Si atoms in the plane of substitution with neighboring atoms gets substantially reduced, hence the value of γ_s for cleavage near these planes is more severely affected than for other planes.

The unstable stacking fault energies for Al-substituted $MoSi_2$ are obtained by replacing half or all of the Si atoms in the operative (013) slip plane. With 100% planar substitution (33% by volume) of aluminum for silicon, γ_{us} is reduced by 29%. These calculations did not include atomic relaxation. To estimate the effects of relaxation, we calculated the energy of a configuration corresponding to the highest energy barrier,

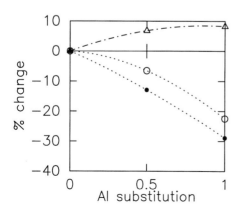

Fig. 10. Percentage change in γ_s (open circles), γ_{us} (closed circles), and D (open triangles) as a function of planar concentration of Al substitution. The lines connecting these points are guides to the eye

Table 2

Effects on γ_s and γ_{us} of full planar substitution of Si by Mg, Al, Ge, and P. D is the disembrittlement parameter and D_0 its value for the ideal crystal ($D_0 = 0.92$)

element	γ_s (J/m^2)	γ_{us} (J/m^2)	$(D - D_0)/D_0$
Si	2.74	2.99	0.00
Mg	1.61	1.45	0.21
Al	2.12	2.13	0.08
Ge	2.32	3.05	−0.17
P	1.78	2.58	−0.26

and find that while the atomic relaxation reduces γ_{us} by 12 to 20%, the percentage reduction in γ_{us} (relative to its value in the ideal crystal) due to Al-substitution remains essentially unchanged. We have also studied how the effects of Al-substitution on γ_s and γ_{us} change with the concentration. We calculated these quantities for 0% (pure MoSi$_2$), 50%, and 100% planar concentrations of Al; the results are shown in Fig. 10. The reduction in γ_{us} (shown as filled circles) varies nearly linearly with Al concentration. The percentage reduction in γ_{us} is larger than that in γ_s for all the concentrations considered and particularly for small concentrations; consequently, the disembrittlement parameter D is enhanced as shown in Fig. 10.

We conducted a more limited investigation into the effects of substitution of Mg, Ge and P for silicon. The results for γ_s and γ_{us} with full planar substitution are summarized in Table 2. It was found that both γ_s and γ_{us} decrease for substitution by Mg and P. For Ge, γ_s decreases but γ_{us} increases slightly. Overall, the disembrittlement parameter D decreases with the number of valence electrons in the substituted element, indicating that alloying with acceptor elements should be better for the enhancement of ductility than alloying with isoelectronic or donor elements. Additional results (Fig. 11) for γ_{us} for substitution with these elements at 50% substitution reveal that, in general, γ_{us} varies to a good approximation linearly with concentration.

Finally we explored the effects of substituting V, Nb, Tc, and Re for Mo. The results for γ_s and γ_{us} are presented in Table 3. As we have shown above, brittle failure breaks Si–Si bonds; therefore substitution for Mo has little effect on γ_s. On the other hand, strong Mo–Si bonds are broken during dislocation nucleation, and hence any substitu-

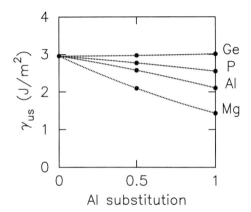

Fig. 11. γ_{us} as a function of planar concentration of Mg, Al, Ge, and P substitution for Si. The lines connecting these points are guides to the eye

Table 3
Effects on γ_s and γ_{us} of full planar substitution for Mo with V, Nb, Tc, and Re

element	γ_s (J/m^2)	γ_{us} (J/m^2)	$(D - D_0)/D_0$
Mo	2.74	2.99	0.00
V	2.49	2.30	0.18
Nb	2.60	2.42	0.18
Tc	2.35	2.34	0.10
Re	2.25	2.55	−0.04

tion for Mo which weakens these bonds should and does lead to a reduction in γ_{us} comparable to that found in the substitution for Si. The disembrittlement parameter D increases with V, Nb, and Tc substitutions, although, as for Si substitution, the effect is weaker for the alloying elements with a higher valence charge. D decreases slightly with substitution of Re for Mo. The effects of V and Nb substitution on D are very similar, because of their same valence.

In Fig. 12, we display charge densities on (110) planes for the cases of 50% substitution of Mo by V and Tc, and compare them to the ideal MoSi$_2$ crystal charge density. In contrast with the substitutions for Si, the effects on charge density seem more extended here. The covalent nature of Mo–Si bonds persists, but is somewhat weakened. This is consistent with the small changes in γ_s we find with these substitutions. The effect of V substitution for Mo on the bonds with nearest neighbor Si atoms is found to be stronger than that of Tc substitution for Mo.

5. Discussion and Conclusions

The interpretation of macroscopic mechanical properties such as ductility and brittleness in terms of fundamental cohesive properties is a complex and difficult task. The present work uses a highly simplified treatment, which borrows heavily from fracture mechanics and dislocation theory, to make the connection between the macroscopic properties and the microscopic structure and composition; first-principles quantum mechanical calculations are used to obtain reliable values for the crucial quantities that enter in the fracture mechanics analysis.

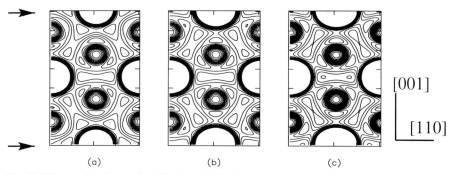

Fig. 12. Electronic charge densities in the (110) plane of a) V-substituted and b) Tc-substituted MoSi$_2$ (V and Tc atoms are substituted for Mo atoms in the top and the bottom layers indicated by arrows). For comparison, we display the charge density of monolithic MoSi$_2$ in c)

There are many approximations and simplifications inherent in this approach. For example, in our analysis of the effects of micro-alloying on the ductility of $MoSi_2$, we assume that the crystal structure of the alloys is the same as that of the matrix material; we have also assumed that the alloying element is readily soluble in the host material. In a similar vein, the sheer numerical complexity of the quantum mechanical calculations, even for this simple model, has forced us to use small supercells which correspond to large concentrations and ordered lattice sites for the alloying elements, even though our stated intention was to look at the *microalloy* regime. We make no attempt to justify these approximations, but rationalize them with the above-mentioned philosophy of providing the most useful and relevant information within the constraints of a real problem. Our results show trends for ductility which can be correlated with electronic structure, and which we expect to provide a cost-effective guide to experiment. We would especially like to emphasize that we were not aware of the recent experimental work [39, 40] confirming our calculated trends with aluminum substitution until after the present calculations were completed. This experimental confirmation is particularly gratifying in light of the drastic approximations employed. We wish to suggest that tests of substitutional microalloys which are predicted to enhance brittleness of $MoSi_2$ (such as Ge or P substituting for Si or Re substituting for Mo) would be particularly useful in providing additional confirmation for the theoretical approach employed here.

In conclusion, through the use of a simple model for ductile versus brittle response, first-principles calculations can be a valuable tool for obtaining microscopic insights into the mechanical behavior of solids. We illustrated this point with an example application to a prototypical brittle solid, silicon. In addition, we illustrated how this approach can be used to provide a guide for experimental efforts to disembrittle a material, molybdenum disilicide, which is of interest for real applications. With ongoing advances in computational resources and algorithms for materials modeling, we expect that the applicability of first-principles methods to modeling and design of materials will become broader.

Acknowledgements This work was performed with support from the Office of Naval Research under SBIR contract no. N00014-97-C-0104. The authors wish to acknowledge significant collaborations with V. V. Bulatov and J. R. Rice, and useful discussions with J. J. Petrovic (Los Alamos National Laboratory) and R. J. Hecht (Pratt & Whitney).

References

[1] C. E. INGLIS, Trans. Inst. Naval Arch. **55**, 219 (1913).
[2] P. HOHENBERG and W. KOHN, Phys. Rev. **136**, B864 (1964).
  W. KOHN and L. J. SHAM, Phys. Rev. **140**, A1133 (1965).
[3] V. VITEK, Phil. Mag. **18**, 773 (1968).
[4] M. S. DUESBERY and V. VITEK, Acta Mater. **46**, 1481 (1998).
[5] J. R. RICE, J. Mech. Phys. Solids **40**, 239 (1992).
[6] R. PEIERLS, Proc. Phys. Soc. London **52**, 34 (1940).
[7] F. R. N. NABARRO, Proc. Phys. Soc. London **59**, 256 (1947).
[8] J. P. HIRTH and J. LOTHE, Theory of Dislocations, Wiley, New York 1982.
[9] V. VITEK and M. YAMAGUCHI, J. Phys. F (Met. Phys.) **3**, 537 (1973).
[10] B. JOOS, Q. REN, and M. S. DUESBERY, Phys. Rev. B **50**, 5890 (1994).
[11] V. BULATOV and E. KAXIRAS, Phys. Rev. Lett. **78**, 4221 (1997).

[12] B. Joos and M. S. Duesbery, Phys. Rev. Lett. **78**, 226 (1997).
[13] A. A. Griffith, Phil. Trans. Roy. Soc. (London) **A221**, 163 (1920).
[14] A. Kelly, W. R. Tyson, and A. H. Cottrell, Phil. Mag. **15**, 567 (1967).
[15] J. Rice and R. Thomson, Phil. Mag. **29**, 73 (1974).
[16] J. R. Rice, in: Topics in Fracture and Fatigue, Ed. A. S. Argon, Springer-Verlag, Berlin 1992; J. Mech. Phys. Solids, **40**, 239 (1992).
J. R. Rice and G. Beltz, J. Mech. Phys. Solids, **42**, 333 (1994).
[17] P. B. Hirsch and S.G. Roberts, Phil. Mag. A **64**, 55 (1991).
[18] M. Khantha, D. P. Pope, and V. Vitek, Phys. Rev. Lett. **73**, 684 (1994).
[19] J. M. Kosterlitz and D.J. Thouless, J. Phys. C (Solid State Phys.) **6**, 1181 (1973).
[20] phys. stat. sol. (b) **217**, No. 1 (2000).
[21] D. M. Ceperly and B. J. Alder, Phys. Rev. Lett. **45**, 566 (1980).
J. P. Perdew and A. Zunger, Phys. Rev. B **23**, 5048 (1981).
[22] J. C. Phillips, Phys. Rev. **112**, 685 (1958).
V. Heine and M. L. Cohen, Solid State Phys. **24** (1970).
[23] A. M. Rappe, K. M. Rabe, E. Kaxiras, and J. D. Joannopoulos, Phys. Rev. B **41**, 1227 (1990).
[24] H. J. Monkhorst and J. D. Pack, Phys. Rev. B **13**, 5188 (1976).
[25] M. P. Teter, M. C. Payne, and D. C. Allan, Phys. Rev. B **40,** 12255 (1989).
[26] G. Kreese and J. Furthmuller, Comput. Mater. Sci. **6**, 15 (1996).
[27] J. H. Rose, J. R. Smith, F. Guinea, and J. Ferrante, Phys. Rev. B **29**, 2963 (1984).
[28] E. Kaxiras and M. S. Duesbery, Phys. Rev. Lett. **70**, 3752 (1993).
[29] G. H. Vineyard, J. Phys. Chem. Solids **3**, 121 (1957).
[30] Y. Juan and E. Kaxiras, Phil. Mag. A **74**, 1367 (1996).
[31] Y.-M. Juan, Y. Sun, and E. Kaxiras, Phil. Mag. Lett. **73**, 233 (1996).
[32] A. K. Bhattacharya and J. J. Petrovic, J. Amer. Ceram. Soc. **74**, 1045 (1992).
[33] J. J. Petrovic, R. E. Honnell, T. E. Mitchell, T. E. Wade, and K. J. McClellan, Ceram. Eng. Sci. Proc. **12**, 1633 (1991).
[34] T. C. Lu, A. G. Evans, R. J. Hecht, and R. Mehrabian, Acta Metall. et Mater. **39**, 1853 (1991).
[35] S. Maloy, A. H. Heuer, J. J. Lewandowski, and J. J. Petrovic, J. Amer. Ceram. Soc. **74**, 2704 (1991).
[36] S. Rawal, Martin Marietta Astronautics, ONR Workshop on MoSi2, Hyannis, June 1995.
[37] K. Ito, H. Inui, Y. Shirai, and M. Yamaguchi, Phil. Mag. A **72**, 1075 (1995).
[38] U. V. Waghmare, V. Bulatov, E. Kaxiras, and M. S. Duesbery, Phil. Mag. A **79**, 655 (1999).
[39] P. Peralta, S. A. Maloy, F. Chu, J. J. Petrovic, and T. E. Mitchell, Scripta Mater. **37**, 1599 (1997).
[40] P. Peralta, F. Chu, S. A. Maloy, P. Santiago, J. J. Petrovic, and T. E. Mitchell, Proc. AFOSR-Sponsored Internat. Conf. Computer-Aided Design of High-Temp. Mater., 30 July to 2 August 1997, Santa Fe (New Mexico).

phys. stat. sol. (b) **217**, 565 (2000)

Subject classification: 71.15.Mb; 71.20.Rv; S12

Comparison of Simulation Methods for Organic Molecular Systems: Porphyrin Stacks

F. Della Sala[1]) (a), J. Widany (a, b), and Th. Frauenheim (b)

(a) INFM and Department of Electronic Engineering,
University of Rome "Tor Vergata", Via di Tor Vergata 110, I-00133 Roma, Italy

(b) Fachbereich Physik, Theoretische Physik, Universität GH Paderborn,
D-33098 Paderborn, Germany

(Received August 10, 1999)

We present the results of density-functional theory based tight-binding (DFTB) and semi-empirical (PM3) Hartree-Fock (HF) calculations on the structural and electronic properties of porphyrin molecular stacks. The geometry of the isolated molecule as well as the molecular stack has been optimized using the DFTB approach. The results for the single molecule are in good agreement with *ab initio* calculations. The DFTB method predicts an alternating charge distribution among the carbon atoms in the porphyrin molecule, which is responsible for the formation of a tilted stacking structure. The electronic band structures obtained with both methods are in very good agreement. The results show a strong and non monotonic dependence of the band dispersion from the tilting angle. A transition from indirect to direct band gap is found, going from the face-to-face to the stable configuration at 25°.

1. Introduction

Organic molecular systems in crystalline, semi-crystalline or amorphous form have recently attracted a strong interest due to their possible use in electronic and opto-electronic devices. Organic Field Effect Transistors (O-FET) [1] and Organic Light Emitting Diodes (O-LED) [2] are presently manufactured and well working.

The current research is focused on finding organic molecular systems which show the best properties in terms of light emission efficiency and/or mobility of charge carriers. The properties of the single molecule do not change considerably in the molecular system. This is due to the weak nature of the inter-molecular forces (van der Waals and/or electrostatic). As a consequence, the characteristics of the molecular system can be controlled by tailoring the single molecules. Moreover, well ordered multi-layer structures can be produced [3] without problems due to strain and interface defects, which are the limiting factors in the quality of inorganic multi-layers.

Up to now there is a considerable gap between experimental progress and theoretical understanding. New organic materials have been synthesized by varying the substituents of target molecules. The success of this tailoring approach is measured by efficiency of the device. However, theoretical insight of the fundamental material properties is the basis for a rational optimization of the device properties. Therefore, a simulation tool is necessary to predict the electronic characteristics of the single molecule as well as the aggregate of molecules.

---
[1]) Phone: +39 06 7259 7366; Fax: +39 06 2020519; e-mail: fabiods@zeus.eln.uniroma2.it

Organic molecules exposing interesting properties for opto-electronic applications contain between 40 and 100 atoms. The computational costs for a first-principle treatment of such systems are too high. Therefore, the methods of choice are tight-binding or semi-empirical schemes. In this paper, we show the applicability of a density-functional based tight-binding (DFTB) and a semi-empirical (PM3) Hartree-Fock (HF) method. We compare the two theoretical approaches on a specific organic system, the porphyrin molecular stack.

2. Porphyrin Molecular Stack

Metallo-porphyrins (Fig. 1) and related tetrapyrrolic macrocycles (for example phthalocyanines) are very promising materials [5, 6] for opto-electronics due to their extended π-conjugate ring system. Moreover, they show a high stability and chemical flexibility. An organic P/N junction has been grown recently contacting two films of different types of porphyrins [7].

Almost all metals can be put in the center of the ring. In this way the properties of the metallo-porphyrins can be tuned with a high degree of freedom. Whereas the phthalocyanines crystallize in a fixed geometry (α and β form), the smaller porphyrins are not rigid: the geometry can change depending on the environment, the substituents or the presence and type of a metal.

A characteristic of these macrocycles is the spontaneous formation of tilted molecular stacks as shown in Fig. 2. The stacking properties are determined by π–π interactions [8] both in solution and in the solid state.

It is known that the exact stacking geometry of the aggregates strongly influences the optical [9], conductivity [10] and photo-conductivity [11] properties.

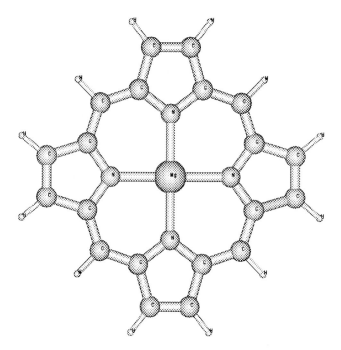

Fig. 1. Metallo-porphyrin

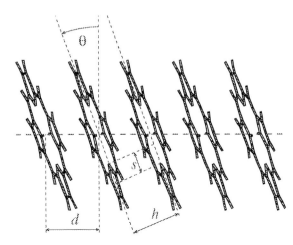

Fig. 2. Schematic view of the one dimensional stack: d is the lattice constant, h the inter-planar distance, s the slipping distance and θ the tilting angle

The presence of a divalent metal does not change the π-ring properties of the porphyrin molecular stacks. Therefore, the free base H_2-porphyrin can be used as a simple model system to study the π–π interaction contribution to the stacking properties. This is also a compelled choice, due to the actual incapability of the DFTB method in treating metals, whereas the PM3 is parameterized for transition metals too.

3. Methods

Density functional based methods allow a very accurate and theoretically well founded description of the electronic density for a wide variety of materials. Recent works show their applicability also to organic molecular crystals and polymers [12, 13]. In this study we use the DFTB method [14, 15] for the geometrical optimization of the single molecule as well as the molecular stack.

Conjugate gradient energy minimization is applied to determine the equilibrium geometries. The band structures of the molecular stacks are obtained in the usual way by setting up a Bloch sum over the pseudo-atom orbitals and calculating the Kohn-Sham eigenvalues for a series of wave vectors along the symmetry line in the Brillouin zone.

The semi-empirical (PM3) Hartree-Fock method [16] is a common approach to predict properties of organic molecules. Applying a finite cluster approach [17] the method can be used to evaluate the band structure of periodic systems. The interaction matrix elements of the whole periodic system derive from the simulation of a finite cluster. However, good results can be obtained only in a mono-dimensional system, as shown in calculations on polymers [18].

As far as the interaction between molecules in a stack is weak, the crystal orbitals $|\psi_{n,k}\rangle$ can be expressed by an LCMO (Linear Combination of Molecular Orbitals) expansion

$$|\psi_{n,k}\rangle = \frac{1}{\sqrt{M}} \sum_{l=-\infty}^{+\infty} e^{ikld} \sum_{\alpha} c_{n\alpha}(k) |\phi_\alpha(l)\rangle, \quad (1)$$

where M is a normalization constant and the sum in α is over the previously determined single molecule orbitals $\phi_\alpha(l)$, located on the lattice site l.

The required matrix elements can be approximated by

$$\langle \phi_\beta(0)| H |\phi_\alpha(l)\rangle \approx \langle \phi_\beta(0)| H_n |\phi_\alpha(l)\rangle, \quad (2)$$

where H_n is the Hartree-Fock operator for a finite cluster of n molecules. For $n = 5$ we have obtained well converged results.

The semi-empirical HF-SCF calculation can be handled for such cluster without high computational demands. Starting the SCF cycle with the electronic density of the isolated molecule the calculation converges very fast.

The band structure can be obtained by solving the eigenvalues problem

$$\sum_{\alpha} c_{n,\alpha}(k) \sum_{l} e^{ikld} H_{\beta,\alpha}(l) = \epsilon_n(k) c_{n,\beta}(k). \tag{3}$$

The sum in (3) can be restricted to $l \in \{0, +1, -1\}$ considering only the first neighbor molecule interaction.

Both methods described above use local minimal basis sets. Bernardini [19] studied the porphyrin stack using a plane wave basis set (25 Ry cut-off). The total energy calculations were performed in DFT-LDA using exchange and correlation potentials as parameterized by Perdew and Zunger [20]. Ion–electron interaction was described using ultra-soft pseudo-potentials [21]. However, the computational costs of this method are very high, a total energy calculation with the DFTB method is approximately 200 times faster.

4. Equilibrium Geometry

The crystallographic structure of porphyrin shows an essentially planar molecule [22]. The bonding situation is dominated by a delocalization of the π-electron density throughout the porphyrin macrocycle. This results in C–C and C–N bond distances which are intermediate between those expected for single and double bonds.

Ab-initio or semi-empirical HF methods cannot be used in general for geometry optimization of porphyrins due to the strong correlation effects. In fact, Foresman et al. [23] obtained a frozen resonance form with alternating single and double bonds at the spin-restricted HF SCF/3-21G level of theory. While restricted HF theory yields a bond alternating structure of D_{3h} symmetry for the annulene ring, inclusion of electron correlation at the MP2/6-31G level favors a delocalized D_{6h} structure in agreement with experimental evidence [24]. Overall, MP2 theory is likely to provide a meaningful description of the bond length alternation in porphyrin while HF-SCF and DFT approaches may predict to much or to little bond alternation, respectively.

We optimized the geometry of the porphyrin molecule with the DFTB method. Despite the use of a minimal basis set, the results are in very good agreement with all-electron *ab initio* MP2 calculations [24]. In Fig. 3 a comparison of the DFTB results with that from calculations at the MP2 level using the DZP2 basis set (Gaussian basis set including d-functions) is shown.

In a second step we optimized the geometry of the molecular stack. The geometry of the isolated molecule remains almost unchanged in the stack. This is due to the stability of porphyrin in the planar configuration and to the relative weakness of inter-molecular interactions. The total energy of the porphyrin molecular stack has been calculated varying both the tilting angle θ and the inter-planar distance h. In Fig. 4 we show the total energy as function of θ for three different inter-planar distances: $h = 3.40$ Å (dotted line), 3.45 Å (solid line), and 3.50 Å (dashed line).

The difference in the curves clearly indicates that the balance between electrostatic repulsion and attraction is the determining parameter for the stability of the stack. In

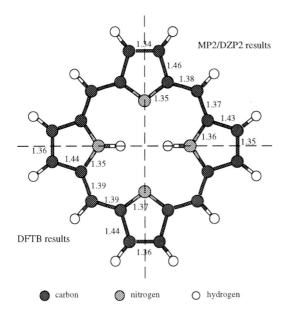

Fig. 3. Bond lengths for H_2-porphyrin: lower-left DFTB, upper-right MP2/DZP2 results from Ref. [24]

fact, by increasing the inter-planar distance both attraction as well as repulsion are reduced. The most stable configuration is found for an inter-planar distance of $h = 3.45$ Å. The importance of electrostatic effects for the stability of porphyrin stacks has already been discussed in literature [8].

The face-to-face stack is highly unstable for all distances. In the range $20° \leq \theta \leq 30°$ the molecular stack is stable with respect to the isolated molecule. The equilibrium geometry is found at $\theta = 25°$ for $h = 3.45$ Å. As shown in the inset of Fig. 4 the minimum of the total energy shifts slightly with the change of the inter-planar distance. This shift is such that the slipping distance $s = h \tan(\theta)$ is kept constant. The result compares well with experimental data [25].

A Mulliken charge analysis of the single molecule shows an inhomogeneous charge distribution, leading to alternating electron excess and electron deficiency on the carbon atoms, as shown in Fig. 5. The same charge alternation is found in the stack. By increasing the tilting angle the slipping distance s increases changing the alignments

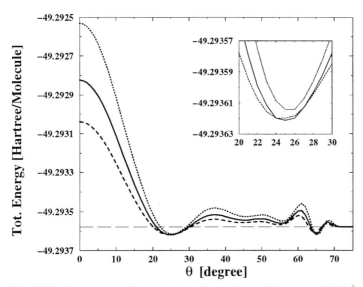

Fig. 4. Total energy vs. tilting angle, for inter-planar distances of 3.40 Å (dotted line), 3.45 Å (solid line), and 3.50 Å (dashed line). The long-dashed line indicates the free molecule energy

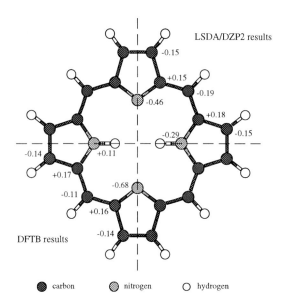

Fig. 5. Mulliken charges for H_2-porphyrin: lower-left DFTB, upper-right LSDA/DZP2 results

between atoms of the different molecules. The equilibrium configuration is characterized by a minimized inter-molecular distance between the carbon atoms with opposite charges. A similar stabilization mechanism has been reported for naphtalene/pyrimidine stackings, where the alignment of oppositely charged atoms lead to face-to-face or face-to-edge stackings, depending on the substituents [26].

We have performed LSDA/DZP2 calculations on the single molecule in order to compare the Mulliken charges with the DFTB results. The VWN potential has been applied for exchange-correlation [27]. There are only very small deviations and the charge alternation is fully retained, as shown in the upper-right part of Fig. 5.

In contrast, the PM3 method fails completely in obtaining this charge alternation. Inclusion of configuration interaction with single and double excitation does not lead to a better situation. Therefore, the PM3 method, being also not parameterized for estimating inter-molecular interactions, is not suitable to predict the correct stacking geometry.

A second minimum is observed at $\theta = 65°$. The stabilization derives from the same electrostatic interaction as described above, but fewer carbons are involved. For tilting angles $\theta \geq 70°$ we obviously obtain a very long slipping distance and therefore, the porphyrin molecules are not interacting. Naturally, the total energy of the stack converges to the free molecule energy (indicated by the long-dashed line in Fig. 4).

A rotation of the porphyrin molecules, as predicted in a previous study [28], is found to be energetic unfavorable.

The plane wave calculations by Bernardini also show the existence of a minimum in the total energy for the titled configuration. However, the exact determination of the absolute minimum is computationally too demanding.

5. Band Structure

The band structure of the molecular stack has been calculated using both approaches as outlined in the Methods section.

In Fig. 6 we report the results of a (PM3) Hartree-Fock LCMO calculation for an H_2-porphyrin stack in a face-to-face geometry ($\theta = 0°$). The DFTB band structure is in good agreement with the presented data. We show the first two valence and the first two conduction bands, other bands being well separated in energy.

The valence band structure is characterized by an anti-bonding interaction at Γ (with a higher energy) and a bonding interaction at X (with a lower energy). The conduction

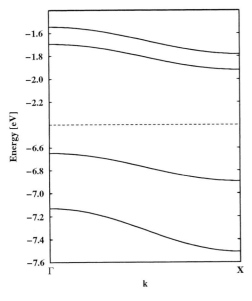

Fig. 6. (PM3) HF band structure for H_2-porphyrin with $\theta = 0°$

band shows a similar shape, leading to the formation of an indirect band gap. With increasing tilting angle we observe a transition from indirect to direct gap. For the minimum configuration we obtain the band structure represented in Fig. 7.

In this free-electron description the conduction band refers to the reduced state whereas the valence band refers to the oxidized state. The band gap gives the energy required for creating a dissociated electron–hole couple. It is not comparable with the absorption spectrum which is dominated by localized excitons on single molecules [9].

For the minimum geometry, the HF gap is 4.85 eV, whereas the DFTB calculation gives a value of 1.76 eV. It is well known that the free-electron band gap is overestimated by HF, due to the neglect of electron–electron correlation, and underestimated by DFT calculations due to the discontinuity of the V_{xc} potential (see for example [29]).

It is worth noting that quasi-particle calculations, which could give more reliable results, cannot be performed on such large systems. The main result of these many-body correction is to shift the position of the bands in energy while the wave functions remain unchanged [30]. Consequently, our results describe correctly the changes in the band structure between stacks with different tilting angle. Up to now there are no experimental data available for the free particle band gap.

In Fig. 8 we report the direct/indirect band gaps as function of the tilting angle. The results obtained by the two methods are very similar.

The structures with a direct band gap have a higher probability of electron–hole

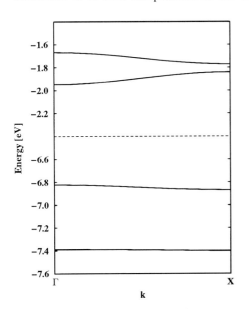

Fig. 7. (PM3) HF band structure for H_2-porphyrin with $\theta = 25°$

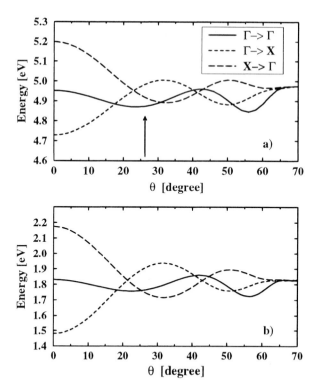

Fig. 8. Band gaps vs. tilting angle: a) (PM3) HF, b) DFTB results. The arrow indicates the stable geometry

recombination in comparison with indirect band gap stacks. Therefore, the electroluminescence and photo-current properties of stacks change with varying tilting angle. Similar behavior has been found experimentally [11].

For all tilting angles the band structures show a simple sinusoidal shape. This means that the crystal orbitals in (1) derive from the interaction of the same orbital on different molecules,

$$|\psi_{n,k}\rangle \approx \frac{1}{\sqrt{M}} \sum_{l} e^{ikld} |\phi_n(l)\rangle . \qquad (4)$$

Our approach is consistent with the excess electron and excess hole band description [31]. From expression (4) we see that the Wannier function for the stack can be very well approximated with isolated molecular orbitals.

The most significant bands originate directly from the Gouterman four orbitals [4] which determine the electronic and optical properties of the porphyrins. The symmetry of these π orbitals is: A_u(HOMO-1), B_{1u}(HOMO), B_{2g}(LUMO), B_{3g}(LUMO + 1). We note that the symmetry of these orbitals within the (PM3) HF approach coincides with the DFTB as well as the DFT ones [12].

Considering only this four orbitals the Hamilton matrix can be written as

$$\begin{bmatrix} E_1 + \beta_1(\theta)\cos(ka) & 0 & 0 & a_1(\theta)\cos(ka) \\ 0 & E_2 + \beta_2(\theta)\cos(ka) & a_2(\theta)\cos(ka) & 0 \\ 0 & a_2(\theta)\cos(ka) & E_3 + \beta_3(\theta)\cos(ka) & 0 \\ a_1(\theta)\cos(ka) & 0 & 0 & E_4 + \beta_4(\theta)\cos(ka) \end{bmatrix}, \qquad (5)$$

where the indices 1, 2, 3, 4 stand for B_{3g}, B_{2g}, B_{1u}, A_u, respectively.

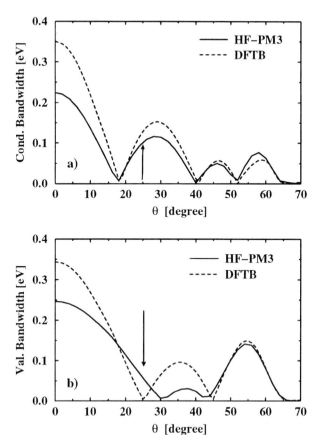

Fig. 9. Band widths vs. tilting angle: a) first conduction, b) first valence band. The arrows indicate the stable geometry

A number of matrix elements are zero due to the symmetry of the stack (C_{2h}, except $\theta = 0°$ with D_{2h}). The $a_i(\theta)$ mixing elements are very small (zero for $\theta = 0°$) compared to the β_i. Therefore, the matrix can be considered as diagonal, thus confirming the validity of (4).

Fig. 9. shows the dependence of the band width as function of the tilting angle. The results of the calculations with the two methods are in very good agreement.

In contrast to common expectations, the band widths do not decrease monotonously with the tilting angle due to the increasing center-to-center distance. They show an oscillating behavior originating from the different overlap of orbitals for different tilting angles.

Neglecting the unstable face-to-face geometry, the following features are worth noting: we find a maximum in the conduction band width for the neighbors of the equilibrium geometry, and also a correspondent minimum for the valence band width. Both band widths are strongly suppressed for $\theta \approx 40°$. The valence band shows a peak at $\theta \approx 55°$. For $\theta \geq 70°$ the slipping distance s is greater than the molecule dimension, so there exists no p_z orbital overlap.

We want to point out that in contrast to the DFTB method in a semi-empirical Hartree-Fock calculation the one-electron contribution to the off-diagonal Fock matrix element is proportional to the calculated overlap of Slater type atomic orbitals (the

Hückel approximation). These off-diagonal matrix elements influence strongly the band structure.

Finally, we want to remark that the presence of substituents and/or divalent metals does not change the HOMO and LUMO orbitals. While the minimum geometry can be influenced by substituents, the results reported in Fig. 9 are valid in general.

6. Summary

In this paper, we report on the structural and electronic properties of porphyrin stacks. The H_2-porphyrin has been chosen as a model system to study the nature of the $\pi-\pi$ interactions between the macrocycles.

The DFTB method determines the equilibrium geometry of the isolated molecule in good agreement with experimental and *ab initio* results. An alternating charge distribution is found among the carbon atoms. The alignment of atoms with opposite charges stabilizes molecular stacks with certain tilting angles. The stability of the molecular stack is determined by electrostatic interaction.

The two applied theoretical approaches lead to consistent results for the band structures. We observe a transition from an indirect to a direct band gap going from the face-to-face stacking to the minimum configuration at 25°. The band widths of the HOMO and LUMO exhibit an oscillating behavior with increasing tilting angle.

In summary, we have shown that the opto-electronic properties of porphyrin stacks are strongly influenced by the tilting angle, which is in turns determined by the charge distribution in the single molecules.

References

[1] Z. Bao, A. Lovinger, and A. Dodabalapur, Appl. Phys. Lett. **69**, 3066 (1996).
[2] Z. Shen, P.E. Burrows, V. Bulovic', S.R. Forrest, and M.A. Thompson, Science **276**, 2009 (1997).
[3] F.F. So and S.R. Forrest, Phys. Rev. Lett. **66**, 2649 (1991).
[4] M. Gouterman, J. Chem. Phys. **30**, 1139 (1959).
[5] K. Yamashita, Y. Harima, and T. Matsubayashi, J. Phys. Chem. **93**, 5311 (1989).
[6] C.J. Schramm, R.P. Scaringe, D.R. Stojakovic, B.M. Hoffman, J.A. Ibers, and T.J. Marks, J. Amer. Chem. Soc. **102**, 6702 (1980).
[7] T.J. Savenije, E. Moons, G.K. Boshloo, A. Goossens, and T.J. Schaafsma, Phys. Rev. B **55**, 9685 (1997).
[8] C.A. Hunter and J.K.M Sanders, J. Amer. Chem. Soc. **112**, 5525 (1990).
[9] G.A. Schick, I.C. Schreiman, R.W. Wagner, J.S. Lindsey, and D.F. Bocian, J. Amer. Chem. Soc. **111**, 1344 (1989).
[10] R. Jones, R.H. Tredgold, A. Hoorfar, and P. Hodge, Thin Solid Films **113**, 115 (1984).
[11] C.-Y. Liu, H. Tang, and A.J. Bard, J. Phys. Chem. **100**, 3587 (1996).
[12] D. Lamoen, P. Ballone, and M. Parrinello, Phys. Rev. B **54**, 5097 (1996).
[13] G. Brocks, Phys. Rev. B **55**, 6816 (1997).
[14] D. Porezag, Th. Frauenheim, Th. Köhler, G. Seifert, and R. Kaschner, Phys. Rev. B **51**, 12947 (1995).
[15] G. Seifert, D. Porezag, and Th. Frauenheim, Internat. J. Quantum Chem. **98**, 85 (1996).
[16] J.J.P. Stewart, J. Comput. Chem. **10**, 209 (1989); **10**, 221 (1989).
[17] P.G. Perkins and J.J.P. Stewart, J. Chem. Soc., Faraday Trans. II **76**, 520 (1979).
[18] A.J.W. Tol, J. Chem. Phys. **100**, 8463 (1994).
[19] F. Bernardini, private communication.
[20] J.P. Perdew and A. Zunger, Phys. Rev. B **23**, 5048 (1981).
[21] D. Vanderbilt, Phys. Rev. B **41**, 7892 (1990).

[22] B.M.L. CHEN and A. TULINSKY, J. Amer. Chem. Soc. **94**, 4151 (1972).
[23] J.B. FORESMAN, M. HEAD-GORDON, J.A. POPLE, and M.J. FRISCH, J. Phys. Chem. **96**, 135 (1992).
[24] J. ALMLÖF, T.H. FISCHER, P.G. GASSMAN, A. GHOSH, and M. HÄSER, J. Phys. Chem. **97**, 10964 (1993).
[25] W.R. SCHEIDT and Y.J. LEE, Structure and Bonding 64, Springer-Verlag, Berlin 1987.
[26] A.V. MUEHLDORF, D. VAN ENGEN, J.C. WARNER, and A.D. HAMILTON, J. Amer. Chem. Soc. **110**, 6561 (1988).
[27] S.H. VOSKO, L. WILK, and M. NUSAIR, Canad. J. Phys. **58**, 1200 (1980).
[28] T.G. GANTCHEV, F. BEAUDRY, J.E. VAN LIER, and A.G. MICHEL, Internat. J. Quantum Chem. **46**, 191 (1993).
[29] R.W. GODBY, M. SCHLÜTER, and L.J. SHAM, Phys. Rev. Lett. **56**, 2415 (1986).
[30] M.S. HYBERTSEN and S.G. LOUIE, Phys. Rev. B **34**, 5390 (1986).
[31] O.H. LEBLANC, JR., J. CHEM. PHYS. **35**, 1275 (1961).

phys. stat. sol. (b) **217**, 577 (2000)

Subject classification: 61.50.Ah; 71.15.Fv; 71.20.Tx; S10.1

Quantum Mechanical Investigations on the Insertion Compounds of Early Transition Metal Oxides

F. Corà and C.R.A. Catlow

Davy-Faraday Research Laboratory, The Royal Institution of Great Britain, 21 Albemarle Street, London W1X 4BS, U.K.
e-mail: furio@ri.ac.uk

(Received August 10, 1999)

We present the results of electronic structure ab initio Hartree-Fock calculations performed on the perovskite, hexagonal, pyrochlore and layered polymorphic structures of MoO_3 and WO_3. We also examine their sodium and potassium insertion compounds, i.e. the molybdenum and tungsten bronzes; the materials examined are at the basis of applications in electrochromic devices and rechargeable batteries. Results of the calculations allow to gain insight into the processes occurring at the atomic level during the insertion process: we examine the relative stability of the host MoO_3 and WO_3 frameworks under addition of electrons to the conduction band, and extraframework ions in the empty interstices of the host lattice, which occur upon electrochemical insertion. The extraframework ions show a templating behaviour for the structure of the host transition metal oxide framework, reminiscent of zeolite synthesis. We exploit the knowledge gained in the latter field of research to examine the feasibility of the synthesis of novel microporous transition metal oxide lattices with composition MoO_3 and WO_3.

1. Introduction

Transition Metal Oxides (TMOs) are the object of growing attention in current science for the potentially unique applications enabled by their physical and chemical properties. While some features of TMOs, such as the colossal magneto-resistance effect of doped $LaMnO_3$ [1], are linked to the electronic configuration of the transition metal ions in the solid, and are specific to a particular material, other properties are found across a wide range of structures and compositions. The ability of TMOs to form stable insertion compounds belongs to the second category, and is displayed by several oxides in which the transition metal ions have a high formal oxidation state, such as TiO_2, MnO_2 or WO_3. The latter property is linked to a common feature in the crystalline structure of the cited oxides: the high oxygen/metal (O/M) ratio, in fact, leaves empty interstices in the oxygen sublattice, which behave as a solid solvent towards the accommodation of external species into the solid framework. Species suitable for insertion are the alkali metals or small organic ammonium ions.

The insertion process is exploited in two technologically important applications of TMOs, namely electrochromic devices [2, 3] and rechargeable batteries [4, 5]; both are based on a simple electrochemical cell design, in which the TMO component forms one of the electrodes, and the insertion of extra species is regulated by the flow of an external electric current. The reaction occurring at the TMO electrode (let us call it MO_n)

can be written as

$$MO_n + x(A^+ + e^-) \rightleftharpoons A_xMO_n. \tag{1}$$

In the electrochemical cell, the species A is inserted in cationic form, balanced by a stoichiometric injection of electrons into the MO_n host via the external circuit, to ensure the charge neutrality of the electrode.

Electrochromic devices exploit the change of properties of the TMO which occurs upon insertion: they employ early TMOs, in which the transition metal ions before insertion have electronic configuration $d^{(0)}$ and are transparent. The introduction of charge carriers upon insertion, gives rise to an intense light absorption in the visible region, which modifies the sample colouration; this change is exploited in displays, smart windows and car manufacturing. The most widely used electrochromic oxides are MoO_3 and WO_3.

In the rechargeable batteries, the electrochemical insertion is instead employed to store energy, under chemical form, to be subsequently released in portable applications, such as mobile phones, portable computers or electric cars. TMOs researched for application in rechargeable batteries include both early TMOs, such as V_2O_5 or V_6O_{13}, and late TMOs such as MnO_2 and CoO_2, in which the transition metal ions have partially filled d-shells.

Electrochromic devices and rechargeable batteries, therefore, exploit the same process at the atomic level, related to the reversible insertion and removal, under applied electric currents, of external ions into the host TMO electrode. Not surprisingly, they have a common set of requirements, among which the structural stability of the TMO host under continuous cycling of charge and discharge, to guarantee long life-time of the devices, and a high mobility of the inserted ions inside the host, to reduce energy losses due to parasitic side reactions. At a fundamental level, electrochromic materials and electrodes for rechargeable batteries can therefore be examined together.

Among the materials researched for the cited applications, early TMOs are common to both electrochromic devices and rechargeable batteries; in the paper we shall limit our attention to the case in which the transition metal ions in the host material have electronic configuration $d^{(0)}$. We examine explicitely the molybdenum and tungsten trioxides; the qualitative features of the insertion process are, however, the same in the whole class of early TMOs.

The insertion compounds of MoO_3 and WO_3 have metallic conductivity and appearance, and are also known as *bronzes*; in the following of the paper, we shall refer to them as composed of an MO_3 *framework*, in whose interstices are located the inserted, or *extraframework*, cations.

Gaining an appropriate understanding of the atomic-level processes occurring in the TMO host during the insertion process, is of fundamental importance to design new materials with improved performance, and is a field suited for modelling work. As clear from the reaction (1), the insertion process has two components: the addition of electrons, which populate the energy levels at the bottom of the conduction band in the TMO framework, and the insertion of extraframework cations in the empty interstices of the TMO host. In the paper we shall examine separately their effect, by means of quantum mechanical, ab initio Hartree-Fock calculations.

The remainder of the paper is organised as follows: in Section 2 we give the details of the calculations performed; the structural features of known polymorphs of Mo and W trioxides and bronzes are described in Section 3. The effect of the insertion process on the TMO host is dealt with in Section 4. In Section 5 we design new framework structures with composition MO_3, and calculate their relative stability with respect to the existing polymorphs of MoO_3 and WO_3; we also examine the possibility of the synthesis for the newly designed polymorphs. In Section 6, finally, we summarise the major conclusions of our work.

2. Computational Details

The computational model that we have employed to study the Mo and W trioxides and bronzes is based on periodic boundary conditions, at the ab initio Hartree-Fock (HF) level of approximation, as implemented in the computer program CRYSTAL [6, 7]. The working structure of the code is described in more detail in reference [8] of the present volume.

The wave-function of the solid is described in terms of crystalline orbitals; these are obtained as linear combination of localised functions, or Atomic Orbitals (AOs), associated with the constituent atoms. The basis sets employed to describe the Mo and W ions are composed of an effective core pseudopotential of the Hay-Wadt type [9], coupled with a split-valence set of basis functions for the valence electrons. Their derivation and performance in describing the stable polymorphs of the binary oxides can be found in references [10] and [11]. The oxygen basis set is a 8-51G, optimised for MgO [12], and already used in several other oxides [13, 14]; finally, the basis sets associated with the alkali ions are derived from previous studies on their oxides [15]. Basis sets of similar quality have proved adequate to describe most crystalline materials [7]; the presence of unoccupied AOs on both the anionic and cationic sites gives flexibility to the basis, and allows the polarizability of both species in their crystalline environment, and the back-donation of electrons from the oxygens to the metal to be taken into account.

Numerical approximations need to be introduced in the implementation of the self-consistent-field equations for infinite systems. In the CRYSTAL code, the accuracy in the evaluation of the infinite Coulomb and exchange series is controlled by a set of "cut-off" tolerances (see reference [7] for details); in the present case we have chosen a set of very "hard" computational tolerances (8, 7, 8, 8, 16), to reduce the effect of numerical inaccuracies to a minimum.

Reciprocal space integrals are performed as a weighted sum at a discrete set of k-points. The k-points employed sample uniformly the first Brillouin zone, and include all the high symmetry positions of the reciprocal lattice; weights are attributed according to the Pack-Monkhorst scheme [16]. The number of k-points is chosen in such a way that the results of the numerical integration are well converged; in the calculations reported in the following sections, we have used an $8 \times 8 \times 8$ sampling for the insulating oxides in the perovskite structure, increased to $12 \times 12 \times 12$ for the metallic bronzes. The k-space sampling in the other polymorphs has been scaled to take into account the different cell dimensions.

Geometry optimisations have been performed with a numerical conjugated-gradient procedure, stopped when the energy difference at subsequent cycles had fallen below 10^{-6} Hartrees.

3. Known Polymorphs of the Mo and W Trioxides and Bronzes

The Mo and W trioxides and bronzes have a complex structural chemistry; among their known structural types, four are of direct relevance to investigate the insertion process at the basis of electrochromic devices and rechargeable batteries; these are the perovskite, layered, hexagonal and pyrochlore structures, shown in Fig. 1. In the following of the discussion, we refer to the four polymorphs as PV (perovskite), L (layered), Hex (hexagonal) and PY (pyrochlore), respectively. The M–O framework in all four polymorphs has stoichiometry MO_3.

The thermodynamically stable phase of WO_3 at ambient conditions is an antiferroelectric distorted PV structure [17 to 22], which is metastable in MoO_3 [23]. In contrast, the stable phase of MoO_3 is the L polymorph, or α-MoO_3 [24 to 26], not displayed by WO_3. A metastable hexagonal form has been synthesized for both WO_3 [27 to 29] and MoO_3 [30 to 32] while a polymorph isostructural to the octahedral backbone of the mineral pyrochlore, $Ca_2Nb_2O_7$, has recently been observed for WO_3 [33 to 35], but not for MoO_3.

The relative order of stability of the four polymorphs in the bronzes is different. If the insertion process is obtained electrochemically, the structure of the host MO_3 frame-

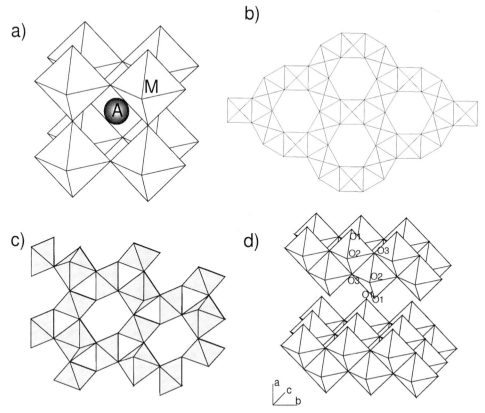

Fig. 1. Connectivity of the MO_6 octahedra in the MO_3 framework of Mo and W trioxides and bronzes: a) PV, b) Hex, c) PY and d) L polymorphs

work is set before the insertion; if instead the extraframework ions are present during the synthesis of the host framework, the type and quantity of extraframework ions influences the structure obtained for the MO_3 framework. In the W-bronzes family, the PV polymorph is stable for small inserted atoms (H, Li, Na) [36 to 38], while larger atoms or groups of atoms (K, Cs, Rb, In, NH_4) favour the Hex structure [39 to 41]. Inserted atoms which are even larger (Tl, Cs, Rb) stabilise the PY structure [33 to 35], depending also on the synthesis conditions [42 to 44]. Examining the structural stability in the MoO_3 family, we see instead that MoO_3 is layered, Na_xMoO_3 is a cubic perovskite [45], while K_xMoO_3 and the bronzes of larger inserted atoms are stable in different layered forms (see [46] and references therein). A high pressure phase of composition $K_{0.92}MoO_3$ has also been reported as cubic PV [47].

To understand the effect of the extraframework ions on the phase stability, it is instructive to examine the structure of the four polymorphs, and the characteristics of their interstices.

The MO_3 framework of PV, Hex and PY structures is based on corner-sharing only of MO_6 octahedra. In PV materials, octahedra form four-membered rings in each of the crystallographic planes; interstices are located in the dodecahedral sites (A) of the O sublattice, in 1:1 ratio with the framework M ions. The Hex structure is instead composed of an alternating of three- and six-membered rings of octahedra in the *ab* plane; this arrangement is repeated along *c*, leaving a set of parallel triangular and hexagonal channels, of which only the hexagonal is large enough to host inserted species. One hexagonal interstice exists in the structure every three MO_3 formula units. The size of the channels is modulated by the stacking of planes along *c*: the minimum diameter is located in the plane at fractional coordinate $z = 0$, which contains the M ions and four of their nearest O ions (equatorial oxygens); the plane at $z = 0.5$ contains only the axial oxygens of the MO_6 octahedra. We refer to the plane at $z = 0$ as to the hexagonal window of the channels.

In the PY structure, six-membered rings of octahedra are oriented along the $\langle 111 \rangle$ directions of the cubic unit cell, creating a three-dimensional network of interconnecting hexagonal channels. The sites where channels intersect are in 1:2 ratio with the framework M ions. Finally, in the L polymorph the MO_6 octahedra connect corner- and edge-sharing in the *bc* crystallographic plane, leaving a layered structure in the perpendicular direction, *a*. Interstices are available in the interlayer region.

It is important to recognise that the four polymorphs have a different number of interstitial sites, and of different dimensions. In the three corner-sharing arrangements (PV, Hex and PY), for instance, the size of the vacant interstices increases in the order of PV < Hex < PY. This is obviously an important feature of the host TMO framework in insertion processes, and has to be taken properly into account to understand the structural stability of the TMO electrodes in the technologic applications examined.

In the following sections, we shall investigate with computer modelling the relative stability and electronic properties of the four framework structures described above, for the pure MO_3 materials and for their Na and K bronzes. The focus of the study will be to understand the factors that stabilise the different polymorphic structures; the choice of Na and K as inserted ions is due to the fact that the host MO_3 framework has different stable structures in the two cases (PV or Hex in the W-bronzes; PV or layered in the Mo oxides), so that comparing the interaction of guest Na and K ions with the host TMO framework is of particular interest.

4. The Insertion Process

In this section, we apply computational modelling to examine the insertion process described in equation (1). The effect of the added electrons and extraframework ions is investigated separately in the following subsections.

4.1 Conduction band electrons

To investigate the effect of a different count of electrons on the properties of the TMO framework, we examine the electronic structure and M–O bonding properties of the PV and L polymorphs. The study of each polymorph is partitioned in two steps: first, we construct an idealised structure, formed by regular MO_6 octahedra, with the M ion on-centre; we refer to this phase as pseudo-cubic. Next, we examine ideal distortions from the pseudo-cubic phase, involving the off-centering of the M ions in their octahedra, to investigate the driving force for structural distortions. The effect of the inserted ions in the bronzes is limited here to the modification of the number of valence electrons in the materials.

In Fig. 2, we report the HF band structure of PV-WO_3, for increasing ferroelectric displacements (in fractional coordinates) of W towards one of its nearest neighbour

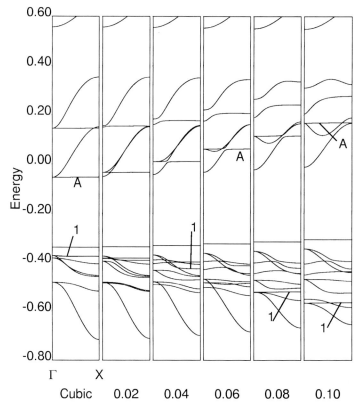

Fig. 2. HF band structure of PV-WO_3, for increasing ferroelectric displacements of the W ions (in fractional coordinates) towards one of their nearest neighbour oxygens

oxygens. The qualitative features of the band structure described for WO_3 are the same in all the other PV-structured TMOs [48]. In the cubic phase, the solution in the Γ point of reciprocal space consists of pure M-d and O-2p AOs; covalent mixing of the AOs on metal and oxygen is symmetry forbidden. The solution along the Γ–X direction of reciprocal space consists instead of π bonding (in the valence band) and antibonding (in the conduction band) combinations of the M-d and O-2p orbitals. For symmetry constraints, the two energy levels labelled "A" and "1" in Fig. 2, remain non k-dependent and M–O non-bonding. When the W ion is displaced off-centre, covalent π interactions between the metal ion and its closest oxygen are no longer symmetry forbidden; the levels A and 1 hybridise in a π M–O bonding level (1, in the valence band) and a π M–O antibonding level (A, in the conduction band).

When the transition metal ion has electronic configuration $d^{(0)}$, as in the binary oxides MoO_3 and WO_3, only the M–O bonding level in the valence band is populated, and the distortion is energetically stable; when the electronic configuration of M is greater than $d^{(0)}$, as in the bronzes, the population of the M–O antibonding level in the conduction band opposes the distortion. The latter result is clear in Fig. 3, where we report the calculated internal energy for WO_3 and $NaWO_3$ following ferro- and antiferro-electric distortions from the cubic phase. While in WO_3 the W off-centering is energetically favourable, the $NaWO_3$ structure is stable in the cubic phase. The topic is examined more extensively in references [48, 49].

The L polymorph has a more complex structure, which contains corner- and edge-sharing between the MO_6 octahedra, and O ions with a different coordination number: the outermost oxygens in each layer are coordinated to only one M ion, oxygens be-

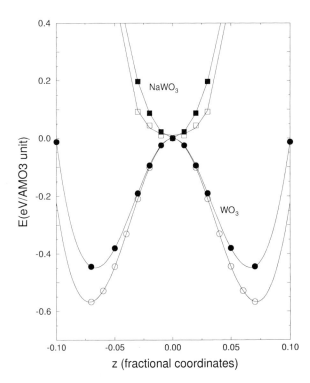

Fig. 3. Calculated internal energy of PV-structured WO_3 (circles) and $NaWO_3$ (squares), as a function of the displacement of the W ions (in fractional coordinates) towards one nearest neighbour oxygen. Full and empty symbols refer to a ferro- and antiferro-electric distortion of the W sublattice, respectively

longing to corner-shared octahedra are two-coordinate, and the oxygens located where the MO_6 octahedra connect by edge-sharing are three-coordinate. In the following discussion, the oxygen atoms which are *n*-coordinate in the L structure are labelled as O_n. In the coordination octahedron around each M ion, one oxygen is of type O_1, two are of kind O_2, and three are O_3's.

Despite the different structure and atomic coordination, the M–O bonding character of the energy levels in the L polymorph is the same as described for the PV structure: levels in the valence band are of M–O bonding character, while the conduction band contains antibonding combinations of metal and oxygen AOs.

In the calculated solution for the L polymorph, even when the MO_6 octahedra are regular by construction as in the pseudocubic phase, the non symmetry-equivalence of the oxygens induces a preferential bonding of the M ion with the one-coordinate O_1 ion, and a more ionic interaction with the three-coordinate oxygens, O_3. The latter result is clear when examining the features of the calculated electronic density; in pseudocubic L-MoO_3, for instance, the Mo–O_n bond populations calculated with a Mulliken partition of charges, are +0.280, −0.029 and −0.187 |e|, respectively for the Mo–O_1, Mo–O_2 and Mo–O_3 bonds.

When the MO_6 octahedra are regular, the L polymorph is highly unstable with respect to the PV structure. The difference in calculated energy is 4.281 eV per formula unit in MoO_3, and 5.471 eV per formula unit in WO_3, indeed very high.

As for the PV structure, we have examined idealised distortions of the L polymorph, by keeping the oxygen sublattice fixed in the coordinates of the pseudocubic phase, and displacing the transition metal ion off-centre in its coordination octahedron. The energy difference that accompanies the displacement of M towards the three non symmetry-

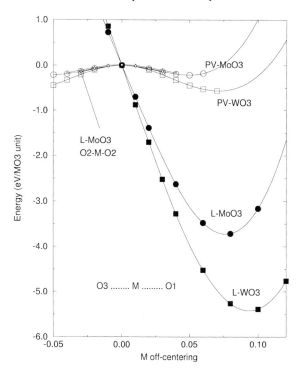

Fig. 4. Energy change, in eV per MO_3 formula unit, for a displacement off-centre of the M ions (in fractional coordinates, referred to the *b* and *c* lattice parameters) in pseudocubic L-MoO_3 and WO_3. The full symbols refer to the movement of M along the axial O_3–M–O_1 direction in L-MoO_3 (circles) and L-WO_3 (squares), from O_3 (negative displacements) to O_1 (positive displacements). Small open triangles refer to the movement of Mo in the equatorial O_2–M–O_2 direction in L-MoO_3. The corresponding energy change for the antiferroelectric distortions of the pseudocubic PV structure of MoO_3 (circles) and WO_3 (squares), are added for comparison as open symbols

equivalent oxygens is reported in Fig. 4. The latter figure confirms that the M ions in the L polymorph are located in a chemically asymmetric crystalline environment: the displacement of M towards O_1 provides a very strong energy gain, while the movement of M towards O_3 is energetically unstable. The comparison makes even more dramatic the difference between the non symmetry-equivalent oxygens of the L polymorph, already suggested by their different bond population with the Mo ion in L-MoO_3.

The energy gained in optimising the distance of the strong M–O_1 bond ($\simeq 4$ eV per formula unit) is crucial in stabilising the L polymorph with respect to the other structures; the calculated energy of the distorted L-MoO_3 polymorph is in fact in the same range as that of the PV structure.

In Fig. 5 we report the calculated density of states (DOS) for the pseudocubic phase of L-WO_3, and for the distorted structure in which W has been displaced at its equilibrium distance towards the O_1 ion. It is important to note that in the conduction band of the distorted structure, very few states are present at lower energy than the π^* M–O_1 antibonding level (labelled as A in the figure). The same qualitative feature is present in the DOS of L-MoO_3; it has a very important consequence for the application of L-structured trioxides as hosts for insertion processes: upon insertion of extraframework atoms, and the corresponding population of the conduction band states of the MO_3 framework, the M–O_1 antibonding states will be populated at very low concentration of the inserted species. As in the PV polymorph, occupation of the antibonding states cancels the driving force for the M off-centering. As shown earlier in Fig. 4, the ener-

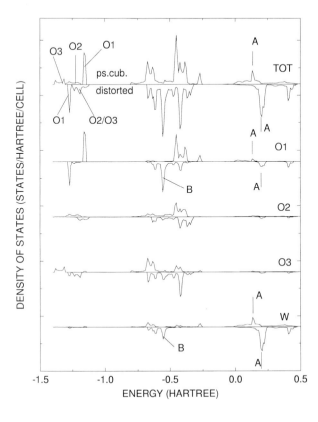

Fig. 5. Calculated density of states of L-WO_3 in the valence region: total value and projections of the DOS onto its components in the basis set of each non symmetry-equivalent ion of the structure, as marked. In each projection, the upper plot refers to the pseudocubic phase, the lower plot to the equilibrium displacement of W towards O_1. The label A indicates the π^* M–O_1 level in the conduction band; B the corresponding M–O_1 bonding level in the valence band

getic importance of the M off-centering is much higher in the L than in the other polymorphs; results of the calculations suggest therefore that the L polymorph of MoO$_3$ might be unstable towards structural modifications upon insertion. The latter result agrees with the experimental structural determination of the molybdenum bronzes (see the review in reference [46]), which may contain layerings in the structure, but never edge-sharing of the MoO$_6$ octahedra. We attribute the instability of the L polymorph to the absence of the M–O$_1$ covalent binding energy, and hence to the instability of the terminal oxygens O$_1$, which occurs upon insertion.

To validate the analysis given above, we have attempted to perform explicit calculations on metallic L-structured oxides. None of the calculations performed has reached convergence; this is very often an indication that the system investigated is energetically very unstable.

The M–O bonding pattern in the valence and conduction bands described here, and its change upon structural distortions, is common to all the polymorphs investigated; a partial population of the conduction band will therefore oppose the off-centering of M in its MO$_6$ octahedron. When we are interested in the structure of the insertion compounds A$_x$MO$_3$, in which the metal ion has configuration d$^{(x)}$, we shall consider only framework structures in which the M ions are on-centre in their coordination octahedra.

Table 1

Calculated energies for the MO$_3$ framework structures examined. ΔE is the relative energy, in eV per MO$_3$ formula unit, with respect to the PV polymorph. The upper part of the table refers to the fully optimised phase of each polymorph, and applies to the relative stability of the binary oxides MoO$_3$ and WO$_3$. The calculated energy of L-MoO$_3$ includes an estimate for electron correlation in the interlayer binding; the fully optimised Hex-WO$_3$ phase includes off-centerings of the W ions along the axial direction, without long-range order in the structure [50]. The bottom part of the table refers to the structures optimised with the M ions on-centre in their coordination octahedra, and applies to the relative stability of the MO$_3$ frameworks in the bronzes

polymorph	ΔE (eV)	
	MoO$_3$	WO$_3$
	fully optimised	
PV	0	0
Hex	–	0.454
PY	0.957	1.524
L	−0.006	–
	optimised with M on-centre	
PV	0	0
Hex	0.239	0.318
PY	0.561	0.725
L	4.281	5.471
PV$^{2\times1}$	–	0.736
PV$^{2\times2}$	0.962	1.634
H$^{2\times1}$	0.524	0.902
H$^{2\times2}$	1.128	1.393

The calculated energies can be employed to define a scale of relative stability for the MO_3 framework of the different polymorphs examined. Since the stability of the structural distortions depends on the number of conduction band electrons, different energy values should be employed, depending on the material of interest. In particular, when examining the stability of the pure MO_3 framework of MoO_3 and WO_3, in which each transition metal ion has electronic configuration $d^{(0)}$, the energy of the fully optimised structures should be used for each polymorph, while in the examination of the bronzes A_xMO_3, chemical distortions which involve the off-centering of M in its octahedron should be neglected.

We have performed both constrained and unconstrained geometry optimisations, for the four polymorphs examined of both MoO_3 and WO_3, and constructed the scale of relative stability for the host MO_3 framework. Results are summarised in Table 1. We note again that the undistorted L polymorph is highly unstable with respect to the other three structures examined, suggesting that bronzes based on L-MoO_3 would not be stable following the cation insertion. The relative stability of the other three polymorphs, parallels instead their relative density, as higher electrostatic energies stabilise the densest structures; in pseudocubic WO_3, for instance, the Madelung field (measured as the difference of electrostatic potential in the W and O sites) is highest in the PV polymorph, while in the Hex and PY it is respectively 0.125 and 0.365 V less favourable. The latter data are roughly proportional to the energy difference between the polymorphs.

4.2 Extraframework ions

In the discussion given in Section 4.1, we have described the inserted ions in the bronzes only as a source of extra electrons; we now try to quantify the steric interaction of the inserted ions with the host MO_3 framework. We do so by making reference to the Na and K insertion compounds of Hex-WO_3.

As a first step in our investigation, we have calculated the electronic distribution, and fully optimised the structure of the host WO_3 framework. As appropriate for the bronzes, the geometry optimisation has been constrained to have the W ions on-centre in their octahedra, in the positions with fractional coordinate $z = 0$. These calculations provide an accurate description of the crystalline environment in which the inserted Na and K ions are hosted. In particular, from the knowledge of the equilibrium geometry and electronic density in the host WO_3 framework, we have calculated the electrostatic potential generated in the hexagonal channels of the structure. The resulting plot is reported in Fig. 6a, in a plane cutting the hexagonal channels along the diameter, and passing through the W (at fractional coordinate $z = 0$) and axial O ions of the structure (located at $z = 0.5$). The potential V is shifted in such a way as to have $V = 0$ in the centre of the hexagonal window.

Let us now examine the potential energy E of a particle with charge +1, moving along the centre of the hexagonal channel, in the c direction. In the following discussion, the energy is shifted to have $E = 0$ for $z = 0.5$. In an idealised description, relative to the insertion of a massless probe charge into the hexagonal channel, its insertion energy can be derived from the knowledge of the electrostatic potential V plotted in Fig. 6a, which represents the Coulomb field acting on the probe charge inside the solid. This value is reported with a continuous line in Fig. 6b. The interstitial position in

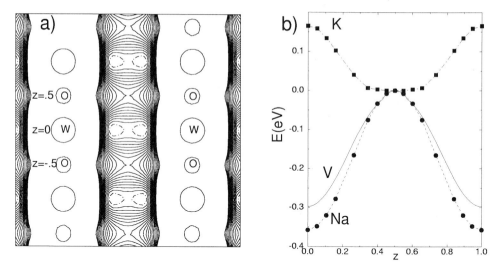

Fig. 6. a) Electrostatic potential V along the hexagonal channels in Hex-WO_3 ($V = 0$ for $z = 0$). b) Insertion energy E (in eV) of a +1 ion in the hexagonal channels of Hex-WO_3 ($E = 0$ for $z = 0.5$). Continuous line: Coulomb energy calculated from the electrostatic potential plotted in a); circles: calculated insertion energy of Na^+ in $Na_{1/3}WO_3$; squares: calculated insertion energy of K^+ in $K_{1/3}WO_3$

which the field is most favourable is located in the hexagonal window, the closest point to the framework oxygens; the migration of the probe particle between the hexagonal windows in two adjacent *ab* planes passes through an energy maximum corresponding to the fractional coordinate 0.5; this position is the furthest away from the framework oxygens.

The other forces acting in the bronzes between the framework and the inserted ions will cause deviations from the ideal Coulomb-only behaviour described above; such forces are represented in the QM Hamiltonian employed, and are therefore included in the solution when performing the calculations on the bronzes. We examined the insertion of Na and K in the Hex-WO_3 framework, corresponding to a stoichiometry $Na_{1/3}WO_3$ and $K_{1/3}WO_3$, in which all the available interstitial sites are occupied. The lattice parameters employed have been optimised for each of the two bronzes examined; we then fixed the lattice parameters, and calculated the relative insertion energy, upon migration of the extraframework ions along the centre of the hexagonal channel. The result is plotted in Fig. 6b. At each fractional coordinate of the inserted ion, the geometry of the surrounding WO_3 framework has been fully reoptimised to accommodate the perturbation introduced by the extraframework ion. To make the calculations feasible, E is calculated without altering the periodicity of the lattice, and all the inserted ions are moved simultaneously. The process described may not be physically relevant to represent the migration of ions in the real materials; this may in fact involve the migration of cation vacancies in the sublattice of the extraframework ions. The energy barrier between maxima and minima as plotted in Fig. 6b would coincide with the activation energy measured experimentally for the migration of the inserted ions, only in the hypothesis of a concerted movement of subdomains of extraframework ions. We note in Fig. 6b that the energy profile for the migration of the Na ion parallels that derived from the electrostatic potential in the cavity; an extra stabilisation in the insertion energy at $z \simeq 0$ is due to the polarisation of the host lattice by the Na ion, excluded when calculating the electrostatic potential generated by the pure host lattice. The hypothesis of an idealised, Coulomb-only interaction between Na^+ and the WO_3 framework is satisfied, suggesting that the Na^+ ion is small enough to have negligible short-range repulsion with the host lattice. The energy profile for the migration of K^+ is instead completely different: its insertion position with lowest energy corresponds to $z = 0.5$, i.e. the position along the channel where the Coulomb field is least favourable. The different behaviour of Na^+ and K^+ ions can be ascribed only to their different ionic sizes, and hence to the steric constraints imposed by the host lattice onto the migration of the larger K^+ ions. The analysis is confirmed by examining the Mulliken population analysis of the calculated electronic distribution: the bond population between an inserted Na ion and each of its nearest framework oxygens equals 0.000 $|e|$ in the hexagonal window, and has a maximum of 0.001 $|e|$ in the position with fractional coordinate $z = 0.5$; in contrast, the same bond population between the inserted K ion and each framework oxygen equals -0.017 $|e|$, indicating appreciable short-range repulsion, when K^+ is in the hexagonal window, and decreases to 0.000 $|e|$ when the K ion is located at $z = 0.5$. The difference in relative energy for the Na and K insertion, as reported in Fig. 6b, gives therefore a direct estimate of the energetics of the steric K^+–O repulsion, at each of the fractional coordinates z examined. From the plot of Fig. 6b, we estimate it as $\simeq 0.52$ eV per K ion, for $z = 0$, in the hexagonal window of the channels.

Let us now return to the data of Table 1. The energy difference between the three corner-sharing (PV, Hex and PY) structures of MoO_3 and WO_3 is in the range of $\simeq 0.5$ eV per MO_3 formula unit. The steric repulsion energy between the framework and the inserted ions, estimated earlier, has the same order of magnitude of the energy difference between the polymorphs; it therefore plays a relevant rôle in the energetics of the bronzes, and may affect the relative stability of different framework structures. This is confirmed by examining the experimentally reported crystal structures, for instance of the W bronzes, as a function of the inserted ions: increasing the size of the extraframework ion we move from the cubic-perovskite form of $NaWO_3$ towards the hexagonal K_xWO_3 and Rb_xWO_3, to the pyrochlore structure obtained by insertion of Cs and of primary and secondary ammonium salts. The latter comparison suggests that if the extraframework ions are present during the synthesis, the energy arising from the short-range repulsion is sufficient to offset the energy difference between the polymorphic structures, and to shift the stability towards the polymorphs with the appropriate interstitial dimensions. We can in such a case identify the extraframework ions as *inorganic templates*, or structure-directing agents, for the forming MO_3 framework.

If the extraframework ions are instead inserted after the MO_3 framework is formed, a phase transformation will require an activation energy, and the framework structure synthesized may have at least a kinetic, if not thermodynamic, stability. The results presented here infer nonetheless that inserted ions incompatible with the interstices of the host TMO framework will induce a strain in the host material; the latter feature may prevent the employability of such host/guest combinations in applications which require a reversible cycling of insertion and disinsertion processes over a long lifetime of the device, such as rechargeable batteries and electrochromic devices.

5. Design of New Polymorphic Structures

The template/host relation between extra-framework ions and TMO structure highlighted in the previous section, recalls similar fields of research; the use of organic cationic templates, for instance, is the standard procedure in the synthesis of microporous alumino-silicates (zeolites), which heavily relies on the presence of organic cations in the synthesis medium, to create molecular-sized interstices in the forming SiO_2 structure. The correlation between the shape of the organic cation and that of the micropores obtained in the zeolitic material is now recognised (see, for instance, references [51 to 54]).

Zeolites are thermodynamically unstable compared to the α-quartz structure of SiO_2; the topic is examined with the same computational technique employed here in a recent paper [55]. The energy difference between the silica polymorphs is in the range of ≃0.1, 0.2 eV per formula unit, much smaller than the one we calculated for MoO_3 and WO_3; excluding the L polymorph, the energy range of the PV, Hex and PY structures is in fact of ≃0.5 eV per formula unit.

We attribute the higher energy difference in the Mo and W trioxides to two effects:

(i) the higher ionic charges, which increases the Madelung field, and hence the energy toll to pay when the density of the solid is decreased by introducing larger interstices; and

(ii) the rigidity of the octahedral framework of TMOs, compared to the more flexible tetrahedral framework of silicates, whose Si–O (or Al–O) backbone is formed only by σ bonds. σ bonds allow in fact an easy rotation of the structure along the M–O direction, with small energy barriers, to respond to local strains. The flexibility of the O–Si–O, and especially of the Si–O–Si angles in zeolites is confirmed by the wide range of values observed, which covers almost all values between 90° and 180°. In comparison, the π frontier orbitals in early TMOs, confer rotational rigidity to the M–O bonding, which can less easily adapt to local strains.

It is important to note, however, that the difference in framework energy between the polymorphs examined of MoO_3 and WO_3 is still in the range of energy provided by the steric repulsion, and can be reversed, as confirmed by the different stable structures of the alkali W bronzes. We believe therefore that a template/host approach, similar to that employed in the synthesis of zeolites, can also be applied to obtain novel TMO frameworks.

Attempts to use larger organic cations, usually quaternary ammonium salts, to template novel TMO frameworks have already been reported in the literature [44, 56]. However, they resulted in the formation of inorganic polyanions, the Keggin structures, rather than extensive M–O frameworks.

By applying computational modelling, we shall now investigate the feasibility for the synthesis of frameworks with a *microporous* architecture also for TMOs. The new structures shall have larger interstices compared to the existing polymorphs. The advantages of TMOs with larger pore structures are manifold: for example, such materials could satisfy the requirement for high ionic conductivity and structural stability upon ion insertion and migration, necessary for the applications in electrochromic devices and rechargeable batteries. Furthermore, if the pore sizes reach the dimension of small organic molecules, and provided the new frameworks have sufficient stability at high temperatures, microporous TMOs would allow heterogeneous catalytic applications. Synthesizing novel materials that could combine the molecular-sieve characteristics of zeolites, with the much higher acidic strength and redox properties of early transition metal cations, would indeed be a major breakthrough in catalysis.

Of course, this is a very challenging task, since microporous structures built up on MO_6 octahedra (the stable coordination polyhedra of transition metal cations) have so far never been reported. The only exception are the OMS sieves based on edge-sharing of $Mn^{IV}O_6$ units [57 to 59], in which the framework has stoichiometry MO_2 and not MO_3 as required for the Mo and W oxides.

Each step in the design process requires the solution of a series of new questions: from the building of the novel structure on paper without having available existing structural references, to testing its relative stability with respect to the existing polymorphs of the material, to providing indications about the synthesis of the newly designed material.

In the following discussion, we shall examine the topic with an analytic approach, exploiting the insight that can be obtained from computer modelling to provide information on the relative stability and on the required synthesis conditions for new, microporous structures of TMOs.

In particular, we employ a computational strategy that has been recently developed for the *de-novo* design of structure-directing agents in zeolites [60], and implemented in a computer code, ZEBEDDE [61]. Given a target microporous architecture, the organic template is gradually grown computationally within the pores of the host, starting from a small molecular fragment used as seed. Optimal space filling is achieved by growing the template molecule within the porous host; at each stage of the tentative template construction, the position, orientation and configuration of the molecule is altered, to maximise its interaction energy with the host. The technique generates, in this way, a set of organic templates which are likely to provide optimal space filling and/or maximum interaction energy with the selected microporous host, thus helping to direct the synthesis towards the target structure.

The method highlighted above requires the knowledge of the porous TMO structure as a starting point; the first problem to solve is therefore that of designing a suitable target structure. The stability requirements, as can be deduced from the discussion of Section 4.2, are that each oxygen ion be shared between at least two neighbouring transition metal ions, avoiding one-coordinate oxygens which would break the framework connectivity, as in the L polymorph. The structure must also contain the maximum extent of corner-sharing of octahedra, which, as shown, is stable upon insertion of extraframework ions and reduction of the host transition metal cations.

The design of new polymorphs can be reformulated as a problem of tiling in space. If we connect the M ions along corner-shared octahedra in the PV, Hex and PY polymorphs, we obtain a tiling based on adjoining triangles, squares and hexagons. Suitable polygons, that allow a complete space-filling under the constraint of periodic boundary conditions necessary to yield a crystalline material, may only have 3, 4 or 6 vertices. The PV, Hex and PY structures exhaust the possibilities based on corner-sharing only of single octahedral units. In the new structures we need therefore to introduce some extent of edge-sharing between octahedra.

The PV and Hex structures can be further imagined as a two-dimensional tiling, repeated by corner-sharing of the axial oxygens along the third (*c*) crystallographic direction. In the design of new structures, we shall retain this two-dimensional approach, as a useful starting point to assess the feasibility of the method.

If we now define a new building block, as a chain of n aligned edge-shared octahedra, a complete series of new structures can be obtained from the same two-dimensional tiling of the PV and Hex phases, using tiles with sides of different length, for instance with polygons whose sides have alternating lengths of m and n units (in the case examined, chains of m and n edge-shared octahedra). The procedure is shown in Fig. 7.

Given the method of construction, we call the new structures $PV^{m \times n}$ and $H^{m \times n}$ (the 1×1 structures reproduce the PV and Hex polymorphs). A schematic representation of the first members of the $H^{m \times n}$ and $PV^{m \times n}$ series is reported in Fig. 8. All the structures obtained with the method highlighted above retain an MO_3 framework stoichiometry.

We have at this point calculated the electronic distribution, geometry and energy of first two members, the 2×1 and 2×2 of the $H^{m \times n}$ and $PV^{m \times n}$ series, for both compositions MoO_3 and WO_3. In all cases, the structure has been optimised under the constraint of keeping the transition metal ions on-centre in their coordination octahedra (i.e. in the equatorial plane, at $z = 0$ in all the structures under investigation). If, as for the zeolites, cationic templates are used in the synthesis, the transition metals will be in a reduced state; we have therefore excluded the chemical distortions of the MO_6 octahedra from the energy minimisation, but allowed the lattice parameters and the internal coordinates of the M and O ions to relax fully under the local Coulomb field generated by the neighbouring ions.

All the new structures examined represent local minima in the potential energy surface; hence, if synthesised, they would have a non-zero activation barrier towards phase transformations, and be kinetically stable.

In Fig. 9, we report the calculated equilibrium electronic density for the four microporous polymorphs examined of WO_3, which is useful to visualise the microporous architecture of the new materials. The size of the plot is different for each polymorph, due to the different values of their lattice parameters and interstitial sizes: the $PV^{2 \times 1}$

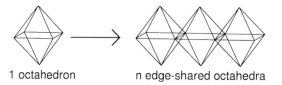

1 octahedron → n edge-shared octahedra

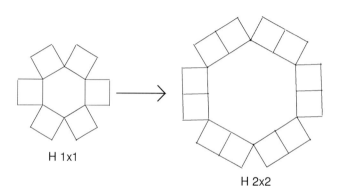

H 1x1 → H 2x2

Fig. 7. Construction of microporous MO_3 structures based on chains of m and n edge-shared octahedra. The example chosen shows the structure $H^{2 \times 2}$

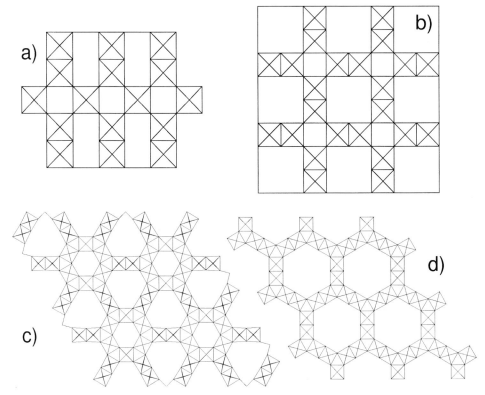

Fig. 8. Disposition of octahedra in the MO$_3$ framework of the a) PV$^{2\times1}$, b) PV$^{2\times2}$, c) H$^{2\times1}$, and d) H$^{2\times2}$ microporous structures

structure is drawn in a 30 × 30 Bohr window; the PV$^{2\times2}$ in a 40 × 40 Bohr window, and the two H$^{2\times n}$ structures in a 50 × 50 Bohr window. In each map, the oxygens are located in the maxima of the electron density (the distorted circles in Fig. 9); the W ions in the lower local maxima, at the centre of four oxygens. W is described with a large-core effective core potential, and hence only the valence electron density is plotted in the W sites. The location of the O and W ions is labelled only in the PV$^{2\times2}$ structure (Fig. 9b).

In Fig. 9a, we further note that the PV$^{2\times1}$ structure can be imagined as derived from the PV polymorph by imposing a regular sequence of stacking faults along the vertical axis of the map. In the PV and PV$^{2\times1}$ lattices, the disposition of the equatorial oxygens is in fact the same, and forms a sequence of square interstices similar to a chess-board. In the PV polymorph, the M ions occupy alternate interstices, forming a pattern similar to the alternating of black and white squares on the chess-board, which may be labelled as *ababab*. In the PV$^{2\times1}$ structure shown in Fig. 9a, instead, the M ions in each third line are shifted horizontally of one interstice, in a pattern equivalent to *aabaabaab*. From the density map of Fig. 9a, it is also clear that the interstices of the PV$^{2\times1}$ polymorph are too small to host any structured organic molecule.

The calculated internal energy of the microporous polymorphs is compared in Table 1 with that of the known MO$_3$ frameworks. The microporous structures are unstable, as

Fig. 9. Calculated equilibrium electronic density for the four microporous polymorphs of WO_3 examined, in the equatorial plane of the structure. The polymorphs are the same displayed in Fig. 8, and are respectively the a) $PV^{2\times1}$, b) $PV^{2\times2}$, c) $H^{2\times1}$, and d) $H^{2\times2}$

expected; the energy difference, however, is still comparable with that among the known polymorphs, and has only a marginally larger magnitude than the steric forces examined in Section 4.2. We consider therefore possible of overcoming this energy difference with steric effects. Given the relative instability of the framework, the template must, however, be selected very carefully, and must fit very tightly in the porous structure to stabilise the new polymorphs; this is probably why previous attempts with *randomly* selected organic templates have failed. The search for templating molecules with the required features may achieve a much higher accuracy by applying the computational method devised for the synthesis of zeolites. As a final step in the computational procedure, we have therefore applied the *de-novo* method to design templating agents specific for the $H^{2\times1}$ and $H^{2\times2}$ structures, those whose interstices are more likely to accommodate organic ions.

The implementation of the method in the code ZEBEDDE is based on interatomic potentials, which are not available to describe the interaction of organic molecules with the transition metal ions Mo and W examined here. For this reason, we have employed a two-step procedure in the template construction. First, we have optimised the host MO_3 framework with QM methods, as described earlier. Second, the framework structure is fixed at its energy-minimised geometry, and the template is grown in the pores of the host. During this second step, the host geometry is not allowed to change; energy-minimised structures have been obtained by relaxing only the template molecule, at each stage of its construction.

The steric interactions between the MO_3 framework and the organic template have been evaluated with the same set of interatomic potentials used for zeolites [62], including the term between framework oxygens and organic molecule, but ignoring the M-template short-range interaction potential. The latter appears, however, a good approximation, as the shape of the pores (see Fig. 9) is given by the oxygens and not by the M ions of the structure. The effect of the transition metal ions was instead included in the electrostatic field, by attributing to them the net charges calculated at the QM level via a Mulliken population analysis of the electronic distribution.

The estimate of interactions via interatomic potentials describes with sufficient accuracy the two-body repulsion forces between the framework oxygen ions and the inserted species, at the basis of the steric, templating effect of the extraframework ions.

Candidate templates are ranked by their calculated binding energy (E_B) with the host lattice: the higher E_B, the more favourable the molecule is considered as a template [52]. The pores of the $H^{2\times1}$ structure have dimensions comparable to un-substituted hydrocarbon chains, suggesting that N-amines or N-diamines, as used in zeolite synthesis may be suitable as templates. The pores of the $H^{2\times2}$ polymorph have instead a size comparable to the twelve-membered rings in zeotypes, and allow the insertion of more structured organic moieties. The calculations reveal that adamantane and di-azobicyclooctane (DABCO) derivatives are of suitable dimensions, the latter being the most promising. A picture of the DABCO template, as located by the code ZEBEDDE

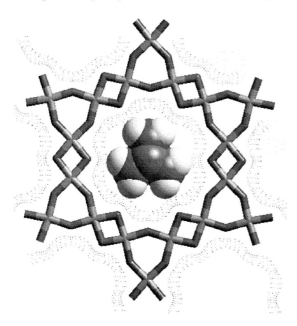

Fig. 10. Di-azobicyclooctane (DABCO) molecule, as located by the code ZEBEDDE inside the pores of the $H^{2\times2}$ polymorph. The darker atoms in the framework are oxygens; the lighter atoms the transition metals. The shaded surface represents the solvent-accessible surface inside the pores, and renders the volume occupied by the framework oxygens

inside the pores of the $H^{2\times 2}$ polymorph of WO_3, is reported in Fig. 10. It is interesting to note there the match between the six-fold symmetry of the porous $H^{2\times 2}$ structure and the six hydrogen atoms of the DABCO organic template. The shaded surface in Fig. 10 represents the solvent-accessible surface inside the pores, and renders the volume occupied by the framework oxygens. Each hydrogen of the template is in close, 1:1, correspondence with one framework oxygen, suggesting that a strong stabilisation from hydrogen bonding may arise in such a configuration.

We believe that the computational study presented here shows the feasibility on energetics grounds of an extensive range of microporous chemistry based on octahedral MoO_3 and WO_3, and suggests that such structures may be synthesised using a host/guest templating approach. Future experimental studies will test the prediction of these calculations. If successful, the new structures would have been derived *completely* on theoretical grounds, from the design stage, up to the 'recipe' for their synthesis, a target that computer modelling should pose itself in future advanced applications.

6. Conclusions

We have employed periodic ab initio HF calculations to examine the relative stability and electronic properties of the Mo and W trioxides, and how they are modified during the insertion process that leads to the formation of the bronzes A_xMO_3. The process is of technological relevance for its application in electrochromic devices and rechargeable batteries.

The insertion of alkali metals or other small organic molecules into the empty interstices of the MO_3 lattice, may induce structural instabilities in the host TMO framework, with two possible origins:

(i) electronic: the population of the energy levels at the bottom of the conduction band of the host TMO, which are of M–O antibonding character, cancels the tendence of the solid towards structural distortions, related to the off-centering of the transition metal ions in their coordination octahedra. When the energy gained during the distortion of the parent material is pronounced, as in L-MoO_3, the TMO framework may be unstable upon insertion;

(ii) steric: when the size of the extraframework ions is not compatible with the interstitial dimensions of the host lattice, the presence of extraframework ions is able to reverse the relative stability of different polymorphic structures of the host oxide.

Extraframework ions show a templating effect towards the stable structure of the host TMO framework, which may be exploited for the synthesis of novel polymorphs. We have examined the latter topic by designing new microporous structures of MoO_3 and WO_3, and exploring the feasibility of their synthesis, employing modelling in advance with respect to experiment.

Acknowledgements We would like to thank Dr. D.W. Lewis for the calculations with the code ZEBEDDE. EPSRC is gratefully acknowledged for funding this research project and for the provision of time on the IBM/SP2 computer at the Daresbury Laboratory.

References

[1] R. von Helmolt, J. Wecker, B. Holzapfel, L. Schultz, and K. Samwer, Phys. Rev. Lett. **71**, 2331 (1993).
Y. Moritomo, A. Asamitsu, H. Kuwahara, and Y. Tokura, Nature **380**, 141 (1996).
[2] M. Green, Chemistry and Industry **17**, 641 (1996).
[3] R.J. Mortimer, Chem. Soc. Rev. **26**, 147 (1997).
[4] J.R. Owen, Chem. Soc. Rev. **26**, 259 (1997).
[5] G. Pistoia (Ed.), Lithium Batteries, Elsevier Publ. Co., Amsterdam 1994.
[6] C. Pisani, R. Dovesi, and C. Roetti, Hartree-Fock ab initio Treatment of Crystalline Systems, Lecture Notes Chem., Vol. 48, Springer-Verlag, Heidelberg 1988.
[7] V.R. Saunders, R. Dovesi, C. Roetti, M. Causà, N.M. Harrison, R. Orlando, and C.M. Zicovich-Wilson, Crystal 98, User's manual, University of Torino, Turin (Italy) 1998.
[8] R. Dovesi et al., phys. stat. sol. (b) **217**, 63 (2000).
[9] P.J. Hay and W.R. Wadt, J. Chem. Phys. **82**, 270, 284, 299 (1985).
[10] F. Corà, A. Patel, N.M. Harrison, C. Roetti, and C.R.A. Catlow, J. Mater. Chem. **7**, 959 (1997).
[11] F. Corà, A. Patel, N.M. Harrison, R. Dovesi, and C.R.A. Catlow, J. Amer. Chem. Soc. **118**, 12174 (1996).
[12] M. Causà, R. Dovesi, C. Pisani, and C. Roetti, Phys. Rev. B **33**, 1308 (1986).
R. Dovesi, Solid State Commun. **54**, 183 (1985).
[13] R. Nada, C.R.A. Catlow, R. Dovesi, and C. Pisani, Phys. Chem. Minerals **17**, 353 (1990).
[14] P. D'Arco, G. Sandrone, R. Dovesi, R. Orlando, and V.R. Saunders, Phys. Chem. Minerals **20**, 407 (1993).
[15] R. Dovesi, C. Roetti, M. Prencipe, C. Freyria Fava, and V.R. Saunders, Chem. Phys. **156**, 11 (1991).
[16] H.J. Monkhorst and J.D. Pack, Phys. Rev. B **13**, 5188 (1976).
[17] W.L. Kehl, R.G. Hay, and D. Wahl, J. Appl. Phys. **23**, 212 (1952).
[18] J. Wyart and M. Foex, C.R. Acad. Sci. (France) **233**, 2459 (1951).
[19] E. Salje, Acta Cryst. B **33**, 574 (1977).
[20] B.O. Loopstra and P. Boldrini, Acta Cryst. **21**, 158 (1966).
[21] E. Salje and K. Viswanathan, Acta Cryst. B **31**, 356 (1975).
[22] R. Diehl, G. Brandt, and E. Salje, Acta Cryst. B **34**, 1105 (1978).
[23] E.M. McCarron, Chem. Commun. 336 (1986).
J.B. Parise, E.M. McCarron, and R. von Dreele, J. Solid State Chem. **93**, 193 (1991).
[24] P.G. Dickens and D.J. Neild, Trans. Faraday Soc. **64**, 13 (1968).
[25] N.C. Stephenson and A.D. Wadsley, Acta Cryst. **19**, 241 (1965).
[26] J. Graham and A.D. Wadsley, Acta Cryst. **20**, 93 (1966).
[27] M. Figlarz, Progr. Solid State Chem. **19**, 1 (1989).
[28] T. Nanba and I. Yasui, J. Solid State Chem. **83**, 304 (1989).
T. Nanba, Y. Nishiyama, and I. Yasui, J. Mater. Res. **6**, 1324 (1991).
T. Nanba, T. Takahashi, S. Takano, J. Takada, A. Osaka, Y. Miura, T. Kudo, and I. Yasui, J. Ceram. Soc. Jpn. **103**, 222 (1995).
[29] B. Gerand, G. Nowogrocki, J. Guenot, and M. Figlarz, J. Solid State Chem. **29**, 429 (1979).
[30] I.P. Olenkova, L.M. Plyasova, and S.D. Kirik, Reaction Kinetics and Catal. Lett. **16**, 81 (1981).
[31] J.D. Guo, P. Zavalij, and M.S. Whittingham, European J. Solid State and Inorg. Chem. **31**, 833 (1994); J. Solid State Chem. **117**, 323 (1995).
[32] N.A. Caiger, S. Crouchbaker, P.G. Dickens, and G.S. James, J. Solid State Chem. **67**, 369 (1987).
[33] C. Gemin, A. Driouiche, B. Gerand, and M. Figlarz, Solid State Ionics **53-6**, 315 (1992).
[34] T. Kudo, J. Oi, A. Kishimoto, and M. Hiratani, Mater. Res. Bull. **26**, 779 (1991).
[35] A. Driouiche, F. Abraham, M. Touboul, and M. Figlarz, Mater. Res. Bull. **26**, 901 (1991).
[36] A.B. Swanson and J.S. Anderson, Mater. Res. Bull. **3**, 149 (1968).
[37] P.J. Wiseman and P.G. Dickens, J. Solid State Chem. **17**, 91 (1976).
[38] P.J. Wiseman and P.G. Dickens, J. Solid State Chem. **6**, 374 (1973).
[39] I. Tsuyumoto, A. Kishimoto, and T. Kudo, Solid State Ionics **59**, 211 (1993).

[40] H. Prinz, U. Müller, and M.L. Haeierdanz, Z. Anorg. Allg. Chem. **609**, 95 (1992).
[41] R.C.T. Slade, P.R. Hirst, and B.C. West, J. Mater. Chem. **1**, 281 (1991).
[42] Q.M. Zhong and W. Colbow, Thin Solid Films **196**, 305 (1991).
[43] J.D. Guo, K.P. Reis, and M.S. Whittingham, Solid State Ionics **53-6**, 305 (1992).
[44] P. Zavalij, J.D. Guo, M.S. Whittingham, R.A. Jacobson, V. Pecharsky, C.K. Bucker, and S.J.H. Wu, J. Solid State Chem. **123**, 83 (1996).
[45] N.C. Stephenson, Acta Cryst. **20**, 59 (1966).
[46] E. Canadell and M.H. Whangbo, Chem. Rev. **91**, 965 (1991).
[47] T.A. Bither, J.L. Gillson, and H.S. Young, Inorg. Chem. **5**, 1569 (1966).
[48] F. Corà and C.R.A. Catlow, Faraday Discuss. **114** (1999), in press.
[49] F. Corà, M. Stachiotti, C.O. Rodriguez, and C.R.A. Catlow, J. Phys. Chem. B **101**, 3945 (1997).
[50] F. Corà, Ph.D Thesis, The Royal Institution of Great Britain and the University of Portsmouth, 1999.
[51] M.E. Davis and R.F. Lobo, Chem. Mater. **4**, 759 (1992).
[52] D.W. Lewis, C.M. Freeman, and C.R.A. Catlow, J. Phys. Chem. **99**, 11194 (1995).
[53] A.P. Stevens, A.M. Gorman, C.M. Freeman, and P.A. Cox, J. Chem. Soc., Farad. Trans. **92**, 2065 (1996).
[54] R.E. Boyett, A.P. Stevens, M.G. Ford, and P.A. Cox, Zeolites **17**, 508 (1996).
[55] B. Civalleri, C.M. Zicovich-Wilson, P. Ugliengo, V.R. Saunders, and R. Dovesi, Chem. Phys. Lett. **292**, 384 (1998).
[56] G.G. Janauer, A. Dobley, J.D. Guo, P. Zavalij, and M.S. Whittingham, Chem. Mater. **8**, 2096 (1996).
[57] Y.F. Shen, S.L. Suib, and C.L. Oyoung, J. Amer. Chem. Soc. **116**, 11020 (1994).
[58] Y.F. Shen, R.N. Deguzman, R.P. Zerger, S.L. Suib, and C.L. Oyoung, Stud. Surf. Sci. and Catal. **83**, 19 (1994).
[59] S.R. Wasserman, K.A. Carrado, S.E. Yuchs, Y.F. Shen, H. Cao, and S.L. Suib, Physica B **209**, 674 (1995).
[60] D.W. Lewis, D.J. Willock, C.R.A. Catlow, J.M. Thomas, and G.J. Hutchings, Nature **382**, 604 (1996).
[61] D.W. Lewis and D.J. Willock, code ZEBEDDE (ZEolites By Evolutionary De-novo DEsign), 1996-1998.
[62] cff91-czeo molecular forcefield and Discover 4.0, Molecular Simulation Inc., San Diego 1997.

phys. stat. sol. (b) **217**, 599 (2000)

Subject classification: 71.20.Nr; 78.20.Bh; 78.20.Ci; S7.14; S7.15

From Band Structures to Linear and Nonlinear Optical Spectra in Semiconductors

W. R. L. LAMBRECHT and S. N. RASHKEEV[1])

Department of Physics, Case Western Reserve University, Cleveland, OH 44106-7079, USA

(Received August 10, 1999)

The theory of the relation between optical properties and electronic band structures is reviewed. In the first part, we describe various approaches to go beyond the long-wavelength limit perturbation theoretical approach in the independent particle approximation for linear optical response functions. In the latter, electron–electron interaction effects are only included in as far as they are implicit in the underlying band structure (typically within the local density functional approximation). We discuss the inclusion of quasiparticle corrections to the band structures, local-field and electron–hole interaction effects. A case study of GaN is used to illustrate the discrepancies between theory and experiment that arise from neglect of these effects. On the other hand, the paper also illustrates that in the process of extracting the band structure information from optical spectra, the independent particle model still plays a central role. In the case of nonlinear optical response, even the independent particle model was only recently fully developed. Recent progress in this area and their implementation within the context of first-principles band structure methods is presented next. Some examples are used to illustrate the potentially richer information contained in NLO spectra in relation to the underlying band structures. Secondly, progress in understanding the trends in NLO coefficients in some classes of materials is illustrated with a study of chalcopyrites.

1. Introduction

The study of optical spectra in solids has formed an integral part of electronic structure calculations since early on in their development. In fact, early semi-empirical band structure methods were based on fitting the pseudopotential form factors so as to reproduce measured interband optical transition energies at selected points in the Brillouin zone. With the advent of density functional theory [1] (DFT) and the local density approximation (LDA), the emphasis in electronic structure calculations has shifted toward predictions of total energy properties. Nevertheless, optical studies should still be considered as an important experimental validating probe of the electronic structure.

Unfortunately, the connection between the band structure and the optical spectra is far less straightforward in the context of the DFT than it was in the naive semi-empirical band structure model. In fact, the eigenvalues of the Kohn-Sham (KS) equations appearing in this theory, strictly speaking have no longer the meaning of a quasiparticle (QP) band structure. The latter have the meaning of individual excitations of electrons

---

[1]) Present address: Department of Physics and Astronomy, Box 1807, Station B. Vanderbilt University, Nashville, TN 37235, USA, Permanent address: P. N. Lebedev Physical Institute, Russian Academy of Sciences, 117924 Moscow, Russian Federation.

or holes, as would be measured in photoemission or inverse photoemission spectroscopy. Optical properties, however, involve the simultaneous excitation of both and thus are further modified by the interactions among these.

In this paper, we review the current viewpoint of how electronic band structure is related to optical properties. First, we discuss briefly how the quasiparticle corrections are included in the band structure. Then, we recall the basics of the formalism for linear optics so that a clear picture is obtained of what approximations are made. We start from the simplest independent particle model. Next, we discuss the topics of local-field and excitonic corrections. This section is mainly meant to provide a guide to the literature on these topics, including the most recent progress. It makes no claims of any original contributions in this respect on the part of the present authors nor a claim of completeness in the description. We then discuss briefly the particular implementation we have used in our own work, which uses the linear muffin-tin orbital (LMTO) basis set. Next, we describe a case study, namely the spectra of wurtzite and zincblende GaN to illustrate how the effects discussed above affect the spectra. At the same time, it illustrates that the independent particle model we have used in our own work still plays a central role in relating features in the spectrum to their specific band structure origin. We then turn to nonlinear optics (NLO). It turns out that until recently, it was not fully understood how to proceed for general NLO processes even in the basic independent particle model. The reason for this is that unlike in linear optics where only interband transitions come into the picture in insulators, an interplay between interband transitions and intraband effects plays a role in nonlinear optics. This was recently clarified by the work of Sipe and coworkers [2 to 5], although a similar approach was also already known in older Russian literature [6]. We will briefly summarize these developments here. We then show some applications of this formalism as implemented within the context of the LMTO method. We show how the rich fine structure of NLO spectra can potentially be exploited to extract band structure information. We also show that in spite of its present shortcomings in predicting accurate absolute values in the static limit, the formalism can provide insights into trends in the efficiency of nonlinear optical effects and has promise to predict novel materials with improved NLO efficiencies.

2. Quasiparticle Corrections

In the Kohn-Sham (KS) version of DFT [1], an auxiliary system of independent particles is introduced which has the same density as the actual interacting system of electrons. The variational principle for the total energy in terms of the density then leads to the condition that these independent particles interact with an effective potential rather than just the external potential. This determines the self-consistency condition on the potential

$$v_{\text{eff}}(\mathbf{r}) = v_{\text{ext}}(\mathbf{r}) + v_{\text{H}}(\mathbf{r}) + v_{\text{xc}}(\mathbf{r}), \qquad (1)$$

in which the Hartree potential $v_{\text{H}}(\mathbf{r}) = e^2 \int n(\mathbf{r}')/|\mathbf{r} - \mathbf{r}'| \, d^3r'$ and the exchange–correlation potential, $v_{\text{xc}}(\mathbf{r}) = \delta E_{\text{xc}}/\delta n(\mathbf{r})$, which appears as a functional derivative of the exchange–correlation total energy, both depend on the density to be determined. The electron density $n(\mathbf{r}) = \sum_i^{\text{occ}} |\psi_i(\mathbf{r})|^2$ is obtained by filling the lowest eigenstates of the one-particle Schrödinger equation,

$$\left[-\frac{\hbar^2}{2m}\nabla^2 + v_{\text{eff}}(\mathbf{r})\right]\psi_i(\mathbf{r}) = \varepsilon_i\psi_i(\mathbf{r}),\tag{2}$$

according to the Pauli principle. Strictly speaking the eigenvalues of this Kohn-Sham Schrödinger equation, which is what one usually calls "the band structure", have nothing to do with the actual quasiparticle excitations of the real many-electron system but are merely an auxiliary device to calculate total energy properties.

In order to study the excitation of the system, a more appropriate starting point is the quasiparticle band structure obtained in the GW approximation (GWA) [7 to 11]. The equation describing the latter is

$$\left[-\frac{\hbar^2}{2m}\nabla^2 + v_{\text{ext}}(\mathbf{r}) + v_{\text{H}}(\mathbf{r})\right]\Psi_i(\mathbf{r}) + \int \Sigma_{\text{xc}}(\mathbf{r}, \mathbf{r}', E_i)\,\Psi_i(\mathbf{r}')\,\mathrm{d}^3r' = E_i\Psi_i(\mathbf{r}).\tag{3}$$

This method derives its name from the approximation for the self-energy operator, $\Sigma_{\text{xc}}(1, 2) = iG(1, 2)\,W(1, 2)$ in terms of the one-electron Green's function $G(1, 2)$ and screened Coulomb interaction $W(1, 2) = \int \mathrm{d}(3)\,v(1, 3)\,\epsilon^{-1}(3, 2)$, which in the static approximation of the dielectric function can be thought of as a screened form of the Hartree-Fock exchange term. Here, 1, 2, ... stand for the position coordinates and time. In Eq. (3) its Fourier transform with respect to time appears, which makes it frequency or energy dependent. It describes single-particle or single-hole excitations of the system and thus relates strictly speaking to inverse or direct photoemission spectroscopies. Unfortunately, the nonlocal character and the energy dependence of the self-energy operator in the GWA severely complicate the method. Calculations in the GWA are still quite time consuming and only possible for systems of moderate size. Thus, one naturally seeks further approximations.

Equation (3) differs from the KS Schrödinger equation, Eq. (2), in that the self-energy operator replaces the exchange–correlation potential. Thus, the KS equation can be viewed as an approximation to the QP equation with $\Sigma_{\text{xc}}(\mathbf{r}, \mathbf{r}', E) \approx v_{\text{xc}}(\mathbf{r})\delta(\mathbf{r} - \mathbf{r}')$ as an approximation to the self-energy operator The most commonly used approximation for $v_{\text{xc}}(\mathbf{r})$ is the local density approximation (LDA), in which $E_{\text{xc}} = \int n(\mathbf{r})\,\varepsilon_{\text{xc}}[n(\mathbf{r})]\,\mathrm{d}^3r$ is given in terms of the exchange–correlation energy per particle in a homogeneous electron gas at the corresponding local density.

Since in most semiconductors and insulators the most important difference between the LDA-KS and QP-GWA band structures consists in a more or less constant upward shift of the conduction band, the LDA-KS band structure still forms a reasonable starting point to understand the origin of spectroscopic features. Actually, the valence bands also require QP self-energy corrections. In particular, the QP self-energy correction increases as one goes down in energy from the valence band maximum and this increase is typically discontinuous as one crosses gaps to more localized atomic-like bands or semicore states. However, the optical response in the visible to near UV is typically dominated by the transitions between the upper valence bands and the lower conduction bands for which the constant shift or "scissor" correction is a good starting point for the discussion. Nevertheless, one should keep in mind that even when the band structure is changed by a scissor shift $\sum_{i\mathbf{k}}^{\text{unocc}} \Delta_{i\mathbf{k}}\,|i\mathbf{k}\rangle\langle i\mathbf{k}|$, one has implicitly changed the

Hamiltonian from which the eigenstates are derived to a nonlocal one. Thus, care will be needed in obtaining the matrix elements and the optical functions. This will be further discussed in Section 3.4.

Recently, new treatments of exchange correlation within DFT have been proposed which provide eigenvalues closer to the QP spectrum. The first of these is the so-called exact-exchange (EEX) approach [12, 13]. In it, one still uses an exchange–correlation *potential*, i.e. a local operator, but it is obtained in a rather nonlocal way from the density. It makes use of the fact that the exchange part of the potential can be written down analytically in terms of one electron orbitals (chosen here to be the solutions of a KS equation rather than a Hartree-Fock equation), which in turn depend implicitly on the density. Thus, one uses the chain rule for functional derivatives,

$$v_x(\mathbf{r}) = \frac{\delta E_x}{\delta n(\mathbf{r})} = \sum_i \left(\frac{\delta E_x}{\delta \psi_i}\right)\left(\frac{\delta \psi_i}{\delta v_{\text{eff}}(\mathbf{r})}\right)\left(\frac{\delta v_{\text{eff}}(\mathbf{r})}{\delta n(\mathbf{r})}\right), \qquad (4)$$

and makes use of the fact that the change in KS orbitals upon a change in effective potential and the change in density upon a change in effective potential can be obtained by first-order perturbation theory. The key is then to invert the last relation to obtain instead the change in effective potential due to a change in density. This method is also known as the *optimized potential method* and was originally introduced by Talman and Shadwick [14] in atomic calculations. Recent implementations indicate that the eigenvalues of the corresponding KS Schrödinger equation are quite close to the QP equations near the gap primarily because the error in self-interaction of the LDA is avoided and that good total energies as well as optical properties can be obtained.

A second promising approach is the so-called generalized Kohn-Sham (GKS) method [15, 16] in which a nonlocal Hartree-Fock type KS equation is used as one-particle equation. Some results of the method and a brief discussion are provided by van Schilfgaarde et al. in [16]. This possibility was already suggested in the original paper by Kohn and Sham [1]. In this case, the exchange–correlation potential is replaced by a nonlocal (but still energy independent) Fock-exchange type operator. The latter is now chosen to be a statically screened exchange operator and the remainder of correlation energy is then approximated for example in a local density fashion. This method makes use of the fact that statically screened exchange is already a good approximation to the GW equation and thus is guaranteed to give eigenvalues close to those of the GW method. On the other hand, it is still formulated within a DFT context so that total energies can also be obtained from the same one-electron states within a self-consistent method.

While clearly a KS or GKS equation with eigenvalues closer to the QP spectrum is desirable, we can still proceed to do useful work within the limitations of the LDA + scissors approach. In the present paper, we examine some case studies to show both the successes and limitations of this approach. In particular, we emphasize throughout that the goal of our approach is not so much to obtain perfect agreement between theoretical and experimental spectra, but rather to understand the correct relation between the two and how to extract band-structure information from it. This will include information on how well for example QP corrections are predicted by current GWA calculations. We also discuss the complexity of relating energy band transitions at specific **k**-points to features in the spectra.

For nonlinear optics (NLO) the QP corrections play an even more important role. As we will see below, some terms in the NLO response functions involve two factors

$(E_i(\mathbf{k}) - E_j(\mathbf{k}) - \hbar\omega)^{-1}$ in which i and j represent conduction and valence bands, respectively. Thus, the gap underestimates of these energy differences have numerically a large effect on the values obtained, which can easily amount to a factor 2 to 3 and in some cases where the uncorrected gap goes to zero can lead to totally meaningless results.

Even when using the best QP energies available, we should be cautious in interpreting optical spectra because of the electron–hole interaction (or excitonic) effects. Inclusion of the electron–hole interactions has only recently been implemented in a first-principles approach [17 to 20]. The excitonic effects shift the transition energies only modestly in most materials – exceptions may be strongly ionic insulators – but lead to important shifts in oscillator strengths, modifying the qualitative shape of the spectra and absolute intensities but leaving the transition energies relating to the band structure intact. Thus, again the underlying band structure remains closely related to the optical spectrum.

3. Basic Formalism for Linear Optics

3.1 Independent particle model

We are concerned with the response of the electronic system to the electric field $\mathbf{E} \propto \exp i(\mathbf{q} \cdot \mathbf{r} - \omega t)$, of an electromagnetic wave in the long-wavelength limit $q \to 0$. The field $\mathbf{E} = \dot{\mathbf{A}}/c$ is given in terms of the vector potential for which the transverse gauge[2]) $\nabla \cdot \mathbf{A} = 0$ is chosen and in our first treatment the $\mathbf{r}$ dependence of $\mathbf{A}$ is neglected from the start. The electric field is here meant to be the macroscopic average electric field inside the material. In that sense, the response to it is not a true response function. The latter would require us to calculate the response to the external electric field. We will come back to this distinction later, when discussing local field corrections. We clarify that this first treatment is, of course, approximate in that it neglects local field effects and is used as starting point of the discussion for reasons of its simplicity.

The problem is then basically to calculate the macroscopic average or expectation value $\langle \mathbf{J}^i \rangle$ of the *induced current operator*,

$$\mathbf{J}^i(t) = \frac{e}{m} \int \psi^\dagger(\mathbf{r}, t) \left[\mathbf{p} - \frac{e}{c} \mathbf{A}(t) \right] \psi(\mathbf{r}, t) \, \mathrm{d}^3 r, \qquad (5)$$

written here in terms of annihilation (creation) field operators $\psi(\mathbf{r}, t)$ ($\psi^\dagger(\mathbf{r}, t)$), and the momentum operator $\mathbf{p} = -i\hbar\nabla$. The Hamiltonian of the system including the interaction with the field is given by:

$$\mathcal{H}(t) = \int \psi^\dagger(\mathbf{r}, t) \left[\frac{1}{2m} \left(\mathbf{p} - \frac{e}{c} \mathbf{A}(t) \right)^2 + V(\mathbf{r}) \right] \psi(\mathbf{r}, t) \, \mathrm{d}^3 r, \qquad (6)$$

with $V(\mathbf{r})$, the effective one-electron (in practice LDA) potential. The unperturbed Hamiltonian $\mathcal{H}_0$ corresponds to leaving out the $\mathbf{A}$ term. Note that the induced current is

---

[2]) The use of the transverse or Coulomb gauge is convenient if the Coulomb interactions are treated as instantaneous and retardation effects are neglected. See G. D. MAHAN, Many-Particle Physics, Plenum Press, New York 1990 (p. 60). The question of gauge invariance of the results is discussed in Ref. [3] in particular with regards to the tranformation to the length-gauge to be discussed later in Section 6.

given by $J^i = e[\partial \mathcal{H}/\partial \mathbf{p}]$, i.e. the charge of the electron e times a velocity, obtained as usual in Hamiltonian theory by the partial derivative of the Hamiltonian with respect to the canonical momentum $\mathbf{p}$.

Note that the Hamiltonian can be written as $\mathcal{H}(t) = \mathcal{H}_0 + \mathcal{H}_1 + \mathcal{H}_2$ with

$$\mathcal{H}_1 = -\frac{e}{mc}\mathbf{A}(t) \cdot \int \psi^\dagger(\mathbf{r}, t)\,\mathbf{p}\psi(\mathbf{r}, t)\,d^3r, \tag{7}$$

$$\mathcal{H}_2 = \frac{e^2}{2mc^2}\mathbf{A}^2(t)\,N. \tag{8}$$

The last equation contains the total number operator $N = \int \psi^\dagger(\mathbf{r}, t)\,\psi(\mathbf{r}, t)\,d^3r$. Because of the $\mathbf{r}$-independence of $\mathbf{A}$, the $\mathbf{A}^2$ term only introduces the same time-dependent phase factor for all eigenstates and can thus be ignored. One now sets up a perturbation expansion for the current expectation value in powers of the electric field.

Using $d\mathbf{P}(t)/dt = \mathbf{J}^i(t)$, or its Fourier transform, $-i\omega \mathbf{P}(\omega) = \mathbf{J}^i(\omega)$, one obtains the *induced polarization* $\mathbf{P}(\omega)$ as a power series in the electric field, which defines the various linear and non-linear optical polarizabilities

$$P_a = \chi^{(1)}_{ab} E_b + \chi^{(2)}_{abc} E_b E_c + \ldots, \tag{9}$$

in which the indices a, etc. indicate Cartesian components. In the NLO case, we here should introduce $\mathbf{E}$ components of different frequencies. For example, the second term is then written more explicitly as $P^{(2)}_a(\omega_3) = \chi^{(2)}(-\omega_3, \omega_1, \omega_2)\,E_b(\omega_1)\,E_c(\omega_2)$. The perturbation calculation can be worked out in various ways. One typically expands the field operators in the basis of the time-independent (Bloch) eigenstates of $\mathcal{H}_0$, $\psi_i(\mathbf{k}, \mathbf{r})$ and uses the density matrix method – see for example the Appendix of Ref. [21]. The resulting well-known expression for the linear response function is

$$\chi^{(1)}_{ab}(-\omega, \omega) = \frac{e^2}{\hbar m^2 \omega^2} \int_{BZ} \frac{d^3k}{(2\pi)^3} \sum_{ij} \frac{f_{ij}(\mathbf{k})\,p^a_{ij}(\mathbf{k})\,p^b_{ji}(\mathbf{k})}{\omega_{ji}(\mathbf{k}) - \omega - i\eta} - \frac{e^2 n_0}{m\omega}\delta_{ab}, \tag{10}$$

in which $\omega_{ji}(\mathbf{k}) = \omega_j(\mathbf{k}) - \omega_i(\mathbf{k})$ and $E_i(\mathbf{k}) = \hbar\omega_i(\mathbf{k})$ are the energy bands and $p^a_{ij}(\mathbf{k})$ are the momentum matrix elements in the Bloch basis, $f_{ij}(\mathbf{k}) = f_i(\mathbf{k}) - f_j(\mathbf{k})$ is a difference of Fermi factors, $f_i(\mathbf{k}) = 1(0)$ for (un)occupied states at zero temperature, n_0 is the average electron density, $\eta \to 0$ is a small positive number and the integral is over the Brillouin zone (BZ). The last term arises in zeroth order in the $\mathcal{H}_1$ perturbation from the $\mathbf{A}$ term in the current operator, the first one from the first-order term in the $\mathcal{H}_1$ perturbation which contains one $\mathbf{p}$ matrix element and one $\mathbf{E}$ factor combined with the $\mathbf{p}$ term in the current operator. A problem is that the expressions obtained in this manner contain a seemingly divergent factor ω^{-2}. Similar divergences plague the higher-order terms. As explained in Sipe and Ghahramani [2] (SG), the divergence in linear response is easily removed by noting that Eq. (10) can be rewritten as

$$\chi^{(1)}_{ab}(-\omega, \omega) = -\frac{e^2 n_0}{\omega^2}\left[\frac{1}{N}\sum_i \int_{BZ}\frac{d^3k}{(2\pi)^3}\,f_i(\mathbf{k})\,[m^*_i(\mathbf{k})]^{-1}_{ab}\right]$$

$$+ \frac{e^2}{\hbar m^2} \int_{BZ}\frac{d^3k}{(2\pi)^3}\sum_{ij}\frac{f_{ij}(\mathbf{k})\,p^a_{ij}(\mathbf{k})\,p^b_{ji}(\mathbf{k})}{\omega^2_{ji}(\mathbf{k})[\omega_{ji}(\mathbf{k}) - \omega]}, \tag{11}$$

in which $[m_i^*(\mathbf{k})]_{ab}^{-1} = \hbar^{-1}\, \partial \omega_i(\mathbf{k})/\partial k_a\, \partial k_b$ is the inverse effective mass tensor of band i at point $\mathbf{k}$ and N is the total number of electrons in the solid. The point is now that in a semiconductor at zero temperature, the $f_i(\mathbf{k})$ are all either 0 or 1 independent of $\mathbf{k}$, so they do not enter the $\mathbf{k}$-integral, which reduces to zero because it is the integral of the second derivatives of a periodic function in $\mathbf{k}$-space over the BZ. Thus, the first term vanishes in a semiconductor at zero temperature. On the other hand, if there are partially filled bands, this divergence is real, but will of course be damped in reality by scattering. It leads to the intraband Drude term. It would for example enter the picture in metals or in a doped semiconductor. In Section 6, we will further pursue this and discuss how to obtain the final form of Eq. (11) directly by separating the interband from the intraband evolution of the electronic system from the start.

For the moment, let us rather reflect on the approximations we have made in deriving Eq. (11). First of all, we have neglected the $\mathbf{r}$ dependence from the start and secondly we have neglected any other terms in the Hamiltonian besides the effective one-electron Hamiltonian and the interaction with the field, and finally, we have assumed that the one-electron states are derived from a Hamiltonian with a simple (i.e. local) potential. We now discuss each of these approximations in turn.

3.2 Local field effects

In the above treatment, $\mathbf{E}$ was assumed to be the *macroscopic* internal electric field in the solid, neglecting the microscopic fluctuations in the density response. Including the effects of the microscopic fluctuations is usually referred to as including "local field" (LF) effects [22 to 25]. Most discussions of this topic are in terms of the longitudinal dielectric function, which describes the response to a scalar potential perturbation $V^{\text{ext}} \propto V \exp(i(\mathbf{q} \cdot \mathbf{r} - \omega t))$. The main point is that the response does not only appear at wave vector $\mathbf{q}$ but includes short-wavelength components $\mathbf{q} + \mathbf{G}$ with $\mathbf{G}$ a reciprocal lattice vector. Thus, the response function becomes a matrix with rows and columns labeled by $\mathbf{G}$. The inverse dielectric constant matrix relates the total internal (self-consistently) screened potential to the external one $V^{\text{scr}} = \epsilon^{-1} V^{\text{ext}}$. According to Adler and Wiser [22, 23], the macroscopic dielectric constant describing the long-wavelength limit response is then given by:

$$\epsilon^{\mathrm{M}}(\omega) = \lim_{\mathbf{q}\to 0} \left[\frac{1}{\epsilon_{\mathbf{G}\mathbf{G}'}^{-1}(\mathbf{q})} \right]_{\mathbf{G}=\mathbf{G}'=0}, \tag{12}$$

which is not the same as simply taking the inverse of the $\mathbf{G} = \mathbf{G}' = 0$ element of the matrix.

For a full discussion of the transverse case, which is the one really applicable to optics, we here follow the treatment of Dolgov and Maksimov [26], which is equivalent to that of Pick [25]. Before discussing the local-field corrections further, it is useful to restate the problem in a more general many-body context than the independent particle perturbation theory used above. First, we need to distinguish between the response to the external field and the total field. Writing it in terms of polarizations rather than currents, we have

$$P_a(\mathbf{r}, \omega) = \int \tilde{\chi}_{ab}(\mathbf{r}, \mathbf{r}', \omega)\, E_b(\mathbf{r}, \omega)\, \mathrm{d}^3 r', \tag{13}$$

$$P_a(\mathbf{r}, \omega) = \int \chi_{ab}(\mathbf{r}, \mathbf{r}', \omega)\, E_b^{\mathrm{e}}(\mathbf{r}, \omega)\, \mathrm{d}^3 r'. \tag{14}$$

Only the latter is a causal response function in the strict sense that $\mathbf{E}^e$ is externally controllable, whereas $\mathbf{E}$ includes the effects produced by the internally induced currents and charges. Here, $\tilde{\chi}_{ab}$ is known as the polarizability and χ_{ab} as the susceptibility. Note that in the independent particle model used above, $\tilde{\chi}_{ab} = \chi^{(1)}$ and is sometimes referred to as the bare polarizability, whereas the susceptibility is then called the "dressed" polarizability. In a many-body framework, the susceptibility is given by

$$\chi_{ab}(\mathbf{r}, \mathbf{r}', \omega) = \frac{e^2}{m} n(\mathbf{r}) \delta(\mathbf{r} - \mathbf{r}') \delta_{ab} + \chi_{ab}^{jj}(\mathbf{r}, \mathbf{r}', \omega) \tag{15}$$

in terms of the *current–current correlation function*, which according to the Kubo formula is given by

$$\chi_{ab}^{jj}(\mathbf{r}, \mathbf{r}', \omega) = -i \int_0^\infty dt\, e^{i(\omega + i\eta)t} \langle [\hat{j}_a(\mathbf{r}, t) \hat{j}_b(\mathbf{r}', 0)] \rangle, \tag{16}$$

in which $\hat{j}_a = (e/2mi) \{\psi^\dagger(\mathbf{r}) \nabla_a \psi(\mathbf{r}) - [\nabla_a \psi^\dagger(\mathbf{r})] \psi(\mathbf{r})\}$ is the standard particle current operator in the absence of the external field.[3] Note that this expression involves expectation values in the exact many-body ground state. In Dolgov and Maksimov's formulation [26] the Hamiltonian used includes in fact the complete electron–electron interaction effects including the effects of retardation. In Kubo's derivation, only the direct instantaneous Coulomb interaction is included explicitly in the Hamiltonian before the perturbation is switched on [26].

In practice, the easier quantity to calculate is the bare polarizability, which, in the independent particle approximation, is also known as the random phase approximation (RPA), using Fourier transforms in both space and time, and is given by

$$\tilde{\chi}_{ab}(\mathbf{q} + \mathbf{G}, \mathbf{q} + \mathbf{G}', \omega) = \frac{e^2}{m} n(\mathbf{G} - \mathbf{G}') \delta_{ab} - \int_{BZ} \frac{d^3k}{(2\pi)^3}$$
$$\times \sum_{ij} \frac{f_{ij}(\mathbf{k}) \langle i\mathbf{k}| j_a(\mathbf{q} + \mathbf{G}) |j\mathbf{k} + \mathbf{q}\rangle \langle j\mathbf{k} + \mathbf{q}| j_b(\mathbf{q} + \mathbf{G}') |i\mathbf{k}\rangle}{\hbar(\omega_{ij}(\mathbf{k}) - \omega - i\eta)}, \tag{17}$$

in which

$$\mathbf{j}(\mathbf{q}) = \frac{e\hbar}{2im} \left(\nabla e^{-i\mathbf{q}\cdot\mathbf{r}} + e^{-i\mathbf{q}\cdot\mathbf{r}} \nabla \right). \tag{18}$$

This equation is obtained by perturbation theory in much the same way as Eq. (10) except that it now involves a more complicated operator in the matrix elements. The susceptibility and the polarizability are related by

$$\chi = \tilde{\chi} - \frac{4\pi}{c^2} \tilde{\chi} O^{-1} \chi, \tag{19}$$

in which an abstract matrix (or operator) notation is used and O^{-1} is the inverse operator of

$$O = (\nabla \times \nabla \times) + \frac{1}{c^2} \frac{\partial^2}{\partial t^2}, \tag{20}$$

---

[3] Strictly speaking, the current and density operators appearing here should be replaced by expressions taking into account screening by magnetic interactions of currents. This may be important in strongly diamagnetic systems but is neglected here. See Ref. [26] for details.

which leads to the wave equation for the optical excitations and follows from the Maxwell equations. Using a Fourier expansion in both time and space,

$$O_{ab}^{-1}(\mathbf{q}+\mathbf{G}) = -\frac{c^2}{\omega^2} \frac{(q_a+G_a)(q_b+G_b)}{|\mathbf{q}+\mathbf{G}|^2} - \frac{c^2}{\omega^2 - c^2|\mathbf{q}+\mathbf{G}|^2}\left(\delta_{ab} - \frac{(q_a+G_a)(q_b+G_b)}{|\mathbf{q}+\mathbf{G}|^2}\right), \quad (21)$$

the transverse part (in the sense of being perpendicular to $\mathbf{q}+\mathbf{G}$) of which — i.e. the second term — is a photon propagator or Green's function and the longitudinal part of which describes the Coulomb interaction between electrons. While these equations describe the full response (including short-wavelength $\mathbf{G} \neq 0$ components), one can obtain a closed expression for the macroscopic component of the total internal electric field in response to an external field, i.e. the term $\mathbf{G}=0$. One starts from the Maxwell (wave) equation for the electric field components $\mathbf{E}(\mathbf{q}+\mathbf{G})$ and splits them into the $\mathbf{G}=0$ and $\mathbf{G}\neq 0$ ones,

$$\begin{pmatrix} \left[O_{ab}(\mathbf{q}) + \frac{4\pi}{c^2}\tilde{\chi}_{ab}(\mathbf{q},\mathbf{q},\omega)\right] & \frac{4\pi}{c^2}\tilde{\chi}_{ab}(\mathbf{q},\mathbf{q}+\mathbf{G},\omega) \\ \frac{4\pi}{c^2}\tilde{\chi}_{ab}(\mathbf{q}+\mathbf{G},\mathbf{q},\omega) & \left[O_{ab}(\mathbf{q}+\mathbf{G})\delta_{\mathbf{G'G}} + \frac{4\pi}{c^2}\tilde{\chi}_{ab}(\mathbf{q}+\mathbf{G'},\mathbf{q}+\mathbf{G},\omega)\right] \end{pmatrix}$$

$$\times \begin{pmatrix} \mathbf{E}_b(\mathbf{q}) \\ \mathbf{E}_b(\mathbf{q}+\mathbf{G}) \end{pmatrix} = \begin{pmatrix} \mathbf{j}_a^e(\mathbf{q}) \\ 0 \end{pmatrix}.$$

Since the external current $\mathbf{j}^e$ (the source of the external field) only involves $\mathbf{G}=0$, one can express the $\mathbf{E}(\mathbf{q}+\mathbf{G})$ for $\mathbf{G}\neq 0$ in terms of the $\mathbf{E}(\mathbf{q})$ component and substitute them in the equation for the $\mathbf{E}(\mathbf{q})$. This is similar to a Löwdin downfolding transformation.

The final transverse macroscopic dielectric function is obtained to be

$$\epsilon_{ab}^{M}(\mathbf{q},\omega) = \delta_{ab} - \frac{4\pi}{\omega^2}\tilde{\tilde{\chi}}_{ab}(\mathbf{q},\omega), \quad (22)$$

with

$$\tilde{\tilde{\chi}}_{ab}(\mathbf{q},\omega) = \tilde{\tilde{\chi}}_{ab}(\mathbf{q}+0,\mathbf{q}+0,\omega),$$

$$= \tilde{\chi}_{ab}(\mathbf{q}+0,\mathbf{q}+0,\omega) - \sum_{\mathbf{G},\mathbf{G'}}{}' \tilde{\chi}_a^{j\varrho}(\mathbf{q}+0,\mathbf{q}+\mathbf{G},\omega)\frac{4\pi}{|\mathbf{q}+\mathbf{G}|^2}$$

$$\times \epsilon^{-1}(\mathbf{q}+\mathbf{G},\mathbf{q}+\mathbf{G'},\omega)\tilde{\chi}_b^{\varrho j}(\mathbf{q}+\mathbf{G'},\mathbf{q}+0,\omega), \quad (23)$$

in which the prime on the summation means excluding the $\mathbf{G}=0$ and $\mathbf{G'}=0$ terms, and in which the current-density correlation function is given by

$$\tilde{\chi}_a^{j\varrho}(\mathbf{q}+\mathbf{G},\mathbf{q}+\mathbf{G'},\omega) = -\omega^{-1}\tilde{\chi}_{ab}(\mathbf{q}+\mathbf{G},\mathbf{q}+\mathbf{G'},\omega)(\mathbf{q}+\mathbf{G'})_b, \quad (24)$$

implying summation over b, and the scalar (or longitudinal) inverse dielectric function is given by

$$\epsilon(\mathbf{q}+\mathbf{G},\mathbf{q}+\mathbf{G'},\omega) = \delta_{\mathbf{G}\mathbf{G'}} - \frac{4\pi}{|\mathbf{q}+\mathbf{G}|^2}\tilde{\chi}^{\varrho\varrho}(\mathbf{q}+\mathbf{G},\mathbf{q}+\mathbf{G'},\omega), \quad (25)$$

or its inverse by,

$$\epsilon^{-1}(\mathbf{q}+\mathbf{G}, \mathbf{q}+\mathbf{G}', \omega) = \delta_{\mathbf{G}\mathbf{G}'} + \frac{4\pi}{|\mathbf{q}+\mathbf{G}|^2} \chi^{\varrho\varrho}(\mathbf{q}+\mathbf{G}, \mathbf{q}+\mathbf{G}', \omega). \quad (26)$$

Similar expressions were also obtained by Pick [25] and Del Sole and Fiorino [27].

The final expressions involve the undressed *current–density* and *density–density* as well as the *current–current* correlation function. Note that the induced charge density and current density are related via the continuity equation, $\nabla \mathbf{J} + d\varrho/dt = 0$, or in terms of the polarization $\nabla \mathbf{P}(\mathbf{r}) = -\varrho(\mathbf{r})$. So, using a Fourier transform for spatial variables, the gradient relating $\mathbf{P}$ to ϱ turns simply into multiplying by a factor $i(\mathbf{q}+\mathbf{G})$. Similarly changing a $\mathbf{J}$ into $\mathbf{P}$ involves dividing by $-i\omega$. Applying factors like this on either side transforms the current–current correlation function into a current–density or density–current correlation function and applying a factor on both sides turns it into a density–density correlation function. For example,

$$\chi^{\varrho\varrho}(\mathbf{r}, \mathbf{r}', \omega) = \omega^{-2} \nabla_a \nabla_b \chi^{jj}_{ab}(\mathbf{r}, \mathbf{r}', \omega),$$

$$\chi^{\varrho\varrho}(\mathbf{q}+\mathbf{G}, \mathbf{q}+\mathbf{G}', \omega) = -\omega^{-2}(\mathbf{q}+\mathbf{G})_a \chi^{jj}_{ab}(\mathbf{q}+\mathbf{G}, \mathbf{q}+\mathbf{G}', \omega) (\mathbf{q}+\mathbf{G}')_b. \quad (27)$$

The same type of relations hold for both the dressed quantities (without tilde) and the undressed ones (with tilde). On the other hand, these dressed correlation functions can also be expressed directly by Kubo-type formulas, as in Eq. (16) replacing simply one or both of the $\hat{j}_a$ operators by the density operator.

The important point in the derivation of Eq. (23) is that for $\omega \ll cG$ the $\mathbf{G} \neq 0$ part of the O^{-1} operator is purely longitudinal. That is the field produced by the induced currents is modified by the longitudinal electron–electron interactions only if retardation effects in the electron–electron interactions carried by photon type interactions are neglected.

A different formulation of LF effects was obtained in a later work of Del Sole and Fiorino [28] by writing the macroscopic dielectric function directly in terms of the long-wavelength components of the susceptibility tensor, (called quasipolarizability in their paper) rather than the polarizability tensor response functions, in other words by working directly with the response to the external field. In the present slightly different notation they are expressed as

$$\epsilon^M_{ab}(\mathbf{q}, \omega) = \delta_{ab} + 4\pi[\chi_{ab}(\mathbf{q}, \omega)]_{\mathbf{0},\mathbf{0}} + (4\pi)^2 \frac{[\hat{q}_a \chi_{ab}(\mathbf{q}, \omega)]_{\mathbf{0},\mathbf{0}} [\chi_{ab}(\mathbf{q}, \omega) \hat{q}_b]_{\mathbf{0},\mathbf{0}}}{1 - 4\pi[\hat{q}_a \chi_{ab}(\mathbf{q}, \omega) \hat{q}_b]_{\mathbf{0},\mathbf{0}}},$$
(28)

involving left and right "longitudinal contractions" obtained by multiplying the tensor with $\hat{q}$ from left or right or both, similar to the way in which current–current response functions are turned into current–density, density–current and density–density response functions. Note that $[\chi_{ab}(\mathbf{q}, \omega)]_{\mathbf{G},\mathbf{G}'} \equiv \chi_{ab}(\mathbf{q}+\mathbf{G}, \mathbf{q}+\mathbf{G}', \omega)$ and that here only $\mathbf{G} = 0$ components appear instead of summations over $\mathbf{G}$. All the many-body physics is now hidden in the χ functions. Del Sole and Fiorino obtained the χ from the two-particle Green's function, as discussed below when we turn to excitonic effects. They showed that in a local basis set the two-particle Green's function can actually be dealt with in practical calculations.

Both of the above formulations take proper account of the mixing of transverse and longitudinal excitations in general crystals. In the case of cubic crystals, these remain separate and it is in fact sufficient to work with the longitudinal response only, because the transverse and longitudinal response functions are the same as long as retardation effects are excluded. The latter is already contained in the above discussion. Instead of a **P** response to **E**, one can simply think in terms of an induced charge density ϱi in response to an external potential V_e or a total potential V because $\mathbf{P} = -\nabla \varrho$ and $\mathbf{E} = -\nabla V$. Thus, only scalar quantities are involved, such as the density–density response function for which we have a similar relation between dressed and undressed versions as Eq. (19)

$$\chi^{\varrho\varrho} = \tilde{\chi}^{\varrho\varrho} + \tilde{\chi}^{\varrho\varrho} v \chi^{\varrho\varrho}, \tag{29}$$

in which we have suppressed the $\mathbf{q} + \mathbf{G}$ and ω dependences, and v stands for a diagonal matrix with elements $v(\mathbf{q} + \mathbf{G}) = 4\pi/|\mathbf{q} + \mathbf{G}|^2$. In this case, the RPA expression becomes

$$\tilde{\chi}^{\varrho\varrho}(\mathbf{q} + \mathbf{G}, \mathbf{q} + \mathbf{G}', \omega)$$
$$= \frac{e^2}{\hbar} \int_{BZ} \frac{d^3k}{(2\pi)^3} \sum_{ij} \frac{f_{ij}(\mathbf{k}) \langle i\mathbf{k}| e^{-i(\mathbf{q}+\mathbf{G})\cdot\mathbf{r}} |j\mathbf{k}+\mathbf{q}\rangle \langle j\mathbf{k}+\mathbf{q}| e^{i(\mathbf{q}+\mathbf{G}')\cdot\mathbf{r}} |i\mathbf{k}\rangle}{\omega_{ij}(\mathbf{k}) - \omega - i\eta}. \tag{30}$$

Within the DFT context, one may now also wish to include the effects of the exchange and correlation in the interactions that modify the total field or potential from the externally applied one, in which case χ becomes

$$\chi = \tilde{\chi}[1 - (v + K_{xc})\tilde{\chi}]^{-1}, \tag{31}$$

in which $K_{xc} = \delta E_{xc}/\delta n(\mathbf{r}) \delta n(\mathbf{r}')$ is the exchange–correlation kernel. Sometimes the inclusion of these exchange–correlation effects is called local-field corrections. For clarity, we will distinguish it from the above discussed local field effects which are present even in the absence of exchange correlation by calling them XC-local field effects.

Gavrilenko and Bechstedt [29] have recently investigated various aspects of local field and exchange-correlation effects on the spectra. The main conclusions of this work are that LF effects in the sense of Eq. (12) decrease the oscillator strength in the region below the main absorption peak and increase it near the main absorption. On the other hand, when the XC-LF effects of Eq. (31) are included, this effect is somewhat reduced. These conclusions agree with earlier work by Hanke and Sham [30] but remove the uncertainties related to an empirical band structure from this earlier work.

Gavrilenko and Bechstedt also studied the effect of including the QP shift of the eigenvalues. They noticed that in order to obtain agreement between theoretical and experimental features such as peak positions and/or the zeros of the real part of the dielectric function, or the value of the static dielectric function, the shifts included need to be smaller than the values obtained in GW calculations. This effect is not well understood but we note that agreement with theory should probably not be expected until excitonic effects are also included; it depends on the accuracy of the GW results used as input.

Although local field effects are usually described in terms of the plane wave expansion used above, this is not the only possibility. More generally, we can for example write the induced charge density as

$$n^i(\mathbf{r}, \omega) = \int \chi(\mathbf{r}, \mathbf{r}', \omega) V(\mathbf{r}', \omega) d^3r', \tag{32}$$

in which χ depends on $\mathbf{r}$ and $\mathbf{r}'$ separately. This means only that $\chi(\mathbf{r}, \mathbf{r}', \omega)$ must be expanded in a double Bloch function expansion

$$\chi(\mathbf{r}, \mathbf{r}', \omega) = \sum_{nn'} \sum_{\mathbf{q}} B^*_{n\mathbf{q}}(\mathbf{r}) \chi_{nn'}(\mathbf{q}, \omega) B_{n'\mathbf{q}}(\mathbf{r}'), \qquad (33)$$

where $B_{n\mathbf{q}}(\mathbf{r})$ can be any suitable basis set of functions satisfying the Bloch condition $B_{n\mathbf{q}}(\mathbf{r} + \mathbf{T}) = \exp(i\mathbf{q} \cdot \mathbf{T}) B_{n\mathbf{q}}(\mathbf{r})$ for any translation vector $\mathbf{T}$. The most frequently used basis set for this purpose is that of plane waves, $B_{\mathbf{G},\mathbf{q}}(\mathbf{r}) \propto \exp i((\mathbf{q} + \mathbf{G}) \cdot \mathbf{r})$. However, other basis sets are possible. The general structure of the theory remains valid if all matrices in the $\mathbf{G}$ indices are simply replaced by whatever new index n labels the basis functions. For example, Aryasetiawan and Gunnarsson [31] recently introduced a product basis set in which suitably orthogonalized products or muffin-tin orbitals are used. The economy in basis size compared to the standard plane waves obtained by this method allowed them to study difficult cases involving highly localized d orbitals in metals/compounds such as Ni and NiO [32]. This follows earlier work by Hanke and Sham who also adopted a local orbital approach [30]. Another approach worth mentioning here is the mixed space approach of Blase et al. [33]. In this case, instead of using a local orbital representation, a real space mesh is introduced but the periodicity of the function in both $\mathbf{r}$ and $\mathbf{r}'$ is still exploited. Both the mixed space and local orbital basis set approaches have important advantages for large systems.

3.3 Excitonic effects

Now we turn to additional terms in the Hamiltonian. Possible terms we should add for a more accurate treatment are: the electron–electron interactions and the electron–phonon interactions. The electron-electron interaction is expected to play an important role beyond what is implicitly included in the effective potential describing our independent particle Hamiltonian, because the excitations produced by the $\mathbf{A}$ terms will introduce electrons and holes which now interact with each other in the perturbed system. Electron–phonon interactions are of course crucial to allow for the description of indirect transitions. In our present formulation because the photon is treated in the long-wavelength approximation has negligible $\mathbf{q} \to 0$, so that only direct vertical transitions occur. The inclusion of electron–phonon interactions has been discussed by Elliott [34], but has to our knowledge only been applied to study the discrete indirect excitons, and has up to now not been carried over to the first-principles calculations of the full continuum part of the spectrum. We will not discuss it further here.

When electron–electron interactions are included in the Hamiltonian, the general Kubo formulation of the problem is still valid, but now we have to calculate the expectation value of the current–current correlation function with the correlated many-body wave functions. The susceptibility can be recognized to be a special case of the more general two-particle Green's function, $\chi(\mathbf{r}, \mathbf{r}', \omega) = -iS(\mathbf{r}, \mathbf{r}, \mathbf{r}', \mathbf{r}'; \omega)$, or $\chi(1, 2) = -iS(1, 1; 2, 2)$, which satisfies the Bethe-Salpeter equation:

$$S(1, 1'; 2, 2') = S_0(1, 1'; 2, 2') + S_0(1, 1'; 3, 3') \Xi(3, 3'; 4, 4') S(4, 4'; 2, 2'), \qquad (34)$$

in which the notation 1 is a shorthand notation for space and time coordinates $(\mathbf{r}_1, t_1)$. The zeroth-order (noninteracting) $S_0(1, 1'; 2, 2') = G(1', 2') G(2, 1)$ corresponds to the polarizability $\Pi(1, 2)$ for $1 = 1'$ and $2 = 2'$. The interaction kernel Ξ consists of a screened direct interaction between electron and hole and an unscreened exchange term.

$$\Xi(1, 1'; 2, 2') = -i\,\delta(1, 1')\,\delta(2, 2')\,v(1, 2) + i\,\delta(1, 2)\,\delta(1', 2')\,W(1, 1'). \tag{35}$$

The reason why the direct interaction needs to be screened while the exchange term should not was given by Sham and Rice [35], who discussed the general many-body theory underpinnings of the effective mass theory approximation to excitons. It amounts to a question of avoiding double counting. The application of this approach to the general continuum part of the spectrum was later developed by Hanke and Sham [30]. As discussed by these authors, the terminology here can get quite confusing. From the point of view of excitons, the direct term is the electron–hole attraction described by the screened Coulomb interaction W and the term involving v is the exciton exchange term. From the point of view of electron–electron interactions or response functions the term involving v arises from relating the susceptibility χ to the polarizability $\tilde{\chi}$ in the RPA. This is also known as the time-dependent Hartree approximation. The term involving W is then viewed as the exchange counterpart to the electron–electron interaction. So, including the v term can be viewed as being equivalent to including v in a density functional based theory by the time dependent Hartree approximation as in Eq. (29). Including the exchange correction to it is approximately treated in DFT by including the K_{xc} in that equation as in Eq. (31). On the other hand, it turns out that the term with W is the stronger one of the two and the most important one to include. It was shown already by Hanke and Sham that it shifts the oscillator strength to lower energy unlike the other XC-local field type terms which tend to do the opposite. Thus, in contrast to previous attempts to include many-body corrections in the susceptibility while staying within the one-particle model, it is now recognized that the better way to go is to explicitly switch to the two-particle formulation.

Recently, new progress has been made in implementing the solution of the Bethe Salpeter equation by introducing the set of single electron–hole pair excited states as basis set [17, 18, 19]. Here we follow Benedict et al. [19] notation defining these states as follows:

$$|f\rangle = \sum_{ij\mathbf{k}} \psi(i, j, \mathbf{k})\, a_{i\mathbf{k}}^\dagger a_{j\mathbf{k}} |0\rangle, \tag{36}$$

in which $a_{i\mathbf{k}}^\dagger$ creates a conduction electron in one of the originally unoccupied single particle eigenstates out of the ground state $|0\rangle$ and $a_{j\mathbf{k}}$ eliminates a valence electron at the same $\mathbf{k}$. The final states are mixtures of the pure band excitations described by the interacting electron–hole wave function $\psi(i, j, \mathbf{k})$ and have energies E_f. An effective two-particle Hamiltonian, $H_{2p-\text{eff}}$ can be written down of which the two-particle Green's function defined above is the resolvent and for which the eigenstates are the above defined $|f\rangle$. In other words diagonalization of this effective two-particle Hamiltonian gives the $\psi(i, j, \mathbf{k})$ eigenvectors. The effective Hamiltonian matrix in this basis consists of a diagonal term containing the differences of the one-particle eigenvalues (corresponding to each electron–hole pair being excited) plus a term describing matrix elements of the interaction kernel Ξ, which mixes different pairs $(i, j, \mathbf{k})$. As one-electron states one uses the best states available, say the ones from a GW calculation. There are now two approaches: either one calculates the $\psi(i, j, \mathbf{k})$ explicitly [17, 18], which then gives information on how much each of the original electron–hole pair states at each $\mathbf{k}$ contributes to the final excitations, or one calculates directly the resolvent [19]. The former has the disadvantage that it involves diagonalization of fairly large matrices because of the coupling of states with different **k**. This apparently requires a very fine mesh in **k**-space [17]. Benedict et al. [19, 20] avoid this problem by making use of a second form of the basis set of single electron–hole pair states written explicitly in real space and utilize the fact that the direct interaction becomes diagonal in this representation if a suitable approximation is made to the screened interaction. In fact, they avoid the problem of the large matrix diagonalization altogether by directly calculating the resolvent which gives the required optical dielectric function. The transformation to a resolvent problem can be done so as to already include the **J** matrix elements. In terms of the states $|f\rangle$, the imaginary part of the dielectric tensor is written as

$$[\epsilon_2(\omega)]_{ab} = \frac{4\pi^2}{\omega^2} \sum_f \langle 0| J_a |f\rangle \langle f| J_b |0\rangle \delta(\hbar\omega - E_f), \qquad (37)$$

but it can be cast into the form

$$[\epsilon_2(\omega)]_{ab} = -4\pi \, \text{Im} \, \langle P_a| \, (\omega - H_{\text{2p-eff}} + i\eta)^{-1} \, |P_b\rangle, \qquad (38)$$

in which $|P_a\rangle = H_{\text{2p-eff}}^{-1} J_a|0\rangle$. The problem thus reduces to calculating the imaginary part of the elements of a resolvent, and it can be solved by means of Haydock's recursion method [36]. Although this approach avoids diagonalizing the large effective two-particle Hamiltonian matrix explicitly, it gives less detailed information on the nature of the transitions because the eigenstates are not calculated explicitly. Nevertheless, a spectral decomposition can still be obtained with some extra work. For more information, we refer to Benedict et al. [19, 20]. A full inclusion of local field effects in the transverse case could possibly be included in these theories following the approach of Del Sole and Fiorino mentioned above [28].

3.4 Nonlocality in potential

One final point needs to be mentioned to wrap up this overview section. So far we have assumed that $V(\mathbf{r})$ in the one-electron Hamiltonian was a simple local potential. If we take as starting point the GWA quasiparticle states, or in a GKS scheme, or even in the case of nonlocal pseudopotentials, the potential contains nonlocal operators, which therefore do not commute with **r**. This means that in calculating the velocity from $i\hbar\mathbf{v} = [\mathbf{r}, H_0]$, terms other than the quasimomentum **p** appear. In Eq. (10), above, the $\mathbf{p}_{ij}$ matrix elements more generally should be replaced by $m\mathbf{v}_{ij}$. This should be clear from Eq. (17), for example, taking the limit for $\mathbf{q} \to 0$. As discussed by Levine and Allan [37], even if we represent the QP corrections simply by means of a scissor's operator $\sum_c |c\rangle \Delta_c \langle c|$, where c stands for the conduction bands, this means that the momentum matrix elements must be renormalized. The specific assumption in their work is that the GW or scissor shifted wave functions do not change appreciably from the LDA wave functions and hence the $\mathbf{r}_{ij}(\mathbf{k})$ matrix elements are assumed to stay unchanged under the scissor shift. Because of the commutation relation $i\hbar\mathbf{v} = [\mathbf{r}, \mathcal{H}]$, we have

$$i\omega_{ij}(\mathbf{k}) \, \mathbf{r}_{ij}(\mathbf{k}) = \mathbf{v}_{ij}(\mathbf{k}) \qquad (39)$$

and $i\tilde{\omega}_{ij}(\mathbf{k}) \, \mathbf{r}_{ij}(\mathbf{k}) = \tilde{\mathbf{v}}_{ij}(\mathbf{k})$, where quantities with tilde indicate those corresponding to the Hamiltonian with scissor or GW shift. Thus, one can obtain the new $\tilde{\mathbf{v}}_{ij}(\mathbf{k})$ or

$\tilde{\mathbf{p}}_{ij}(\mathbf{k}) = m\tilde{\mathbf{v}}_{ij}(\mathbf{k})$ matrix elements needed in Eq. (11) by the renormalization

$$\tilde{\mathbf{v}}_{ij}(\mathbf{k}) = \mathbf{v}_{ij}(\mathbf{k}) \frac{\tilde{\omega}_{ij}(\mathbf{k})}{\omega_{ij}(\mathbf{k})}. \tag{40}$$

This means effectively that in the denominator in Eq. (11) only the $\omega_{ij}(\mathbf{k})$ in the resonance factor (the one including $-\omega$) should be replaced by the shifted $\tilde{\omega}_{ij}(\mathbf{k})$ energy differences if we still use the matrix elements calculated in the LDA. The use of LDA wave functions for the optical matrix elements is consistent with the usual practice of calculating the GW corrections to the eigenvalues in first-order perturbation theory, justifications of which can be found in Ref. [10]. If we write the expressions for $\chi^{(1)}$ directly in terms of $\mathbf{r}_{ij}(\mathbf{k})$ matrix elements, as will be discussed in Section 6, it is also clear that if we finally calculate these in terms of $\mathbf{p}_{ij}(\mathbf{k})$ using the above relationships and using LDA wave functions, we should also use LDA energy differences in Eq. (39).

We note here that inclusion of a constant scissor shift and at the same time renormalizing the matrix elements is not entirely consistent as was discussed in Ref. [38]. In fact, when renormalized matrix elements as well as energy denominators are substituted in the $\mathbf{k} \cdot \mathbf{p}$ effective mass expression,

$$\left[\frac{m}{m_i^*}\right]_{ab} = \delta_{ab} + \frac{1}{\hbar m} \sum_{i \neq j} \frac{p_{ij}^a(\mathbf{k}) p_{ji}^b(\mathbf{k}) + p_{ij}^b(\mathbf{k}) p_{ji}^a(\mathbf{k})}{\omega_{ij}(\mathbf{k})}, \tag{41}$$

the second term will be increased, meaning that the effective mass would decrease. On the other hand, with a constant scissor shift the mass should stay the same and generally speaking one would rather expect the masses to increase if the gap increases. Thus, it appears that neither assuming $\mathbf{p}_{ij}(\mathbf{k})$ nor $\mathbf{r}_{ij}(\mathbf{k})$ to stay the same when changing from LDA to QP wave functions is entirely correct. In some sense the correct link between matrix elements and eigenvalues is lost when the eigenfunctions used in the calculation of the matrix elements are not obtained from the same Hamiltonian. In Ref. [38] we used a somewhat empirical adjustment to the Hamiltonian instead, and then recalculated the matrix elements with the new eigenvectors. Unfortunately, this empirical adjustment only works well for the lowest conduction band. It gave superior results for GaAs because there the optical functions of interest are dominated by transitions to the lowest conduction band in the region of the main features in $\epsilon_2(\omega)$. However, this is not generally the case. Nevertheless, this indicates that schemes such as the GKS method in which a nonlocal exchange correlation potential is used within the Kohn-Sham equation are promising. In fact, they give eigenvalues in closer agreement to the dynamical QP equations as a starting point, and their wave functions are explicitly obtained so that one does no longer need to decouple the eigenvalue differences used in the optics calculation from the eigenfunctions used in the matrix elements.

The importance of including nonlocality corrections explicitly was also emphasized by Adolph et al. [39]. They pointed to the nonlocalities arising from the nonlocal pseudopotentials as well as the QP shift related effect. Further work on this topic was done recently by Del Sole and Girlanda [40] and Bechstedt et al. [41]. They investigated how taking into account the dynamic (energy dependent) effects in the GW method affects the optical spectra. It was found [40] that when the energy dependence of the self-energy contribution to the $i\mathbf{k}$ QP energy $E_i(\mathbf{k})$, which can be quantified by $\beta_i(\mathbf{k}) = -\langle i\mathbf{k}| \partial \Sigma(\mathbf{r}, \mathbf{r}', E)/\partial E |i\mathbf{k}\rangle\, E = \varepsilon_i(\mathbf{k})$, with $\varepsilon_i(\mathbf{k})$ the LDA eigenvalue and $|i\mathbf{k}\rangle$ the LDA Bloch state, is included in the renormalizations of the matrix elements, considerable spectral strength is shifted away from the QP peaks into satellite structures. The overall reduction of the main peaks' spectral strength considerably worsens agreement with experiment. On the other hand [41], this effect was found to be almost completely cancelled by corresponding corrections which must then be made to the vertex (in other words the Ξ of Eq. (34)) of the polarizability calculation. To isolate specifically these effects, the authors of Ref. [41] still neglect the direct electron–hole interaction or excitonic effect. In other words, their approach is a further generalization of Eq. (31), in the sense that the kernel K_{xc} appearing here now includes a first-order approximation to the dynamical effects in the screened Coulomb interaction. The cancellation found in this work seems to justify why approaches based on using GW eigenvalues in the spectral functions but not including the dynamical effects in the matrix element calculation (as for example in Benedict et al. [19, 20]) work well.

We have taken some detour explaining the essentials of the various approaches to go beyond the "standard" approach, which we define here to consist in calculating the polarizability in perturbation theory starting from an effective one-particle Hamiltonian, and neglecting local field and excitonic effects completely. The practical results of inclusion of LF effects have already been discussed. The results for excitonic effects will be considered in detail in the next section using a specific case study.

4. Band Structure Method Aspects

Most implementations of the above methods have been carried out in the framework of pseudopotential plane wave basis set band structure methods. The advantage of this basis set is that calculating the momentum matrix elements is trivial. The disadvantages are that the nonlocal pseudopotentials that are required for accurate calculations complicate the matrix elements, as was mentioned at the end of the previous section, and that a large number of basis functions is required to calculate the handful of bands of interest. Thus, it is costly to diagonalize the matrices and sometimes compromises are made in terms of the number of **k**-points included in the integrals over the BZ. Nevertheless, it is using this method and also using plane waves for the expansion of reponse functions that most of the calculations which go beyond the standard approach have been carried out.

In our own work, we have used the linear muffin-tin orbital method[4]) in its atomic sphere approximation [42, 43]. The momentum matrix elements are calculated from the so-called one-center partial wave expansions of the LMTO method. As is well known, within the LMTO method, one can either expand the total wave functions in multi-centered muffin-tin orbitals

$$\chi_{RL}^{\mathbf{k}} = \phi_{RL}^{\mathbf{k}} - \sum_{R'L'} \dot{\phi}_{R'L'}^{\mathbf{k}} S_{R'L',RL}(\mathbf{k}), \tag{42}$$

in which $S_{R'L',RL}(\mathbf{k})$ are structure constants and $\phi_{RL}^{\mathbf{k}}$ and $\dot{\phi}_{RL}^{\mathbf{k}}$ are atom-centered partial waves (i.e. radial functions times spherical harmonics) at energy E_v and their energy derivatives respectively, or in a one-center expansion in terms of the $\phi_{RL}^{\mathbf{k}}(E_{n\mathbf{k}}) = \phi_{RL}^{\mathbf{k}} + (E_{n\mathbf{k}} - E_v) \dot{\phi}_{RL}^{\mathbf{k}}$. Here, R labels the atoms in the unit cell and Bloch summation has already been carried out, and $L = l, m$ labels the spherical harmonic components. Within the so-called nearly orthogonal representation [43] the expansion

---

[4]) The actual version of the LMTO used in our work is now known as the "second generation LMTO" and is described in [43].

coefficients of the Bloch eigenstates defined by

$$\psi_{n\mathbf{k}} = \sum_{RL} C_{RL}^{n\mathbf{k}} \phi_{RL}^{\mathbf{k}}(E_{n\mathbf{k}}) = \sum_{RL} C_{RL}^{n\mathbf{k}} \chi_{RL}^{\mathbf{k}}, \qquad (43)$$

coincide. The matrix elements of the momentum operator between Bloch states thus reduce straightforwardly to internal products of the eigenvectors of the LMTO secular equation and matrix elements between the atom-centered partial waves. By expanding the momentum operator $-i\hbar\nabla$ in its spherical (rank-1) tensor components $-i\hbar\nabla_m$ with $m = 0, \pm1$, given by $\nabla_0 = \partial/\partial_z$, $\nabla_{\pm1} = \mp(1/\sqrt{2})(\partial/\partial_x \pm i\partial/\partial_y)$, the latter matrix elements in turn reduce to radial matrix elements and Clebsch-Gordan coefficients. The Wigner-Eckart theorem is used so that only the radial matrix element for ∇_0 needs to be calculated explicitly. The final radial matrix elements appearing are between $\phi_{Rl'}$ and ϕ'_{Rl}, the radial derivative of ϕ_{Rl} with $l' = l \pm 1$ and the corresponding ones with ϕ replaced by $\dot{\phi}$. All these quantities are readily available in the LMTO codes. We note that even though the equations appearing in Section 6 are written in terms of $\mathbf{r}$ matrix elements (to be properly defined below) these are ultimately calculated from $\mathbf{p}$ or $\mathbf{v}$ velocity matrix elements using the relation Eq. (39).

There may be some concern about the accuracy of the matrix elements because of the ASA. A correction may be applied to the matrix elements for the contributions of the interstitial region and the overlap of the spheres. This is usually known as the combined correction term. Its effects on matrix elements were discussed by Alouani et al. [44]. The convergence with respect to the angular momentum quantum number was discussed by Uspenski et al. [45]. In practice, we find it often is necessary to go to slightly higher angular momentum cut-off for obtaining matrix elements than one would normally use for the band structure for the same material. Further examples are given in Rashkeev et al. [38]. To put these concerns to rest for the time being, that paper showed excellent agreement with FLAPW (full-potential linear augmented plane-wave) calculations [5] for the few cases studied by both this method and ASA-LMTO. Furthermore, these results concerned second harmonic generation (SHG), which is more sensitive to matrix elements than linear optics.

It is perhaps worthwhile pointing out that while the ASA approximation to the charge density is a poor approximation for total energy and hence structural properties, the ASA approximation to the potential (even if obtained self-consistently from an ASA charge density) is quite accurate and the band structures are generally speaking in good agreement with those obtained from full-potential methods. The reasons for this are among others discussed in the article on the LMTO method in this volume [46]. Of course, the quality of the ASA results is sensitive to an appropriate choice of sphere radii. While the general guidelines for choosing sphere radii are clear, this sometimes may add some uncertainty in the study of systems with unusual crystal structures. Our own practice has been to cross check the accuracy of the electronic structure with FP-LMTO calculations before undertaking a study of the optical properties using the ASA.

A full-potential calculation of the optical matrix elements has also been developed by Alouani on the basis of the FP-LMTO codes by Wills, see for example Ref. [47]. In this approach, the interstitial region is treated by extrapolating the smooth LMTO tail inside the spheres, thus defining a smooth pseudofunction in all space. The optical matrix elements are then calculated by using this dual representation of the LMTO. The matrix elements of the pseudofunction are easily calculated in all space by Fourier transformation, i.e. like in a plane wave method. The spherical harmonic expansions of the pseudofunction inside the spheres are treated as before in the ASA and the contributions inside the spheres are subtracted and replaced by the corresponding contributions of the actual LMTO inside the sphere. This method is thus closer in spirit to the FLAPW method in that it uses plane wave expansions in treating the interstitial region. The approach is exactly the same as the "combined correction" approach described in Andersen's 1975 paper [42] but now applied to non-overlapping muffin-tin spheres. The disadvantage of this FP-LMTO implementation is that it looses the simplicity and minimal basis set character of the ASA-LMTO method.

In comparing the results from various calculations, care should be taken to consider case by case. For example, calculations using lattice constants obtained from the minimization of the LDA total energy, which are as a rule slightly smaller than the experimental values, give rise to shifts in eigenvalues (due the band deformation potentials) and thus slightly different optical spectra as well. While in the purist's point of view, one should use theoretical lattice constants to stay consistent, one might argue that there is no need to use a wrong lattice constant when trying to make predictions of the optical properties. Unlike the calculations of phonons where it is essential to perform the expansions of the total energy about the theoretical equilibrium positions, one here is after all concerned with providing the best possible description of the electronic system and its optical response given the positions of the atoms.

5. Case Study: GaN

In this section we examine the optical reponse functions of GaN as an example which was investigated already in considerable detail in our previous work [48 to 50]. Nevertheless, some new aspects, in particular relating to polarization dependence are dis-

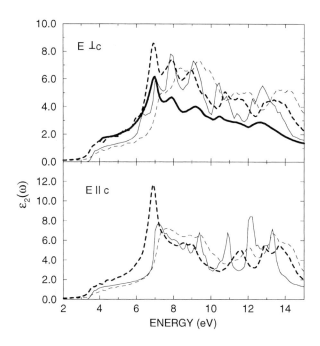

Fig. 1. Imaginary part of the dielectric function $\epsilon_2(\omega)$ in wurtzite GaN: top panel for $\mathbf{E} \perp \mathbf{c}$, bottom panel for $\mathbf{E} \parallel \mathbf{c}$: thick solid line, experiment from Ref. [48], thin solid line: ASA-LMTO results in independent-particle approximation, thick (thin) dashed lines: results from Ref. [20], including (excluding) electron–hole interaction effects

cusssed here for the first time. GaN was recently also studied by Benedict and Shirley [20] including excitonic effects so that we can examine how important these effects are.

The top panel of Fig. 1 shows the imaginary part of the dielectric function (for $\mathbf{E} \perp \mathbf{c}$) of wurtzite GaN compared to experiment. The latter, shown by the heavy solid line, was obtained by Kramers-Kronig analysis of normal incidence reflectivity on basal plane GaN, and is in agreement with earlier data by Olson et al. [51]. The first theoretical curve (thin solid line) is obtained in the LDA but shifted by a constant scissor's correction of ≈ 1 eV, so as to give the correct minimum band gap and was calculated using the LMTO method [48]. One may note that there is good agreement between theory and experiment, as far as the energies of the main features are concerned. This is why the standard model band structure approach as defined in this paper is still useful to interpret optical spectra, or vice versa why optical spectra are useful to obtain band structure information. The present results give support to the approximate validity of a constant scissor shift difference between the LDA and QP band structures. In contrast, the available QP band structure results for GaN [52, 53] indicate a trend of increasing QP shift with increasing energy in the conduction band [50]. We caution to add that the trend is not completely uniform. There are also variations from $\mathbf{k}$-point to $\mathbf{k}$-point, indicating that the wave function character plays a role. Furthermore, the trend is stronger in the results of Rubio et al. [52] than in Palummo et al. [53]. In any case, the trend is not supported by the above comparison between theory and experiment and illustrates how the comparison with optical data provides valuable information on checking the status of current GW calculations. The effect may in part be due to the use of first-order perturbation theory in the calculation of the GW corrections.

Next, we examine the discrepancies in peak intensity. First of all, one may notice an increasing overestimate with increasing energy of the present theoretical intensity compared to the experiment. In Ref. [48] it was argued that in fact, this can in large part be explained by an experimental error: the measured reflectivity is lower than the specular reflectivity needed to derive $\epsilon_2(\omega)$ because of diffuse scattering. This problem becomes worse with higher and higher energy because the wavelength becomes comparable to the lateral length scale of the surface roughness. However, there is also a genuine error in the theory which was not appreciated at the time of writing Ref. [48]. In fact, when examining the first three peaks, one may notice that their relative intensities are exactly in opposite order in theory and experiment. Anticipating the discussion below, we note that this problem is essentially solved when the electron–hole interaction is included as indicated by the thick dashed line curve in this figure representing the recent results by Benedict and Shirley [20] including electron–hole interaction effects. We note, however, that their independent-particle result indicated in the thin dashed line differs from ours. In particular, the peaks at higher energy are shifted to higher energy. This ultimately leads to a remaining discrepancy with experiment when the electron–hole interaction is switched on. This is in part due to the manner in which QP effects are included. In fact, in their calculation, the conduction and valence bands were "stretched" so as to mimic the GW results of Rubio et al. [52]. The discrepancy at higher energies again indicates that this increasing shift of the conduction band states in GW is not supported by experiment.

Further indications that something is wrong with the independent-particle model are obtained by considering zincblende GaN, for which $\epsilon_2(\omega)$ is given in Fig. 2. In this case, the experimental data are from spectroscopic ellipsometry [54]. The thin solid line, the theory curve, again corresponds to our LMTO calculation [49]. Here, there appears to

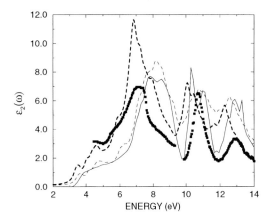

Fig. 2. Imaginary part of the dielectric function $\epsilon_2(\omega)$ in zincblende GaN: square dots line, experiment from [54], thin solid line: ASA-LMTO results in independent-particle approximation, thick (thin) dashed lines: results from Ref. [20], including (excluding) electron–hole interaction effects

be an important difference in peak position between that theory and experiment for the first main absorption peak, usually associated with the so-called E_1, E_2 transitions in zincblende crystals. In fact, the theory overestimates this peak position by about 1 eV. Note that the same 1 eV scissor shift has been applied to the LDA position of the spectrum as for wurtzite. In fact, this shift is so large that other authors decided that better agreement is obtained for this peak using the unshifted LDA results [47], their explanation being that perhaps QP corrections are smaller at this energy than at the minimum gap. In that case the higher peaks do no longer coincide but this was ignored because there was some doubt about the validity of these higher experimental peaks. In fact, in a later paper [55], the peaks between 10 and 12 eV seem to have disappeared! The reason for this is that these peaks were deemed to be spurious features due to second order diffraction of the light from the spectrometer and were filtered out. We believe the authors may have "thrown the baby out with the bathwater", so to speak and that this feature does exist. In any case, if one lets the peaks at 7 eV coincide with the unshifted LDA theory, then the higher peaks at 10 and 13 eV are underestimated by the theory by 1 eV. So, this then requires a QP correction which first goes down from 1 eV at the gap to about 0 in the region of the E_1, E_2 peaks and then back up to 1 eV. This is hard to understand within GW theory.

To shed further light on the question, we must now examine the origin of the features in terms of specific band-to-band transitions and consider the various types of corrections beyond the "standard model" discussed in Section 3.

First, how do we decompose the optical function into its contributions? Traditionally, this is done more or less by inspection and emphasis is placed on transitions at critical points, typically occurring at high symmetry points in the BZ. This approach has been further developed to include analytical line shapes corresponding to different types of critical points by Cardona et al. [56] and is used for the present material by Logothetidis et al. [57] and Petalas et al. [47]. On the other hand, we can of course break down our calculated $\epsilon_2(\omega)$ function back into the components from which it was calculated. First, we can decompose the function in contributions from each pair of valence i, and conduction j bands (i, j), ordered simply in order of increasing energy. This is shown for zincblende GaN in Fig. 3. It becomes obvious then that this feature mainly results from transitions between the top two valence bands and the lowest conduction band. A difficulty here is that at certain symmetry points, these bands become degenerate and so there are

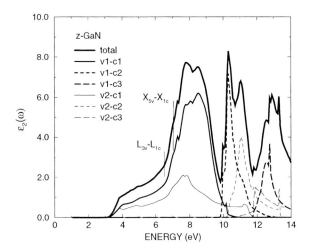

Fig. 3. Decomposition of $\epsilon_2(\omega)$ in individual band-to-band contributions

also slight contributions from the next lower valence bands to the first conduction band arising from the points where these are degenerate to the topmost valence band.

Next, which regions of **k**-space contribute most? Assuming that the matrix elements do not vary strongly with **k**, this is determined by the joint density of states (JDOS). The latter is strong in regions where the bands are parallel. We therefore find it useful to plot the band-pair energy differences in the Brillouin zone in the same way as we would plot the band structure itself. This is shown in Fig. 4, in which energy differences to various conduction bands from valence band 1, 2, 3 (counting down from the top) are indicated by different symbols on the curves. Regions where the energy differences are constant then correspond to regions of high density of states and extrema correspond to van Hove singularities. The latter can either be minima, saddle points of type 1 or type 2 or maxima. The corresponding expected features in the spectral function are well known and for instance discussed in Jones and March [58]. Now, it becomes apparent that the peak in question near 7 eV arises from the flat regions near X and L, these points marking singularities. Further inspection shows, however, that the transitions at the points L and X themselves, which in the traditional interpretation would

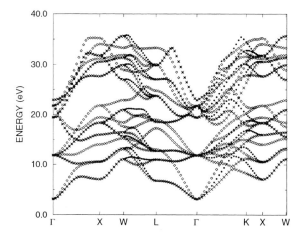

Fig. 4. Transition energy band structure: crosses, transitions originating from topmost valence band v_1, full circles, transitions from v_2, open diamonds, transitions from v_3

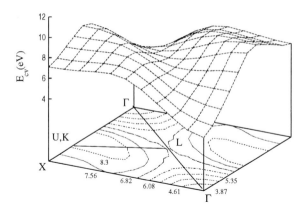

Fig. 5. Energy band difference between lowest conduction band and topmost valence band in z-GaN shown in $\Gamma - X - L$ plane

have been interpreted as critical points lie in fact just below the main peak! They correspond to small changes in slope on the partial curves shown in the decomposition of Fig. 3. The energy band difference can be further examined by making a plot of the $E_{c1}(\mathbf{k}) - E_{v1}(\mathbf{k})$ surface for $\mathbf{k}$ in the entire $\Gamma - X - L$ plane. This is shown in Fig. 5. Now, one recognizes more dramatically that there is in fact a fairly large range of $\mathbf{k}$-space in which this band difference is almost constant and is thus strongly contributing to this peak. Thus, the detailed calculations do not support the emphasis that is traditionally placed on high symmetry critical points. It would clearly be wrong to take the peak position to coincide with the exact extremum points at L and X. Rather it corresponds to the average energy position of this large more or less flat region.

A similar analysis has been carried out in great detail for wurtzite GaN in Ref. [48]. In fact, in this case, because the Brillouin zone is smaller and hence there are more bands, one finds that many features are made up from several band-to-band contributions for each of which a JDOS study has to be made to identify the prime $\mathbf{k}$-space locations. This part of the analysis is unfortunately still somewhat ad hoc and based on visual inspection of correlations between peak positions and regions of $\mathbf{k}$-space with flat energy band differences. This explains why Christensen and Gorczyca [59] obtained assignments slightly different from ours for some features in wurtzite GaN, even though the method for the calculation and the band-to-band decomposition were performed in an almost identical manner. It would be desirable to obtain an unbiased automatic procedure for doing this, which ideally would also take into account the strength of the matrix elements and its variation with $\mathbf{k}$.

Nevertheless, this as an aside, let us now return to our problem of the zincblende spectrum. Clearly, what happens here is that a large region of $\mathbf{k}$-space contributes to the theoretical peak while the traditional assignments of E_1 to the $\Gamma - L$ line and E_2 to the neighborhood of the X point are somewhat off because these occur at lower energies on the slope of the peak. Can one now explain the discrepancy by differences in QP corrections by inspecting the gap corrections at X and L compared to those of the minimum gap which occurs at Γ? Apparently not. While QP shifts obtained from GW calculations in many III–V semiconductors are indeed slightly larger at Γ than at X [9], the opposite is found for GaN [52]. Again, there are discrepancies here between the two GW calculations. Palummo et al. [53] do find a smaller gap correction for the X_{1c} state than for the Γ_{1c} state. Nevertheless the direct gap at X increases more than the direct gap at Γ because of the valence band GW shift. In Rubio et al. results, the direct correction is in fact 0.7 eV larger at X (and 1.3 eV at L) than at Γ. On the other hand, GW calculations by Palummo et al. [53] give only slightly larger (0.1 to 0.2 eV) GW shifts at X and L than at Γ. In either case taking into account the **k**-dependence of the QP correction goes in the wrong direction concerning a reconciliation of theory and experiment! This is not to say that these calculations are wrong, but only that additional effects must play a role. Also, the GW calculations by Palummo et al. [53] are more in line with expectations from other III–V semiconductors than Rubio et al. [52].

Next, what about LF and XC-LF effects? From the studies of Gavrilenko and Bechstedt [29] (admittedly on other materials), it would appear that these corrections are small and furthermore also go in the wrong direction. In fact, they find the oscillator strength to move towards the main peaks from the energy region below them, whereas, here we need exactly the opposite.

Thus, we arrived in Ref. [49] at the conclusion that excitonic effects were the most likely culprit for the discrepancy. We argued that these might be expected to be particularly strong because of the extent of the region of the parallel bands and because of the high ionicity of GaN. Subsequently, this explanation was confirmed by the work of Benedict and Shirley [20].

In fact, we have already included their results including and excluding electron–hole interaction effects in Fig. 2. Clearly, in zincblende GaN the shift of oscillator strength from the main first main peak to lower energy is so strong that the whole peak appears to shift. Similarly, in wurtzite GaN, the inclusion of electron–hole interactions also shifts the oscillator strength to lower energies, resulting in fact in a relative order of peak intensities in agreement with experiment. This example clearly illustrates that inclusion of the excitonic effects, in particular the direct electron–hole interaction is one of the most important corrections to the independent-particle model needed to bring theory and experiment into agreement. While other corrections should be examined carefully in a complete theory, for example, the energy dependence of GW shifts and vertex corrections, there are partial cancellations between such effects, which may allow one to neglect them in an approximate theory, if the approximations are done in a judicious manner. But the excitonic effect clearly must be included for a quantitative agreement with experiment.

What is interesting is that, at least in the case study examined of GaN, the excitonic effects shift back the oscillator strength towards the high-symmetry critical points X and L. This perhaps explains the success of the empirical models which placed emphasis on these points. If this is a generally occurring phenomenon, which remains to be seen, but seems worthwhile of further exploration, then the reason for the strong correlation of optical features with high-symmetry points is not so much that they have the highest joint density of states but rather that the electron–hole interaction increases the oscillator strength near such points. Intuitively, this could be understood in the sense that if the bands are really parallel there (because they correspond to extrema) then the electron and hole move with the same velocities. Thus, the electron and hole would then have a better chance of interacting with each other which strengthens the oscillator strength of these transitions, because the overlap of their correlated wave functions is higher. Apparently, this increase in oscillator strength would vary much more rapidly with deviations from the parallel bands than the JDOS, thus emphasizing the critical point. Thus, the electron–hole interaction model may finally reconcile the two opposing views which have emerged over the years regarding the optical spectra: in the empirical viewpoint these are dominated by high-symmetry critical points, whereas in the calculations resulting from first-principles calculations in the independent-particle model, peaks are correlated with much wider regions of **k**-space.

It is important to recognize here that the continuum excitonic effects shift oscillator strengths but do not shift the energies of the transitions (the energies E_f in Eq. (37)) from their independent-particle prediction very much. These shifts in fact are probably comparable to the binding energies of the discrete excitons appearing below the band as bound states and which are known to be typically only of the order of 0.01 eV even in wide band gap semiconductors, unless very different band masses are involved. An alternative viewpoint on these excitonic effects is to treat them within an effective mass approach as resonances in the continuum [60 to 62]. Thus, with the exception of very ionic materials, the spectra even when corrected for these excitonic effects still show a strong correlation with the ones obtained in the independent-particle model, and thus to the original band structure. This is why after all, these spectra are still so useful to determine band strucure energies and this is the most important message of this paper.

Now, we turn to the effects of matrix elements. For wurtzite GaN, the optical tensor in fact has two independent components, one for $\mathbf{E} \perp \mathbf{c}$ and one for $\mathbf{E} \parallel \mathbf{c}$. While only the former has been measured, it is of interest to consider their difference because this will yield information on the matrix elements, the JDOS underlying both being the same obviously. In the bottom panel of Fig. 1 we compare the $\epsilon_2(\omega)$ functions for the two polarizations and compare with the results of Benedict and Shirley. Clearly the most important difference between the two polarizations is a strong shift of oscillator strength from the higher two peaks in the 6 to 10 eV region to the lower energy region for the $\mathbf{E} \parallel \mathbf{c}$ spectrum compared to the older spectrum. Also a smaller feature at 6.5 eV present for $\mathbf{E} \perp \mathbf{c}$ is missing for $\mathbf{E} \parallel \mathbf{c}$. Similar effects appear in the calculations including electron–hole interaction effects by Benedict and Shirley [20]. However, the oscillator shift towards the first strong peak is even stronger. To analyze the origin of this, we again consider the decompositions into partial contributions in Fig. 6. Only a few relevant to our discussion are shown. First of all, we note that the peak in question for $\mathbf{E} \parallel \mathbf{c}$ is almost entirely reproduced by just the contributions of the two topmost valence bands to the first conduction band, denoted in the figure as $v_1 \to c_1$ and $v_2 \to c_1$. In contrast, the first few peaks in the other polarization include contributions from bands

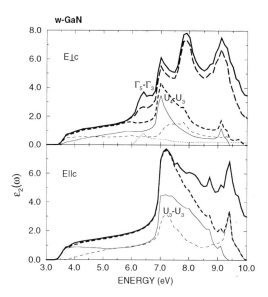

Fig. 6. Polarization dependence of $\epsilon_2(\omega)$ in w-GaN: top panel $\mathbf{E} \perp \mathbf{c}$, bottom panel $\mathbf{E} \parallel \mathbf{c}$, both containing various partial decompositions: thin solid line $v_1 \to c_1$, thin dashed line $v_2 \to c_1$, thick dashed: line sum of previous two, dotted line in top panel: $v_4 \to c_2$, long dashed line $(v_1 + v_2 + v_3) \to (c_1 + c_2)$.

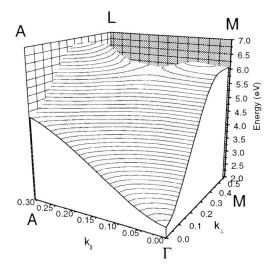

Fig. 7. Energy band difference $E_{c1}(\mathbf{k}) - E_{v1}(\mathbf{k})$ in MLAΓ plane

$v_1 - v_3$ to $c_1 - c_2$. Even then, the feature at 6.5 eV is not included. The latter clearly arises from the $(v_4, v_5) \to c_2$ contribution, only the first of which is shown. Inspection of the band structure difference indicates that this is related to the $\Gamma_5 \to \Gamma_3$ transition, which is only allowed for $\mathbf{E} \perp \mathbf{c}$. This is, however, a rather small feature in GaN because of the small joint density of states associated with the neighborhood of the Γ-point. The conduction bands have in fact a much smaller effective mass than the valence band, so the bands are not very parallel.

Returning now to the main feature just above 7 eV, we note that its strong enhancement for $\mathbf{E} \parallel \mathbf{c}$ is related to the $v_2 \to c_2$ contribution. Inspection of the band differences given in Ref. [48] shows that at 7 eV there is an extremum along the ML line. This corresponds to a transition $U_3 \to U_3$ which is allowed only for $\mathbf{E} \parallel \mathbf{c}$. In contrast for $\mathbf{E} \perp \mathbf{c}$, the allowed transition is $U_4 \to U_3$. Thus, for the perpendicular polarization it is primarily the transitions from the topmost valence band that contribute here, whereas for the parallel polarization the transitions from the next lower energy valence band are the ones that contribute. We note that along ML the two topmost valence bands are of symmetry U_4 and U_3, respectively, and lie almost on top of each other between L and the point at 2/3 of the ML axis measured from M. We note that the extrema in these band differences along U are both saddle points because they are a minimum in the ML direction but a maximum in the perpendicular direction. The energy-difference surfaces $E_{c1}(\mathbf{k}) - E_{v1}(\mathbf{k})$ and $E_{c1}(\mathbf{k}) - E_{v2}(\mathbf{k})$ are shown in the MLAΓ plane in Figs. 7 and 8, respectively. Note that the two lowest conduction bands cross be-

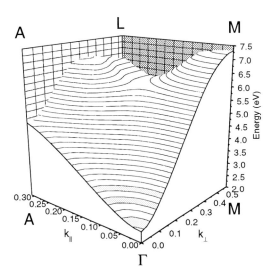

Fig. 8. Energy band difference $E_{c1}(\mathbf{k}) - E_{v2}(\mathbf{k})$ in MLAΓ plane

tween M and L and Fig. 7 shows the lowest band for any **k**. Thus, at the crossing there appears a cusp. This can be seen along ML at $k_\parallel \approx 0.09(2\pi)/a$. Between 0.15 and 0.2 along this axis one may see the saddle point we are discussing here. Clearly, a flat region extends from this point in the $k_\perp$ direction and wraps itself around the two hills at M and L. A similar discussion applies to Fig. 8. These saddle point structures are expected to lead to peaks. It is of interest to note that the point at 2/3 from M to L corresponds to the point on which both the zincblende L_{zb} and X_{zb} are folded when converting to a wurtzite Brillouin zone. As is well known in zincblende the E_1 transition in this energy range corresponds to nearly parallel bands along $\Gamma - L$ near L. Thus, this occurrence of nearly parallel bands apparently survives when going to wurtzite. So, the first strong peak in **E** $\parallel$ **c** is predominantly of $U_3 \to U_3$ character right near this saddle point. Note that this saddle point corresponds to the L_{zb} points which are not parallel to the wz **c** axes, of which there are six. The fact that there are six of these contributes to its high intensity.

The higher peaks for **E** $\perp$ **c** at 7.9 and 9.1 eV have various contributions as was discussed in detail in Ref. [48]. For the first one of these, a strong contribution is related to transitions $M_4 \to M_3(2)$ to the second M_3 conduction band, which is only allowed for **E** $\perp$ **c** and closely related transitions between these same bands $v_1 \to c_2$ near the L point. The peak at 9.1 eV has as one of its main contributions a transition $U_2 \to U_1$ ($v_3 \to c_1$) which again is only allowed for **E** $\perp$ **c**. Thus, the suppression of both of these peaks for **E** $\parallel$ **c** can be understood. The polarization effects in GaN closely resemble those of other wurtzite semiconductors. For example, they were already in 1965 identified in CdSe by Cardona [63]. Our present interpretation, however, is different.

In summary of this section, the example of GaN illustrates how the interpretation using the standard model interplays with including excitonic effects.

1. Excitonic effects produce strong oscillator shifts in the peaks which may sometimes give the impression of a peak shift and hence may confuse assignment to band structure features.

2. The peaks in the independent-particle picture (standard model) do usually not correspond to high-symmetry points but to rather extensive regions of **k**-space where the bands happen to be nearly parallel. In contrast the excitonic effects appear to shift the oscillator strength towards the high-symmetry critical points.

3. The polarization effects can then be well understood on the basis of the symmetry labeled single-particle bands and selection rules at the high symmetry critical points.

6. Basic Formalism for Nonlinear Optics

Having seen that in spite of its shortcomings there is still some virtue to the independent-particle model in linear optics, we now turn to nonlinear optics. In this case, there have been only a few attempts to go beyond the independent-particle picture and those were only for the static values or the region below the absorption threshold [64, 65]. They both considered response to a longitudinal perturbation. In fact, as mentioned already in the introduction, in the nonlinear optics case there were until recently problems even to understand the independent-particle picture. So, we now restrict ourselves to that model and examine what the problem is.

We return to our discussion in Section 3, Eq. (11), on the removal of apparent divergences. This was a serious problem in nonlinear optics, because different sum rules needed to be invoked. The original proof by Aspnes [66] that the divergences disappear in semiconductors was valid only for cubic crystals. The key to understanding this problem is found in a systematic separation of intraband and interband motion from the start.

Returning to Eq. (6), and following a treatment by Sipe et al. [4] one can achieve this separation by incorporating the intraband part of the evolution of the field operator in a new one defined by

$$\tilde{\psi}(\mathbf{r}, t) = \psi(\mathbf{r}, t) \, e^{-i\mathbf{K}(t)\cdot\mathbf{r}}, \tag{44}$$

with $\mathbf{K}(t) = -\mathbf{A}(t)\,e/c$, and which satisfies the following equation of motion:

$$i\hbar \frac{d\tilde{\psi}}{dt} = [\tilde{\psi}, \mathcal{H}_{\text{eff}}], \tag{45}$$

with the effective Hamiltonian given by

$$\mathcal{H}_{\text{eff}}(t) = \int \tilde{\psi}^\dagger(\mathbf{r}, t) \left[H_0 - e\mathbf{r}\cdot\mathbf{E}(t)\right] \tilde{\psi}(\mathbf{r}, t) \, d^3r. \tag{46}$$

As one may notice here, the perturbation now has the so-called "length-gauge" form $e\mathbf{r}\cdot\mathbf{E}(t)$.[5]

The new field operators can now be expanded in the Bloch wave functions of the original unperturbed Hamiltonian using fermion creation and annihilation operators $a_i^\dagger(\mathbf{k})$ and $a_i(\mathbf{k})$,

$$\tilde{\psi}(\mathbf{r}, t) = \sum_i \int_{\text{BZ}} \frac{d^3k}{(2\pi)^3} \, a_i(\mathbf{k}) \, \psi_i(\mathbf{k}, \mathbf{r}). \tag{47}$$

To rewrite the effective Hamiltonian in terms of these operators, one needs to express matrix elements of $\mathbf{r}$ in the basis set of Bloch functions. The problem now is that these are ill-defined if one looks at it in a naive way because the Bloch functions extend over an infinite crystal. The solution to this problem is to turn to the so-called crystal momentum representation which was discussed extensively by Blount [67] but is also given by Lifschitz and Kaganov [68] and can even be found in the well-known *Course on Theoretical Physics* by Landau and Lifshitz [69]. The proper definition of these matrix elements is

$$\langle i\mathbf{k}|\,\mathbf{r}\,|j\mathbf{k}'\rangle = \langle i\mathbf{k}|\,\mathbf{r}_{\text{inter}}\,|j\mathbf{k}'\rangle + \langle i\mathbf{k}|\,\mathbf{r}_{\text{intra}}\,|j\mathbf{k}'\rangle ,$$
$$\left[1 - \delta_{ij}\right]\delta(\mathbf{k}-\mathbf{k}')\,\xi_{ij}(\mathbf{k}) + \delta_{ij}[\delta(\mathbf{k}-\mathbf{k}')\,\xi_{ii}(\mathbf{k}) + i\delta(\mathbf{k}-\mathbf{k}')\,\nabla_{\mathbf{k}}], \tag{48}$$

in which

$$\xi_{ij}(\mathbf{k}) = \frac{i(2\pi)^3}{\Omega_c} \int_{\Omega_c} u_i^*(\mathbf{k}, \mathbf{r})\, \nabla_{\mathbf{k}} u_j(\mathbf{k}, \mathbf{r})\, d^3r, \tag{49}$$

---

[5] The transformation from the Hamiltonian of Eq. (6) to the effective Hamiltonian involves the unitary transformation of Eq. (44). In a Lagrangian formulation, one may view this as adding the total derivative with respect to time of a function, which in turn can be viewed as arising from a gauge transformation of the electromagnetic vector and scalar potentials. For further details, see C. COHEN-TANNOUDJI, J. DUPONT-ROC, and G. GRYNBERG, Photons and Atoms, Introduction to Quantum Electrodynamics, Wiley, New York 1989 (p. 269). This transformation is possible only in the long-wavelength limit.

in which $u_i(\mathbf{k}, \mathbf{r})$ is the periodic part of the Bloch function, and Ω_c is the volume of the unit cell. The $\xi_{ij}(\mathbf{k})$ are closely related to the velocity matrix elements,

$$\xi_{ij}(\mathbf{k}) = \frac{\mathbf{v}_{ij}(\mathbf{k})}{i\omega_{ij}(\mathbf{k})}.\tag{50}$$

As discussed by Blount, $\xi_{ii}(\mathbf{k})$ in itself is not uniquely defined but the combination between square brackets in Eq. (48) is. The $\mathbf{r}_{\text{intra}}$ and $\mathbf{r}_{\text{inter}}$ are defined as the terms with $\delta_{ij} = 0$ and $\delta_{ij} \ne 0$, respectively, and are called the *intraband* and *interband* contributions to the position operator, respectively. The result of the appearance of the $\nabla_{\mathbf{k}}$ term in the definition of the $\mathbf{r}$ matrix elements is that $\mathbf{k}$-derivatives appear whenever $\mathbf{r}$ is combined with other operators. However, it turns out that in practice $\mathbf{r}$ only appears in commutators (for example the current is defined in terms of the velocity which is given by $i\hbar \mathbf{v} = [\mathbf{r}, \mathcal{H}_{\text{eff}}]$) and it can be shown that in that case the derivatives never appear isolated but always in conjunction with ξ factors. Neither of these are uniquely defined by themselves, but their combination is. Thus, it makes sense to define a so-called *generalized derivative*. For any simple operator Q – that is one for which the matrix elements only involve a $\delta(\mathbf{k} - \mathbf{k}')$ function but no derivative term – one has for the matrix elements of the commutator with $\mathbf{r}_{\text{intra}}$,

$$\langle i\mathbf{k}| [\mathbf{r}_{\text{intra}}, Q] |j\mathbf{k}'\rangle = i\delta(\mathbf{k} - \mathbf{k}') [Q_{ij}]_{;\mathbf{k}},\tag{51}$$

in which the generalized derivative, denoted by ;$\mathbf{k}$, is defined as follows:

$$[Q_{ij}(\mathbf{k})]_{;k_b} = \nabla_{k_b} Q_{ij}(\mathbf{k}) - i[\xi_{ii}^b(\mathbf{k}) - \xi_{jj}^b(\mathbf{k})] Q_{ij}(\mathbf{k}),\tag{52}$$

in which we have explicitly indicated the Cartesian component in question to clarify that the ξ Cartesian component should be taken along the same direction as the $\nabla_{\mathbf{k}}$.

Applying the expansion of the new field operator, the effective Hamiltonian becomes

$$\mathcal{H}_{\text{eff}}(t) = \mathcal{H}_0 - [\mathbf{P}_{\text{intra}}(t) + \mathbf{P}_{\text{inter}}(t)] \cdot \mathbf{E}(t),\tag{53}$$

with

$$\mathbf{P}_{\text{inter}}(t) = e \sum_{ij} \int_{\text{BZ}} \frac{d^3k}{(2\pi)^3} \mathbf{r}_{ij}(\mathbf{k}) a_i^\dagger(\mathbf{k}) a_j(\mathbf{k}),\tag{54}$$

$$\mathbf{P}_{\text{intra}}(t) = ie \sum_i \int_{\text{BZ}} \frac{d^3k}{(2\pi)^3} \left\{ \frac{1}{2} [a_i^\dagger(\mathbf{k}) \nabla_{\mathbf{k}} a_i(\mathbf{k}) - (\nabla_{\mathbf{k}} a_i^\dagger(\mathbf{k})) a_i(\mathbf{k})] \right.$$
$$\left. - i\xi_{ii}(\mathbf{k}) a_i^\dagger(\mathbf{k}) a_i(\mathbf{k}) \right\},\tag{55}$$

in which $\mathbf{r}_{ij}(\mathbf{k}) = \xi_{ij}(\mathbf{k})$ for $i \ne j$ and $= 0$ otherwise.

We now calculate the macroscopic average of the induced current in terms of the density matrix $\varrho_{ij}(\mathbf{k})$.

$$\langle \mathbf{J}^i \rangle = \text{Tr}(\mathbf{J}^i \varrho),$$
$$= \frac{1}{i\hbar} \text{Tr}([\mathbf{P}_{\text{intra}}, \mathcal{H}_{\text{eff}}] \varrho) + \frac{d}{dt} \text{Tr}(\mathbf{P}_{\text{inter}} \varrho),$$
$$= \langle \mathbf{J}_{\text{intra}} \rangle + \langle \mathbf{J}_{\text{inter}} \rangle.\tag{56}$$

The density matrix itself $\varrho_{ij}(\mathbf{k}) = \langle a_i^\dagger(\mathbf{k}) a_j(\mathbf{k}) \rangle$ satisfies

$$i\hbar \frac{d\varrho_{ij}(\mathbf{k})}{dt} = \hbar\omega_{ij}(\mathbf{k}) \varrho_{ij}(\mathbf{k}) - e\mathbf{E}(t) \\ \times \sum_l (\mathbf{r}_{il}(\mathbf{k}) \varrho_{lj}(\mathbf{k}) - \varrho_{il}(\mathbf{k}) \mathbf{r}_{lj}(\mathbf{k})) - ie\mathbf{E}(t) \cdot [\varrho_{ij}(\mathbf{k})]_{;\mathbf{k}}, \quad (57)$$

which follows from the usual equation of motion of the density operator,

$$i\hbar \frac{d\varrho}{dt} = [\mathcal{H}_{\text{eff}}, \varrho]. \quad (58)$$

Instead of solving the dynamical equations Eq. (57) directly, one uses perturbation theory. Starting from the zeroth order $\varrho_{ij}^{(0)}(\mathbf{k}) = f_i(\mathbf{k}) \delta_{ii}$, which assumes an insulator (or clean semiconductor) at zero temperature, one calculates its contributions to various orders in $\mathbf{E}(t)$ resulting from the perturbation term ($\neq H_0$, in Eq. (46) or Eq. (53)), and then combines them with the various terms in Eq. (56) to finally identify contributions to $\mathbf{P}$ to the various orders in $\mathbf{E}$, as defined in Eq. (9).

Using $\mathbf{E}(t) = \sum_n \mathbf{E}(\omega_n) \exp(-i\omega_n t)$, the first order correction to the density matrix is obtained by putting the zeroth order of the density matrix on the r.h.s. of Eq. (57), and the first order on the left, expanded also as $\varrho_{ij}^{(1)}(\mathbf{k}, t) = \sum_n \varrho_{ij}^{(1)}(\mathbf{k}, \omega_n) \exp(-i\omega_n t)$. Identifying terms in the same ω_n, one obtains

$$\varrho_{ij}^{(1)}(\mathbf{k}, t) = \sum_n \mathbf{R}_{ij}(\mathbf{k}, \omega_n) \cdot \mathbf{E}(\omega_n) e^{-i\omega_n t}, \quad (59)$$

with

$$\mathbf{R}_{ij}(\mathbf{k}, \omega_n) = \frac{e}{\hbar} \frac{f_{ji}(\mathbf{k}) \mathbf{r}_{ij}(\mathbf{k})}{(\omega_{ij}(\mathbf{k}) - \omega_n)}, \quad (60)$$

because $[\varrho_{ij}^{(0)}(\mathbf{k})]_{;\mathbf{k}} = \mathbf{0}$. Iterating Eq. (57) with this substituted in the r.h.s., one obtains the second order contribution on the left

$$\varrho_{ij}^{(2)}(\mathbf{k}, \omega_3) = \frac{e}{i\hbar(\omega_{ij}(\mathbf{k}) - \omega_3)} \\ \times \left\{ -R_{ij;c}^b(\mathbf{k}, \omega_1) + i \sum_l [r_{il}^c(\mathbf{k}) R_{lj}^b(\mathbf{k}, \omega_1) - R_{il}^b(\mathbf{k}, \omega_1) r_{lj}^c(\mathbf{k})] \right\} \\ \times E^b(\omega_1) E^c(\omega_2) e^{-i(\omega_1 + \omega_2) t}, \quad (61)$$

with $\omega_3 = \omega_1 + \omega_2$, and a summation is implied over Cartesian indices b and c and over frequencies ω_1 and ω_2. Here we have introduced the shorthand notation $R_{ij;c}^b(\mathbf{k}, \omega_1) = [R_{ij}^b(\mathbf{k}, \omega_1)]_{;k_c}$.

It is important to note that

$$\langle i\mathbf{k} | [\mathbf{r}_{\text{intra}}, \varrho^{(0)}] | i\mathbf{k}' \rangle = i\delta(\mathbf{k} - \mathbf{k}') \nabla_\mathbf{k} f_i(\mathbf{k}), \quad (62)$$

which vanishes for any completely filled or completely empty band. Thus, for example $\varrho_{ii}^{(1)}(\mathbf{k}) = 0$, and consequently there is no first order contribution to $\langle \mathbf{P}_{\text{intra}} \rangle$. It also means, however, that in higher order terms of the perturbation theory, combinations involving $[\mathbf{r}_{\text{intra}}, \varrho^{(0)}]$ will fall out and it is these simplifications that avoid the undesirable divergences.

Now, using $\langle i\mathbf{k}|\mathbf{P}_{\text{inter}}|j\mathbf{k}'\rangle = \delta(\mathbf{k}-\mathbf{k}')\, er_{ij}(\mathbf{k})$ one recovers the result of Eq. (11) with the first term vanishing, because we have already assumed an insulator.

In second order, there will be several contributions: 1. a term arising from $\mathbf{P}_{\text{inter}}$ involving $\varrho_{ij}^{(2)}$,

$$\langle \mathbf{P}_{\text{inter}}^{(2)}\rangle = e \int_{\text{BZ}} \frac{d^3k}{(2\pi)^3} \sum_{ij} \varrho_{ij}^{(2)}(\mathbf{k}, t)\, \mathbf{r}_{ji}, \tag{63}$$

2. a term arising from the $\mathbf{J}_{\text{intra}}$ term in Eq. (56) which contains one electric field factor with the first order $\varrho_{ij}^{(1)}$,

$$\langle J_{\text{intra}}^a(t)^{(2,1)}\rangle = -\frac{e^2}{i\hbar}\int_{\text{BZ}} \frac{d^3k}{(2\pi)^3}\sum_{ij} \varrho_{ij}^{(1)}(\mathbf{k},t)\, r_{ji;a}^b(\mathbf{k})\, E^b(t), \tag{64}$$

and 3. a term arising from $\mathbf{J}_{\text{intra}}$ with the second order $\varrho_{ij}^{(2)}$,

$$\langle J_{\text{intra}}^a(t)^{(2,2)}\rangle = e\int_{\text{BZ}} \frac{d^3k}{(2\pi)^3}\sum_{i} \varrho_{ii}^{(2)}(\mathbf{k},t)\, v_{ii}^a(\mathbf{k}). \tag{65}$$

When substituting Eq. (61) in here, it should be noted that for the diagonal term the $R_{ii;c}(\mathbf{k}) = 0$, and Eq. (65) can be rewritten as

$$\langle J_{\text{intra}}^a(t)^{(2,2)}\rangle = \frac{e^2}{\hbar\omega_3}\left[\int_{\text{BZ}} \frac{d^3k}{(2\pi)^3}\sum_{ij} \Delta_{ij}^a(\mathbf{k})\, r_{ji}^c(\mathbf{k})\, R_{ij}^b(\mathbf{k},\omega_1)\right]$$
$$\times E^b(\omega_1)\, E^c(\omega_2)\, e^{-i(\omega_1+\omega_2)t}, \tag{66}$$

in which $\Delta_{ij}^a(\mathbf{k}) = v_{ii}^a(\mathbf{k}) - v_{jj}^a(\mathbf{k})$.

We note that even in Eq. (63), the second order density matrix of Eq. (61) contains an intraband contribution $R_{ij;c}^b$ involving in fact a generalized derivative in the sense of Eq. (52). In this sense, the interband contribution to $\chi^{(2)}$ is modified by intraband motion. If one separates out this contribution, one can eventually isolate a pure interband contribution,

$$\left[\chi_{abc}^{(2)}(-\omega_3, \omega_1, \omega_2)\right]_{\text{inter}_1} = \frac{e^3}{\hbar^2}\int_{\text{BZ}}\frac{d^3k}{(2\pi)^3}\sum_{ij}\frac{r_{ji}^a(\mathbf{k})}{\omega_{ij}(\mathbf{k})-\omega_3}$$
$$\times \sum_l \left[\frac{f_{jl}(\mathbf{k})\, r_{il}^c(\mathbf{k})\, r_{lj}^b(\mathbf{k})}{\omega_{lj}(\mathbf{k})-\omega_1} - \frac{f_{li}(\mathbf{k})\, r_{il}^b(\mathbf{k})\, r_{lj}^c(\mathbf{k})}{\omega_{il}(\mathbf{k})-\omega_1}\right]. \tag{67}$$

The other term arising from the interband part reads

$$\left[\chi_{abc}^{(2)}(-\omega_3, \omega_1, \omega_2)\right]_{\text{inter}_2} = i\frac{e^3}{\hbar^2}\int_{\text{BZ}}\frac{d^3k}{(2\pi)^3}\sum_{ij}\frac{r_{ji}^a(\mathbf{k})}{\omega_{ij}(\mathbf{k})-\omega_3}\left[\frac{f_{ji}(\mathbf{k})\, r_{ij}^b(\mathbf{k})}{\omega_{ij}(\mathbf{k})-\omega_1}\right]_{;k_c}. \tag{68}$$

The first intraband contribution leads to

$$\left[\chi_{abc}^{(2)}(-\omega_3, \omega_1, \omega_2)\right]_{\text{intra}_1} = \frac{e^3}{i\hbar^2\omega_3}\int_{\text{BZ}}\frac{d^3k}{(2\pi)^3}\sum_{ij} r_{ji;a}^b(\mathbf{k})\, \frac{f_{ji}(\mathbf{k})\, r_{ij}^c(\mathbf{k})}{\omega_{ij}(\mathbf{k})-\omega_1}, \tag{69}$$

while the second one becomes

$$\left[\chi^{(2)}_{abc}(-\omega_3, \omega_1, \omega_2)\right]_{\text{intra}_2} = \frac{e^3}{i\hbar^2\omega_3^2} \int_{\text{BZ}} \frac{d^3k}{(2\pi)^3} \sum_{ij} \Delta^a_{ji}(\mathbf{k}) \, r^c_{ji}(\mathbf{k}) \, \frac{f_{ji}(\mathbf{k}) \, r^b_{ij}(\mathbf{k})}{\omega_{ij}(\mathbf{k}) - \omega_1}. \quad (70)$$

In summary, whereas in first order, there is only a pure interband contribution to $\chi^{(1)}$, in second order the intra and interband contributions get mixed up. The interband part of the polarization as defined in Eq. (54) is modified by an intraband part in the second order density matrix. The intraband part of the polarization is modified by interband transitions in the sense that it contains contributions from $r^a_{ij;b}$ as in Eq. (64).

The pure interband term of Eq. (67) resembles the equation that would be obtained for a finite system [70]. It appears as if each $\mathbf{k}$ point corresponded to an individual molecule. It can be further modified by replacing indices $a \to b$ and simultaneously $\omega_1 \to \omega_2$, adding the two equations and dividing by two so as to make the final expression look more symmetric. In a naive generalization of the equations for a molecular system, these would be the only terms appearing.

The terms resulting from $\mathbf{J}_{\text{intra}}$ still have factors ω_3^{-1} (Eq. (69)) and ω_3^{-2} (Eq. (70)) in front and would thus seem to diverge as $\omega_3 \to 0$. However, it can be shown that if all frequencies are below the resonances $\omega_{ij}(\mathbf{k})$, these can be combined into an expression that is manifestly divergence free [3],

$$\left[\chi^{(2)}_{cba}(-\omega_3, \omega_1, \omega_2)\right]_{\text{intra}} = \frac{e^3}{2i\hbar^2} \int_{\text{BZ}} \frac{d^3k}{(2\pi)^3} \sum_{ij} \left[\frac{r^b_{ji}(\mathbf{k})}{\omega_{ij}(\mathbf{k}) + \omega_2}\right]_{;k_c} \frac{f_{ji}(\mathbf{k}) \, r^a_{ij}(\mathbf{k})}{\omega_{ij}(\mathbf{k}) - \omega_1}, \quad (71)$$

in the limit that ω_1 and ω_2 both go to zero. This last transformation involves essentially a partial integration but no new sum rules.

Before the expressions become convenient for numerical calculations, some further rewritings are required. First of all, the intrinsic permutation symmetry of the expressions with respect to $\mathbf{E}(\omega_1)$ and $\mathbf{E}(\omega_2)$ is utilized to symmetrize the expressions. Secondly, time reversal symmetry,

$$\mathbf{r}_{ij}(\mathbf{k}) = \mathbf{r}_{ji}(-\mathbf{k}),$$
$$\mathbf{v}_{ij}(\mathbf{k}) = -\mathbf{v}_{ji}(-\mathbf{k}), \quad (72)$$

can be invoked to simplify some expressions. Some of the expressions obtained so far contain products of different factors $(\omega_{ij}(\mathbf{k}) - \omega_i)^{-1}$. These can be expanded in partial fractions so as to generate expressions involving only one such factor. A small imaginary part is then introduced in each ω factor and in the limit of this imaginary part going to zero, a delta function is obtained for the imaginary part resulting from that term, while the corresponding real part is obtained from it by a Kramers-Kronig transformation.

$$r^b_{ji;a}(\mathbf{k}) = \frac{r^a_{ji}(\mathbf{k}) \, \Delta^b_{ij}(\mathbf{k}) + r^b_{ji}(\mathbf{k}) \, \Delta^a_{ij}(\mathbf{k})}{\omega_{ji}(\mathbf{k})}$$
$$+ \frac{i}{\omega_{ji}(\mathbf{k})} \sum_l [\omega_{li}(\mathbf{k}) \, r^a_{jl}(\mathbf{k}) \, r^b_{li}(\mathbf{k}) - \omega_{jl}(\mathbf{k}) \, r^b_{jl}(\mathbf{k}) \, r^a_{li}(\mathbf{k})], \quad (73)$$

rather than from the direct definition in Eq. (52). The final expressions become too lengthy to include them here in full generality. Expressions pertaining to second harmonic generation, in which case $\omega_1 = \omega_2 = \omega$ and $\omega_3 = 2\omega$ were given in Ref. [38]. It is noteworthy that the final expressions for the static limit,

$$\left[\chi^{(2)}_{cba}\right]_{\text{inter}} = \frac{e^3}{\hbar^2} \int_{BZ} \frac{d^3k}{(2\pi)^3} \sum_{ijl} \frac{r^c_{ij}(\mathbf{k}) \{r^b_{jl}(\mathbf{k}) r^a_{li}(\mathbf{k})\}}{\omega_{ij}(\mathbf{k})\omega_{jl}(\mathbf{k})\omega_{li}(\mathbf{k})}$$
$$\times \left[\omega_i(\mathbf{k}) f_{jl}(\mathbf{k}) + \omega_j(\mathbf{k}) f_{li}(\mathbf{k}) + \omega_l(\mathbf{k}) f_{ij}(\mathbf{k})\right], \tag{74}$$

with $\{r^b_{jl}(\mathbf{k}) r^a_{li}(\mathbf{k})\} = (r^b_{jl}(\mathbf{k}) r^a_{li}(\mathbf{k}) + r^a_{jl}(\mathbf{k}) r^b_{li}(\mathbf{k}))/2$ and

$$\left[\chi^{(2)}_{cba}\right]_{\text{intra}} = \frac{ie^3}{4\hbar^2} \int_{BZ} \frac{d^3k}{(2\pi)^3} \sum_{ij} \frac{f_{ij}(\mathbf{k})}{\omega_{ji}(\mathbf{k})^2}$$
$$\times \left[r^a_{ij}(\mathbf{k}) (r^b_{ji;c}(\mathbf{k}) + r^c_{ji;b}(\mathbf{k})) + r^b_{ij}(\mathbf{k}) (r^a_{ji;c}(\mathbf{k}) + r^c_{ji;a}(\mathbf{k}))\right.$$
$$\left. + r^c_{ij}(\mathbf{k}) (r^a_{ji;b}(\mathbf{k}) + r^b_{ji;a}(\mathbf{k}))\right] \tag{75}$$

are symmetric under all possible permutations of the a, b, c indices, a symmetry which is known as Kleinman symmetry [71], and they are purely real.

The resulting equations are equivalent to previous formulations such as that of Moss et al. [72] or Aspnes [66] in the case of cubic crystals, except that here no approximations were made of neglecting so-called virtual hole processes, i.e. terms in which the three bands appearing are respectively valence, valence and conduction band. The main advantages of the new expressions are that they are valid for non-cubic crystals whereas the formulation by Aspnes [66] made use of the cubic symmetry in order to cast the equation into its final form. A generalization to non-cubic crystals was previously given by Ghahramani et al. [73] but involved the introduction of a new sum rule. The present formulation of Sipe and Ghahramani [2] and by Aversa and Sipe [3] provides a more general framework for discussing general second as well as third order response functions. Furthermore, as will become clear in the next section, the separation in intraband and interband processes will be found to provide new insights into the relation between band structures and the magnitude of the nonlinear response functions.

The results obtained in the length-gauge formalism [3] are, of course, equivalent to those obtainable within the velocity-gauge formalism [4]. As is discussed by Aversa and Sipe [3], however, this gauge invariance rests on the validity of the basic commutator relationship $[r_a, p_b] = i\hbar\delta_{ab}$. In practical implementations of the electronic structure calculations, this relationship among operators may not be exactly fulfilled, for example when finite basis sets are introduced or cut-offs on the number of states are included in the representation of these operators. In this sense the length-gauge formalism appears to have certain advantages in that it is more forgiving because it is already manifestly divergence free.

This completes our discussion of the equations for second-order susceptibilities. Suffice it to say that this procedure has already been generalized to third order [3]. Let us just point out once more that what made it possible to arrive at divergence-free expressions in the static limit is the systematic inclusion of intraband contributions. This was previously realized by Genkin and Mednis [6], although it is only after the work of Sipe et al. [3, 2] that numerical implementations in terms of first-principles band structure became possible [5].

Perhaps even more important, however, the expressions for $\mathbf{J}_{\text{intra}}$ have real divergencies in the case that $\omega_3 =\to +0$ but ω_1 and $\omega_2 = -\omega_1 + \omega_3$ are in the absorbing energy region [4]. These contributions were already pointed out in the past [3, 74]. The terms in ω_3^{-1} lead to a direct current, while the terms in ω_3^{-2} have the meaning of a $d\mathbf{J}/dt$, or injection of current, introduced but the interaction with the light. They are thus nonlinear types of photoconductivity. The $d\mathbf{J}/dt$ contribution for example can be written

$$\frac{d}{dt}\langle J_a^{(2)}\rangle = \eta_{abc}(0,\omega,-\omega)\, E_b(\omega)\, E_c(-\omega), \tag{76}$$

with

$$\eta_{abc}(0,\omega,-\omega) = \frac{\pi e^3}{2\hbar^2} \int_{\text{BZ}} \frac{d^3 k}{(2\pi)^3} \sum_{ij} \Delta_{ij}^a(\mathbf{k})\, f_{ji}(\mathbf{k}) [r_{ji}^c(\mathbf{k}), r_{ij}^b(\mathbf{k})]\, \delta(\omega_{ij}(\mathbf{k}) - \omega), \tag{77}$$

in which $[r_{ji}^c(\mathbf{k}), r_{ij}^b(\mathbf{k})] = r_{ji}^c(\mathbf{k})\, r_{ij}^b(\mathbf{k}) - r_{ji}^b(\mathbf{k})\, r_{ij}^c(\mathbf{k})$. This effect occurs in the case of the interaction of circularly (or elliptically) polarized light. It can be understood as an interference effect between the amplitude for absorption for two orthogonal polarizations. Note that Eq. (76) still implies a summation over b, c, ω and $-\omega$. If an electric field with both ω and $-\omega$ is present, the summation over both of these amounts to the same except for a factor of two. So one can also multiply this equation with two and then sum only once over ω. If only one particular frequency ω is present, then only a sum over Cartesian components is left. The interesting thing here is that Eq. (77) involves a δ-function factor exactly the same as in the JDOS or $\epsilon_2(\omega)$. However, the matrix elements in front of it are different from those in the linear optical response function. Since they arise from a third rank tensor like all other second order NLO coefficients, they are only nonvanishing for crystals without an inversion center. Further restrictions apply because the function changes sign if b and c are interchanged. Thus, nonvanishing components only survive in a limited set of crystal systems. Their identification as a special case of the nonlinear susceptibilities was recently emphasized by Sipe et al. [4]. Their full discussion goes beyond the present review. We merely note that this effect may provide an interesting alternative to linear optics for probing band structure effects.

In conclusion, even in the independent-particle model without inclusion of local field effects, the theory of nonlinear optical properties is quite complex. The most important recent breakthrough has been to systematically sort out the intraband contributions to these susceptibilities, as sketched above. The role of local field effects and excitonic effects on these processes remains to be studied.

7. Applications of NLO Calculations

In this section, we present some examples to illustrate how nonlinear optics is related to the underlying band structure.

7.1 Frequency dependent $\chi^{(2)}$

First of all, we note that the equations for second harmonic generation consist of a number of resonant terms. In this sense the imaginary part, $\text{Im}\,\chi^{(2)}(-2\omega,\omega,\omega)$ resembles the $\epsilon_2(\omega)$, and provides a link to the band structures. The difference, however, is

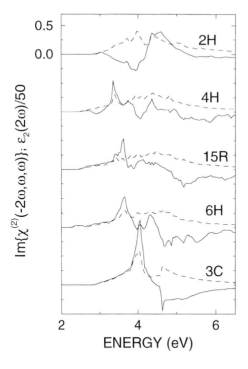

Fig. 9. 2ω resonance terms in $\operatorname{Im}\chi^{(2)}(-2\omega,\omega,\omega)$ in units 10^{-6} esu (solid lines), compared to $\epsilon_2(2\omega)$ (dashed lines) for a number of SiC polytypes

that whereas in $\epsilon_2(\omega)$ only the absolute value of the matrix elements squared enters, the matrix elements entering the various terms in $\chi^{(2)}$ are more varied. They are in general complex and can have any sign. Thus, $\operatorname{Im}\chi^{(2)}(-2\omega,\omega,\omega)$ can be both positive and negative. Secondly, there appear both resonances when 2ω equals a interband energy and when ω equals an interband energy. Fig. 9 shows the 2ω resonances and Fig. 10 the single ω resonance contributions to $\operatorname{Im}\chi^{(2)}(-2\omega,\omega,\omega)$ compared to $\epsilon_2(\omega)$ for a number of SiC polytypes. They clearly show a greater variation from polytype to polytype than the linear optics function. In some sense they resemble a modulated spectrum. Secondly, we note that the 2ω resonances occur at half the frequency corresponding to the interband transition. Thus, the incoming light need not be as high in the UV to detect these higher lying interband transitions. This may be important for wide band gap semiconductors where laser light sources reaching the higher interband transitions are not available. Nevertheless, one still needs to be able to detect the corresponding 2ω signal in the UV.

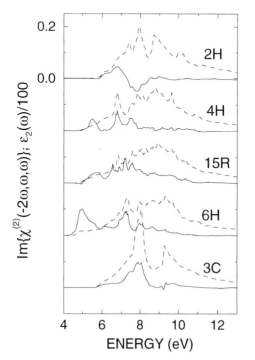

Fig. 10. ω resonance terms in $\operatorname{Im}\chi^{(2)}(-2\omega,\omega,\omega)$ in units 10^{-6} esu (solid lines), compared to $\epsilon_2(\omega)$ (dashed lines) for a number of SiC polytypes

Unfortunately, the intrinsic richness of $\chi^{(2)}$ spectra remains largely to be explored experimentally. We are not aware of any attempts to measure both the real and imaginary parts of these spectral functions as one standardly does in linear optics. The difficulty is undoubtedly that it is fairly difficult already to obtain the $\chi^{(2)}$ magnitude at a few frequencies. Kramers-Kronig analysis would require knowing the function for a large range of ω up into the UV, for which no adequate lasers are available. One might wonder whether it would be possible to obtain both real and imaginary parts of $\chi^{(2)}$ for a limited frequency range using measurements of $I(2\omega)$ at different angles, in a manner similar to spectroscopic ellipsometry. The difficulty is that the angular variation of $\chi^{(2)}$ is strong due to the third rank tensorial character and the geometry is thus usually chosen so as to maximize the conversion coefficient under study. Also, the angles strongly affect the efficiency of power transfer from a beam of one frequency to another via the degree of phase matching. Only for one particular set of angles, efficient pumping occurs. Therefore these phase matching effects may well overwhelm the intrinsic angular effects. Before one could attempt gathering a sufficiently complete set of intrinsic data to unravel the real and imaginary parts, one would need to correct for these phase matching effects.

There are to the best of our knowledge only a few measurements of the total absolute value of $\chi^{(2)}$ above the absorbing region. The agreement between theory and experiment was discussed for GaAs in Ref. [38]. Even the absolute values of $\chi^{(2)}$ contain much sharper variations with frequency than the $\epsilon_2(\omega)$ function and could be useful to extract band structure information. In particular, in the range $E_g/2 < \omega < E_g$, only 2ω resonances can occur and may provide valuable information about critical points in the region $E_g < E < 2E_g$. In Ref. [75] we have given an example of how the features in $\chi^{(2)}$ in this region may be related to the transitions between a small set of close lying bands near the gap. This example involved the 4H and 6H polytypes of SiC in which several bands occur near the gap because of the large unit cell. This situation is similar to that in semiconductor superlattices in which the conduction band and valence bands of the original semiconductor of the quantum well split into so-called minibands.

Another recognized opportunity for exploiting SHG is that of surfaces or interfaces. In systems in which inversion symmetry forbids $\chi^{(2)}$, the symmetry may be broken at an interface or surface. Hence a measurement of $\chi^{(2)}$ spectra then specifically measures the interface band structure. Comparison between theory and experiment for such cases would be of interest and could further enhance such spectroscopies but remains a largely unexplored area.

7.2 Static $\chi^{(2)}$

For lack of experimental data to compare with for the dynamic $\chi^{(2)}$, most work has focused on the static values. Our own emphasis in this area has been to try to understand the trends in some classes of semiconductors. The situation here is difficult because comparing absolute values is more subject to error than comparing spectroscopic features, as was already discussed at length for linear optics in the first part of this paper.

Thus, the neglect of local field effects and the band gap problem become more serious concerns. Nevertheless, the theoretical developments outlined in Section 6 have allowed us to gain some understanding of the trends of $\chi^{(2)}$ in chalcopyrites [76]. The latter class of materials was chosen because these materials have relatively high $\chi^{(2)}$

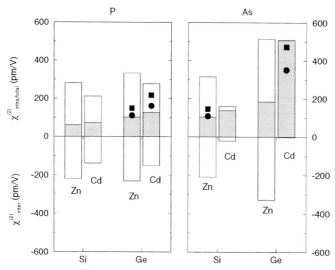

Fig. 11. Static $\chi^{(2)}$ values for II–IV–V$_2$ compounds, top panels total value and intraband part, bottom panels interband contribution, experimental data indicated by circles are from Ref. [78], by squares are from Ref. [79]

values, and in addition have adequate birefringence to allow for angular tuning phase matching. This makes them attractive for practical applications of frequency conversion, in particular in the infrared range. They are used both to frequency double CO$_2$ laser lines (lying between 9 and 11 μm) and in so-called optical parametric oscillators (OPOs) and related devices. In these the difference frequency between a pump probe at frequency ω_1 and an idle beam at ω_2 is generated as output. The system is set up such that optimal energy transfer between these beams occurs. For a discussion of these practical aspects of NLO we refer the reader to Ref. [77].

Here we are concerned with the trends in $\chi^{(2)}$ in a series of II–IV–V$_2$ chalcopyrites, with II = Zn or Cd, IV = Si or Ge and V = As or P. The calculated values for $\chi^{(2)}$ from Ref. [76] are given in Fig. 11 as the filled bar graphs. Also given is their decomposition into intra- and interband contributions. They are arranged so as to make the Si–Ge, Zn–Cd and P–As trends obvious. For example $\chi^{(2)}$ obviously increases when going from Si to Ge and from Zn to Cd and from P to As, keeping the other components constant each time. The few available experimental values are indicated by circles [78] and squares [79]. Although the agreement between theory and experiment is by no means perfect, the values obtained are within the ballpark and it appears that at least the trends with atomic species are correctly reproduced.

Note that the interband parts are negative in all cases and in most cases largely compensate the intraband part. The exceptions are the Cd–As compounds in both cases of which the interband part is much smaller in magnitude than the intraband part. This is quite interesting because unexpected. It raises the question what features in the band structure of these two compounds distinguish them from the other compounds.

We investigated the reasons for the cancellation of intra- and interband parts by inspecting the corresponding frequency dependent imaginary parts of the

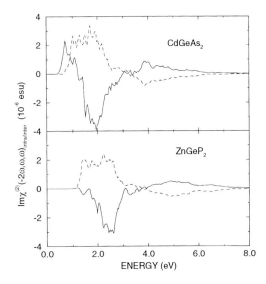

Fig. 12. Inter- (solid lines) and intraband (dashed lines) contributions to Im $\chi^{(2)}(-2\omega, \omega, \omega)$ for CdGeAs$_2$ (top) and ZnGeP$_2$ (bottom)

$\chi^{(2)}(-2\omega, \omega, \omega)$. These are shown in Fig. 12 for an example of a "normal" compound, ZnGeP$_2$, and an anomalous one, CdGeAs$_2$. First of all, one now sees that the opposite sign of intra- and interband parts not only occurs in the static value but occurs almost energy by energy. This is true over the entire energy range in ZnGeP$_2$ and over most of the range ($E > 1$ eV) for CdGeAs$_2$. However, below about 1 eV, we now see that CdGeAs$_2$ has a positive interband contribution. The signs of the inter- and intraband part are difficult to understand *a-priori* because a variety of matrix element products comes into play and both ω and 2ω resonances occur in both the pure interband, Eq. (67), and the interband contribution modified by intraband motion, Eq. (68), when these are further worked out into separate resonance terms. Nevertheless, it is clear that upon taking the Kramers-Kronig (KK) transform,

$$\text{Re}\left\{\chi^{(2)}(0, 0, 0)\right\} = \frac{2}{\pi} \mathcal{P} \int \frac{\text{Im}\left\{\chi^{(2)}(-2\omega, \omega, \omega)\right\}}{\omega} d\omega, \qquad (78)$$

the real part of the $\chi^{(2)}$ at zero energy, the static value will in the case of ZnGeP$_2$ be negative for the interband part because the KK weights the low energy part of the spectral function more heavily. In the case of CdGeAs$_2$, however, a strong compensation appears between the first positive region and the second negative region for $\chi^{(2)}_{\text{inter}}$. Thus, the anomalously low interband value in CdGeAs$_2$ is in some sense due to the positive value of the interband imaginary part in the low energy region $E < 1$ eV. The positive peak at 1 eV correlates with a peak in the $\epsilon_2(\omega)$ function at about 2 eV, suggesting a 2ω resonance is responsible for it. The peak would then seem to be related to regions of parallel bands between the upmost valence and lowest conduction bands near the points T, H and N in the Brillouin zone, as can be seen in Fig. 13. Unfortunately, it is not clear at this point how to understand the signs of the matrix elements in simple terms.

The only apparent difference in the band structures of ZnGeP$_2$ and CdGeAs$_2$ is that CdGeAs$_2$ is a direct gap semiconductor whereas ZnGeP$_2$ is indirect [80]. In ZnGeP$_2$ the lower two conduction bands at Γ lie much closer to each other and the conduction band minimum is about at the same energy at T, N and Γ. Further analysis is necessary to understand the effect of these band structure differences on the matrix elements and resonances in $\chi^{(2)}$.

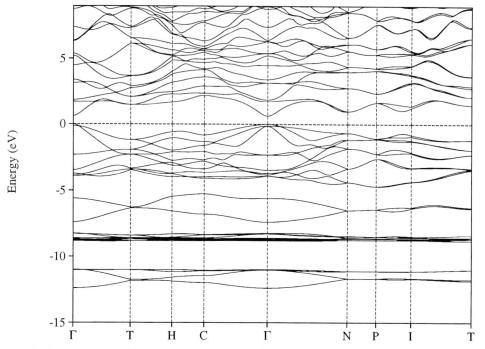

Fig. 13. Band structure of CdGeAs$_2$ obtained by the FP-LMTO-LDA method with empirically corrected band gap

8. Conclusions

In this paper, we have given a survey of the state of the art in calculation of optical properties from the underlying electronic structure in crystalline solids and in particular in semiconductors or insulators. While the relation between the band structure and the optical properties is far from simple if one applies the current insights into the manybody nature of the electronic structure, the underlying independent-particle picture still plays a central role. This is important because that is what makes optical studies relevant as an experimental check on the band structure.

In particular, one first must clearly define what band structure one is talking about. We have discussed the relation between the Kohn-Sham density functional theory band structure and the quasiparticle band structure. The most recent advances in density functional theory, such as exact exchange and generalized Kohn-Sham schemes attempt to bring these two closer together than is the case in the local density approximation, which may be hoped to make studies of optical properties more feasible for larger systems because the full complications of dynamic corrections of the GW approximation may not be required.

We discussed the local field effects, which are due to the fact that the internal field in the material is not the same as the externally applied one because even a long-wavelength excitation produces short-wavelength screening responses which modify the effective field. While most practical investigations have been limited to the longitudinal case, the transverse case [25 to 28] has been worked out in principle but will require further work to be applied in practical calculations. It may be less important than thought for optical properties because the theory reveals that the essential screening effects are in fact longitudinal if retardation effects are neglected, which is a good approximation in the visible and UV, the region that probes the important bands near the band gap. Also, it appears from recent calculations that the longitudinal LF effects modify the redistribution of oscillator strength only in a modest way, if exchange–correlation effects are included in their evaluation [29]. Far more important corrections to the spectra seem to appear from the direct electron–hole interaction. While this contribution has taken the longest time to include in the first-principles calculations, because of its complexity, there has been important recent progress in this area. Over the years, there has been quite some confusion as to which effects were more important and all too often going beyond a previous approximation led actually to worse agreement with experiment. This situation now seems to become clearer thanks to the careful work of various authors to investigate one by one the various complicating factors, such as including or not including nonlocality, dynamic corrections to spectral strength and so on. To arrive at a practical approximate and yet accurate scheme, it now appears that

– QP corrections to the bands primarily involved in the optical spectra can to a fairly good approximation (although not exactly) be calculated with static screening, which makes new DFT schemes such as screened exchange attractive;

– they should essentially only be included in the energy denominators and not in the matrix-element factors, unless one is willing to pay the price of additional correction factors in the vertex corrections of the polarizability [40, 41];

– the direct electron–hole interaction effect [17 to 19] plays a significant role in shifting oscillator strength and seems to weight the critical point extrema more heavily than the independent-particle spectra;

– local basis set expansions or mixed real and reciprocal representations are promising to make the calculations more tractable for large systems [31, 33].

The third point above may be the explanation for the discrepancy that has emerged over the years between older semi-empirical theories emphasizing critical points at high symmetry points in the BZ in the analysis of optical spectral functions in terms of the underlying band structure and the results found by practitioners of first-principles band structure calculations, who find generally fairly large regions of **k**-space contributing to their calculated spectral functions.

This new understanding of the special role played by the critical points may facilitate the detailed analysis of polarization dependences. The latter ultimately go back to selection rules and symmetry analysis in the underlying independent particle approximation and emphasize the continuing importance of a full symmetry analysis of the band structure. This was illustrated by the example of GaN.

As far as nonlinear optics is concerned, there still is a lot of work to do to put it on a level equal to that of linear optics. Local field and excitonic effects have essentially not yet been addressed within the first-principles framework, with the exception of the work of Dal Corso et al. [65] and Levine et al. [64] in the non-absorbing region of the spectrum only. The recent progress has been in developing an understanding of the role played by intraband motion in the nonlinear optical properties within the independent-particle approximation in the long-wavelength limit, and neglecting local field effects. Not only has this made the expressions suitable for numerical calculations within a first-principles band-structure approach, by making them explicitly divergence free and displaying all the underlying symmetries, such as Kleinman permutation symmetry in the static limit, explicitly, it also has led to unifying various nonlinear optical phenomena, such as second harmonic generation (SHG) $\chi^{(2)}(-2\omega, \omega, \omega)$, sum (SFG) $\chi^{(2)}(\omega_1 + \omega_2, \omega_1, \omega_2)$, and difference (DFG) frequency generation $\chi^{(2)}(\omega_1 - \omega_2, \omega_1, -\omega_2)$, the linear electro-optical effect (LEO) $\chi^{(2)}(\omega, \omega, 0)$, and optical rectification, $\chi^{(2)}(0, \omega, -\omega)$. Among these, perhaps the most interesting ones are the photogalvanic effects or current injection phenomena [4]. The new insights into the role of the intraband contribution also start to provide some insights into the trends of NLO coefficients in classes of materials for which these properties are of practical relevance. Finally, the spectroscopic aspects of NLO response functions were shown to be intrinsically much richer in structure, because they involve the matrix elements in more interesting combinations. However, this feature remains so far elusive to experiments because of the intrinsic difficulties in determining the full real and imaginary parts of these response functions.

Finally, the authors apologize to all those whose work was not included here. This survey no doubt only describes a small cross section of the literature on the subject with which we became familiar in the process of our research and the writing of this article. It is not meant to be comprehensive or giving an overview but rather a hopefully somewhat didactic comment on the current state of this field.

Acknowledgements We would like to thank Dr. John Sipe for inspiring discussions on nonlinear optics, Dr. Lorin Benedict for discussions on the excitonic effects and for bringing the problem of the polarization dependence in GaN to our attention, Drs. Friedhelm Bechstedt and Vladimir I. Gavrilenko for keeping us up to date on their results, and Dr. S. Limpijumnong for help with some of the graphics and calculations. This work was supported by the National Science Foundation under grant No. DMR95-29376.

References

[1] P. HOHENBERG and W. KOHN, Phys. Rev. **136**, B864 (1964).
W. KOHN and L. J. SHAM, Phys. Rev. **140**, A1133 (1965).
[2] J. E. SIPE and E. GHAHRAMANI, Phys. Rev. B **48**, 11705 (1993).
[3] C. AVERSA and J. E. SIPE, Phys. Rev. B **52**, 14636 (1995).
[4] J. E. SIPE and A. I. SHKREBTII, private communication.
J. E. SIPE, A. I. SHKREBTII, and O. PULCI, phys. stat. sol. (a) **170**, 431 (1998).
[5] J. L. P. HUGHES and J. E. SIPE, Phys. Rev. B **57**, 13630 (1997); Phys. Rev. B **53**, 10751 (1996).
[6] V. M. GENKIN and P. MEDNIS, Zh. Eksper. Teor. Fiz. **54**, 1137 (1968) [Soviet Phys. − J. Exper. Theor. Phys. **27**, 609 (1968)]; Fiz. Tverd. Tela **10**, 1 (1968) [Soviet Phys. − Solid State **10**, 1 (1968)].
[7] L. HEDIN and S. LUNDQVIST, in: Solid State Physics, Avances in Research and Applications, Eds. F. SEITZ, D. TURNBULL, and H. EHRENREICH, Vol. 23, Academic Press, New York 1969 (p. 1).
[8] M. S. HYBERTSEN and S. G. LOUIE, Phys. Rev. Lett. **55**, 1418 (1985); Phys. Rev. B **34**, 5390 (1986).
[9] X. ZHU and S. G. LOUIE, Phys. Rev. B **43**, 14142 (1991).
[10] R. W. GODBY, M. SCHLÜTER, and L. J. SHAM, Phys. Rev. Lett. **56**, 2415 (1986); Phys. Rev. B **37**, 10159 (1988).
[11] F. ARYASETIAWAN and O. GUNNARSSON, Rep. Progr. Phys. **61**, 237 (1998).
[12] M. STAEDELE, J. A. MAJEWSKI, P. VOGL, and A. GÖRLING, Phys. Rev. Lett. **79**, 2089 (1997).
M. STAEDELE, M. MOUKARA, J. A. MAJEWSKI, P. VOGL, and A. GÖRLING, Phys. Rev. B **59**, 10031 (1999).
[13] T. KOTANI, J. Phys.: Condensed Matter **10**, 9241 (1998).
T. KOTANI and H. AKAI, Phys. Rev. B **54**, 16502 (1996).
T. KOTANI, Phys. Rev. B **50**, 14816 (1994).

[14] J. D. Talman and W. F. Shadwick, Phys. Rev. A **14**, 36 (1976).
[15] A. Seidl, A. Görling, P. Volg, J. A. Majewski, and M. Levy, Phys. Rev. B **53**, 3764 (1996).
[16] H. Rücker, unpublished.
M. van Schilfgaarde, A. Sher, and A.-B. Chen, J. Cryst. Growth **178**, 8 (1997).
A. Sher, M. van Schilfgaarde, M. A. Berding, S. Krishnamurty, and A.-B. Chen, MRS Internet J. Nitride Res. **4S1**, G5.1 (1999).
[17] M. Rohlfing and S. G. Louie, Phys. Rev. Lett. **81**, 2312, 3320 (1998).
[18] S. Albrecht, L. Reining, R. Del Sole, and G. Onida, Phys. Rev. Lett. **80**, 4510 (1998).
[19] L. X. Benedict, E. L. Shirley, and R. B. Bohn, Phys. Rev. Lett. **80**, 4514 (1998).
[20] L. X. Benedict and E. L. Shirley, Phys. Rev. B **59**, 5441 (1999).
[21] D. J. Moss, E. Ghahramani, J. E. Sipe, and H. M. van Driel, Phys. Rev. B **41**, 1542 (1990).
[22] S. L. Adler, Phys. Rev. **126**, 413 (1962).
[23] N. Wiser, Phys. Rev. **129**, 62 (1963).
[24] R. M. Pick, M. H. Cohen, and R. M. Martin, Phys. Rev. B **1**, 910 (1976).
[25] R. M. Pick, Adv. Phys. **19**, 269 (1970).
[26] O. V. Dolgov and E. G. Maksimov, in: The Dielectric Function of Condensed Systems, Eds. L. V. Keldysh, D. A. Kirzhnitz, and A. A. Maradudin, Elsevier Publ. Co., Amsterdam 1989 (p. 221).
[27] R. Del Sole and E. Fiorino, Solid State Commun. **38**, 169 (1981).
[28] R. Del Sole and E. Fiorino, Phys. Rev. B **29**, 4631 (1984).
[29] V. I. Gavrilenko and F. Bechstedt, Phys. Rev. B **54**, 13416 (1996); **55**, 4343 (1997).
[30] W. Hanke and L. J. Sham, Phys. Rev. B **21**, 4656 (1980).
[31] F. Aryasetiawan and O. Gunnarsson, Phys. Rev. B **49**, 16214 (1994).
[32] F. Aryasetiawan, O. Gunnarsson, M. Knupfer, and J. Fink, Phys. Rev. B **50**, 7311 (1994).
[33] X. Blase, A. Rubio, S. G. Louie, and M. L. Cohen, Phys. Rev. B **52**, R2225 (1995).
[34] R. J. Elliott, Phys. Rev. **108**, 1384 (1957).
[35] L. J. Sham and T. M. Rice, Phys. Rev. **144**, 708 (1966).
[36] R. Haydock, in: Solid State Physics, Advances in Research and Applications, Eds. F. Seitz, D. Turnbull, and H. Ehrenreich, Vol. 35, Academic Press, New York 1980 (p. 215).
[37] Z. H. Levine and D. C. Allan, Phys. Rev. Lett. **63**, 1719 (1989).
[38] S. N. Rashkeev, W. R. L. Lambrecht, and B. Segall, Phys. Rev. B **57**, 3905 (1998).
[39] B. Adolph, V. I. Gavrilenko, K. Tenelsen, F. Bechstedt, and R. Del Sole, Phys. Rev. B **53**, 9797 (1996).
[40] R. Del Sole and R. Girlanda, Phys. Rev. B **54**, 14376 (1996).
[41] F. Bechstedt, K. Tenelsen, B. Adolph, and R. Del Sole, Phys. Rev. Lett. **74**, 1528 (1997).
[42] O. K. Andersen, Phys. Rev. B **12**, 3060 (1975).
[43] O. K. Andersen, O. Jepsen, and D. Glötzel, in: Highlights of Condensed Matter Theory, Eds. F. Bassani, F. Fumi, and M. P. Tosi, North-Holland Publ. Co., New York 1985 (p. 59).
O. K. Andersen, O. Jepsen, and M. Šob, in: Lecture Notes in Physics: Electronic Band Structure and its Applications, Ed. M. Yussouff, Springer-Verlag, Berlin 1987.
[44] M. Alouani, J. M. Koch, and M. A. Khan, J. Phys. F **16**, 473 (1986).
[45] Yu. A. Uspenski, E. G. Maksimov, S. N. Rashkeev, and I. I. Mazin, Z. Phys. B **53**, 263 (1983).
[46] R. W. Tank and C. Arcangeli, phys. stat. sol. (b) **217**, 89 (2000).
[47] J. Petalas, S. Logothetidis, S. Boultadakis, M. Alouani, and J. M. Wills, Phys. Rev. B **52**, 8082 (1995).
[48] W. R. L. Lambrecht, B. Segall, J. Rife, W. R. Hunter, and D. K. Wickenden, Phys. Rev. B **51**, 13516 (1995).
[49] W. R. L. Lambrecht, K. Kim, S. N. Rashkeev, and B. Segall, Mater. Res. Soc. Symp. Proc. **395**, 455 (1996).
[50] W. R. L. Lambrecht, in: Gallium Nitride (GaN) I, Eds. J. I. Pankove and T. D. Moustakas, Series Semiconductors and Semimetals, Vol. 50, Eds. R. K. Willardson and E. R. Weber, Academic Press, San Diego 1998 (p. 369).
[51] C. G. Olson, D. W. Lynch, and A. Zehe, Phys. Rev. B **24**, 4629 (1981).
[52] A. Rubio, J. L. Corkill, M. L. Cohen, E. L. Shirley, and S. G. Louie, Phys. Rev. B **48**, 11810 (1993).
[53] M. Palummo, L. Reining, R. W. Godby, C. M. Bertoni, and N. Börnsen, Europhys. Lett. **26**, 607 (1994).

M. Palummo, R. Del Sole, L. Reining, F. Bechstedt, and G. Cappellini, Solid State Commun. **95**, 393 (1995).
[54] C. Janowitz, M. Cardona, R. L. Johnson, T. Cheng, T. Foxon, O. Günther, and G. Jungk, BESSY Jahresbericht 1994 (p. 230).
[55] T. Wethkamp, K. Wilmers, N. Esser, W. Richeter, O. Ambacher, H. Angerer, G. Junk, R. L. Johnson, and M. Cardona, Thin Solid Films **314**, 745 (1998).
[56] M. Cardona, Modulation Spectroscopy, Academic Press, New York 1969.
[57] S. Logothetidis, J. Petalas, M. Cardona, and T. D. Moustakas, Phys. Rev. B **50**, 18017 (1994).
[58] W. Jones and N. H. March, Theoretical Solid State Physics, Vol. 1, Wiley, New York 1973 (p. 28).
[59] N. E. Christensen and I. Gorczyca, Phys. Rev. B **50**, 4397 (1994).
[60] B. Velický and J. Sak, phys. stat. sol. **16**, 157 (1966).
[61] E. O. Kane, Phys. Rev. **180**, 852 (1969).
[62] S. Adachi, Phys. Rev. B **41**, 1003 (1990).
[63] M. Cardona, Solid State Commun. **1**, 109 (1963).
[64] Z. H. Levine, Phys. Rev. B **42**, 3567 (1990).
Z. H. Levine and D. C. Allan, Phys. Rev. B **44**, 12781 (1991).
Z. H. Levine, Phys. Rev. B **49**, 4532 (1994).
J. Chen, L. Jönsson, J. W. Wilkins, and Z. H. Levine, Phys. Rev. B **56**, 1787 (1997).
[65] A. Dal Corso, F. Mauri, and A. Rubio, Phys. Rev. B **53**, 15638 (1996).
A. Dal Corso and F. Mauri, Phys. Rev. B **50**, 5756 (1994).
[66] D. E. Aspnes, Phys. Rev. B **6**, 4648 (1972).
[67] E. I. Blount in: Solid State Physics, Advances in Research and Applications, Vol. 13, Eds. F. Seitz and D. Turnbull, Academic Press, New York 1962 (p. 305).
[68] I. M. Lifschitz and M. I. Kaganov, Uspekhi Fiz. Nauk **69**, 419 (1960) [Soviet Phys. – Uspekhi **2**, 831 (1960)].
[69] E. M. Lifschitz and L. P. Pitaevskii, in: Statistical Physics, Part 2, Landau and Lifschitz, Course of Theoretical Physics, Vol. 9, Pergamon Press, Oxford 1980 (Chapter VI, p. 223).
[70] R. W. Boyd, Nonlinear Optics, Academic Press, Boston 1992 (p. 136).
[71] D. A. Kleinman, Phys. Rev. **126**, 1977 (1962).
[72] D. J. Moss, J. E. Sipe, and H. M. van Driel, Phys. Rev. B **36**, 9708 (1987).
[73] E. Ghahraramani, D. J. Moss, and J. E. Sipe, Phys. Rev. B **41**, 1542 (1991).
[74] V. I. Belincher and B. I. Sturman, Uspekhi Fiz. Nauk **130**, 415 (1980) [Soviet Phys. – Uspekhi **23** 199 (1980)].
V. I. Belincher et al., Zh. Eksper. Teor. Fiz. **83**, 649 (1982) [Soviet Phys. – J. Exper. Theor. Phys. **56**, 359 (1982)].
[75] S. N. Rashkeev, W. R. L. Lambrecht, and B. Segall, Phys. Rev. B **57**, 9705 (1998).
[76] S. N. Rashkeev, S. Limpijumnong, and W. R. L. Lambrecht, Phys. Rev. B **59**, 2737 (1999).
[77] R. L. Sutherland, Handbook of Nonlinear Optics, Marcel Dekker Inc., New York 1996.
[78] S. K. Kurtz, J. Jerphagnon, and M. M. Choy, in: Landolt-Börnstein, Numerical Data and Functional Relationships in Science and Technology, New Series, Vol. III/11, Eds. K.-H. Hellwege and A. M. Hellwege, Springer-Verlag, Berlin 1979 (p. 671).
[79] V. G. Dmitriev, G. G. Gurzadyan, and D. N. Nikogosyan, Handbook of Nonlinear Optical Crystals, Springer Series in Optical Sciences, Vol. 64, Springer-Verlag, Berlin 1991.
[80] S. Limpijumnong, W. R. L. Lambrecht, and B. Segall, Phys. Rev. B **60**, 8087 (1999).

phys. stat. sol. (b) **217**, 641 (2000)

Subject classification: 61.46.+w; 71.15.Fv; 71.24.+q; 73.20.Hb; 78.30.Am; 78.66.Jg; S5.11

Si Nanoparticles as a Model for Porous Si

M.J. CALDAS

Instituto de Física, Universidade de São Paulo, 05508-900 Cid. Universitária, São Paulo, SP, Brasil
e-mail: mjcaldas@usp.br

(Received August 10, 1999)

I describe here our investigation of the structural and optical properties of small hydrogenated Si nanoparticles, including the effect of weak surface oxidation. The investigation was conducted through self-consistent semi-empirical Hartree-Fock techniques, using the Configuration Interaction extension to include correlation effects, and allowing full relaxation of all atoms. Two techniques were specially reparametrized to treat semiconductor systems, MNDO/AM1 and ZINDO/CI. For clean Si:H particles, with diameters of up to ≈ 15 Å, we found that the first optical transition is size dependent, confirming the effect of quantum confinement, and is strongly blue-shifted with respect to bulk Si. The actual value of the transition energy is also affected by structural relaxation. The character of the transition can be described as excitonic, in that it develops in the crystalline-like core of the particle, and involves an almost pure one-electron excitation. Decay however can occur through a highly localised surface state, a transient Si–H–Si bridge defect, that is pinned in energy in the red. Initial oxidation of the surface, as Si–OH units, or as backbond Si–O–Si incorporation, does not affect the energy of the first absorption transition, or the decay through the bridge defect. However, more complete oxidation should kill the red-luminescence though hardening of the surface. Incorporation of a Si=O silanone unit perturbs both absorption and emission properties, introducing a first transition in the red-orange, localised over the defect. We propose that the bridge defect can explain the emission properties of freshly prepared porous Si. We also gained insight in the effect of oxidation of these systems, and propose that is not a simple process, and that the emission properties cannot be understood through one single mechanism. This picture could only be obtained through techniques that allow for investigation of excited states.

1. Scenario

Although known for a long time, porous Si (po-Si) made the first appearance as a highly promising material for opto-electronic devices in 1991. Two independent and almost simultaneous papers [1, 2] reported optical activity from po-Si, with efficient photoluminescence (PL) in the red-orange region, that is, above the bulk Si band gap. Since then there has been continued interest in this material, resulting in a large number of papers published yearly. As can be seen from Table 1, more than one paper a day has been published touching on po-Si in the past six years, including several reviews per year. Porous Si is usually obtained by electrochemical etching of Si wafers in hydrofluoric acid HF (or other etchant combinations), but it can also be fabricated by stain etching. A review of experimental data up to 1996 can be found in Ref. [3]. A large amount of work has been devoted to the study of the mode of growth of pores, including theoretical simulations [3, 4]. Pores develop preferentially in the (001) direction from the surface, and the residual Si matrix is crystalline to a high degree, but the resulting structure is not ordered because pores interconnect, sometimes growing in a

Table 1

Number of articles touching on porous Si published in indexed journals: numbers obtained through the Institute for Scientific Information search engine, using the keywords [porous silicon OR porous Si], in the timespan indicated. The papers for 1999 cover until August

year	articles (total)	review articles	year	articles (total)	review articles
1988	8	–	1994	389	3
1989	14	–	1995	458	9
1990	10	–	1996	495	7
1991	46	–	1997	497	12
1992	153	3	1998	463	4
1993	257	3	1999	354	6

dendritic fashion [3]. Larger pore-diameters can be obtained from type n^+ and p^+ wafers (10 to 100 nm pore diameter) than from type p^- wafers, in which a finer structure of pores and interconnected crystalline Si nanostructures is detected. It is generally agreed that pore surface in freshly prepared po-Si is hydrogen terminated. Porous layers can be obtained with porosities of up to 95%, porosity decreasing with depth from the surface. Around 80% porosity is needed for efficient luminescence, and there is general agreement [3] that there is direct correlation between the presence of Si nanostructures and the luminescent properties of highly porous Si. po-Si is a very open structure, so the escape probability for the emitted light is high, however, the luminescence efficiency ($\approx 3\%$) calls also for an explanation of its very occurrence: in the case of bulk Si the indirect band gap forbids the radiative recombination for electrons and holes and there is no light emission. A very plausible explanation for the luminescence is then that, as an effect of the porosity, carriers are confined to wire or dot-like regions, of very reduced effective dimensions (of the order of nanometers). This quantum confinement at once relaxes the symmetry selection rules that forbid the band gap recombination in bulk Si, and opens the gap, producing the visible light emission. Increasing porosity would decrease the dimensions of the residual Si crystalline material, and in agreement with the confinement model, the optical absorption of po-Si has been shown to blue-shift with porosity. Another very convincing argument for the confinement is provided by work on Si nanoparticles [5]. Si:H nanoparticulate material is also optically active, and a strong blueshift in absorption is seen with decreasing particle diameter.

Depending on the excitation light, po-Si can emit light in a wide range of wavelengths, from around 350 to 1500 nm, and it has been shown [3] that different energies correspond to different mechanisms (either for excitation, or for recombination and decay). A puzzling characteristic is, however, that the photoluminescence does not necessarily follow the behavior of absorption with porosity or particle diameter [5]. In fact the visible band (red-orange), the only band active also in electro- and cathodoluminescence, is virtually pinned at around ≤ 1.5 to $1.8\,\text{eV}$ while the absorption moves up to 2.8 to $3.0\,\text{eV}$. This led several authors to invoke localised states, mostly surface states, as the origin of this band. Thus, if on the one hand confinement seems to be a necessary condition, the radiative decay could be caused by other mechanism (or mechanisms). Some of the proposed mechanisms rely on the existence of surface defects, while others suggest the existence of a hydrogenated amorphous Si phase [6] at the surface of the pores, which would then trap carriers and provide recombination channels. Another point that has been raised by several authors relates to the oxidation state of the surface. It is known that ageing in air can drastically affect the optical properties of samples. A layer of oxide eventually forms in aged samples, or can be intentionally formed by special procedures. It has then been proposed [7, 8] that the radiative centers would be defects at the SiO_2/Si interface, or at the SiO_2 layer itself. The fact that luminescence exists in fresh po-Si samples, on the other hand, would seem to imply that either oxygen is not needed at all, or that at least very little oxygen can be sufficient. Through analysis of Raman and infrared (IR) spectra, it was suggested that the PL originated from siloxenic-like surface coating [9] of the pores. Further studies [10] showed this promising hypothesis to be incorrect, but the role of oxygen at the surface is not yet clarified.

Theoretical modeling could be expected to help unravel the several hypothesis suggested by experiment. However, in spite of the impressive progress in theory in the past decades, a reliable description of these systems is still very difficult. It is in this case necessary not only to study structural relaxations at the microscopic level, but also to model these disordered structures at quantum level, that is, taking account of charge-transfer processes, and giving a reasonable description of localised and excited electronic states. On the other hand, the claim that the crystallinity of the original Si survives to some extent in the nanostructures calls for a scheme that is capable of describing also the relevant properties of extended crystalline systems. In a few words, it is necessary to treat the optical properties consistently from the smallest cluster limit, where correlation effects are expected to be important, up to large crystallites or wires.

The effective-mass approximation (EMA) has been used extensively [11] to study optical properties of Si wires and dots, however it cannot treat deviations [12] from the perfect crystalline conditions other than the idealised confinement. On the other extreme of sophistication are the ab-initio local-density appoximation (LDA) techniques, which can give quantitative answers to questions on the ground-state properties [13] of unit cells with up to 10^2 atoms, but the study of optical properties is hindered by the well-known shortcomings [13, 14] of LDA when treating excited states. Very special work [15] has been devoted to small Si:H particles within density functional theory beyond LDA, however the size of the particles has to be kept to a minimum. Turning to Hartree-Fock theory, which might – with extensions such as Configuration Interaction or perturbation approximations – give accurate answers to excited state properties, ab initio investigations with suitable basis-sets are also prohibitively demanding for any but the smallest clusters.

Empirical LDA (ELDA) can easily go to much larger cells [12, 16] and together with empirical tight-binding techniques (ETB) can give a fair description of the electronic structure for systems close to the bulk environment. Empirical techniques have been divided [17] into those which give good fits to the conduction band of bulk Si, and those which do not. I remark that even if the fit to the extended states is good, this does not guarantee a good description of the localised states one expects to find in po-Si. Also, geometries in this case have to be guessed, from bulk and bulk–surface relaxations.

Structural properties could be studied through tight-binding molecular Dynamics [18] since it may be applied to relatively large cells. However, it is not easily applied to atoms such as oxygen, which would be important to investigate, and cannot in the standard forms [19] (since a new self-consistent method, based on LDA, has been proposed to treat Si, O and H) treat charge tranfer accurately. Atomistic molecular dynamics, which may be sucessfully applied to geometries of unit cells of the order of 10^4 to 10^6 atoms, gives no clue to the electronics of the system. Both techniques take the parameters from bulk situations, and are thus well adapted to treat semiconductor systems. Semi-empirical Hartree-Fock techniques, on the other hand, are usually parametrized for small organic molecules, where the sp^3 hybridization is not relevant, and so the straigthforward application to semiconductor systems might yield poor results.

It is thus very difficult to choose a tool for the theoretical study of po-Si, and the discussion can become quite technical (see, e.g., critique of the work of Ref. [14] in Ref. [17]) and little enlightening. To complicate things results are often conflicting, as is apparent from the data shown in Table 2. The data are for the energy of the first optical transition of Si particles, with perfectly hydrogenated surfaces, obtained from different models; we see that for any of the sizes studied, the transition energy can differ by more than 1.5 eV from one model to another. Even accounting for the difference between models, this discrepancy is unexpected, and arises from the fact that we are dealing with structures at the limit between molecules and solids, where both localization and delocalization effects can play an important role. I group in Table 2 only work on pure hydrogenated systems, but certainly we need also to look at properties of oxidized material. Again in this respect we have somewhat conflicting results from different techniques, as we will see later.

We see by the amount of literature already devoted to the subject (see Table 1) that there has been a widespread effort to elucidate the behavior of po-Si material. I will not here try to make a thorough or even fair survey of the field, which would be next

Table 2

Gap energy of H-saturated Si nanoparticles calculated through different models. The number of atoms in the particle is indicated in the first column. TB[a]): semi-empirical tight-binding, third neighbors; LDA-PW[b]): supercell, pseudopotentials within Local Density Approximation, plane-waves basis set; LDA-LCAO[c]): LCAO basis set, LDA, constant shift of +0.6 eV in gap energy; MNDO-PM3[d]): semi-empirical Hartree-Fock, MNDO standard parametrization PM3; ETB[e]): empirical TB, nearest neighbors only. Estimates can also be made, from LDA-PW[f]) calculations for a particle with around 70 Si atoms, gap $\approx$2.9 eV: empirical pseudopotentials, LDA-PW with excitonic corrections; from ab initio LDA-PW[g]) calculations, for a particle with 10 Si atoms, gap $\approx$6.3 eV: relaxed geometry, GW corrections plus excitonic correlation

number of atoms	transition energy (eV)				
	TB[a])	LDA-PW[b])	LDA-LCAO[c])	MNDO-PM3[d])	ETB[e])
10	–	4.62	5.10	3.43	–
17	4.29	–	5.05	3.10	–
29	–	3.32	4.95	2.76	–
35	–	–	4.90	2.74	2.95
66	–	2.95	3.98	2.62	–
87	3.10	–	4.05	3.10	–
123	–	2.45	3.60	2.45	–
239	–	–	3.50	–	1.68

[a]) Ref. [55]; [b]) Ref. [56]; [c]) Ref. [57]; [d]) Ref. [38]; [e]) Ref. [36]; [f]) Ref. [16]; [g]) Ref. [15].

to impossible. I will rather concentrate on some of the theoretical models advanced for the luminescence, and focus on the studies [20 to 24] performed by our group.

We found that the luminescence in the red can be generated at the perfectly hydrogenated surface of Si nanoparticles, by light-induced transient defects. There is no need for pre-existing saturation defects (dangling bonds) or extraneous species. Another finding is that oxygen in the most stable forms (hydroxyl and back-bonded) does not induce radiative decay, on the contrary, it may well quench the luminescence. However, a specific surface attachment (as a silanone, Si=O) may indeed cause luminescent decay. Both mechanisms can explain the luminescence efficiency, and the Stokes shift between absorption and emission.

Our studies were conducted within the Hartree-Fock formalism, with the Configuration Interaction extension. In principle, the formalism would allow us to obtain reliable transition energies, and also reliable excited state geometries. Due to the relatively large size of the particles involved, we resort to semi-empirical techniques, for which, as already mentioned, there were no optimised parametrizations for semiconductor systems. This led us to reparametrize two techniques for the needed atoms, Si, O and H. In the next section, I briefly review the approximations involved in the theoretical treatment. I discuss in Sections 3 and 4 results for nanocrystals and hydrogenated nanoparticles, in Section 5 the effect of oxidation, and summarize our results in Section 6.

2. NDO/Crystal Approach: Reparametrization of Quantum Chemistry Techniques

The semiempirical Hartree-Fock techniques, developed mostly for Quantum Chemistry (QC) applications, have the advantage over ETB schemes that calculations can be carried through to self-consistency in the charge density. One can take into account charge rearrangements, which allows us to determine stable geometries. More important is that electronic correlation is routinely included in these QC techniques, through the method of Configuration Interaction (CI), which allows us to analyse excited states and understand optical properties.

QC techniques are designed to furnish, with the best possible accuracy, chosen specific properties (target properties) of molecules. This must be done without sacrificing theoretical consistency with Hartree-Fock and ease of calculation. In this spirit, the Hamiltonian is simplified through the introduction of as few as possible semi-empirical parameters, which must satisfy the requirement of transferability (contrary to ETB, the same set of parameters must describe a particular atom in different bonding environments) and are adjusted to fit the target properties. Starting from the usual Hartree-Fock variational procedure, the single-particle wavefunctions (Molecular Orbitals, MO) are expanded in Linear Combinations of Atomic Orbitals ϕ_μ (LCAO). The most relevant simplifications are: (a) the frozen-core approximation, so that only valence-shell electrons are treated selfconsistently; (b) the LCAO expansion is restricted to a minimum-basis set, e.g. for Si or O only valence s and p orbitals are included; (c) all electronic and nuclear energies are written in parametric form, but as said above parameters should be associated with particular elements, not bonding environments, and be transferable from one environment to another; (d) finally, to keep the number of two-electron two-center integrals manageable, the overlap element is neglected ($\phi_\mu \phi_\nu \, dr = 0$, where μ and ν are centered on different atoms). This last approximation is known as Neglect of Differential/Diatomic Overlap (NDO), but it is important to note that the actual overlap between orbitals, ultimately responsible for bonding, is calculated and brought back through the one-electron *resonance integrals*. I again stress that the electronic structure is obtained through a full self-consistent iterative procedure. Most parameters are taken already from atomic properties. A suitable set of systems is then chosen to complement the parametrization procedure (parametrization ensemble), which should cover the most useful bonding situations. There are different approximations to the Hamiltonian (to study different properties), which distinguish one technique from another, and different parametrizations within the same technique. The transferability property for the parameters holds, however, within a given set of molecular systems that are similar in some way to the set of sample systems included in the parametrization ensemble. The problem is, here, that the sp^3 hybridization of tetrahedral semiconductors is not included with the required weight in the standard ensemble systems.

We introduced a procedure [23] to adapt a semi-empirical Hartree-Fock technique to the study of complex semiconductor structures, comprising compatible use of a full Bloch-periodic Supercell Model [25], and a Cluster Model with adequate treatment of surface dangling bonds. An alternative to **k**-vector periodicity is the cyclic-cluster model, equivalent in our case to the use of $\mathbf{k} = \mathbf{0}$ only, known as Γ-point Brillouin-zone sampling (see [25]); with large enough supercells/clusters, the results should be convergent, see Ref. [23]. The parametrization is meant for use with finite systems (nanoparticles, or simulation of defects in extended systems through the cluster model). However, if we want to simulate perturbations in an infinite medium, we must be able to simulate the crystal itself. On the other hand, to obtain tranferable parameters one should include in the parametrization ensemble as many representative systems as possible. In the ensemble should enter systems for which there exist reliable data: in the case of semiconductors there are not as many different compounds built from Si, O and H as, for example, organic compounds built from C, O and H. We included in the parametrization ensemble a mix of crystalline systems, small molecules, and localised defects, all involving the Si, O and H atoms.

We have chosen to reproduce the structural and optical properties at low temperatures of bulk Si, and one of the low temperature SiO_2 isomorphs, α-quartz; the isolated oxygen interstitial defect in Si; and small molecules where the tetrahedral environment is relevant (silane, disilane, disiloxane). In the set of properties we included the lattice parameter for Si and SiO_2; the bulk modulus for Si, geometries and vibrational frequencies for the small molecules and for the interstitial O defect in Si. We also fit the phonon spectrum for bulk Si: in this way, we avoid including the higher-energy structures for the crystal, that are in general too far from equilibrium at low temperatures. For optical properties we included the gap for Si and SiO_2, and the optical spectra for the small molecules. We then recheck the parameters by looking at different but still well-known systems, such as the Si-A center, and more complex siloxenic molecules.

The techniques we have chosen are the Modified Neglect of Diatomic Overlap in the Austin Model version (MNDO/AM1), designed to describe accurately the geometrical features of molecules, including formation energies and vibrational frequencies. We will refer to the new set of parameters as AM1/Crystal, to distinguish it from former parametrizations (the original [26] AM1 and the more recent [27] PM3). For optical properties we have parametrized ZINDO/CI, the spectroscopic version [28] of the Intermediate Neglect of Differential Overlap method (we will refer to the new parametrization as ZINDO/Crystal). The reader is referred to Ref. [23] for technical details.

3. Nanocrystals and Quantum-Confinement Effects

I present here a small summary of the quality of our fit for the ensemble systems. In particular for the AM1/Crystal parametrization, a detailed description can be found in Ref. [23]. We obtain best fits, within 4% of the experimental value, (a) to the geometries and vibrational properties of the molecules, with the standard molecular code and the new parametrization, and (b) for the lattice parameters of the crystals, with the periodic supercell NDO code. For the Si bulk crystal we obtain not only the bulk modulus (within 12% error, compared to 87% with the usual parametrization), but also phonon energies for the high-symmetry Γ, X and L points, using a frozen-phonon [29, 30] approach. Our fit is within 8% error, the largest relative errors appearing for the transverse acoustic frequencies, as expected. The frequencies and charge distributions for the O defects in Si, studied within the nanocrystal approach (described in the following) are also within few percent of experiment. *We do not scale any of our frequencies when comparing to experiment*, as is common practice. Relative stabilities of different configurations are also checked for siloxane compounds, and are found to be in good agreement with LDA results.

For ZINDO/Crystal we obtained best fits to the optical spectra and first ionization potentials of the small molecules, always using the geometries obtained with AM1/Crystal. At the same time, using the periodic code, we adjusted parameters to describe the valence band properties and optical gap character for the crystals. We obtain the correct symmetries, and fair valence bandwidth for both Si and α-quartz (we are focusing on the first optical transitions, so the lower valence states were not optimised). We also obtain the correct indirect gap character for Si, with however the expected overestimation of HOMO-LUMO (Highest Occupied − Lowest Unoccupied MO) gap energy, and underestimation of conduction bandwidth, that are recognised to be due to lack of correlation effects. In fact, unoccupied eigenvalues in Hartree-Fock theory carry no precise physical meaning, so that the HOMO-LUMO eigenvalue difference in general cannot be taken to relate to a "gap energy" or optical transition. To obtain these energies correlation must be included: in the Configuration Interaction method, the first correction comes from calculating the total energy of a "pure" excited configuration. A pure excited configuration is associated with a Slater determinantal function, with a "hole" in the formerly occupied states and an "electron" in one of the formerly unoccupied states (single excitation), or two holes and two electrons, and so on. Full correlation can then be achieved if we allow interaction between the ground and all such excited configurations: a way of accounting for the need of electrons to avoid each other, which is not taken into account in simple Hartree-Fock. Clearly, CI correlation effects gain importance with confinement of the electrons. Therefore, a method that simulates the bulk crystal is not guaranteed to simulate small particles equally well, since the degree of localisation varies too much from one case to the other. This is why we are careful to check that both our methods reproduce also the properties of silane, disilane and disiloxane.

The quality of the fit for the small molecules, already including CI, can be seen to be good from Table 3. The inclusion of correlation effects through CI is not straightfor-

Table 3

Ionization potentials for silane, disilane and disiloxane, and optical transition energies for disilane and disiloxane, as calculated by ZINDO/CI, standard parametrization, and as calculated by the ZINDO/Crystal parametrization, compared to experiment. The optical transitions for silane carry always a strong contribution from Rydberg states, which should not be describable with minimum basis sets

molecule	ZINDO	ZINDO/Crystal	exp.[a]	molecule	ZINDO	ZINDO/Crystal	exp.[a]
	ionization potentials (eV)				optical transitions (eV)		
SiH_4	13.96	13.56	12.36	Si_2H_6	8.35	7.67	7.78
Si_2H_6	11.49	11.16	10.53		9.23	8.54	8.39
	12.00	12.94	12.0	OSi_2H_6	6.42	7.79	7.90
OSi_2H_6	14.76	11.39	11.19		7.10	8.16	8.27
	14.89	11.76	11.37		7.86	9.68	8.92
					9.73	10.01	9.54

[a]) Refs. [58], [59] and [60].

ward for the infinite periodic solid. It is, however, crucial to our parametrization that we have an estimate of these effects: to do that we study *nanocrystals*, simulated by clusters in the perfect (theoretical) lattice geometry, saturated by pseudo-atoms [23, 31, 32].

We generate pseudo-host-atoms (Si′) which have an s-type valence orbital only, and is capable in principle of saturating one or at most two directional bonds, if placed at correct crystalline positions. The parameters for these pseudo-atoms are the same as for the replaced atom, with the exception of those that depend on the orbital type and valence charge. These are adjusted to furnish, in the internal shells of the cluster, certain desired characteristics: (a) the correct hybridization for the internal host atoms; (b) minimum possible charge inequivalence for atoms in shells that, although inequivalent in the nanocrystal cluster, would be equivalent in the crystal; (c) minimum possible forces on internal host atoms (for AM1/Crystal). No effort is done at this point to obtain a given value for the HOMO-LUMO energy difference. CI is then performed for these nanocrystals, and the gap is "extrapolated" to infinite radius.

We studied "spherical" nanocrystals in T_d symmetry ($Si_{35}Si'_{36}$, $Si_{71}Si'_{60}$) and "box-like" nanocrystals in D_{3d} symmetry ($Si_{42}Si'_{44}$). The results for the three largest Si nanocrystals are shown in Fig. 1a, and we see that our extrapolated value for the gap of bulk Si is extremely close to the experimental value. It is also relevant that the gap transition is essentially HOMO-LUMO, and that these orbitals preserve the character of the (indirect) gap-edge orbitals obtained for the bulk. These results should be similar to results from EMA, in that we simply truncate the crystal, which is surrounded by vacuum. There are no dangling bonds, but we have no hydrogen bonds either, as in ETB or ELDA models. Of course, our extrapolation involves a choice of the representative radius [12] and other considerations [15], however, we find that the dependence of the gap energy with size, at these quite small sizes and taking pure quantum confinement effects into account, should be close to R^{-1}, as found by several other authors, and not to R^{-2} as predicted by EMA.

Tigelis and coworkers [11] used EMA for a "bumpy" wire, that is, a wire with different diameters along the length. They find that the "bumpiness" induces a ladder of

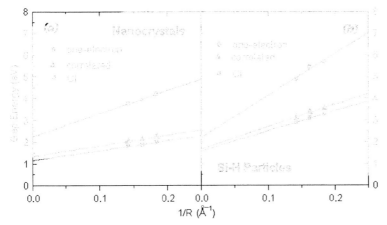

Fig. 1. Energy of the first optical transition for a) Si nanocrystals and b) nanoparticles, plotted against the inverse radius $1/R$ (R is a characteristic length, given by the distance to the center of the structure of the outer Si shell). We show the Hartree-Fock one-electron value, $\epsilon_f - \epsilon_i$, (always HOMO-LUMO for these transitions), the value including virtual orbital correction ($\epsilon_f - \epsilon_i - J_{if} + 2K_{if}$), and the CI value

eigenvalues, with the lower eigenfunctions localised at regions of the wire with large diameter. As the energy goes up, the states become less localised until at sufficiently high energies, the states are delocalised along the length of the wire. At this point, the energy is compatible with the smaller diameter along the wire. The explanation [11] for the Stokes shift between absorption and emission, already suggested by earlier experimental analysis [33], would follow simply: absorption occurs at the inner core of the wire, whereas luminescence occurs after trapping of the electron–hole pair at one of the larger wire sections. This very interesting hypothesis would have no parallel for nanoparticle material, where the same puzzling behavior is detected [5]. Looking at our results, the sizes consistent with the red-orange emission would be around 2 to 4 nm diameter. The absorption threshold, however, could not be explained at all, since even the smallest crystals would absorb well below blue.

4. Si:H Nanoparticles and Surface Effects

Our next step is the simulation of Si nanoparticles, with the surface perfectly saturated by hydrogen atoms. We studied small hydrogenated Si particles in T_d symmetry (spherical), centered either on a Si atom (Si_5H_{12}, $Si_{17}H_{36}$, $Si_{29}H_{36}$, $Si_{35}H_{36}$, $Si_{71}H_{60}$) or on a T_d interstitial site ($Si_{10}H_{16}$); and box-like bond-centered clusters in D_{3d} symmetry ($Si_{26}H_{30}$, $Si_{44}H_{42}$). Now the geometries are allowed to relax, through AM1/Crystal, before studying the optical properties. The first finding is that even for particles with as few as 29 Si atoms, in tetrahedral arrangement, the vibrational spectrum reproduces quite well the phonon dispersion of the bulk. This particle has just five "bulk-like" perfectly coordinated Si atoms. We show in Fig. 2 the vibrational spectrum for two particles, with 35 and 44 Si atoms, where we see clearly also the Si–H surface modes. There occurs for all particles a non-linear relaxation that is responsible for a split-off Si–Si mode, with energy slightly higher than the optical phonon at Γ. This is due to a compression of the first shell towards the central atom. On the other hand, the overall

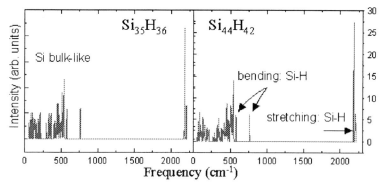

Fig. 2. Vibrational spectra obtained through AM1/Crystal for the particles $Si_{35}H_{36}$ and $Si_{44}H_{42}$. We indicate the Si bulk-like modes and the surface Si–H modes

effect of relaxation is an expansion of the Si core of the particles, relative to nanocrystals (such non-linear relaxations are also found [23] to a lesser extent in hydrogenated Si(111) surfaces). This results are relevant to the debate on the dimensionality of the luminescent structures in po-Si, since it is shown that particles of this size may cause the phonon lines seen [34] experimentally.

The optical spectra are obtained for the optimum geometries of each particle with ZINDO/Crystal, with single-excitations CI. We consider only singlet excited states, since singlet–triplet splitting should not be so large as to affect the validity of our results (as inferred by the results of Ref. [35], the splittings should not exceed 0.1 eV for these small particles). The spectra are always rich, as we show in Fig. 3 for two different-symmetry particles, but with no transitions in the red-orange region. The average diameter of the larger Si structures studied here is of the order of 10 to 15 Å, and correspond to gaps in the region ≈ 2.5 to 3.5 eV (the $Si_{35}H_{36}$ cluster with a 3.25 eV transition has an average diameter of ≈ 11 Å). Extrapolation of these values to infinite

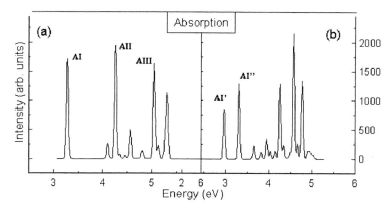

Fig. 3. Optical spectra for the particles a) $Si_{35}H_{36}$ and b) $Si_{44}H_{42}$ calculated through ZINDO/Crystal, in the *relaxed ground state geometry*; the lines are gaussian-broadened by 0.01 eV. The first peaks can be traced in more than 85% to the HOMO-LUMO one-electron excitation; the LUMO symmetry is a_1 in both cases, but the HOMO symmetry is lowered from t_2 of T_d in a) to $a_1 + e$ of C_{3v} in b), hence the splitting of AI to AI' and AI''

diameter would yield a gap of ≈1.5 eV, as seen in Fig. 1b, which underlines that, besides quantum confinement effects, also structural relaxations play a role in defining the gap of such small particles. Since relaxation effects in the core of the particles must disappear after some critical size, it is not reasonable to extract from our results a range of diameters for particles that could absorb in the red-orange region.

Early theoretical work on hydrogenated small Si clusters usually neglected correlation altogether, optical properties being extracted from HOMO-LUMO differences [36, 37]. Exceptions are the works of Kumar and coworkers [38] and Wang and Zunger [39]. The first authors used, however, the MNDO Hamiltonian in the PM3 parametrization [27]: apart from the fact that PM3 is still a molecular parametrization, the Hamiltonian itself is better adapted for the calculation of geometries. They obtain absorption energies of ≈3.4 eV, and a Stokes shift of the order of 0.6 eV for the smaller clusters. Wang and Zunger work within LDA, which includes and usually overestimates correlation in isotropic form; a point worth noting [14] is that the dieletric function was extrapolated from the bulk crystal to the small particles. More recently, Rohlfing and Louie [15] studied small Si:H particles, and went beyond LDA (with the GW formalism) in order to include correlation in the calculation of the first optical excitations; as stated in Table 2, they obtain optical gaps even higher than our results. Our main results [20] seem, however, to agree qualitatively with the predictions of all these authors, in that the absorption appears to be too high to account for all the optical properties of porous silicon.

I show in Fig. 4 an analysis of the origin of the low-energy optical transitions for one of the particles (results are very much the same for any of the larger ones). The three

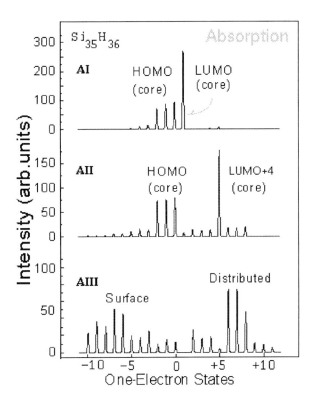

Fig. 4. The origin of the absorption peaks of the $Si_{35}H_{36}$ cluster (≈3.25, 4.25 and 5.05 eV) is analysed by plotting the weight of the single-particle states contributing to the transition (see text). The single-particle levels in the abscissa are numbered by integers indicative of the energy ordering, from the Highest Occupied Molecular Orbital (HOMO). The HOMO is triply degenerate (states $-2, -1, 0$), while the Lowest Unnocupied MO (LUMO, state $+1$) is singly degenerate. The HOMO-LUMO single-particle excitation is therefore clearly the dominating contribution to the lowest transition

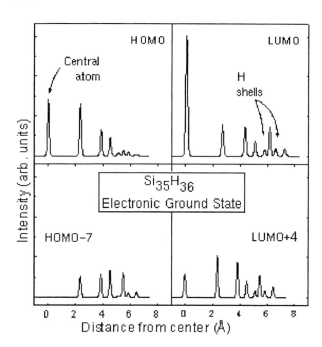

Fig. 5. Spatial localization of the HOMO and LUMO for the $Si_{35}H_{36}$ cluster: The probability density associated to a typical atom in a given shell is plotted as a function of its distance r from the center of the cluster. All the peaks correspond to Si shells, except those marked by arrows corresponding to the hydrogen shells at the surface of the cluster. Clearly the most important contribution comes from the internal crystalline-like Si shells. We show also (lower panels) two other states, occupied (state HOMO-7) and unoccupied (state LUMO+4), that represent more distributed probability densities, and that contribute to the second AII and third AIII absorption transitions; we see that even for these states the surface is not dominant

main peaks for the optical absorption spectrum obtained for the $Si_{35}H_{36}$ particle (Fig. 3a) are resolved into one-electron-excitation contributions in Fig. 4, where we project the determinantal states appearing in the CI in terms of single "hole" or "electron" states. Note that the horizontal axis carries eigenvalue numbers, not energies, and that eigenvalues are numbered from the "valence band top" (HOMO). It is clearly seen that the lowest one-electron excitation (HOMO-LUMO) dominates the first peak: we may thus identify the optical transition as mainly derived from the HOMO-LUMO single particle excitation. In the case of box-like particles, there is a symmetry splitting of the single-particle states defining the top of the valence band, so the optical transitions are also split, as seen in Fig. 3b: everything applying to peak AI, applies almost identically to peaks AI' and AI''. We now deconvolute the combination of atomic orbitals for the HOMO and LUMO, in contributions coming from different atoms of the cluster (Fig. 5). This is obtained by simply summing the probability densities associated to the orbitals of all the atoms which belong to a given shell in the cluster (sitting at a given distance from the center); the result is then divided by the number of atoms in the shell for normalization. (For example, in this $Si_{35}H_{36}$ cluster there is a central atom, a first shell of 4 Si atoms, then the second and third shells have 12 Si atoms, and the fifth shell has 6 Si atoms; all the hydrogens form 12-atom shells). From the spatial localization of the HOMO it results that the internal shells of Si atoms account for almost 80% of the orbital distribution, while the most important individual contribution comes from the central atom (Fig. 5, HOMO). The same analysis applies to the LUMO spatial distribution (Fig. 5, LUMO). The contribution from the hydrogen atoms is very small, and this happens consistently for all states close to the gap. In accordance with previous theoretical results, both on small hydrogenated particles [16, 38] and on hydrogenated Si surfaces [23], we find the hydrogen-related one-electron states around 4 eV below the valence band maximum.

For the higher-lying states there is already a significant configuration mixing, as I exemplify also in Fig. 4; however, as indicated, the first contributions from surface states (Fig. 5, lower panels) appear only at the third main peak, that is, higher than 5 eV: As a result, we find that all optical absorption below 5 eV has crystalline character.

Summarizing our results so far, we find that the optical gap of Si:H particles, in the stable relaxed geometry, shows a blue shift with respect to bulk Si that is due to both confinement and relaxation effects. In agreement with most previous theoretical results, for particles in the diameter range 1.5 to 2 nm the first optical transition lies around 3 eV, too high to account for the red PL in po-Si; also, this first transitions are crystalline in character, which is compatible with the size dependence but seems conflicting with the data on po-Si, where we see a large dependence of the red PL on the surface [5] conditions. We obtained, however, another relevant information: for all studied particles, the first optical transition is essentially a HOMO-LUMO one-electron excitation. This explains why different treatments of correlation give virtually similar results: there is none or extremely little configuration mixing for these particular transitions. Even more interesting, this allows us to study the excited state reached under illumination [21, 22]: we simply study through AM1/Crystal the properties of the particle with one electron taken from the HOMO to the LUMO. With the geometry optimised for this electronic excited state, we again study the optical properties, through ZINDO/Crystal.

I show in Fig. 6 the spectra so obtained for the particles $Si_{35}H_{36}$ and $Si_{44}H_{42}$, which must be compared to Fig. 3. The difference is striking: the first transition is drastically red shifted, and the region from 1 to 3 eV is now completely filled with lines. On the other hand, the similarity of the effect for the two particles is also remarkable.

It must be noted that these are the spectra more properly associated with the luminescence; the spectra in Fig. 3 should correspond to absorption only. However, these optical spectra are obtained for the relaxed geometry of the first excited electronic state of each particle (the state reached though the HOMO-LUMO excitation); thus, only the lower lines in these spectra are likely to be seen in luminescence.

The luminescent process in po-Si has been called "exciton trapping". What our study indicates is that indeed the absorption process might be said to create an exciton, since

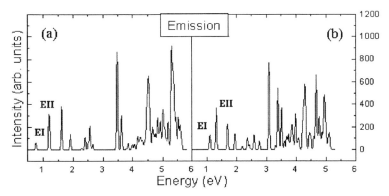

Fig. 6. Optical spectra for the a) $Si_{35}H_{36}$ and b) $Si_{44}H_{42}$ calculated through ZINDO/Crystal, in the *relaxed excited state geometry*; the lines are gaussian-broadened by 0.01 eV. The lower peaks should be seen in emission

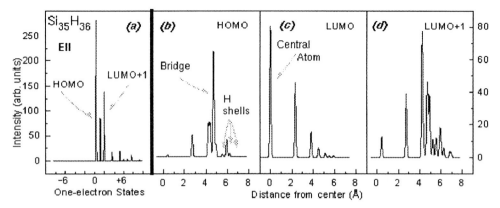

Fig. 7. Deconvolution of the emission peak for particle $Si_{35}H_{36}$ showing the contribution from the molecular orbitals; we see it is now a mixed transition, involving mostly the HOMO, LUMO and LUMO+1. The spatial localisation for the relevant orbitals is shown in parts b) HOMO; c) LUMO; and d) LUMO+1; all orbitals are now singly degenerate: note the localisation of the HOMO and LUMO+1

the transition clearly involves just the two one-electron levels and could be described as creating an electron-hole pair. The situation after the "exciton trapping" is, however, more complex. We again investigate the origin of the lower peaks, by deconvolution in one-electron contributions, as shown in Fig. 7 for the particle $Si_{35}H_{36}$ (as expected from the similarity of the spectra, results are very similar for the two particles). It is seen that there is now a clear configuration mixing effect, already for the lowest two transitions. In emission, the transitions mainly decay to the HOMO, but originate both from the LUMO and from a second virtual state. Figure 7 depicts the analysis of the emission for the particle $Si_{35}H_{36}$: we note that the LUMO is still crystalline in character, but the HOMO is very different from the HOMO of the ground state. It is no longer degenerate, and shows a very large density concentration over just two Si atoms at the surface of the cluster. The other virtual state involved, LUMO+1, also shows this localised character.

What we find in fact is that the particle in the excited state undergoes a strong symmetry-lowering spontaneous distortion, extremely localised on two neighboring surface Si atoms, each bonded to just one H atom. In the distortion the Si atoms bend and stretch their bond by 0.26 Å, keeping the distance to their internal neighbors almost unchanged. At the same time, one of the H atoms moves to bridge the bond, so that a surface configurational defect is created, with no complete rupture of any bonds in the particle. We show in Fig. 8 a scheme of the defect, focusing on a six-fold Si ring that reaches from the surface to the central core of the particle. If we analyse the bond order in this bridge-defect configuration we find that the H atom is bonded almost

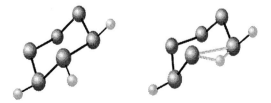

Fig. 8. Scheme of the bridge defect, showing a six-fold Si ring that reaches to the center of the particle, in the normal ground state (left) and in the excited state bridge configuration (right); visualization for the box-like $Si_{44}H_{42}$ particle, Fig. 9a

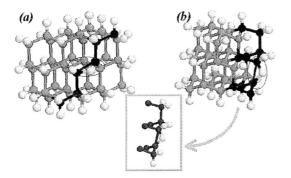

Fig. 9. Visualization of the particles a) $Si_{44}H_{42}$ and b) $Si_{35}H_{36}$ in the excited state geometries. We draw attention to the bridge defect, and to the line of surface atoms that accomodate the distortion. In the inset, we focus on the line relevant to the $Si_{35}H_{36}$ particle

equally to both Si atoms, that maintain 40% of the regular bond strength to each other. We show in Fig. 9 representations of the defect in the $Si_{35}H_{36}$ (spherical) and $Si_{44}H_{42}$ (box-like) particles: we can see that the defect is basically the same, and occurs also basically at the same spot, in one of the (110)-like surface lines of Si atoms. In the case of the spherical particle, the bridge occurs next to a corner (SiH_2 group), but in the "roomier" box-like particle it moves more to the center of the line. However, except for that, the defect is essentially unchanged.

This bridge defect has a considerable oscillator strength for optical decay, so we might expect efficient luminescence from such a center. Furthermore, the arrangement is highly unstable relative to the normal surface configuration, and there is no evidence for a metastable (local minimum) distorted configuration in the ground state. So once the center optically decays, it should decay non-radiatively back to the normal surface configuration. In other words, we do not expect dangling-bond creation [40] or optical bleaching [41] in this case. An earlier suggestion by Allan and coworkers [42] to explain the luminescence in po-Si invoked a surface defect: by removal of the H atoms saturating neighboring Si surface atoms, a surface dimer could be created, and the authors show that by stretching such a dimer one could create an efficient optical decay center, similar to the bridge defect; the decay energy would be, by the authors' calculation, around 0.7 eV, lower than our estimate for the bridge. Our results show that there is no need for actual built-in defects, since this bridge is photocreated by the absorption process on perfectly H-saturated nanoparticles. It requires, however, a "soft" environment, ideally provided by the surface of such particles, since there is considerable distortion of the atoms involved.

We find another distortion (breathing) mode for the excited state, from which there is also efficient optical activity; this symmetric relaxed configuration is, however, much higher in energy than the bridge defect state. Our results allow then for a consistent explanation of the photoluminescence processes in Si:H nanoparticles, which should hold also for freshly prepared po-Si. For a given size distribution in a sample, we have a distribution of absorption thresholds that result in an exponential tail; for each particle, the first optical absorption originates in the crystalline core and the gap is size dependent, showing quantum confinement effects. Once in the excited state it can decay through a symmetric state, Stokes shifted by ≈ 0.2 to 0.3 eV, that is, in the blue-green region for particle diameters around 2 nm; due to the (symmetric) mode of relaxation, it should be a fast channel. The H surface atoms are not involved in either absorption, or blue-green luminescence.

The red-PL, on the other hand, is associated with the photocreated Bridge defect and, due to the extremely localised character of the defect wavefunction, is "pinned" in energy. We do not have an estimate for the decay rate: since the distortion is very large the Jahn-Teller coupling term could slow the process down, or it could be that the decay is not directly to this luminescent excited state, but passes through some intermediate state. The presence of H atoms is essential to the red luminescence, as is the softness of the environment: the intensity of luminescence should increase with decreasing particle size.

The capture of the exciton at surface resonators has also been proposed by several authors, to explain the PL behavior with temperature, and has been simulated explicitly by Létant and Fishman [43]. The authors use a very simplified model, so comparison is difficult, however, the bridge defect would seem to fit very well in this picture.

Our results so far dit not touch on the question of oxidation of the surface, however, indicate clearly that the effect can be dramatic. We discuss it in the next section.

5. The Role of Oxygen

Porous Si was used at first to obtain oxide layers for Silicon-on-Insulator technoloy, so the oxidation has been carefully Studied [44] even before the discovery of its optical properties. It is certain that po-Si is easily oxidized, even through dry oxidation: this is not the expected behavior for hydrogen-saturated Si surfaces, which indeed have been widely used in the industry exactly because of its stability [45] against dry air oxidation. However, the much larger and less ordered surface area for po-Si, which probably carry also much more saturation defects, makes it easier to oxidize. Notwithstanding the abundance of experimental studies, it is difficult to have a clear picture. If some authors are very positive in stating that the samples under investigation are "freshly prepared" and carry clean, hydrogen-terminated surfaces, there have been statements that there are usually no oxygen-free samples, since oxidation would start in the first minutes once the sample is exposed to air. To exemplify and citing only a few papers, we comment next on IR spectroscopy, electron spectroscopy (ES) and PL experiments.

Kato and coworkers [44] studied po-Si oxidized in dry air, by conventional IR spectroscopy focused on the Si–H stretching band. The authors find the band, *prior to oxidation*, at 2090 to 2140 cm^{-1}; they follow the oxidation process by the shift in energy of this band, which is known to occur for every hydrogenated Si surface under oxidation: the O atoms enter in the back-bonds, introducing broader bands at higher energies (from ≈ 2120 to $2260\ cm^{-1}$ in that case). They find that the Si–H surface species (albeit back-bonded to O) survives oxidation as long as the oxidation temperature is kept below around 500 K. While these authors pointed to the existence of a non-oxidized phase, the similarity between the IR spectra of po-Si and siloxene led Brandt and coworkers [9] to propose that the PL originated from a siloxene coverage of the pores: soon after Tischler and Collins [10] showed that PL could be obtained from oxygen-free samples. The initial stages of oxidation of po-Si (by atomic O^+) were also studied by Koizumi and coworkers [46], through ES and PL measurements. They found for the fresh samples only the signatures of the four-fold Si–Si coordination, and of Si–H and Si–H_2 surface terminations. Si–O signals, not visible at first, were seen to develop with oxygen dose. Their samples emitted at ≈ 1.8 eV, and they found that the PL band did not change energy, or shape, with oxidation. However, there was a clear inverse correlation of oxygen dose and PL intensity. In fact, the PL almost disappeared after 6.0×10^{18} cm^{-2} oxygen dose, which was attributed by the authors either to a decrease in the number of luminescent centers, or to an increase of non-radiative surface states. It is specially noted by Koizumi and coworkers that the incident beam in IR experiments can by itself cause O incorporation. From the opposite standpoint, Wolkin and collaborators [47] stated recently that oxidation starts after only 3 min in air at room temperature, and eliminates the Si–H coverage. They base their argument on IR measurements focused at the Si–H bands at 2100 and 664 cm^{-1}. Indeed, the Si–H stretching bands are seen to decrease in intensity, and "disappear" after a day of exposure to air, but as discussed above this is to be expected, since the band will be shifted to higher energies (we follow the shift until it can be seen, but unfortunately the figure in Ref. [47] does not reach to high enough energies); in the lower energy region the directly O-related bands are superimposed on the bending and wagging Si–H resonators [44], so the analysis is not clear.

There are several other papers on the subject, but the situation is actually controversial, and the truth may be that the PL in po-Si can originate from more than one center or family of centers. We might again expect some help from theory. There has been work on the oxidation of hydrogenated Si surfaces (see Ref. [23]), on small related molecules [48, 35], and related systems [49 to 51]. The first issue to be addressed is the mode of incorporation of O into the structure, and relative stabilities between different incorporation sites. If we restrict our attention to the initial oxidation, well before formation of a complete oxide layer, we might distinguish (see Fig. 10) different incorporation modes: *additive*, as hydroxyl (Si–OH) surface units, or interstitially (Si–O–Si) in the first surface layer backbonds, or saturating dangling bonds as a silanone (Si=O) unit; *corrosive*, substituting for a Si–H$_2$ unit. Most early studies are dedicated to backbond versus hydroxyl incorporation, either in molecules or infinite (siloxene-like) systems. It is generally agreed that backbond is preferred over hydroxyl incorporation, by ≈ 0.5 eV per O atom, as is also our result, and in agreement with experiment (see Ref. [24] for a compilation).

We studied several oxidation modes imposed on the spherical Si$_{35}$H$_{36}$ particle, starting from a single O atom, up to 12 O atoms in different configurations, but mostly on

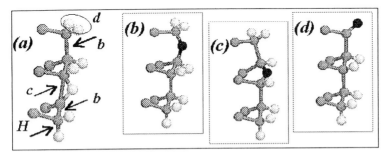

Fig. 10. Visualization of the sites for oxygen incorporation. The oxidation was imposed on the particle (Fig. 9b), at different sites, indicated in a) with arrows; the line of surface atoms is extracted form the inset of Fig. 9. Some geometries are also visualized: in b) single backbond oxidation of the outermost shell or "corner", in c) back oxidation of the (110)-like surface bond, in (c) silanone incorporation at the corner; hydroxyl incorporation is only indicated (H) in a), and we also studied corrosion of the corner (substitition of the SiH$_2$ group by one oxygen atom) and multiple backbond or hydroxyl oxidation

the external surface. This particle has 5 bulk-like Si core atoms; in the next shell each of the 12 Si atoms is already bonded to a H atom, so oxidation comes anyhow very close to the core. We calculated optimal geometries and vibrational frequencies through AM1/Crystal. Our results indicate [24] that incorporation as a silanone is highly unstable relative to almost any other form, which seems in agreement with the fact that the group SiO is easily etched out from regular Si surfaces [45, 52]. We also found that island formation (that is, clustering of Si–O–Si groups sharing a surface Si atom) is favored, even if not as much as in the crystalline (111)Si:H surface: ≈ 0.7 eV per pair, compared to 1.25 eV per pair in the surface. In this soft environment, stress relaxation is not so relevant as in the perfect surface (in fact, the number for po-Si may vary depending on the incorporation site).

Our results for the Si–H resonator frequencies on the particle show the same trend found for the (111)Si:H surface [23] in that there is a shift to higher energies with backbond oxidation. The magnitude of the shift depends on the particular site, and (as is the case for the Si:H surface) on the O clustering. In the particle, the Si–H stretch mode can be found at energies already slightly higher, and the response to oxidation is stronger, so the displacements can be larger than in the surface by around 5 to 10 meV. The IR lines are already broadened by the initial disorder of the surface. Under oxidation, disorder enhances both the shift and the broadening.

We then studied the optical properties of these weakly oxidized particles, through ZINDO/Crystal. The results for three cases are depicted in Figs. 11 (absorption) and 12 (emission). Focusing first on the absorption, we note the overall similarity of the spectra. The first absorption peak loses intensity, but is not shifted relative to the clean particle. These results are a little unexpected, since several authors [50, 51, 53] find that OH saturation of Si structures usually brings about a closing of the gap, affecting mainly the conduction band minimum. Indeed, this is also what we find for siloxenic molecules [24]. However, here the effect is not felt because the absorption involves

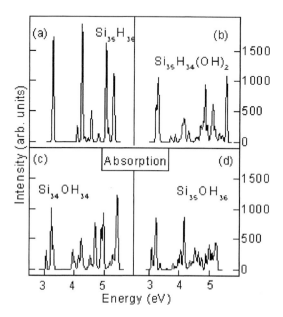

Fig. 11. Optical absorption spectra for some of the particles obtained by weak oxidation of the $Si_{35}H_{36}$ clean particle, calculated through ZINDO/Crystal, in the relaxed ground state geometries; the lines are gaussian-broadened by 0.01 eV. The spectrum for the clean particle is shown again in a). In b) we show the result for double-hydroxylation of a corner (H in Fig. 10), in c) for the silanone incorporation (d in Fig. 10), and in d) for backbond oxidation (b in Fig. 10)

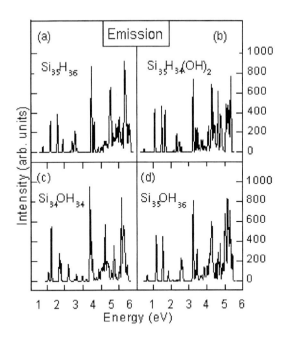

Fig. 12. Optical emission spectra for the particles obtained by weak oxidation of the $Si_{35}H_{36}$ clean particle, calculated through ZINDO/Crystal, in the relaxed excited state geometries; the lines are gaussian-broadened by 0.01 eV. The emission spectrum for the clean particle is shown again in a). In b) we show the result for double-hydroxylation of a corner (H in Fig. 10), in c) for the silanone incorporation (d in Fig. 10), and in d) for backbond oxidation (b in Fig. 10). The strong similarity of d) and b) must be noted

almost exclusively the "crystalline core" region. For initial oxidation by OH saturation, or by backbond incorporation (Fig. 11b, d), the first optical transition is dominated by the HOMO-LUMO one eletron excitation, which still shows charge concentration in the core region. This is the case even for the particle which has been etched in one corner (Fig. 11c). More affected is the second main absorption peak, which is seen to red-shift strongly and lose intensity already with the extremely weak oxidation simulated in the cases in Fig. 11. This is due to the composition of the transition, involving other one-electron excitations (LUMO+1, HOMO−1) that are spread to the surface and feel the presence of the impurity. For stronger oxidation patterns (not shown) it can happen that the order of the one-electron levels is affected, so that the HOMO or LUMO carry contributions from the surface shells: even then, in the cases studied, the first optical transition comes mainly from just one one-electron excitation, originating in the core. One may speculate that for larger particles, where the proportion of core to surface Si atoms is higher, the absorption will be even less affected by initial stages of oxidation.

I discuss next the emission from these oxidized particles, which we simulated in the same way as for clean particles. The results for the spectra are in Fig. 12. Interestingly, there is no strong effect on the emission either, compared to the emission from the clean particle, apart from a small shift of the first, less intense peak (the second peak is virtually not affected). As indicated in Fig. 10, some of the sites could interfere directly with the formation of the transient Si–H–Si bridge. We found in fact that *the presence of the O atoms attracts the bridge defect*, that is, the already existing distortion promoted by the hydroxyl radical or by backbond incorporation (in this case, in the site labeled as b in Fig. 10), or even the corrosion of a corner, causes the bridge-like distortion of the excited state to occur in the close vicinity. The resulting emission is almost identical to the normal one, and in particular occurs in the same energy range. The spectrum is also identical for backbond oxidation in the site labeled as c in Fig. 10, indicating the decay is through a bridge defect. However, *in that case, the defect is strongly repelled from the oxidation site.* We may then speculate that as oxidation proceeds, still in the first surface layer, occupying these backbonds with O atoms will decrease the number of available sites for transient defect formation, thus decreasing the number of effective luminescent centers.

It remains to be discussed the case of silanone, which is indeed completely different from the others [24]. I show in Fig. 13 the optical absorption and emission spectra for the particle $Si_{34}H_{34}Si=O$, that is, the same particle I have been discussing, with one SiH_2 surface group substituted by the Si=O silanone group. As can be seen, in this case the absorption spectrum is already affected, and a first transition is inserted in the red-orange region. This first peak is also mainly composed of one one-electron HOMO-LUMO excitation, only in this case the HOMO (or "hole state") is highly concentrated over the O atom, as obtained also by Wolkin and coworkers [47] through semi-empirical TB calculations; the LUMO (or "electron state") is again crystalline core-like. We then simulated the excitation, and the resulting emission spectrum (also shown in Fig. 13) is markedly different from the others, not only in energy, but particularly in the origin of the first peaks. In this case, the transient bridge defect is not formed, because the excitation is already concentrated over the Si=O group. The first peaks in emission concentrate in the region of the silanone, but over the neighbor Si atoms. The silanone incorporation promotes in the ground state a distortion of the Si–Si backbonds, that increase slightly to 2.39 Å, while the Si–O bond length is 1.58 Å. In the excited state, the distortion is enhanced, with one of the Si–Si bonds stressed to 2.59 Å, while the Si–O bond shortens slightly. Gole and Dixon [35] performed ab initio calculations for the ground state (singlet) and excited state (triplet) of small silanone molecules, and predict a lengthening of the Si–O bond on excitation (of 0.12 Å for the singlet → triplet excitation in the small $(SiH_3)_2Si=O$ molecule); in the larger environment of the particles this is reverted [24], because in the excited state charge is trapped at the stressed Si–Si bond.

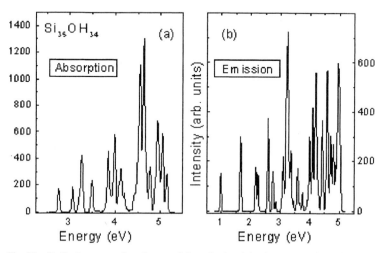

Fig. 13. Optical spectra for the particles obtained by incorporation of a silanone unit (Si=O) substituting for a SiH_2 unit at a "corner" of the $Si_{35}H_{36}$ particle; a) absorption spectrum, calculated through ZINDO/Crystal, in the relaxed ground state geometries; b) emission spectrum, calculated in the relaxed excited state geometries. The lines are gaussian-broadened by 0.01 eV

We learn that oxidation of Si:H nanoparticles is a complex process, and further work is certainly needed. It may well be that several different mechanisms [54] are involved. In particular, we found that oxidation may cause opposite effects in the emission lines. Normal backbond oxidation hardens the surface and eventually kills the PL by decreasing the number of luminescent centers. Silanone incorporation, on the other hand, introduces a new recombination center that is active (and should be seen) also in absorption.

6. Conclusions

In summary, we have studied the structural and optical properties of Si:H nanoparticles, also under weak oxidation. In order to do this, we have developed parametrizations for two semi-empirical techniques within the Hartree-Fock formalism, based on a definite procedure designed for semiconductor systems. The techniques chosen were AM1, for geometries and vibrational frequencies, and ZINDO, for optical properties. It is specially relevant to the case under study that correlation effects be included in these techniques: we chose the Configuration Interaction extension to Hartree-Fock, because of the possibility of analysis of the spectra in physically-intuitive one-electron states.

Our results allow us to derive a consistent description for the optical processes in porous Si. In agreement with the majority of experimental data, and also most theoretical models, the absorption can be explained by quantum confinement at crystalline Si regions of nanometer dimensions. However, we show that structural relaxations also play an important role in the actual value of the gap energy. The absorption occurs at the "crystalline" core of the particles, or regions, and is exciton-like in character, that is, involves an almost pure one-electron excitation. Optical emission can occur though different channels, depending on the chemical state of the surface. For perfectly hydrogenated regions, there should be a fast channel, Stokes shifted in energy from the absorption by only ≈ 0.2 to 0.3 eV, which is also tied to the core of the particle. There is, however, also a slow channel, strongly red-shifted and pinned in energy, that originates from coupling of the electronic excitation to the surface modes, with generation of a transient Si–H–Si bridge-like defect. We propose this state to be responsible for the red-PL in freshly prepared po-Si samples.

The bridge-defect center should survive initial oxidation of the surface, but further oxidation is found to harden the surface and decrease the number of effective sites for optical decay. Thus, in heavily oxidized po-Si other centers should be responsible for the luminescence. In particular, O incorporation as silanone (Si=O) units is found to be optically active in the red-orange region. However, one cannot exclude the occurrence of other O-related centers.

I stress that our findings, in particular the occurrence of the transient surface bridge, could not be investigated by a theoretical model that did not take correlation effects into account, and did not open the possibility of investigating excited electronic states. This shows how the careful use of empirical or semi-empirical models can still be of value when studying complex, disordered systems.

Acknowledgements I wish to thank my collaborators in this long study, E. Molinari, S. Ossicini and especially R. J. Baierle. This work has been supported by Fundaçaõ de Amparo à Pesquisa do Estado de São Paulo FAPESP and Conselho Nacional de Desenvolvimento Científico e Tecnológico CNPQ, Brazil, and Consiglio Nazionale della Ricerca CNR and Istituto Nazionale di Fisica della Materia INFM, Italy. Calculations were mostly done at Laboratório de Computacão Científica Avancada, LCCA-USP.

References

[1] V. LEHMANN and U. GÖSELE, Appl. Phys. Lett. **58**, 856 (1991).
[2] L. T. CANHAM, Appl. Phys. Lett. **57**, 1004 (1991).
[3] A. CULLIS, L. CANHAM, and P. CALCOTT, J. Appl. Phys. **82**, 909 (1997).
[4] O. TESCHKE et al., J. Appl. Phys. **78**, 590 (1995).
[5] Y. KANEMITSU, Phys. Rep. **263**, 1 (1995).
[6] R. P. VASQUEZ et al., Appl. Phys. Lett. **60**, 1004 (1992).
[7] S. M. PROKES and O. J. GLEMBOCKI, Phys. Rev. B **49**, 2238 (1994).
[8] G. QIN and G. G. QIN, J. Appl. Phys. **82**, 2572 (1997).
[9] M. S. BRANDT et al., Solid State Commun. **81**, 307 (1992).
[10] M. A. TISCHLER and R. T. COLLINS, Solid State Commun. **84**, 819 (1992).
[11] I. G. TIGELIS, J. P. XANTHAKIS, and J. L. VOMVORIDIS, phys. stat. sol. (a) **165**, 125 (1998).
[12] A. ZUNGER and L.-W. WANG, Appl. Surf. Sci. **102**, 350 (1996).
[13] F. BUDA, J. KOHANOFF, and M. PARRINELLO, Phys. Rev. Lett. **69**, 1272 (1992).
[14] S. ÖGÜT, J. R. CHELIKOWSKI, and S. G. LOUIE, Phys. Rev. Lett. **79**, 1770 (1997).
[15] M. ROHLFING and S. G. LOUIE, Phys. Rev. Lett. **80**, 3320 (1998).
[16] L. WANG and A. ZUNGER, J. Phys. Chem. **98**, 2158 (1994).
[17] C. DELERUE, G. ALLAN, and M. LANNOO, J. Lum. **80**, 65 (1999).
[18] M. TANG, L. COLOMBO, J. ZHU, and T. D. DE LA RUBIA, Phys. Rev. B **55**, 14279 (1997).
[19] R. KASCHNER, T. FRAUENHEIM, T. KÖLER, and G. SEIFERT, J. Comp.-Aided Mater. Design **4**, 53 (1997).
[20] R. J. BAIERLE, M. J. CALDAS, E. MOLINARI, and S. OSSICINI, Braz. J. Phys. **26**, 631 (1996).
[21] R. J. BAIERLE, M. J. CALDAS, E. MOLINARI, and S. OSSICINI, Solid State Commun. **102**, 545 (1997).
[22] R. J. BAIERLE, M. J. CALDAS, E. MOLINARI, and S. OSSICINI, Mater. Sci. Forum **258–263**, 11 (1998).
[23] R. J. BAIERLE and M. J. CALDAS, Quantum Chemistry Study of Semiconductor Systems: Initial Oxidation of the Si–H (111) surface, J. Mod. Phys. B (1999), to be published.
[24] R. J. BAIERLE and M. J. CALDAS, Optical Properties of Weakly Oxidized Si:H Nanoparticles, to be published.
[25] P. DEÁK and L. C. SNYDER, Phys. Rev. B **36**, 9619 (1987)
[26] M. J. S. DEWAR, E. G. ZOEBISH, E. F. HEALY, and J. J. P. STEWART, J. Amer. Chem. Soc. **107**, 3902 (1985).
[27] J. J. P. STEWART, J. Comput. Chem. **10**, 209 (1989).
[28] J. RIDLEY and M. ZERNER, Theoret. Chem. Acta **32**, 111 (1973).
[29] K. KUNC and R. M. MARTIN, Phys. Rev. Lett. **48**, 406 (1982).
[30] M. T. YIN and M. L. COHEN, Phys. Rev. B **25**, 4317 (1982).
[31] M. J. CALDAS, C. W. RODRIGUES, and P. L. SOUZA, Mater. Sci. Forum **83–87**, 1015 (1991).
[32] C. K. ONG, A. H. HARKER, and A. M. STONEHAM, Interface Sci. **1**, 139 (1993).
[33] P. J. VENTURA, M. C. DO CARMO, and K. P. O'DONNELL, J. Appl. Phys. **77**, 323 (1995).
[34] P. D. J. CALCOTT et al., J. Phys.: Condensed Matter **5**, L91 (1993).
[35] J. L. GOLE and D. A. DIXON, Phys. Rev. B **57**, 12002 (1998).
[36] T. HUAXIANG, Y. LING, and X. XIDE, Phys. Rev. B **48**, 10978 (1993).
[37] C. DELERUE, G. ALLAN, and M. LANNOO, Phys. Rev. B **48**, 11024 (1993).
[38] R. KUMAR, Y. KITOH, K. SHEGEMATSU, and K. HARA, Jpn. J. Appl. Phys. **33**, 909 (1994).
[39] L. WANG and A. ZUNGER, Phys. Rev. Lett. **73**, 1039 (1994).
[40] X. ZHOU et al., Phys. Rev. B **54**, 7881 (1996).
[41] M. J. CALDAS and E. MOLINARI, Solid State Commun. **89**, 499 (1994).
[42] G. ALLAN, C. DELERUE, and M. LANNOO, Phys. Rev. Lett. **76**, 2961 (1996).
[43] S. LÉTANT and G. FISHMAN, Phys. Rev. B **58**, 15344 (1998).
[44] Y. KATO, T. ITO, and A. HIRAKI, Jpn. J. Appl. Phys. **27**, L1406 (1988).

[45] B. S. Meyerson, F. J. Himpsel, and K. J. Uram, Appl. Phys. Lett. **57**, 1034 (1990).
[46] T. Koizumi et al., Jpn. J. Appl. Phys. **35**, L803 (1996).
[47] M. V. Wolkin et al., Phys. Rev. Lett. **82**, 197 (1999).
[48] K. Sakata, A. Tachibana, S. Zaima, and Y. Yasuda, Jpn. J. Appl. Phys. **37**, 4962 (1998).
[49] P. Deák et al., Phys. Rev. Lett. **69**, 2531 (1992).
[50] M. R. Pederson, W. E. Pickett, and S. C. Erwin, Phys. Rev. B **48**, 17400 (1993).
[51] C. G. Van de Walle and J. E. Nortrhup, Phys. Rev. Lett. **70**, 1116 (1993).
[52] A. Stesmans and V. V. Afanasev, Appl. Phys. Lett. **72**, 2271 (1998).
[53] S. Ossicini, phys. stat. sol. (a) **170**, 377 (1998).
[54] Z. Hajnal and P. Deák, J. Non-Cryst. Solids **230**, 1053 (1998).
[55] L. Vervoort, A. Saúl, F. Bassani, and F. A. D'Avitaya, Thin Solid Films **297**, 163 (1997).
[56] M. Hirao and T. Uda, Surf. Sci. **306**, 87 (1994).
[57] B. Delley and E. F. Steigmeier, Phys. Rev. B **47**, 1397 (1993).
[58] D. R. Lidie (Ed.), Handbook of Chemistry and Physics, 78 ed., CRC Press, New York 1997/1998.
[59] U. Itoh, Y. Toyoshima, and H. Onuki, J. Chem. Phys. **85**, 4867 (1986).
[60] M. Gelize and A. Dargelos, Comput. Phys. **49**, 333 (1991).

phys. stat. sol. (b) **217**, 665 (2000)

Subject classification: 71.15.La; 71.15.Mb; 76.30.–v

Paramagnetic Defects

U. GERSTMANN, M. AMKREUTZ, and H. OVERHOF[1])

*AG Theoretische Physik, Fachbereich Physik der Universität-GH Paderborn,
Warburger Str. 100, D-33098 Paderborn, Germany*

(Received August 10, 1999)

Ab-initio calculations of paramagnetic hyperfine interactions for deep defects in semiconductors provide information about the magnetization density distribution in space. A comparison of theoretical results with corresponding data from magnetic resonance experiments allows to estimate the accuracy of the magnetization densities obtained theoretically. In many cases this comparison is decisive in establishing the defect's atomistic structure.

1. Introduction

In the local spin density approximation (LSDA) of the density functional theory (DFT) the spin density distribution in real space is the fundamental quantity. Total energies, derived from these spin densities, are used for the determination of lattice relaxations, ionization levels, defect formation energies, etc. which can be compared with corresponding experimental quantities. Unfortunately, there are no experimental results that can be directly compared with the spin density distributions, on which all theory is based.

There is an exception for *paramagnetic* defect states: here the magnetization density, i.e. the difference of the spin-up and the spin-down densities can in fortunate cases be directly converted into paramagnetic hyperfine interaction (HFI) parameters for the interactions with the defect nucleus and with the ligand nuclei. These data can be directly compared with experimental data that are obtained from magnetic resonance experiments.

From this comparison we can derive interesting information both about the theoretical description of defects in solids and about the defect system under study:
– The accuracy of the magnetization density distribution as far as this quantity enters the HFI parameters.
– The range of applicability and also the limitations of empirical models like the vacancy model or the Ludwig and Woodbury (LW) model. This discussion is useful only if the theoretical spin densities have proven to be reliable.
– The determination of the atomistic structure of a given defect. Experiments can determine several defect properties including the point group symmetry of a defect state, but usually cannot provide firm information about the lattice site (substitutional or interstitial, e.g.). The theoretical treatment is complementary starting from an atomistic model and determining the properties for this defect model. In many cases one can find a unique defect model with HFI properties that quantitatively agree with those of mag-

---

[1]) Tel.: (+495251)-602334; Fax: (+495251)-603435; e-mail: h.overhof@phys.upb.de

netic resonance experiments while for *all* reasonable alternative models not even a qualitative agreement is found.

The paper is organized as follows: in the next section we discuss in which cases HFI parameters can be computed from spin density distributions. In the following sections we discuss some examples of s-like, p-like and d-like defects states. We conclude the paper by a short discussion, in which we try to find out what the next steps should be in an attempt to a more rigorous understanding both of the theoretical methods and of the defect systems to which these methods are applied.

2. Calculation of HFI Parameters

The introduction of a single deep defect into a perfect crystal destroys the translational symmetry. In the supercell method one restores the translational symmetry replacing the isolated defect by a three-dimensional periodic array of defects at equivalent sites. In this method the computation of the HFI with the ligand nuclei is difficult, because a given ligand interacts with the defects in all the different supercells.

In the cluster method one abandons the translational symmetry and treats a defect embedded in a finite crystallite, the cluster. The method, although simple in principle, suffers from the difficulty that the band gap of the host and therewith the position of the defect-induced states is not well defined. The accuracy of calculated HFI values for ligands which necessarily are close to the cluster boundary is questionable but has not been tested yet.

For most numerical calculations of the defect HFI and ligand HFI a Green's function technique has been used. Here the isolated defect and its first few shells of neighbors form the "perturbed region", which is embedded into an otherwise perfect crystal. Although the incorporation of a lattice relaxation around the isolated point defect is possible in principle, the computations presented in this paper have been performed for unrelaxed host lattices using the Linear Muffin-Tin Orbital method in the Atomic-Spheres Approximation (LMTO-ASA) [1].

Usually, the many-body ground state for a given charge state of a defect must be constructed from several different configurations of defect-induced single-particle states. In fortunate cases different configurations lead to many-body states of different spin and/or of different point group symmetries. In this case, the configuration that leads to the lowest total energy is identified with the ground state of the defect system.

If, however, a state with given spin and point group symmetry can be constructed from several configurations which are close in energy to the configuration with the lowest energy, then the many-body ground state has to be calculated in a configuration interaction scheme. Such a calculation requires the knowledge of the corresponding many-body wave functions, which usually cannot be constructed from the LSDA spin densities without further assumptions.

For a given defect state (we shall always deal with defects on tetrahedral sites in tetrahedral hosts) a degenerate single-particle gap state may be partially occupied. In this event the ground state exhibits orbital degeneracy which causes a static Jahn-Teller distortion. This distortion lowers the symmetry of the system, thereby lifting the degeneracy, until the relaxed defect is in an orbital singlet state. The splitting introduced by the Jahn-Teller distortion can be small enough that interactions like the spin–orbit interaction are comparable. In this case it is again necessary to construct the many-particle wave functions from the spin densities, a procedure which is not unique except in fortunate cases [2].

This paper shall be concerned exclusively with orbital singlet ground states, which can be treated in a single configuration, because all competing configurations have much higher total energies. This ground state manifold (which still has spin degeneracy) with gyromagnetic ratio g_s is split by the hyperfine interactions with the nuclei at $\mathbf{R}_k$ with nuclear spin $\mathbf{I}_k$ and gyromagnetic ratio g_{N_k}, ($\mathbf{r}_k = \mathbf{r} - \mathbf{R}_k$),

$$\mathcal{H}_{\mathrm{HF}} = -\frac{\mu_0}{4\pi} \sum_k g_s g_{N_k} \mu_B \mu_N \left(\frac{8\pi}{3} \delta(\mathbf{r}_k) \mathbf{S} + \frac{\mathbf{L}_k}{r_k^3} + \frac{(3\mathbf{r}_k \cdot \mathbf{S}) \mathbf{r}_k - r_k^2 \mathbf{S}}{r_k^5} \right) \cdot \mathbf{I}_k. \quad (1)$$

For orbital singlet ground states there is no first order orbital angular momentum contribution to the hyperfine interactions [3].

Experimentally, the HFI splitting is analyzed in terms of the hyperfine tensors $\mathbf{A}_k$ as

$$\mathcal{H}_{\mathrm{HF}} = \sum_k \mathbf{I}_k \mathbf{A}_k \mathbf{S}, \quad (2)$$

where the HFI is split into an isotropic part, $a_k \mathbf{1}$, and the traceless anisotropic contribution, $\mathbf{B}_k$, which, in the principle axis system of $\mathbf{A}_k$ is given by

$$\mathbf{A}_k = (a_k \mathbf{1} + \mathbf{B}_k) \quad \text{with} \quad \mathbf{B}_k = \begin{pmatrix} -b_k + b'_k & 0 & 0 \\ 0 & -b_k - b'_k & 0 \\ 0 & 0 & 2b_k \end{pmatrix}. \quad (3)$$

These tensors are given in terms of the three HFI parameters a_k, b_k, and b'_k. In many cases the HFI with the defect nucleus and its first few shells of crystallographically equivalent neighbors can be resolved by EPR or by ENDOR. For those ligand shells for which the anisotropic HFI can be resolved, the symmetry class of this shell can be determined from the angular dependence of $\mathbf{B}_k$. There is, however, no direct clue to the relative distances of the ligands to the defect center.

If in an LSDA calculation for an orbital singlet state the spins of the Kohn-Sham spin-orbitals are aligned parallel (or anti-parallel) to the z-axis, then the resulting magnetization density can be directly identified with the magnetization density of the $|S, M_S = S\rangle$ many-particle spin state, which is unique. We can therefore calculate the hyperfine interactions in this case directly from the magnetization density obtained from the LSDA calculation

$$a_k = \frac{1}{2S} \frac{8\pi}{3} g_s g_{N_k} \mu_B \mu_N \int m(\mathbf{r}_k) \, \delta_{\mathrm{Th}}(\mathbf{r}_k) \, \mathrm{d}^3 r_k, \quad (4)$$

$$\mathbf{B}_k = \frac{1}{2S} g_s g_{N_k} \mu_B \mu_N \int m(\mathbf{r}_j) \frac{3\mathbf{r}_k \otimes \mathbf{r}_k - r_k^2 \mathbf{1}}{r_k^5} \, \mathrm{d}^3 r_k. \quad (5)$$

In a nonrelativistic treatment, $\delta_{\mathrm{Th}}(\mathbf{r}_k)$ reduces to the Delta function. For all but the lightest nuclei, however, a relativistic theory should be used [4], in which case

$$\delta_{\mathrm{Th}}(\mathbf{r}_k) = \frac{1}{4\pi r_k^2} \frac{r_{\mathrm{Th}}/2}{[r_k + r_{\mathrm{Th}}/2]^2}, \quad (6)$$

i.e. for the isotropic HFI the magnetization density is averaged over a sphere with Thomson radius

$$r_{\text{Th}} = \frac{Ze^2}{mc^2}, \qquad (7)$$

which is about ten times the nuclear radius.

3. s-Like Defect States

Many deep donor point defects exhibit a gap state composed from s-like single-particle orbitals. These defects are in a paramagnetic state if the s state is singly occupied giving rise to a $^2A_1$ orbital singlet defect state. For this defect state the HFI with the defect nucleus can be calculated directly from the magnetization density. Results compiled in a recent review [5] show that the calculated (isotropic) HFI with the defect nucleus compares well with the corresponding experimental value with differences in the range of a few percent. For very heavy nuclei (Sn, Pb) the agreement is somewhat poorer, it appears that the (scalar relativistic) calculation tends to overestimate the HFI for these systems. The HFI with the ligand nuclei has both isotropic and anisotropic contributions. Here the agreement between theoretical and experimental data is somewhat poorer with deviations of the order of 10 to 20% for the first few ligand shells. Note however, that in most cases the ligand HFI corresponds to s-like and p-like magnetic moments at the ligands that are in the 10^{-3} to 10^{-4} range.

According to EPR and ENDOR, most deep defects with s-like defect states in a cubic host are located on a tetrahedral site. Experimentally, substitutional sites hardly can be distinguished from tetrahedral interstitial sites. Combining the experimental data with results from theoretical calculations provides a unique identification. For S_{Si}^+, e.g., the substitutional site has a single $^2A_1$ paramagnetic state. From symmetry reasons the HFI with the defect nucleus for this state is isotropic. In contrast, the sulfur defect on the interstitial has no tetrahedral $^2A_1$ state. The close quantitative agreement of experimental and theoretical data for the substitutional defect model combined with qualitative disagreement for all alternative models is the best conceivable identification of a atomistic defect model. Furthermore, the close agreement between experimental data and theoretical results [5] shows that the calculated magnetization densities are fairly accurate, at least in the vicinity of the nuclei (which is the region probed by Eqs. (4) and (5)).

4. p-Like Defect States

The lattice vacancies in tetrahedral elemental semiconductors are the p-like defect systems best investigated both experimentally and theoretically. Besides the vacancies, there are very few deep paramagnetic p-like defect states, Cd_{Si}^+ being one of the rare exceptions.

4.1 Lattice vacancies

Naively one might consider vacancies in elemental semiconductors to be the simplest deep defects. Removing one atom from the lattice leaves four dangling bonds which after symmetrization to dangling bond orbitals transform according to the a_1 and the t_2

irreducible representations of T_d, respectively. The crystal field lowers the a_1 states with respect to the t_2 states. For the neutral charge state four of the dangling bond states are occupied, leading to the $a_{1\uparrow}^1 a_{1\downarrow}^1 t_{2\uparrow}^2$ ground state configuration that transforms according to $^3T_1$. For the vacancy in Si this orbitally degenerate state is subject to a Jahn-Teller distortion (see e.g. Watkins [6]). The computation of the resulting lattice distortion within the LDA turned out to be quite difficult (see e.g. Mercer et al. [7]) because the energy surface is rather flat with several local minima.

For the neutral Si vacancy in 3C-SiC and for the neutral vacancy in diamond the situation is even worse: The ground state is not a $^3T_1$ orbital triplet and spin triplet as predicted by the LSDA for the tetrahedral vacancy. Instead a ^1E diamagnetic ground state [8,9] is found which requires a calculation including configuration interactions to treat the electron correlation effects properly. These correlations become important, not because of being extremly large, but because there are several configurations which happen to have similar energies.

For the negative charge state of the Si vacancy V_{Si}^- in SiC there is a single $a_{1\uparrow}^1 a_{1\downarrow}^1 t_{2\uparrow}^2$ configuration that can give rise to the $^4A_2$ orbital singlet high-spin ground state that is confirmed experimentally [13,14] by ENDOR. The HFI with the first C and Si ligand shells have been resolved for 3C-, 4H-, and 6H-SiC polytypes. The HFI parameters are virtually identical for all three polytypes and in fair agreement with calculated data for 3C-SiC [14], thus completing the defect identification.

For the neutral vacancy in diamond there are several calculations for the lattice relaxation. Whereas the results agree that the relaxation is small and essentially of the breathing type, the values for the relaxation are contradictory: Li and Lowther [10] report a 10% inward relaxation, Breuer and Briddon [11] find a 13% outward relaxation, whereas according to Zywietz et al. [12] the outward relaxation is 7.3%.

For the negative charge state of the lattice vacancy V_C^- in diamond, the S1 center, the observed ground state is also an $^4A_2$ orbital singlet state. Furthermore, for the *neutral* vacancy there is a single $a_{1\uparrow}^1 a_{1\downarrow}^0 t_{2\uparrow}^3$ configuration which results in an excited $^5A_2$ orbital singlet state. For both states correlation effects can be ignored, because there are no further $a_1^{n_1} t_2^{n_2}$ configurations that could give rise to an A_2 spin quartet (quintet) state.

While the LSDA calculation for the $^4A_2$ ground state of the negatively charged vacancy is straightforward, the $^5A_2$ excited state of the neutral vacancy requires that the $a_{1\downarrow}$ orbital is left unoccupied. In the LSDA this orbital gives rise to a valence band resonance rather than to a gap state [15]. While for localized states in the gap we can just prescribe the occupancy, the DOS distribution of a hole in the valence band is not as well-defined in general. However, Fig. 1 displaying the vacancy-induced change of the DOS distribution shows that vacancy-induced $a_{1\uparrow}$ and $a_{1\downarrow}$ resonances are reasonably sharp [15]. In a self-consistent calculation for the $^5A_2$ excited state it is therefore possible to leave the $a_{1\downarrow}$ states in the upper valence band region unoccupied both for the calculation of the spin densities and for the total energies. This choice fairly well represents a single hole of symmetry a_1.

The $^5A_2$ state of the neutral vacancy is observed under illumination only: The ND1 absorption band with a 3.149 eV zero phonon line (ZPL) excites a negatively charged vacancy $^4A_2$ ground state into a $^4T_1$ excited state. For this excited state with one of the $a_{1\uparrow}^1 a_{1\downarrow}^0 t_{2\uparrow}^3 t_{2\downarrow}^1$ configurations we find an excitation energy of 3.77 eV, in fair agreement with the experimental value for ND1. The $^4T_1$ excited state is auto-ionized into the neutral $^5A_2$ state with orbital singlet $a_{1\uparrow}^1 a_{1\downarrow}^0 t_{2\uparrow}^3$ configuration. From this high-spin state

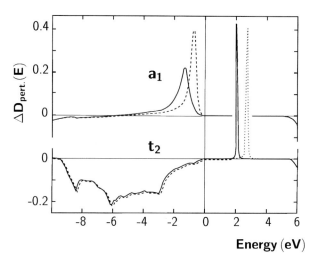

Fig. 1. Induced density of states $\Delta D_{\text{pert.}}(E)$ for the negatively charged V_C^- vacancy broken up into a_1 states (upper panel) and t_2 states (lower panel). Full lines mark spin-up states, broken and dotted lines denote occupied and unoccupied spin-down states, respectively

there are no allowed optical transitions into any of the lower-lying states of the neutral vacancy and, therefore, the excited $^5A_2$ state is extremely long-lived [16]. It has been argued [17] that the upper $a_{1\downarrow}$ state should be a localized gap state rather than a resonance as calculated using the LSDA, because otherwise the optical ZPL could not be as sharp as observed. This is certainly not true. While the LSDA expands the electron spin density into localized states and into resonance states, the resulting defect states always have a sharp energy.

A comparison of the calculated ligand HFI data with the experimental data (see Table 1) shows a fair agreement for the dipolar terms and also for the isotropic HFI with the nuclei of the first shell of ligands. The isotropic HFI with the ^{27}Si nearest neighbor ligands does not have the correct sign, but the modulus is very small corresponding to 10^{-4} of the HFI of a C 2s electron only.

In Table 2 the isotropic HFI is broken up into the contributions from the gap state, the valence band polarization (which for the excited state of V_C^0 includes the a_1 hole state), and the core polarization which for the C ligands is due to the 1s states only.

We have plotted in Fig. 2 the calculated total magnetization densities, the contribution of the valence bands to the magnetization density, and the change of the particle density as a contour plot in the $(1\bar{1}0)$ plane. While the contribution of the valence band

Table 1

Calculated ligand hyperfine interaction constants (in MHz) for the unrelaxed $^5A_2$ excited state of V_C^0 compared with experimental data from van Wyk et al. [16] and for the $^4A_2$ state of V_C^- compared with experimental data from Isoya et al. [39]

			(1,1,1) ligand		(2,2,0) ligand			(1,1,$\bar{3}$) ligand		
			a	b	a	b	b'	a	b	b'
$^5A_2$	V_C^0	this work	88.0	16.4	−4.7	1.2	0.2	−0.5	0.2	0.1
		exp.	53.73	18.70	6.36	1.2	−	−	−	−
$^4A_2$	V_C^-	this work	126.0	18.1	−3.2	1.2	0.3	−0.4	0.2	0.2
		exp.	101.7	20.0	10.7	1.37	0.085	−	−	−

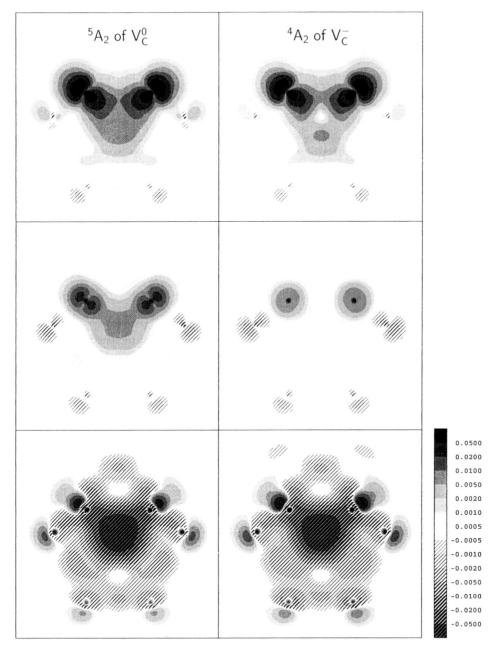

Fig. 2. Contour plot of the total magnetization densities (upper panel), the magnetization densities induced in the valence bands (middle panel) and the total induced particle density (bottom) for the excited $^5A_2$ state of V_C^0 (left) and for the $^4A_2$ ground state of V_C^- (right), respectively

Table 2

Contributions to the isotropic ligand hyperfine interaction constants (in MHz) from the gap state, the polarization of the valence band states, and from the 1s core state calculated for the unrelaxed $^5A_2$ excited state of V_C^0 and for the $^4A_2$ state of V_C^- compared with experimental data from van Wyk et al. [16] and from Isoya et al. [39], respectively

		$^5A_2$ excited state of V^0			$^4A_2$ ground state of V^-		
		(1,1,1)	(2,2,0)	(1,1,$\bar{3}$)	(1,1,1)	(2,2,0)	(1,1,$\bar{3}$)
this work	gap state	72.0	2.8	1.1	106.0	4.4	1.7
	valence band	43.0	−7.0	−1.4	49.0	−7.3	−2.0
	1s core	−27.0	−0.5	−0.2	−33.0	−0.3	0.2
	total	88.0	−4.7	−0.5	126.0	−3.2	−0.4
exp.		53.73	6.36	–	101.7	10.7	–

spin polarization is small for V_C^-, the contribution for the V_C^0 excited state is much larger, since there is a $a_{1\downarrow}$ hole in the valence band DOS.

There is a striking difference between the localization length of the magnetization density and that of the induced particle density. While the former density is confined to the nearest neighbors, as already anticipated in the defect molecule approach of Coulson and Kearsley [8], the latter extends over a much larger volume. One has to be careful, therefore, when translating experimental HFI data into LCAO atomic orbitals. The HFI data represent the magnetization density while the spatial extent of a LCAO orbital corresponds to the particle density distribution. Quite obviously, the magnetic polarization is a more local phenomenon than the dielectric polarization of the particle density.

4.2 The Cd_{Si}^+ donor state

For the Cd defect in Si a $^4A_2$ ground state with tetrahedral symmetry has been observed by EPR [18]. The gyromagnetic ratio $g \simeq 2$ excludes that this is an effective-mass-like state of Cd_{Si}^- with $J = 3/2$. The involvement of Cd could be proven by doping the crystal with ^{111}Cd and ^{113}Cd isotopes. Both have $I = 1/2$, with a slightly different gyromagnetic ratio g_N which is reflected by slightly different HFI values.

A tetrahedral Cd-related defect could either be due to Cd on a lattice position or to Cd on a tetrahedral interstitial site. Self-consistent LMTO-ASA calculations [19] show that the interstitial Cd_i defect has a gap state that transforms according to a_1. This defect can give rise to a $^2A_1$ paramagnetic defect with $S = 1/2$ only which with an isotropic HFI of $a = 3.9$ GHz cannot be mistaken for a $^4A_2$ defect.

Hence the Cd_{Si}^+ donor state is the only possible candidate for a $^4A_2$ tetrahedral defect state involving Cd. For this defect state the calculated isotropic HFI is $a = 4.2$ MHz only, a factor of 15 smaller than the $a = 64$ MHz value observed experimentally [18]. The reason for this apparent failure of the computation is not clear at the present time, but we suspect the use of the LSDA and/or the neglect of lattice relaxations as primary sources for this error.

The isotropic HFI for a p-like defect state is an indirect effect, caused by the spin polarization of valence band and core states, analogous to 3d transition metal (TM) defects with a very small and essentially p-like magnetic moment in the Cd case. Still the discrepancy between calculated and experimental values for a is rather unexpected and the reasons for this apparent failure of our calculations have to be investigated in the near future.

5. 3d Transition Metal Defects

3d transition metal (TM) point defects have been extensively investigated experimentally since the pioneering paper of Ludwig and Woodbury [20]. In elemental semiconductors these defects are built in both as interstitial and as substitutional defects and are often observed in several charge states acting as donors and acceptors.

In the tetrahedral crystal field the d-electron states are split by Δ_{cr} into states transforming according to the e and to the t_2 irreducible representations, respectively. For the interstitial point defects the e-states are higher in energy than the t_2-states while for the substitutional defects the order is reversed. The states are further split by the exchange interaction Δ_x into $e_\uparrow$, $e_\downarrow$, $t_{2\uparrow}$, and $t_{2\downarrow}$ states. If $\Delta_x > |\Delta_{cr}|$ we will have high-spin ground states, otherwise low-spin states will result. If the d-electrons of a defect occupy a single subshell (e.g. as in a $t_{2\uparrow}^2 t_{2\downarrow}^0 e_\uparrow^0 e_\downarrow^0$ configuration) or if all states are occupied except for a single subshell (e.g. $t_{2\uparrow}^3 t_{2\downarrow}^3 e_\uparrow^1 e_\downarrow^0$) the distinction between high-spin and low-spin states cannot be made based on the resulting total spin, i.e. it must be left to the theoretical treatment. According to the famous Ludwig and Woodbury (LW) model *all TM point defects in all charge states* should have high-spin ground states.

Since for the first 25 years the high-spin prediction of the LW model has not failed for any of the 3d TM defect states in Si *for which experimental EPR data are observed*, it is widely used for the interpretation of EPR spectra. There are, however, experimental [28] and theoretical [27,21] proofs that substitutional Ni_{Si}^- is in a low-spin state as are several other 3d TM defects for which no EPR has been observed to date.

5.1 3d transition metal defects in Si

As has been noted by Zunger [21], these defects exhibit a dual nature: the observed magnetic moment shows 3d orbitals that are localized as in an atom, but for such orbitals the ionization energies for different charge states must differ by about 10 eV. It would therefore be quite unlikely to have more than two different charge states within the 1.2 eV band gap of Si.

The resolution of this apparent contradiction can be seen from Table 3. Here the occupations of the d-like states are listed for the substitutional Mn_{Si} point defect in its four different charge states. Going from Mn_{Si}^+ with $S = 2/2$ to Mn_{Si}^{2-} with $S = 5/2$ the magnetic moment of the d states changes from 2.9 to 3.8 only, while the total d occupancy is even slightly reduced. This demonstrates the effectiveness of the *self-regulating response* [21]: increasing the d occupancy of the gap state decreases the occupancy of the valence band resonances. The same effect is demonstrated in the contour plots of the densities in Fig. 3. The increase of the magnetization densities occurs predominantly at the ligand atoms. The moderate increase of the induced particle density at the Mn atom and its nearest neighbors is followed by a significant decrease in the more distant neighborhood.

In Table 4 we compare calculated and experimental HFI data for all the orbital singlet 3d TM defect states observed in Si on tetrahedral interstitial and substitutional sites. Quite generally the calculated values are by about a factor of two smaller in modulus than

Table 3

Occupation number $N^{\Gamma\alpha}_{o,d,\text{TM}}$ contributed from gap state and valence band resonances for differently charged Mn_{Si} point defects in Si

spin		$\text{Mn}^{+}_{\text{Si}}$ 2/2	$\text{Mn}^{0}_{\text{Si}}$ 3/2	$\text{Mn}^{-}_{\text{Si}}$ 4/2	$\text{Mn}^{2-}_{\text{Si}}$ 5/2
gap	$d^{\uparrow}$	–	$t_{2,\uparrow}$ 0.1656	$t_{2,\uparrow\uparrow}$ 0.3652	$t_{2,\uparrow\uparrow\uparrow}$ 0.6105
valence band		$e_{\uparrow\uparrow}$	$e_{\uparrow\uparrow}$	$e_{\uparrow\uparrow}$	$e_{\uparrow\uparrow}$
e	$d_{\uparrow}$	1.8227	1.8241	1.8044	1.7538
e	$d_{\downarrow}$	0.1021	0.0698	0.0450	0.0355
t_2	$d_{\uparrow}$	2.2472	2.2164	2.1909	2.0854
t_2	$d_{\downarrow}$	1.1009	0.9367	0.7571	0.6561
total		$e_{\uparrow\uparrow}$	$e_{\uparrow\uparrow}t_{2,\uparrow}$	$e_{\uparrow\uparrow}t_{2,\uparrow\uparrow}$	$e_{\uparrow\uparrow}t_{2,\uparrow\uparrow\uparrow}$
	$d_{\uparrow} + d_{\downarrow}$	5.2729	5.2026	5.1626	5.1413
	$d_{\uparrow} - d_{\downarrow}$	2.8669	3.1004	3.5584	3.7581

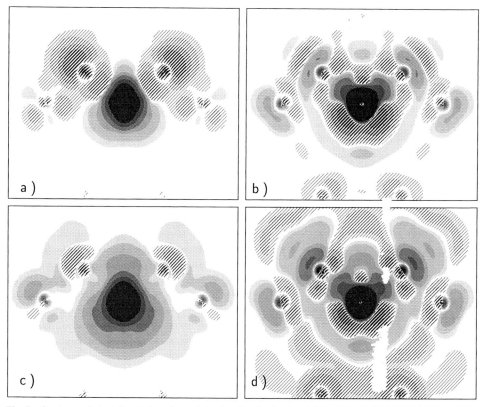

Fig. 3. Contour plot of the induced particle density for isolated Mn_{Si}. Part a) shows the total magnetization density and b) the induced particle density for $\text{Mn}^{+}_{\text{Si}}$, c) and d) show the corresponding densities for the $\text{Mn}^{2-}_{\text{Si}}$ charge state

Table 4
Experimental electron hyperfine interaction constants (in MHz) for 3d transition metals in Si compared with theoretical data. The theoretical data have been broken up in contributions from the polarization of the valence band a_{vb} and from the core states a_{core}. The magnetic moment μ_{TM} within the TM ASA sphere is also given

	S		TM	g_{exp}	a_{exp}	$a_{theor.}$	a_{vb}	a_{core}	μ_{TM}
interstitial TMs									
d^3	3/2	$^4A_2$	^{47}Ti$^+$	1.9986	± 15.65[a]	− 10.0	+ 10.5	− 20.5	1.37
			^{51}V^{2+}	1.9892	−126.21[b]	− 58.6	+ 64.8	−123.5	1.81
d^5	5/2	$^6A_1$	^{53}Cr$^+$	1.9978	+ 31.99[b]	+ 15.8	+ 16.1	− 32.0	3.36
			^{55}Mn^{2+}	2.0066	−160.30[b]	− 66.2[c]	+ 73.9	−139.0	3.05
d^8	2/2	$^3A_2$	^{55}Mn$^-$	2.0104	−213.64[b]	−126.0[c]	+102.0	−228.0	2.01
			^{57}Fe^0	2.0699	± 20.93[b]	− 15.2[d]	+ 12.5	− 27.7	2.06
substitutional TMs									
d^2	2/2	$^3A_2$	^{53}Cr^0	1.9962	± 37.62[b]	+ 23.4	− 36.1	+ 59.5	2.51
			^{55}Mn$^+$	2.0259	−189.30[b]	−116.5[c]	186.0	−302.5	2.85
d^5	5/2	$^6A_1$	^{55}Mn^{2-}	2.0058	−121.5[b]	− 61.5[c]	+107.1	−168.5	4.03

[a] van Wezep and Ammerlaan, 1985 [22].
[b] Ludwig and Woodbury, 1962 [20].
[c] Overhof, 1995 [23].
[d] Weihrich and Overhof, 1996 [24].

the corresponding experimental values. As in the case of Cd$_{Si}^+$, the isotropic HFI is due to the polarization of the core states by the magnetic moment of the d-like states: The exchange interaction is attractive for the majority spin (spin up), and therefore, the very localized 1s$_\uparrow$ state is slightly dragged outwards compared with the 1s$_\downarrow$ state, thus giving rise to a negative magnetization at the nuclear site. This polarization is rather small (of the order of 10^{-6} for the 1s state of 3d TM defects), since the 1s electrons are very tightly bound and have little overlap with the 3d states. The same effect is with 10^{-3} much stronger for the somewhat more extended 2s$_\uparrow$ states, which still are very tightly bound. The 3s$_\uparrow$ and the 4s$_\uparrow$ (valence band) states are less localized than the 3d-like magnetic moment and therefore, the up-spin density is slightly pushed into the nuclear region.

The resulting magnetization density at the nucleus has a positive contribution from the valence band polarization. The negative contribution arising from the core states exceeds the former contribution by a factor two in modulus. The reason for the discrepancy by about a factor of two between experimental and theoretical values is unclear at the moment, about the same discrepancy is observed for 3d TM point defects in II–VI semiconductor compounds [25]. Battocletti et al. [26] have shown that a similar discrepancy observed for ferromagnetic metals can be relieved in part, if the LSDA is augmented by generalized gradient corrections (GGA).

5.2 Nickel defects in diamond

In contradiction to the famous LW model [20] the late 3d TM defects in Si exhibit low-spin ground states both for the interstitial and for the substitutional site [27]. According to our calculations the same is true for all charge states of the tetrahedral Ni point defects in diamond.

5.2.1 Ni_C^- in diamond

For the substitutional site our results confirm the vacancy model [29,21] when modified to reflect the electronic structure of the vacancy in diamond: instead of a C_{2v}-distorted $S = 1/2$ ground state observed for the V_{Si}^- in Si, the V_C^- vacancy in diamond has tetrahedral symmetry and an $S = 3/2$ ground state. The vacancy model predicts for Ni_C^- that the interaction of the 3d states of Ni with the electronic states of V_C^- leads to the same $^4A_2$ ground state that is observed for V_C^-. For Ni the 3d states are considerably lower in energy than the t_2 gap states of V_C^-. The occupied $t_{2\uparrow}$ gap states are therefore predominantly dangling-bond-like. As can be seen from Fig. 4, the gap state of the Ni_C^- defect in fact resembles the gap state of a vacancy with a rather small (21%) admixture of Ni 3d states. This small admixture is the reason for the 0.4 eV exchange splitting for Ni_C^- which is much smaller than the 3.4 eV crystal-field splitting of the e and t_2 single particle states.

The t_2 single particle gap states contribute the dominant part of the magnetization density. There is practically no contribution from the fully occupied resonant $e_\uparrow$ and $e_\downarrow$ states, nor from the valence band resonances. Comparing the calculated HFI data for Ni_C^- with experimental data from Isoya et al. [30], we find again the discrepancy between calculated and experimental value for the isotropic HFI with the 3d TM nucleus, this time slightly less than a factor of two. The calculated ligand HFI values are in fair agreement with the experimental data which shows the calculation to be able to correctly describe the magnetization density distribution. Note in particular that the magnetization density at the C(111) ligands is much smaller for the Ni_C^- defect than for the V_C^- vacancy in diamond. Hence our calculations show in agreement with the experimental HFI data that the vacancy model must not be considered to be a quantitative description of 3d TM defects.

According to our calculations, Ni_C can exist in the 2+, +, neutral, and single negatively charged states with an ionization energy of $E^{0/-} = E_v + 3.85$ eV, in general agreement with the 3.0 eV value deduced from photo-EPR [32]. All charge states are paramagnetic with the exception of the Ni_C^{2+} charge state.

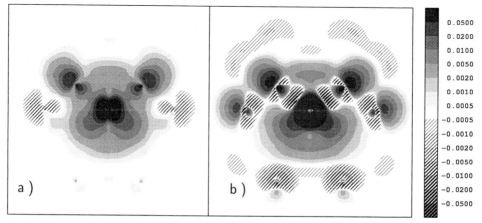

Fig. 4. Contour plots of the induced particle densities for isolated Ni_C^-. Part a) shows the total magnetization density and b) the induced particle density in a (110) plane

5.2.2 The NIRIM-1 point defect

The nickel-related NIRIM-1 $S = 1/2$ point defect observed by Isoya et al. [31] exhibits a symmetry-lowering lattice relaxation at low temperatures (below 4 K) and appears as a defect with tetrahedral symmetry at elevated temperatures. In either spectra no HFI with the ^{13}C ligands nor with the ^{61}Ni central nucleus was observed. For diamond samples that are heavily p-doped by boron the NIRIM-1 EPR signal disappears.

For the positive charge state of Ni_C^+ the LW model predicts an $S = 5/2$ high-spin ground state. In contrast, our calculations result in a low-spin $^2T_1$ ground state, an orbital triplet that would be subject to a Jahn-Teller distortion. According to our results, not even an $S = 5/2$ excited state can be constructed for Ni_C^+. However, we can exclude that NIRIM-1 is to be identified with Ni_C^+: according to Table 5 the HFI of the Ni_C^+ state with the nearest neighbor ^{13}C(1, 1, 1) ligands is much larger than the EPR linewidth and should not be unresolved.

Instead we tentatively identify NIRIM-1 with the positive charge state Ni_i^+ of the interstitial Ni. This defect with an ^2E ground state has according to our calculations a HFI with the nearest neighbor ^{13}C(1, 1, 1) ligand shell that is smaller than that for Ni_C^+ by more than one order of magnitude. This interaction may well be hidden under the broader NIRIM-1 line.

With the identification of the Ni_i^+ defect as origin of the NIRIM-1 EPR spectrum we would expect that in heavily B-doped diamond the $^3A_2$ ground state of Ni_i^{2+} should be observed in EPR, while the NIRIM-1 EPR spectrum disappears. In the experiments, no new EPR spectrum was found upon heavy B-doping. We have investigated whether pairing of Ni with B acceptors offers an explanation [35]. In particular, we have investigated the trigonal $(Ni_i - B_C)$ pair. In p-type diamond this pair is tightly bound with a pair formation energy of 1.3 eV for the $(Ni_i - B_C)^+$ charge state (for the $(Ni_i - B_C)^0$ charge state the pair formation energy amounts to 0.95 eV while for the $(Ni_i - B_C)^-$ no pairing is predicted). However, the electronic ground state of $(Ni_i - B_C)^+$ is again $^3A_2$ (of the group C_{3v}) as is expected for a Ni_i^{2+} point defect interacting with a B_C^- negative charged center.

Alternatively we have investigated the stability of trigonal $(Ni_C - B_C)$ pairs. For p-type material we find a 1.3 eV pair formation energy for the $(Ni_C - B_C)^+$ charge state of the pair which is diamagnetic. The formation of this pair could explain the absence of a Ni_C^+ EPR signal, but not of a Ni_i^+ signal in p-type diamond.

Table 5

Comparison of the calculated hyperfine interactions (in MHz) for the unrelaxed $^4A_2$ state of Ni_C^- with experimental data from Isoya et al. [39]. Theoretical hyperfine data for the $^2T_2$ state of Ni_C^+ and for the ^2E state of Ni_i^+ are also given

			^{61}Ni	nearest neighbor		next nearest neighbor		
			a	a	b	a	b	b'
$^1A_2$	Ni_C^-	this work	13.5	31.9	7.67	−4.12	0.38	0.14
		exp	17.1	18.85	9.32	8.63	1.04	0.05
$^2T_2$	Ni_C^+	this work	12.7	35.3	10.5	0.82	0.45	0.41
$^4A_2$	Ni_i^+	this work	25.2	0.51	−0.90	5.99	3.4	0.43

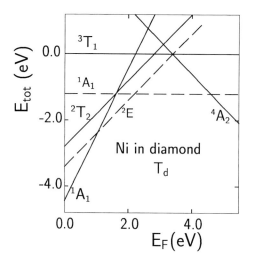

Fig. 5. Total energies $E_{tot}(E_F)$ for Ni_C (full lines) and Ni_i (dashed lines) in the different charge states

The most probable explanation is given by a comparison of the total energies of isolated Ni_C point defects as compared to isolated Ni_i sketched in Fig. 5: For n-type samples the substitutional Ni_C^- in the negative charge state has the lowest total energy (there is no Ni_i^- charge state), while at intermediate p-type doping levels the Ni_i^+ is the most stable defect state. For highly doped p-type material, however, the interstitial Ni_C^{2+} defect is energetically more favourable than Ni_i^+. Hence the absence of the NIRIM-1 spectrum in highly p-doped diamond simply reflects that in thermal equilibrium the Ni_i^+ is no longer stable.

The simultaneous observation of the Ni_C^- and NIRIM-1 defects in a single sample presents a problem: Since NIRIM-1 shows tetrahedral symmetry above helium temperatures, it is very likely an isolated point defect. Both for Ni_i and for Ni_C the $E^{+/0}$ donor level is well below the $E^{0/-}$ acceptor level of Ni_C at $E_v + 3.6$ eV. Isoya et al. [31] report that the color distribution of the samples is quite inhomogeneous, which might explain why both defects can be observed simultaneously in a dark EPR experiment.

5.2.3 Comparison with cluster calculations

There are three different theoretical cluster calculations that can be compared with our Green's function results. The CNDO parametrized cluster calculations of Paslovsky and Lowther [33,34] disagree with our results: While in our calculations the gap state of Ni_C transforms according to t_2, the cluster calculations result in an additional a_1 state that is lower in energy than the t_2 gap state. For Ni_C^- the ground state would be a spin doublet. In order to explain the experimentally observed quartet of the EPR experiment [30], Lowther [17] has considered a strong spin-orbit interaction that could give rise to a $J = 3/2$ ground state that transforms according to Γ_8. For such a state the Landé-factor must be close to 4/3, in contrast to the observed value of 2.0319.

Our results are similar to the results of a cluster calculation by Goss et al. [36]. In this calculation it was shown that the lattice relaxation preserves the T_d symmetry of both the Ni_i^+ and the Ni_C^- point defects. In contrast, the results of the cluster calculation by Jinlong et al. [37] for Ni_i^+ predict a 0.1 Å trigonal distortion.

In a recent DV-X_α cluster calculation by Ohashi et al. [38] the ligand HFI with the $^{13}C(1, 1, 1)$ ligands was calculated for the $^4A_2$ ground state of Ni_C^-. These authors obtain a value for the isotropic HFI that is about twice the experimental value while their value for b is slightly below our calculated value. The authors show that an outward lattice relaxation by 4.5% would reduce the value of a to its experimental value with little effect on the value for b.

5.3 3d transition metal defects in group-III nitrides

The large direct bandgap of the group-III nitrides (GaN, AlN, and InN) makes this class of semiconductors attractive for optoelectronic devices. Until now, only a few centers have been investigated by EPR in the nitrides: besides shallow donors the TM defects Fe, Ni, and Mn in GaN [41 to 44], and Cr in AlN [45] have been identified. Unfortunately, the HFI was resolved for the last two impurities only. The large difference in electronegativity between nitrogen (3.04) and gallium (1.81) indicates that GaN is quite ionic. As is customary in ionic compounds, the oxidation state of the 3d TM defect is given rather than the charge state. To cope with the semi-core character of the Ga 3d-derived bands [46 to 48] we have treated these d bands like ordinary valence bands in the LMTO-ASA formalism.

5.3.1 Manganese defects in GaN

The first observation of EPR in GaN *with resolved hyperfine structure* was reported by Baranov et al. [43,44] for Mn_{Ga}^{2+} (note that this is the singly negative charge state of the defect). The EPR spectrum of an $S = 5/2$ state interacting with one nucleus with nuclear spin $I = 5/2$ and 100% natural abundancy proves that ^{55}Mn is involved and that it is in a $3d^5$ high-spin configuration. Assuming that the defect substitutes for Ga, the defect state can be identified as Mn_{Ga}^{2+}.

Usually, 3d TM defects are incorporated on the gallium site in GaN [49]. Therefore, their nearest neighbourhood is given by four nitrogen atoms arranged in a tetrahedral configuration, both for cubic and wurtzite GaN. Because of the strongly localized magnetization density of 3d defects (see also Fig. 6), it can be assumed that the HFI parameters are quite similar in both polytypes. The same was approximately true for the V_{Si} vacancies in SiC.

For Mn_{Ga} we find three different charge states, going from the positive Mn_{Ga}^{4+} with $S = 3/2$ to the negatively charged Mn_{Ga}^{2+} with $S = 5/2$. Thus, in its ground state Mn_{Ga} is paramagnetic for all possible charge states confirming the high-spin prediction of the

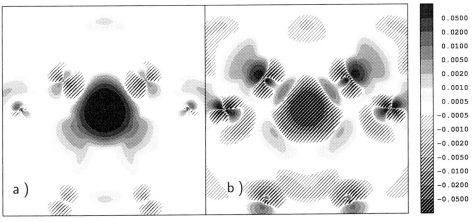

Fig. 6. Contour plots of the induced particle densities for isolated Mn_{Ga}^{2+}. Part a) shows the strongly localized magnetization density and b) the induced particle density in a (110) plane

Table 6

Calculated hyperfine interactions (in MHz) for the possible charge states of substitutional incorporated Mn_{Ga}

		S		^{55}Mn a	^{14}N(1,1,1) a	b	^{69}Ga(2,2,0) a	b	b'
Mn_{Ga}^{2+}	$3d^5$	5/2	$^6A_1$	−98.9	1.4	0.24	43.0	2.00	0.62
Mn_{Ga}^{3+}	$3d^4$	4/2	$^5T_2$	−104.3	2.2	0.18	19.7	1.59	0.03
Mn_{Ga}^{4+}	$3d^3$	3/2	$^4T_1$	−108.4	1.6	−0.51	−3.9	0.45	0.22
Mn_{Ga}^{5+}	$3d^2$	2/2	$^3A_2$	−116.6	0.7	−2.18	−24.7	−2.54	0.54

model of Ludwig and Woodbury [20]. The other charge states have high-spin ground states.

According to our results, for Mn_{Ga}^{2+} with $S = 5/2$ the calculated value of the isotropic HFI with the ^{55}Mn nucleus is also smaller by a factor of about two than the experimental value of 210 MHz (see Table 6).

5.3.2 Iron defects in GaN

No HF structure was observed for iron defects in GaN. It is difficult to detect the HFI of iron since the natural abundancy of 2.15% of the magnetic ^{57}Fe isotope is quite small. For Fe in GaN, we do not believe that isotope enrichment experiments would be successful, since the calculated HF splitting with ^{57}Fe is only 13.5 MHz for the Fe_{Ga}^{3+} ($S = 5/2$) state (see Table 7). This HF splitting would be within the linewidth of the central EPR line [43,44], even if we allow for the factor-of-two correction. Similar arguments are valid for the ligand HFI constants. Thus, no resolved HF structure is expected for Fe_{Ga}.

ESR, ODMR, Zeeman and photoluminescence (PL) investigations of the 1.299 eV zero phonon line (ZPL) in GaN have shown [50,41] that it is a ($^4T_1$–$^6A_1$) transition of a $3d^5$ electron system with $S = 5/2$ ground state of Fe. This high-spin ground state is confirmed by our calculations (see also Table 8). Furthermore, if we approximate the $^4T_1$ excited state by an $e_\uparrow^1 e_\downarrow^1 t_{2\uparrow}^3$ configuration (this is the excited state configuration with the lowest total energy), we obtain a ($^4T_1$–$^6A_1$) transition energy of 0.97 eV according to our total energy LSDA calculations.

Baur et al. [51,52] locate the $Fe_{Ga}^{(3+/2+)}$ acceptor level at $E_v + 2.5$ eV by an analysis of the spectral dependence of this PL excitation. This result has been challenged recently

Table 7

Calculated hyperfine interactions (in MHz) for the possible charge states of substitutional incorporated Fe_{Ga}. Note: In the case of Fe_{Ga}^{4+} the low-spin configuration is energetically favoured by about 0.5 eV. Thus, the ground state is a spinless $^1A_1$ state without hyperfine interaction

		S		^{57}Fe a	^{14}N(1,1,1) a	b	^{69}Ga(2,2,0) a	b	b'
Fe_{Ga}^{2+}	$3d^6$	4/2	$^5T_2$	−13.5	3.4	0.43	29.9	1.72	0.09
Fe_{Ga}^{3+}	$3d^5$	5/2	$^6A_1$	−12.9	4.8	0.83	−16.9	0.52	0.22
Fe_{Ga}^{4+}	$3d^4$	0	$^1A_1$						

Table 8
ZPL energies of TM emissions in GaN and AlN taken from [49]. Possible common identifications are also given

GaN energy (eV)	AlN energy (eV)	identification	3d TM defect
1.299	1.297	$3d^5$, $(^4T_1-^6A_1)$	Fe^{3+}
1.193	1.201	$3d^2$, $(^1E-^3A_2)$	Cr^{4+}
1.047	1.043	$3d^7$, $(^4T_2-^4A_2)$	Ni^0, Co^{2+}
0.931	0.943	$3d^2$	V^{3+}
0.82	0.797		TM complex with V

by Heitz et al. [53] who claim that the $Fe_{Ga}^{(3+/2+)}$ level should be at $E_v + 3.17$ eV, considerably higher than the value of Baur et al. Our self-consistent total energy calculations give $E^{3+/2+} = E_v + 3.14$ eV, a value that is in better agreement with the results of Heitz et al.

5.3.3 Chromium defects in AlN

Since the neighbourhood of strongly localized TM_{Ga} defects in all group-III nitrides is formed by a $TM-N_4$ complex, some physical properties are expected to be quite similar for different nitrides. In fact, as can be seen in Table 9 the ZPL energies in GaN and AlN are very similar. The close analogy of other PL characteristics suggests a common assignment of the ZPLs to TM defects in both materials [49].

The identification of the ZPL near 1.2 eV with a $(^1E-^3A_2)$ transition of a $3d^2$ configuration is out of question [54]. However the assignment to Cr^{4+} [45,56] or Ti^{2+} [53,55] impurities is discussed controversially. The fair agreement of calculated and experimental acceptor levels for iron give us some confidence to assume that the levels for other 3d TM defects in the nitrides will also be realistic. For Ti_{Ga} the calculated acceptor level in GaN turns out to be high in the conduction band. In contrast, for Ti_{Al} in AlN the first acceptor level is located about 0.2 eV below the conduction band edge. Nevertheless, in both materials the excited states are resonances that hardly can give rise to a ZPL with a linewidth that is below 1 meV. Thus, the identification of the 1.2 eV ZPL with Ti defects as assumed by several authors [53,55] appears to be quite unlikely.

If we tentatively approximate the ^1E excited state by the ground state of the $e_\uparrow^{n_1} e_\downarrow^{n_2} t_{2\uparrow}^{n_3} t_{2\downarrow}^{n_4}$ subspace with $S = 0$, we end up with a $e_\uparrow^1 e_\downarrow^1$ configuration for the ^1E excited

Table 9
Calculated hyperfine interactions (in MHz) for the possible charge states of substitutional incorporated Cr_{Al}

		S		^{53}Cr	^{14}N(1,1,1)		^{27}Al(2,2,0)		
				a	a	b	a	b	b'
Cr_{Al}^+	$3d^5$	5/2	$^6A_1$	20.2	−1.1	0.70	16.5	0.75	0.003
Cr_{Al}^{2+}	$3d^4$	4/2	$^5T_2$	23.8	−0.5	0.51	15.5	0.98	0.21
Cr_{Al}^{3+}	$3d^3$	3/2	$^4T_1$	27.5	−0.7	−0.13	10.9	1.09	0.14
Cr_{Al}^{4+}	$3d^2$	2/2	$^3A_2$	29.3	−2.8	−1.36	4.5	1.21	0.18
Cr_{Al}^{5+}	$3d^1$	1/2	^2E	37.1	−2.4	−3.15	−2.7	−1.96	1.54

state for which we obtain the transition energy $E_{(^1E-^3A_2)} = 0.78$ eV. Note that our results for the transition energies of the excited states differ from the experimental data by about 0.4 eV for Fe_{Ga}^{3+} and Cr_{Al}^{4+}, and by about 0.6 eV for the neutral vacancy in diamond. We suspect that this discrepancy could be relieved in part by gradient corrections to the LSDA and/or the inclusion of lattice relaxation effects.

For AlN an EPR spectrum with resolved hyperfine structure was presented by Baur et al. [45] who observed an isotropic EPR signal with $g = 1.997$ and resolved HFI structure. At least four lines with a HFI splitting of 53.51 MHz where observed besides the central line. Their relative intensities are consistent with Cr only, for which the isotope ^{53}Cr with $I = 3/2$ has 9.5% natural abundance. The defect state was therefore identified to be the ^2E state of Cr_{Al}^{5+}. For this charge state, the calculated isotropic hyperfine interaction (37.1 MHz, see Table 9) is about 70% of the experimental value.

After neutron irradiation the Cr-related EPR signal disappears while simultaneously the 1.201 eV ZPL increases. Neutron irradiation is known to create nitrogen vacancies in AlN [57], thus raising the Fermi level. The ^2E ground state of the Cr_{Al}^{5+} giving rise to the observed EPR signal is thus transformed into the $^3A_1$ ground state of Cr^{4+} which is the final state of the 1.201eV ZPL.

According to our calculations, Cr_{Al} defects in AlN exist in five different charge states, from Cr^{5+} with $3d^1$ electronic configuration to Cr^+ in a $3d^5$ high-spin configuration. In particular, there is no evidence for an existing spinless charge state. Thus, all ground states should be paramagnetic. Yet a paramagnetic EPR signal for the $^3A_2$ ground state of Cr_{Al}^{4+} in AlN is still awaiting detection.

6. Conclusions

Ab-initio calculations for deep defects in semiconductors yield magnetization densities which in fortunate cases (orbital singlet states) can be directly converted into magnetic hyperfine interactions. Comparison with corresponding data from magnetic resonance experiments shows that in particular for s-like states observed for deep donors the magnetization densities as derived from the LSDA are very accurate for the defect atom and also for several ligand shells. For the p-like vacancy states the calculated ligand HFI values compare well with experimental data. The isotropic HFI with the defect nucleus of 3d TM defects is caused by the spin polarization of the core shells and the valence band resonant states. Here the calculated values turn out to be too small by a factor of about two if compared with experimental data. The reason for this rather universal factor-of-two error is unclear at the moment. In most cases the magnetic moment within the defect ASA shell is comparable with the total magnetic moment of the defect, thus an error in the localization of the magnetic moment is rather unlikely. Two possible sources of error can be imagined: the use of the LSDA and the neglect of lattice relaxations. For ferromagnetic metals a similar discrepancy is known. In an attempt to solve the problem going beyond the LSDA, Battocletti et al. [26] have used several different parametrizations of a generalized gradient approximation (GGA). The resulting HFI data compared better in general with the experimental data than the LSDA results, but unfortunately, in a statistical sense only. Further, more, the results for different GGA parametrizations are by no means identical.

Experience has shown that the LSDA calculations are very reliable in predicting the spin state of the ground states of a defect and in predicting the different charge states as well as the corresponding ionization energies. This sets several strong limitations to the identification of a given defect with suitable defect models. Furthermore, as the donor and acceptor levels are predicted quite accurately, the theoretical results can be used to discount some defect assignment: If the ground state of the defect is represented by a resonant state, it is quite unlikely that this state should be the final state of a PL transition.

Although the LSDA is a theory for ground states only, it can be used for the calculation of excited states, if these happen to transform according to a different irreducible representation for the many-body wave function (or likewise, if it belongs to a different spin quantum number) and if this wave function can be approximated by a *single* configuration. For the V_C in diamond we have seen that the magnetization density for the excited state is given as accurately as the magnetization density of the ground state. We have shown for three examples (V_C in diamond, Fe^{3+}_{Ga} in GaN, and Cr^{4+}_{Al} in AlN) that the calculated transition energies turn out to be somewhat too small. It must be left to future investigations to find out, whether and how these results can be improved.

Acknowledgement One of the authors (U.G.) is indebted to the Deutsche Forschungsgemeinschaft for financial support.

References

[1] O. Gunnarsson, O. Jepsen, and O. K. Andersen, Phys. Rev. B **27**, 7144 (1983).
[2] H. Weihrich and H. Overhof, Semicond. Sci. Technol. **13**, 1374 (1998).
[3] C. P. Slichter, Principles of Magnetic Resonance with Examples from Solid State Physics, Harper & Row, New York 1963 (reprinted in Springer Ser. Solid State Sci., Vol. 1, Springer-Verlag, Heidelberg 1980).
[4] S. Blügel, H. Akai, R. Zeller, and P. H. Dederichs, Phys. Rev. B **35**, 3271 (1987).
[5] H. Overhof, in: Special Defects in Semiconducting Materials, Ed. R. P. Agarwala, Solid State Phenomena **71**, 93 (2000).
[6] G. D. Watkins, in: Deep Defects in Semiconductors, Ed. S. T Pantelides, Gordon & Breach, New York 1986 (p. 177).
[7] J. L. Mercer, J. S. Nelson, A. F. Wright, and E. B. Stechel, Modelling Simul. Mater. Sci. Engng. **6**, 1 (1998).
[8] C. A. Coulson and M. J. Kearsley, Proc. Roy. Soc. A **241**, 433 (1957).
[9] M. Lannoo and J. Bourgoin, Point Defects in Semiconductors I, Theoretical Aspects, Springer Ser. Solid State Sci. Vol. 22, Springer-Verlag, Berlin/Heidelberg 1981.
[10] L. H. Li and J. E. Lowther, Phys. Rev. B **53**, 11 (1996).
[11] S. J. Beuer and P. R. Briddon, Phys. Rev. B **51**, 984 1995).
[12] A. Zywieth, J. Furthmüller, and F. Bechstedt, phys. stat. sol. (b) **210**, 13 (1998).
[13] T. Wimbauer, B. K. Meyer, A. Hofstaetter, and A. Scharmann, in: Proc. 23rd Internat. Conf. Physics of Semiconductors, Ed. M. Scheffler and R. Zimmermann, World Scientific Publ. Co., Singapore 1996 (p. 2645).
[14] T. Wimbauer, B. K. Meyer, A. Hofstaetter, A. Scharmann, and H. Overhof, Phys. Rev. B **56**, 7384 (1997).
[15] G. B. Bachelet, G. A. Baraff, and M. Schlüter, Phys. Rev. B **24**, 4736 (1981).
[16] J. A. van Wyk, O. D. Tucker, M. E. Newton, J. M. Baker, G. S. Woods, and P. Spear, Phys. Rev. B **52**, 12657 (1995).
[17] J. E. Lowther, Phys. Rev. B **51**, 91 (1995).
[18] W. Gehlhoff, A. Näser, M. Lang, and G. Pensl, Mater. Sci. Forum **258**, 423 (1997).
[19] A. Näser, W. Gehlhoff, H. Overhof, and R. A. Yankov, phys. stat. sol. (b) **210**, 753 (1998).
[20] G. W. Ludwig and H. H. Woodbury, Solid State Phys. **13**, 223 (1962).
[21] A. Zunger, Phys. Rev. B **28** 3628 (1983); Solid State Phys. **39**, 276 (1986); Phys. Rev. B **11**, 849 (1975).

[22] D. A. van Wezep, R. van Kemp, E. G. Sieverts, and C. A. J. Ammerlaan, Phys. Rev. B **32**, 7129 (1985).
[23] H. Overhof, Mater. Sci. Forum **196**, 1363 (1996).
[24] H. Weihrich and H. Overhof, Phys. Rev. B **54**, 4680 (1996).
[25] M. Illgner and H. Overhof, Semicond. Sci. Technol. **11**, 977 (1996).
[26] M. Battocletti, H. Ebert, and H. Akai, Phys. Rev. B **53**, 9776 (1998).
[27] F. Beeler, O. K. Andersen, and M. Scheffler, Phys. Rev. Lett. **55**, 1498 (1985); Phys. Rev. B **41**, 1603 (1990).
[28] L. S. Vlasenko, N. T. Son, A. B. van Oosten, C. A. J. Ammerlaan, A. A. Lebedev, E. S. Taptygov, and V. A. Khramtsov, Solid State Commun. **73**, 292 (1990).
[29] G. W. Watkins, Physica B **117/118**, 9 (1983).
[30] J. Isoya, H. Kanda, J. R. Norris, J. Tang, and M. K. Bowman, Phys. Rev. B **41**, 3905 (1990).
[31] J. Isoya, H. Kanda, and Y. Uchida, Phys. Rev. B **42**, 9843 (1990).
[32] D. Hofmann, M. Ludwig, P. Christmann, B. K. Meyer, L. Pereira, L. Santos, and E. Pereira, Phys. Rev. B **50**, 17618 (1994).
[33] S. Paslovsky and J. E. Lowther, J. Phys.: Condensed Matter **4**, 775 (1992).
[34] S. Paslovsky and J. E. Lowther, J. Phys. Chem. Solids **54**, 243 (1993).
[35] U. Gerstmann, M. Amkreutz, and H. Overhof, Proc. ICDS-20 (Internat. Conf. Defects in Semiconductors), in press.
[36] J. Goss, A. Resende, R. Jones, S. Öberg, and P. R. Briddon, Mater. Sci. Forum **196/201**, 67 (1995).
[37] Y. Jinlong, Z. Manhong, and W. Kelin, Phys. Rev. B **49**, 15525 (1994).
[38] S. Ohashi, O. Fukunaga, J. Isoya, and J. Tanaka, Jpn. J. Appl. Phys. **36**, 1126 (1997).
[39] J. Isoya, H. Kanda, Y. Uchida, S. C. Lawson, S. Yamasaki, H. Itoh, and Y. Morita, Phys. Rev. B **45**, 1436 (1992).
[40] K. Atobe, M. Honda, N. Fukuoka, M. Okada, and M. Nakagawa, Jpn. J. Appl. Phys. **29**, 150 (1990).
[41] R. Heitz, P. Thurian, I. Loa, L. Eckey, A. Hoffmann, I. Broser, K. Pressel, B. K. Meyer, and E. N. Mokhov, Appl. Phys. Lett. **67**, 2822 (1995).
[42] P. G. Baranov, I. V. Ilyin, and E. N. Mokhov, Solid State Commun. **101**, 611 (1997).
[43] P. G. Baranov, I. V. Ilyin, E. N. Mokhov, and A. D. Roenkov, Semicond. Sci. Technol. **11**, 1843 (1996).
[44] P. G. Baranov, I. V. Ilyin, and E. N. Mokhov, Mater. Sci. Forum **258/263**, 1167 (1997).
[45] J. Baur, U. Kaufmann, M. Kunzer, J. Schneider, H. Amano, I. Akasaki, T. Detchprohm, and K. Hiramatsu, Mater. Sci. Forum **196/201**, 55 (1995).
[46] V. Fiorentini, M. Methfessel, and M. Scheffler, Phys. Rev. B **47**, 13353 (1993).
[47] A. F. Wright and J. S. Nelson, Phys. Rev. B **50**, 2159 (1994).
[48] W. R. L. Lambrecht, B. Segall, S. Strite, G. Martin, A. Agarwal, H. Morkoc, and A. Rockett, Phys. Rev. B **50**, 14155 (1994).
[49] B. K. Meyer, A. Hoffmann, and P. Thurian, Group III Nitride Semiconductor Compounds, Ed. B. Gil, Physics and Applications, Oxford University Press, 1998.
[50] K. Maier, M. Kunzer, U. Kaufmann, J. Schneider, B. Monemar, I. Akasaki, and H. Amano, Mater. Sci. Forum **143/147**, 93 (1994).
[51] J. Baur, K. Maier, M. Kunzer, U. Kaufmann, and J. Schneider, Appl. Phys. Lett. **65**, 2211 (1994).
[52] J. Baur, K. Maier, M. Kunzer, U. Kaufmann, and J. Schneider, Mater. Sci. Engng. B **29**, 61 (1995).
[53] R. Heitz, P. Maxim, L. Eckey, P. Thurian, A. Hoffmann, I. Broser, K. Pressl, and B. K. Meyer, Phys. Rev. B **55**, 4382 (1997).
[54] R. Heitz, P. Thurian, K. Pressel, I. Loa, L. Eckey, A. Hoffmann, I. Broser, B. K. Meyer, and E. N. Mokhov, Phys. Rev. B **52**, 16508 (1995).
[55] K. Pressel, R. Heitz, L. Eckey, I. Loa, P. Thurian, A. Hoffmann, B. K. Meyer, S. Fischer, C. Wetzel, and E. E. Haller, MRS Symp. Proc. **395**, 491 (1995).
[56] J. Baur, K. Maier, M. Kunzer, U. Kaufmann, J. Schneider, H. Amano, I. Akasaki T. Detchprohm, and K. Hiramatsu, Appl. Phys. Lett. **64**, 857 (1994).

phys. stat. sol. (b) **217**, 685 (2000)

Subject classification: 71.15.Mb; 71.15.Pd; 73.25.+i; S5; S5.11; S7.14

Large-Scale Applications of Real-Space Multigrid Methods to Surfaces, Nanotubes, and Quantum Transport

J. Bernholc, E. L. Briggs, C. Bungaro, M. Buongiorno Nardelli, J.-L. Fattebert, K. Rapcewicz, C. Roland, W. G. Schmidt, and Q. Zhao

North Carolina State University, Raleigh, NC 27695-8202, USA

(Received August 10, 1999)

The development and applications of real-space multigrid methods are discussed. Multigrid techniques provide preconditioning and convergence acceleration at all length scales, and therefore lead to particularly efficient algorithms. When using localization regions and optimized, non-orthogonal orbitals, calculations involving over 1000 atoms become practical on massively parallel computers. The applications discussed in this chapter include: (i) dopant incorporation and ordering effects during surface incorporation of boron, which lead to the formation of ordered domains at half-monolayer coverage; (ii) incorporation of Mg into GaN during growth, and in particular the conditions that would lead to maximum p-type doping; (iii) optical fingerprints of surface structures for use in real-time feedback control of growth; and (iv) mechanisms of stress release and quantum transport properties of carbon nanotubes.

1. Introduction

Electronic structure calculations based on density functional theory have been enormously successful in predicting and explaining the properties of materials. Pseudopotentials provide a further simplification, since they eliminate the need to explicitly consider the core electrons. Nevertheless, the computational demands of calculations necessary to study complex or artificially structured materials are still very large. This has stimulated the interest in new "real-space" methods, which can enable even larger calculations and/or reduce the computational cost.

The real-space methods avoid the use of plane waves, which extend throughout the entire system, so that the vast majority of operations are "local" in real-space. This has several advantages. Advanced mathematical techniques can be used, which automatically separate the various length scales present in the problem, substantially accelerating convergence (see below). Parallelization becomes much easier, because each processor can be assigned a given region of space. Finally, a real-space formulation is needed to develop and use $O(N)$ techniques, which overcome the $O(N^3)$ scaling of the computational cost of traditional electronic structure techniques.

Real-space methods that employ atom-centered functions, e.g., gaussians or atomic orbitals, have long been used in electronic structure calculations. However, large numerical bases are also very interesting since they allow one to attain any desired numerical accuracy by systematically increasing the number of degrees of freedom. Furthermore, the multiscale aspects of real-space calculations are easiest to exploit with methods based on techniques of applied mathematics: multigrids, wavelets and finite elements. The numerical bases used in recent calculations include finite elements [1, 2, 4, 5], grids

[6 to 14] and wavelets [15 to 17]. These bases allow for natural implementation of mesh refinement [6, 8, 10, 13], cluster boundary conditions, and efficient domain decomposition approaches on massively parallel computers [9, 10].

The computational effort in traditional electronic structure calculations must ultimately scale as $O(N^3)$, where N is the number of atoms. This is because the wavefunction of each electron can in general extend over the whole solid, and therefore computing one wavefunction will take at least $O(N)$ operations. Since the number of electrons grows linearly with the number of atoms, the computational effort must grow at least as $O(N^2)$. Furthermore, the individual wavefunctions must be orthogonal to each other and the orthogonalization or diagonalization effort will ultimately dominate, since they scale as $O(N^3)$. Recently, a number of ingenious methods have been proposed for evaluating the total energy in $O(N)$ operations, see Refs. [18, 19] for recent reviews. The main idea in most methods is to rewrite the total energy expression with help of optimized *localized* functions. Each function is confined to a given region of space, but the various confinement regions overlap so that there is little loss of generality. Clearly, overlaps between functions localized in regions sufficiently far away will vanish, which results in the number of non-zero overlaps being $O(N)$. By exploring this sparsity while avoiding the calculations of the individual eigenfunctions it is possible to evaluate the total energy in $O(N)$ operations. The $O(N)$ methods have already been successful for tight-binding Hamiltonians, where interactions between the various atoms are described with short-range parameters. For true density-functional calculations, however, the advancement has been much slower, due to convergence problems in optimizing the many degrees of freedom.

The present chapter reviews the development and applications of multigrid-based methods for electronic structure calculations, focusing mainly on the work performed at North Carolina State University. In the current implementation, a real-space grid is used as a basis. The grid is fine enough to accurately describe the wavefunctions, the density, and the first-principles ionic pseudopotentials that represent the atoms. An important ingredient of the method is a compact, accurate and efficient discretization scheme, which also aids in parallelization. Multigrid methods [20], which provide automatic preconditioning on all length scales, are used to greatly reduce the number of iterations needed to converge the electronic wavefunctions. In particular, the advantage of multigrids became apparent for systems that are difficult to converge, due to, e.g., charge sloshing or multiple length scales. Quantum molecular dynamics, which must conserve total energy to very high accuracy, works very well and allows for large time steps. Parallelization has also been very effective, and the computational speed on the massively parallel Cray T3E supercomputer has scaled linearly with the number of processors, up to the maximum number we had access to (1024). Our method was already tested and applied to a large number of systems. These include the C_{60} molecule, diamond, Si and GaN supercells, and ab initio MD simulations. Further applications include simulations of surface melting of Si and of strain-induced deformations in carbon nanotubes, properties of Si, GaN, and InP surfaces, $In_xGa_{1-x}N$ quantum wells, and initial studies of a large amyloid β peptide ($C_{146}O_{45}N_{42}H_{210}$) implicated in Alzheimer's disease. The accuracy of the method was also tested by direct comparison with plane-wave calculations, when possible, and the results were in excellent agreement in all cases.

Very recently, we have investigated the performance of $O(N)$ techniques in the context of grid-based DFT calculations. By developing a generalization of the Galli-Parrinello method [21], we are able to use non-orthogonal, grid-based localized orbitals, while accelerating convergence through multigrid techniques and maintaining a high accuracy discretization. The localization of the orbitals makes the computation of the wavefunctions and the electron density scale as $O(N)$. These are the most expensive parts of the calculations. However, in order to minimize the number of iterations, we found it advantageous to also include partially occupied and unoccupied orbitals in the calculations, which requires an explicit diagonalization. The unoccupied orbitals accelerate the convergence even further, leading to the overall number of iterations that is not much greater then when orthonormal, fully delocalized orbitals are used. Since the codes are fully parallel, efficient SCALAPACK routines are used for the diagonalization and matrix operation, so that the $O(N^3)$ part is less than 20% even for calculations involving more than a thousand atoms.

The optimized localized orbitals are also very useful apart from facilitating exceedingly large calculations. They enable efficient calculations of quantum transport properties, which rely on the expansion of Hamiltonian and Green's function matrices in terms of localized functions [22]. Since one can use very few localized functions per atom, e.g., four for the case of carbon atoms, the dimensions of the relevant matrices are minimized, which has enabled ab initio calculations of quantum conductance for large nanotube structures [23].

The rest of this paper is organized as follows. In Section 2 the methodology of real-space multigrid calculations [9] is briefly summarized, together with its generalization to the case of localized, non-orthogonal orbitals [24]. This is followed by several examples of recent applications of multigrid methods. Impurity and dopant incorporation at Si(100) surfaces is discussed in Section 3. Section 4 focuses on wide gap nitride semiconductors, and in particular on the incorporation modes of Mg, the dominant p-type dopant, into GaN(0001) surfaces. Section 5 describes the calculations of optical signatures of surface reconstructions, for in situ monitoring and control of growth. Section 6 focuses on defect formation mechanisms in carbon nanotubes under severe tensile stress and the electronic effects of such defects on quantum conductance.

2. Summary of Real-Space Multigrid Methodology

2.1 Grid calculations with multigrid acceleration

In the method described in Ref. [9], a real-space mesh is used to represent the wavefunctions, the charge density, and the ionic pseudopotentials. The density functional equations are discretized using a generalized eigenvalue form:

$$\mathbf{H}_{\mathrm{mehr}}[\psi_n] = \tfrac{1}{2}\mathbf{A}_{\mathrm{mehr}}[\psi_n] + \mathbf{B}_{\mathrm{mehr}}[\mathbf{V}_{\mathrm{eff}}\psi_n] = \epsilon_n \mathbf{B}_{\mathrm{mehr}}[\psi_n],$$

where $\mathbf{A}_{\mathrm{mehr}}$ and $\mathbf{B}_{\mathrm{mehr}}$ are the components of the *Mehrstellen* procedure [25], which is based on Hermite's generalization of Taylor's theorem. It uses a weighted sum of the wavefunction and potential values to improve the accuracy of the discretization of the *entire* differential equation, not just the kinetic energy operator. Only nearest and next-nearest neighbor points are used in the discretization, and the short-ranged representation of the Hamiltonian leads to an efficient domain-decomposition-based implementation on massively parallel computers.

To efficiently solve the discretized equations, we use multigrid iteration techniques that accelerate convergence by employing a sequence of grids of varying resolutions.

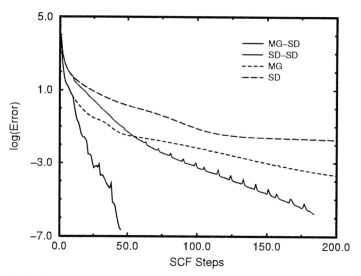

Fig. 1. Convergence rates from a random start for a 64-atom diamond cell with a N impurity: SD steepest descents, MG multigrid, SD-SD steepest descents with subspace diagonalizations, and MG-SD multigrid with subspace diagonalizations

The solution is obtained on a grid fine enough to accurately represent the pseudopotentials and the electronic wavefunctions. However, the iterations are accelerated by solving both the Poisson and the density functional equations on coarser grids and transferring the resulting corrections to the fine grid. This procedure provides excellent

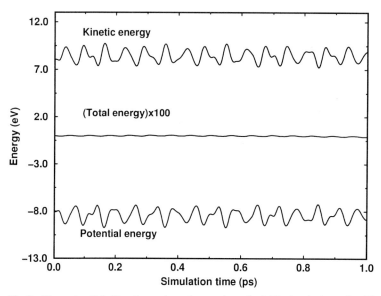

Fig. 2. The potential, kinetic, and total energies of a MD simulation of a 64-atom Si cell at 1100 K with an 80 a.u. time step. The total energy curve is multiplied by a factor of 100. The potential and total energies have been shifted by 251.171 a.u.

preconditioning for all length scales present in a system and leads to very fast convergence rates, see Fig. 1. The operation count to converge one wavefunction with a fixed potential is $O(N_{\text{grid}})$.

The real space methodology is also very suitable for high quality quantum molecular dynamics simulations. For example, a 1 ps test simulation of bulk Si at 1100 K conserved the total energy to within 27 μeV per atom, see Fig. 2. The details of the current multigrid implementation, which include new compact discretization schemes in real space for systems with cubic, orthorhombic, and hexagonal symmetries, and multilevel algorithms for the iterative solution of density functional equations are described in Ref. [9].

2.2 Towards linear scaling in ab initio electronic structure calculations

Real-space methods are inherently local, and therefore appropriate for imposing localization constraints on the basis representing the orbitals or for local mesh refinements [6, 8 to 10, 13]. In the Galli-Parrinello method [21], the electron density is expanded in terms of localized, non-orthogonal orbitals ϕ_i. The conjugate orbitals, defined by $\bar{\phi}_i = \sum S_{ji}^{-1} \phi_j$, where S is the overlap matrix, satisfy the relation $\langle \bar{\phi}_i | \phi_j \rangle = \delta_{ij}$. The total energy can then be written in a form where the relevant quantities depend on the product $\bar{\phi}_i^* \phi_i$ and the corresponding Euler equations are derived. If the orbitals ϕ_i are constrained to be zero outside of fixed localization radii, one obtains a variational $O(N)$ expression for the total energy whose accuracy, compared to the full density-functional solution, depends on the size of the localization radii. One should stress that the shapes of the orbitals inside their respective localization regions are not predetermined, but are optimized variationally in an iterative fashion. Our initial tests of this approach on 64- and 216-atom cells of Si have demonstrated that while good accuracy can be obtained for sufficiently large localization radii, the convergence rate in optimizing the localized functions on a grid was unacceptably slow. However, by including unoccupied orbitals and explicit multigrid preconditioning, the number of iterations was reduced to a level not greatly different from that of the standard multigrid method without any localization approximation. An outline of the essential steps of our approach is given below. A full discussion will be published elsewhere [24].

Consider the Mehrstellen finite difference discretization of the Kohn-Sham equations, which leads to a generalized eigenvalue problem

$$H\psi = \epsilon B\psi,$$

where H and B are real $M \times M$ sparse matrices. It is convenient to write the eigenvectors ψ as the columns of a $M \times N$ matrix denoted Ψ. In a non-orthogonal basis given by

$$\chi = (\varphi_1, \ldots, \varphi_N), \qquad (1)$$

the matrix steepest-descent direction, D, is

$$D = B\chi\Theta - H\chi, \qquad (2)$$

where

$$\Theta = (\chi^\dagger B \chi)^{-1} (\chi^\dagger H \chi). \qquad (3)$$

In the particular case of B being the identity matrix, the result (2) is equivalent to that given in [21]. However, eq. (2) provides an invariant expression for the true steepest descent (SD) direction for any non-orthogonal basis. Without preconditioning, the step along the SD direction must be very small for numerical stability reasons and the convergence of the process can be very slow, especially if $M \gg N$. To prevent this slowdown, we use an iterative multigrid operator as a preconditioner that reduces the high frequency components of the steepest-descent vectors.

When working with non-orthogonal functions χ instead of the eigenfunctions Ψ, localization constraints can easily be imposed on χ to reduce the cost of the calculation. At this time we are using spherical localization regions around each atom, forcing the functions to be zero outside. In particular, this truncation linearizes the cost of computing updates to the orbitals and forming the electron density — the most expensive parts of the algorithm.

As mentioned above, inclusion of some unoccupied and/or partially occupied orbitals in the N computed orbitals substantially improves the convergence rate and the range of applicability of the method. For a chemical potential μ, we define the matrix Υ by its matrix elements:

$$\Upsilon_{ij} = \delta_{ij} f((\epsilon_i - \mu)/kT),$$

where f is the Fermi-Dirac distribution. The density matrix is then given by:

$$\varrho = \Psi \Upsilon \Psi^*.$$

For $T = 0$, the density matrix ϱ is a projector onto the states with eigenvalues lower than μ. Its dimension is given by the number of degrees of freedom in the discretization, i.e., the number of nodes in the real-space grid. In the finite differences approach, this number is far too large to apply iterative methods that require $\varrho(r, r')$. However, one can express ϱ in the basis χ:

$$\varrho_\chi = \chi^* \varrho \chi.$$

Introducing the matrix $\bar{\varrho}_\chi = S^{-1} \varrho_\chi S^{-1}$ as in [26], the expectation value $\bar{A}$ of an operator A is then given by

$$\bar{A} = \text{tr}(\bar{\varrho}_\chi A_\chi).$$

The matrix $\bar{\varrho}_\chi$ can be computed by operations on submatrices of dimension N. The knowledge of $\bar{\varrho}_\chi$ allows for the computation of the electronic density ϱ and the total energy from the non-orthogonal wavefunctions χ.

The above method still contains parts that scale as $O(N^3)$, although with a much reduced prefactor. These include the inversion of the overlap matrix and a submatrix diagonalization. However, by using standard parallel linear algebra packages [27], even without taking advantage of the sparsity of either the overlap or the Hamiltonian matrices, the cost of $O(N^3)$ operations remained below 20% for calculations involving over 1100 atoms.

The localized functions obtained in the process of minimizing the total energy are also quite useful in their own right. Fig. 3 shows one such function for a carbon nanotube. The optimization, which was started from random numbers, led to a chemically very intuitive result: a function strongly localized on a carbon–carbon bond. The quantum-mechanical basis that automatically localizes in bonding regions should lead to ac-

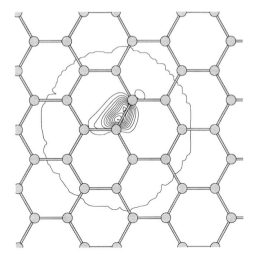

Fig. 3. An optimized electron-density function for a carbon nanotube. Note that although the allowed localization region extends over 6 Å, this function is largely confined to one carbon–carbon bond. The plotting plane is along the surface of the nanotube

curate results while keeping the number of basis functions to a minimum. The latter fact has already been very useful in performing efficient ab initio calculations of quantum transport, see Section 6.

The (2×1) reconstructed Si(100) surface, which is preferred for growth of Si-based devices, has complex structural and electronic properties. It is covered by rows of dimerized three-fold-coordinated Si atoms, with a high density of dangling bonds and anisotropic stress-fields. Correspondingly, the adsorption, incorporation and segregation of impurities at this surface is very complex. In an extensive set of studies, we have investigated the adsorption, incorporation and segregation of a number of common impurities at this surface. Some of our results are summarized below, while a full account is given in Refs. [28 to 30].

Among the considered impurities, two qualitatively distinct classes of behavior were uncovered. Al, Ga, In and Sn exhibit similar behavior. They prefer to adsorb in the trenches between surface dimer rows and their incorporation into the surface is highly unfavorable. Impurities in the other class (which include C, B, P, As and Sb) prefer to adsorb on top of surface dimer rows. Their incorporation into the surface is either energetically favorable or only marginally unfavorable. Within each group, atomic size differences result in quantitative shifts of the incorporation energies. The majority of these impurities, which include donors, acceptors and isoelectronic impurities, are surfactants, in that they prefer the adlayer or surface layer. Boron and carbon alone, which are first-row elements, prefer subsurface incorporation. Furthermore, we found that kinetic effects play an important role in determining the distribution of incorporated configurations.

The above results explain a variety of experimental investigations. Indeed, Al, Ga and In of group III are found in the adlayer. In group V, P and As readily incorporate into the surface [31, 32], while Sb does not incorporate to any significant extent [33]. For B, different models of subsurface B incorporation into the second [34] or the third layer [35] of the surface have been proposed in the literature. Our calculations support the former model over the latter. The results for C and Sn are also consistent with published theory and experiment [33, 36, 37].

Two of the impurities, B and As, have been selected for in-depth investigations [29, 30]. Due to space limitations, we only discuss B here, which is known to form ordered structures at 0.5 monolayer coverage. This effect is important, because an ordered dopant layer could substantially reduce electron scattering phenomena.

We found that the segregation and ordering of boron impurities at high concentration levels near the Si(100) surface is due to a complex interplay between chemical and

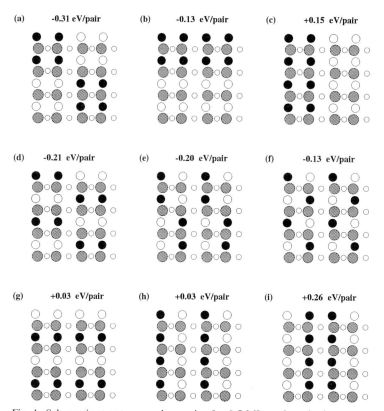

Fig. 4. Schematic structures and energies for 0.5 ML surface doping

strain effects. Like the other acceptors in group III (Al, Ga, In), boron is energetically favored to segregate to the surface, in order to saturate surface dangling bonds and to relieve the strain arising from its large atomic size mismatch with host Si atoms. However, unlike other group III acceptors, it prefers substitutional configurations in the first and second layers of the surface. With increasing doping, aggregation of boron impurities in the second layer of the surface becomes more and more favorable, up to a critical level corresponding to 0.5 ML of the impurities. The low-energy modes of ordering correspond to zigzag patterns of subsurface pairs, see Fig. 4. Above this doping level, the binding energies of ordered impurity configurations decrease sharply, and they become highly repulsive for a complete surface or subsurface impurity monolayer.

A number of experimental observations can be explained on the basis of our results. The equilibrium structure found at 0.5 ML doping has $c(4 \times 4)$ symmetry. However, both at low and high doping levels, several distinct modes of ordering are found to be essentially degenerate. This favors the formation of domain structures over long-range ordering of any one structure. Therefore, even at 0.5 ML coverage only (2×1) LEED patterns are observed [38 to 40], while domains with $c(4 \times 4)$ and (4×4) periodicities are observed in STM [34]. If some of the metastable ordered structures can be stabilized, there would be a potential for engineering complex impurity patterns at the nanoscale.

One common feature of all the surfaces with subsurface boron atoms is the pronounced flattening and lowering of surface Si dimers with B neighbors, of up to 40...50% of the interlayer spacing. The lowered Si dimers are not visible in simulated STM images. However, the removal of such boron-bonded Si dimers to create surface divacancies is energetically very expensive, much more so than on the clean Si(100) surface [28]. Therefore, the strain fields induced by heavy boron doping are not relieved by the creation of ordered patterns of surface Si divacancies. These results provide an alternative and more consistent explanation of the dark features extensively observed on B-doped Si(100) in STM investigations [34, 35, 41].

3. Mg Incorporation at GaN Surfaces

Controlled incorporation of impurities is critical to semiconductor science and technology. Specific dopant concentrations are necessary to achieve desired conductivities, recombination rates in light emitting devices or other electrical characteristics. In general, doping of bulk materials can be performed in three different ways: (i) by in-diffusion following growth, (ii) by ion implantation, and (iii) during growth. The in-diffusion of impurities is limited by the maximum solubility, which is a thermodynamic quantity that can be determined from either experiments or calculations. Ion implantation has no such limit, but in many materials the implantation damage is severe and cannot be annealed out, limiting the usefulness of this technique. During growth, the incorporation of an impurity occurs at the surface and depends sensitively upon the surface and its environment. For example, impurity energetics and consequently its solubility in the near-surface region can be significantly different from that in the bulk. This is because under typical growth conditions diffusion at the surface is much faster than in the bulk and the impurity density of the near-surface region can be frozen in as the film grows, leading to a concentration in the grown film different from that expected on the basis of its bulk solubility. Indeed, under these conditions it is the surface properties, instead of the bulk properties, that most strongly influence the impurity concentration. This is particularly true in a compound semiconductor, where many variables affect impurity incorporation during growth. Surface reconstruction patterns are often complex; in a given growth direction there are at least two possible surface terminations, and the surface structure varies as a function of the chemical potentials of the atomic constituents. The interplay of these effects leads to a complicated doping behavior, which, however, can be used to achieve a desired impurity concentration [42].

We have examined the incorporation of Mg in wurtzite GaN, which is an important paradigmatic case. There is a growing awareness that the two polarities of the (0001) GaN surface, nominally corresponding to the Ga and N faces, exhibit very different behavior. Indeed, striking differences have been observed in the morphology of the (0001) and (000$\bar{1}$) surfaces [43 to 49], and the strongly ionic nature of the GaN bond together with the low symmetry of the surface results in unusual reconstruction patterns that differ significantly from those observed in other III–V semiconductors. Since as-grown GaN exhibits unintentional n-type conductivity and controlled doping of GaN is necessary for optoelectronic devices, the achievement of good p-type conductivity has been a priority. However, Mg, which is the preferred p-type dopant, has a relatively large ionization energy. Consequently, high Mg concentrations are required to achieve the desired hole densities. In practice, the refined control of doping required to obtain

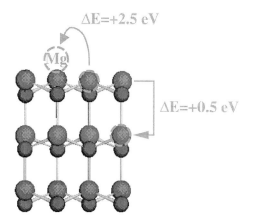

Fig. 5. The energetics of Mg incorporation at the Ga-terminated surface

reproducibly the particularly high Mg concentrations needed has been difficult to accomplish. Our calculations show that the interplay of surface orientation, reconstruction patterns and the availability of the species involved (as measured by their chemical potentials) determines the incorporation characteristics, and that intuition derived from bulk solubility considerations may be misleading [42]. For example, superior incorporation of Mg at the Ga-substitutional site (Mg_{Ga}) is expected under N-rich conditions, because of the decreased competition between Mg and Ga for the Ga site. However, this reasoning is a poor guide for determining the conditions for efficient incorporation. Indeed, we found that the growth conditions for optimal incorporation are reversed when the Mg is incorporated through a surface of different orientation (in this case, the film has different polarity). For the (0001) Ga-surface (see Fig. 5), we find that best incorporation occurs at N-rich and moderate Ga-rich conditions. On the contrary, for the (000$\bar{1}$) N-surface, in N-rich conditions Mg displays a strong tendency to segregate and superior incorporation occurs in a Ga-rich environment, when the surface is gallium terminated. Growth procedures may therefore need to be tailored in a nontrivial fashion to the properties of the particular growth surface(s) in order to achieve effective doping with specific impurities.

Our other work in this area examined the reconstruction patterns at GaN surfaces [48], band offsets at heterointerfaces [50], the effects of pyroelectric fields on recombination and lasing [51], and defect formation in wide gap nitride alloys [52].

4. Reflectance Anisotropy Spectra for Large Systems

The techniques of optical spectroscopy are evolving from fundamental semiconductor-surface studies to the control of semiconductor processing in situ and with real-time feedback. In particular, Reflectance Difference Spectroscopy (RDS), also known as Reflectance Anisotropy Spectroscopy (RAS), is a reliable in situ monitoring tool, applicable to ultrahigh vacuum as well as to gas-phase environments [53 to 56]. It is particularly useful for the wide range of materials that are grown by metal-organic vapor phase epitaxy (MOVPE), where the well-developed electron- or ion-beam techniques cannot be used. However, first-principles calculations of RDS spectra are very expensive, because thick slabs and a large number of excited states must be used in computing the surface optical response. We have incorporated the formalism [57, 58] for calculating RDS into our parallel multigrid code. This has enabled us to perform, to our knowledge for the first time, a well-converged study of the optical response of realistic III–V(001) surface structures [59]. In particular, we were able to resolve the controversies regarding the InP(001)(2 × 4) surface structure [60, 61].

The comparison between theory and experiment, although hindered by the computational shortcomings of the DFT-LDA (see below), allowed for clear identification of the dominant surface features, or structural fingerprints, present under various growth conditions. This characterization should be very useful in real-time feedback and control of InP growth by MOVPE. However, the theoretical energy scale had to be "stretched" in order to match the experimental data. This is due to the well-known deficiencies of DFT-LDA in computing excited state properties.

In contrast to InP, the precise GaP(001) surface structure was essentially unknown prior to our study [62]. On the basis of comprehensive total energy results we established the surface phase diagram. The cation-rich surface turned out to contain mixed dimers, in analogy to the corresponding InP surface. Our theoretical finding is corroborated by recent core-level results [63].

The RA spectra calculated for the energetically favored structures are strongly dependent on the surface geometry: The structures featuring cation–cation surface bonds show a pronounced negative anisotropy in the low energy region, with minima between 2.0 and 2.3 eV. The strength of that anisotropy is directly correlated to the number of Ga–Ga bonds along the [110] direction. On the other hand, we find that P–P dimers give rise to a relatively broad positive anisotropy between about 2.4 and 4.4 eV. These relations between the calculated RA and the surface stoichiometry agree very well with the experimental observations [62]. Furthermore, the results obtained for GaP are in line with the earlier findings for InP, supporting the existence of structural "fingerprints" in the optical spectra. This is an important outcome with respect to the technological application of optical spectroscopy for semiconductor growth.

While a good qualitative description of the measured optical anisotropy was achieved, in particular with respect to the chemical trends, the calculated spectra still deviate significantly from experiment, in particular concerning the energetic positions of the peaks and the line shape. In Fig. 6 we show the DFT-LDA calculation for the Ga-rich GaP surface in comparison with the measured curve. Highlighted in the figure is the pronounced negative anisotropy at low energy (denoted by S). This feature is mainly related to optical transitions between filled Ga–Ga bonding states and empty dangling

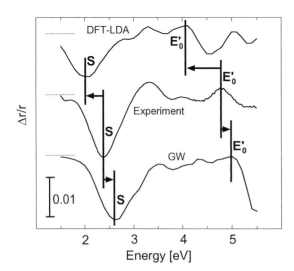

Fig. 6. Calculated and measured RA spectra for the Ga-rich GaP surface (see text)

bond states located at surface Ga atoms. Its energetic position is underestimated in the DFT-LDA calculation by about 0.3 eV with respect to the measured spectrum [62]. A much stronger underestimation of about 0.7 eV occurs for the essentially bulk-related features at higher photon energies. This non-uniform shift of the calculated spectra cannot be remedied by a simple scissor-operator approach. We applied an approximate GW-approach, based on a model dielectric function in conjunction with an approximate treatment of local-field effects and the dynamical screening [64, 65]. The resulting RA spectrum is also presented in Fig. 6. Compared to the DFT-LDA results we observe considerable improvements: The most important effect of the quasiparticle corrections is a non-uniform shift of the spectrum towards higher photon energies. We also observe a stronger shift for the bulk-related features than of the surface signatures.

5. Nanotubes

5.1 Axial tension and plastic transformations

Nanotubes are extremely unusual in their ability to withstand huge deformations, and are thus particularly suitable to mechanical applications. We have previously shown that nanotubes can be reversibly bent to very large angles [66] and this prediction has already been confirmed experimentally. Furthermore, our early classical molecular dynamics simulations indicated exceptional strength, which needed to be investigated further. However, strength measurements on nanotubes are extremely difficult, due to their very short lengths. In lack of any experimental data we performed extensive ab initio simulations to determine the dominant mechanism of strain release. Since the longest ab initio simulations correspond to real times of the order of only tens of picoseconds, simulations at 10% tensile strain and 1800 K were performed, using multiple Nose-Hoover thermostats [67]. Under those conditions the dynamical processes are greatly accelerated and a lengthy simulation has identified the key transformation that releases tension in highly strained carbon nanotubes. This transformation rotates a carbon–carbon bond, converting four hexagons into two pentagons and two hexagons coupled in pairs, the so-called (5-7-7-5) defect. Depending on tube symmetry, temperature and strain rate, it can lead to either breakage or plastic behavior in carbon nanotubes. Once the defect transformation is identified, the thermodynamic stability of nanotubes under strain can be determined by computing defect formation energies as a function of strain. Fig. 7 shows that beyond about 5% tension, a nanotube of "armchair" symmetry, in which some carbon–carbon bonds are oriented perpendicular to its axis, can release its excess strain via spontaneous formation of such defects. The predicted strength is indeed the highest by far of any material to date, and there should be no shortage of exciting applications once nanotubes are grown in sufficient quantity and quality. Furthermore, the activation energy for the bond rotation is quite high (see Fig. 8) so that the observed nanotube strengths at ambient temperatures could be substantially higher.

Detailed ab initio simulations for many tube geometries are too expensive at present, but much qualitative and semiquantitative information has been obtained through classical and tight-binding molecular dynamics [67, 68]. It turns out that the appearance of a (5-7-7-5) defect can be interpreted as a nucleation of a degenerate dislocation loop in the hexagonal network of the graphite sheet. The configuration of this primary dipole is a (5-7) core attached to an inverted (7-5) core. The (5-7) defect behaves thus as a

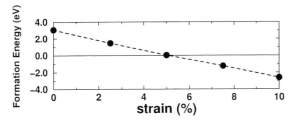

Fig. 7. Formation energy of the (5-7-7-5) defect for the (5,5) tube at different strains, obtained in ab initio calculations

single edge dislocation in the graphitic plane. Once nucleated, the (5-7-7-5) dislocation loop can ease further relaxation by separating the two dislocation cores, which glide through successive bond rotations [67]. This corresponds to a plastic flow of dislocations and gives rise to ductile behavior. In contrast, in the case of nanotubes of the so-called zig-zag symmetry, the same C–C bond is parallel to the applied tension, which is already the minimum energy configuration for the strained bond. The formation of a defect is then limited to rotation of the bonds oriented 120° with respect to the tube axis. Our calculations show that the formation energies of these defects are strongly dependent on curvature, i.e., on the diameter of the tube. Simulations show that nanotubes with a diameter smaller 1.1 nm can deform plastically regardless of their symmetry, while wider tubes must eventually break in a brittle fashion.

As mentioned above, activation energies for the formation defects in perfect nanotubes are very high (cf. Fig. 8). However, due to the high temperature at which the growth of nanotubes occurs, defects will form for thermodynamic reasons and then remain frozen in. For a material at thermal equilibrium, the number of defects of a particular type is given by $N_{sites} e^{(-G_F/kT)}$, where N_{sites} is the number of potential sites and G_F is the Gibbs free energy of formation of the defect. The entropic contributions are usually a small part of G_F; one can thus use the energy of formation to obtain a lower bound. For single-walled tubes, which are grown at ~1500 K, the above formula suggests that even at zero strain there might be point defects in nanotubes every few tenths of a mm. The presence of the frozen-in defects will certainly limit the ultimate strength of nanotubes and thus some of their proposed uses.

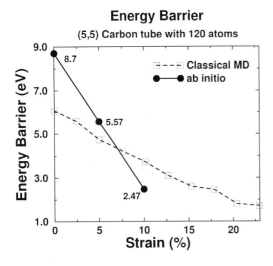

Fig. 8. Preliminary results for the activation energy for bond rotation for the (5,5) tube computed by the multigrid method. The corresponding values obtained by molecular statics using the Tersoff-Brenner potential are also shown. Note that while the qualitative trends are similar between the two approaches, there are substantial quantitative differences

5.2 Quantum transport

Nanotubes can be either metallic or semiconducting, depending on their symmetry (or index). The plastic transformations described above change nanotube index between the two (5-7) pairs. This raises the prospect of easily fabricating metal–semiconductor junctions [67], and thus all-carbon electronic devices [69]. This possibility has stimulated our interest in quantum transport in such structures.

The electronic and transmission properties of carbon nanotubes have already been extensively studied both experimentally and theoretically [70 to 80]. The general problem of calculating the conductance in carbon nanotubes has been addressed with a variety of techniques that reflect the various approaches in the theory of quantum transport in ballistic systems. Most of the existing calculations derive the electronic structure of carbon nanotubes from a simple π-orbital tight-binding Hamiltonian that describes the bands of the graphitic network via a single nearest-neighbor hopping parameter. Since the electronic properties of carbon nanotubes are basically determined by the sp^2 π-orbitals, the model gives a reasonably good qualitative description of their behavior and, given its simplicity, it has become the model of choice in a number of theoretical investigations. However, although qualitatively useful to interpret experimental results, this simple Hamiltonian lacks the accuracy that more sophisticated methods are able to provide. In particular, the π-orbital model is limited to studies of topological changes, and cannot address the effects of topology-preserving distortions, such as bending, on conductance.

Recently, we have developed a general scheme for calculating quantum conductance that is particularly suitable for realistic calculations of electronic transport properties in extended systems [22]. This method is based on the Surface Green's Function Matching formalism and efficiently combines the iterative calculation of transfer matrices with the Landauer formula for coherent conductance. It is very flexible and applicable to any system described by a Hamiltonian with a localized orbital basis. The only quantities that enter are the matrix elements of the Hamiltonian operator, with no need for the explicit knowledge of the electron wavefunctions for the multichannel expansion. The last fact makes the numerical calculations efficient also for systems described by multi-orbital Hamiltonians. In order to obtain the highly-localized ab initio orbitals and potentials necessary for the efficient application of this scheme, we use the localized-orbi-

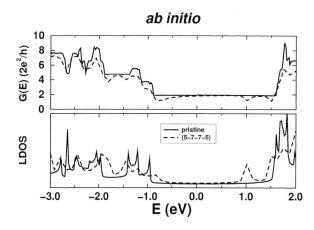

Fig. 9. Ab initio conductance and local density of states for a pristine (5,5) nanotube and for one with a (5-7-7-5) defect. The Fermi energy is used as reference

tal method described in Section 2.2. This allows for DFT calculations of quantum conductance to be carried out for systems with complicated topology, contacts, etc.

As the first application, we evaluated the conductance of a (5,5) tube, pristine and with a (5-7-7-5) defect, see Fig. 9. It is clear from the results that a single defect does not affect the conductance near the Fermi level, but there are significant effects near the valence bands maximum and in the upper part of the gap. Our results for the (5,5) tube with a single (5-7-7-5) defect compare very well with a recent ab initio calculation [79] for a larger (10,10) tube, and with tight-binding calculations for the same system. Further applications of this formalism to multi-component nanotube systems are in progress.

Acknowledgement This work was supported in part by ONR, NSF, NASA, and DoE.

References

[1] B. Hermansson and D. Yevick, Phys. Rev. B **33**, 7241 (1986).
[2] S. R. White, J. W. Wilkins, and M. P. Teter, Phys. Rev. B **39**, 5819 (1989).
[3] H. Murakami, V. Sonnad, and E. Clementi, Internat. J. Quant. Chem. **42**, 785 (1992).
[4] E. Tsuchida and M. Tsukada, Phys. Rev. B **52**, 5573 (1995).
[5] J. E. Pask, B. M. Klein, C. Y. Fong, and P. A. Sterne, Phys. Rev. B **59**, 12352 (1999).
[6] J. Bernholc, J.-Y. Yi, and D. J. Sullivan, Faraday Disc. Chem. Soc. **92**, 217 (1991).
[7] J. R. Chelikowsky, N. Troullier, and Y. Saad, Phys. Rev. Lett. **72**, 1240 (1994).
[8] F. Gygi and G. Galli, Phys. Rev. B **52**, R2229 (1995).
[9] E. L. Briggs, D. J. Sullivan, and J. Bernholc, Phys. Rev. B **52**, R5471 (1995); **54**, 14362 (1996).
[10] G. Zumbach, N. A. Modine, and E. Kaxiras, Solid State Commun. **99**, 57 (1996).
[11] K. A. Iyer, M. P. Merrick, and T. L. Beck, J. Chem. Phys. **103**, 227 (1995).
[12] A. P. Seitsonen, M. J. Puska, and R. M. Nieminen, Phys. Rev. B **51**, 14057 (1995).
[13] J.-L. Fattebert, J. Comput. Phys. **149**, 75 (1999).
[14] F. Ancilotto, P. Blandin, and F. Toigo, Phys. Rev. B **12**, 7868 (1999).
[15] K. Cho, T. A. Arias, J. D. Joannopoulos, and P. K. Lam, Phys. Rev. Lett. **71**, 1808 (1993).
[16] S. Wei and M. Y. Chou, Phys. Rev. Lett. **76**, 2650 (1996).
[17] C. J. Tymczak and X.-Q. Wang, Phys. Rev. Lett. **78**, 3654 (1997).
[18] G. Galli, Curr. Opin. Solid State Mater. Sci. **1**, 864 (1996).
[19] S. Goedecker, Rev. Mod. Phys., in press (1999).
[20] A. Brandt, Math. Comp. **31**, 333 (1977); GMD Studien **85**, 1 (1984).
[21] G. Galli and M. Parrinello, Phys. Rev. Lett. **69**, 3547 (1992).
[22] M. Buongiorno Nardelli, Phys. Rev. B **60**, 7828 (1999).
[23] M. Buongiorno Nardelli, J.-L. Fattebert, and J. Bernholc, to be published (1999).
[24] J.-L. Fattebert and J. Bernholc, to be published (1999).
[25] L. Collatz, The Numerical Treatment of Differential Equations, Springer-Verlag, Berlin 1966.
[26] R. Nunes and D. Vanderbilt, Phys. Rev. B **50**, 17611 (1994).
[27] L. S. Backford, J. Choi, A. Cleary, E. D'Azevedo, J. Demmel, I. Dhillon, J. Dongarra, S. Hammarling, G. Henry, A. Petitet, K. Stanley, D. Walker, and R. C. Whaley, SCALAPACK User's Guide, SIAM, Philadelphia 1997.
[28] M. Ramamoorthy, E. L. Briggs, and J. Bernholc, Phys. Rev. Lett. **81**, 1642 (1998).
[29] M. Ramamoorthy, E. L. Briggs, and J. Bernholc, Phys. Rev. B **59**, 4813 (1999).
[30] M. Ramamoorthy, E. L. Briggs, and J. Bernholc, Phys. Rev. B **60**, 8178 (1999).
[31] R. D. Bringans, D. K. Biegelsen, and L. E. Swartz, Phys. Rev. B **44**, 3054 (1991).
[32] Y. Wang, X. Chen, and R. J. Hamers, Phys. Rev. B **50**, 4534 (1994).
[33] J. Nogami, A. A. Baski, and C. F. Quate, J. Vac. Sci. Technol. A **8**, 245 (1990); Phys. Rev. B **43**, 9316 (1991); **44**, 1415 (1991); **44**, 11167 (1991); Appl. Phys. Lett. **53**, 2086 (1991); **58**, 475 (1991).
[34] Y. Wang, R. J. Hamers, and E. Kaxiras, Phys. Rev. Lett. **74**, 403 (1995).
[35] Z. Zhang et al., J. Vac. Sci. Technol. B **14**, 2684 (1996).

[36] J. Tersoff, Phys. Rev. Lett. **74**, 5080 (1995).
[37] P. C. Kelires and E. Kaxiras, Phys. Rev. Lett. **78**, 3479 (1997).
[38] R. L. Headrick, B. E. Weir, A. F. J. Levi, D. J. Eaglesham, and L. C. Feldman, Phys. Rev. Lett. **57**, 2779 (1990).
[39] B. E. Weir, R. L. Headrick, Q. Shen, L. C. Feldman, M. S. Hybertsen, M. Needels, M. Schluter, and T. R. Hart, Phys. Rev. B **46**, 12861 (1992).
[40] R. Cao, X. Yang, and P. Pianetta, J. Vac. Sci. Technol. B **11**, 1455 (1993).
[41] T. Komeda and Y. Nishioka, Appl. Phys. Lett. **71**, 2277 (1997).
[42] C. Bungaro, K. Rapcewicz, and J. Bernholc, Phys. Rev. B **59**, 9771 (1999).
[43] F. A. Ponce, D. P. Bour, W. T. Young, M. Saunders, and J. W. Steeds, Appl. Phys. Lett. **69**, 337 (1996).
[44] B. Daudin, J. L. Roviére, and M. Arlery, Appl. Phys. Lett. **69**, 2480 (1996).
[45] Z. Liliental-Weber, Y. Chen, S. Ruvimov, and J. Washburn, Phys. Rev. Lett. **79**, 2835 (1997).
[46] M. M. Sung, J. Ahn, V. Bykov, J. W. Rabalais, D. D. Koleske, and A. E. Wickenden, Phys. Rev. B **54**, 14652 (1996).
[47] E. S. Hellman, D. N. E. Buchanan, D. Wiesmann, and I. Brener, MRS Internet J. Nitride Semicond. Res. **1**, 16 (1996).
[48] K. Rapcewicz, M. Buongiorno Nardelli, and J. Bernholc, Phys. Rev. B **56**, R12725 (1997).
[49] A. R. Smith, R. M. Feenstra, D. W. Greve, J. Neugebauer, and J. E. Northrup, Phys. Rev. Lett. **79**, 3934 (1997).
A. R. Smith, R. M. Feenstra, D. W. Greve, M.-S. Shin, M. Skowronski, J. Neugebauer, and J. E. Nothrup, Appl. Phys. Lett. **72**, 2114 (1998).
[50] M. Buongiorno Nardelli, K. Rapcewicz, and J. Bernholc, Phys. Rev. B **55**, R7323 (1997).
[51] M. Buongiorno Nardelli, K. Rapcewicz, and J. Bernholc, Appl. Phys. Lett. **71**, 3135 (1997).
[52] P. Boguslawski and J. Bernholc, Phys. Rev. B **59**, 1567 (1999).
[53] D. E. Aspnes and A. A. Studna, Phys. Rev. Lett. **54**, 1956 (1985).
[54] D. E. Aspnes, Surf. Sci. **307/309**, 1017 (1994).
[55] W. Richter and J. T. Zettler, Appl. Surf. Sci. **101**, 465 (1996).
[56] J. F. McGilp, D. Weaire, and C. H. Patterson, Epioptics. Linear and Nonlinear Optical Spectroscopy of Surfaces and Interfaces, Springer-Verlag, Berlin 1995.
[57] R. Del Sole, Solid State Commun. **37**, 537 (1981).
[58] F. Manghi, R. Del Sole, A. Selloni, and E. Molinari, Phys. Rev. B **41**, 9935 (1990).
[59] W. G. Schmidt, E. L. Briggs, J. Bernholc, and F. Bechstedt, Phys. Rev. B **59**, 2234 (1999).
[60] D. Pahlke, J. Kinsky, C. Schultz, M. Pristovsek, M. Zorn, N. Esser, and W. Richter, Phys. Rev. B **56**, R1661 (1997).
K. B. Ozanyan, P. J. Parbrook, M. Hopkinson, C. R. Whitehouse, Z. Sobiesierski, and D.I. Westwood, J. Appl. Phys. **82**, 474 (1997).
M. Zorn, T. Trepk, J. T. Zettler, B. Junno, C. Meyne, K. Knorr, T. Wethkamp, M. Klein, M. Miller, W. Richter, and L. Samuelson, Appl. Phys. A **65**, 333 (1997).
J. Kinsky, C. Schultz, D. Pahlke, A. M. Frisch, T. Herrmann, N. Esser, and W. Richter, Appl. Surf. Sci. **123**, 228 (1998).
[61] C. D. MacPherson, R. A. Wolkow, C. E. J. Mitchell, and A. B. McLean, Phys. Rev. Lett. **77**, 691 (1996).
M. Shimomura, N. Sanada, Y. Fukuda, and P. J. Moller, Surf. Sci. **359**, L451 (1996).
[62] A. M. Frisch, W. G. Schmidt, J. Bernholc, M. Pristovsek, N. Esser, and W. Richter, Phys. Rev. B **60**, 2488 (1999).
[63] N. Sanada et al., Surf. Sci. **419**, 120 (1999).
[64] F. Bechstedt et al., Solid State Commun. **84**, 765 (1992).
[65] W. G. Schmidt et al., in preparation.
[66] S. Iijima, C. Brabec, A. Maiti, and J. Bernholc, J. Chem. Phys. **104**, 2089 (1996).
[67] M. Buongiorno Nardelli, B. I. Yakobson, and J. Bernholc, Phys. Rev. B **57**, R4277 (1998).
[68] M. Buongiorno Nardelli, B. I. Yakobson, and J. Bernholc, Phys. Rev. Lett. **81**, 4656 (1998).
[69] B. I. Dunlap, Phys. Rev. B **46**, 1933 (1992).
L. Chico, V. H. Crespi, L. X. Benedict, S. G. Louie, and M. L. Cohen, Phys. Rev. Lett. **76**, 971 (1996).
[70] P. G. Collins, A. Zettl, H. Bando, A. Thess, and R. Smalley, Science **278**, 100 (1996).
S. N. Song, X. K. Wang, R. P. H. Chang, and J. B. Ketterson, Phys. Rev. Lett. **72**, 697 (1994).

L. Langer, L. Stockman, J. P. Heremans, V. Bayot, C. H. Olk, C. Van Haesendonck, Y. Bruynseraede, and J.-P. Issi, J. Mater. Res. **9**, 927 (1994).

L. Langer, V. Bayot, E. Grivei, J.-P. Issi, J. P. Heremans, C. H. Olk, L. Stockman, C. Van Haesendonck, and Y. Bruynseraede, Phys. Rev. Lett. **76**, 479 (1996).

S. J. Tans, M. H. Devoret, H. Dai, A. Thess, R. E. Smalley, L. J. Georliga, and C. Dekker, Nature **386**, 474 (1997).

A. Bachtold, C. Strunk, J.-P. Salvetat, J.-M. Bonnard, L. Forró, T. Nussbaumer, and C. Schönenberger, Nature **397**, 673 (1999).

[71] A. Bezryadin, A. R. M. Verschueren, S. J. Tans, and C. Dekker, Phys. Rev. Lett. **80**, 4036 (1998).

[72] S. Paulson, M. R. Falvo, N. Snider, A. Helser, T. Hudson, A. Seeger, R. M. Taylor II, R. Superfine, and S. Washburn, http://xxx.lanl.gov/abs/cond-mat/9905304, preprint (1999).

[73] W. Tian and S. Datta, Phys. Rev. B **49**, 5097 (1994).

[74] R. Saito, G. Dresselhaus, and M. S. Dresselhaus, Phys. Rev. B **53**, 2044 (1996).

[75] L. Chico, L. X. Benedict, S. G. Louie, and M. L. Cohen, Phys. Rev. B **54**, 2600 (1996).

[76] R. Tamura and M. Tsukada, Phys. Rev. B **55**, 4991 (1997); **58**, 8120 (1998).

[77] M. P. Anantran and T. R. Govindan, Phys. Rev. B **58**, 4882 (1998).

[78] A. A. Farajian, K. Esfarjani, and Y. Kawazoe, Phys. Rev. Lett. **82**, 5084 (1999).

[79] H. J. Choi and J. Ihm, Phys. Rev. B **59**, 2267 (1999).

[80] A. Rochefort, F. Lesage, D. R. Salahub, and P. Avouris, http://xxx.lanl.gov/abs/cond-mat/9904083, preprint (1999).

phys. stat. sol. (b) **217**, 703 (2000)

Subject classification: 73.20.Dx; 73.40.Kp; 73.61.Tm; S7.15

Semiconductor Nanostructures

A. Di Carlo

INFM and Dipartimento di Ingegneria Elettronica, Università di Roma "Tor Vergata", I-00133 Roma, Italy

(Received August 10, 1999)

Empirical tight-binding approaches (TB) are applied to study semiconductor nanostructures. We will show how tight-binding relaxes all the limitations of simplified approach based on envelope function approximations, maintaining, at the same time, the computational cost low. We will present TB calculations for semiconductor nanostructures ranging from single quantum well up to nanostructured device where an electron current flow is present.

1. Introduction

Nanostructures based on semiconductor heterojunctions are nowadays used in electronic and optoelectronic devices. As a matter of example, long-wavelength lasers for telecommunications have active regions formed by a sequence of quantum wells obtained from the heterojunction of two or more semiconductors with different band gaps [1]. On the other hand, physical phenomena related to semiconductor nanostructures such as the confinement of electron in zero-, one-, two-dimensions are of great interest and have contributed to define new concepts in condensed matter physics [2].

The theoretical investigation of semiconductor nanostructures is of crucial importance since it allows both to investigate fundamental physics and to optimize nanostructured devices. The capability of theoretical techniques to predict or investigate physical phenomena concerning nanostructures is essentially related to the possibility of applicating these techniques to treat the nanostructures which are usually composed by a large number of atoms. The theoretical tool widely used to study the nanostructure problem is based on the $\mathbf{k} \cdot \mathbf{p}$ approach within the envelope function approximation (EFA) context [3]. In this case, only the envelope of nanostructure wavefunction is described regardless to atomic details. Despite the fact that EFA is a very simplified description of the system, based on several approximations which have been debated in the literature [4], envelope functions approaches have obtained great success mainly due to the fair compromise between simplicity of the method and reliability of the results.

Modern applications, however, push nanostructures to dimension and uses where EFA may not be as accurate as one would expect. This is for example the case of scaling down of silicon MOSFET technology [5], where oxide dimension, channel thickness and channel length are such that the application of EFA is highly questionable. On the other hand, however, as soon as we relax envelope function approximations going to more ab-initio approaches, the complexity of the problem becomes intractable. Thus, the use of an intermediate level approach which improves the description of the system (leading to ab-initio results), but with a complexity similar to the $\mathbf{k} \cdot \mathbf{p}$ EFA, is highly required.

Empirical tight-binding methods (TB) [6,7] are very suited for the study of semiconductor nanostructures [8 to 23]. TB relaxes all the approximations upon which the EFA is constructed and allows to define atomic details, detailed band structure in the whole Brillouin zone [24], strain, charge self-consistency [17]. Moreover, the complexity of empirical TB approaches is not very different from $\mathbf{k} \cdot \mathbf{p}$ EFA approaches.

In the following we will show, by performing characteristic calculations, how TB can be efficiently used to approach the nanostructure problem. We will discuss examples ranging from the single quantum well system up to electronic devices where an electron current flow is present. This paper is divided as follows: in Section 2 we will review the theory the empirical-TB approach applied to nanostructure is based on, Section 3 will present results for nanostructures where the translational symmetry is broken in one dimension, while in Section 4 we will show calculations for nanostructures confined in two-dimensions.

2. Theory

2.1 Definition of the TB Hamiltonian for the heterostructure

In this section, we discuss the theory of tight-binding models applied to systems where the translational symmetry is broken in one direction (z), usually referred as the *growth axis*. This is the case of several electron devices such as diodes, BJT transistors as well as heterojunction resonant tunnel diodes or other 1D confined quantum structures. The case where translational symmetry is broken in two dimensions will be discussed in Section 4. Perpendicular to the growth direction there is a two-dimensional atomic plane where the periodicity is preserved. In order to set our notation, the growth direction will be labeled by "⊥" and vectors lying on the two-dimensional plane by "∥". The lattice vector can be written as sum of two terms,

$$\mathbf{R} = m\mathbf{d} + \mathbf{R}_\parallel, \tag{1}$$

with m being an integer. Here, $\mathbf{d}$ is a vector parallel to the growth direction with module equal to the distance between two atomic planes and $\mathbf{R}_\parallel$ is a vector on the m-th atomic plane. In other words, $m\mathbf{d}$ is the projection of $\mathbf{R}$ on the growth direction and $\mathbf{R}_\parallel$ the projection on the m-th atomic plane. In general $\mathbf{d}$ will depend also on the atomic plane index m. This is the case, for example, of particular growth direction or strained structures. To reduce the complexity of notations, the lattice vector $\mathbf{R}$ includes also the basis atom displacement, if present.

The wave function of the system, $|E, \mathbf{k}_\parallel\rangle$ can be written as linear combination of planar Bloch sums $|\alpha, m, \mathbf{k}_\parallel\rangle$ [25,12,18],

$$|E, \mathbf{k}_\parallel\rangle = \sum_{\alpha,m} C_{\alpha,m}(E, \mathbf{k}_\parallel) |\alpha, m, \mathbf{k}_\parallel\rangle \tag{2}$$

with

$$|\alpha, m, \mathbf{k}_\parallel\rangle = \frac{1}{\sqrt{N}} \sum_{\mathbf{R}_\parallel} e^{i\mathbf{k}_\parallel \cdot \mathbf{R}_\parallel} |\alpha, \mathbf{R}\rangle, \tag{3}$$

where $|\alpha, \mathbf{R}\rangle$ is the localized orbital basis function, $\mathbf{k}_\parallel$ is the in-plane wave vector and N is the number of unit cells in the atomic plane. The subindex α refers both to the basis atom index and to the atomic orbital index.

For a given $\mathbf{k}_\parallel$, the eigenstates E are calculated by solving the secular equation

$$H\,|E,\mathbf{k}_\parallel\rangle = (H^S + V^H)\,|E,\mathbf{k}_\parallel\rangle = E\,|E,\mathbf{k}_\parallel\rangle, \qquad (4)$$

where H^S is the system tight-binding Hamiltonian and V^H is the Hartree potential (see Section 2.3). In the Bloch sum expansion, considering orthonormalized basis function we have

$$\sum_{m',\alpha'} H_{m,\alpha;m',\alpha'} C_{m',\alpha'} = E C_{m,\alpha}. \qquad (5)$$

2.2 Definition of the system: boundary conditions

There are mainly three different ways to define boundary conditions for the secular equation (5). Open chain (cluster) boundary conditions, periodic boundary conditions and transmitting boundary conditions. The choice will depend on the physical system we are interested in.

2.2.1 Open chain boundary conditions

In this case we consider a limited number of atomic planes in an empty space. Far away from the cluster of planes the wave function goes to zero. In the case of nearest-neighbor interaction, the Hamiltonian matrix takes a tridiagonal band form well suited for diagonalization procedures. The drawback of this method is the appearance of dangling bonds which will induce the presence of spurious surface states. These states can be, however, removed either with hydrogenating the dangling bonds or by direct inspection of the associated wave function which has a different character (localization at the surface) with respect to real quantized states.

2.2.2 Periodic boundary condition

When the translational symmetry is periodically broken (for example in superlattices) we can limit the calculation to the supercell defining the periodic structure and applying the Bloch theorem to define the wave function boundary condition. The Hamiltonian matrix which will result has a more complicated structure with respect to the open chain boundary condition case. Indeed, for a superlattice and with nearest-neighbor interaction the Hamiltonian matrix will have a band form plus two other contributions in the top-right corner and bottom-left corner. This matrix form is not suitable for typical fast diagonalization routines, however the sparsity of the matrix will still reduce the computational effort.

2.2.3 Transmitting boundary condition

As soon as the system allows current flow, we need to consider appropriate boundary conditions to the Eq. (5) in order to allow such flux of particles. This can be achieved by applying the transmitting boundary conditions.

In the nearest-neighbor interaction case and for a [001] growth direction (see Schulman and Chang Ref. [26] for the general case), the secular equation Eq. (5) is

$$\mathbf{H}_{m,m+1}\mathbf{C}(m+1) + \mathbf{H}_{m,m-1}\mathbf{C}(m-1) + (\mathbf{H}_{m,m} - \mathbf{E})\,\mathbf{C}(m) = 0. \qquad (6)$$

Here $\mathbf{H}_{m,m'}$ is the Hamiltonian matrix with elements $H_{\alpha,\alpha'}$, where α belongs to the atomic plane m and α' to the atomic plane m'. $\mathbf{C}(m)$ are the expansion coefficients $C_{\alpha,m}$ and $\mathbf{E}$ is the diagonal energy matrix. Considering two successive atomic planes (anion and cation plane), called layer (l), we arrive to define, via Eq. (6), the transfer matrix Γ that connects two successive layers. The recursive relation between successive layer coefficients is

$$[\mathbf{C}^a \mathbf{C}^c](l) = \Gamma(l) [\mathbf{C}^a \mathbf{C}^c](l-1). \tag{7}$$

We subdivide the system into three regions, namely the left (L) and right (R) semi-infinite bulk-like regions, and a middle (M) region where the translational symmetry is broken.

In semi-bulk regions (L, R), the wave function is expressed either by Eq. (2) or in terms of complex bulk states [27], i.e. by a linear combination of propagating Bloch states and evanescent states [18],

$$\langle z \mid \mathbf{k}_\parallel, E \rangle = \sum_i [h_{\mathrm{L},i} \langle z \mid \phi_{\mathrm{L},i}(\mathbf{k}_\parallel, E) \rangle] \quad \text{for} \quad z \to \mathrm{L},$$

$$\langle z \mid \mathbf{k}_\parallel, E \rangle = \sum_i [h_{\mathrm{R},i} \langle z \mid \phi_{\mathrm{R},i}(\mathbf{k}_\parallel, E) \rangle] \quad \text{for} \quad z \to \mathrm{R}, \tag{8}$$

where $\phi_{\mathrm{L(R)},i}$ are the complex bulk states of the left (right) region and $h_{\mathrm{L(R)},i}$ are the expansion coefficients

Comparing Eq. (2) and Eq. (8), we can relate the complex bulk expansion in regions L and R through the transfer matrix of region M,

$$\mathbf{h}_{\mathrm{R}} = \mathbf{U}_{\mathrm{R}}^{-1} \prod_{l \in M} \Gamma(l) \, \mathbf{U}_{\mathrm{L}} \mathbf{h}_{\mathrm{L}}, \tag{9}$$

where $(\mathbf{h}_{\mathrm{R(L)}})_i = h_{\mathrm{R(L)},i}$. The $\mathbf{U}_{\mathrm{R(L)}}$ matrices transform the complex bulk states of region L (R) into the in-plane Bloch sums. Equation (9), solved for each incoming state, allows us to calculate the total transmission coefficient. Indeed, the transmission coefficient for a given incoming state (i) from the left region into a given outgoing state (j) in the right region can be written as

$$T_{(\mathrm{L},i \to \mathrm{R},j)}(\mathbf{k}_\parallel, E) = |h_{\mathrm{R},j}|^2 \left| \frac{\mathbf{v}_{\mathrm{R},j}(\mathbf{k}_\parallel, E)}{\mathbf{v}_{\mathrm{L},i}(\mathbf{k}_\parallel, E)} \right|, \tag{10}$$

where $\mathbf{v}$ is the group velocity. Thus, the total current flowing through the device is given by

$$j = \frac{(-e)}{(2\pi)^3 \hbar} \int_{BZ_\parallel} d\mathbf{k}_\parallel \int_{-\infty}^{+\infty} dE \sum_{i,j} T_{(\mathrm{L},i \to \mathrm{R},j)}(\mathbf{k}_\parallel, E) \, [f_{\mathrm{R}}(E) - f_{\mathrm{L}}(E)], \tag{11}$$

where $f_{\mathrm{R(L)}}$ are the Fermi distribution functions in the right and left doped region, respectively.

2.3 Self-consistency

The influence of the electronic free-charge rearrangement can be included at a Hartree level by solving the Poisson equation, $d^2 V^{\mathrm{H}}/dz^2 = -\varrho(z)/\varepsilon$, where ε is the static dielectric constant [17]. Here we consider only the contribution to the Hartree potential that comes from the free-charge, that is the electron in the conduction band and holes in the valence band, and we do not consider the valence electrons which are accounted (not self-consistently) in the dielectric constant. The charge density in the m-th plane $\varrho(m)$ is defined, for an orthonormal basis, by

$$\varrho(m) = -\frac{e}{(2\pi)^2} \int_{BZ_\parallel} d\mathbf{k}_\parallel \sum_{n,\alpha} |C_{\alpha,n}(E_n, \mathbf{k}_\parallel)|^2 \tilde{f}(E_n, E_F), \qquad (12)$$

where e is the electron charge and n labels the energy levels for a given $\mathbf{k}_\parallel$. The function $\tilde{f}(E_n, E_F)$ is defined as follows:

$$\tilde{f}(E_n, E_F) = \begin{cases} f(E_n, E_F) & \text{for the conduction states,} \\ 1 - f(E_n, E_F) & \text{for the valence states,} \end{cases} \qquad (13)$$

where $f(E_n, E_F)$ is the Fermi distribution function with a given Fermi level E_F. $\tilde{f}(E_n, E_F)$ is a well behaved function which is different from zero only in the proximity of the valence and conduction band edges.

Poisson and Schrödinger equations in the TB representation are iteratively solved until convergence is reached. All of the boundary condition discussed in Section 2.2 can be used to account self-consistency.

2.4 Parameterization of TB matrix elements

The central issue of the empirical TB approach is the proper parameterization of Hamiltonian matrix elements. Typically the sp^3s^* nearest-neighbor model [7] is used for III–V semiconductors, while the inclusion of second-neighbors is often considered to model Silicon based heterostructures. However, very recently, the definition of the $sp^3d^5s^*$ Hamiltonian model [24] has improved the TB description of bulk bands. In fact, the sp^3s^* model is able to reproduce band dispersion along principal axes but it will fail along a general direction of the Brillouin zone. This is especially true for the conduction band. In the $sp^3d^5s^*$ model such limitations are eliminated and comparisons with ab-initio results are in really good agreement.

2.5 Alloys

The tight-binding parameterization for the alloys such as AlGaAs, InGaAsP etc. are obtained from those of the binary materials by applying the Virtual Crystal Approximation (VCA) to Hamiltonian matrix elements, lattice constants and elastic constants. Sometimes, however, a nonlinear interpolation for the Hamiltonian matrix elements is needed to account the disorder-induced nonlinear variation of the band structure parameters such as energy gap, masses etc [19]. The virtual crystal approximation is also applied at the interface between two materials, which is considered as a monolayer with the average tight-binding parameters of the two adjacent materials.

The virtual crystal approximation can be relaxed by applying other high-order methods such as Coherent Potential Approximation (CPA) or T-matrix approaches [28].

2.6 Optical properties

When optical properties are of interest, one can make use of the Kubo formula to define the susceptibility tensor which is related to the current–current response function of the electromagnetic perturbation. This can be easily calculated within the tight-binding scheme *without* introducing new fitting parameters [29]. If we consider a σ-polarized light, the absorption coefficient can be written as [29,17]

$$a(\omega) = \frac{4\pi^2}{nc\omega S} \sum_{E, E', \mathbf{k}_\|} [f(E) - f(E')] \, \delta(\hbar\omega + E - E') \, |\sigma \cdot \langle E, \mathbf{k}_\| | \mathbf{J} | E', \mathbf{k}_\| \rangle|^2. \tag{14}$$

Here, S is the transverse area of the primitive cell, n is the refractive index c the speed of light, $\mathbf{J}$ the current operator and $f(E)$ the distribution function. The matrix elements of the current operator can be expressed as

$$\langle E, \mathbf{k}_\| | \mathbf{J} | E', \mathbf{k}_\| \rangle$$
$$= \frac{ie}{\hbar dN} \sum_{\substack{\mathbf{R}_\|, m, \alpha \\ \mathbf{R}'_\|, m', \alpha'}} C^*_{\alpha', m'}(E, \mathbf{k}_\|) \, C_{\alpha, m}(E', \mathbf{k}_\|) \, e^{i\mathbf{k}_\| \cdot (\mathbf{R}'_\| - \mathbf{R}_\|)} \, [\mathbf{R}' - \mathbf{R}] \, H_{\alpha', \alpha}(\mathbf{R}' - \mathbf{R}). \tag{15}$$

Here $H_{\alpha', \alpha}(\mathbf{R}' - \mathbf{R}) = \langle \alpha', \mathbf{R}' | H | \alpha, \mathbf{R} \rangle$.

2.7 Strained layers

When a crystal is strained, the translation vectors become

$$\mathbf{R}' = (\hat{\mathbf{1}} + \hat{\boldsymbol{\epsilon}}) \, \mathbf{R}, \tag{16}$$

where $\hat{\boldsymbol{\epsilon}}$ is the symmetric strain tensor.

In ZB crystals and for [001] growth direction Eq. (16) is sufficient to describe the strain for each atom in the unit cell. However, if the growth direction is along the [111] or [110] directions, there is an additional internal displacement. This is usually described by the *internal-strain parameter* ξ which ranges from 0 to 1. In this case, Eq. (16) must be replaced by

$$\mathbf{R}' = (\hat{\mathbf{1}} + \hat{\boldsymbol{\epsilon}}) \, \mathbf{R} - \frac{a_0}{2} \, \epsilon_{xy} \xi \, [111]$$

for [111] growth direction and

$$\mathbf{R}' = (\hat{\mathbf{1}} + \hat{\boldsymbol{\epsilon}}) \, \mathbf{R} - \frac{a_0}{2} \, \epsilon_{xy} \xi \, [001]$$

for [110] growth direction.

Let us focus on the [001] growth direction and on the biaxial strain which is the typical situation we have in the pseudomorphic growth of heterojunctions. For this direction, the biaxial strain can be decomposed into the sum of the strain tensor for a pure hydrostatic pressure and a uniaxial stress in the growth direction. Thus, the strain tensor can be written as follows:

$$\hat{\boldsymbol{\epsilon}} = \begin{pmatrix} \epsilon_\| & 0 & 0 \\ 0 & \epsilon_\| & 0 \\ 0 & 0 & \epsilon_\perp \end{pmatrix}.$$

According to Eq. (16) we have

$$\epsilon_\perp = \frac{a_\perp}{a_0} - 1, \qquad \epsilon_\| = \frac{a_\|}{a_0} - 1, \tag{17}$$

where a_0 is the unstrained lattice constant. The strain tensor components are not independent but are related each other,

$$\epsilon_\perp = -2\frac{C_{12}}{C_{11}}\epsilon_\parallel \Rightarrow a_\perp = a_0\left[1 - 2\frac{C_{12}}{C_{11}}\left(\frac{a_\parallel}{a_0} - 1\right)\right],$$

where C_{11}, C_{12} are the elastic stiffness constants.

The effect of the strain induces changes of two parameters in the empirical tight-binding: the geometrical factors and the two-center matrix elements. The scaling properties of the matrix element has been extensively discussed by Harrison and coworkers [30,31,6] who showed that a τ^{-2} dependence on the nearest-neighbor distance (τ) between two atoms reasonably reproduces the chemical properties of several materials. Although this law provides a very reasonable general trend, it has been shown [32,33,34] that a more refined power dependence can provided better results. Usually, the modify Harrison scaling law is defined by

$$H_{llm}^{(s)} = H_{llm}\left(\frac{\tau}{\tau^{(s)}}\right)^{\eta_{llm}}, \tag{18}$$

where the matrix elements are in the "molecular" notation [30] and τ is the distance between the two atoms with ($\tau^{(s)}$) and without (τ) strain.

2.8 Numerical implementation

The numerical implementation of the TB approach is of crucial importance. By itself, the method is computationally quite heavy since the diagonalization of very large matrices is needed. In the following we describe efficient numerical implementation to reduce the computational cost of the algorithm.

When it is possible, the use of open-chain boundary conditions is preferred to the other boundary conditions since the Hamiltonian matrix which results has a band form. This implies that very efficient diagonalization methods, suited for this matrix form, can be applied. In order to speed up the calculations, an hybrid diagonalization method can be used to solve Eq. (5). This uses a standard (LAPACK [35]) routine to calculate eigenvalues and an inverse iteration scheme [36] to calculate eigenvectors. The advantage of this procedure over others relies on the fact that only few eigenvectors are needed, namely those closed to the energy band gap.

The $\mathbf{k}_\parallel$ integration needed to calculate the charge density Eq. (12), the absorption/gain coefficient Eq. (14) or the current density Eq. (11) is performed in the 2D Brillouin zone by using a uniform k-point grid in the irreducible wedge, which is obtained by considering the symmetry properties of the Hamiltonian. Only few portion of the irreducible wedge will contribute to the calculated properties. For example, if we are interested in the absorption/gain coefficient close to the energy gap, we can limit the integration to the region $|\mathbf{k}_\parallel| < 0.1\, 2\pi/a$, whereas if we calculate the self-consistent charge density in a typical situation we obtain converged results by using $5\mathbf{k}_\parallel$ points with $|\mathbf{k}_\parallel| < 0.06\, 2\pi/a$ for a direct band gap material and $8\mathbf{k}_\parallel$ points with $|\mathbf{k}_\parallel| < 0.2\, 2\pi/a$ for an indirect band gap material.

In the case of optical transition, two nearly spin-degenerate valence and conduction subbands are involved. Each squared optical matrix element is summed over the two final conduction states and averaged over the two initial valence states [37].

In our procedure we first calculate (and store) the energy levels and the squared optical matrix elements for each $\mathbf{k}_\parallel$, then we evaluate the absorption/gain coefficient by performing the sums over the carrier distribution functions in Eq. (14). To reduce the numerical fluctuations induced by the finite number of $\mathbf{k}_\parallel$ points considered in this sum (≈ 1600), we sum over a much finer $\mathbf{k}_\parallel$ grid (≈ 160000 points). The energy levels and squared matrix elements at these new $\mathbf{k}_\parallel$ points are obtained by using a bilinear interpolation of the calculated quantities. This is allowed since the variations of both energy levels and squared matrix elements in the irreducible wedge are quite smooth.

3. Nanostructures Confined in One-Dimension: Quantum Wells

In the following calculations we will apply the empirical tight-binding approach to study several nanostructures where the translational symmetry is broken in one direction. This means that in two directions the system is perfectly periodic, while in the other direction the presence of heterojunction or/and confining potential will destroy the periodicity.

3.1 Valence band mixing

Let us apply the theory developed in the previous section to study the valence band mixing (VBM), that is the mixing of valence states due to the presence of heterojunctions. To investigate this topic the TB approach is well suited since several contributions to mixing of bands arise from microscopic details (for example anion or cation terminated wells present different behaviors, etc.).

To study the VBM we will calculate the difference in the polarization of the light which arises from a pump and probe experiment [38]. In a quantum well, heavy-hole (HH) and light-hole (LH) subbands are coupled, which implies that the third component of the angular momentum, m_J (what in the following will be called the spin) is not a good quantum number. In other words, any electronic valence band (VB) state has a certain probability of its spin being $+1/2$, $-1/2$, $+3/2$ or $-3/2$. This is usually known as the VBM.

In a quantum well system and for a σ-polarized electromagnetic perturbation the squared optical matrix element can be written as (see Eq. (14))

$$M_{cv}^2(\mathbf{k}_\parallel) = |\langle c, \mathbf{k}_\parallel | \boldsymbol{\sigma} \cdot \mathbf{J} | v, \mathbf{k}_\parallel \rangle|^2. \tag{19}$$

Here, c(v) labels the conduction (valence) subband including spin indices.

Let us consider a typical experiment where a quantum well is optically excited with σ^+ circularly polarized light and the emitted light is analyzed in its σ^+ and σ^- components. The excitation with circularly polarized light allows the selection of the spin of the photoexcited carriers because the population of the excited electrons and holes with a certain spin depends on the excitation energy E. If we assume that the spin relaxation of holes is very fast, the intensity of the σ^+ (σ^-) polarized light, I^+ (I^-), at short times, before the electron spin relaxation takes place, will reflect the initial population of photoexcited electrons with a certain spin. Thus, the study of the polarization degree of the PL at $t = 0$, $P(0)$ as a function of E, will give relevant information about the dependence of the VBM on k.

For a given excitation energy (E) we calculate the excitation probability for the σ^+ polarization

$$\mathcal{P}_x^+(c, v, E) = \int d\mathbf{k}_\| \, |\langle c, \mathbf{k}_\| | \boldsymbol{\sigma}^+ \cdot \mathbf{J} | v, \mathbf{k}_\| \rangle|^2 \, \delta(E - E_c(\mathbf{k}_\|) - E_v(\mathbf{k}_\|)), \qquad (20)$$

where $E_n(\mathbf{k}_\|)$ is the n-th subband dispersion.

The recombination probability is calculated considering that all the electrons are relaxed to $\mathbf{k}_\| \approx 0$,

$$\mathcal{P}_r^+(c, v) = |\langle c, \mathbf{0} | \boldsymbol{\sigma}^+ \cdot \mathbf{J} | v, \mathbf{0} \rangle|^2, \qquad (21)$$

$$\mathcal{P}_r^-(c, v) = |\langle c, \mathbf{0} | \boldsymbol{\sigma}^- \cdot \mathbf{J} | v, \mathbf{0} \rangle|^2. \qquad (22)$$

The luminescence intensity for each polarization and for an excitation energy E is given by

$$I^+(E) = \sum_{c,v,v'} \mathcal{P}_x^+(c, v, E) \, \mathcal{P}_r^+(c, v'), \qquad (23)$$

$$I^-(E) = \sum_{c,v,v'} \mathcal{P}_x^+(c, v, E) \, \mathcal{P}_r^-(c, v'). \qquad (24)$$

In our calculation the excitation process involves the first two valence subbands and the first conduction subband, while the recombination process occurs between the first conduction subband and the first valence subband.

Finally, the degree of polarization of the luminescence is calculated as

$$\mathbf{P}(t=0, E) = \frac{I^+(E) - I^-(E)}{I^+(E) + I^-(E)}. \qquad (25)$$

For comparison with published experiments, the calculated value of $\mathbf{P}(t=0)$ has to be reduced to $\mathbf{P}(t_0)$, where $t_0 = 3$ ps is the half-width of the laser pulse; for this purpose, it has been considered that $\mathbf{P}(t)$ decays exponentially with a characteristic time of 30 ps, as obtained from experiments.

We have investigated a GaAs$_{1-x}$P$_x$ QWs with well widths of 120 Å and phosphorus compositions (x) of 5%. The barriers are made with Ga$_{0.65}$Al$_{0.35}$As barriers. This choice of the composition induces a HH–LH degenerate ground state. Fig. 1 shows the experimental [38] and theoretical results for the structure. Fig. 1 reflects a strong VBM for $\Delta E > 40$ meV since $P(t_0)$ becomes as low as 0.15. The experimental results are in good agreement with the theory, which predicts a polarization degree of ≈ 0.2 to 0.15 above the HH–LH subband edges. We should point out that $P(t_0)$ never reaches values close to unity, even for low excitation energy. This is mainly due to the presence of VBM even at $\mathbf{k}_\| = 0$ for this degenerate condition. Such behavior cannot be accounted in a $k \cdot p$-EFA theory which usually neglects these contributions. However, recent extensions to the $\mathbf{k} \cdot \mathbf{p}$ EFA theory [39] did relax such limitation by introducing a mixing parameter obtained with the TB theory.

3.2 Gain in semiconductor optical amplifiers

In the previous section it has been shown how valence-band mixing can be studied by means of spin-relaxation experiment and how tight-binding can account for this effect. A typical optoelectronic device where mixing effects are very important, expecially in the degenerate configuration described in the previous section, is the Semiconductor

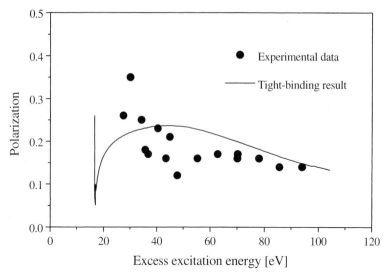

Fig. 1. The solid points indicate the polarization degree of the luminescence at 3 ps as a function of the excess excitation energy above the band gap for the GaAlAs/GaAsP QW. The solid line shows the result of tight-binding calculations

Optical Amplifier (SOA). Such device allows to amplify optical beams by means of stimulated emissions. In other words, this device is similar to a semiconductor laser except for the fact that cavity mirrors are absent. Beside its application as an optical amplifier, SOA can be used as a wavelength converter by using either gain saturation or four-wave mixing effects.

One requisite which should be satisfied by the SOA is the independence of the optical gain from the polarization of the light. This is an important requirement since the light that should be amplified, has a random polarization. However, in a quantum well SOA (QW-SOA), the optical gain has a larger TE-polarization contribution with respect to TM-polarization. This is mainly due to the fact that the fundament transition, (between first conduction level and fist heavy-hole level) occurs only for TE-polarized photons.

To overcome this intrinsic limitation of QW-SOA, strained quantum wells are generally used. By applying strain, one can vary the relative position of heavy and light holes [40]. For a degenerate energy level condition between heavy and light hole (as that of Section 3.1), the transition probability from conduction band to these levels, becomes equal for TE and TM photon polarization, allowing a perfect polarization independence of the QW-SOA. However, in a real device, one should account also for the difference in the confinement factor of the two TE and TM modes. One can find that TM polarized mode has a smaller confinement factor with respect to the TE polarized mode, thus to have a real polarization independence one should amplify more the TM mode to overcome the reduction in the confinement factor.

The tight-binding model is one of the more indicated tools to investigate the physical factors conditioning gain in SOAs, especially for its capability to exhaustively describe the properties of strained epitaxial, the effect of thin layer perturbations and valence band mixing at a microscopic level. It has been shown that TB methods are very useful,

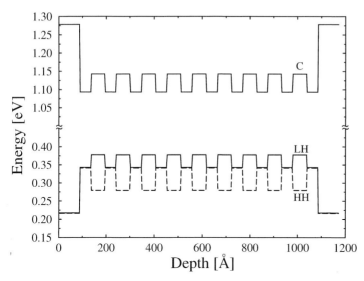

Fig. 2. Conduction and valence band edges of the Semiconductor Optical Amplifier (SOA)

for example, to calculate the polarization dependence of the amplified light in delta-strained InP/InGaAs SOAs [19].

In order to show typical tight-binding calculations in the field of semiconductor optical amplifiers we have calculated the optical gain in InP based multi-quantum well SOA [41]. The active layer consisted of a stack of ten 4.8 nm, unstrained $In_{0.53}Ga_{0.47}As$ wells separated by 6.0 nm, $In_{0.4}Ga_{0.6}As$ barriers. This was cladded by two separate-confinement, 0.13 mm thick, $In_{0.74}Ga_{0.26}As_{0.56}P_{0.44}$ layers (both undoped). The tight-binding approach was used to describe the band structure and to calculate the gain coefficient. Parameterizations are those used in Ref. [19].

Fig. 2 shows the conduction and valence band edges in the SOA. Heavy and light holes split due to the presence of tensile strain in the barrier. Minimum energy is for the LH in the barrier, while HH minimum energy is in the well. Moreover, LH levels are split to lower energy with respect to the situation without strain. These variations induced by the barrier strain affect considerably the optical properties of the structure. In Fig. 3 we show the gain coefficient calculated form Eq. (14). We observe an enhancement of the TM gain with respect to the TE gain, a prerequisite to achieve an optical independence to the light polarization. The gain spectra show a step structure typical of quantum well absorption, however, band structure effects are evidenced, for example, from the non-constant value of the gain coefficient in each step.

3.3 Si/SiGe interfaces

In the above calculation we have analyzed the optical properties of an SOA where free-charge rearrangement was not present. However, self-consistent device simulations can be obtained directly from the self-consistent tight-binding (SC-TB) [42]. As an example we have calculated the self-consistent band edge profile of a Si/SiGe heterojunction. Such structure is the building block of p-type modulation doped field effect transistor (MODFET). This calculation confirms that self-consistency is achieved regardless

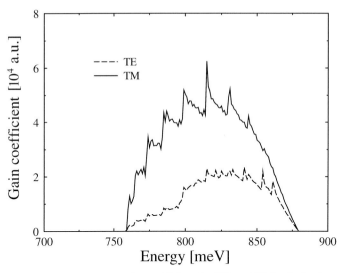

Fig. 3. Calculated SOA gain coefficient for TE and TM polarizations

of the type of the energy gap, and this can be applied to indirect band gap material such as Si an SiGe.

The calculate valence band profile is shown in Fig. 4 for a p δ-doped SiGe/Si. Strain in the SiGe well is responsible for the splitting of heavy and light hole bands.

3.4 Nitride based quantum wells

Very recently we have assisted to the development of LED and lasers based on nitride semiconductor such as GaN, AlN, InN and related alloys [43]. These devices emit blue

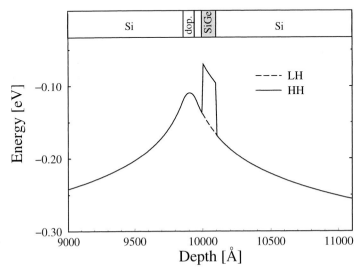

Fig. 4. Self-consistent tight-binding result for the valence band edge of a $Si/Si_{0.7}Ge_{0.3}$ δ-doped ($p = 10^{18}$ cm^{-3}) heterojunction

radiation and are attractive for consumer electronics and other spectroscopical applications. Despite the impressive achievements, several puzzling issues remain unsettled, such as the exact mechanism responsible for laser action and the unusually high threshold densities required, the red shift of the transition energy observed for increasing well width etc. These phenomena are related to the peculiarity of wurtzite nitrides in having a non-zero macroscopic polarization, consisting of both spontaneous (pyroelectric) [44] and piezoelectric components. These polarizations produce an internal electric field as soon as heterojunctions between different nitride-based semiconductors are formed. Such field, usually in the MV/cm range, induces large band bending of valence and conduction bands (Stark effect), altering the usual behavior of quantum wells. Obviously, polarization fields may be screened by free charge during the operating condition of nitride-based quantum well laser or LEDs. This issue can be investigated by means of the self-consistent TB as soon as we include polarization fields in the Poisson equation

$$\frac{d}{dz} D = \frac{d}{dz}\left(-\varepsilon \frac{d}{dz} V^H + P_T\right) = e(p - n). \qquad (26)$$

The (position-dependent) quantities D, ε, and V, are the displacement field, dielectric constant, and potential, respectively. The (position-dependent) transverse polarization P_T is the sum of the spontaneous component P_s, calculated ab initio [44], and of the piezoelectric component $P_{pz} = 2e_{31}\epsilon_{xx} + e_{33}\epsilon_{zz}$, involving the calculated [44] piezoelectric constants e_{31} and e_{33}, and the strain tensor components ϵ_{xx} and ϵ_{zz}. These are obtained via elasticity theory in terms of the barrier (a_b) and well (a_w) lattice parameters [44] and elastic constants as $\epsilon_{xx} = (a_b - a_w)/a_w$ and $\epsilon_{zz} = -2\epsilon_{xx}C_{13}/C_{33}$. TB parameters have been determined fitting the DFT-LDA band structure as outlined in Ref. [24].

Here we consider a $Al_{0.15}Ga_{0.85}N/GaN$ superlattice (SL) with a AlGaN barrier width of 100 Å and several well widths. Similar calculations for InGaN/GaN single quantum wells can be found in Ref. [21].

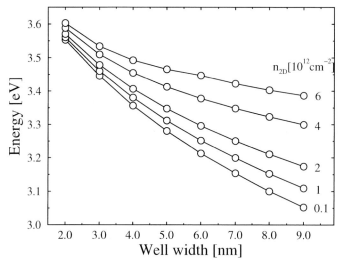

Fig. 5. Self-consistent tight-binding results for $Al_{0.15}Ga_{0.85}N/GaN$ C1 → V1 transition as a function of the well width

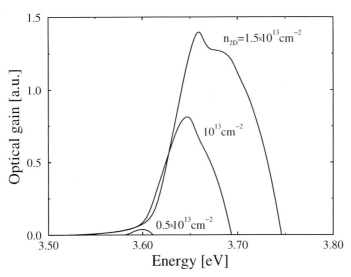

Fig. 6. Self-consistent tight-binding results for $Al_{0.15}Ga_{0.85}N/GaN$ gain spectra with 4 nm well width

Fig. 5 shows the calculated photoluminescence (PL) peak position (that is the energy of the transition between first conduction band level and first valence band level, C1→V1) as a function of the superlattice well width for several values of the free-charge density in the well.

As seen in Fig. 5, upon increasing the density we observe a blue shift in the transition energy. This is due to the progressive recovery of flat band conditions upon increasing the density. This blue shift effect is also visible in the calculated gain spectra of the SL shown in Fig. 6. Here we plot the gain spectra for several free-charge densities. The structure begins to gain as soon as the density is higher than the transparent density (n_t) which in our structure is of the order of 5×10^{12} cm^{-2}. For densities higher than n_t the structure shows a gain with a maximum that blue shifts as the density increases. Such blue shift is much larger than the usual band filling blue shift typical of conventional nanostructure. Band-filling effects are, however, clearly shown from the calculated spectra, since the curve becomes larger as soon as the density increases.

The above calculation shown hows SC-TB can be used to model nitride-based heterojunction laser an LED, where conventional methods fail due to the presence of high polarization fields.

3.5 Zener tunneling in multi-quantum wells

In the following calculation we will show how empirical TB can be applied to study nanostructures where an electron current flow is present.

Wannier-Stark ladders, [45,18] i.e. the energy levels arising from the presence of an electric filed in a semiconductor, have been observed optically in superlattices and multiple quantum wells (MQWs) [46,47]. In the following, we examine the formation of Wannier-Stark ladders in MQW and how they show-up in the Zener current of a reversely biased diode [22,23].

Let us consider a p–i–n structure with an intrinsic region that contains a coupled multiple quantum wells (p–MQW–n). To be concrete, we assume the intrinsic region to extend over ten periods of a superlattice with a unit cell of 2ML AlAs/5ML GaAs.

The n- and p-regions are heavily doped and we assume that the applied reverse potential drops only in the intrinsic region. A left and right spacer layer of 10 nm between the doped regions and the MQW region has also been included in order to consider a typical p–MQW–n nanostructure. We calculate the stationary current flowing in this structure by employing the transmitting boundary conditions of Section 2.2.3.

Let us focus on the electronic states associated with the intrinsic region. If the electric field in the MQW region is uniform, the subbands form localized and equidistantly spaced Wannier-Stark resonances. Since, however, the extent of the electric field in the intrinsic region is limited, only 10 WS resonances form. Specifically, it has been shown [18] that a prerequisite for the formation of WS resonances is that the total potential drop eV_D across the field region exceeds the total zero field miniband width Δ. This condition can be easily satisfied in superlattices because the band width of the minibands is much smaller than that of bulk solids. The condition suffices for the observation of WS ladders in an optical experiment where one can populate the WS resonances in the conduction band by exciting electrons directly from the valence bands. In transport, on the other hand, to populate these resonances by a tunneling mechanism one needs an emitter band where the electron can enter the field region via tunneling, and a collector band where the electrons can leave the device having resonantly tunneled into the WS ladders. Therefore, in order to observe a WS ladder in the Zener tunneling regime one needs an applied potential greater than the miniband width and such that the WS, valence band (emitter) and conduction band (collector) are aligned.

When this condition is achieved, the presence of such quasi-localized levels induce a resonant Zener tunneling between the valence band and the conduction band. This is reflected in the transmission coefficient as well as in the calculated Zener current. In Fig. 7 we show the transmission coefficient for $k_\parallel = 0$ and a potential drop of 3.1 V, as

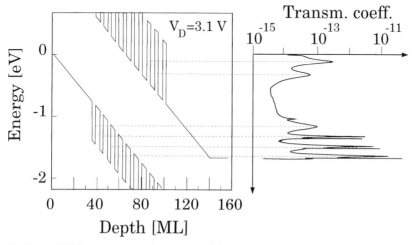

Fig. 7. p–MWQ–n structure for a potential drop of 3.1 V. Right part: transmission coefficient of the structure. The dashed lines relate the transmission coefficient peaks to the Wannier-Stark ladders of the p–MWQ–n diode

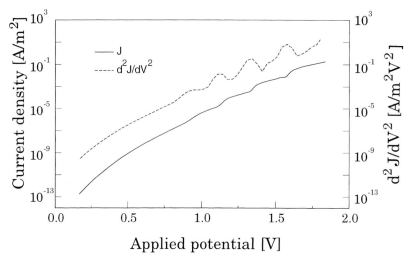

Fig. 8. Calculated Zener current density and its second derivative vs. the applied potential for the p–MWQ–n diode

obtained from Eq. (10). The crucial point is that the tunneling coefficient from the p to the n doped regions exhibits highly pronounced resonances whenever the energy of the incoming and outgoing electron is aligned with one of the WS resonances associated with the multiple quantum well structure in the intrinsic zone. At low energy, these peaks correspond to the resonances induced by the Wannier-Stark ladders of the valence band, whereas the high energy portion corresponds to resonances in the conduction WS states. The resonance peaks in the valence band reveal a fine structure due to the presence of heavy and light hole WS ladders. The energy spacing between Wannier-Stark resonance of the same miniband is eFa, where a is the unit cell length of the multiple quantum well. Such resonant peaks will affect the Zener current density.

The onset of each WS ladder induces an abrupt increase of the tunneling current as shown in Fig. 8. In fact, by increasing the reverse bias, we increase also the number of WS ladders involved in the tunneling process. The abruptness of such a process depends on the linearity of the field in the intrinsic region as well as on the doping density of the n- and p-regions. If the field extends also in the doped region because of low doping, the onset of the WS resonance gets smeared out. Nevertheless, the derivative of the current density still exhibits clear resonances.

In Fig. 8, the oscillation in the current comes from the WS of the valence band. Nevertheless, the current resonances induced by WS ladders of the conduction band can be observed if we use a p–MQW–n structure where the left spacer layer (between p and MQW) is greater than the right one.

The injection of current by Zener tunneling into WS ladders in a p–MQW–n diode can be used as a mechanism to invert the population of the adjacent WS states in order to produce a laser in the THz frequencies.

4. Nanostructures Confined in Two Dimensions: Quantum Wires

Quasi one-dimensional electron confinement has been achieved in a variety of ways, based on direct etching, strain-induced quantum confinement, growth on vicinal substrates and so on [48,49]. Beside all, the growth of V-shaped quantum wires have attracted most of the attention. V-shaped quantum wires are grown on a (100) GaAs substrate patterned with [$\bar{1}$10]-oriented V grooves [49]. In this case, the GaAs wire is embedded in an AlAs/GaAs superlattice which constitutes the confining barrier for the electrons [50,51]. The study of such structures by means of envelope function schemes is very critical. In fact, the restrictions of the EFA become very serious when indirect gap semiconductors are used either in the wire (e.g., in Si wires) or in the barrier (AlAs or AlAs/GaAs superlattices) regions. The use of *effective parameters*, e.g. effective barrier height, is then highly questionable. Theoretical study of V-shaped QWRs based on a tight-binding model [52,20] relax the restriction of EFA, maintaining at the same time the possibility to treat realistic structures with more than 10^4 atoms in the unit cell. This has been made feasible by recent advances in diagonalization algorithms [53,54,55,17].

The unit cell of the considered V-shaped quantum wires structures is 550×220 Å^2 (see Ref. [20]). The wire is oriented along the [$\bar{1}$10] direction, while the directions of confinement are fixed to be [110] and [001], respectively, named x-axis and z-axis.

The full wave function $|E, \mathbf{k}\rangle$ for the quantum wire is made up of a linear combination of wire Bloch sums $|\alpha, k, \mathbf{R}_\perp\rangle$,

$$|E, k\rangle = \sum_{\alpha, \mathbf{R}_\perp} C_{\alpha, \mathbf{R}_\perp}(E, k) |\alpha, k, \mathbf{R}_\perp\rangle \tag{27}$$

with

$$|\alpha, k, \mathbf{R}_\perp\rangle = \sum_m e^{ikmd} |\alpha, \mathbf{R}\rangle . \tag{28}$$

With respect to Eqs. (2) and (3), here k is a one-dimensional wave number and $\mathbf{R}$ is broken up into two terms as $\mathbf{R} = \mathbf{R}_\parallel + \mathbf{R}_\perp$, where $\mathbf{R}_\parallel = m\mathbf{d}_\parallel$ is the vector parallel to the quantum wire direction [$\bar{1}$10] ($|\mathbf{d}_\parallel| = (\sqrt{2}/2)\,a$, m is an integer) and $\mathbf{R}_\perp = \mathbf{R}_z + \mathbf{R}_x$ is a vector on the quantization plane with components $\mathbf{R}_z$ and $\mathbf{R}_x$ along the directions [001] and [110], respectively. Periodic supercell boundary conditions are used for the Schrödinger's equation.

The large Hamiltonian matrix H which results from such realistic dimensions of the unit cell, can been diagonalized by using a Lanczos algorithm without reorthogonalization [56]. Following to the Folded Spectrum Method [53,54], we do not directly diagonalize the Hamiltonian H but rather the matrix $A = (H - \lambda I)^2$, where λ (energy offset) defines an energy in the vicinity of which the eigenvalues are searched. With respect to the standard Lanczos algorithm, we have found that the lowest eigenstates of the associated tridiagonal matrix visit all the excited states of A before collapsing (Lanczos phenomena) on the lowest eigenvalue of A. This allows us to identify high energy states during the Lanczos iteration. Eigenvectors (wavefunctions) can be obtained from the knowledge of the eigenvalues by using the inverse iteration technique [17,36]. The overall efficiency of the Lanczos/inverse iteration approach is very high and a typical band structure for realistic wire geometries can be carried out in a few hours of calculation on a workstation. Details of the calculation can be found in Ref. [20].

Fig. 9 shows the energy of the first three quantized levels of the conduction band for $k = 0$ as a function of the dimension (number of atoms) of the unit supercell. We can see that to obtain a a converged result, the number of atoms in the supercell should be of the order of 2×10^4. We should noticed that, the higher the quantized level energy is

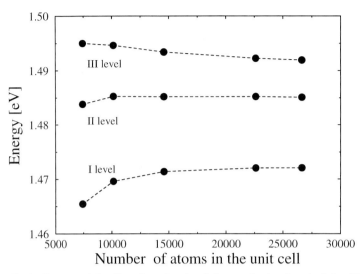

Fig. 9. Energy of the first three levels of the conduction band of the V-shaped wire as a function of the number of atoms in the supercell. Here $k = 0$

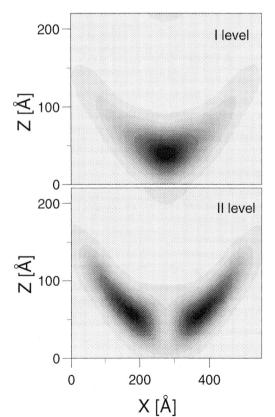

the more the number of atoms required to achieve a converged result increases. This is due to the wavefunction which is more extended as the quantized level energy increases, thus the supercell should be large enough to account for the extension of the wavefunction of the highest energy level we are interested to calculate.

This effect is clearly shown in Fig. 10 where we plot the squared wavefunction of the first two levels of the valence band (which are very similar to those of the conduction band) in the supercell. An average between anion and cation contribution is made

Fig. 10. Contour plot of the squared wavefunction for the first two levels of the valence band in the V-shaped quantum wire. Darker gray represents higher values

in order to avoid the oscillation of the wave function due to the charge transfer between anion and cation [12]. The shape of the squared wavefunction reflects the V-shape of the wire and the extension of the second level wavefunction is larger than that of the first level.

5. Conclusion

We have shown how empirical tight-binding is well suited to describe nanostructure going beyond the usual enevelope function approximations. This step is required to describe the behavior of modern semiconductor nanostructures. The theory presented can be easily extended to include ab-initio tight-binding methods. This is of great interest for the study of nanostructured devices from an ab-initio point of view. This work have been partially supported by the EC Community under "Ultrafst" TMR project and by Italian MURST fundings.

References

[1] A. YARIV, Optical Electronics in Modern Communications, Oxford University Press, New York 1997.
[2] Y. S. TANG, Physics of Semiconductor nanostructures, World Scintific Publ. Co., Singapore 1999.
[3] G. BASTARD, Wave Mechanics Applied to Semiconductor Heterostructures, Les Edition de Physique, Les Ulis 1988.
[4] M. G. BURT, J. Phys.: Condensed Matter **4**, 6651 (1992).
[5] H. S. P. WONG, D. J. FRANK, P. M. SOLOMON, C. H. J. WANN, and J. J. WELSER, in: Special Issue on Quantum Devices and Their Applications, Eds. A. C. Seanaugh and P. Mazumder, Proc. IEEE **87** 537 (1999).
[6] J. C. SLATER and G. F. KOSTER, Phys. Rev. **94**, 1498 (1954).
 D. W. BULLET, Solid State Physics, **35**, 129 (1980).
 J. A. MAJEWSKI and P. VOGL, The Structure of Binary Compounds, Eds. F. R. DE BOER and D. G. PETTIFOR, Elsevier Publ. Co., Amsterdam 1989.
[7] P. VOGL, H. P. HJALMARSON, and J. D. DOW, J. Phys. Chem. Solids **44**, 365 (1983).
[8] D. L. SMITH and C. MAILHIOT, Rev. Mod. Phys. **62**, 173, (1990).
[9] T. B. BOYKIN, J. P. A. VAN DER WAGT, and J. S. HARRIS, Phys. Rev. B **43**, 4777 (1991).
 T. B. BOYKIN, Phys. Rev. B **51**, 4289 (1995).
[10] S. K. KIRBY, D. Z.-Y. TING, and T. C. MCGILL, Phys. Rev. B **48**, 15237 (1993).
[11] M. S. KILEDJIAN, J. N. SCHULMAN, K. L. WANG, and K. V. ROUSSEAU, Phys. Rev. B **46**, 16012 (1992).
[12] J. N. SCHULMAN and Y. C. CHANG, Phys. Rev. B **31**, 2056 (1985).
[13] G. ARMELLES and V. R. VELASCO, Phys. Rev. B **54**, 16428 (1996).
[14] I. A. PAPADOGONAS, A. N. ANDRIOTIS, and E. N. ECONOMOU, Phys. Rev. B **55**, 10760 (1997).
[15] M. DI VENTRA and A. BALDARESCHI, Phys. Rev. B **57**, 3733 (1998).
[16] A. DI CARLO and P. LUGLI, Semicon. Sci. Technol. **10**, 1673 (1995).
[17] A. DI CARLO, S. PESCETELLI, M. PACIOTTI, P. LUGLI, and M. GRAF, Solid State Commun. **98**, 803 (1996).
[18] A. DI CARLO, P. VOGL, and W. PÖTZ, Phys. Rev. B **50**, 8358 (1994).
[19] A. DI CARLO, A. REALE, L. TOCCA, and P. LUGLI, IEEE J. Quantum Electronics **34**, 1730 (1998).
[20] S. PESCETELLI, A. DI CARLO, and P. LUGLI, Phys. Rev. B **56**, R1668 (1997).
 A. DI CARLO, S. PESCETELLI, A. KAVOKIN, M. VLADIMIROVA, and P. LUGLI, Phys. Rev. B **57**, 9770 (1998).
[21] F. DELLA SALA, A. DI CARLO, P. LUGLI, F. BERNARDINI, V. FIORENTINI, R. SCHOLTZ, and J.-M. JANCU, Appl. Phys. Lett. **74**, 2002 (1999).
[22] C. HAMAGUCHI, M. YAMAGUCHI, H. NAGASAWA, M. MORIFUJI, A. DI CARLO, P. VOGL, G. BÖHM, G. WEIMANN, Y. NISHIKAWA, and S. MUTO, Jpn. J. Appl. Phys. **34**, 4519 (1995).
[23] K. MURAYAMA, M. MORIFUJI, C. HAMAGUCHI, A. DI CARLO, P. VOGL, G. BÖHM, and M. SEXL, phys. stat. sol. (b) **204**, 368 (1997).

[24] J.-M. JANCU, R. SCHOLZ, F. BELTRAM, and F. BASSANI, Phys. Rev. B **57**, 6493 (1998).
R. SCHOLZ, J.-M. JANCU, and F. Bassani, MRS Proc. **491** 383 (1998).
[25] G. C. OSBOURN and D. L. SMITH, Phys. Rev. B **19**. 2124 (1979).
[26] J. N. SCHULMAN and YIA-CHUNG CHANG, Phys. Rev. B **27**, 2346 (1983).
[27] YIA-CHUNG CHANG and J. N. SCHULMAN, Phys. Rev. B **25**, 3975 (1982).
[28] E. N. ECONOMOU, Green's Functions in Quantum Theory, 2nd ed., Springer-Verlag, Heidelberg/Berlin 1993.
[29] M. GRAF and P. VOGL, Phys. Rev. B **51**, 4940 (1995).
[30] W. A. HARRISON, Electronic Structure and the Properties of Solids, Freeman, San Francisco (CA) 1980.
[31] S. FROYEN and W. A. HARRISON, Phys. Rev. B **20**, 2420 (1979).
[32] S. Y. REN, J. D. DOW, and D. J. Wolford, Phys. Rev. B **25**, 7661 (1982).
[33] C. PRIESTER, G. ALLAN, and M. LANNOO, Phys. Rev. B **37**, 8519 (1988); **38**, 9870 (1988); **38**, 13451 (1988).
Y. FOULON and C. PRIESTER, Phys. Rev. B **44** 5889 (1991).
[34] Q. M. MA, K. L. WANG, and J. L. SCHULMAN, Phys. Rev. B **47**, 1936 (1993).
[35] E. ANDERSON, Z. BAI, C. BISCHOF, J. DEMMEL, J. DONGARRA, J. DU CROZ, A. GREENBAUM, S. HAMMARLING, A. MCKENNEY, S. OSTROUCHOV, and D. SORENSEN, LAPACK User's Guide, SIAM, Philadelphia, 1992.
[36] W. H. PRESS, B. P. FLANNERY, S. A. TEUKOLSKY, and W. T. VETTERLING, Numerical recipes, Cambrige University Press, 1986.
[37] YIA-CHUNG CHANG and J. N. SCHULMAN, Phys. Rev. B **31**, 2069 (1985).
[38] E. PERZ, L. MUÑOZ, L. VIÑA, E. S. KOTELES, and K. M. LAU, in: Proc. 23rd Internat. Conf. Physics of Semiconductors, Eds. M. SCHEFFLER and R. ZIMMERMAN, World Scientific Publ. Co., Singapore 1996 (p. 1975)..
[39] E. L. IVCHENKO, A. YU. KAMINSKI, and U. RÖSSLER, Phys. Rev. B **54**, 5852 (1996).
[40] S. L. CHUANG, Physics of Optoelectronic Devices, Wiley, New York 1995.
[41] K. MAGARI, M. OKAMOTO, and Y. NOGUCHI, IEEE Photon. Technol. Lett. **3**, 998 (1991).
[42] A. REALE, A. DI CARLO, S. PESCETELLI, M. PACIOTTI, and P. LUGLI, VLSI Design **8**, 469 (1998).
[43] S. NAKAMURA and G. FASOL, The Blue Laser Diode, Spinger-Verlag, Berlin 1997.
[44] F. BERNARDINI, V. FIORENTINI, and D. VANDERBILT, Phys. Rev. B **56**, R10024 (1997).
F. BERNARDINI and V. FIORENTINI, Phys. Rev. B **57**, R9427 (1997).
[45] G. H. WANNIER, Phys. Rev. **100**, 1227 (1955); Phys. Rev. **101**, 1835 (1956); Phys. Rev. **117**, 432 (1960); Rev. Mod. Phys. **34**, 645 (1962).
[46] E. E. MENDEZ, F. AGULLÓ-RUEDA, and J. M. HONG, Phys. Rev. Lett. **60**, 2426 (1988); Phys. Rev. B **40** 1357 (1989).
[47] P. VOISIN, J. BLEUSE, C. BOUCHE, S. GAILLARD, C. ALIBER, and A. REGRENY, Phys. Rev. Lett. **61**, 1639 (1988).
[48] K. KASH, A. SCHERER, J. M. WORLOCK, H. G. CRAIGHEAD, and M. C. TAMARGO, Appl. Phys. Lett. **49**, 1043 (1986).
M. KOHL, D. HEITMANN, P. GRAMBOW, and K. PLOOG, Phys. Rev. Lett. **63**, 2124 (1989).
L. N. PFEIFFER, K. W. WEST, H. L. STÖRMER, J. P. EISENSTEIN, K. W. BALDWIN, D. GERSHONI, and J. SPECTOR, Appl. Phys. Lett. **56**, 1697 (1990).
P. M. PETROFF, J. GAINES, M. TSUCHIYA, R. SIMES, L. GOLDREN, H. KROEMER, J. ENGLISH, and A. C. GOSSARD, J. Cryst. Growth **95**, 260 (1989).
[49] E. KAPON, D. M. HWANG, and R. BHAT, Phys. Rev. Lett. **63**, 430 (1989).
R. BHAT, E. KAPON, D. M. HWANG, M. A. KOZA, and C. P. YUN, J. Cryst. Growth **93**, 850 (1988).
[50] R. RINALDI, M. FERRARA, and R. CINGOLANI, Phys. Rev. B **50**, 11795 (1994).
[51] R. RINALDI, R. CINGOLANI, M. LEPORE, M. FERRARA, I. M. CATALANO, F. ROSSI, L. ROTA, E. MOLINARI, P. LUGLI, U. MARTI, D. MARTIN, F. MORIER-GEMOUD, P. RUTERANA, and F. K. REINHART, Phys. Rev. Lett. **73**, 2899 (1994).
[52] Y. ARAKAWA, T. YAMAUCHI, and J. N. SCHULMAN, Phys. Rev. B **43**, 4732 (1991).
[53] L.-W. WANG and A. ZUNGER, J. Chem. Phys. **100**, 2394 (1994); Comput. Mater. Sci. **2**, 326 (1994).
[54] G. GROSSO, L. MARTINELLI, and G. PASTORI PARRAVICINI, Phys. Rev. B **51**, 13033 (1995).
[55] L. COLOMBO, Annu. Rev. Computational Physics IV, Ed. D. STAUFFER, World Scientific Publ. Co., Singapore 1996 (pp. 147 to 183).
[56] C. LANCZOS, J. Res. Nat. Bur. Standards **45**, 225 (1950); H. Q. LIN and J. E. GUBERNATIS, Computer in Phys. **7**, 400 (1993).

Subject Index

ab initio molecular dynamics 42, 389
ab initio potential energy surface 396
Abell-Tersoff potential 24
adsorption of carbon-trioxide 430
adsorption on Si (111), (001) 350
AIMPRO (Ab Initio Modelling PROgram) 131, 434
alkali halide clusters 323, 329
all-electron methods 219
alpha (α)-helices 357, 366
alpha (α)-quartz 539
alumina (0001), alpha (α) 377
alumina (0001), surface energy 377
amino acids 366
amorphous semiconductors 352
amorphous silicon 461
Andersen's pressure control 464
atomic orbital basis 335

band structure 117, 599
band structure, GaN 616
band structure, method 614
band structure, porphyrines 571
band structure, III–V semiconductors 452
basis sets 135, 450
Bethe lattice 20
Bethe-Salpeter equation 610
binding energy 326
biomolecules 345, 357
Bloch waves (states) 10, 704
BN-fullerenes 338
bond energy 29
bond-angle distribution 468
bond-order formalism 24, 25, 30, 31, 35
bond-order potentials 289
boundary conditions 11, 18, 98, 705
bridge defect 654, 659
Brillouin zone (BZ) 10, 13, 18
Brillouin-zone, integral 11, 14
brittle fracture 283
brittleness 545
buffer layer 458

CaO, alkali doping (Li, Na) 78
carbon diffusion in GaAs 473, 488
carbon nitride (C_3N_4) 535
carbon trioxide (CO_3) 430
Car-Parrinello MD 165
chalcopyrites 633
charge density 157, 482
chemical hardness 52
chemical potential 480
chalcogenide glasses 294
clusters 9, 522
clusters, anionic chlorides 329
clusters, Fe 301

clusters, Ge 183
clusters, halides 329
clusters, Li 315
clusters, Si 293
CO_3 on graphite 442
coarse-grained molecular dynamics 258
coarse-grained potential 278, 286
conductance 698, 699
configuration interaction 645
configurational heat capacity 314
conjugated bond 36
continuum mechanics 266
contracted atomic orbitals 474
core polarization 670
Coulomb interaction 474
Coulomb potential 206
coupling of length scales 251, 267
crack propagation 252, 267, 284
crambin 349
CRYSTAL-program 63, 579
crystal momentum representation 625
crystals 9
Cu impurities in Si 525
current–current correlation 604
current–current response 707
CVD silicon growth 390
cyclic cluster model 18, 19

DNA 346, 365, 373, 374
deep donors 668
defect molecule model 17
deformation potential 453
density matrix 627
density matrix formulation 233, 235
density of states 29, 393
devices 703, 704
DFT (density-functional theory) 26, 42, 46, 90, 174, 202, 219, 293, 335, 391, 515, 551, 358, 665, 685
DFTB (density-functional tight-binding) 44, 48
diamond lattice 13
diatomics 228
dielectric tensor 607, 612
dielectric function, GaN 616
dielectric constant, a-Si 469
dihedral angle rotation 37
Dirac equation 95
dislocation dynamics 547
dislocation loop 696, 697
dislocations, GaN 473
disordered solids 20
dispersion interaction 273, 357
dissociative adsorption of H_2 401
dissociative chemisorption 301
divide-and-conquer approach 245

domain decomposition   282
doping   685, 691
double-counting terms   474
ductility   545
dynamical load balancing   198
dynamical matrix   294

edge dislocation, GaN   502
effective mass   419, 451, 613
effective-mass approximation   643
effective pair potential   34
eigenvalue problem   113
electric dipole moment   295
electron affinity   293, 324, 326, 331
electron counting rule (ECR)   479
electron–hole recombination   454
electron–phonon coupling parameters, Li, S   423, 425
electron–phonon interactions   419
electron spectroscopy   656
electron temperature   462
electronic density of states, a-Si   469
electronic polarizability   294
Eliashberg spectral function   421
embedded cluster model (ECM)   14
embedding matrix   16
empirical potentials   23, 263
empirical tight binding   703, 704
ENDOR   668
energy functional   233
enthalpy, generalized   465, 467
EPR   668
equilibrium shape crystals   410
erosion rate   429
exchange correlation energy   152
excited states   449, 655, 669
exciton corrections   454
excitonic effects   621, 624, 637
excitons   610

FCD (full charge density method)   406
F-centres   83
Fe-clusters   301
fictious Lagrangian   265
finite difference pseudopotential method   178
finite elements   183, 255, 262, 266
Finnis-Sinclair potential   30
Fock matrix   73
formation energies, defects in GaAs   484
Franck-Condon factors   59
Friedel model   408
fullerenes   337, 350

GaAs, defect formation energies   484
GaAs, elastical constants   128
GaAs, line defects   168
GaAs, point defects   167

GaAs, surface energy   480
GaAs, surface reconstruction   479
GaAs, vacancy   483
GaAs, voids   473
GaN, defects   493
GaN, dielectric function   616
X-point approximation   475
Gaussian basis (orbitals)   137, 202, 685
Gaussian functions   435
Ga-vacancy–oxygen complex   504
Ge, clusters   183, 185
general eigenvalue problem   164, 475
GeS_2 308
$GeSe_2$ 304
GGA   220, 224, 293, 325, 390, 675
Gibbs free energy   697
glasses   293
gold, nanoclusters, nanowires   342
graphite   435
Green function   15, 16, 666
group III-nitrides   679
GW approximation   601
GW method   617

hafnia   541
Hamiltonian matrix   207
handshaking atoms   269, 275
harmonic approximation   294
Harris-Foulkes functional   27, 28, 45
Hartree energy   148
Hartree-Fock method   645, 515
Hartree-Fock theory   65
Hellmann-Feynman forces   476
Hellmann-Feynman theorem   160, 265, 463
heterojunction   703
hexavacancy in GaAs   522
HOMO-LUMO gap   300, 437
Hubbard U 45, 52
Hückel theory   435
Hunds rule   28
hybrid simulations   259, 261
hydrocarbon potential   31, 32
hydrogen bonding   346, 357, 362
hydrogen dissociation   393, 396
hyperfine interaction   665, 670
hyperfine coupling constants   81

impurities   525, 691, 692, 693
infrared absorption (IR)   371
InGaAs alloy   455
intersubband transition   456
ion mobility measurement   299
ionic bonding   329
ionization energies   293
ionization potential   326
IR spectroscopy   293, 301, 656
island shapes   414

Subject Index

isobaric quench   465
isomerization, thermal   333

joint density of states   332, 619, 620

Keating model   14, 20
kink formation energy   415
KKR equation   106
Kohn-Sham energy   474
Kohn-Sham equation   132, 175, 220, 294, 325
Kohn-Sham potential   90
$\mathbf{k} \cdot \mathbf{p}$ model   455
$\mathbf{k} \cdot \mathbf{p}$ theory   703, 704, 711
$\mathbf{k}$-point sampling   380
Kramers-Kronig transformation   629

Langevin dynamics   184
large scale electronic structure   231
LCAO (linear combination of atomic orbitals)   49, 265, 325, 359, 645
LCMO (linear combination of molecular orbitals)   567
line defects   168
linear optics   603
linear response   192, 420
linear scaling (methods)   231, 234, 335
liquid silicon   465
lithium clusters   312, 315
lithium fluoride   83, 227
LMTO, ASA   91, 121, 406, 615, 666
LMTO, basis sets   108
LMTO, density matrix   120
LMTO, Hamiltonian   112
LMTO method   89, 614
local field effects   605
localization constraints   241
localization region   232
localized orbitals   704
London dispersion formula   273
L(S)DA   44, 91, 125, 133, 293, 380, 665
luminescence   711, 715
lumped mass approximation   268

magnetic anisotropy energies   212
magnetic molecules   212
magnetization density distribution   665, 667, 670
metal surfaces   405
metallic nanostructures   342
Metropolis Monte Carlo   313
micro-electromechanical systems (MEMS)   253
molecular cluster model (MCM)   17, 18
molecular dynamics   184, 255, 263, 311, 389, 461
molecular stacking interactions   360, 373
molecules, CO   57
molecules, CO_2 56

molybdenum bronzes   580
molybdenum disilicides   555
moment method   282
mono-atomic steps   412
Monte Carlo method   311
muffin tin approximation   91
muffin tin eigenfunction   107
muffin tin potentials   121
Mulliken analysis   53
Mulliken-charge   474, 475
multiscale modeling   251

nanoclusters, gold   342
nanocrystals   648
nanopipes, GaN   473, 493, 499
nanoscale materials   335
nanostructure   703, 704
nanotubes   337, 685, 696
nanotubes, bending, oxidation, opening   340
nanotubes, vibrational properties   339
nanowires, gold   343
neglect of differential overlap   646
NiO band structure   75
NiO spin densities (FM/AFM)   77
nonlinear core corrections   222
nonlinear optics   599, 624, 631, 637
non-orthogonality   43
norm conservation   221
normal modes   296
Nosé-Hoover thermostat   289, 464
NRLMOL-program   57, 197
numerical grids   205, 685

optical absorption   59
optical activity   458
optical gain   713
optical gap, a-Si   469
optical response   694
optical spectra   192, 647, 650
optical tensor   622
optimized potential method (OPM)   602
order-N 46, 335, 685
organic field effect transistor (OFET)   565
organic light emitting diodes (OLED)   565
organic molecules, PTCDA   56
orthonormalization constraints   237
oscillator strength   621, 624
overlap matrix   143
overlapping subsystems   245
oxidation of Si   656
oxides, silica   538
oxygen migration   433, 441

pair correlation function   468
paladium surface   393
parallel implementation   179, 197, 399, 473, 476, 685

partial waves 93, 95, 104
Pd(100), hydrogenation 391
Pd(100), dissociation of H$_2$ 391
periodic Hartree-Fock method 63, 68
perturbed crystal approach 15, 16
phase transition liquid–amorphous 466
phonon spectra, solid argon 286
phonon spectrum, liquid silicon 465
phonons, lithium 424
photodissociation spectroscopy 301
photoelectron spectra 323, 331
photoemission spectra 185
photoluminescence 641, 655
pi(π)-bond 32, 36
point defects, C, Si, SiC, GaAs 166, 513, 668
Poisson equation 121, 125, 149, 706
polarizability 191, 295, 604
polarization 624, 629
polarization functions 140
polycyclic aromatic hydrocarbons 437
porous Silicon 641
porphyrines 566, 568, 571
positron annihilation spectroscopy (PAS) 484
power spectrum 190
precipitation 524
pressure control in tight-binding 461
protein 349
pseudoatoms 17
pseudopotentials 134, 177, 219, 336, 380, 685

QM/MM 360
quantum confinement 648
quantum molecular dynamics 244
quantum size effect 455
quantum transport 698
quantum well 455, 703, 710, 716
quantum wire 718
quasiparticle corrections 600
quasiparticle-effects 617, 637
quenching 461

radical bonds 32, 36
radiolytic oxidation of graphite 429
Rahman-Parrinello approach 464
Raman spectroscopy 293, 304
reaction dynamics 398
real space method 176, 685
resonant Raman 57
resonant spin tunneling 210
restricted Hartree-Fock 65, 67
rippled facets 416

scaling 182, 208, 234, 478
SCC-DFTB 50, 357, 473
scissors operator 617
screening sphere 98, 102, 143
screening radii 102

screw dislocation GaN 483, 495
second-moment approximation 30
second-order susceptibilities 630
self-consistent-charge approximation (SCC) 46
self-interaction relaxation corrections (SIRC) 449
semiconducting optical amplifier (SOA) 712, 713
semiempirical Hartree-Fock 645
semiempirical methods 358
shape transition 300
Si, amorphous 461
Si, band structure 113, 119
Si, brittleness of 553
Si, charge density 124
Si, clusters 294
Si, ductility 553
Si, elastic constants 128
Si, liquid phase 461, 465
Si, liquid surface 352
Si, point defects 167, 513, 521, 673
Si, total energy 127
Si, vacancy clusters 522
Si(001) CVD growth 390
Si(111), (001) surface adsorption 350
Si:H nanoparticles 649
SIESTA-program 335
sigma(σ)-bond 32
silanone 660
silica 539
silicides 555
silicon carbon nitride 536
silicon nitride (Si$_3$N$_4$) 534
simulated annealing 183
SiO$_2$ 646
slab model 479
solids crystalline 9
special **k**-points 11, 12
specific heat 468
spin density 206
spin–orbit coupling 213
spin polarization 54
spin polarization energy 224
split interstitial Si 519
stacking energies, DNA 374
stacking faults 547
sticking probability 399
Stilling-Weber potential 282
stishovite 539
strain tensor 709
supercell model (SCM) 13
superhard materials 533
superlattice(s) 453, 715
surface energy, alumina (0001) 377
surface energy, anisotropy 409
surface energy, GaAs 480
surface energy, metals 405

Subject Index

surface reactions   389
surface reconstruction, GaAs   473
surface reconstruction, Si   390
surfaces and interfaces   19
surfactants   691

Tauc plot   469
temperature control in tight-binding   465
thermal isomerization   325
thermodynamic stability   379
thermodynamic variables   462
III–V semiconductors   473
tight-binding, parallel implementation   473
tight-binding parameters   450
tight-binding theory   27, 43, 234, 255, 264, 357, 449, 703
time-dependent (TD)-LDA   192
total energies   89, 124, 141, 233, 379, 557
transfer matrix method   706
transition metal compounds, NiO   74
transition metal defects   673
transition metal oxides   577

unrestricted Hartree-Fock   67

vacancies in Si   521
vacancy aggregates in GaAs   483
vacancy clusters in Si   522
valence band mixing   710
valence band offset   453
valence-force model   25
Van der Waals interaction   361, 373
Van Hove singularities   619
velocity autocorrelation function   465
Verlet algorithm   264, 269
vibrational frequencies   293, 333
vibrational modes   188, 208
vibrational power spectrum   469
virtual crystal approximation   455, 707

Wannier functions   241

yellow luminescence   473

Zener tunneling   716
zirconia   539